Chinese HERBAL MEDICINE

MODERN APPLICATIONS OF TRADITIONAL FORMULAS

Chinese HERBAL MEDICINE

MODERN APPLICATIONS OF TRADITIONAL FORMULAS

CHONGYUN LIU
and
ANGELA TSENG
with
Sue Yang

CRC PRESS
Boca Raton London New York Washington, D.C.

Library of Congress Cataloging-in-Publication Data

Liu, Chongyun.
Chinese herbal medicine : modern applications of traditional formulas / by Chongyun Liu and Angela Tseng with Sue Yang.
p. cm.
Includes bibliographical references and index.
ISBN 0-8493-1568-9 (alk. paper)
1. Herbs—Therapeutic use. 2. Medicine, Chinese. I. Tseng, Angela, II. Yang, Sue, III. Title.

RM666.H33L559 2003
615'.321—dc22

2003055711

Direct all inquiries to CRC Press LLC, 2000 N.W. Corporate Blvd., Boca Raton, 33431.

Visit the CRC Press Web site at www.crcpress.com

International Standard Book Number 0-8493-1568-9
Library of Congress Card Number 2003055711
Printed in the United States of America 1 2 3 4 5 6 7 8 9 0
Printed on acid-free paper

Contents

Part III Single Herbs Classification

Appendices

Foreword

Traditional Chinese medicine is rapidly advancing onto the world stage as a medicine that has endured through time and history to assuage the ills of humankind and assist in the prevention of further diseases and imbalances of health. Chinese herbal medicine is notable for its sophistication in addressing clinical concerns while remedying an individual's particular needs. Thousands of years old, the herbal medicine tradition remains a vibrant player in today's health field.

Chinese Herbal Medicine: Modern Applications of Traditional Formulas, written by Dr. Chongyun Liu and Angela Tseng, is the newest addition to the busy clinician's repertoire. With over 20 years of clinical experience, Dr. Liu supplies more than 860 formulas with descriptions of appropriate customization that will enable practitioners to tailor herbal formulas to meet patients' needs. In addition, a listing of over 640 individual herbs provides quick access to clinical modifications. Of significant note is the extensive review of 190 allopathic medical conditions with corresponding traditional Chinese medicine diagnoses and formulations.

The publication of this text is another step forward in bringing Chinese medicine to the forefront of the integration between Eastern and Western medicine. May this book guide and assist practitioners and students of traditional Chinese medicine in their lifetime pursuit of excellence on the journey of clinical study.

Terry Courtney, M.P.H., L.Ac.

About the Authors

Chongyun Liu, O.M.D., L.Ac.

A prominent authoritative figure in the field of Chinese medicine, Dr. Chongyun Liu is a graduate of Luzhou Medical School and Chengdu University of Chinese Medicine. With an impressive list of accomplishments, Dr. Liu has over 20 years of experience in research, clinical practices and as an educator in China, Europe, and the United States. In addition, numberous articles and books written by Dr. Liu have been published in both China and the United States. Currently, he is a professor in the Acupunture and Oriental Medicine Department at Bastyr University and a clinical faculty member at the Bastyr Center for Natural Health.

Angela Tseng, L.Ac, Dipl.Ac., Dipl.C.H.

A respected educator of traditional Chinese medicine, Ms. Tseng is a graduate from the University of Toronto with a degree in chemistry and Bastyr University with a master's of science in acupuncture and a certificate in Chinese herbal medicine. She is a licensed practitioner in the State of Washington with certificates in both acupunture and Chinese herbology from the National Certification Commission for Acupuncture and Oriental Medicine (NCCAOM). She also received advanced hospital training in China. Currently, she is a faculty member in the Acupuncture and Oriental Medicine Department at Bastyr University and a supervisor at the Bastyr Center for Natural Health.

Acknowledgments

The authors are grateful for the assistance of Sue Yang, L.Ac., for her hard work in organizing the information in Part III and for the layout design; to Hugo Lau, L.Ac., for his translation of the Standardized Syndrome Differentiation of Traditional Chinese Medicine (Appendix 1); and to David Plotnik, L.Ac., and Kelly Neu, L.Ac., for their compilation and organization of the Glossary of Traditional Chinese Medical Terminology (Appendix 2). Our sincere thanks to Terry Courtney, M.P.H., L.Ac., for her encouragement in writing this book and to Steve Given, L.Ac., Matt Ferguson, L.Ac., and Ying Wang, O.M.D., L.Ac., for their suggestions for its title.

We also express special thanks to Sonia Xiaosong Wang and Mary Liu for their utmost support and constructive recommendations, and for their patience and understanding when Chongyun Liu was not able to spend quality time with them during the book-writing process.

Angela Tseng, L.Ac. and Sue Yang, L.Ac. would especially like to extend their appreciation to Dr. Chongyun Liu for giving them the opportunity to collaborate on this project, which helped reinforce their knowledge of Chinese herbal medicine in clinical applications.

Introduction

The clinical practice of traditional Chinese medicine (TCM) includes differential diagnosis and treatments based on the analysis of signs and symptoms and differentiation of syndromes. The art of selecting the appropriate base herbal formula is dependent on this analysis and the modifications made according to the presenting signs and symptoms of the patient. Upon completing the study of the fundamental theories of TCM, diagnosis, pathology, and materia medica, senior students and new practitioners become concerned with how to apply their knowledge in clinical settings, with the selection of appropriate formulae, and with the modification process according to the manifestations or sometimes to the conventional diagnoses. Therefore, this book was designed, edited, and written with the goal of solving these concerns in the hope that this book will serve as a bridge between the study of TCM fundamental theories and clinical applications.

The most common syndrome differentiation methods in clinical practice emphasized by TCM schools today are based on the theory of Zang Fu organs, Qi, blood, and body fluids, Yin, Yang, and meridian systems. Knowledge of these enable, for example, the diagnosis of Spleen and Kidney Yang deficiency, Lung Qi deficiency, Stomach Yin deficiency, and Liver Yang hyperactivity, etc. The organization of the herbal formulas in this book is based on the systems of the Five Zang organs. In Part I, under each chapter of an organ system, the formulas are discussed according to their syndrome differentiations. Formulas related to the Five Zang organs are each individually discussed in their designated chapter and are followed by formulas used to treat multiple organ systems, for a combined total of ten chapters. Under each core formula, related formulas are discussed. For example, under Gui Zhi Tang [Cinnamon Twig Decoction], Gui Zhi Jia Fu Zi Tang [Cinnamon Twig Decoction plus Prepared Aconite], Gui Zhi Jia Gui Tang [Cinnamon Twig Decoction with Increased Dosage of Cinnamon], and other related formulas are discussed. Also, modifications of the core formula are listed according to the presenting and concomitant signs and symptoms. After determining the syndrome pattern, practitioners and students can consider different formulas under the same categories with common modifications and related formulas according to a patient's signs and symptoms. In addition, the majority of the sections within each chapter contain a summarized comparison of all the formulas discussed to assist readers in making an appropriate choice of the numerous similar formulas. The benefit of this layout makes syndrome differentiation and formula selection an easier process.

Part II of this book is convenient for readers to reference information when patients have been given a Western diagnosis or if a practitioner needs to look into different syndrome patterns and treatments for a particular condition or disorder. Under each allopathic medical system, such as digestive, respiratory, and gynecological, common disorders are included with their TCM syndrome patterns, treatment principles, indicated formulas, and modifications. This section, then, also serves the purpose of bridging the gap between conventional medicine and traditional Chinese medicine. In this book there are over190 conditions and disorders included to assist readers who are familiar with conventional systems to find the appropriate traditional Chinese formulas.

Part III of this book is a table layout of over 640 single herbs with their Pinyin and common English names, temperatures, tastes, channels entered, functions, indications, dosages, and cautions.

Practitioners can easily refer to this section for quick access to the properties of single herbs and for assistance in precise selection of herbs in formula modifications. Thus, it is easier to understand the composition of the formulas and to modify a formula.

There were two avenues of thought developed while writing this book. First, it should follow the TCM order of approach of (1) diagnosis, (2) formulas, (3) single herbs, and (4) modifications of formulas and, second, it should incorporate conventional medical diagnosis with TCM syndrome differentiations, providing formulas, single herbs, and their modifications as remedies. The first had as its goal directly combining accumulated academic knowledge with clinical applications in the selection of herbal formula and herbal additions and subtractions. The second had the objective of connecting conventional medicine and traditional Chinese medicine. This book serves as a desktop reference guide for medical doctors and other health care providers to utilize and learn about the Chinese herbal formulas their patients are taking. These two approaches are like two bridges connecting the theories and clinical practice of traditional medicine with modern medicine.

For each differentiation of conventional medical conditions and disorders, please refer to the Standard Syndrome Differentiation of TCM (Appendix I) for the signs and symptoms purpose. The glossary is provided to accurately explain ancient TCM terminologies. All formulas and single herbs are organized in Pinyin to English and English to Pinyin cross-reference formats to enable convenient access for users. Toxic herbs are divided into three categories for clinical use that are listed at the beginning of Part III.

Findings from modern research in terms of the application of traditional formulas in modern practice or the clinical application of the pharmacological actions of the single herbs are also included in this book. Thus, ancient formulas are applied to modern disorders. The English translations of formula names follow the commonly used terminologies in this field in the United States. The authors used the Pinyin system to translate single herb names and formula titles. The names of classical formulas remain the same in this book as they appeared in classical text, such as Xi Jiao Di Huang Tang [Rhinoceros Horn and Rehmannia Decoction] and Xi Huang Wan [Rhinoceros Gallstone Pill]. However, endangered animal products in these formulas are removed and substituted with other products such as Shui Niu Jiao substituting for Xi Jiao, Shan Yang Jiao substituting for Ling Yang Jiao, and bobcat bone substituting for Hu Gu.

If there are mistakes or missing information in the content, we would be grateful for any errors pointed out for future revised editions since English is not the authors' native tongue. Thank you.

PART I

Formulas According to TCM Zang Fu Syndrome Differentiation

CHAPTER 1

Lung Syndromes and Formulas

WIND–COLD INVASION

The main symptoms of wind–cold invasion include chills and fever, presence or absence of sweating, headache, body aches, floating tight pulse, and white tongue coating. Some other symptoms include nasal congestion, clear nasal discharge, cough, watery sputum, or stifling sensation at chest which relates to Lung Qi stagnation. Based on the treatment principle that when the problem is at the skin level, the sweating method is used to release the exterior, this type of syndrome often uses releasing-exterior and promoting-sweating herbs such as Ma Huang, Gui Zhi, Xi Xin, Zi Su Ye, Jing Jie, and Fang Feng as the chief herbs in a formula. These formulas are warm, acrid, and releasing-exterior, which can promote the dispersing function of the Lung and expel evils.

Ma Huang Tang [Ephedra Decoction] *Shang Han Lun*

Ma Huang	9 g
Gui Zhi	9 g
Xing Ren	9 g
Gan Cao	3 g

Main Indication: Wind–cold invasion, chills and fever, headache, body ache, absence of sweating, wheezing, decreased taste sensitivity, floating, and tight pulse.

Preparation: Decoction. Take the warm decoction in three doses until mild sweating is induced.

Treatment Principles: Acrid–warm and promoting-sweating method. This can be used for wind–cold wheezing and Bi syndrome due to wind–cold–damp.

Related Formulas:

Formula	Herbs	Indication
San Ao Tang [Three-Unbinding Decoction]	Core formula – Gui Zhi	Wind–cold invasion, headache, body aches, wheezing, coughing, fullness sensation at chest, and clear, watery sputum This formula is good for stopping wheezing Note: Its sweat promoting function is less than that of Ma Huang Tang

(*continued*)

Formula	Herbs	Indication
Hua Gai San [Canopy Powder]	Core formula – Gui Zhi + Sang Bai Pi Zi Su Zi Fu Ling Chen Pi	Wind–cold invasion, cough with rebellious Lung Qi, fullness at chest and diaphragm, tightness at neck and back, nasal congestion, and body heaviness Note: This formula is stronger than Ma Huang Tang in its stop cough, expel phlegm, direct Lung Qi-downward, and stop wheezing functions
Ma Huang Jia Zhu Tang [Ephedra Decoction plus Atractylodes]	Core formula + Bai Zhu	Cold–damp at the exterior with whole body aches for which it is appropriate to use sweating method This formula releases the exterior and expels dampness
Ma Xing Yi Gan Tang [Ephedra, Apricot Kernel, Coicis, and Licorice Decoction]	Core formula – Gui Zhi + Yi Yi Ren	Wind–damp with whole body aches and fever, which is worse during the day The treatment principle of this formula is to release the exterior and expel dampness. However, this formula is more appropriate for patient with some heat symptoms Note: This is different from Ma Huang Jia Zhu Tang which is for cold type syndromes
Gu Jin Lu Yan Xu Ming Tang [Prolong Life Decoction from *Gu Jin Lu Yan*]	Core formula + Dang Gui Chuan Xiong Ren Shen Shi Gao Gan Jiang	Wind stroke with heat rashes, inability to control body movements and speech, inability to distinguish pain location or body stiffness, and inability to turn This formula also treats patient who prefers to lie down but is unable to lie in a flat position; cough with rebellious Lung Qi; and puffy face and eyes

Da Qing Long Tang [Major Bluegreen Dragon Decoction] *Shang Han Lun*

Ma Huang	10 g
Gui Zhi	6 g
Xing Ren	9 g
Gan Cao	6 g
Shi Gao	30 g
Sheng Jiang	9 g
Da Zao	12 g

Main Indication: Wind–cold with severe chills and fever, whole body aches, restless agitation, absence of sweating, and tight and floating pulse.

Preparation: Decoction.

Treatment Principles: Release exterior and clear internal heat method.

Comparison** to Ma Huang Tang**:

1. This formula treats restlessness and agitation due to internal heat, in addition to wind–cold symptoms.
2. It is indicated for more severe body heat symptoms.
3. The patient often has thirst.

Jin Gui Yao Lue uses this formula for Yi Yin syndrome (overflowing beverage syndrome). Swelling at the four limbs should be treated by sweating method. (Yi Yin syndrome: Water flows into the four limbs. Patient should sweat but cannot sweat, thus experiences heaviness and pain in the body.)

Gui Zhi Tang [Cinnamon Twig Decoction] *Shang Han Lun*

Gui Zhi	10 g
Bai Shao Yao	10 g
Sheng Jiang	10 g
Gan Cao	6 g
Da Zao	4 pieces

Main Indication: Ying and Wei disharmony and unregulated Yin and Yang. Headache, stiff neck, fever, presence of sweating, chills, nasal noise, dry heaves, and floating pulse.

This formula is indicated when patient has any one of these three symptoms: headache and stiff neck, chills with sweating, or clear nasal discharge.

Preparation: Decoction, taken warm. It can also be taken with hot congee to promote mild sweating.

Treatment Principle: Release the exterior and muscle layer, and harmonize Ying and Wei level.

Related Formulas:

Formula	Herbs	Indication
Gui Zhi Jia Fu Zi Tang [Cinnamon Twig Decoction plus Prepared Aconite]	Core formula + Fu Zi	This formula is good for excessive sweating that cannot be stopped, chills, urinary difficulty, and stiffness of the four limbs
Gui Zhi Jia Gui Tang [Cinnamon Twig Decoction with Increased Dosage of Cinnamon Twig]	Core formula with increased dose of Gui Zhi	This formula treats cold induced Running Piglet Syndrome. Patient experiences rushing sensation from lower abdomen to Heart and abdominal pain Note: Treatment principle is to harmonize Liver and Spleen, warm the Middle Jiao and direct the rebellious Qi downward
Gui Zhi Jia Hou Po Xing Zi Tang [Cinnamon Twig Decoction plus Magnolia Bark and Apricot Kernel]	Core formula + Hou Po Xing Ren	This formula treats chronic bronchitis with symptoms indicated in Gui Zhi Tang [Cinnamon Twig Decoction]
Gui Zhi Jia Ge Gen Tang [Cinnamon Twig Decoction plus Kudzu]	Core formula + Ge Gen	This formula treats stiff neck and back, and chills with spontaneous sweating
Ge Gen Tang [Kudzu Decoction]	Core formula + Ge Gen Ma Huang	This formula treats stiff neck and back, and absence of sweating
Huang Qi Gui Zhi Wu Wu Tang [Astragalus and Cinnamon Twig Five-Substance Decoction]	Core formula – Gan Cao + double Sheng Jiang, Huang Qi	This formula treats Ying, Wei, Qi, and blood deficiencies, evil pathogens entering blood level leading to Blood Bi, stubborn numbness in muscles, and painful Bi
Gui Zhi Jia Huang Qi Tang [Cinnamon Twig Decoction plus Astragalus]	Core Formula + Huang Qi	This formula treats yellow sweats, and sweating above the waist or under armpit. The sweat stains cloth the color of Huang Bai. Patient also has coldness in lower legs, heaviness, and pain in the body
Gui Zhi Jia Shao Yao Tang [Cinnamon Twig Decoction plus Peony]	Core formula with Double the dosage of Shao Yao	This formula treats postpartum mastitis. It also treats spasmodic pain in any part of the body
Gui Zhi Fu Zi Tang [Cinnamon Twig and Prepared Aconite Decoction]	Core formula – Shao Yao + Fu Zi	This formula treats contending wind and damp leading to body pain and restlessness; absence of nausea and thirst, and floating, empty, and choppy pulse
Gui Zhi Jia Cang Er Xin Yi Tang [Cinnamon Twig Decoction plus Xanthium and Magnolia Flower]	Core formula + Cang Er Zi Xin Yi Hua	This formula is for Gui Zhi Tang [Cinnamon Twig Decoction] syndrome with severe clear nasal discharge
Gui Zhi Shao Yao Zhi Mu Tang [Cinnamon Twig, Peony, and Anemarrhena Decoction]	Core formula – Da Zao + Bai Zhu Fu Zi Ma Huang Fang Feng Zhi Mu	This formula treats wind cold–damp Bi syndrome which has transformed into heat. Thus patient shows red, painful and swollen joints with lingering wind cold–damp evils

(continued)

Formula	Herbs	Indication
Jia Wei Gui Zhi Tang [Augmented Cinnamon Twig Decoction]	Core formula + Jiang Ban Xia Zi Shi Ying Wu Zhu Yu Huai Xiao Mai	This formula is for neurotic nausea and vomiting
Qian Jin Gui Zhi Qu Shao Yao Jia Zao Jia Tang [Cinnamon Decoction minus Peony plus Gleditsiae from Thousand Ducat Formula]	Core formula – Shao Yao + 1 burned, skinless Zao Jia	This formula is good for Lung atrophy with vomiting and thin saliva. It can also treat excess saliva in children and rhinitis with redness between nose and mouth and on cheeks
Gui Zhi Qu Shao Yao Jia Fu Zi Tang [Cinnamon Twig Decoction minus Peony with Prepared Aconite] *Shang Han Lun*	Gui Zhi 9 g Zhi Gan Cao 9 g Sheng Jiang 9 g Da Zao 3 pieces Zhi Fu Zi 3 g	Harmonize Ying and Wei Treats Tai Yang syndrome. Wind–cold, exterior deficiency, chest fullness, slight aversion to cold and chills, and rapid pulse

Ren Shen Bai Du San [Ginseng Powder to Overcome Pathogenic Influences] *Xiao Er Yao Zheng Zhi Jue*

Qiang Huo	10 g
Du Huo	10 g
Chuan Xiong	10 g
Chai Hu	10 g
Qian Hu	10 g
Zhi Ke	10 g
Jie Geng	10 g
Dang Shen (or Ren Shen 3 g)	10 g
Fu Ling	10 g
Gan Cao	3 g

Main Indication: Exterior wind cold–damp with constitutional deficiency. The symptoms include chills, fever, headache, absence of sweating, sore and painful limbs and joints, and cough with sputum.

Preparation: Add some Bo He, Sheng Jiang. Decoction: take the warm decoction in three doses.

Treatment Principle: Tonify Qi and release exterior.

Modification:

Addition	Indication
Di Yu Hai Zhu Gen	This modification is effective in treating tetanus and rabies

Related Formulas:

Formula	Herbs	Indication
Bai Du San [Overcome Pathogenic Influences Powder]	Core formula – Dang Shen	The same exterior syndrome as indicated in Ren Shen Bai Du San [Ginseng Powder to Overcome Pathogenic Influences] but without Zheng Qi deficiency
Jing Fang Bai Du San [Schizonepeta and Ledebouriella Powder to Overcome Pathogenic Influences]	Core formula – Dang Shen + Jing Jie Fang Feng	Treats more severe exterior cold

(continued)

Formula	Herbs	Indication
Yin Qiao Bai Du San [Honeysuckle and Forsythia Powder to Overcome Pathogenic Influences]	Core formula – Dang Shen + Jin Yin Hua Lian Qiao	Treats influenza and beginning stage of sore toxins
Xiao Huang Bai Du San [Mirabilitum and Rhubarb Powder to Overcome Pathogenic Influences]	Core formula – Dang Shen + Mang Xiao Da Huang	Treats the same exterior syndrome with constipation and excess accumulation of heat toxins

Shen Su Yin [Ginseng and Perilla Leaf Decoction] *He Ji Ju Fang*

Ren Shen	24 g
Zi Su Ye	24 g
Ge Gen	24 g
Qian Hu	24 g
Ban Xia (fried with ginger juice)	24 g
Fu Ling	24 g
Chen Pi	15 g
Gan Cao	15 g
Zhi Ke	15 g
Mu Xiang	15 g
Jie Geng	15 g

(In *Jian Yi Fang*, Shen Su Yin [Ginseng and Perilla Leaf Decoction] does not contain Mu Xiang.)

Main Indication: For patient with common cold with deficient constitution. Symptoms include chills, fever, headache, sticky nasal discharge and sputum, cough with sticky sputum, fullness sensation at epigastrium, and floating and soft pulse.

Preparation: Powder the herbs. Take 12 g of the powder and decoct with ginger and date water. Take the decoction warm.

Treatment Principle: Tonify Qi and release exterior, regulate Qi flow, and transform phlegm.

Comparison: The common characteristic of Shen Su Yin [Ginseng and Perilla Leaf Decoction] and Ren Shen Bai Du San [Ginseng Powder to Overcome Pathogenic Influences] is the base formula. The base formula is Si Jun Zi Tang [Four-Gentleman Decoction] without Bai Zhu.

The difference between the two formulas:

Ren Shen Bai Du San [Ginseng Powder to Overcome Pathogenic Influences]	Shen Su Yin [Ginseng and Perilla Leaf Decoction]
More exterior syndrome than interior, thus use Qiang Huo and Du Huo	More interior syndrome than exterior, thus use Mu Xiang, Chen Pi, and Ban Xia
Wind–damp at the exterior; patient experiences pain and soreness at limbs and joints, thus using Qiang Huo, Du Huo, and Chuan Xiong.	Evil pathogens at Tai Yin which causes fullness at chest and epigastrium due to Spleen Qi not transporting, thus using Mu Xiang, Chen Pi, and Ban Xia

Related Formula:

Formula	Herbs	Indication
Cong Bai Qi Wei Yin [Scallion Decoction with Seven Ingredients]	Cong Bai 9 g Ge Gen 9 g Dan Dou Chi 6 g Sheng Jiang 6 g Mai Men Dong 9 g Sheng Di Huang 9 g Water 3 cups	This formula nourishes blood and releases exterior Treats exterior syndrome with blood deficiency

Xiang Su San [Cyperum and Perilla Leaf Decoction] *He Ji Ju Fang*

Xiang Fu (fried, hair removed)	12 g
Zi Su Ye	12 g
Chen Pi	60 g
Zhi Gan Cao	30 g

Main Indication: Exterior wind–cold with interior Qi stagnation. Patient has chills, but skin is warm to the touch. Other symptoms include headache without sweating, fullness and stifling sensation at chest and epigastrium, poor appetite, and thin, white tongue coating.

This formula can be used for any exterior wind–cold with inhibited Qi Ji (functional activities of vital energy) or constitutional Liver Qi stagnation, disharmony between Liver and Stomach with exterior wind–cold.

Preparation: Powder the herbs. Take 6 g three times a day. It can also be taken as decoction.

Treatment Principle: Regulate Qi and release exterior.

Related Formula:

Formula	Herbs	Indication
Xiang Su Cong Chi Tang [Cyperum, Perilla Leaf, Scallion, and Prepared Soybean Decoction]	Core formula + Cong Chi Tang [Scallion and Prepared Soybean Decoction]	Increased strength of releasing exterior

Cong Chi Tang [Scallion and Prepared Soybean Decoction] *Zhou Hou Fang*

Cong Bai	5 stalks
Dou Chi	10 g

Main Indication: Beginning stage of a common cold, fever, chills, absence of sweating, headache, nasal congestion, and mild evil pathological conditions.

This is suitable for the beginning stage of a common cold in which evil pathogens have not pass the surface and are still very shallow at the skin layer.

Preparation: Decoction.

Treatment Principle: Acrid neutral to release exterior.

Related Formulas:

Formula	Herbs	Indication
Huo Ren Cong Chi Tang [Scallion and Prepared Soybean Decoction from the Book to Safeguard Life]	Core formula + Ma Huang Gan Ge Gen	Treats first 1 or 2 days of a wind–cold condition with headache, neck, back and body pain, chills without sweating, and tight pulse
Cong Chi Jie Geng Tang [Scallion, Prepared Soybean, and Platycodon Decoction]	Core formula + Shan Zhi Zi Jie Geng Bo He Lian Qiao Gan Cao Dan Zhu Ye	Treats beginning stage of wind–warmth and wind–heat condition
Cong Chi Bo Bang Tang [Scallion, Prepared Soybean, Peppermint, and Arctii Decoction]	Core formula + Bo He Niu Bang Zi	Treats beginning stage of wind–heat condition
Cong Chi Bai Hu Tang [Scallion, Prepared Soybean, and White Tiger Decoction]	Core formula + Shi Gao Zhi Mu Gan Cao Powdered Jie Geng	Treats wind–cold constricting exterior with internal excess heat accumulation

Ma Huang Fu Zi Xi Xin Tang [Ephedra, Asarum, and Prepared Aconite Decoction] *Shang Han Lun*

Ma Huang	6 g
Fu Zi	15 g
Xi Xin	3 g

Main Indication: Exterior syndrome with Yang deficiency. Symptoms include fever and severe chills where the patient still feels cold wearing heavy clothing; fatigue, and deep and faint pulse.

This formula can be used to treat Yang deficient, cold type, with sudden loss of voice (aphonia), severe sore throat or as if throat is obstructed by something (as in chronic pharyngitis). To treat myocardial infarction due to conductive obstruction at the ventricle, this formula is often added to the core formula.

Preparation: Decoct Fu Zi first, 40 minutes to 1 hour, until it does not cause numbness in the mouth, then add other herbs. Take the warm decoction in three doses.

Treatment Principle: Warm Yang to release exterior.

Related Formula:

Formula	Herbs	Indication
Ma Huang Fu Zi Gan Cao Tang [Ephedra, Prepared Aconite, and Licorice Decoction]	Core formula – Xi Xin + Gan Cao	Treat similar syndrome as Ma Huang Fu Zi Xi Xin Tang [Ephedra, Asarum, and Prepared Aconite Decoction], but this formula has less sweat promoting function In *Jin Gui Yao Lue*, this formula is called Ma Huang Fu Zi Tang [Ephedra and Prepared Aconite Decoction], which treats edema syndrome with deep pulse. This formula promotes Yang and sweating, diffuses Lung, and promotes water flow as urination

Jiu Wei Qiang Huo Tang [Nine-Herb Decoction with Notopterygium]
Ci Shi Nan Zhi

Qiang Huo	9 g
Cang Zhu	9 g
Fang Feng	9 g
Xi Xin	3 g
Bai Zhi	9 g
Chuan Xiong	6 g
Sheng Di Huang	9 g
Huang Qin	9 g
Gan Cao	3 g

Main Indication: Exterior wind cold–damp syndrome. Symptoms include chills, fever, no sweating, headache, stiff neck, sore and painful body and limbs, and bitter taste with mild thirst.

Preparation: Decoction.

Treatment Principle: Release exterior and expel dampness.

This formula is used for chills and fever, more cold than heat, headache, and sore, painful, limbs and body.

Modification:

Minus	Addition	Indication
Sheng Di Huang	Huo Xiang	Same presentation with damp accumulation in Middle Jiao. Symptoms include epigastric discomfort, nausea, and greasy coat
	Xing Ren Qian Hu	Same presentation with cough and sticky sputum
	Niu Bang Zi Bo He	Same presentation with dry throat or sore throat This formula is especially effective in treating toothache with exterior syndrome
	Jin Yin Hua Lian Qiao	Same presentation with heat

Related Formulas:

Formula	Herbs	Indication
Shen Zhu San [Marvelous Atractylodis Powder]	Qiang Huo 30 g Xi Xin 30 g Bai Zhi 30 g Chuan Xiong 30 g Gao Ben 30 g Gan Cao 30 g Cang Zhu 90 g	Cold and damp exterior syndrome. Symptoms include headache, stiff neck, fever, chills, and sore and painful body and limbs Exterior wind condition with nasal congestion, heavy body, cough, and dizziness Note 1: Powder the herbs; each time take 10 g and decoct with ginger, green onion, and water. Take the warm decoction Note 2: This formula works strongest to scatter cold and expel damp. Thus, it is good for exterior syndrome without internal heat
Bao Zhen Tang [Preserve-True Decoction]	Chuan Xiong 6 g Gao Ben 6 g Cang Zhu 24 g Gan Cao 3 g	Exterior cold with damp, causing severe pain

Qiang Huo Sheng Shi Tang [Notopterygium Decoction to Overcome Dampness] *Nei Wai Shang Bian Huo Lun*

Qiang Huo	9 g
Du Huo	9 g
Fang Feng	6 g
Gao Ben	6 g
Chuan Xiong	6 g
Man Jing Zi	6 g
Gan Cao	3 g

Main Indication: Exterior cold–damp, headache or heaviness sensation at head, heaviness and pain at lumbar area and spine, or whole body ache with inability to turn; chills, mild fever, and floating pulse.

Preparation: Decoction.

Treatment Principle: Release exterior and expel dampness.

This formula scatters the wind (cold) and damp at the upper part of the body, especially good for syndromes with heaviness and pain at head and neck.

Modification:

Addition	Indication
Fang Ji 1.5 g (alcohol washed) Fu Zi 1.5 g (mild condition) Chuan Wu 1.5 g (severe condition)	These additions are specifically for heaviness at the body and waist area, which is a result of cold–damp in the channels. This formula warms the channel, scatters cold, and assists Yang to transform damp
Fang Ji Yi Yi Ren Gui Zhi Shi Gao Huang Bai Cang Zhu	For the similar presentation with damp–heat heaviness in body and burning pain at joints. The additional herbs help to clear heat and expel dampness

Related Formulas:

Formula	Herbs	Indication
Juan Bi Tang [Remove Painful Obstruction Decoction]	Qiang Huo 9 g Fang Feng 9 g Dang Gui 9 g Huang Qi 9 g Chi Shao Yao 9 g Jiang Huang 9 g Gan Cao 3 g Sheng Jiang 5 slices	This formula can benefit Qi, harmonize Ying, expel wind, and eliminate damp Deficiency of Ying and Wei level with wind–damp Bi pain, and numbness and pain at shoulder, arm, neck, hand, and feet
San Bi Tang [Three-Painful Obstruction Decoction]	Zhi Huang Qi 3 g Ren Shen 3 g Bai Zhu 3 g Gan Cao 1.5 g Dang Gui 1.5 g Chuan Xiong 1.5 g Bai Shao Yao 1.5 g Rou Gui 1.5 g Fang Ji 1.5 g Fang Feng 1.5 g Zhi Wu Tou 1.5 g Xi Xin 1 g Sheng Jiang 3 slices Hong Zao 2 pieces	Benefit Qi and blood, and expel wind–damp. Treat wind–cold–damp Bi, Qi stagnation, blood stasis, and spasms of hands and feet

Chuan Xiong Cha Tiao San [Ligusticum Chuanxiong Powder to be Taken with Green Tea] *He Ji Ju Fang*

Qiang Huo	6 g
Fang Feng	6 g
Bai Zhi	6 g
Gan Cao	6 g
Jing Jie	12 g
Chuan Xiong	12 g
Xi Xin	3 g
Bo He Ye	3 g

Main Indication: Exterior wind syndrome with frontal, temporal or vertex headache, chills, fever, blurred vision, dizziness, nasal congestion, thin, white tongue coating, and floating and slippery pulse.

Preparation: Powder the herbs. Each time take 3 g with tea after meal. This formula can also be taken as decoction.

Treatment Principle: Expel wind and scatter evil pathogens.

The herbal ingredients of this formula all release exterior and scatter wind which is based on the idea that "illness of the top (head) can only be reached by wind herbs."

This formula can also be used for the wind syndrome of headache caused by chronic rhinitis.

Related Formula:

Formula	Herbs	Indication
Ju Hua Cha Tiao San [Chrysanthemum Powder to be Taken with Green Tea]	Core formula + Ju Hua Jiang Can	This formula can expel wind and clear heat. Thus, it is for syndrome with heat presentation

Xiang Ru San [Elsholtzia Powder] *He Ji Ju Fang*

Xiang Ru	9 g
Bai Bian Dou	9 g
Hou Po	12 g

Main Indication: This formula is for treating cold–damp invasion during summer months. This can be a result of drinking cold beverage or cool wind invasion in the evening time. In this syndrome, Yang Qi is obstructed by Yin pathogens and patient experiences aversion to cold but skin is warm to touch; heaviness and pain at head, absence of sweating, abdominal pain, vomiting, and diarrhea.

Preparation: Decoction. Take the decoction cold.

Treatment Principle: Expel dampness and release exterior.

This formula is for cold–damp invasion during summer time. This syndrome includes both exterior and interior syndromes. Thus, it is not suitable for summer heat stroke.

Related Formulas:

Formula	Herbs	Indication
Huang Lian Xiang Ru Yin [Elsholtzia Decoction with Coptis]	Core formula – Bian Dou + Huang Lian	Summer heat stroke and heat excess with thirst and restlessness Note: For summer heat syndrome
Si Wei Xiang Ru Yin [Four-Ingredient Elsholtzia Decoction]	Core formula + Huang Lian	Treats the same presentation as the original formula with heat, severe thirst, yellow greasy tongue coating Note: For summer heat syndrome
Xin Jia Xiang Ru Yin [Newly-Augmented Elsholtzia Decoction]	Core formula – Bian Dou + Jin Yin Hua Lian Qiao Fresh Bian Dou Hua	Summer heat invades Lung with a presentation-like cold. The pulse is flooding and big at right, but small at left; red face, thirst, and absence of sweating
Jia Wei Xiang Ru Tang [Augmented Elsholtzia Decoction]	Core formula + Qing Hao Jin Yin Hua Lian Qiao Hua Shi Gan Cao	Common cold due to summer–damp. This formula can be modified with fresh He Ye, Pei Lan, or Huo Xiang
Shi Wei Xiang Ru Yin [Ten-Ingredient Elsholtzia Decoction]	Core formula + Fu Ling Gan Cao Ren Shen Huang Qi Bai Zhu Ju Pi Mu Gua	Treats the same presentation as the original formula with Middle Jiao Qi deficiency and excessive sweating

Xiao Xu Ming Tang [Minor Prolong Life Decoction] *Qian Jin Fang*

Gui Zhi	9 g
Fu Zi	15 g
Chuan Xiong	9 g
Ma Huang	6 g
Ren Shen	6 g
Bai Shao Yao	12 g
Xing Ren	6 g
Fang Feng	9 g
Huang Qin	9 g
Fang Ji	9 g
Gan Cao	5 g
Sheng Jiang	9 g

Main Indication: Wind evils invading channels, leading to spasm and tightness of muscle, tendons, and channels; hemiplegia; and speech difficulty (slow and slurred); deviation of mouth and eyes; stiff neck and headache. This formula also treats wind–damp Bi syndrome.

Preparation: Decoction. Decoct Fu Zi first, then add all other herbs. Take the warm decoction in three doses.

Treatment Principle: Warm the channel, unblock Yang, strengthen body Zheng Qi, and expel wind.

This formula often treats stroke due to exterior wind with a presentation of hemiplegia and painful wind–damp Bi. For mouth and eyes deviation caused by wind evil invading channel, one should add Wu Gong.

This formula is contraindicated for internal Liver wind syndrome. When it is used inappropriately, it is like adding wood to fire.

Related Formula:

Formula	Herbs	Indication
Zhu Li Yin Zi [Bamboo Sap Decoction]	Core formula + Shi Gao Ling Yang Jiao (Shan Yang Jiao) Zhu Li	Wind-stroke with paralysis, absence of body ache, inability to move the four limbs, minimal consciousness, but not severe

Xiao Feng San [Eliminate Wind Powder] *He Ji Ju Fang*

Jing Jie Sui	9 g
Bo He	9 g
Qiang Huo	9 g
Fang Feng	9 g
Chuan Xiong	9 g
Jiang Can	9 g
Chan Tui (fried)	9 g
Fu Ling	9 g
Chen Pi	6 g
Hou Po	6 g
Dang Shen	20 g

Main Indication: Wind evil attacking the upper part of the body leading to dizziness and nasal congestion. Or wind–damp at the channel with numbness at skin and muscle, faint rash, and itchiness.

Preparation: Powder. Take 5 to 9 g with tea. For boil and ringworm, take powder with warm alcohol. It can also be taken as decoction.

Treatment Principle: Expel wind and scatter evil.

Related Formula:

Formula	Herbs	Indication
Xiao Feng San [Eliminate Wind Powder from *Zhang Shi Yi Tong*]	Core formula – Fu Ling + Huo Xiang	Feng Dan and itchiness

Comparison of Formulas Treating Wind–Cold Invasion

There are 14 main formulas under this category primarily composed of warm acrid releasing exterior herbs to treat wind–cold evil invading the body surface. The first few formulas all treat exterior symptoms such as chills, fever, headache, and body aches, which are the common characteristics of these formulas. The differences of the formulas are discussed below.

1. The main indications for Ma Huang Tang [Ephedra Decoction] are absence of sweating and presence of wheezing which is a result of cold evil constricting the exterior leading to Lung Qi rebellion. Thus, Ma Huang is used as a chief herb to promote sweating and calm wheezing which follows the principle of using acrid warm herbs and promoting sweating.
2. The main indications for Gui Zhi Tang [Cinnamon Twig Decoction] are fever, spontaneous sweating, and aversion to wind. This is a result of wind evil attacking Wei Qi, leading to Ying and Wei disharmony. Thus, Gui Zhi and Bai Shao Yao are used to harmonize Ying Wei. This is the main formula for exterior deficiency and exterior syndrome with spontaneous sweating.
3. Da Qing Long Tang [Major Bluegreen Dragon Decoction] treats both Ma Huang Tang [Ephedra Decoction] syndrome presentations which are wind–cold constricting surface, exterior excess without sweating, and restlessness agitation caused by interior heat. Thus, Shi Gao is added to Ma Huang Tang [Ephedra Decoction] to clear heat. This formula now not only releases exterior but also clears internal heat, demonstrating the principle of treating exterior and interior at the same time.
4. Ren Shen Bai Du San [Ginseng Powder to Overcome Pathogenic Influences] treats constitution deficient patient with cough and phlegm, and sore and painful limbs and joints. Thus, Qiang Huo and Du Huo which expel wind and dispel dampness are used together with Ren Shen, tonifying the body Zheng Qi. This demonstrates the method of tonifying Qi and releasing exterior together. The primary function of Shen Su Yin [Ginseng and Perilla Leaf Decoction] is expelling phlegm, and the secondary function is to tonify Qi and release exterior. When this formula is compared to Ren Shen Bai Du San [Ginseng Powder to Overcome Pathogenic Influences], it is seen that Shen Su Yin treats more exterior syndromes and Ren Shen Bai Du San, more interior syndromes. This is the main difference between the two.
5. The main indications of Xiang Su San [Cyperus and Perilla Leaf Powder] are poor appetite, distention, and a stifling sensation at chest and epigastrium which indicate stagnation of Qi dynamics. Thus, Zi Su Ye, which can release exterior and unblock Qi stagnation, is used as the main herb. It is accompanied by Xiang Fu which smoothes Liver and regulates Qi and Chen Pi which awakens Spleen and regulates Qi. This formula demonstrates the principles of releasing exterior and regulating Qi.
6. Ma Huang Fu Zi Xi Xin Tang [Ephedra, Prepared Aconite, and Asarum Decoction] treats cold in both exterior and interior. The main manifestations include severe aversion to cold which cannot be relieved by thick clothing, and a deep and minute pulse. Thus, the formula is composed of Ma Huang and Xi Xin to release exterior, and is assisted by Fu Zi to tonify Yang. This demonstrates the principle of tonifying Yang and releasing exterior.
7. Jiu Wei Qiang Huo Tang [Nine-Herb Decoction with Notopterygium] and Qiang Huo Sheng Shi Tang [Notopterygium Decoction to Overcome Dampness] both treat exterior evil with dampness. The manifestations include headache, sore and painful body, or heavy pain at limbs and inability to turn. Thus, Qiang Huo and Fang Feng are main herbs in the formula which correspond to the principle of expelling wind and dampness.
8. The characteristic of Xiang Ru San [Cyperus and Perilla Leaf Powder] is treating concurring symptoms of chills, fever, heavy and painful limbs, abdominal pain, and vomiting and diarrhea. This is a result of cold–damp accumulation leading to both exterior and interior pain. Thus, the formula combines Xiang Ru, Bai Bian Dou, and Hou Po, which demonstrates the principle of treating exterior (releasing exterior) and interior (expelling dampness) together.
9. The main indication of Chuan Xiong Cha Tiao San [Ligusticum Chuanxiong Powder to be Taken with Green Tea] is headache. Thus, all herbs in this formula can expel wind and stop pain. However, one should still carefully differentiate the syndrome and only use this formula when exterior syndrome is presenting.
10. The main indication for Xiao Xu Ming Tang [Minor Prolong-Life Decoction] is external wind invading the body, leading to deviation of mouth and eyes, and paralysis. Xiao Feng San [Eliminate Wind Powder] treats wind evil lodged at meridians and collaterals, causing stubborn numbness and itchiness. They are both unique in their different ways of treating wind.

WIND–HEAT INVASION

This is the beginning stage of wind–warm diseases where the evil pathogens are at the Wei level, leading to exterior heat syndromes. The warm evil pathogens attack the lungs causing fever, slight aversion to wind and cold, thirst, preference for cold drinks, cough, sore throat, red tongue tip, and floating, rapid pulse. The acrid, cool, releasing exterior herbs that are commonly used are Jin Yin Hua, Lian Qiao, Bo He, Niu Bang Zi, Sang Ye, and Ju Hua. They are often the main herbs in formulas such as Yin Qiao San [Honeysuckle and Forsythia Powder] and Sang Ju Yin [Mulberry Leaf and Chrysanthemum Decoction]. Furthermore, there are also acrid, cool, releasing-exterior formulas that are formed by acrid, warm, releasing-exterior herbs and acrid, cold, heat-clearing herbs such as in Ma Xing Shi Gan Tang [Ephedra, Apricot Kernel, Gypsum, and Licorice Decoction], Jia Wei Cang Er Zi San [Augmented Xanthium Powder], and Xiao Feng San [Eliminating Wind Powder].

For patient with Yin-deficient constitution or postpartum wind–cold transforming to heat or dryness, Yin-nourishing herbs such as Yu Zhu and Mai Men Dong, and blood-nourishing herbs such as Sheng Di Huang, and Bai Wei should be added to the acrid, cool, releasing-exterior formulas. This combination is referred to as "nourishing Yin to release exterior" and "nourishing blood to promote sweating" methods. The representing formulas are Jia Jian Wei Rui Tang [Modified Polygonatum Odoratum] and Qi Wei Cong Bai Yin [Seven-Ingredient Scallion Decoction].

Formula that vents rashes and releases exterior is also included in this category. It is composed of Sheng Ma, Ge Gen, and Xi He Liu to vent rashes and other heat clearing, releasing toxin herbs as in Zhu Ye Liu Bang Tang [Lophatherus, Willow, and Burdock Seeds Decoction]. This type of formula is often used at the beginning stage of measles where rashes are difficult to surface.

Yin Qiao San [Honeysuckle and Forsythia Powder] *Wen Bing Tiao Bian*

Jin Yin Hua	30 g
Lian Qiao	30 g
Jie Geng	6 g
Niu Bang Zi	18 g
Jing Jie Sui	18 g
Bo He	18 g
Dan Dou Chi	15 g
Dan Zhu Ye	12 g
Gan Cao	9 g
Lu Gen	30 g

Main Indication: Beginning stage of warm disease; fever but not feeling cold, or with mild chills, headache, thirst, cough, sore throat, red tongue tip, thin white or thin yellow tongue coating, and floating and rapid pulse. This belongs to Upper Jiao wind–heat.

Preparation: Decoction. To be taken four times a day.

Treatment Principle: Acrid, cool, releasing-exterior method.

This formula has wide applications. It can be used to treat common cold and flu, measles, epidemic encephalitis, encephalitis B, and other beginning stages of febrile diseases. When the patient mainly manifests as Upper Jiao wind–heat syndrome, then this formula with modifications can be prescribed. However, this formula can only be used for heat-type exterior syndrome without dampness.

Modification:

Minus	Addition	Indication
	Huo Xiang Yu Jin	Stifling sensation at the chest and epigastric region
	Tian Hua Fen	Severe thirst
	Ma Bo Xuan Shen	Swollen neck and painful throat
Jing Jie Sui Dan Dou Chi	Bai Mao Gen Ce Bai Tan Shan Zhi Zi	Epistaxis
	Xing Ren	Cough
	Sheng Di Huang Mai Men Dong	Heat condition moving toward interior
	Huang Qin Zhi Mu Shan Zhi Zi	Heat condition moving toward interior not relieved by adding Sheng Di Huang and Mai Men Dong; scanty urination

Related Formulas:

Formula	Herbs	Indication
Jia Jian Yin Qiao San [Modified Honeysuckle and Forsythia Powder]	Lian Qiao 30 g Jin Yin Hua 24 g Xuan Shen 15 g Xi Jiao 6 g (Shui Niu Jiao) Mai Men Dong 15 g Dan Zhu Ye 9 g	Heat abundance, delirium, slurred speech, restlessness, thirst, red tongue with yellow coat at the center, weak and rapid pulse Note: He Ye Juice (2 to 3 teaspoons) is added to decoction. To be taken three times a day
Jia Jian Yin Qiao San [Modified Honeysuckle and Forsythia Powder] *Guang Wen Re Lun*	Jin Yin Hua 15 g Lian Qiao 9 g Niu Bang Zi 6 g Xing Ren 3 g Xian Zhu Ye 30 pieces Mu Zei 3 g Gua Lou Pi 6 g Chuan Bei Mu 6 g Zi Cao 9 g Mu Dan Pi 6 g	Excess fire syndrome The evil pathogens are at the muscle layer causing faint rashes or spots. This formula is used in the draining Wei and venting Ying method

Sang Ju Yin [Mulberry Leaf and Chrysanthemum Decoction] *Wen Bing Tiao Bian*

Sang Ye	9 g
Ju Hua	12 g
Lian Qiao	6 g
Jie Geng	9 g
Xing Ren	9 g
Bo He	3 g
Lu Gen	15 g
Gan Cao	3 g

Main Indication: Beginning stage of wind–warm disease. Cough, slight fever, slight thirst; thin, white tongue coating, and floating, rapid pulse. This is a formula for Upper Jiao wind–heat.

Preparation: Decoction.

Treatment Principle: Acrid, cool, releasing-exterior, method; diffuses Lung and stops cough.

This formula is not only used for beginning stage of wind–warm diseases, in *Wen Bing Tiao Bian* it is also used to treat "cough due to fall dryness." It is effective in treating dry cough without phlegm. It is also used to treat upper respiratory tract infection, early stage of measles, wind rash, mild type of bronchitis, and syndromes with Upper Jiao wind–heat.

Jia Wei Cang Er Zi San [Augmented Xanthium Powder] (Jing Yan Fang)

Cang Er Zi	15 g
Xin Yi Hua	9 g
Bo He	6 g
Bai Zhi	9 g
Ju Hua	9 g
Jin Yin Hua	18 g
Lian Qiao	12 g

Main Indication: Sinusitis, thick turbid nasal discharges, frontal headache, and heat manifestations. This belongs to heat accumulation in the Lung syndrome.

Preparation: Decoction. To be taken continuously for 5 to 10 bags.

Treatment Principle: Clear and diffuse wind–heat.

Related Formulas:

Formula	Herbs	Indication
Cang Er Zi San [Xanthium Powder] (Ancient Formula)	Cang Er Zi Xin Yi Hua Bo He Bai Zhi	Cold type of sinusitis with nasal congestion, clear nasal discharge, headache or head tightness, or with wind–cold syndrome
Xiong Zhi Shi Gao Tang [Ligusticum Chuanxiong, Angelica Dahurica, and Gypsum Decoction]	Chuan Xiong 10 g Bai Zhi 10 g Shi Gao 20 g Ju Hua 10 g Gao Ben 10 g Qiang Huo 10 g	Wind–heat headache

Jia Jian Wei Rui Tang [Modified Polygonatum Odoratum Decoction] *Tong Su Shang Han Lun*

Yu Zhu (Wei Rui)	9 g
Bai Wei	6 g
Cong Bai	3 stalks
Dan Dou Chi	12 g
Bo He	6 g
Jie Geng	6 g
Zhi Gan Cao	3 g
Hong Zao	3 pieces

Main Indication: Constitution Yin deficiency with exterior evil invasion. The symptoms include headache, fever, slight aversion to wind, mild sweating or absence of sweating, dry throat, cough, restlessness, thirst, red tongue, and rapid pulse.

Preparation: Decoction. To be taken warm.

Treatment Principle: Nourish Yin and release exterior.

When the exterior syndrome is not released, it is not appropriate to use Yin nourishing herbs too early to prevent trapping the evil pathogens in the interior. However, when Jin Ye (body fluids) are deficient with exterior syndrome, herbs that promote sweating not only can relieve the exterior syndrome but may also exhaust the Yin fluids. Therefore, in order to achieve both goals, Yin nourishing herbs and sweat promoting herbs are used together. This is the main formulating principle of this formula.

Qi Wei Cong Bai Yin [Seven-Ingredient Scallion Decoction] *Wai Tai Mi Yao*

Sheng Di Huang	15 g
Mai Men Dong	12 g
Ge Gen	12 g
Cong Bai	3 stalks
Dan Dou Chi	6 g
Sheng Jiang	6 g
Water	3 cups

Main Indication: Blood deficiency with external evil invasions. This formula is for external evil invasion with headache, fever, slight chills, and absence of sweating, after a long-term illness or blood loss.

Preparation: Decoction. To be taken warm.

Treatment Principle: Nourish blood and release exterior.

This formula is used for exterior syndrome following blood loss or a long-term illness. Both Yin and blood are exhausted, leading to tidal fever, night sweats, and cough with blood. It is also used for postpartum exterior syndrome.

Xiao Feng San [Eliminating Wind Powder] *Yi Zong Jin Jian*

Jing Jie	5 g
Fang Feng	9 g
Chan Tui	9 g
Niu Bang Zi	6 g
Cang Zhu	12 g
Ku Shen	9 g
Mu Tong	12 g
Huo Ma Ren	12 g
Shi Gao	18 g
Zhi Mu	9 g
Sheng Di Huang	12 g
Dang Gui	6 g
Gan Cao	3 g

Main Indication: Wind–heat with dampness accumulate at the muscle and skin layers. This can be seen in eczema, wind rashes, itching, exudation after scratching, yellow tongue coating, and floating, rapid, forceful pulse.

Preparation: Decoction. To be taken on an empty stomach.

Treatment Principle: Expel wind and damp, clear heat, and cool blood.

Modification:

Addition	Indication
Jin Yin Hua Lian Qiao Ye Ju Hua Pu Gong Ying Tian Kui Zi	Excess heat toxins These herbs clear heat and detox
Mu Dan Pi Zi Cao	Excess blood–heat These herbs cool the blood
Yi Yi Ren Tu Fu Ling Di Fu Zi Bai Xian Pi	Severe dampness These herbs drain dampness and stop itching
Bai Hua She Quan Xie Wu Gong	Excess wind These herbs expel wind
Fu Ping	Severe itching
Ci Wei Pi Chuan Shan Jia Di Long Jiang Can	Pain due to channel obstructions These herbs open the channels and stop pain Also for severe itching due to white lichen on scalp

Related Formula:

Formula	Herbs	Indication
Pi Pa Qing Fei Yin [Loquat Leaf Decoction to Clear the Lung]	Pi Pia Ye 6 g Huang Lian 3 g Sang Bai Pi 6 g Huang Bai 3 g Gan Cao 3 g Ren Shen 1 g	Disperse wind and clear Lung Treat wind–heat attacking Lung, leading to acne and rashes

Zhu Ye Liu Bang Tang [Lophatherus, Willow, and Burdock Seeds Decoction] *Xian Xing Zhai Yi Xue Guang Bi Ji*

Xi He Liu (Cheng Liu)	15 g
Jing Jie	3 g
Ge Gen	6 g
Chan Tui	3 g
Bo He Ye	5 g
Fried Niu Bang Zi	5 g
Zhi Mu	3 g
Xuan Shen	6 g
Mai Men Dong	9 g
Gan Cao	3 g
Dan Zhu Ye	30 pieces

Main Indication: Lung Qi unable to circulate and rashes are not vented completely. It is seen with wheezing, cough, and irritability and restlessness.

Preparation: Decoction.

Treatment Principle: With mildly diffused Lung Qi, vent rashes, clear heat, and nourish Yin.

Modification:

Addition	Indication
Shi Gao 15 g	Excess heat and high fever
Mai Men Dong 30 g	with restlessness and thirst

Comparison of Formulas Treating Wind–Heat Invading Lung

Yin Qiao San [Honeysuckle and Forsythia Powder] is the representative formula in treating the beginning stage of wind–heat and wind–warmth. It focuses on dispersing wind–heat of the Upper Jiao and relieving toxicity. Sang Ju Yin [Mulberry Leaf and Chrysanthemum Decoction] treats cough due to wind–heat attacking Lung. The stop-coughing function of this formula is stronger, but dispersing wind–heat and relieving toxicity functions are weaker. Cang Er Zi San [Xanthium Powder] and Jia Wei Cang Er Zi San [Augmented Xanthium Powder] focuses on treating nasal congestion and discharge due to invasion of external pathogens. Qi Wei Cong Bai Yin [Seven-Ingredient Scallion Decoction] treats external pathogen invasion with blood-deficient constitution. Jia Jian Wei Rui Tang [Modified Polygonatum Odoratum Decoction] treats external pathogenic invasion with Yin deficiency. Zhu Ye Liu Bang Tang [Lophatherus, Willow, and Burdock Seeds Decoction] treats rash that is difficult to surface or Lung Qi failing to disperse. Xiao Feng San [Eliminate Wind Powder] treats wind rash and eczema due to wind–heat with damp accumulating at muscles and tendons.

EXTERIOR DEFICIENCY WITH SPONTANEOUS SWEATING

This is exterior deficiency syndrome due to Lung Qi deficiency and Wei Qi not consolidating the exterior leading to spontaneous sweating. Huang Qi, Dang Shen, and Bai Zhu are often used to treat the root by tonifying Qi, solidifying Wei, and consolidating the exterior. Mu Li, Ma Huang Gen, Wu Wei Zi, and Fu Xiao Mai are used to treat the manifestation by astringing and stopping the sweat. This type of herbal combination, treating both the root and manifestation, is suitable for spontaneous sweating due to exterior deficiency. However, this method is not effective in treating spontaneous sweating due to Ying and Wei disharmony.

Yu Ping Feng San [Jade Windscreen Powder] *Shi Yi De Xiao Fang*

Huang Qi	24 g
Bai Zhu	9 g
Fang Feng	9 g

Main Indication: Spontaneous sweating due to exterior deficiency. It is also good for deficient person who suffers from recurrent colds.

Preparation: Decoction.

Treatment Principle: Tonify Qi and consolidate exterior method.

Related Formula:

Formula	Herbs	Indication
Bai Zhu San [Atractylodes Powder]	Core formula + Mu Li	Especially for severe, spontaneous sweat

Mu Li San [Oyster Shell Powder] *He Ji Ju Fang*

Ma Huang Gen	30 g
Huang Qi	30 g
Mu Li	30 g
Fu Xiao Mai	30 g

Main Indication: All types of deficiency that lead to unstable exterior or sudden deficiency due to illness which lead to spontaneous sweating, worse at night, palpitation, being easily frightened, weakness, and deficiency and shortness of breath.

Preparation: Coarsely grind the first three herbs. Take 9 g of the powder and decoct with Fu Xiao Mai. Decoction to be taken warm.

Treatment Principle: Tonify Qi, consolidate exterior, astringe sweating, and anchor Yang.

Mu Li is the chief herb in this formula which astringes sweat and anchors Yang. Thus, this formula can also be used for night sweating.

Related Formula:

Formula	Herbs	Indication
Mu Li San [Oyster Shell Powder] *Zheng Zhi Zhun Sheng*	Core formula – Huang Qi + Chi Shi Zhi Nuo Mi Fen (Glutenous Rice Powder) Bing Pian	This is an external formula to astringe sweating and consolidate the exterior Note: Grind the ingredients into powder. Apply topically to sweating area

WHEEZING AND COUGH DUE TO LUNG COLD

This indicates for Lung cold and rebellious Qi manifesting as wheezing, breathlessness, cough, thin white phlegm, absence of thirst or thirst with preference of hot drinks, thin slippery tongue coating, and floating tight pulse. It can also be used for the above symptoms with headache, fever, and chills. Ma Huang, Gui Zhi, Gan Jiang, and Xi Xin are often used to warm and scatter Lung cold. Ban Xia, Hou Po, Zi Su Zi, Xing Ren, She Gan, Xuan Fu Hua, and Wu Wei Zi are used to direct the rebellious Lung Qi downward to smooth wheezing. Lung cold and rebellious Qi can also manifest as cough, thus herbs that stop cough and expel phlegm such as Zi Wan, Kuan Dong Hua, Qian Hu, and Ban Xia are often added. These herbs together with Ma Huang, Gui Zhi, Gan Jiang, and Xi Xin demonstrate the treatment principle of warming the Lung to stop cough.

Xiao Qing Long Tang [Minor Bluegreen Dragon Decoction] *Shang Han Lun*

Ma Huang	9 g
Gui Zhi	9 g
Ban Xia	12 g
Gan Jiang	9 g
Xi Xin	3 g
Wu Wei Zi	6 g
Bai Shao Yao	9 g
Gan Cao	6 g

Main Indication: Exterior wind–cold with internal accumulation of water Yin (congested fluid). Symptoms include fever, chills, absence of sweating, cough, wheezing, copious thin phlegm, moist slippery tongue coating, absence of thirst, and floating and tight pulse. This can also treat the cold type of phlegm in Yin wheezing and cough, or heavy painful body and entire swelling of skin and muscle.

Preparation: Decoction. Take the warm decoction in three doses.

Treatment Principle: Release exterior, cleanse the congested fluid, warm the Lung, and direct the rebellious Qi downward.

This formula can also be used to treat phlegm Yin accumulation in lungs with exterior cold type of chronic bronchitis and chronic wheezing and breathlessness.

Related Formula:

Formula	Herbs	Indication
Xiao Qing Long Jia Shi Gao Tang [Minor Bluegreen Dragon Decoction with Gypsum]	Core formula + Shi Gao	This formula treats patients with water accumulation below the Heart, cough, wheezing, fever, and restlessness Note: Gypsum not only treats internal heat with restlessness, but it also inhibits the sweating function of Ma Huang and reinforces promoting urination function of Ma Huang

Su Zi Jiang Qi Tang [Perilla Fruit Decoction for Directing Qi Downward]
He Ji Ju Fang

Ban Xia	9 g
Zi Su Zi	9 g
Gan Cao	3 g
Rou Gui	3 g
Qian Hu	9 g
Hou Po	12 g
Chen Pi	6 g
Dang Gui	6 g
Sheng Jiang	3 slices

Main Indication: Yang deficiency at Middle Jiao with phlegm accumulation and rebellious Qi. Symptoms often include wheezing with shortness of breath, cough with thin phlegm, throat discomfort, fullness and stifling sensation at the chest and epigastrium, poor appetite and apathy, and thin white or slippery, greasy tongue coating.

Preparation: Decoction. Take the warm decoction in three doses.

Treatment Principle: Warm and transform cold phlegm, regulate Qi, and direct the rebellious Qi downward.

This formula is only suitable for Yang deficiency with aversion-to-cold syndrome without exterior condition. This can be seen in chronic bronchitis and bronchial asthma which manifests as cough and wheezing, difficulty in breathing, and swelling in the head. The formula can also be used to treat late-stage, deficient, cold type pediatric whooping cough.

She Gan Ma Huang Tang [Belamcanda and Ephedra Decoction] *Jin Gui Yao Lue*

She Gan	9 g
Ma Huang	9 g
Zi Wan	9 g
Kuan Dong Hua	12 g
Ban Xia	12 g
Sheng Jiang	9 g
Xi Xin	6 g
Wu Wei Zi	6 g
Da Zao	9 g

Main Indication: Cold beverage accumulates internally which disturbs the dispersing and descending function of Lung Qi. Symptoms include cough, wheezing, rattling phlegm sounds in throat, and white sticky tongue coating.

Preparation: Decoction. To be taken warm.

Treatment Principle: Warm Lung, direct the rebellious Lung Qi downward, stop cough, and expel phlegm.

Hou Po Ma Huang Tang [Magnolia Bark and Ephedra Decoction] *Jin Gui Yao Lue*

Hou Po	15 g
Ma Huang	9 g
Shi Gao	24 g
Xing Ren	9 g
Ban Xia	12 g
Gan Jiang	6 g
Xi Xin	3 g
Wu Wei Zi	6 g
Fu Xiao Mai	9 g

Main Indication: The evil beverage attacks the Lung, which disturbs the dispersing and descending functions. Symptoms include wheezing, cough, fullness at chest, difficulty breathing, rattling phlegm sound in the throat, and floating pulse.

Preparation: Decoction. Take the warm decoction in three doses.

Treatment Principle: Disperse Lung and cleanse the congested fluid.

This formula is effective in treating wheezing and coughing with phlegm syndromes either due to Lung cold or Lung heat. However, in treating Lung cold with wheezing, coughing, and excess phlegm, one should increase the dosage of Gan Jiang and decrease the dosage of Shi Gao.

Comparison of Formulas Treating Wheezing and Cough Due to Lung Cold

1. Xiao Qing Long Tang [Minor Bluegreen Dragon Decoction], She Gan Ma Huang Tang [Belamcanda and Ephedra Decoction], Su Zi Jiang Qi Tang [Perilla Fruit Decoction for Directing Qi Downward], and Hou Po Ma Huang Tang [Magnolia Bark and Ephedra Decoction] can all treat syndromes of internal cold beverage accumulation, rebellious Lung Qi. However, in comparing the warming function to the Lung, Xiao Qing Long

Tang [Minor Bluegreen Dragon Decoction] is strongest, followed by Su Zi Jiang Qi Tang [Perilla Fruit Decoction for Directing Qi Downward], She Gan Ma Huang Tang [Belamcanda and Ephedra Decoction], and Hou Po Ma Huang Tang [Magnolia Bark and Ephedra Decoction], the least effective. In terms of the directing Lung Qi downward and smoothing wheezing, Xiao Qing Long Tang [Minor Bluegreen Dragon Decoction] is still the strongest, followed by She Gan Ma Huang Tang [Belamcanda and Ephedra Decoction], Hou Po Ma Huang Tang [Magnolia Bark and Ephedra Decoction], and Su Zi Jiang Qi Tang [Perilla Fruit Decoction for Directing Qi Downward], the weakest. In terms of the stopping-cough function, She Gan Ma Huang Tang [Belamcanda and Ephedra Decoction] is the best. In terms of expelling phlegm function, Su Zi Jiang Qi Tang [Perilla Fruit Decoction for Directing Qi Downward] is better than other formulas.

2. Hou Po Ma Huang Tang [Magnolia Bark and Ephedra Decoction] is not a typical formula for warming the Lung and directing the rebellious Qi downward. It is a formula combining both cold and heat herbs. It can be used to treat wheezing due to cold or due to heat, depending on the dosage of Gan Jiang and Shi Gao. In the formula, the function of Shi Gao controls the sweat-promoting function of Ma Huang, thus this formula is categorized as the method of warming the Lung and directing rebellious Qi downward.
3. The reason that Xiao Qing Long Tang [Minor Bluegreen Dragon Decoction] can also release the exterior is because of the usage of Ma Huang and Gui Zhi together. The reason that Su Zi Jiang Qi Tang [Perilla Fruit Decoction that Direct Qi Downward] can also enter the blood level is due to the presence of Dang Gui.

WHEEZING AND COUGH DUE TO LUNG HEAT

This indicates Lung heat causing wheezing and cough. The symptoms include coarse breathing, cough with yellow sticky thick phlegm, red face, spontaneous sweating, thirst with preference of cold drinks, red tongue, yellow tongue coating, and slippery and rapid pulse. The herbs that are commonly used include Shi Gao, Zhi Mu, Sang Bai Pi, Di Gu Pi, Huang Qin, Huang Lian, Jin Yin Hua, and Lian Qiao. If wheezing is the main manifestation, then Ma Huang, Xing Ren, Zi Su Zi, Ban Xia, and/or Hou Po are added to form Ma Xing Shi Gan Tang [Ephedra, Apricot Kernel, Gypsum, and Licorice Decoction], Xie Bai San [Drain the White Powder], and Ding Chuan Tang [Arrest Wheezing Decoction]. If cough is the main manifestation, then Zi Wan, Kuan Dong Hua, Ma Dou Ling, Bai Bu, Bai Qian, Bei Mu, Gua Lou, and/or Pi Pa Ye are added to form Qing Jin Hua Tan Tang [Clear the Metal and Transform Phlegm Decoction] and Qing Qi Hua Tan Wan [Clear the Qi and Transform Phlegm Pill].

Ma Xing Shi Gan Tang [Ephedra, Apricot Kernel, Gypsum, and Licorice Decoction] *Shang Han Lun*

Ma Huang	9 g
Xing Ren	9 g
Shi Gao	30 g
Gan Cao	3 g

Main Indication: Heat evils congest the Lung. Symptoms include fever, thirst, wheezing, cough, coarse breathing, absence or presence of sweating, yellow tongue coating, and slippery, rapid pulse.

Preparation: Decoction.

Treatment Principle: Diffuse Lung Qi, clear heat, direct the rebellious Qi downward, and arrest wheezing.

This formula is effective in treating pediatric frequent urination disorder with wheezing and cough due to Lung Qi failing to circulate. This is based on the theory that Lung is the upper source of water which dominates and regulates the waterways. This is also an example of treating the above for disease below.

Modification:

Addition	Indication
Xie Bai San [Drain the White Powder]	Severe wheezing
Jin Yin Hua Lian Qiao Shan Zhi Zi Huang Qin Huang Lian Yu Xing Cao	Severe heat

Related Formulas:

Formula	Herbs	Indication
Yin Qiao Ma Xing Shi Gan Tang [Honeysuckle, Forsythia, Ephedra, Apricot Kernel, Gypsum, and Licorice Decoction]	Core formula + Jin Yin Hua Lian Qiao	To treat symptoms as indicated in the main formula. However, this formula has a stronger detox function
Su Ting Ma Xing Shi Gan Tang [Perilla Fruit, Lepidii Seeds, Ephedra, Apricot Kernel, Gypsum, and Licorice Decoction]	Core formula + Zi Su Zi Ting Li Zi	To treat wheezing, cough with copious phlegm. The functions of expelling phlegm and directing the rebellious Qi downward are enhanced

Xie Bai San [Drain the White Powder]
Xiao Er Yao Zheng Zhi Jue

Di Gu Pi	12 g
Sang Bai Pi	15 g
Gan Cao	3 g
Geng Mi	9 g

Main Indication: Lung heat wheezing and cough, steaming heat at the skin and muscle, especially worse in the late afternoon; thin rapid pulse, and red tongue.

Preparation: Decoction.

Treatment Principle: Clear heat and smooth wheezing.

This formula is for the patient with cough that is clear, high pitched, and without phlegm. It is contraindicated for exterior syndrome.

Related Formulas:

Formula	Herbs	Indication
Sang Dan Xie Bai San [Mulberry Leaf, Moutan Powder to Drain the White]	Core formula – Geng Mi + Sang Ye Ju Hua Mu Dan Pi Xing Ren Bei Mu Jin Yin Hua	This formula treats warm toxins in throat Sha (eruptive disease). It also treats wheezing and cough due to Lung heat
Sang Dan Xie Bai Tang [Mulberry Leaf, Moutan Decoction to Drain the White]	Core formula + Sang Ye Zhu Ru Mu Dan Pi Chuan Bei Mu Jin Ju Da Zao	It treats Liver fire scorching Lung. Cough that leads to flank pain which leads to inability to turn the body. In severe cases, there can be hemoptysis. This is referred to as "the wood confines the sound of metal." Thus, this formula clears the Liver to protect the Lung For severe heat cases, add Shan Zhi Zi and Qing Dai

Ding Chuan Tang [Arrest Wheezing Decoction]
She Sheng Zhong Miao Fang

Bai Guo	30 pieces
Ma Huang	6 g
Zi Su Zi	9 g
Gan Cao	3 g
Kuan Dong Hua	9 g
Xing Ren	6 g
Sang Bai Pi	12 g
Huang Qin	9 g
Ban Xia	9 g

Main Indication: Wind–cold constricts the exterior, and phlegm heat accumulates internally. This is a syndrome of cold wrapping the heat. The symptoms include copious thick phlegm, cough, wheezing; yellow, greasy tongue coating, and slippery rapid pulse.

Preparation: Decoction. To be taken in three doses.

Treatment Principle: Diffuse Lung Qi, clear heat, direct the rebellious Qi downward, and transform phlegm.

This formula can be used to treat chronic bronchitis, bronchiole asthma, and cough and wheezing as a result of a common cold.

Qing Jin Hua Tan Tang [Clear the Metal and Transform Phlegm Decoction] *Tong Zhi Fang*

Huang Qin	12 g
Shan Zhi Zi	12 g
Zhi Mu	15 g
Sang Bai Pi	15 g
Gua Lou Ren	15 g
Bei Mu	9 g
Mai Men Dong	9 g
Ju Hong	9 g
Fu Ling	9 g
Jie Geng	9 g
Gan Cao	3 g

Main Indication: Heat phlegm congests the Lung. The symptoms include yellow, thick phlegm; yellow, greasy tongue coating, and soggy, rapid pulse.

Preparation: Decoction.

Treatment Principle: Clear heat and transform phlegm.

Qing Qi Hua Tan Wan [Clear the Qi and Transform Phlegm Pill] *Yi Fang Kao*

Dan Nan Xing	45 g
Zhi Ban Xia	45 g
Ju Hong	30 g
Fu Ling	30 g
Xing Ren	30 g
Gua Lou Ren	30 g
Huang Qin	30 g

Main Indication: Phlegm heat knotted internally. Symptoms include cough; yellow, thick, sticky phlegm, rapid breathing with nausea in severe cases, chest and epigastric fullness and distention, maybe with fever, maybe with fright and palpitation; insomnia, scanty dark urine, red tongue with yellow, greasy tongue coating, and slippery, rapid pulse.

Preparation: Grind the herbs into powder. Form the pills with ginger juice. Take 6 to 9 g each time. The dosage should be decreased if it is prepared as decoction.

Treatment Principle: Clear heat, transform phlegm.

Bei Mu Gua Lou San [Fritillaria and Trichosanthes Fruit Powder] *Yi Xue Xin Wu*

Bei Mu	5 g
Gua Lou	6 g
Tian Hua Fen	3 g
Fu Ling	3 g
Ju Hong	3 g
Jie Geng	3 g

Main Indication: Dryness attacks the Lung damaging Jin Ye (body fluids). Symptoms include hacking cough, unproductive phlegm, choking sensation and pain at the throat, rapid wheezing, rebellious Qi, thin white or thin yellow dry tongue coating, and slippery rapid or thin rapid pulse.

Preparation: Decoction.

Treatment Principle: Moisten the Lung, and transform phlegm.

This formula is used to treat dry cough due to fall dryness or Lung dryness, and dry cough at the late stage of warm diseases, as well as the late stage of the chronic exhaustive type of diseases.

Comparison of Formulas Treating Wheezing and Cough Due to Lung Heat

1. Ma Xing Shi Gan Tang [Ephedra, Apricot Kernel, Gypsum, and Licorice Decoction], Xie Bai San [Drain the White Powder], and Ding Chuan Tang [Arrest Wheezing Decoction] are all effective in directing the rebellious Qi downward and smoothing wheezing. However, Ma Xing Shi Gan Tang [Ephedra, Apricot Kernel, Gypsum, and Licorice Decoction] treats excess heat type of wheezing. Xie Bai San [Drain the White Powder] is indicated for deficient heat type of wheezing. Ding Chuan Tang [Arrest Wheezing Decoction] is for wheezing and cough due to phlegm heat accumulation internally which includes expelling phlegm and stop coughing herbs, in comparison to the two former formulas.
2. Qing Jin Hua Tan Tang [Clear the Metal and Transform Phlegm Decoction], Qing Qi Hua Tan Wan [Clear the Qi and Transform Phlegm Pill], and Bei Mu Gua Lou San [Fritillaria and Trichosanthes Fruit Powder] are all effective in transforming phlegm. They all stop cough and smooth wheezing by clearing heat and transforming phlegm.

UPPER JIAO HEAT TOXINS

This indicates warm–heat pathogenic evils invading Upper Jiao, leading to swollen, painful, ulcerated throat and cheeks, or heat toxins congest the Lung, leading to swelling, pus, wheezing, cough, and chest pain. Herbs that clear heat and relive toxins, such as Jin Yin Hua, Lian Qiao, Da Qing Ye, Ban Lan Gen, Yu Xing Cao, Huang Qin, Ma Bo, She Gan, and Wei Jing, are often selected to form formulas as in Jia Jian Pu Ji Xiao Du Yin [Modified Universal Benefit Decoction to Eliminate Toxin].

Jia Jian Pu Ji Xiao Du Yin [Modified Universal Benefit Decoction to Eliminate Toxin] *Wen Bing Tiao Bian*

Jin Yin Hua	30 g
Lian Qiao	30 g
Bo He	9 g
Jie Geng	30 g
Niu Bang Zi	18 g
Jiang Can	15 g
Ma Bo	12 g
Xuan Shen	30 g
Ban Lan Gen	15 g
Jing Jie Sui	9 g
Gan Cao	15 g

Main Indication: Warm toxins congest and stagnate the Upper Jiao. Symptoms include swollen painful throat, swelling in front and back of the ear, swollen cheeks, red face, or throat not painful but swollen externally, and deafness in severe cases. Tongue is red with yellow, dry coating, and rapid, wiry pulse.

Preparation: Grind into coarse powder. Take 20 to 25 g each time with fresh Wei Jing decoction. To be taken every 4 hours.

Treatment Principle: Clear heat, relieve toxins, scatter wind, and benefit the throat.

This formula is used to treat epidemic parotitis, tonsillitis, heat toxins at face. This formula is not appropriate for syndromes with dampness.

Modification:

Minus	Addition	Indication
Jing Jie Sui	Shi Gao Sheng Di Huang Huang Qin	Severe heat type
	Da Qing Ye Cao He Che	Severe toxin type

Related Formulas:

Formula	Herbs	Indication
Pu Ji Xiao Du Yin [Universal Benefit Decoction to Eliminate Toxin]	Huang Qin (wine fried) 15 g Huang Lian (wine fried) 15 g Chen Pi 12 g Lian Qiao 12 g Ban Lan Gen 20 g Ma Bo 6 g Niu Bang Zi 6 g Bo He 6 g Gan Cao 30 g Xuan Shen 30 g Jiang Can 2 g Sheng Ma 2 g Chai Hu 6 g Jie Geng 6 g	This treats Da Tou Wen, [toxins at the face, similar to mumps] with fever, chills, red swollen face, throat discomfort, dry tongue, dry throat, and floating, rapid, forceful pulse. This formula can scatter wind and evil, clear heat, and relieve toxins
Jia Jian Pu Ji Xiao Du Yin #1 [Modified Universal Benefit Decoction to Eliminate Toxin #1]	Above formula – Chen Pi Sheng Ma Chai Hu Jie Geng + Long Dan Cao Xia Ku Cao Pu Gong Ying	This treats severe heat-toxin type of parotitis If after taking the decoction, fever and pain decreases but the hardness is not gone yet, add Hai Zao and Kun Bu to be used with Gan Cao. This will attack that which is knotted and soften the hardness
Jia Jian Pu Ji Xiao Du Yin #2 [Modified Universal Benefit Decoction to Eliminate Toxin #2]	Pu Ji Xiao Du Yin [Universal Benefit Decoction to Eliminate Toxin] – Chen Pi Ma Bo Niu Bang Zi Sheng Ma Chai Hu Jie Geng + Quan Xie Ju Hua Gou Teng Pu Gong Ying Da Qing Ye	This formula can treat evil toxins attacking internally (meningitis), leading to fever, stiff neck, headache, vomiting, somnolence, unconsciousness, fright, red lips, deep red tongue (with or without tongue coating), purple finger prints in children, and thin, rapid pulse For severe nausea and vomiting, add Dai Zhe Shi, and Zhu Ru For somnolence and unconsciousness, take Zhi Bao Dan [Greatest Treasure Special Pill] at the same time For spasm, add Bai Shao Yao and Gan Cao and take with Zi Xue Dan [Purple Snow Special Pill]

Yin Qiao Ma Bo San [Honeysuckle, Forsythia, and Puffball Powder] *Wen Bing Tiao Bian*

Lian Qiao	30 g
Niu Bang Zi	18 g
Jin Yin Hua	15 g
She Gan	9 g
Ma Bo	9 g

Main Indication: Warm–dampness. Warm–heat accumulates in Upper Jiao, obstructing the throat causing painful throat.

Preparation: Pound the above ingredients into powder. Take 18 g of the powder and cook with fresh Wei Jing Tang. The decoction is ready when the water turns to a big boil.

Treatment Principle: Clear heat, transform damp, benefit the throat, and relieve toxin.

This formula can treat high fever, swollen red painful throat due to tonsillitis and scarlet fever, or skin rashes and spots, red tongue, white or yellow tongue coating, and floating rapid pulse. It can also treat suppurative stomatitis and ulcerations in tongue and mouth by combining with Dao Chi San [Guide Out the Red Powder].

Modification:

Addition	Indication
Dao Chi San [Guide Out the Red Powder]	Pustule type of stomatitis, ulceration of the mouth and tongue
Jie Geng Hua Shi	This modification is for obstructed throat without pain, which is an indication of more severe damp evil pathogen. The two herbs open above and leach below
Da Qing Ye Ban Lan Gen	Severe heat toxin

Related Formula:

Formula	Herbs	Indication
Jie Geng Gan Cao Tang [Platycodon and Licorice Decoction] *Shang Han Lun* also known as Jie Geng Tang [Platycodon Decoction]	Jie Geng 4 g Sheng Gan Cao 6 g	Clear heat, relieve toxicity, benefit the throat, and stop pain Treats sore throat due to heat in Shao Yin. This formula can also be used in lung abscess syndrome

Qing Fei Jie Du Tang [Clear the Lung and Relieve Toxicity Decoction] *Quan Guo Zhong Cao Yao Zi Liao Hui Bian*

Da Qing Ye	15 g
Ban Lan Gen	15 g
Zi Cao Gen	15 g
Cao He Che	15 g
Bai Bu	15 g
Guan Zhong	9 g
Yin Chen Hao	9 g
Jie Geng	6 g
Gan Cao	6 g
Shan Dou Gen	9 g

Main Indication: Warm evil attacks the Lung. Symptoms include fever, headache, sore throat, cough, rash, red tongue, yellow tongue coating, and floating, rapid pulse or slippery rapid pulse.

Preparation: Decoction. To be taken 1 dose every 4 hours. For children, the dose decreases to half.

Treatment Principle: Clear heat, relieve toxin, drain the Lung, and benefit throat.

This formula can treat warm–heat virus congestion in Upper Jiao type of viral infection of the respiratory tract, such as epidemic parotitis, viral pneumonia, shingles, and chicken pox.

Wei Jing Tang [Reed Decoction] *Qian Jin Fang*

Wei Jing	60–120 g
Yi Yi Ren	30 g
Dong Gua Ren	24 g
Tao Ren	9 g

Main Indication: Lung abscess. Symptoms include cough, slight fever, coughing or vomiting foul phlegm, pus, blood in severe cases, dull pain at chest, scaly skin, and slippery, rapid pulse. These belong to the syndrome of wind–heat congesting the Upper Jiao, forming blood, rotten flesh, and pus.

Preparation: Decoction. One bag per day. For severe condition, two bags per day.

Treatment Principle: Clear heat, scatter accumulation, drive out stasis, and expel pus.

Modification 1:

Minus	Addition	Indication
	Jin Yin Hua Lian Qiao Huang Qin	Excess heat
	Huang Bai Yu Xing Cao Qian Li Guang	Pus obstruction leading to Lung Bi syndrome
	He Huan Pi Bi Ji Dang Shen Huang Qi	Pus already expelled, but there is unhealed Lung damage

Modification 2:

Addition	Indication
Pu Gong Ying Zi Hua Di Ding Jin Yin Hua Lian Qiao	Lung carbuncles not yet formed These herbs can promote elimination and scattering
Jie Geng Gan Cao Zhe Bei Mu Yu Xing Cao or Jie Geng Bai San [Platycodon White Powder]	Presence of abscess These herbs promote the draining of pus and eliminating abscess Jie Geng Bai San includes Jie Geng, Zhe Bei Mu, and Ba Dou

Modification 3:

Minus	Addition	Indication
Tao Ren	Sang Bai Pi Huang Qin Zhe Bei Mu	Middle stage of measles where the rash has already surfaced, and the symptoms include fever, cough, copious phlegm, thirst, and rapid breathing which indicates severe Lung heat
	Gua Lou Pi Pi Pa Ye	Late stage of heat febrile disease in which the heat is not totally cleared, with cough with copious phlegm

Related Formula:

Formula	Herbs	Indication
Qian Jin Wei Jing Tang Jia Xing Ren Hua Shi Tang [Thousand Gold Reed Decoction with Apricot Kernel and Talcum Decoction] *Wen Bing Tiao Bian*	Core formula + Xing Ren Hua Shi	This formula treats heat–damp congesting the Lung, leading to rapid wheezing If heat is excessive, add Huang Qin If phlegm is copious with severe wheezing, add Ting Li Zi, Zi Su Zi, Ban Xia

Comparison of Formulas Treating Heat Toxins in Upper Jiao

1. The formulas focus on clearing heat and relieving toxins. The application of clearing the Lung and relieving toxins is very wide. One can modify Qing Fei Jie Du Tang [Clear the Lung and Relieve Toxin Decoction] to treat any respiratory tract infections. Jia Jian Pu Ji Xiao Du Yin [Modified Universal Benefit Decoction to Eliminate Toxin] is most effective in treating parotitis. Yin Qiao Ma Bo San [Honeysuckle, Forsythia, and Puffball Powder] is the best in treating throat complaints. Wei Jing Tang [Reed Decoction] not only can treat Lung abscess, but can also combine with herbs that clear heat and relieve toxins to treat lobar pneumonia.
2. Comparing their strength of relieving toxins, Qing Fei Jie Du Tang [Clear the Lung and Relieve Toxin Decoction] is the strongest, followed by Jia Jian Pu Ji Xiao Du Yin [Modified Universal Benefit Decoction to Eliminate Toxin], then Yin Qiao Ma Bo San [Honeysuckle, Forsythia, and Puffball Powder]; Wei Jing Tang [Reed Decoction] is the weakest.

COUGH AND WHEEZING DUE TO LUNG DEFICIENCY

There are two types of cough and wheezing due to Lung deficiency. They are Qi deficiency and Yin deficiency. Lung Qi deficiency type often presents with chronic cough, low voice, shortness of breath, apathy, scanty speech, wheezing with cough, spontaneous sweating, and weak constitution. Herbs that tonify Lung Qi and stop cough, such as Ren Shen, Fu Ling, Ge Jie, Zi Wan, Kuan Dong Hua, and Dong Chong Xia Cao, are used in Ren Shen Ge Jie San. Yin deficiency type is often Lung Yin deficiency with Liver and Kidney Yin deficiency. Symptoms include emaciation, apathy, malar flush, tidal fever, night sweats, insomnia, excessive dreaming, spermatorrhea in male and irregular menstruation in female, red tongue, and thin rapid pulse. Other symptoms are cough with sticky phlegm or phlegm with tinge of blood, and chest pain. Herbs from different categories are combined in the formula: Yin nourishing herbs, such as Sheng Di Huang, E Jiao, Feng Mi (honey), Shan Yao, Tian Men Dong, Mai Men Dong, Bei Sha Shen, Shi Hu, and Yu Zhu; heat clearing herbs such as Mu Dan Pi, Di Gu Pi, Zhi Mu, Xuan Shen, and Bai Wei; stop cough, transform phlegm herbs, such as Bai Bu, Bai He, Zi Wan, Kuan Dong Hua, Bei Mu, Xing Ren, and Hai Ge Ke; and stop bleeding herbs such as Bai Ji, San Qi, Xue Yu Tan, Qian Cao Gen, Han Lian Cao, and Xian He Cao. These formulas include Yue Hua Wan [Moonlight Pill] and Qing Hao Bie Jia Tang [Artemisia Annua and Soft-Shelled Turtle Shell Decoction]. For both Qi and Yin deficiencies, one can use Sheng Mai San [Generate the Pulse Powder] to tonify Qi and nourish Yin.

Ren Shen Ge Jie San [Ginseng and Gecko Powder] *Wei Sheng Bao Jian*

Ge Jie	1 pair
Xing Ren	150 g
Zhi Gan Cao	150 g
Ren Shen	60 g
Fu Ling	60 g
Bei Mu	60 g
Sang Bai Pi	60 g
Zhi Mu	60 g

Main Indication: Lung deficiency with heat. Symptoms often include chronic cough, wheezing, breathlessness, cough up saliva or white pus, restlessness and heat at chest, body emaciation, and floating and deficient pulse.

Preparation: Grind the above ingredients into fine powder. 2 to 3 g each time and twice a day.

Treatment Principle: Tonify the deficiency, clear heat, transform phlegm, and stop wheezing.

This formula originally treated "highly contagious Lung tuberculosis," and it uses Ge Jie as the chief herb to treat long-term illness with cough and wheezing due to Lung Qi deficiency. It can also arrest wheezing and stop cough which demonstrate the ability in treating both the root and the branch at the same time. In clinical practice, it can be prescribed for chronic bronchitis and chronic asthma, and for the resting and remittence period of tuberculosis.

Bu Fei E Jiao Tang [Tonify the Lungs Decoction with Ass-Hide Gelatin] (Originally called E Jiao San) *Xiao Er Yao Zheng Zhi Jue*

E Jiao	9 g
Ma Dou Ling	6 g
Niu Bang Zi	6 g
Gan Cao	3 g
Xing Ren	9 g
Nuo Mi (Glutenous Rice)	15 g

Main Indication: Lung Yin deficiency. Symptoms include coarse breathing, rapid wheezing, dry throat, cough with scanty phlegm, or phlegm with tinge of blood; red tongue, scanty coat, and floating rapid pulse.

Preparation: Decoction.

Treatment Principle: Nourish Yin, tonify Qi, clear heat, and stop cough.

Modification:

Addition	Indication
Qing Dai Shan Zhi Zi	When there is blood in phlegm and severe heat, this indicates Liver fire attacking the Lung. Thus add these two herbs to clear Liver and calm the Lung

Sheng Mai San [Generate the Pulse Powder] *Qian Jin Yao Fang*

Ren Shen	6 g
Mai Men Dong	9 g
Wu Wei Zi	3 g

Main Indication: Qi and Yin both deficient. Symptoms include Qi and Yin deficiency, leading to fatigue, shortness of breath, too lazy to talk, dizziness, vertigo, apathy, dry mouth, thirst, and deficient and rapid pulse. Or it can treat Lung deficiency type of chronic cough, cough with scanty phlegm, shortness of breath, spontaneous sweating, dry throat, dry tongue, and deficient pulse.

Preparation: Decoction.

Treatment Principle: Tonify Qi and generate fluids.

Modification:

Minus	Addition	Indication
Ren Shen	Xi Yang Shen	For severe fluid damage Ren Shen is used for severe Qi damage
	Huang Qi Dang Gui	For excessive sweating and scanty hesitant urination. Promote urinating herbs are contraindicated
Wu Wei Zi	Large dosage of Zi Wan	Only right cun pulse is rapid with dribbling urination. This is an indication of Lung Yin deficiency. This modification will increase the Qi transformation function

Related Formulas:

Formula	Herbs	Indication
Wu Wei Mai Men Dong Tang [Five-Ingredient Ophiopogonis Decoction] *Wai Tai Mi Yao*	Shi Gao Gan Cao	This formula treats restlessness, thirst, and preference of drinks due to residual internal heat from Qi and Yin, both deficient This formula can also treat syndromes of exhaustion of Jin Ye (body fluids) with presence of heat, deficient rapid pulse, dry mouth, sweating, or cough, and other rebellious symptoms
Jia Jian Sheng Mai San [Modified Generate the Pulse Powder] *Wen Bing Tiao Bian*	Bei Sha Shen Mai Men Dong Wu Wei Zi Mu Dan Pi Sheng Di Huang	The formula is prepared as a decoction and divided into two doses Treats hidden summer heat in Tai Yin, red tongue, thirst, and excessive sweating This formula is sour and sweet to transform Yin. It is stronger in nourishing Yin and clearing heat function When Bei Sha Shen is substituted with Ren Shen, then it treats fatigue and weakness at late stage of febrile disease with symptoms that worsen at night, deep red tongue, and scanty tongue coating

Yue Hua Wan [Moonlight Pill] *Yi Xue Xin Wu*

Tian Men Dong	30 g
Mai Men Dong	30 g
Sheng Di Huang	60 g
Shu Di	60 g
Shan Yao	60 g
Bai Bu	60 g
Bei Sha Shen	60 g
Fu Ling	60 g
E Jiao	45 g
Chuan Bei Mu	21 g
San Qi	15 g
Ta Gan	30 g

Main Indication: Lung and Kidney Yin deficiency. Symptoms include dry cough with scanty phlegm or phlegm with tinge of blood, or with tidal fever, burning palms and feet, chest pain, dry throat, dry mouth, red tongue tip and sides, and thin, rapid pulse.

Preparation: Decoct Sang Ye into thick juice, then dissolve E Jiao into the decoction. Grind the above herbs into powder and add the decoction and honey to make pills with a diameter of 1.5 cm. Taken one pill three times a day.

Treatment Principle: Nourish Yin, moisten the Lung, stop cough, and stop bleeding.

In *Zhou Hou Fang*, Ta Gan San [Otter Liver Powder] is composed of only Otter Liver to treat cold consumptive disease (chronic tuberculosis). Bai Bu is also a herb to treat highly contagious consumptive disease with steaming bone.

Comparison of Formulas Treating Cough and Wheezing Due to Lung Deficiency

1. Ren Shen Ge Jie San [Ginseng and Gecko Decoction] treats Lung Qi deficiency, thus ginseng and gecko are chief herbs to tonify Lung Qi. Sheng Mai San [Generate the Pulse Decoction] treats both Qi and Yin deficiency, and so use Ren Shen and Mai Men Dong to tonify Qi and Yin. Yue Hua Wan [Moonlight Pill] treats Lung and Kidney Yin deficiency, so use Tian Men Dong, Mai Men Dong, Sheng Di Huang, Shu Di, sweating, and cold herbs to nourish Yin and regulate both Kidney and Lung.
2. The main indication for Ren Shen Ge Jie San [Ginseng and Gecko Decoction] is wheezing and rebellious Qi; thus, Ge Jie, Xing Ren, and Sang Bai Pi are used to direct the Qi downward and smooth wheezing. The main indication for Yue Hua Wan [Moonlight Pill] is coughing up phlegm with blood, thus Bai Bu and Ta Gan are used to stop cough. The main indications for Sheng Mai San [Generate the Pulse Decoction] are shortness of breath, spontaneous sweating, dry mouth, and thirst, thus Ren Shen and Mai Men Dong are used to tonify Qi, and Yin and Wu Wei Zi are used to astringe Lung Qi.
3. Sheng Mai San [Generate the Pulse Decoction] not only treats both Lung Qi and Yin deficiency, but it is also a formula strengthening the Heart to prevent collapse. Yue Hua Wan [Moonlight Pill] contains Ta Gan and Bai Bu, thus it can be used to treat Lung consumptive disease.

CHRONIC COUGH DUE TO LUNG DEFICIENCY

This indicates chronic cough leading to scattered and exhausted Lung Qi. Wu Wei Zi, Wu Mei, and Ying Su Ke are often used as chief herbs, assisted by Kuan Dong Hua and Bei Mu to stop cough and transform phlegm. The representing formulas are Jiu Xian San [Nine-Immortal Powder], Wu Wei Zi Tang [Schisandra Decoction], and Bu Fei Tang [Tonify the Lung Decoction]. If the symptoms include chronic cough with coughing up blood, this is an indication of Lung Yin deficiency. Thus, herbs such as Bai Ji, E Jiao, Wu Wei Zi, and Hai Ge Ke Powder are used to astringe the lung and stop bleeding. It is then assisted by Bai He, Bai Bu, and Bei Mu to moisten the Lung, stop cough, and transform phlegm. This is Bai Ji Pi Pa Wan [Bletilla and Loquat Leaf Pill].

This treatment is only appropriate for chronic cough due to Lung deficiency without exterior syndromes. If these formulas are used with the presence of exterior symptoms, then the exterior pathogenic factors may be trapped inside.

Jiu Xian San [Nine-Immortal Powder] *Yi Xue Zheng Zhuan*

Ren Shen	9 g
E Jiao	12 g
Kuan Dong Hua	12 g
Jie Geng	9 g
Sang Bai Pi	15 g
Bei Mu	9 g
Wu Wei Zi	9 g
Wu Mei	15 g
Ying Su Ke	9 g

Main Indication: Lung Qi deficiency and inability to astringe. Symptoms include chronic cough, wheezing, and spontaneous sweat with severe cough and deficient and rapid pulse.

Preparation: Powder the above ingredients. Take 10 g of powder three times a day with warm water.

Treatment Principle: Astringe the Lung and stop cough.

This formula is used for chronic cough, exhaustion of Qi and Yin, cough, wheezing, and spontaneous sweating. It is not appropriate for excess phlegm, damp accumulation, or with the presence of exterior pathogenic factors.

Wu Wei Zi Tang [Schisandra Decoction] *Lei Zheng Huo Ren Shu*

Ren Shen	6 g
Mai Men Dong	9 g
Wu Wei Zi	9 g
Xing Ren	6 g
Ju Pi	6 g

Main Indication: Lung Qi and Yin both deficient. Symptoms include cough with scanty phlegm, wheezing leading to spontaneous sweat, dry mouth, dry tongue, and deficient and rapid pulse.

Preparation: Decoct with ginger and red date. Take the decoction in two doses.

Treatment Principle: Benefit the Qi and generate fluid, astringe Lung, and stop cough.

This formula astringes (Wu Wei Zi), disperses (Xing Ren), tonifies (Ren Shen and Mai Men Dong), and also moves (Ju Pi). It is a formula that treats both the root and the branch with its graceful formulation.

Bu Fei Tang [Tonify the Lung Decoction] *Yong Lei Ling Fang*

Ren Shen	9 g
Huang Qi	24 g
Shu Di	24 g
Wu Wei Zi	6 g
Zi Wan	9 g
Sang Bai Pi	12 g

Main Indication: Lung Qi deficiency. Symptoms include shortness of breath, spontaneous sweating, sometimes hot and sometimes cold, cough, wheezing, pale tongue, and weak pulse.

Preparation: Decoction. If the formula is made into pills, dosage should be increased.

Treatment Principle: Tonify Qi and stabilize the exterior.

This formula not only tonifies the Lung Qi above, but can also tonify the Kidney and essence below. Thus it can also treat syndromes of deficient Lung and Kidney, with Kidney not grasping the Qi.

Bai Ji Pi Pa Wan [Bletilla and Loquat Leaf Pill] *Zheng Zhi Zhun Sheng*

Bai Ji	30 g
E Jiao (fried with Hai Ge Ke powder)	15 g
Ou Jie	15 g
Pi Pa Ye	15 g
Sheng Di Huang (fresh juice)	30 g

Main Indication: Liver and Kidney Yin deficiency with deficient fire attacking Lung. Symptoms include cough with blood or phlegm with tinge of blood; tidal fever, night sweat, red tongue with scanty coating, and rapid, thin pulse.

Preparation: Grind Bai Ji and Pi Pa Ye into powder. Decoct the rest of the herbs, then mix with the powder to form pills. Take 5 to 15 g each time, two to three times a day. This can also be prepared as decoction.

Treatment Principle: Nourish Yin, clear heat, astringe Lung, and stop bleeding.

Modification:

Addition	Indication
Qing Dai	Severe heat
Shan Zhi Zi Huang Qin	This modification clears more Liver heat and is more effective in stopping bleeding

Related Formula:

Formula	Herbs	Indication
Bai He Pian [Lily Bulb Tablet]	Bai He Bai Ji Bai Bu Hai Ge Ke powder Qing Dai	This treats Lung Yin deficiency with deficient heat as in bronchiectasis with bleeding and hemoptysis in Lung tuberculosis

LUNG QI UNABLE TO CIRCULATE

This indicates cough, wheezing, and breathlessness from Lung Qi's being unable to circulate due to various reasons. As described in previous methods, when there is cold restricting the exterior, then Xin Jia San Ao Tang [Newly-Augmented Three-Unbinding Decoction] is added to diffuse Lung Qi, release the exterior, and scatter the cold. When there is heat accumulation in the Lung, then Xin Jia Xiang Ru Yin [Newly-Augmented Elsholtzia Decoction] is added to circulate the stagnation and scatter heat. When there is wheezing and coughing due to rebellious Lung Qi, then Ma Xing Shi Gan Tang [Ephedra, Apricot Kernel, Licorice, and Gypsum Decoction] is added to diffuse Lung Qi and direct the rebellious Qi downward.

The treatment focuses on cough due to stagnant Lung Qi with phlegm damp congestion. Thus, it is suitable to diffuse Lung Qi and transform phlegm mainly with Jie Geng, Xing Ren, Pi Pa Ye, Sang Ye, Jing Jie, and Bo He to open and diffuse Lung Qi. It is often combined with Qian Hu, Zi Wan, Kuan Dong Hua, Bai Bu, Bai Qian, Ban Xia, and other phlegm transforming and stopping cough herbs to form Zhi Sou San [Stop Coughing Powder] and Xing Su San [Apricot Kernel and Perilla Leaf Powder].

Furthermore, since Lung and Large Intestine are internally and externally related, addition of Jie Geng, Xing Ren, and other herbs that open and diffuse Lung Qi can treat constipation. This is the method that is referred to as "lift the pot and open the lid" which is based on Lung Qi opening the Large Intestinal Qi and regulating the bowel. This is another method of treating constipation.

Zhi Sou San [Stop Coughing Powder] *Yi Xue Xin Wu*

Jie Geng	10 g
Jing Jie	10 g
Zi Wan	10 g
Bai Bu	10 g
Bai Qian	10 g
Chen Pi	6 g
Gan Cao	3 g

Main Indication: Lung Qi unable to circulate due to external pathogenic factors. Symptoms include cough, unproductive cough, and floating and moderate pulse.

Preparation: Decoction.

Treatment Principle: Diffuse Lung Qi and stop coughing.

This formula treats cough due to external pathogenic invasion. It is effective in treating excess syndrome.

Modification:

Addition	Indication
Fang Feng Zi Su Ye Sheng Jiang	More symptoms of cold. These herbs help to scatter cold
Sang Ye Ju Hua Niu Bang Zi Bo He Huang Qin Lu Gen	More symptoms of heat These herbs can clear heat and stop cough
Ban Xia Fu Ling Zi Su Zi	Phlegm damp obstructing the middle manifesting as cough with copious sputum. Herbs are added to expel phlegm and stop cough

Xing Su San [Apricot Kernel and Perilla Leaf Decoction] *Wen Bing Tiao Bian*

Xing Ren	10 g
Zi Su Ye	6 g
Ban Xia	10 g
Ju Pi	6 g
Fu Ling	12 g
Gan Cao	2 g
Qian Hu	6 g
Jie Geng	6 g
Zhi Ke	6 g
Sheng Jiang	6 g
Da Zao (without pit)	3 pieces

Main Indication: Cool and dryness attack the Lung, leading to mild headache, chills, aversion to cold, absence of sweating, nasal congestion, dry throat, cough with watery phlegm, and floating wiry pulse.

Preparation: Decoction.

Treatment Principle: Diffuse Lung Qi and transform phlegm.

This formula treats mild exterior cool dryness. It follows the treatment principle from *Huang Di Nei Jing*, "to treat dryness invading the interior, one should use bitter and warm (herbs) and assist with sweet and acrid (herbs)." This formula and Sang Ju Yin both treat cough due to exterior syndrome. The former is warm and the latter is cool.

WARM DRYNESS DAMAGING LUNG

This indicates exterior heat dryness causing dry cough without phlegm or with scanty phlegm that is hard to expel; nasal dryness, dry mouth, sore throat, chest pain with severe cough, red tongue tip, thin yellow tongue coating, and small, rapid pulse. Zhi Mu, Lu Gen, Shi Gao, Sang Ye, Pi Pa Ye, and Xing Ren are appropriate herbs to clear and disperse the dryness and heat. Bei Sha Shen, Mai Men Dong, Tian Men Dong, and Yu Zhu can moisten the Lung and generate fluids.

Sang Xing Tang [Mulberry Leaf and Apricot Kernel Decoction]
Wen Bing Tiao Bian

Sang Ye	15 g
Xing Ren	10 g
Bei Sha Shen	15 g
Zhe Bei Mu	10 g
Dan Dou Chi	10 g
Shan Zhi Zi	12 g
Li Pi (Asian Pear Peel)	30 g

Main Indication: Warm dryness damages Lung. Symptoms include headache, fever, thirst, dry cough without phlegm, red tongue, thin white dry tongue coating, and rapid, big pulse on the right side.

Preparation: Decoction.

Treatment Principle: Clear, circulate, and moisten the dryness.

This method can be used for fever and dry cough without phlegm in upper respiratory infections.

Modification:

Minus	Addition	Indication
Dan Dou Chi	Qing Dai Gua Lou Pi He Zi	This treats hemoptysis due to bronchiectasis
	Shan Zhi Zi (heavy dose) Qing Dai Qing Hao Mu Dan Pi	This modification is used when there is nasal bleeding. These herbs clear Liver and cool blood
	Niu Ban Zi Xuan Shen	For dry and painful throat. These two herbs benefit the throat and clears heat
	Gua Lou Pi	Cough with yellow sticky phlegm. Gua Lou Pi clears heat and transforms phlegm

Sha Shen Mai Dong Tang [Glenhnia and Ophiopogonis Decoction] *Wen Bing Tiao Bian*

Bei Sha Shen	10 g
Mai Men Dong	10 g
Yu Zhu	6 g
Tian Hua Fen	6 g
Sheng Bian Dou	5 g
Sheng Gan Cao	3 g
Dong Sang Ye	6 g

Main Indication: Dryness damages Lung Yin. Symptoms include dry throat, thirst, fever, or dry cough without phlegm, dry tongue coating and lack of fluids, and thin pulse at cun position.

Preparation: Decoction. Take the decoction in two doses.

Treatment Principle: Sweet cold generates fluids.

This formula originally treats Lung Yin deficiency. However, dry throat and thirst are also symptoms of Stomach Yin deficiency, thus this formula is good for fluid damages in both Lung and Stomach.

Related Formula:

Formula	Herbs	Indication
Ye Shi Yi Wei Tang [Ye's Benefit the Stomach Decoction]	Core formula – Tian Hua Fen	Same indication as the main formula

Qing Zao Jiu Fei Tang [Eliminate Dryness and Rescue the Lungs Decoction] *Yi Men Fa Lu*

Shi Gao	15 g
Dong Sang Ye	10 g
Xing Ren	10 g
Pi Pa Ye (without the fine hair)	3 pieces
Ren Shen	10 g
Gan Cao	3 g
E Jiao (Melted)	6 g
Mai Men Dong	6 g
Huo Ma Ren (Fried, ground)	3 g

Main Indication: Warm dryness damages Lung causing both Qi and Yin deficient. Symptoms include headache, fever, dry cough without phlegm, rebellious Qi and wheezing, dry throat, dry nose, fullness at chest, hypochondriac pain, restlessness, thirst, dry tongue without coating, and thin rapid pulse.

Preparation: Decoction. Taken warm and frequently.

Treatment Principle: Clear dryness and moisten the Lungs.

This is the main formula to treat when warm dryness damages Lung and damages both Qi and Yin with evil heat still presenting.

Modification:

Addition	Indication
Chuan Bei Mu Sheng Gua Lou	Excess phlegm
Sheng Di Huang	Blood dryness
Xi Jiao (Shui Niu Jiao) Ling Yang Jiao (Shan Yang Jiao) Niu Huang	Excess Heart and Liver fire upward, blazing, and attack the Lung

Related Formula:

Formula	Herbs	Indication
Jia Jian Yu Shi Qing Zao Jiu Fei Tang [Modified Yu's Eliminate Dryness and Rescue the Lung Decoction]	Core formula – Huo Ma Ren E Jiao + Jin Yin Hua juice Jin Zhi (purified feces)	This formula treats the beginning stage of diphtheria with headache, fever, and chills; slight rapid and choppy of the right cun pulse, dry throat without phlegm, white spots at the throat, presence or absence of cough, but with a stuck sensation at the throat and normal eating

Yang Yin Qing Fei Tang [Nourish Yin and Clear the Lungs Decoction] *Chong Lou Yu Yao*

Sheng Di Huang	30 g
Xuan Shen	24 g
Mai Men Dong	20 g
Sheng Gan Cao	6 g
Bei Mu	10 g
Bai Shao Yao (Fried)	12 g
Mu Dan Pi	12 g
Bo He	6 g

Main Indication: Lung and Kidney Yin deficiency with contagious toxin infection. Symptoms include painful throat, appearance of white curdy deposits in the throat which are difficult to scrape off, fever at the onset, dry nose and dry lips, heavy breathing which sounds like wheezing but is not wheezing, red tongue, scanty coating on tongue, and floating, rapid pulse.

Preparation: Decoction. One bag of herbs per day. In severe cases, two bags per day.

Treatment Principle: Nourish Yin, clear heat, cool blood, and relieve toxin.

This is an effective formula to treat beginning stage of diphtheria. If this formula is taken early enough, the mucous membrane can be cleared in 1 to 2 days. The more severe case will take five to six bags. After the disappearance of the white membrane, the patient should continue to take one to two bags of herbs to prevent the recurrence. This formula can also be used for Lung Yin deficiency causing general throat complaints.

Modification:

Addition	Indication
Sang Ye Qing Hao Ge Gen	With fever and chills
Shan Zhi Zi (higher dose) Huang Qin Long Dan Cao Di Gu Pi	Severe internal heat

Jin Yin Hua Lian Qiao Pu Gong Ying Ban Lan Gen Zi Hua Di Ding Shi Gao Zhi Mu	Excess toxins
Gua Lou Zhu Ru	Copious phlegm
Da Huang Mang Xiao	Constipation

Related Formula:

Formula	Herbs	Indication
Kang Bai Hou He Ji [Anti-Diphtheria Mixture]	Sheng Di Huang 30 g Mai Men Dong 9 g Xuan Shen 9 g Huang Qin 18 g Lian Qiao 18 g	This treats mild type of diphtheria, laryngitis, pharyngitis, and tonsillitis

Comparison of Formulas that Treat Warm Dryness Damaging Lung

1. The strengths of the formulas to clear dryness and heat, from the strongest to weakest, are as follows: Kang Bai Hou He Ji [Anti-Diphtheria Mixture], Yang Yin Qing Fei Tang [Nourish the Yin and Clear the Lungs Decoction], Qing Zao Jiu Fei Tang [Eliminate Dryness and Rescue the Lung Decoction], Sha Shen Mai Dong Tang [Glenhnia and Ophiopogonis Decoction], and Sang Xing Tang [Mulberry Leaf and Apricot Kernel Decoction].
2. The moisten dryness and generating fluid strengths of the formula from the strongest to the weakest are Yang Yin Qing Fei Tang [Nourish the Yin and Clear the Lungs Decoction], Kang Bai Hou He Ji [Anti-Diphtheria Mixture], Qing Zao Jiu Fei Tang [Eliminate Dryness and Rescue the Lung Decoction], Sha Shen Mai Dong Tang [Glenhnia and Ophiopogonis Decoction], and Sang Xing Tang [Mulberry Leaf and Apricot Kernel Decoction].
3. Sang Xing Tang [Mulberry Leaf and Apricot Kernel Decoction], Sha Shen Mai Dong Tang [Glenhnia and Ophiopogonis Decoction], and Qing Zao Jiu Fei Tang [Eliminate Dryness and Rescue the Lung Decoction] treat dryness damage to both Qi level and Lung fluids. Thus, Dong Sang Ye are used in all of these formulas to moisten the Lung and direct the rebellious Qi downward. It is assisted by Xing Ren and Shan Zhi Zi to recover the dispersing and descending function of the Lung. The other two formulas, Kang Bai Hou He Ji [Anti-Diphtheria Mixture] and Yang Yin Qing Fei Tang [Nourish the Yin and Clear the Lungs Decoction], treat dryness and heat damage to both Ying and blood, thus Sheng Di Huang, Xuan Shen, and Mu Dan Pi are used to clear heat in the Ying level and nourish Yin and blood.
4. Sang Xing Tang [Mulberry Leaf and Apricot Kernel Decoction], Sha Shen Mai Men Dong Tang [Glenhnia and Ophiopogonis Decoction], and Qing Zao Jiu Fei Tang [Eliminate Dryness and Rescue the Lung Decoction] do not have a relieving toxins function. They treat general dryness and heat damaging the Lung. The other two formulas have a relieving toxins function so they can be used to treat diphtheria.

PHLEGM AND WATER CONGESTING THE LUNGS

This indicates Lung Qi unable to circulate, Jin Ye unable to distribute, Upper Jiao congestion which leads to phlegm, and water obstruction with the symptoms of wheezing, inability to lie flat, and chest and hypochondriac pain. It also treats Lung Qi failing to descend and inability to open the waterway, thus there is difficult and hesitant urination or even edema. Commonly used herbs include Jie Geng, Xing Ren, Pi Pa Ye, Ma Huang, and other acrid herbs to open the Lung and circulate Lung Qi; thus, the movement of Qi leads to transformation of dampness. Furthermore, Ting Li Zi, Sang Bai Pi, Mu Tong, Fu Ling, and Wei Jing are used to open the waterway and regulate the San Jiao. Together, these form the method of bitter acrid to unblock and direct Lung Qi downward, open the source, and guide the flow. Thus, diseases can be cured by clearing the flow of water.

Yue Bi Tang [Maidservant from Yue Decoction] *Jin Gui Yao Lue*

Ma Huang	15 g
Shi Gao	50 g
Sheng Jiang	9 g
Da Zao	10 g
Gan Cao	6 g

Main Indication: Feng Shui (wind edema) due to Lung being unable to disperse and descend. Symptoms include aversion to wind, whole body swelling, absence of thirst, presence of continuous spontaneous sweat, absence of big fever, and floating and tight pulse.

Preparation: Decoction. Taken warm in three doses. If patient also complains of aversion to wind, add one whole Zhi Fu Zi.

Treatment Principle: Circulate Lung Qi and promote water.

This formula uses Ma Huang in a heavy dose to disperse and direct Lung Qi downward, and to open and adjust the waterway to reduce swelling. However, the presenting symptoms also include spontaneous sweating; thus, Shi Gao is used to counteract the promoting sweat function of Ma Huang. In the original text, the two symptoms "presence of continuous spontaneous sweat, without big fever" indicate that the main function of Ma Huang in this formula is not promoting sweat, and the main function of Shi Gao in this formula is not clearing heat.

Yue Bi Jia Ban Xia Tang [Maidservant from Yue Decoction plus Pinellia] *Jin Gui Yao Lue*

Ma Huang	10 g
Shi Gao	30 g
Sheng Jiang	6 g
Da Zao	12 g
Gan Cao	6 g
Ban Xia	15 g

Main Indication: Exterior cold with interior Yin syndrome, Lung Qi unable to circulate. Symptoms include cough, wheezing, protruding of the eyes, and floating, big pulse.

Preparation: Decoction. Taken warm in three doses.

Treatment Principle: Disperse Lung Qi and cleanse Yin (congested fluid).

The original dose of Ma Huang in this formula is 18 g which is larger than all other stop wheezing formulas. Thus, the key point of this formula is to treat wheezing with a high dose of Ma Huang to direct the rebellious Qi downward and stop wheezing. However, the dosage of Shi Gao needs to be large to counteract the promoting sweat function of Ma Huang.

Yue Bi Jia Zhu Tang [Maidservant from Yue Decoction plus Atractylodes] *Jin Gui Yao Lue*

Ma Huang	10 g
Shi Gao	40 g
Sheng Jiang	10 g
Gan Cao	3 g
Da Zao	12 g
Bai Zhu	10 g

Main Indication: Edema due to Lung unable to disperse and descend. Early symptoms include fever, cough, scanty urination, puffiness, and swelling which is more commonly seen on face, eyelids, and lower limbs. Pulse is minute, floating, and tight.

Preparation: Decoction.

Treatment Principle: Disperse Lung Qi and promote urination.

This formula is used to treat the beginning stage of acute glomerular nephritis which promotes sweating, urination, and decreases swelling.

Referring to various formulas of Zhang Zhong Jing, other formulas beside this one that are composed of Ma Huang and Shi Gao include Ma Xing Shi Gan Tang [Ephedra, Apricot Kernel, Gypsum, and Licorice Decoction], Yue Bi Tang [Maidservant from Yue Decoction], Yue Bi Jia Ban Xia Tang [Maidservant from Yue Decoction plus Pinellia], Yue Bi Jia Zhu Tang [Maidservant from Yue Decoction plus Atractylodes], Hou Po Ma Huang Tang [Magnolia Bark with Ephedra Decoction], Da Qing Long Tang [Major Bluegreen Dragon Decoction], Xiao Qing Long Jia Shi Gao Tang [Minor Bluegreen Dragon with Gypsum Decoction], and Gui Zhi Er Yue Bi Yi Tang [Two Parts of Cinnamon Twig Decoction plus One Part of Maidservant from Yue Decoction]. Among these formulas, some use Ma Huang to promote sweating and release exterior, and some use Ma Huang to disperse and direct Lung Qi downward and stop wheezing and coughing. Some use Ma Huang to disperse Lung Qi and promote urination to treat Yin evil and edema. The combination of Ma Huang and Shi Gao has two functions. One is to fully utilize the dispersing, descending, and promoting urination functions of Ma Huang without promoting sweating. Thus, Shi Gao is added to counteract this promoting sweat function, such as in Ma Xing Shi Gan Tang [Ephedra, Apricot Kernel, Gypsum, and Licorice Decoction], Yue Bi Tang [Maidservant from Yue Decoction], Yue Bi Jia Ban Xia Tang [Maidservant from Yue with Pinellia Decoction], and Yue Bi Jia Zhu Tang [Maidservant from Yue Decoction plus Atractylodes]. Another purpose is to promote sweating and release exterior with Ma Huang and clear heat with Shi Gao. In this case, the dosage of Shi Gao is reduced or combined with Gui Zhi which can enhance the promoting sweat function without counteracting by Shi Gao, such as in Da Qing Long Tang [Major Bluegreen Dragon Decoction] and Gui Zhi Er Yue Bi Yi Tang [Two Parts of Cinnamon Twig Decoction plus One Part of Maidservant from Yue Decoction].

All of the above formulas can promote sweating, release exterior, and promote urination to treat edema. The key is the dosage ratio of Ma Huang and Shi Gao. If the dosage of Shi Gao is three times bigger than Ma Huang, then the formula can only disperse Lung Qi and promote urination but not promote sweating. If the dosage of Shi Gao is not bigger than three times Ma Huang, then the formula can promote sweating, release exterior, and also promote urination to reduce swelling.

Related Formulas:

Formula	Herbs	Indication
Yue Bi Jia Zhu Fu Tang [Maidservant from Yue Decoction plus Atractylodes and Prepared Aconite]	Core formula + Fu Zi	Treats beriberi, whole body swelling and fullness, and difficulty urinating, or aversion to wind, numbness, and lack of sensation in the feet
He Xiong Huang San [Combined Ligusticum and Rhubarb Powder]	Core formula + Chuan Xiong Da Huang	This formula treats burning, swelling, painful eye balls; swelling eyelids or ulcerated sty; itching pain with light sensitivity; excess discharges and tears. This formula is effective in treating eye disorders due to congested heat in Lung and Liver, Qi, blood, and Jin and Ye congestion

Shi Zao Tang [Ten-Jujube Decoction] *Shang Han Lun*

Da Zao	10 pieces
Yuan Hua	equal amount
Gan Sui	
Da Ji	

Main Indication: Water evil congested in the chest; symptoms include cough, thin saliva with chest and hypochondriac pain, hard focal distention at the epigastrium, dry heaves, shortness of breath, headache, and vertigo, or patient presents with significant sweats, slippery tongue coating, and deep and wiry pulse. This formula is also indicated for edema with abdominal swelling, fullness at the hypochondrium which causes pain with palpation, or even pain radiating to the shoulder and back.

Preparation: Grind the latter three herbs into powder. Taken 1 g of the powder with warm decoction made from 10 pieces of Da Zao.

Treatment Principle: Purge and drive out congested fluid.

This formula is used for patient with congested fluid in chest and abdomen and with a strong constitution. For patient with a weak constitution, a tonification formula should be used alternately with this one. Gan Cao is contraindicated to be used with this formula. It is also contraindicated for pregnant women.

Related Formula:

Formula	Herbs	Indication
Kong Xian Dan [Control Mucous Special Pill] *San Yin Fang*, also called Miao Ying Wan [Beautiful Voice Pill], Zi Long Wan [Purple Dragon Pill]	Gan Sui Da Ji Bai Jie Zi (equal amounts)	Grind the equal amount of herbs into powder and form into pills the size of Firmiana seeds. Take 5 to 10 pills with light ginger tea before bedtime. Indication: All pathological conditions resulting from phlegm and mucous congested above and below the diaphragm; or dull unbearable pain, or pulling, tightening, and moving pain at neck, chest, back, lower back, abdomen, hypochondrium, hands, feet, and hip joints; or numbness at the skin surface like paralysis. Or unclear mind, sleepiness, and fatigue; or no taste in mouth, and thick sticky phlegm and saliva. Or drooling during sleep; or numbness, vertigo, and dizziness, and fullness and stifling sensation with a phlegm constitution. This formula is Shi Zao Tang [Ten-Jujube Decoction] minus Yuan Hua, Da Zao, and added Bai Jie Zi. It has the function of purging and driving out phlegm

Qing Fei Yin [Clear the Lung Decoction] *Zheng Zhi Hui Bu*

Shan Zhi Zi	equal amounts
Huang Qin	
Mai Men Dong	
Sang Bai Pi	
Che Qian Zi	
Mu Tong	
Fu Ling	

Main Indication: Lung heat with Qi obstruction, Lung fails to disperse and descend. Symptoms include short and rapid breathing, difficult urination, dry mouth and thirst; thin, white tongue coating, and rapid pulse.

Preparation: Decoction.

Treatment Principle: Clear heat and promote urination (open the source and guide the flow).

This formula is used to treat interior heat in lungs with obstructed urinary pathway. The key that indicates this formula is both short and rapid breathing and difficulty urinating. If short and rapid breathing are absent, then the diagnosis should be Middle Jiao excess heat.

Shang Jiao Xuan Bi Tang [Disperse Painful Obstruction at the Upper Jiao Decoction] *Wen Bing Tiao Bian*

Pi Pa Ye (remove fur)	10 g
Yu Jin	10 g
She Gan	6 g
Bai Tong Cao	3 g
Dan Dou Chi	6 g

Main Indication: Upper Jiao damp–heat, Lung Qi obstruction and congestion. Symptoms include Qi rebellion with dry heaves or distention and stifling sensation at the chest, steaming heat at the skin, thin sticky tongue coating, and soggy pulse.

Preparation: Decoction. Taken in two doses.

Treatment Principle: Gently disperse painful obstruction at Lung. (Open the source and guide the flow.)

Modification:

Addition	Indication
Mu Zei Ban Xia Huang Qin Qing Pi Bing Lang Lu Gen	Treats beginning stage of damp–heat with distending pain at both sides of the head, fullness, stifling sensation, or pain at chest and hypochondrium, painful throat, and sticky yellow tongue coating. This simultaneously treats Lung and San Jiao
San Xiang Tang [Three Fragrant Decoction] (see Liver Syndrome)	Painful obstruction at the Qi level with hiccups and severe stifling sensation at chest. This combination strengthens the dispersing and descending functions

Mu Fang Ji Tang [Cocculus Decoction] *Jin Gui Yao Lue*

Mu Fang Ji	12 g
Shi Gao	60 g
Gui Zhi	10 g
Ren Shen	12 g

Main Indication: Zhi Yin, congested fluid at the chest and diaphragm. Symptoms include wheezing, fullness sensation, hard and fullness at the epigastrium, dark complexion, and deep tight pulse. It is for patients who have had the symptoms for a few weeks and have been treated with vomiting and the downward draining method without any effect.

Preparation: Decoction. Taken warm.

Treatment Principle: Drain heat and drive out congested fluid.

Modification:

Minus	Addition	Indication
	Fu Ling	To strengthen the promoting-urination function
Shi Gao	Mang Xiao	When the patient is in a severe condition and symptoms are only relieved for a few days with the original formula, one can use the original formula again. If still not seeing any results, then use this modification to treat the stubborn condition

Related Formula:

Formula	Herbs	Indication
Mu Fang Ji Jia Fu Ling Mang Xiao Tang [Cocculus with Mirabilitum and Poria Decoction]	Core formula – Shi Gao + Fu Ling Mang Xiao	Treats the syndrome that is indicated by Mu Fang Ji Tang with difficult urination, severe hard fullness, and absence of restlessness and thirst

Ting Li Da Zao Xie Fei Tang [Descurainia and Jujube Decoction to Drain the Lungs] *Jin Gui Yao Lue*

Ting Li Zi	12 g
Da Zao	5 pieces

Main Indication: Phlegm and water congesting the Lungs. Symptoms include wheezing with inability to lie down or inability to rest due to Zhi Yin; sticky or slippery tongue coating and deep wiry pulse. This formula is also indicated for puffiness and edema of whole body, face and eyes, cough, Qi rebellion, wheezing, rapid breathing, breathlessness, and fullness and distention at the chest.

Preparation: Decoction. Taken as one dose.

Treatment Principle: Drain Lung, promote urination.

Related Formula:

Formula	Herbs	Indication
Jia Wei Ting Li Da Zao Xie Fei Tang [Augmented Descurainia and Jujube Decoction to Drain the Lungs]	Core formula + Gan Sui 1 g Bai Zhu 9 g Bai Shao Yao 10 g Sang Bai Pi 12 g Huang Qin 12 g	Powder Gan Sui, decoct the rest of the herbs. One dose per day and ten doses for one treatment Treat fluid accumulation in the chest cavity due to tuberculosis

Ma Huang Lian Qiao Chi Xiao Dou Tang [Ephedra, Forsythia, and Aduki Bean Decoction] *Shang Han Lun*

Ma Huang	9 g
Lian Qiao	15 g
Xing Ren	9 g
Chi Xiao Dou	15 g
Zi Bai Pi	15 g
Sheng Jiang	9 g
Gan Cao	6 g
Da Zao	4 pieces

Main Indication: Damp–heat obstructing the skin, between the muscles and San Jiao. Symptoms include whole body turning bright yellow (Yang Huang); edema (Yang Shui) begins with puffiness at the face and eyes, then spreads throughout the body, difficulty urinating, and sudden onset. It is also used for damp–heat obstructing the collaterals, leading to Feng Dan, and faint skin rash with itching.

Preparation: Decoction.

Treatment Principle: Disperse Lung, vent the evils, promote urination, and clear damp–heat.

This formula uses sweating and two methods of promoting urination to treat edema. It is effective in treating urticaria, glomerular nephritis, and hepatitis with jaundice.

Zi Bai Pi (Cortex catalpae) clear heat and resolve toxins. If it cannot be found, Sang Bai Pi can be used as a replacement.

Comparison of Formulas Treating Phlegm and Water Congesting the Lungs

1. The nine formulas discussed in this category all can disperse Lung phlegm and water or drain Lung and promote urination. All can be used in Lung congestion failing to disperse and descend, damp water congestion, and obstruction.
2. Yue Bi Tang [Maidservant from Yue Decoction] and Yue Bi Jia Zhu Tang [Maidservant from Yue Decoction plus Atractylodes] are focused on treating edema. Shi Zao Tang [Ten-Jujube Decoction] and Mu Fang Ji Tang [Cocculus Decoction] focus on treating water accumulation at the chest and abdomen. Yue Bi Jia Ban Xia Tang [Maidservant from Yue Decoction plus Pinellia] and Ting Li Da Zao Xie Fei Tang [Descurainia and Jujube Decoction to Drain the Lungs] focus on treating wheezing and Qi rebellion due to internal Yin evil obstruction. Ma Huang Lian Qiao Chi Xiao Dou Tang [Ephedra, Forsythia, and Aduki Bean Decoction] is effective in treating Feng Dan faint skin rash due to damp–heat stagnating at skin and muscle. Shang Jiao Xuan Bi Tang [Disperse Painful Obstruction at the Upper Jiao Decoction] is designed for hiccups due to damp–heat obstruction. Qing Fei Yin [Clear the Lung Decoction] treats Lung heat congesting and obstructing Lung, water ways not regulated, and obstruction of urination.

3. Yue Bi Tang [Maidservant from Yue Decoction], Yue Bi Jia Ban Xia Tang [Maidservant from Yue Decoction plus Pinellia], Ma Huang Lian Qiao Chi Xiao Dou Tang [Ephedra, Forsythia, and Aduki Bean Decoction], and Shang Jiao Xuan Bi Tang [Disperse Painful Obstruction at the Upper Jiao Decoction] demonstrate the method of dispersing Lung and promoting urination. All other formulas in this category demonstrate the method of draining the Lung and promoting urination.

Jin Fei Cao — Damp - cold phlegm in Lu → cough

Xuan Fu Hua
Ban Xia
Jing Jie
Bo He
Chen Pi
Fu Ling
Gan Cao

COUGH

Excess vs Defi

strong voice
strong cough
young
strong constitution

CHAPTER 2

Lung and Heart or Spleen Syndromes and Formulas

LUNG AND HEART SYNDROMES

Intense Heat in Both Qi and Blood

Lung dominates Qi and Wei, and Heart dominates blood and Ying. At the beginning stage of a warm febrile disease when the pathogens reach Ying level, a deep red tongue with a white or yellow coating indicates that the evil pathogens at the Qi level are not yet resolved. Thus, one should drain Wei and vent Ying with Qing Ying Tang [Clear the Nutritive Level Decoction] (see Chapter 7 on Heart syndromes). In these formulas, Jin Yin Hua, Lian Qiao, and other herbs are acrid and cool to release the exterior. Xi Jiao (Shui Niu Jiao), Sheng Di Huang, and other herbs can clear the Ying level and drain heat so they function to vent heat and transform Qi. If symptoms of the Qi level such as high fever, restlessness, thirst, and sweating are present with blood heat symptoms such as rashes, bleeding, deep red tongue, or even spasm and syncope, then there is severe heat in both blood and Qi. In this case, one should use clearing heat from the Qi level herbs such as Shi Gao, Zhi Mu, Jin Yin Hua, Shan Zhi Zi, and Huang Lian with cooling blood and relieving toxicity herbs such as Xi Jiao (Shui Niu Jiao), Sheng Di Huang, Mu Dan Pi, Zi Cao, Da Qing Ye, and Ban Lan Gen to make up the formulas that clear both heat and blood, as in Qing Wen Bai Du Yin [Clear Epidemics and Overcome Toxin Decoction] and Liu Nao He Ji [Compound Mixture for Epidemic Encephalitis].

Qing Wen Bai Du Yin [Clear Epidemics and Overcome Toxin Decoction] *Yi Zhen Yi De*

Shi Gao	Large dose 180–240 g Medium dose 60–120 g Small dose 30–60 g
Sheng Di Huang	Large dose 18–24 g Medium dose 10–15 g Small dose 6–12 g
Huang Lian	Large dose 12–18 g Medium dose 6–12 g Small dose 3–6 g
Xi Jiao (Substitute with 5 to 10 times Shui Niu Jiao)	Large dose 18–24 g Medium dose 10–15 g Small dose 6–10g
Shan Zhi Zi	6–15 g

(continued)

Huang Qin	6–15 g
Lian Qiao	6–15 g
Zhi Mu	6–15 g
Mu Dan Pi	6–15 g
Chi Shao Yao	6–15 g
Xuan Shen	6–15 g
Dan Zhu Ye	6–15 g
Gan Cao	6–15 g
Jie Geng	6–15 g

Main Indication: Severe heat in Qi and blood. Symptoms include intense fever, body heat, strong thirst, vomiting, splitting headache, restlessness, and irritability leading to delirium and delirious speech; in more severe cases, there are also skin rashes and nosebleed. Other signs include red tongue, scorched lips; deep, thin, and rapid pulses at all six positions; or pulse either deep and rapid or floating, big, and rapid.

Preparation: Boil Shi Gao for 2 to 3 min, then add other herbs to make the decoction. Grind Xi Jiao (Shui Niu Jiao) into powder and take with the decoction.

Treatment Principle: Clear heat, relieve toxins, cool blood, and rescue Yin.

This formula is modified from Huang Lian Jie Du Tang [Coptis Decoction to Relieve Toxicity], Bai Hu Tang [White Tiger Decoction], and Xi Jiao Di Huang Tang [Rhinoceros Horn and Rehmannia Decoction]. Thus, this formula has a strong function to clear heat, relieve toxicity, and clear both Qi and blood. Warm febrile diseases with the above described symptoms such as epidemic meningitis, encephalitis B, and septicemia can be treated with this formula.

Modification:

Addition	Indication
Da Qing Ye Ban Lan Gen	Strengthens the effectiveness of the formula
Zi Hua Di Ding Sheng Da Huang	Swollen head and face
Jin Yin Hua Qing Dai	Swollen cheeks, neck, and parotid gland
Ling Yang Jiao (Shan Yang Jiao) Long Dan Cao Ju Hua Zang Hong Hua	Red capillaries congested in eyes and blurry vision
Da Qing Ye Zi Hua Di Ding	Swelling and pain behind the ear
Mu Tong Tong Bian (Boy Urine)	Making noises and playing with the tongue
Li Zhi (Asian Pear Juice) Zhu Li Xi Gua Zhi (Watermelon juice)	Sticky powder type of tongue coating and slurred speech
Fresh Zhu Ru Pi Pa Ye Chen Xiang Qing Pi Yu Jin Zhi Shi	Coarse breathing, hiccups, and rebellious Qi
Shan Dou Gen	Swollen and painful throat
Ling Yang Jiao (Shan Yang Jiao) Ju Hua Long Dan Cao Sang Zhi Si Gua Luo	Spasms and tremors of tendons and unconscious movement of the four limbs in severe cases

(continued)

Addition	Indication
Huang Bai Mu Tong	Irritable pain at the joints and lumbar pain as if the person had been hit
Pei Lan Ye Qiang Wei (Wild Rose) Sap	Foul smell, offending people
Hua Shi Hu Po Bai Mao Gen Che Qian Cao Niu Xi	Scanty, dark, cloudy, hesitant urination or, in severe cases, with blood and pain (Blood Lin Syndrome)
Mang Xiao Fan Xie Ye Bai Mi (Honey)	Tenesmus or defecating purple blood with foul, turbid, or dysentery-like stools.

Related Formulas:

Formula	Herbs	Indication
Liang Ying Qing Qi Tang [Cool Ying and Clear the Qi Decoction]	Xi Jiao (Shui Niu Jiao) Sheng Di Huang Mu Dan Pi Chi Shao Yao Huang Lian Shan Zhi Zi Xuan Shen Lian Qiao Dan Zhu Ye Shi Gao Shi Hu Lu Gen Bai Mao Gen Bo He Gan Cao	Decoction. Treats epidemic throat Sha (Scarlet Fever), severe heat in both Qi and blood, with high fever, thirst, irritability, restlessness, and delirium; delirious speech in severe cases; swollen red, severely painful throat, and ulceration in severe cases; rashes and deep red tongue with yellow dry coating
Jia Jian Qing Wen Bai Du Yin [Modified Clear Epidemics and Overcome Toxin Decoction]	Sheng Di Huang Chi Shao Yao Xi Jiao (Shui Niu Jiao) Xuan Shen Shi Gao Zhi Mu Da Qing Ye Lian Qiao Huang Qin Huang Lian Niu Huang	Decoction. Taken with Shen Xi Dan [Miraculous Rhinoceros Horn Pill] Treats epidemic meningitis, encephalitis B In severe heat cases, add Ling Yang Jiao (Shan Yang Jiao) For tremor, add Jiang Can, Di Long, Quan Xie, and Wu Gong and take with Zi Xue Dan [Purple Snow Pill] For excess phlegm, add Tian Zhu Huang, Long Dan Cao, Ban Xia, and Hou Zao (0.5–1 g) and take with An Gong Niu Huang Wan [Calm the Palace Pill with Cattle Gallstone]; for internal lock-up and external collapse, take with Du Shen Tang [Unaccompanied Ginseng Decoction]

Liu Nao He Ji [Compound Mixture for Epidemic Encephalitis] *Jiang Xi Zhong Yi Xue Yuan Fang*

Sheng Shi Gao	60 g
Zhi Mu	15 g
Da Qing Ye	30 g
Fresh Sheng Di Huang	60 g
Chi Shao Yao	9 g
Mu Dan Pi	12 g
Huang Lian	12 g
Huang Qin	12 g
Lian Qiao	15 g
Dan Zhu Ye	9 g
Gan Cao	9 g
Jie Geng	6 g
Shui Niu Jiao Decoct first to get 200 mL of decoction	120 g

Main Indication: Scorched Qi and blood. High fever, severe headache, nausea, vomiting, painful limbs, stiffness of the neck, painful or swollen, red throat, clear red bleeding spots on skin, deep red tongue, and rapid pulse.

Preparation: Decoct Shi Gao and Da Qing Ye first, then combine with Shui Niu Jiao decoction, then add rest of the herbs to make a final decoction of 200 to 400 mL. Divide into three doses. Take two to four doses per night. If the decoction is hard to swallow due to severe vomiting, then take Shi Chang Pu and Zhu Li first to stop vomiting before taking this formula.

Treatment Principle: Clear heat, relieve toxins, cool blood, and rescue Yin.

This formula is designed to treat epidemic meningitis. However, this formula can also be used to treat various warm febrile diseases with scorched Qi and blood syndromes.

LUNG AND SPLEEN STOMACH SYNDROMES

Spleen and Lung Qi Deficiency

Lung and Spleen have the functions that mutually support each other. They are both the source of postnatal vital Qi (clear air from the nature and essence Qi from food). If both Spleen and Lung are deficient, symptoms of pale face, shortness of breath, too deficient to talk, fatigue, poor appetite, and loose stools can be seen. In this case, both Spleen and Lung need to be tonified by using Si Jun Zi Tang [Four-Gentleman Decoction]. (Please refer to the section on tonifying Qi to benefit Spleen.)

If both Middle Jiao and Lung have an excess of exterior and interior symptoms, Fang Feng Tong Sheng San [Ledebouriella Powder that Sagely Unblocks] is used to release exterior and clear heat from both exterior and interior.

Fang Feng Tong Sheng San [Ledebouriella Powder That Sagely Unblocks] *Huang Di Su Wen Xuan Ming Lun Fang*

Fang Feng	15 g
Ma Huang	15 g
Da Huang (prepared with wine)	15 g
Mang Xiao	15 g
Jing Jie	15 g
Bo He	15 g
Shan Zhi Zi	15 g
Hua Shi	90 g
Shi Gao	30 g
Lian Qiao	15 g
Huang Qin	30 g
Jie Geng	30 g
Chuan Xiong	15 g
Dang Gui	15 g
Bai Shao Yao	15 g
Bai Zhu	15 g
Gan Cao	60 g

Main Indication: Interior and exterior excess heat. Symptoms include strong fever and chills, headache, dizziness, red and sore eyes, difficulty swallowing, nasal congestion with thick and sticky nasal discharge, bitter taste in mouth, focal distention and stifling sensation at chest and diaphragm;

constipation, dark yellow urine, red tongue, yellow, greasy tongue coating, and flooding, wiry, rapid, and slippery pulse.

Preparation: Grind all ingredients into powder, take 6 to 9 g as draft with 3 pieces of Sheng Jiang. With appropriate reduction in dosage, this formula can also be prepared as decoction.

Treatment Principles: Disperse wind, clear heat, and drain downward.

CHAPTER 3

Spleen and Stomach Syndromes and Formulas

DEFICIENT COLD IN THE MIDDLE JIAO

For deficient cold in the Middle Jiao, warming the Middle Jiao and strengthening the Spleen are the main treatment methods. They treat Spleen Yang deficiency leading to vomiting, diarrhea, epigastric and abdominal pain alleviated by warmth and worsened by cold, white tongue coating, absence of thirst, and deep slow pulse or tight wiry pulse. Gan Jiang, Chuan Jiao, Ding Xiang, Wu Zhu Yu, Gao Liang Jiang, and Rou Gui are often used to warm the Middle Jiao and to scatter the cold. The formula is often assisted by Dang Shen, Bai Zhu, Gan Cao, and Sha Ren to tonify Qi and strengthen the Spleen, and to recover the Yang in Middle Jiao, thus scattering the evil cold, as does Li Zhong Wan [Regulate the Middle Pill].

Li Zhong Wan [Regulate the Middle Pill] *Shang Han Lun*

Ren Shen	90 g each
Gan Jiang	
Zhi Gan Cao	
Bai Zhu	

Main Indication: Deficient cold in the Middle Jiao. Symptoms include diarrhea, vomiting, abdominal pain, absence of thirst, pale tongue, white coating, or black wet tongue coating with tender tongue proper, and deep weak pulse.

Preparation: Make into honey pills. Take 10 g each time, with warm water. If taken as decoction, then the dosage should be modified accordingly.

Treatment Principles: Warm the Middle Jiao and strengthen the Spleen.

This formula warms, tonifies, and strengthens the Spleen. It can be used in deficient cold type ulceration, chronic gastroenteritis leading to poor digestion, borborygmus, and diarrhea, worsening with cold. It can also be used to treat gastromalacia, gastroptosis, and neurofunctional disorders of the Stomach and intestine with above described symptoms. It is also effective in treating Spleen that fails to control the blood.

Modification:

Addition	Indication
Yi Zhi Ren	Increase of saliva after an illness is an indication of Spleen Yang not yet fully recovered and cannot control Jin Ye
Quan Xie Jiang Can	Chronic fright and convulsion in children leading to vomiting and diarrhea that damages Spleen
	In *Jin Gui Yao Lue*, this formula is used to treat "chest Bi" syndrome with deficient cold. It is called Ren Shen Tang [Ginseng Decoction]

Related Formulas:

Formula	Herbs	Indication
Fu Zi Li Zhong Wan [Prepared Aconite Pill to Regulate the Middle Jiao]	Fu Zi Li Zhong Tang	Treats severe cases of deficient cold with four cold extremities
Gui Fu Li Zhong Tang [Cinnamon and Prepared Aconite Decoction to Regulate the Middle Jiao]	Rou Gui Fu Zi Li Zhong Tang	Treats deficient cold in Spleen and Stomach with abdominal pain, vomiting, diarrhea, and cold hands and feet
Ding Yu Li Zhong Tang [Clove and Evodia Decoction to Regulate the Middle Jiao]	Ding Xiang Wu Zhu Yu Li Zhong Tang	Treats duodenal ulcer due to Yang deficiency and cold; symptoms include vomiting, abdominal pain, worsens with cold and alleviates with warmth. The treatment method warms the Middle Jiao and descends the rebellious Qi
Sha Ban Li Zhong Tang [Amomum and Pinellia Decoction to Regulate the Middle Jiao]	Sha Ren Ban Xia Li Zhong Tang	Treats above syndrome with severe vomiting This method warms the Middle Jiao and descends the rebellious Qi
Zhi Shi Li Zhong Tang [Immature Bitter Orange Decoction to Regulate the Middle Jiao]	Zhi Shi Fu Ling Li Zhong Tang	Treats distention and fullness of the epigastrium and abdomen, and deficient cold type of abdominal distention and pain
Zhi Zhong Tang [Treats the Middle Jiao Decoction]	Qing Pi Chen Pi Li Zhong Tang	Treats epigastric and abdominal distention due to cold food accumulation and obstruction
Li Ling Tang [Regulate the Middle Jiao Decoction with Five-Ingredient Powder with Poria]	Wu Ling San Li Zhong Tang	Treats Yang deficient patient with food accumulation, abdominal distention, wheezing, puffiness and swelling, difficult urination, and urinary retention.
Lian Li Tang [Coptis Decoction to Regulate the Middle Jiao]	Huang Lian Li Zhong Tang	Treats Spleen deficient type of chronic enteritis with the key symptom of vomiting sour water It is a formula with both cold and heat herbs
Lian Yu Li Zhong Tang [Coptis and Evodia Decoction to Regulate the Middle Jiao]	Huang Lian Wu Zhu Yu Li Zhong Tang	Treats cold and heat mixed type of digestive ulcer It is a formula with both cold and heat herbs

Da Jian Zhong Tang [Major Construct the Middle Decoction] *Jin Gui Yao Lue*

Chuan Jiao (fried)	9 g
Gan Jiang	15 g
Ren Shen	6 g
Yi Tang	60 g

Main Indication: Extreme exhaustion of Yang in the Middle Jiao and excess internal Yin and cold. Symptoms include excruciating pain at epigastrium and abdomen, and visible abdominal spasm during which patient cannot tolerate being touched; vomiting to the point of not being able to eat; greasy white tongue coating, and wiry slow pulse or deep weak pulse.

Preparation: Decoct the first three herbs, strain the decoction, and then add Yi Tang. To be taken warm.

Treatment Principles: Warm the Middle Jiao, scatter cold, alleviate, and stop the acute pain.

This formula can also treat deficient cold in the Middle Jiao and abdominal pain caused by roundworms because Chuan Jiao can warm the Middle Jiao and also expel parasites.

Xiao Jian Zhong Tang [Minor Construct the Middle Decoction] *Shang Han Lun*

Gui Zhi	15 g
Bai Shao Yao	30 g
Sheng Jiang	15 g
Gan Cao	10 g
Da Zao	10 g
Yi Tang	60 g

Main Indication: Deficient cold in the Middle Jiao and Liver overacting on Spleen. Symptoms include spasmic abdominal pain, a preference for warmth, pressure, palpitation, restlessness, lackluster face, pale tender tongue, and wiry and choppy pulse.

Preparation: Decoct the first five herbs, strain the decoction, and then add Yi Tang. To be taken warm, three times a day.

Treatment Principles: Warm the Middle Jiao, tonify the deficiency, soften the Liver, and moderate the spasm.

This is the first formula that uses the principle of "warm and sweet to clear heat."

This formula modifies from Gui Zhi Tang [Cinnamon Twig Decoction] with double the dosage of Shao Yao and addition of Yi Tang. Double dose of Shao Yao and Yi Tang can nourish Yin and soften Liver to moderate pain, which shows the classic method of "benefit the wood and support the earth." It can also regulate and tonify Yin and Yang to treat fever and palpitation due to exhaustion and overexertion.

This formula is effective in treating Middle Jiao deficient cold type of digestive ulceration, neurasthenia, chronic fatigue syndrome, and aplastic anemia without greasy tongue coating.

Related Formulas:

Formula	Herbs	Indication
Huang Qi Jian Zhong Tang [Astragalus Decoction to Construct the Middle]	Core Formula + Huang Qi	Treats digestive ulceration especially duodenal ulcers, chronic peritonitis, and chronic fatigue syndrome
Gui Qi Jian Zhong Tang [Tangkuei and Astragalus Decoction to Construct the Middle]	Core Formula + Dang Gui Huang Qi	Treats hunger pain, postpartum abdominal pain and tightness at the lower abdomen leading to pain at lower back due to Middle Jiao deficient cold
Dang Gui Jian Zhong Tang [Tangkuei Decoction to Construct the Middle]	Core Formula + Dang Gui	Treats deficient constitution after giving birth with continuous abdominal pain

Comparison of Formulas Treating Deficient Cold in Middle Jiao

All three formulas in this category treat abdominal pain due to Middle Jiao deficient cold. However, the key indications for Li Zhong Wan [Regulate the Middle Pill] are vomiting, diarrhea, and abdominal pain. It focuses on warming the Middle Jiao to restore the ascending and descending function in the Middle Jiao. The key indication for Da Jian Zhong Tang [Major Construct the Middle Decoction] is abdominal pain with visible spasm in which the abdomen refuses palpation or pressure, especially at the middle of the abdomen. It focuses on warming and scattering the congealed cold, and moderating and stopping severe pain. Xiao Jian Zhong Tang [Minor Construct the Middle Decoction] treats abdominal pain that prefers warmth and pressure. The pain is predominant at epigastrium and lower abdomen. It focuses on softening the Liver, moderating pain, and harmonizing Liver and Spleen.

EXCESS HEAT IN QI LEVEL

The "acrid cold to clear heat" method is used to treat Yang Ming channel syndrome or warm febrile disease of the Qi level. Symptoms include high fever, presence of sweating, irritability, thirst, and flooding big pulse which are referred to as the "Four Big Syndrome." Often, Shi Gao, Zhi Mu, and Dan Zhu Ye are used to form the formulas. Heat pathological condition easily damages body fluid; therefore, acrid cold herbs that clear heat are used. This is because acrid relaxes the muscles, and cold clears the heat, and when the heat is gone, the fluid returns. In clinical practice, one should modify the formula by combining acrid cool releasing exterior herbs, clearing Ying level and cooling blood level herbs, or Yin nourishing and fluid promoting herbs, according to the patient's presentation.

Bai Hu Tang [White Tiger Decoction] *Shang Han Lun*

Shi Gao	15–60 g
Zhi Mu	12 g
Gan Cao	6 g
Geng Mi	9 g

Main Indication: Yang Ming channel syndrome or warm febrile disease of the Qi level. Symptoms include high fever, sweating, thirst and restlessness, dry red tongue, flooding big pulse, or slippery rapid pulse.

Preparation: Decoction.

Treatment Principles: Acrid cold to clear heat.

This method is an effective heat clearing, antiinflammatory formula. When it is modified appropriately, it can treat the "Big-Four Syndrome" seen in pneumonia, meningitis, encephalitis B, common cold and flu, heat stroke, influenza, and measles.

This formula is contraindicated for fever due to Yin deficiency and true cold false heat patterns.

Related Formulas:

Formula	Herbs	Indication
Bai Hu Jia Gui Zhi Tang [White Tiger plus Cinnamon Twig Decoction]	Core Formula + Gui Zhi	Treats exterior syndrome with symptoms indicated for Bai Hu Tang [White Tiger Decoction] and pain at the joints. It also treats warm malaria with only heat presentation without cold. Symptoms also include joint pain, occasionally vomiting, and even pulse. It can also treat internal warm febrile patterns with exterior wind cold syndrome
Chai Hu Bai Hu Tang [Bupleurum White Tiger Decoction]	Core Formula + Chai Hu	Treats alternating fever and chills, more heat than cold
Bai Hu He Huang Lian Jie Du Tang [White Tiger with Coptis Decoction to Relieve Toxicity]	Core Formula + Huang Qin Huang Lian Huang Bai Shan Zhi Zi	Treats rashes due to warm toxins, irritation/restlessness, heat sensation, sensory aphasia and insomnia

(continued)

Formula	Herbs	Indication
Bai Hu Cheng Qi Tang [White Tiger and Order the Qi Decoction]	Core Formula + Da Huang Mang Xiao	Treats high fever, thirst, sweating, and delirium with delirious speech, constipation, and scanty hesitant urination
Zhen Ni Bai Hu Tang [White Tiger Decoction to Suppress Rebellion]	Ban Xia Zhu Ru Shi Gao Zhi Mu	Treats the indicating syndrome of Bai Hu Tang [White Tiger Decoction] with rebellious Qi and distention and stifling sensation at the epigastrium
Bai Hu Jia Ren Shen Tang [White Tiger plus Ginseng Decoction]	Core Formula + Ren Shen or Bei Sha Shen	Treats indicating syndrome of Bai Hu Tang [White Tiger Decoction] with excess sweating, but empty big pulse
Cang Zhu Bai Hu Tang [Atractylodes White Tiger Decoction]	Core Formula + Cang Zhu	Treats damp warm febrile diseases, aversion to cold, high fever, thirst, whole body ache. It also treats damp–heat with heat excess
Yin Qiao Bai Hu Tang [Honeysuckle, Forsythia, and White Tiger Decoction]	Core Formula + Jin Yin Hua Lian Qiao	Treats both Wei level and Qi level heat pattern, such as in encephalitis B
Qing Zhen Tang [Clear the Rash Decoction]	Ling Yang Jiao (Shan Yang Jiao) He Ye Lian Qiao Chan Tui Jiang Can Jin Xian Chong Lou Shi Gao Zhi Mu	Treats pediatric rashes or blotches on skin, which are not fully surfaced, excess heat both exterior and interior, restlessness, thirst, sore throat, and hoarseness
Xin Jia Bai Hu Tang [Newly Augmented White Tiger Decoction]	Core Formula + Bo He He Ye Dan Zhu Ye Sang Zhi Hua Shi Zhu Sha	Treats fever, sweating, restlessness, and thirst, faint skin rash, scanty dark burning urination, and hemoptysis and delirium in severe cases
Hua Ban Tang [Transform Blotches Decoction]	Core Formula + Xi Jiao (Shui Niu Jiao) Xuan Shen	Treats heat conditions with delirium, delirious speech and dark red blotches on skin
Xin Jia Yu Nu Jian [Newly Augmented Jade Woman Decoction]	Sheng Di Huang Xuan Shen Mai Men Dong Shi Gao Zhi Mu	Treats heat conditions with scorched Qi and blood
	Modification 1: Xin Jia Yu Nu Jian [Newly Augmented Jade Woman Decoction] + Niu Xi Ci Shi Mu Li Bai Shao Yao	Treats insufficient Shao Yin and excess Yang Ming leading to toothache, loose teeth, and swollen painful gum
	Modification 2: Xin Jia Yu Nu Jian [Newly Augmented Jade Woman Decoction] + Xi Xin Rou Gui	Treats toothache due to external evil pathogens

(continued)

Formula	Herbs	Indication
Xi Ling Bai Hu Tang [Rhinoceros Horn, Antelope, and White Tiger Decoction]	Core Formula + Xi Jiao (Shui Niu Jiao) Ling Yang Jiao (Shan Yang Jiao) Gou Teng Ju Hua	Treats pediatric warm febrile diseases that transform into dryness, exhaustion of fluids leading to wind. Symptoms include absence of nasal discharge, dry eyes without tear, sallow dark dry complexion, delirium, spasm, and convulsion Da Qing Ye and Ban Lan Gen can be added to strengthen the clearing heat and relieving toxin effect
Ling Ma Bai Hu Tang [Antelope, Gastrodia, White Tiger Decoction]	Core Formula + Ling Yang Jiao (Shan Yang Jiao) Tian Ma Gua Lou Ren Chuan Bei Mu	Treats wind warm febrile diseases leading to spasm, and damaged Yin fluids. Symptoms include internal movements of Liver wind causing upward movement of phlegm, blood not nourishing the tendons leading to spasms
Jia Jian Yin Qiao Bai Hu Tang [Modified Honeysuckle, Forsythia, and White Tiger Decoction]	Shi Gao 60 g Zhi Mu 12 g Dan Zhu Ye 12 g Qing Hao 15 g Jin Yin Hua 30 g Lian Qiao 30 g Da Qing Ye 15 g Mu Dan Pi 12 g Huang Qin 9 g Fresh Shi Hu 15 g Fresh Bai Mao Gen 60 g	Treats encephalitis B with severe heat. Symptoms include high fever, restlessness, irritability, red face, short and coarse breathing, severe thirst, red tongue, yellow greasy tongue coating, and rapid and forceful pulse

Zhu Ye Shi Gao Tang [Lophatherus and Gypsum Decoction] *Shang Han Lun*

Dan Zhu Ye	10 g
Sheng Shi Gao	30 g
Ban Xia	12 g
Ren Shen	6 g
Mai Men Dong	30 g
Gan Cao	6 g
Geng Mi	10 g

Main Indication: Late stage of febrile disease with lingering heat signs, damaged Qi and blood. Symptoms include emaciation, shortness of breath, nausea, vomiting, dry throat, thirst, red tongue with scanty coating with deficient and rapid pulse. It also treats restlessness, insomnia due to deficiency with deficient rapid pulse.

Preparation: Decoction.

Treatment Principles: Clear heat, descend rebellion, tonify Qi and generate fluid.

In *Shang Han Lun*, this formula is indicated for damaged Qi and Yin due to "the residual fire in the ashes after the fire is put out."

Related Formula:

Formula	Herbs	Indication
Jia Jian Zhu Ye Shi Gao Tang [Modified Lophatherus and Gypsum Decoction] *Guang Wen Re Lun*	Xi Yang Shen 3 g Sheng Shi Gao 10 g Sheng Gan Cao 2 g Mai Men Dong 6 g Ban Xia 3 g Sugar cane juice 1 cup Fresh Ginger juice 2 drops	It is for emaciation, weakness, shortness of breath, rebellious Stomach Qi, nausea and vomiting, after rashes or blotches have surfaced on skin One should also drink Zhu Ru 10 g and fresh Bai Mao Gen 30 g decoction instead of water

(continued)

Formula	Herbs	Indication
Jian Wei Zhu Ye Shi Gao Tang [Reduced Lophatherus and Gypsum Decoction] *Wen Bing Tiao Bian*	Dan Zhu Ye 15 g Shi Gao 24 g Mai Men Dong 18 g Gan Cao 10 g	Treats warm febrile disease of Yang Ming with floating hurried pulse

Comparison of the Two Formulas Treating Excess Heat in Qi Level

Bai Hu Tang [White Tiger Decoction] treats excess heat syndrome with the "Big Four." Zhu Ye Shi Gao Tang [Lophatherus and Gypsum Decoction] treats residual heat after heat–febrile diseases with damaged Qi and body fluid leading to shortness of breath, nausea, vomiting, dry mouth, thirst, red tongue, scanty coating, and deficient rapid pulse. Bai Hu Tang directly attacks excess heat from the Qi level and Zhu Ye Shi Gao Tang clears heat at the same time it tonifies.

HEAT ACCUMULATION IN THE STOMACH AND INTESTINE

Bitter herbs are used to treat heat accumulation in the Stomach leading to mouth ulcers, bad breath, irritability, thirst, increased hunger, or red swollen or ulcerations at gum, swollen lips, mouth, and cheeks, red tongue, dry yellow tongue coating, and slippery rapid pulse. Herbs such as Shan Zhi Zi, Huang Lian, and Mu Dan Pi are often used as in Qing Wei San [Clear the Stomach Powder]. It is also used to treat acute pancreatitis with high fever, severe sharp pain at the epigastrium, nausea and vomiting due to excess heat at the Middle Jiao or heat accumulation in Stomach and Gallbladder, such as Qing Yi Tang [Clear the Pancreas Decoction] and Da Chai Hu Tang [Major Bupleurum Decoction].

The heat toxins accumulating in Large Intestine lead to Qi and blood stagnation creating intestinal abscess (appendicitis). This should be treated following the principle of keeping the six Fu organs unobstructed. Thus, herbs like Yan Hu Suo, Mu Dan Pi, Jin Yin Hua, Lian Qiao, Huang Qin, Tao Ren, Hong Teng, Bai Jiang Cao, Pu Gong Ying, Zi Hua Di Ding, Da Huang, and other herbs that can regulate Qi, activate blood, cool blood, and relieve toxins can be used. Representing formulas are Da Huang Mu Dan Pi Tang [Rhubarb and Moutan Decoction] and Lan Wei Qing Jie Tang [Clear and Relieve Appendix Decoction].

Zhi Zi Chi Tang [Gardenia and Prepared Soybean Decoction] *Shang Han Lun*

Shan Zhi Zi	10 g
Xiang Dou Chi	10 g

Main Indication: Heat accumulation in the chest. Symptoms include body heat, irritability, restlessness, insomnia with tossing and turning in bed, feeling remorseful, or stifling and restless heat sensation at the chest, if pressing the chest area, then epigastrium showing softness and noises as if the person is hungry, slight yellow tongue coating, excess pulse at cun position.

Preparation: Decoction.

Treatment Principles: Clear heat and alleviate restlessness and irritability.

In this formula, Dou Chi ascends and Shan Zhi Zi descends. Thus, this formula can be used for bleeding at either upper or lower parts of the body. Since Shan Zhi Zi can be used to treat jaundice, in *Sheng Ji Zong Lu* this formula is used to treat symptoms of "toad yellow, distention with blue stripes on tongue, and inability to sleep day and night."

Related Formulas:

Formula	Herbs	Indication
Zhi Zi Gan Cao Chi Tang [Gardenia, Licorice, and Prepared Soybean Decoction]	Shan Zhi Zi Dan Dou Chi Gan Cao	Heat accumulating in the chest with shortness of breath
Zhi Zi Sheng Jiang Chi Tang [Gardenia, Fresh Ginger, and Prepared Soybean Decoction]	Shan Zhi Zi Dan Dou Chi Sheng Jiang	Heat accumulating in the chest with nausea and rebellious Qi
Zhi Zi Hou Po Tang [Gardenia and Magnolia Bark Decoction]	Shan Zhi Zi Hou Po Zhi Shi	Treats restlessness, abdominal fullness, restlessness and insomnia after purging is done to treat a cold induced disease
Zhi Zi Gan Jiang Tang [Gardenia and Ginger Decoction]	Shan Zhi Zi Gan Jiang	Treats slight restlessness, persistent fever after a strong purging to treat a cold induced disease
Jia Wei Zhi Zi Chi Tang [Augmented Gardenia and Prepared Soybean Decoction]	Shan Zhi Zi Dan Dou Chi Zhi Ke Zhu Ru Ban Xia Mu Xiang Huang Lian Su Ye	Treats nausea and/or vomiting clear or sour water, bitter mouth, distending sensation at heat with dizziness, vertigo, stifling sensation at chest, hunger and rumbling, scanty urination, red tongue, yellow tongue coating, wiry, slippery, and rapid pulse

Qing Wei San [Clear the Stomach Powder] *Lan Shi Mi Cang*

Dang Gui	5 g
Huang Lian	6 g
Sheng Di Huang	10 g
Mu Dan Pi	12 g
Sheng Ma	3 g

Main Indication: Damp–heat accumulating in Yang Ming channel. Symptoms include toothache causing headache, fever, preference for cold than heat, red swollen gum or with ulceration, bleeding gums, bad breath, swollen painful cheeks and lips, dry mouth, dry tongue, red tongue with scanty coating, and slippery, big and rapid pulse.

Preparation: Decoction.

Treatment Principles: Clear and disperse Stomach heat, cool blood, and nourish Yin.

This formula treats damp–heat type of gum bleeding. If the bleeding is caused by deficient heat of the Kidney, one should use Liu Wei Di Huang Tang [Six-Ingredient Decoction with Rehmannia]. If the toothache is caused by Yin deficiency with residual heat at the Yang Ming channel, then one should use Yu Nu Jian [Jade Woman Decoction].

Modification:

Addition	Indication
Ge Gen	Treats swollen cheeks and chin due to overdrinking
Yin Chen Huo Xiang Bai Dou Kou	Treats deep gum ulceration and bleeding due to over-indulgence of fatty greasy food and alcohol
Xi Jiao (Shui Niu Jiao) Lian Qiao Huang Qin Shi Gao	Treats acute periodontitis
Da Huang Mang Xiao	Treats bleeding due to any dental pathology with constipation

Related Formulas:

Formula	Herbs	Indication
Jin Jian Qing Wei Yin [Clear the Stomach Decoction from *Yi Zong Jin Jian*]	Huang Lian Sheng Di Huang Mu Dan Pi Sheng Ma Shi Gao Huang Qin	Treats severe damp–heat in the Stomach
Jia Wei Qing Wei San [Augmented Clear the Stomach Powder]	Core Formula + Xi Jiao (Shui Niu Jiao) Lian Qiao Gan Cao	Treats the same pattern as Qing Wei San, but with excess gum bleeding
Mi Yan Qing Wei San [Secret Clear the Stomach Powder from Experiences]	Core Formula + Shi Gao Shan Zhi Zi Huang Qin Bai Shao Yao Cang Zhu Qing Pi Xi Xin Huo Xiang Jing Jie Sui Gan Cao	Treats all types of wind–heat–damp phlegm causing toothache, swollen bleeding gums, and loose teeth

Qing Yi Tang [Clear the Pancreas Decoction] (Tian Jin Nan Kai Yi Yuan Fang)

Chai Hu	15 g
Huang Qin	9 g
Hu Huang Lian	9 g
Bai Shao Yao	15 g
Mu Xiang	9 g
Yan Hu Suo	9 g
Fresh Da Huang (decoct later)	15 g
Mang Xiao (add to final decoction)	9 g

Main Indication: Heat accumulating and obstructing in Gallbladder and Liver as in acute pancreatitis. Symptoms include constant severe epigastric pain, with intermittent increased in pain, abdominal pain upon palpation, tight abdominal muscle, nausea, vomiting, fever, yellow greasy tongue coating with wiry and rapid pulse.

Preparation: Decoction. One bag a day, taken in two doses. For severe condition, use two bags per day and take in four doses.

Treatment Principles: Pacify the Liver and benefit the Gallbladder, drain heat, and open the Fu organs.

This is an effective formula in treating acute pancreatitis. It can be used in any syndromes with heat accumulating in Gallbladder and Stomach with obstruction.

Modification:

Minus	Addition	Indication
	Yin Chen Hao Shan Zhi Zi Long Dan Cao	With severe dampness and/or jaundice
	Ban Xia Dai Zhe Shi	With severe vomiting

(continued)

Minus	Addition	Indication
	Chuan Lian Zi Yan Hu Suo (increase dosage)	With severe pain
	Lai Fu Zi Shen Qu Mai Ya Shan Zha	With food accumulation
	Hou Po Zhi Shi	With chest fullness
	Quan Gua Lou Xie Bai Fang Feng Qin Jiao	With back pain
	Bing Lang Shi Jun Zi Ku Lian Gen Pi	With roundworms in the bile duct
Da Huang Mang Xiao	Fu Zi Gan Jiang	With deficient constitution and coldness in the Middle Jiao

Da Huang Mu Dan Tang [Rhubarb and Moutan Decoction]
Jin Gui Yao Lue

Da Huang (add near end)	12 g
Mu Dan Pi	9 g
Tao Ren	9 g
Dong Gua Ren	15 g
Mang Xiao (add to final decoction)	9 g

Main Indication: Beginning stage of intestinal abscess. Symptoms include lower right quadrant pain that is aggravated and sharp with palpation, abdominal muscle tightness, and normal urination, or aggravating pain on extending the leg straight, preference for flexing the hip and knee, or with fever and constipation, greasy yellow thin tongue coating, slippery rapid and wiry pulse.

Preparation: Decoction.

Treatment Principles: Drain heat and expel stasis.

This formula is used to treat beginning stage of heat toxin accumulation with obstruction of Qi and blood. It can also be used for unsmooth lochial discharge, urinary difficulties, congestion of the blood and water, lower abdominal fullness, and pain and constipation. This can also be used for congested Qi and blood in Lower Jiao causing irregular menstruation, with red and white leucorrhoea syndromes.

Related Formulas:

Formula	Herbs	Indication
Chi Fu Ling Tang [Red Poria Decoction]	Core Formula + Chi Fu Ling	Treats the same indication as Da Huang Mu Dan Tang [Rhubarb and Moutan Decoction]
Mei Ren Tang [Prune Seed Decoction]	Da Huang Mu Dan Pi Mei Ren (Japanese Plum) Dong Gua Ren Mang Xiao Xi Jiao (Shui Niu Jiao)	Treats the same indication as Da Huang Mu Dan Tang [Rhubarb and Moutan Decoction]

(continued)

Formula	Herbs	Indication
Fu Fang Da Huang Mu Dan Tang [Revised Rhubarb and Moutan Decoction]	Da Huang 6 g Yi Yi Ren 6 g Dong Gua Ren 15 g Mu Dan Pi 9 g Jin Yin Hua 15 g Huang Qin 21 g Pu Gong Ying 15 g Lian Qiao 30 g Tao Ren 6 g Dang Gui 9 g Chen Pi 3 g	Decoction Treats acute appendicitis
Fu Fang Hong Teng Jian [Revised Millettia Decoction]	Da Huang 5 g Mu Dan Pi 9 g Jin Yin Hua 12 g Hong Teng 60 g Pu Gong Ying 30 g Lian Qiao 12 g Mo Yao 9 g Ru Xiang 9 g Yan Hu Suo 6 g Gan Cao 3 g	Decoction Treats intestinal abscess pain
Hong Teng Yu Li Ren Jian [Millettia and Prune Seed Decoction]	Yu Li Ren 12 g Yi Yi Ren 30 g Dong Gua Ren 30 g Mu Dan Pi 9 g Bai Jiang Cao 15 g Hong Teng 30 g Pu Gong Ying 15 g Tao Ren 5 g	Decoction Treats acute appendicitis with abdominal pain due to stagnation and stasis or mild syndrome with pus formation

Lan Wei Hua Yu Tang [Transform Blood Stasis Decoction for Appendicitis] (Tian Jin Nan Kai Yi Yuan Fang)

Da Huang	10 g
Mu Dan Pi	10 g
Jin Yin Hua	15 g
Chuan Lian Zi	15 g
Mu Xiang	10 g
Yan Hu Suo	10 g
Tao Ren	10 g

Main Indication: Acute appendicitis, during the Qi stagnation and blood stasis stage but without severe heat. Symptoms include continuous dull or distending lower right quadrant abdominal pain, aversion to palpation and pressure, fullness and stifling sensation at the epigastrium, poor appetite, nausea, normal bowel movement or constipation, clear or yellow urination, white tongue coating with normal tongue proper or with purple spots, and wiry and tight pulse.

Preparation: Decoction. One bag a day.

Treatment Principles: Move Qi, activate blood, drain heat, and unblock the Fu organs.

Modification:

Addition	Indication
Hong Teng 30–60 g	Severe blood stagnation that forms into mass

Lan Wei Qing Hua Tang [Clear and Transform the Appendicitis Decoction] (Tian Jin Nan Tai Yi Yuan Fang)

Jin Yin Hua	30 g
Pu Gong Ying	30 g
Mu Dan Pi	15 g
Da Huang	15 g
Chuan Lian Zi	9 g
Chi Shao Yao	12 g
Tao Ren	9 g
Sheng Gan Cao	9 g

Main Indication: Acute appendicitis, during the heat accumulating stage with congested heat toxins. Symptoms include low-grade fever or afternoon fever, thirst, severe pain at lower right quadrant, constipation, dark urine, red tongue, yellow tongue coating, and slippery rapid pulse or rapid wiry pulse.

Preparation: Decoction. Two bags a day.

Treatment Principles: Clear heat, relieve toxins, move Qi, and activate blood.

Modification:

Addition	Indication
Huang Lian Huang Qin	For patient with more damp and heat
Pei Lan Bai Dou Kou Huo Xiang Geng Mu Tong	Patient with severe dampness

Lan Wei Qing Jie Tang [Clear and Relieve the Appendicitis Decoction] (Tian Jin Nan Kai Yi Yuan Fang)

Jin Yin Hua	60 g
Pu Gong Ying	30 g
Dong Gua Ren	30 g
Da Huang	24 g
Mu Dan Pi	15 g
Mu Xiang	9 g
Chuan Lian Zi	9 g
Sheng Gan Cao	9 g

Main Indication: Acute appendicitis, excess heat and toxin during the heat toxin stage. Symptoms include high fever, thirst, red face, red eyes, dry lips, dry throat, nausea causing inability to eat, lower right quadrant abdominal pain, refusing pressure or palpation, or with hard abdominal muscles in severe cases, constipation, scanty hesitant urination; yellow, greasy tongue coating, deep red tongue, flooding slippery rapid big pulse, or wiry rapid forceful pulse.

Preparation: Decoction. Two bags per day or four bags per day in severe cases, take both day and night.

Treatment Principles: Clear heat, relieve toxins, move Qi, and activate blood.

Modification:

Addition	Indication
Sheng Shi Gao 30 g Tian Hua Fen 15 g	Presenting with high fever and big thirst

Comparison of Formulas Treating Heat Accumulation in Stomach and Intestines

1. Zhi Zi Chi Tang [Gardenia and Prepared Soybean Decoction] treats beginning stage of heat accumulation in Yang Ming. The disease location is at the chest, with the main symptoms of restlessness, irritability, and feeling remorseful.
2. Qing Wei San [Clear the Stomach Powder] treats heat accumulation in the Yang Ming traveling along the channel up to the head, face, and cheeks area. Main symptoms are swelling and pain sensation in head, face, and gum areas.
3. Qing Yi Tang [Clear the Pancreas Decoction] treats congestion and stagnation in Gallbladder and Stomach type of acute pancreatitis. The obstruction leading to epigastric pain.
4. Da Huang Mu Dan Tang [Rhubarb and Moutan Decoction] treats beginning stage of acute appendicitis with the main symptoms of lower right quadrant abdominal pain. Lan Wei Hua Yu Tang [Transform Blood Stasis Decoction for Appendicitis], Lan Wei Qing Hua Tang [Clear and Transform the Appendicitis Decoction], and Lan Wei Qing Jie Tang [Clear and Relieve the Appendicitis Decoction] all treat acute appendicitis. However, Lan Wei Hua Yu Tang [Transform Blood Stasis Decoction for Appendicitis] is best for stagnation and stasis stage without too much heat. Lan Wei Qing Hua Tang [Clear and Transform the Appendicitis Decoction] is most suitable for heat accumulating stage with excess heat and toxins. Lan Wei Qing Jie Tang [Clear and Relieve the Appendicitis Decoction] is most suitable in treating severe stage of excess toxins with pus formation and the spreading of heat and toxins throughout the body.
5. Please consult your medical doctor first in case of diagnosis as acute pancreatitis or acute appendicitis.

FOOD ACCUMULATION AND STAGNATION

When food accumulates in the Stomach or stagnates the Middle Jiao, the commonly seen symptoms include abdominal distention, abdominal pain, fullness and focal distention at the chest and epigastrium, sour regurgitation, aversion to food, or nausea, diarrhea, or not smooth bowel movements with foul smell, yellow or greasy tongue coating, and slippery, excess and forceful pulse. This is an excess syndrome, thus herbs that reduce the accumulation and guide the stagnation to recover the transformation and transportation functions of Spleen and Stomach are often used. These herbs include Shan Zha, Shen Qu, Gu Ya, Mai Ya, Lai Fu Zi, and Ji Nei Jin. Food accumulation and Qi stagnation can transform into dampness with time, thus formulas in this category often combine with Bai Zhu, Fu Ling to tonify Spleen and drain damp, and Zhi Shi and Chen Pi to move Qi.

Severe accumulation and stagnation can be treated with downward draining herbs, such as Da Huang, Qian Niu, and Bing Lang to purge downward.

Long-term accumulation and stagnation with weak Spleen and Stomach is a pattern of both excess and deficiency. Thus, one should use both: tonifying the Spleen and guiding the stagnation treatment principles.

In general, if food accumulation is mainly due to carbohydrates, then Mai Ya and Gu Ya are often used. If it is caused by meat product, then Shan Zha is the main herb used. If caused by fruits and melons, then Ding Xiang and Rou Gui are added in the formula. If food accumulation and poor appetite are caused by imbalanced intestinal flora due to antibiotic use, then Shen Qu, fried Bai Zhu, and Mu Xiang are often used.

Bao He Wan [Preserve Harmony Pill] *Dan Xi Xin Fa*

Shan Zha	10 g
Shen Qu	12 g
Lai Fu Zi	10 g
Chen Pi	10 g
Ban Xia	10 g
Fu Ling	10 g
Lian Qiao	10 g

Main Indication: Food accumulation in the Middle Jiao. Symptoms include fullness and focal distention at the chest and epigastrium, abdominal distention sometimes with pain, belching, sour regurgitation, nausea, vomiting, diarrhea, thick greasy or yellow tongue coating, and slippery pulse.

Preparation: Decoction or make into pill form.

Treatment Principles: Reduce accumulation and guide stagnation.

This is an effective formula in treating food stagnation. In clinical practice, it can be effectively used in treating diarrhea and abdominal pain that is relieved with bowel movements or belching with foul, rotten egg smell without abdominal pain.

Modification:

Addition	Indication
Huang Lian fried with ginger juice	Food accumulation in the Middle Jiao with heat steaming upward, leading to sweating at head

Related Formulas:

Formula	Herbs	Indication
Jia Wei Bao He Wan [Augmented Preserve the Harmony Pill]	Core Formula + Mai Ya	Same indications
Da An Wan [Great Peace Pill]	Core Formula + Bai Zhu	Same indications

Zhi Shi Dao Zhi Wan [Immature Bitter Orange Pill to Guide the Stagnation] *Nei Wai Shang Bian Huo Lun*

Zhi Shi	10 g
Da Huang	10 g
Huang Qin	10 g
Huang Lian	10 g
Fu Ling	10 g
Ze Xie	6 g
Bai Zhu	10 g
Shen Qu	12 g

Main Indication: Damp–heat accumulation and stagnation obstructing the Middle Jiao. Symptoms include focal distention and stifling sensation at chest and epigastrium, or diarrhea with abdominal pain, heavy sensation at the anus, or constipation with yellow urine, red tongue greasy tongue coating, and deep excess pulse.

Preparation: Decoction. It can also be made into pills.

Treatment Principles: Clear heat, eliminate dampness, reduce accumulation, and guide stagnation.

This formula mainly treats excess type of damp–heat obstructing the Middle Jiao leading to diarrhea, uncontrolled diarrhea, or constipation.

Mu Xiang Bing Lang Wan [Aucklandia and Betel Nut Pill] *Ru Men Shi Qin*

Qian Niu Zi	60 g
Bing Lang	15 g
Da Huang	30 g
Huang Lian	15 g
Huang Bai	15 g
Mu Xiang	15 g
Chen Pi	15 g
Qing Pi	15 g
E Zhu	15 g
Xiang Fu	60 g

Main Indication: Internal heat with water and damp accumulation and Qi obstructing the Middle Jiao. Symptoms include epigastric fullness, focal distention and distending pain, difficulty urinating, constipation, or diarrhea with blood, mucous and tenesmus; thick, greasy tongue coating, flabby tongue or with scalloped edges, deep forceful or wiry pulse.

Preparation: Grind the ingredients into powder and form into pills with water at a size of 5 to 10 mm in diameter. Take three times a day.

Treatment Principles: Clear heat, eliminate dampness, guide the stagnation, and unblock the Fu organs.

This formula mainly treats severe cases of damp water accumulation and stagnation that obstruct the Qi in the Stomach and intestines, leading to urinary and bowel dysfunction. These are excess syndromes.

Related Formula:

Formula	Herbs	Indication
Mu Xiang Bing Lang Wan [Aucklandia and Betel Nut Pill]	Core Formula + Zhi Ke San Leng	Severe Qi stagnation

Zhi Shi Xiao Pi Wan [Immature Bitter Orange Pill to Reduce Focal Distention] *Lan Shi Mi Cang*

Dang Shen	10 g
Bai Zhu	10 g
Fu Ling	10 g
Zhi Gan Cao	3 g
Gan Jiang	10 g
Ban Xia	12 g
Zhi Shi	10 g
Hou Po	12 g
Mai Ya	10 g
Huang Lian	6 g

Main Indication: Spleen Qi deficiency, transformation and transportation dysfunction leading to internal accumulation and obstruction. Symptoms include fullness and focal distention at epigastrium, fatigue, apathy; or focal distention at the chest and abdomen, indigestion, irregular bowel movements, flabby tongue with scalloped edges, thick or greasy white tongue coating, soggy pulse.

Preparation: Decoction.

Treatment Principles: Move Qi, reduce focal distention, tonify Qi, and strengthen the Spleen.

This formula treats Spleen dysfunction leading to damp phlegm and food accumulation and obstruction. The key symptoms include focal distention and poor appetite.

Related Formulas:

Formula	Herbs	Indication
Xiao Pi Wan [Reduce the Focal Distention Pill]	Core Formula – Fu Ling Mai Ya + Shen Qu Sha Ren Chen Pi Zhu Ling Ze Xie Huang Qin Jiang Huang	Treats chronic stifling and focal distention at the epigastrium
Huang Lian Xiao Pi Wan [Reduce the Focal Distention Pill with Coptis]	Core Formula – Dang Shen Mai Ya Hou Po + Jiang Huang Ze Xie Chen Pi Huang Qin	Treats epigastric fullness and focal distention, stagnation and congestion, irritability, heat sensation, wheezing and restlessness that show more signs of damp–heat

Zhi Zhu Wan [Immature Bitter Orange and Atractylodes Macrocephala Pill] *Pi Wei Lun*

Zhi Shi	15 g
Bai Zhu	15 g

Main Indication: Spleen deficiency with food stagnation. Symptoms include abdominal distention, chest fullness, and focal distention. If it is in decoction form, it treats the plate-size area of hardness at the epigastrium that is due to Shui Yin fluid accumulation. Pale tongue with white or greasy tongue coating and deep pulse.

Preparation: Grind the ingredients into powder and form into pills with rice that are cooked in He Ye. Take 5 g per day. It can also be taken as a decoction.

Treatment Principles: Tonify Spleen and guide stagnation.

This formula is called Zhi Zhu Tang [Immature Bitter Orange and Atractylodes Macrocephala Decoction] in *Jin Gui Yao Lue*. It treats the plate-size area of hardness in the epigastrium that is caused by Shui Yin. This is because Spleen deficiency leading to water accumulation. Decoction will be more effective in treating this presentation. Pill has a slower action, thus it treats chronic distention and fullness due to Spleen deficiency.

Related Formulas:

Formula	Herbs	Indication
Qu Mai Zhi Zhu Wan [Medicated Leaven, Barley Sprout, Immature Bitter Orange, and Atractylodes Macrocephala Pill]	Core Formula + Shen Qu Mai Ya	Treats overeating leading to fullness and distending sensation at the chest and abdomen
Ju Ban Zhi Zhu Wan [Orange Peel, Pinellia, Immature Bitter Orange, and Atractylodes Macrocephala Pill]	Core Formula + Chen Pi Ban Xia	This formula is stronger in strengthening Spleen and reducing the focal distention function
Xiang Sha Zhi Zhu Wan [Aucklandia, Amomum, Immature Bitter Orange, and Atractylodes Macrocephala Pill]	Core Formula + Mu Xiang Sha Ren	Treats severe cases of Qi obstruction and food stagnation
San Huang Zhi Zhu Wan [Three-Yellow, Immature Bitter Orange, and Atractylodes Macrocephala Pill]	Core Formula + Da Huang Huang Qin Huang Lian Shen Qu Chen Pi	Treats fullness, stifling, distending discomfort due to overindulgence of meat, carbohydrates, spicy, hot and greasy foods which belong to excess heat syndrome

Jian Pi Wan [Strengthen the Spleen Pill] *Zheng Zhi Zhun Sheng*

Dang Shen	15 g
Bai Zhu	9 g
Fu Ling	12 g
Gan Cao	3 g
Shan Yao	15 g
Chen Pi	9 g
Sha Ren	6 g
Mu Xiang	6 g
Shan Zha	9 g
Shen Qu	12 g
Mai Ya	15 g
Huang Lian	3 g
Rou Dou Kou	9 g

Main Indication: Spleen deficiency with food stagnation. Symptoms include indigestion, epigastric and abdominal bloating and focal distention, loose and watery stool; greasy, yellow tongue coating, and weak and forceless pulse.

Preparation: Decoction. It can also be taken as pills.

Treatment Principles: Tonify the Spleen and guide stagnation.

Even though this formula is indicated for syndrome of both excess and deficiency, there is more deficiency than excess and the cause is the deficiency. Thus, deficiency is the root and excess is the branch.

Modification:

Minus	Addition	Indication
Huang Lian	Gan Jiang	Above presentation with coldness
Rou Dou Kou		Absence of watery and loose stool
	Ban Xia	With phlegm and damp

Comparison of Reduce Accumulation and Guide Stagnation Formulas

1. Bao He Wan [Preserve the Harmony Pill] is indicated for food stagnation and damp–heat obstructing the Middle Jiao leading to fullness, bloating, focal distention, and pain. Zhi Shi Dao Zhi Wan [Immature Bitter Orange to Guide the Stagnation Pill] is indicated for more severe cases of accumulation and stagnation, when other reducing food stagnation formulas are not effective. Mu Xiang Bing Lang Wan [Aucklandia and Betel Nut Pill] treats water and damp accumulation and stagnation leading to Qi obstruction in the Stomach and intestines. All three formulas focus on reducing accumulation.
2. Zhi Shi Xiao Pi Wan [Immature Bitter Orange to Reduce the Focal Distention Pill] treats epigastric focal distention and fullness and poor appetite due to Spleen deficiency with damp obstruction. Zhi Zhu Wan [Immature Bitter Orange and Atractylodes Macrocephala Pill] is a mild formula that treats chronic Spleen deficiency leading to chest focal distention and abdominal distention. Jian Pi Wan [Strengthen the Spleen Pill] treats Spleen deficiency leading to food stagnation, poor appetite, and loose stool. All three formulas are tonifying with some reduction.

SPLEEN QI DEFICIENCY

Spleen Qi deficiency often manifests as lethargy, fatigue, shortness of breath, breathlessness especially with movement or exercise, pale face and complexion, laconic speech, poor appetite, loose stool, and weak pulse especially at the guan position. The herbs that are commonly used are Ren Shen, Dang Shen, Bai Zhu, Fu Ling, Gan Cao, Shan Yao, Huang Qi, Sha Ren, and Bai Dou Kou to tonify Qi and strengthen the Spleen. One of the characteristics of Spleen is preference to dryness and aversion to dampness. Spleen deficiency often leads to damp accumulation and Qi obstruction, thus herbs like Chen Pi, Mu Xiang, Yi Yi Ren, and Sha Ren are often added to move Qi and dry dampness.

If Spleen deficiency leads to excess dampness pouring down and manifesting as leucorrhea, then Chun Gen Pi, Yin Xing, Qian Shi, and Lian Xu can be added to eliminate dampness and stop leucorrhea as in Wan Dai Tang [End Discharge Decoction].

Si Jun Zi Tang [Four-Gentleman Decoction] *He Ji Ju Fang*

Ren Shen	15 g
Bai Zhu	10 g
Fu Ling	12 g
Gan Cao	3 g

Main Indication: Spleen Qi deficiency. Symptoms include poor appetite, loose stool, pale sallow complexion, shortness of breath, laconic speech, weakness of the four limbs; thin, white coat with a slightly flabby tongue, and thin and soft pulse.

Preparation: Decoction.

Treatment Principles: Tonify Qi and strengthen the Spleen.

Ren Shen in this formula is often substituted with Dang Shen for mild case of Spleen deficiency. In more severe cases, Ren Shen should be used, especially if there is bleeding due to Qi not being able to control the blood. Ren Shen will be more effective than Dang Shen.

This formula can be used to treat chronic gastritis, neurasthenia, or weak constitution after illness with low energy, poor appetite, poor digestion, abdominal bloating, and fullness which indicates Spleen Qi deficiency.

Related Formulas:

Formula	Herbs	Indication
Qi Wei Bai Zhu San [Seven-Ingredient Actractylodes Macrocephala Decocotion]	Core Formula + Mu Xiang Huo Xiang Ge Gen	Spleen deficiency with heat sensation at the muscle and diarrhea This formula is also called Qian Shi Bai Zhu San [Qian Family's Atractylodes Macrocephala Decoction]
Yi Gong San [Extraordinary Merit Powder]	Core Formula + Chen Pi	Treats nausea, vomiting, diarrhea, and very low appetite
Liu Jun Zi Tang [Six-Gentleman Decoction]	Core Formula + Chen Pi Ban Xia	Treats Stomach deficiency, scanty food intake, nausea, vomiting or diarrhea, or cough with thin watery phlegm
Xiang Sha Liu Jun Zi Tang [Six-Gentleman Decoction with Aucklandia and Amomum]	Core Formula + Chen Pi Ban Xia Mu Xiang Sha Ren	Treats focal distention and fullness due to Qi deficiency and phlegm, food, and Qi stagnation, or nausea and vomiting during pregnancy due to Spleen deficiency
Si Shou Yin [Four Beasts Decoction]	Core Formula + Chen Pi Ban Xia Cao Guo Wu Mei	Treats malaria with phlegm and food stagnation due to Middle Jiao deficiency
Xing Fu Liu Jun Zi Tang [Six-Gentleman Decoction with Arisaemae and Typhonii]	Core Formula + Chen Pi Ban Xia Zhi Nan Xing Bai Fu Zi	Treats wind phlegm migraine headache at Yang Ming
Wu Xie Si Jun Zi Tang [Four-Gentleman Decoction with Aconite and Scorpion]	Core Formula + Chuan Wu (or Wu Gong) Quan Xie Tian Ma	Treats chronic vomiting and diarrhea in kids turning into chronic convulsion due to weak Spleen
Wu Xie Liu Jun Zi Tang [Six-Gentleman Decoction with Aconite and Scorpion]	Core Formula + Chen Pi Ban Xia Chuan Wu Quan Xie	Treats Spleen deficiency wind stroke with excess phlegm, deviating mouth and eyes, and slurred speech Chuan Wu can be substituted with Wu Gong
Zhu Mai Liu Jun Zi Tang [Six-Gentleman Decoction with Lophatherus and Ophiopogonis]	Core Formula + Chen Pi Ban Xia Mai Men Dong Zhu Li (Lophatherus Sap)	Treats Spleen deficiency with phlegm obstruction leading to weakness and flaccidity of the four limbs
Wen Pi San [Warm the Spleen Powder]	Core Formula + Huang Qi Huo Xiang He Zi Ju Pi Jie Geng Yi Yi Ren Sha Ren	If the formula is taken in powder form, then take 10 g per day. Otherwise decoct with ginger and Da Zao Treats Spleen and Heart deficiency in kids, manifesting as playing with the tongue
Sheng Chai Liu Jun Zi Tang [Six-Gentleman Decoction with Cimicifuga and Bupleurum]	Core Formula + Sheng Ma Chai Hu Chen Pi Ban Xia	Treats bloody urination in female due to overindulgence of sexual activity

(continued)

Formula	Herbs	Indication
Zha Qu Liu Jun Zi Tang [Six-Gentleman Decoction with Hawthorn Berry and Medicated Leaven]	Core Formula + Chen Pi Ban Xia Shan Zha Shen Qu Mai Ya	Treats fatigue, apathy, and sleepiness after meal due to Spleen deficiency
Huang Lian Liu Jun Zi Tang [Six-Gentleman Decoction with Coptis]	Core Formula + Chen Pi Ban Xia Huang Lian fried with ginger juice	Treats patient who is hungry but cannot eat due to Stomach deficiency If Wu Zhu Yu is added, then this formula can treat recurrent rumbling sounds that are decreased with food intake and acid sensation due to excess heat
E Jiao Yi Gong San [Extraordinary Merit Powder with Ass-Hide Gelatin]	Core Formula + Chen Pi E Jiao	Treats consumptive symptoms due to chronic coughing up blood, emaciation, fatigue, weakness of the four limbs, five center heat, restlessness, dry throat, red cheeks, malar flush, and night sweats Huang Qi, Bie Jia, Mai Men Dong, and Wu Wei Zi can be added to strengthen the effect
Jia Wei Liu Jun Zi Tang [Augmented Six-Gentleman Decoction]	Core Formula + Cang Zhu Chen Pi Ban Xia Sheng Ma Chai Hu Sheng Jiang	Treats weak Stomach with phlegm, poor appetite, and frequent leucorrhoea
Chai Shao Liu Jun Zi Tang [Six-Gentleman Decoction with Bupleurum and Peony]	Core Formula + Chen Pi Ban Xia Chai Hu Bai Shao Yao Dang Gui	Treats abdominal pain or dysmenorrhea due to Spleen deficiency. It is effective in treating diarrhea with menstruation, or chronic low-grade fever due to deficient Spleen and depressed Liver

Shen Ling Bai Zhu San [Ginseng, Poria, and Atractylodes Macrocephala Powder] *He Ji Ju Fang*

Dang Shen	15 g
Bai Zhu	10 g
Fu Ling	15 g
Gan Cao	3 g
Shan Yao	15 g
Bai Bian Dou	12 g
Lian Zi	15 g
Yi Yi Ren	24 g
Sha Ren	9 g
Jie Geng	6 g
Chen Pi (is added in *Yi Fang Ji Jie*)	6 g

Main Indication: Spleen deficiency, damp obstruction, ascending and descending dysfunction. Symptoms include indigestion, focal distention at chest, stifling sensation in the epigastric, or vomiting or diarrhea, weakness of the four limbs, emaciation, pale flabby tongue with white tongue coating, and deficient pulse.

Preparation: Grind herbs into powder, taken with water. It can also be taken as decoction. The powder can be steamed with lean meat for children to eat.

Treatment Principles: Tonify Qi, strengthen Spleen, ascend the clear and descend the turbid.

This formula is an elegant formula in its herbal combination. It tonifies deficiency, eliminates dampness, with some ascends and some descends. It is especially suitable for pediatric malnutrition due to Spleen deficiency. It is also effective in treating Spleen deficiency with severe dampness in females with symptoms of white vaginal discharge, pale face, obesity, loose stool, swelling of feet, or diarrhea during menstruation. It can also treat chronic nephritis and chronic proteinuria due to Spleen deficiency failing to astringe essence.

Related Formulas:

Formula	Herbs	Indication
Jia Jian Shen Ling Bai Zhu San [Modified Ginseng, Poria, and Atractylodes Macrocephala Powder]	Pao Shen 10 g Fu Shen 6 g Bai Zhu 10 g Gan Cao 6 g Mu Xiang 6 g Sha Ren 3 g Shan Yao 12 g Bai Bian Dou 12 g	Treats irregular menstruation with scanty, thin light red menstrual blood, sometimes with sticky discharge, white moist tongue coating, pale tongue, and soggy pulse Add burned Ai Ye 9 g for abdominal pain Add Du Zhong 12 g and Xu Duan 9 g for lower back pain
Jie Geng Gan Cao Tang [Platycodon and Licorice Decoction] *Shang Han Lun* also known as Jie Geng Tang [Platycodon Decoction]	Jie Geng 4 g Sheng Gan Cao 6 g	Clear heat, relieve toxicity, benefit the throat and stop pain Treats sore throat due to heat in Shao Yin This formula can also be used in Lung abscess syndrome

Wan Dai Tang [End Discharge Decoction] *Fu Qing Zhu Nu Ke*

Ren Shen	6 g
Shan Yao	30 g
Bai Zhu (fried)	30 g
Gan Cao (fried)	3 g
Chen Pi	6 g
Cang Zhu	12 g
Che Qian Zi	10 g
Bai Shao Yao (fried with wine)	15 g
Hei Jie Sui (fried with wine)	2 g
Chai Hu	3 g

Main Indication: Spleen deficiency, excess damp, downward pouring of damp turbid. Symptoms include profuse white vaginal discharge that is sometimes sticky thick and sometimes thin, no foul smell, continuous in severe cases, pale complexion, overweight or obesity, loose stool or swollen feet, white tongue coating, pale flabby tongue or with scalloped edges, and soggy pulse.

Preparation: Decoction.

Treatment Principles: Tonify the Spleen and transform dampness.

This classic formula can tonify the Spleen, smooth the Liver, and also ascend and clear Yang. When Spleen is strengthened, damp will be eliminated and then the vaginal discharge will be cleared.

Modification:

Addition	Indication
Wu Zei Gu Qian Shi Mu Li (Calcined)	Profuse vaginal discharge

Related Formula:

Formula	Herbs	Indication
Jia Jian Wan Dai Tang [Modified End Discharge Decoction]	Pao Shen 12 g Bai Zhu 9 g Bai Shao Yao 6 g Cang Zhu 9 g Gan Cao 3 g Jing Jie (fried) 3 g Chai Hu 2 g Shan Zhi Zi 6 g Huang Bai 6 g Huang Lian 3 g	Decoction Treats vaginal discharge that is yellow, sticky, foul smelling, with occasional abdominal distending pain, regular bowel movements and urination, white or slightly yellow greasy tongue coating, and soggy and rapid pulse If there is also vaginal itching, omit Bai Zhu and add She Chuang Zi 6 g and Jin Yin Hua 6 g

Yi Huang Tang [Change Yellow Decoction] *Fu Qing Zhu Nu Ke*

Shan Yao	30 g
Qian Shi (fried)	30 g
Huang Bai (fried)	6 g
Che Qian Zi (fried with salt water)	3 g
Bai Guo (fried with wine)	10 pieces, crushed

Main Indication: Spleen deficiency, damp obstruction, downward pouring of damp–heat. Symptoms include long–term, unremitting yellow vaginal discharge, sallow complexion, dizziness, vertigo, decreased appetite, delayed menstruation with pale menstrual blood, occasional loose stool, pale yellow urination, and soft and slippery pulse.

Preparation: Decoction.

Treatment Principles: Tonify the Spleen, leach out dampness, clear heat, and stop vaginal discharge.

This formula is designed to treat Spleen deficiency with damp accumulation transforming into heat draining downward, manifesting as yellow vaginal discharge.

Modification:

Addition	Indication
Zhu Ling Chi Shao Yao Mu Dan Pi	Treats severe cases of damp–heat, yellow vaginal discharge

Comparison of Formulas Treating Spleen Qi Deficiency

Si Jun Zi Tang [Four-Gentleman Decoction] is neutral in temperature and is a basic formula to tonify Qi. Shen Ling Bai Zhu San [Ginseng, Poria, and Atractylodes Macrocephala Powder] treats Spleen deficiency and excess damp leading to diarrhea and vaginal discharge. It is especially effective in treating kids. Wan Dai Tang [End Discharge Decoction] treats Spleen deficiency with damp excess pouring downward leading to damp–cold type vaginal discharge. Yi Huang Tang [Change Yellow Decoction] treats Spleen deficiency with damp–heat pouring downward leading to yellow vaginal discharge.

SPLEEN DEFICIENCY WITH QI SINKING

In this pattern, symptoms of the upper body include shortness of breath, laconic speech, and breathlessness. Symptoms of the lower body include uterine prolapse, heavy or prolonged uterine bleeding, lochiorrhea, Beng Lou, leucorrhea, urinary incontinence, enuresis, or urinary difficulties, constipation, rectal prolapse and blood in stool. On the basis of tonifying Qi and strengthening the Spleen, Yang lifting herbs such as Sheng Ma and Chai Hu are added. Huang Qi is almost a must-use herb in Qi deficiency and sinking. It not only can tonify Qi, but also lift Yang Qi.

Furthermore, sinking clear Yang and internal stagnation of the Yang Qi can manifest as fever, spontaneous sweating, thirst with preference of hot drinks, and big but empty pulse which indicates false heat syndrome. When the method of lifting the sunken Yang is used, it can raise the clear Yang and spread the Yang Qi to the surface, thus clear heat. This is often referred to as "sweet warm to clear heat."

"Raising Yang and lifting the sunken" is also a method to correct inappropriate treatment of damp–heat. Damp–heat in the Middle Jiao should be treated with acrid herbs to open, bitter herbs to drain, aromatic herbs to transform, and bland herbs to leach. However, if downward draining method is used by mistake, then it can cause sinking of Yang in the Middle Jiao leading to diarrhea or watery stool. Then, at this time, one should add Chai Hu, Huang Qi, He Ye, Qing Hao, and other herbs to raise the clear Yang and clear damp. However, in formulation, one should use more dry dampness and transform turbid herbs and less drain damp herbs. Higher dosage of drain damp herbs can hinder the function of raising Yang.

Bu Zhong Yi Qi Tang [Tonify the Middle and Augment the Qi Decoction] *Pi Wei Lun*

Huang Qi	24 g
Ren Shen	15 g
Bai Zhu	10 g
Gan Cao	6 g
Chen Pi	9 g
Dang Gui	10 g
Sheng Ma	6 g
Chai Hu	6 g

Main Indication: Spleen deficiency and Qi sinking. Symptoms include:

1. Poor appetite, no taste, laconic speech, fatigue, weak limbs, shortness of breath, and breathlessness with exertion, and empty, big and weak pulse.
2. Fever, spontaneous sweating, and thirst with preference for hot drinks, headache, aversion to cold, laconic speech, poor appetite, and deficient, big pulse.
3. Beng Lou, rectal prolapse, uterine prolapse, blood in stool, chronic dysentery, chronic malarial disorder, difficult, frequent, painful or incontinent urination, and other clear Yang sinking syndromes.
4. Low blood pressure, gastroptosis, myasthenia gravis, neurasthenia, insomnia, and nocturnal emission all belong to deficient and sunken Qi.

Preparation: Decoction. If the formula is made into honey pill, then the dosage should be increased.

Treatment Principles: Tonify the middle, augment the Qi, raise the Yang, and lift the sunken.

The formula can be used in a variety of conditions. The effectiveness is strengthened with appropriate modification.

Modification:

Minus	Addition	Indication
	Double dosage of Huang Qi and Sheng Ma	To treat uterine prolapse and all other prolapse of internal organs
	Mu Xiang	For excessive worrying, Spleen Yang sinks into Lower Jiao, causing diarrhea
	He Zi Bai Dou Kou Wu Wei Zi Wu Mei	Treat chronic diarrhea, weakness of the anus, fecal incontinence, and rectal prolapse
Dang Gui	Mu Xiang	Dysentery leading to Qi collapse, rectal prolapse, internal urgency, and fecal incontinence
	Fang Feng	Treat Qi deficiency and sinking type of constipation, tenesmus, or with scanty, white pus after having dysentery
	Pao Jiang Chi Shi Zhi	Treats stools with pus and blood followed by stools with white foam and white pus
	Bai Dou Kou Mu Xiang Zhu Che Wan [Vessel and Vehicle Pill]	Treat intermittent dysentery
	Double dosage of Sheng Ma Chai Hu Dang Gui	Decoct until becomes a thick paste. Add sesame seed oil and Ma Zi Ren Wan [Hemp Seed Pill] to treat constipation due to Spleen deficiency, fatigue, and laconic speech
	Mu Tong Che Qian Zi	Treat sinking Yang Qi and blocked urination
	Che Qian Zi Ze Xie	Treat Spleen deficient type of Lao Lin (Exhausted Lin) syndrome that is recurrent with over exertion
	Mai Men Dong Wu Wei Zi Liu Wei Di Huang Wan [Six-Ingredient Pill with Rehmannia]	Treat Kidney deficient type of Lao Lin syndrome
	Mu Tong Ze Xie	Treat deficient and sinking Qi leading to Lin syndrome
	Fu Zi	For urinary incontinence worse during the day
	Shan Yao Wu Wei Zi	For deficient Spleen, weak Qi leading to frequent urination worse with exertion
	Suo Quan Wan [Shut the Sluice Pill]	Treat profuse urination and seminal emission after bowel movement
	Pao Jiang	Treat premenstrual diarrhea due to Spleen deficiency Add Huang Lian if there is heat
	Bai Shao Yao (fried with wine)	Treat Beng Lou due to Spleen failing to hold Add Huang Qin if there is heat in Liver
	Wu Zei Gu	Fry Wu Zei Gu first, then grind into powder. Take the powder with vinegar then follow by the core formula Treat chest pain after uterine bleeding, which is Spleen and Heart blood deficiency
Bai Zhu Dang Gui Chen Pi	Bai Shao Yao Wu Wei Zi	Treat tingling and numbness of the face and ten fingers
	Sheng Di Huang Shao Yao	Treat weakness, numbness and tingling of limbs, itching with white flakes
	Yi Zhi Ren	Treat wind stroke urinary incontinence
	Double dosage of Huang Qi	Treat bleeding of the nose and mouth

Related Formula:

Formula	Herbs	Indication	Notes
Sheng Xian Tang [Raise the Sinking Decoction]	Huang Qi 20 g Zhi Mu 10 g Chai Hu 5 g Sheng Ma 3 g Jie Geng 5 g Taken as decoction	For sinking Qi of the chest with shortness of breath, or such difficulty breathing that breathing almost stops, deep slow minute and weak pulse, especially between cun and guan position, or incomplete presence of the six pulses	For severe Qi deficiency with Qi sinking, add Ren Shen, Shan Zhu Yu to astringe the scattered Qi If there is also sinking sensation or pain at lower abdomen, increase Sheng Ma to 6 g

Jia Jian Bu Zhong Yi Qi Tang [Modified Tonify the Middle and Augment Qi Decoction] *Zhong Yi Fu Ke Zhi Liao Xue*

Huang Qi	10 g
Dang Shen	10 g
Bai Zhu	6 g
Chen Pi	6 g
Sheng Ma	3 g
Chai Hu	3 g
E Jiao	6 g
Ai Ye (burned)	6 g
Gan Cao	3 g

Main Indication: Restless fetus due to Qi deficiency and sinking. Symptoms include weak constitution, 4 to 5 months of pregnancy with lower back soreness, abdominal distention or with sinking sensation, fatigue, apathy, restless fetus, scanty vaginal spotting, flabby tongue, white tongue coating, and slippery and forceless pulse.

Preparation: Decoction.

Treatment Principles: Tonify Qi, calm the fetus, raise Yang, and lift the sunken.

This formula can be used for spotting during pregnancy due to Qi deficiency and to treat or prevent habitual miscarriage.

Modification:

Addition	Indication
Du Zhong Xu Duan Sang Ji Sheng Wu Zei Gu	Kidney deficiency. These herbs tonify Kidney and stop bleeding

Related Formulas:

Formula	Herbs	Indication
Jia Jian Bu Zhong Yi Qi Tang [Modified Tonify the Middle and Augment Qi Decoction]	Bu Zhong Yi Qi Tang [Tonify the Middle and Augment Qi Decoction] + Zhi Ke Yi Mu Cao Taken as decoction on empty stomach	Treat uterine prolapse, profuse clear urination, aversion to cold, apathy, sinking and distending sensation of the lower abdomen, lower back pain and soreness, thin white tongue coating, and weak pulse
Jia Jian Bu Zhong Yi Qi Tang [Modified Tonify the Middle and Augment the Qi Decoction]	Above formula – Sheng Ma Chai Hu Zhi Ke	Treat Qi deficiency, lochial retention, shortness of breath, apathy, abdominal distention, and excess lochial discharge

Ju Yuan Jian [Lift the Source Decoction] *Jing Yue Quan Shi*

Huang Qi	24 g
Ren Shen	9 g
Bai Zhu	9 g
Zhi Gan Cao	3 g
Sheng Ma	6 g

Main Indication: Deficient and sinking Qi, failing to control blood. Symptoms include heavy or prolonged menstruation, pale thin watery menstrual blood, pale sallow face, shortness of breath, laconic speech, fright, palpitation, aversion to cold, empty and sinking sensation of the lower abdomen, weakness of the four limbs, pale tongue and weak pulse. It can also treat Beng Lou due to Qi deficiency, sudden heavy bleeding, or prolonged spotting.

Preparation: Decoction.

Treatment Principles: Tonify Qi and raise Yang.

Modification:

Addition	Indication
Wu Zei Gu Qian Cao Wu Mei	Excessive menstrual flow
Huang Qin	Liver heat

Sheng Ma Huang Qi Tang [Cimicifuga and Astragalus Decoction] *Yi Xue Zhong Zhong Can Xi Lu*

Huang Qi	15 g
Dang Gui	12 g
Sheng Ma	6 g
Chai Hu	6 g

Main Indication: Sinking of the Qi mechanism. Symptoms include difficulty or dribbling urination. Sometimes nausea, vomiting or cough, or lying curled up on one side can help with the urination. Tongue is pale with thin, white coat, and pulse is deep and weak, especially the Middle Jiao.

Preparation: Decoction.

Treatment Principles: Tonify Qi and raise the sinking.

This formula is used to treat dribbling or difficulty in urination due to Qi deficiency and sinking. When ascending function of the Qi mechanism is affected, then Qi cannot descend either. Thus, raising the sinking can unblock and descend the Qi.

Comparison of Tonify Qi and Raise the Sinking Formulas

1. All four formulas use Ren Shen, Huang Qi, Sheng Ma, and Chai Hu to tonify Qi and raise the sinking functions.
2. Bu Zhong Yi Qi Tang [Tonify the Middle and Augment Qi Decoction] can be used in variety of conditions. It is indicated for Qi sinking syndromes in upper and lower part of the body or urinary and rectal conditions. Jia Jian Bu Zhong Yi Qi Tang [Modified Tonify the Middle and Augment Qi Decoction] is suitable for restless fetus due to Qi deficiency with vaginal spotting. Thus, Dang Gui, warm and acrid, is eliminated. E Jiao and Ai Ye, Yin and moistening are added. Ju Yuan Jian [Lift the Source Decoction] is

indicated for heavy menstruation, or Beng Lou due to Qi deficiency failing to control blood. Sheng Ma Huang Qi Tang [Cimicifuga and Astragalus Decoction] is designed for difficulty in urination due to sinking of the Qi mechanism.

STOMACH QI REBELLION

The normal Stomach Qi flow should descend. When it does not, then Stomach Qi becomes rebellious leading to nausea, vomiting, and hiccup. Herbs such as Ju Pi, Zhu Ru, Hou Po, Huo Xiang, Ding Xiang, Dai Zhe Shi, Wu Zhu Yu, and Shi Di can descend rebellious Qi and regulate the Stomach.

Stomach Qi rebellion can be further divided into cold and heat type or deficient and excess type. For cold syndromes, Wu Zhu Yu and Ding Xiang can warm the middle and descend the rebellious Qi. It is often combined with Gan Jiang and Rou Gui. For heat syndromes, Zhu Ru and Dai Zhe Shi are used to clear heat and descend the rebellious Qi. It is often combined with Huang Qin and Huang Lian. In deficient cases, one should combine with Ren Shen and Gan Cao. In excess cases, one should combine with Fan Xie Ye, Da Huang, and Zhi Shi. Shi Di and Dao Dou are used in hiccups.

If phlegm damp accumulates in Stomach causing Qi rebellious nausea, vomiting, and belching with focal distention of the chest, excess phlegm, and thick greasy tongue coating, then it is suitable to combine with Ban Xia and Sheng Jiang.

If Liver overacts on Spleen, or Gallbladder and Stomach are not harmonized leading to nausea, vomiting, and hiccup, then it is often combined with Bai Shao Yao, Bai Zhu, Huang Qin, and Zhu Ru. If the symptoms also include cough and wheezing due to Lung Qi rebellion, then Pi Pa Ye and Su Ye are added to unblock and circulate Lung Qi to assist the descending of the Stomach Qi.

As for unblocking the bowel and promoting urination to treat nausea, vomiting, and hiccups there is a method to open and drain the Lower Jiao, and a method to treat below for above pathological condition.

Ju Pi Zhu Ru Tang [Tangerine Peel and Bamboo Shaving Decoction] *Jin Gui Yao Lue*

Ju Pi	20 g
Zhu Ru	20 g
Sheng Jiang	9 g
Ren Shen	6 g
Gan Cao	10 g
Da Zao	10 g

Main Indication: Stomach deficient, Qi rebellion leading to hiccup. More often there is heat accumulation in the Stomach with Stomach deficiency resulting in rebellious Qi. Tongue coating is thin, yellow, and there is a slight rapid pulse at guan position.

Preparation: Decoction. Taken three doses a day.

Treatment Principles: Regulate the middle and descend rebellious Qi.

This formula can be used for acute or chronic gastritis, functional disorder of the Stomach, morning sickness, nausea, vomiting, and hiccup during pregnancy.

Modification:

Addition	Indication
Ban Xia Hou Po Shi Di	Severe nausea, vomiting, and hiccup

Related Formulas:

Formula	Herbs	Indication
Huo Ren Ju Pi Zhu Ru Tang [Tangerine Peel and Bamboo Shaving Decoction from the Book to Safeguard Life]	Core Formula + Ban Xia	Treats the same indication as the core formula. Also treats pediatric vomiting of milk and whooping cough
Zhu Ru Tang [Bamboo Shaving Decoction]	Core Formula – Ren Shen Da Zao + Ban Xia	Treats phlegm-damp type of hiccup and Qi rebellion
Ren Shen Zhu Ru Tang [Ginseng and Bamboo Shaving Decoction]	Core Formula – Da Zao + Ban Xia	For retching and Qi rebellion
Ji Sheng Ju Pi Zhu Ru Tang [Tangerine Peel and Bamboo Shaving Decoction from Formulas that Aid the Living]	Core Formula + Fu Ling Ban Xia Mai Men Dong Pi Pa Ye	Treats Stomach heat, frequent thirst, dry heaves, and inability to eat food
Wai Tai Da Ju Pi Tang [Big Tangerine Peel Decoction from Wai Tai Mi Yao]	Core Formula – Zhu Ru Da Zao	Treats dry heaves, chest fullness, and restlessness due to common cold and flu
Ren Shen Ju Pi Zhu Ru Tang [Ginseng, Tangerine Peel, and Bamboo Shaving Decoction]	Core Formula – Da Zao + Hou Po Ban Xia Huo Xiang	Treats nausea and Qi rebellion due to Stomach deficiency and damp obstruction
Ju Pi Zhu Ru Tang [Tangerine Peel and Bamboo Shaving Decoction]	Core Formula – Ren Shen Da Zao Sheng Jiang + Ban Xia Fu Ling Huang Lian Ge Gen	Treats heat congested in the Stomach, nausea, and dry heaves
Jin Gui Ju Pi Tang [Tangerine Peel from the Golden Cabinet]	Ju Pi Sheng Jiang	Treats slight cold type of dry heaves
Xin Zhi Ju Pi Zhu Ru Tang [Newly Formulated Tangerine Peel and Bamboo Shaving Decoction]	Core Formula – Ren Shen Gan Cao Da Zao + Shi Di	Treats excess type Stomach heat leading to hiccup and Qi rebellion

Jia Wei Huang Lian Su Ye Tang [Augmented Coptis and Perilla Decoction] *Xia Men Yi Xue Za Zhi*

Huang Lian	2 g
Su Ye	3 g
Ban Xia	9 g
Fu Ling	9 g
Zhu Ru	9 g
Pi Pa Ye	9 g
Shi Di	9 g

Main Indication: Disharmony of the Lung and Stomach, Stomach Qi rebellion. Symptoms include frequent nausea and vomiting, or cough with phlegm; yellow or greasy tongue coating, with slippery, rapid, and forceful pulse, especially at the guan position.

Preparation: Decoction. Taken the decoction cold and frequently.

Treatment Principles: Clear heat and descend the rebellion.

This formula clears Stomach heat, unblocks Lung and Stomach, eliminates dampness from the Stomach, disperses Lung Qi, and stops nausea and vomiting. Thus, it is suitable for concurrent symptoms of nausea and cough, or nausea leading to cough, or cough leading to nausea. It can also be used for Stomach heat and Lung Qi obstruction type of nausea and morning sickness during pregnancy.

Qin Lian Ju Ru Tang [Scutellaria, Coptis, Tangerine Peel, and Bamboo Shaving Decoction] (Shang Hai Di Yi Yi Xue Yuan Fang)

Huang Qin	9 g
Huang Lian	3 g
Shan Zhi Zi (burned)	9 g
Zhu Ru	12 g
Chen Pi	12 g
Ban Xia	9 g
Wu Zhu Yu	3 g
Shi Hu	12 g
Bai Shao Yao	12 g

Main Indication: Disharmony of Liver and Stomach, Stomach Qi rebellion. Symptoms include nausea and vomiting during pregnancy, dizziness, vertigo, yellow tongue coating, and slippery rapid pulse.

Preparation: Decoction. Take frequently.

Treatment Principles: Clear heat, unblock Liver, harmonize the Middle Jiao, and descend the rebellion.

The strength of this formula lies in treating Liver Qi stagnation and depression overacting on the Stomach leading to nausea, vomiting, and poor appetite during pregnancy.

Modification:

Minus	Indication
Shi Hu	Above symptoms with excess dampness and thick greasy tongue coating

Related Formula:

Formula	Herbs	Indication
Yi Qing Wan [Supress the Green Pill]	Huang Lian	Ground Huang Lian into powder; add water to make pills the size of Ma Zi Ren. 20 to 30 pills each time Treats Liver stagnation, overacting on Stomach, and Stomach heat leading to nausea and vomiting

Ban Xia Fu Ling Wan [Pinellia and Poria Pill] *Qian Jin Yao Fang*

Ban Xia	12 g
Fu Ling	12 g
Chen Pi	9 g
Gan Cao	3 g
Sheng Jiang	6 g
Xuan Fu Hua (wrapped)	10 g
Zhi Xi Xin	3 g
Jie Geng	6 g
Ren Shen	6 g
Sheng Di Huang	12 g
Bai Shao Yao	18 g
Chuan Xiong	6 g

Main Indication: Spleen deficiency with excess damp accumulation, and disharmony of Liver and Stomach. Symptoms include morning sickness, restlessness, stifling sensation, nausea, vomiting, aversion to smell of food, dizziness, heaviness of the body, heavy and painful limbs and joints, preference for sleeping rather than getting up, yellow, thin or greasy tongue coating, and wiry slippery pulse.

Preparation: Decoction.

Treatment Principles: Nourish blood, soften the Liver, expel phlegm, and descend rebellious Qi.

This formula is indicated for morning sickness of a woman with weak constitution and Liver blood deficiency, leading to damp accumulation and disharmony of Liver and Stomach.

Modification:

Minus	Addition	Indication
Ju Pi Xi Xin	Qian Hu Zhi Mu	With restlessness and thirst, ulcerations on tongue and mouth
Sheng Di Huang	Gui Xin	Slight cold type of diarrhea
Sheng Di Huang	Da Huang Huang Qin	Poor appetite, scanty urination, and stool retention, and deficient heat in the Stomach

Ding Xiang Shi Di Tang [Clove and Persimmon Calyx Decoction] *Zheng Yin Mai Zhi*

Ding Xiang	10 g
Shi Di	10 g
Sheng Jiang	12 g
Ren Shen	10 g

Main Indication: Middle Jiao deficient cold. Symptoms include hiccup and Qi rebellion due to internal injury or chronic sickness. Symptoms are better with heat or lying down. Other symptoms include distending sensation at chest, pale tongue with thin white tongue coating, and slow pulse.

Preparation: Decoction.

Treatment Principles: Warm the Middle Jiao and descend the rebellion.

This formula is often used to treat nausea and vomiting due to chronic gastritis, duodenum ulcer, and chronic cholecystitis.

Related Formulas:

Formula	Herbs	Indication
Shi Di Tang [Persimmon Calyx Decoction]	Core Formula – Ren Shen	Treat chest fullness with continuous hiccup. Ren Shen is eliminated to prevent Qi stagnation
Shi Qian San [Persimmon Coin Powder]	Core Formula – Sheng Jiang	Grind into powder Treat mild cold type of hiccups
Bian Que Ding Xiang San [Clove Powder of Bian Que]	Core Formula – Sheng Jiang + Gan Jiang	Treats severe cold type of hiccups
Ding Xiang Shi Di Zhu Ru Tang [Clove, Persimmon Calyx, and Bamboo Shaving Decoction]	Core Formula – Ren Shen Sheng Jiang + Ju Pi Zhu Ru	Treats severe nausea due to excess phlegm

Xiao Ban Xia Tang [Minor Pinellia Decoction] *Jin Gui Yao Lue*

Ban Xia	15 g
Sheng Jiang	9 g

Main Indication: Rebellious turbid phlegm. Symptoms include phlegm damp constitution, nausea vomiting with copious white clear watery phlegm, distending sensation at chest, difficulty in swallowing, absence of thirst, greasy tongue coating, and soggy pulse.

Preparation: Decoction.

Treatment Principles: Expel phlegm and descend the rebellion.

Related Formulas:

Formula	Herbs	Indication
Ban Xia Gan Jiang San [Pinellia and Dry Ginger Powder]	Ban Xia Gan Jiang	Treat dry heaves, nausea, vomiting, vomiting saliva, and cold in Stomach
Sheng Jiang Ban Xia Tang [Fresh Ginger and Pinellia Decoction]	Ban Xia Fresh Ginger Juice	Treat Tan Yin obstructing at epigastrium or sometimes with visible protrusion
Xiao Ban Xia Jia Fu Ling Tang [Minor Pinellia Decoction with Poria]	Core Formula + Fu Ling	Treat sudden nausea, vomiting, focal distention at epigastrium, accumulation of water around diaphragm, vertigo and dizziness or thirst followed by nausea with water accumulating below the Heart
Gan Jiang Ren Shen Ban Xia Wan [Ginger, Ginseng, and Pinellia Pill]	Gan Jiang Ren Shen Ban Xia	Grind into powder, then form pills Treats continuous nausea and vomiting during pregnancy due to deficient cold of Middle Jiao

Da Ban Xia Tang [Major Pinellia Decoction] *Jin Gui Yao Lue*

Ban Xia	15–30 g
Ren Shen	10 g
Honey	30 g

Main Indication: Middle Jiao deficient cold. Symptoms include vomiting a few hours after meals (such as vomiting in the afternoon after having breakfast or vomiting in the morning after having dinner), hardness and focal distention at the epigastrium, fatigue, apathy, weak constitution, pale tongue with thin, white coating, and weak and deficient pulse.

Preparation: Decoct the first two herbs, strain the decoction, and then add honey.

Treatment Principles: Harmonize the middle and descend the rebellion.

This formula treats nausea and vomiting due to obstruction or narrowing of the cardiac or pyloric sphincter. It can also be used for chronic nausea and vomiting leading to hardness and focal distention at the epigastrium.

Modification:

Addition	Indication
Zhi Ke Chen Pi	With Qi stagnation

Xuan Fu Dai Zhe Tang [Inula and Haematite Decoction] *Shang Han Lun*

Xuan Fu Hua (wrapped)	15 g
Dai Zhe Shi (decoct first)	30 g
Ban Xia	12 g
Sheng Jiang	9 g
Ren Shen	9 g
Gan Cao	6 g
Da Zao	4 pieces

Main Indication: Middle Jiao deficiency with congealed phlegm and Qi rebellion. Symptoms include focal distention in the epigastrium, belching, and nausea with an aversion to food, white greasy tongue coating, and wiry pulse. It also treats Stomach turbidity failing to descend, causing nausea with aversion to food and constipation.

Preparation: Decoction. Taken warm, three times per day.

Treatment Principles: Harmonize Stomach and descend the turbid. (Xuan Fu Hua descends rebellion, dissolves phlegm, and is especially effective in treating belching.)

This formula can be used for nausea, vomiting, and constipation due to acute or chronic gastritis, and partial obstruction of the pyloric sphincter. It can also treat phlegm turbidity obstructing the Middle Jiao type of neurotic nausea; vomiting; and nausea and vomiting during pregnancy.

Jiang Ni Zhi E Tang [Direct the Rebellious Qi Downward and Stop Hiccup Decoction] *Jiang Su Zhong Yi Za Zhi*

Dai Zhe Shi (decoct first)	24 g
Xuan Fu Hua (wrapped)	12 g
Ju Pi	15 g
Zhu Ru	12 g
Ding Xiang	9 g
Shi Di	9 g
Tai Zi Shen	12 g
Gan Cao	9 g
Tian Men Dong	9 g
Mai Men Dong	9 g
Pi Pa Ye (without hair)	9 g

Main Indication: Heat above, cold below, Stomach Qi rebellion. Symptoms include hiccup with soft and low sound, coldness of lower limb, dry red tongue with thin coating, and thin pulse.

Preparation: Decoction. Take warm, three times per day.

Treatment Principles: Descend rebellion, stop hiccup.

This formula is a modification of Xuan Fu Dai Zhe Tang [Inula and Haematite Decoction], Ju Pi Zhu Ru Tang [Tangerine Peel and Bamboo Shaving Decoction], and Ding Xiang Shi Di Tang [Clove and Persimmon Calyx Decoction]. Thus, it can treat hiccups due to a mixed pattern of heat and cold. The functions of descending the rebellion and stopping hiccup are stronger.

Modification:

Addition	Indication
Bai Shao Yao	Liver Qi stagnation with severe spasms

Comparison of Regulate Stomach, Descend Rebellious Qi Formulas and Syndrome Differentiation and Application to Nausea, Vomiting, and Hiccups

All of the above nine formulas can treat nausea, hiccups, and Qi rebellion. In the clinical practice, one does not have to be limited by these nine formulas. According to the etiology of the condition, various formulas can be used. For Gallbladder invading Stomach causing Qi rebellion and nausea, one can use Hao Qin Qing Dan Tang [Artemisia Annua and Scutellaria Decoction to Clear the Gallbladder], Da Chai Hu Tang [Major Bupleurum Decoction], and Xiao Chai Hu Tang [Minor Bupleurum Decoction]. For disharmony of Lung and Stomach leading to nausea, Jia Wei Huang Lian Su Ye Tang [Augmented Coptis and Perilla Decoction] is indicated. For Middle Jiao deficient cold with Stomach Qi rebellion causing nausea and vomiting, one can use Ding Yu Li Zhong Tang [Clove and Evodia Decoction to Regulate the Middle], Da Ban Xia Tang [Major Pinellia Decoction], or Sha Ban Li Zhong Tang [Glehnia and Pinellia Decoction to Regulate the Middle]. For nausea due to excess heat in Yang Ming and Qi rebellion, one can use Zhen Ni Bai Hu Tang [White Tiger Decoction to Suppress Rebellion] or Zhu Ye Shi Gao Tang [Lophatherus and Gypsum Decoction]. For nausea due to Stomach and Spleen Qi deficiency, one can use Xiang Sha Liu Jun Zi Tang [Six-Gentleman Decoction with Aucklandia and Amomum] or Ju Pi Zhu Ru Tang [Tangerine Peel and Bamboo Shaving Decoction]. For Spleen failing to transform dampness with turbid phlegm rebellion causing nausea, one can use Xuan Fu Dai Zhe Tang [Inula and Haematite Decoction] or Xiao Ban Xia Tang [Minor Pinellia Decoction]. For accumulation and obstruction in the Stomach and intestine with turbid Yin failing to descend, causing nausea, one can use Bao He Wan [Preserve Harmony Pill] or modifications of Cheng Qi Tang [Order the Qi Decoction]. For Spleen failing to transport fluids, internal water accumulation leading to nausea and thirst, one can use Wu Ling San [Five-Ingredient Powder with Poria] or Fu Ling Ze Xie Tang [Poria and Alismatis Decoction] (Wu Ling San [Five-Ingredient Powder with Poria] minus Zhu Ling, plus Sheng Jiang and Gan Cao). For nausea due to disharmony of Liver and Stomach, one can use Qin Lian Ju Ru Tang [Scutellaria, Coptis, Tangerine Peel, and Bamboo Shaving Decoction] to treat heat type and Ban Xia Fu Ling Tang [Pinellia and Poria Decoction] for cold type. For cold type of hiccups, one can use Ding Xiang Shi Di Tang [Clove and Persimmon Calyx Decoction]. For heat type of hiccups, one can use Ju Pi Zhu Ru Tang [Tangerine Peel and Bamboo Shaving Decoction]. For mixed type of hiccups, one can use Jiang Ni Zhi E Tang [Direct the Rebellious Qi Downward and Stop Hiccups Decoction].

QI STAGNATION IN THE MIDDLE JIAO

This syndrome is often seen with symptoms of fullness and distention of the epigastrium and abdominal area or Qi stagnation with congealed phlegm obstructing the throat. Treatment should focus on moving Qi and guiding the stagnation. Commonly used herbs include Hou Po, Chen Pi, Zhi Ke, Mu Xiang, Cao Dou Ko, Sha Ren, Gao Liang Jiang, Zi Su Ye, and Da Fu Pi to smooth and unblock the Qi mechanism. Other herbs are selected according to the presenting syndromes.

Excess type of Qi stagnation is often treated with Bing Lang and Lu Hui. For fullness and distention of the abdomen leading to pain, Wu Yao and Tan Xiang are often added. If Qi and food stagnation occur concurrently then Shan Zha, Shen Qu, Mai Ya, or Gu Ya are added. If phlegm is present, then Ban Xia and Fu Ling are added.

Hou Po Wen Zhong Tang [Magnolia Bark Decoction to Warm the Middle] *Nei Wai Shang Bian Huo Lun*

Hou Po	12 g
Chen Pi	9 g
Cao Dou Kou	9 g
Mu Xiang	6 g
Gan Jiang	9 g
Fu Ling	12 g
Gan Cao	3 g

Main Indication: Cold–damp in Spleen and Stomach, obstructed Qi mechanism. Symptoms include fullness and distention at the epigastrium and abdomen, loss of appetite, and fatigue in the extremities; or cold invading Stomach causing episodic pain. The tongue coating is white or greasy with minute wiry pulse.

Preparation: Decoction. Taken warm before meal.

Treatment Principles: Warm the middle, transform dampness, move Qi, and stop pain.

This formula mainly treats syndromes of cold–damp injuring Spleen and Stomach leading to fullness, distention, and pain.

Related Formula:

Formula	Herbs	Indication
Liang Fu Wan [Galangal and Cyperus Pill]	Zhi Gao Liang Jiang Zhi Xiang Fu	Equal amount of the herbs grind into powder. Mix with water to form pill. Take 6 g, twice a day. Treats Liver Qi or cold invading the Stomach. Symptoms include epigastric pain, nausea, vomiting, or with distending pain at the chest and hypochondrium. This formula warms the Stomach, moves Qi, and soothes the Liver

Ban Xia Hou Po Tang [Pinellia and Magnolia Bark Decoction] *Jin Gui Yao Lue*

Ban Xia	12 g
Hou Po	12 g
Fu Ling	12 g
Sheng Jiang	9 g
Zi Su Ye	12 g

Main Indication: Mei He Qi (Plum Pit Qi) is caused by Qi stagnation with congealed phlegm. Symptoms include depressed emotions, a feeling of something obstructing the throat that can neither be ejected nor swallowed, fullness at chest, rapid breathing, or with cough, or with nausea, or with distending sensation and pain at chest and hypochondrium. Tongue is dark and stagnant with purple, dark, or distending sublingual vein, and wiry pulse.

Preparation: Decoction. Taken warm, three times during the day and once at night.

Treatment Principles: Move Qi, dissipate clumps, descend the rebellion, and transform phlegm.

This is an effective formula in treating Mei He Qi, but it is only suitable for knotted Qi and phlegm without heat type. In *Jin Gui Yao Lue*, this formula treats "women with clumps or obstructed sensation in the throat." In later times, the formula treats syndromes of Qi stagnation with phlegm obstruction.

Related Formulas:

Formula	Herbs	Indication
Si Qi Tang [Four Seven Decoction] *Zhi Zhi Fang*	Core Formula + Xiang Fu Gan Cao Hu Po	Treats urinary difficulty, with vaginal pain in severe cases
Si Qi Tang [Four Seven Decoction] *Ju Fang*	Core Formula + Da Zao	Same indication as the main formula
Jia Wei Si Qi Tang [Augmented Four Seven Decoction]	Core Formula – Sheng Jiang + Bai Zhi Mu Xiang Shi Chang Pu	Treats sticky thick leucorrhoea, focal distention, and stifling sensation at the epigastrium, copious phlegm, or with wheezing, nausea, and vomiting
Zi Su San [Perilla Powder] *Sheng Hui Fang*	Core Formula + Zhi Ke Chai Hu Bing Lang Gui Xin	Treats Qi stagnation, distention, and stifling sensation at the chest and diaphragm, and phlegm congestion with inability to swallow

Comparison of Qi Stagnation in Middle Jiao Formulas

Hou Po Wen Zhong Tang [Magnolia Bark Decoction to Warm the Middle] treats cold–damp in the Middle Jiao obstructing the Qi mechanism. It treats Spleen and Stomach at the same time. Liang Fu Wan [Galangal and Cyperus Pill] treats stagnant Liver Qi or cold invading Stomach. Ban Xia Hou Po Tang [Pinellia and Magnolia Bark Decoction] treats emotional upset, Liver Qi stagnation causing Lung and Stomach Qi failing to descend, knotted Qi, and phlegm accumulating in the throat, which is referred to as Mei He Qi.

UNREGULATED ASCENDING AND DESCENDING FUNCTION

When both nausea due to turbid Yin failing to descend in the Stomach and diarrhea due to clear Yang of Spleen failing to ascend happen concurrently, then one should ascend the clear and descend the turbid to regulate the ascending–descending function of the Middle Jiao. Harmonizing the Spleen and Stomach can relieve nausea, vomiting, and diarrhea. Huo Xiang, Zi Su Ye, and Bai Zhi can ascend the clear while Fu Ling, Tong Cao, Yi Yi Ren, Ban Xia, and Hou Po can descend the rebellion and eliminate dampness.

If unregulated ascending and descending function of Spleen and Stomach causing focal distention and congelation in the Middle Jiao occur with unsmooth Qi mechanism, and mixed cold and heat, then one should combine both heat and cold herbs in the formula. Cold bitter herbs such as Huang Qin and Huang Lian are used with acrid warm herbs such as ginger and Ban Xia. Thus, this combination shows the characteristics that Stomach prefers cool and Spleen prefers warm.

Huo Xiang Zheng Qi San [Agastache Powder to Rectify the Qi]
He Ji Ju Fang

Huo Xiang	9 g
Zi Su Ye	9 g
Bai Zhi	9 g
Jie Geng	6 g
Chen Pi	9 g
Hou Po	12 g
Da Fu Pi	15 g
Ban Xia	12 g
Bai Zhu	9 g
Fu Ling	12 g
Gan Cao	3 g

Main Indication: External wind–cold invasion, internal injury with damp obstruction. Symptoms include aversion to cold, fever, heaviness at head, headache, fullness and stifling sensation at chest and diaphragm, painful epigstrium and abdomen, or nausea, vomiting, diarrhea; white, greasy tongue coating, and soggy or floating, rapid pulse. This can also treat vomiting and diarrhea due to cholera.

Preparation: Grind into powder and take 6 g, two to three times a day. Or take as decoction.

Treatment Principles: Release exterior, transform dampness, ascend clear, and descend turbid.

This formula focuses on aromatic herbs to transform dampness and harmonize the Middle Jiao with moderate strength of releasing exterior and scattering cold. It is most suitable for external cold invasion with internal injury of dampness in the summer, and disharmony of Spleen and Stomach. Other sudden illness without known etiology with manifestations of vomiting, diarrhea, fever, and absence of sweat can also be treated with this formula. Recently, this formula has been used for damp injuring Spleen and Stomach type of acute gastroenteritis, external wind–cold invasion, common cold, and flu in the summer, heatstroke, headache, nausea, vomiting, abdominal pain, and diarrhea.

Related Formulas:

Formula	Herbs	Indication
Yi Jia Jian Zheng Qi San [First Modification of Rectify the Qi Powder]	Huo Xiang Geng 9 g Hou Po 9 g Fu Ling Pi 6 g Chen Pi 3 g Xing Ren 6 g Shen Qu 5 g Mai Ya 5 g Yin Chen Hao 6 g Da Fu Pi 3 g	Treats fullness and distention of epigastrium and abdomen with unsmooth bowel movement
Er Jia Jian Zheng Qi San [Second Modification of Rectify the Qi Powder]	Huo Xiang Geng 9 g Hou Po 9 g Fu Ling Pi 9 g Chen Pi 6 g Fang Ji 9 g Da Huang Dou Juan 6 g Tong Cao 5 g Yi Yi Ren 9 g	Treats stifling sensation at the epigastrium, loose stool, body aches, and pale tongue
San Jia Jian Zheng Qi San [Third Modification of Rectify the Qi Powder]	Huo Xiang 9 g Hou Po 9 g Fu Ling Pi 9 g Chen Pi 5 g Xing Ren 9 g Hua Shi 15 g	Treats turbid accumulating internally, stifling sensation at epigastrium, and yellow tongue coating
Si Jia Jian Zheng Qi San [Fourth Modification of Rectify the Qi Powder]	Huo Xiang 9 g Hou Po 6 g Fu Ling 9 g Chen Pi 5 g Cao Guo 3 g Shan Zha 15 g Shen Qu 6 g	Treats turbid accumulating internally, evil obstructing the Qi level; white, slippery tongue, and moderate right pulse
Wu Jia Jian Zheng Qi San [Fifth Modification of Rectify the Qi Powder]	Huo Xiang 6 g Hou Po 6 g Fu Ling 9 g Chen Pi 5 g Cang Zhu 6 g Gu Ya 3 g Da Fu Pi 5 g	Treats turbid accumulating internally and stifling sensation at the epigastrium with loose stool

Comparison of the Five Modifications of Zheng Qi San [Rectify the Qi Powder]

All five formulas treat stifling sensation or distention at the epigastrium due to damp obstructing the Middle Jiao. Thus, Huo Xiang, Hou Po, Chen Pi, and Fu Ling are the basis for these formulas with the aromatic function to transform dampness, bland taste to leach out dampness and promote urination. However, each modification has its strength. The First Modification treats damp obstruction in the three Jiaos, unregulated ascending and descending Qi mechanism. Thus, Xing Ren is added to disperse Upper Jiao; Shen Qu and Mai Ya guide the Middle Jiao; and Yin Chen Hao and Da Fu Pi unblock and regulate the Lower Jiao. The Second Modification treats damp obstruction in the three Jiaos, channels, and collaterals. Thus, Fang Ji and Yi Yi Ren are added to unblock the channel, disperse Bi pain, and stop body aches. Dou Juan and Tong Cao promote urination to solidify the stool. The Third Modification treats turbidity accumulating internally, unsmooth Qi mechanism, and long-term internal constraint transforming into heat. Thus Xing Ren is added to unblock and disperse Lung Qi; Hua Shi is added to clear and drain damp–heat. The Fourth

Modification treats cold–damp syndrome of turbidity obstructing the Qi level. Thus, acrid, hot Cao Guo is used to transform turbidity with its aromatic property and to warm and promote Spleen Yang. Shan Zha and Shen Qu are added to eliminate accumulation and guide stagnation. The Fifth Modification treats severe turbidity. Thus, Cang Zhu is used to dry Spleen dampness, Da Fu Pi to eliminate damp and fullness, and Gu Ya to ascend Stomach Qi.

Liu He Tang [Harmonize the Six Decoction] *He Ji Ju Fang*

Ren Shen	6 g
Bai Zhu	9 g
Chi Fu Ling	9 g
Gan Cao	2 g
Bian Dou	9 g
Huo Xiang	9 g
Sha Ren	6 g
Xing Ren	6 g
Ban Xia	12 g
Hou Po	12 g
Mu Gua	10 g

Main Indication: Dampness injures the Spleen and Stomach ascending and descending dysfunction. Symptoms include vomiting and diarrhea in cholera, focal distention and fullness at chest and diaphragmatic region, fatigue, lethargy, somnolence, white, slippery tongue coating, and soggy or minute slippery pulse.

Preparation: Decoction.

Treatment Principles: Strengthen the Spleen, transform dampness, ascend the clear, and descend the turbid.

This formula is suitable for improper diet during the summer and dampness injuring the Spleen and Stomach, leading to vomiting and diarrhea.

Lian Po Yin [Coptis and Magnolia Bark Decoction] *Huo Luan Lun*

Huang Lian	12 g
Shan Zhi Zi	9 g
Hou Po	15 g
Zhi Ban Xia	12 g
Shi Chang Pu	6 g
Xiang Dou Chi	9 g
Lu Gen	30 g

Main Indication: Damp–heat accumulating internally, ascending and descending dysfunction. Symptoms include vomiting and diarrhea in cholera, focal distention and stifling sensation at chest and epigastrium, scanty dark urine, yellow, greasy tongue coating, slippery and rapid pulse. Symptoms can also be fever, restlessness, irritability, focal distention, and stifling sensation at chest and epigastrium; thirst, dry heaves, loose stool, and yellow, greasy tongue coating.

Preparation: Decoction. One bag per day. For severe cases, two bags per day, taken every 4 hours.

Treatment Principles: Clear heat, transform dampness, regulate Qi, and harmonize the middle.

This formula is used in damp, warm, febrile conditions for syndromes of vomiting and diarrhea that are severe in both heat and dampness with Middle Jiao dysfunction.

Recently, it is used to treat internal accumulation of heat and damp type of bacillary dysentery, common typhoid, and paratyphoid with vomiting and diarrhea as its main symptoms.

Related Formula:

Formula	Herbs	Indication
Ran Zhao Tang [Lighting Decoction] *Chong Ding Huo Luan Lun*	Hua Shi 12 g Dan Dou Chi 9 g Jiao Shan Zhi Zi 6 g Huang Qin 4.5 g Xing Tou Cao 4.5 g Zhi Hou Po 3 g Zhi Ban Xia 3 g Bai Kou Ren 3 g	Decoction. This formula clears heat, drains damp, harmonizes stomach, and awakens the Spleen Treat heat-type cholera and acute vomiting and nausea due to damp–heat

Can Shi Tang [Silk Worm Excrement Decoction] *Huo Luan Lun*

Wan Can Sha (wrapped)	24 g
Ban Xia (fried with vinegar)	12 g
Yi Yi Ren	24 g
Da Dou Huang Juan	12 g
Tong Cao	3 g
Huang Lian	9 g
Huang Qin	12 g
Shan Zhi Zi	6 g
Mu Gua	12 g
Wu Zhu Yu	2 g

Main Indication: Internal accumulation of damp and heat, ascending and descending dysfunction. Symptoms include vomiting and diarrhea in cholera, abdominal pain with spasms of the gastrocnemius muscle, thirst, irritability and restlessness; yellow, thick, dry tongue coating and soggy, rapid, or deep, thin pulse.

Preparation: Decoction.

Treatment Principles: Clear heat, drain damp, ascend the clear, and descend turbid.

This formula treats damp–heat obstruction in the Middle Jiao with severe heat which causes vomiting, diarrhea, and fluid damage, leading to spasms of gastrocnemius muscle.

Ban Xia Xie Xin Tang [Pinellia Decoction to Drain the Epigastrium] *Shang Han Lun*

Ban Xia	15 g
Huang Lian	6 g
Huang Qin	9 g
Gan Jiang	9 g
Ren Shen	9 g
Zhi Gan Cao	3 g
Da Zao	9 g

Main Indication: Cold and heat knotted in Middle Jiao, ascending and descending dysfunction. Symptoms include focal distention and fullness at the epigastrium, dry heaves or nausea, vomiting, borborygmus with diarrhea, and thin, yellow or greasy tongue coating, with wiry and rapid pulse.

Preparation: Decoction.

Treatment Principles: Harmonize the Middle Jiao, descend the rebellion, disperse the clumping, and eliminate focal distention.

The formula originally treats improper purging of a syndrome that is indicated for Xiao Chai Hu Tang [Minor Bupleurum Decoction]. With the damaged middle Yang, external evil invades the body leading to cold and heat knotted in the epigastrium with the manifestation of focal distention (Pi). Focal distention refers to the inability of Qi to ascend and descend, fullness without pain, and soggy upon palpation. Knotted cold and heat and dysfunction of Qi mechanism manifest as dry heaves or nausea, vomiting at the upper part of the body, and abdominal pain, borborygmus, and diarrhea at the lower part of the body. This formula uses both cold and heat herbs, bitter herbs to descend, and acrid herbs to unblock. It tonifies Qi, harmonizes the Middle Jiao, eliminates evils, recovers the body Zheng Qi, and regulates the ascending and descending function, thus eliminating all symptoms. This formula can also be used in acute, chronic gastroenteritis, indigestion in children, and food poisoning with the manifestations of focal distention at chest, fullness epigastrium, and greasy tongue coating.

Related Formulas:

Formula	Herbs	Indication
Gan Cao Xie Xin Tang [Licorice Decoction to Drain the Epigastrium]	Core Formula – Ren Shen + Zhi Gan Cao 9 g Another formula is adding Zhi Gan Cao without eliminating Ren Shen	Treats weak Stomach Qi, indigestion, borborygmus with diarrhea, focal distention and hardness at the epigastrium, dry heaves, restlessness, and unsound sleep
Sheng Jiang Xie Xin Tang [Fresh Ginger Decoction to Drain the Epigastrium]	Core Formula – Gan Jiang + Sheng Jiang	Treats knotted water and heat, epigastrium, focal distention and hardness, dry belching with foul smell, abdominal borborygmus, and diarrhea
Huang Lian Tang [Coptis Decoction]	Core Formula – Huang Qin + Gui Zhi	Treats heat in the chest and cold in the Stomach. Irritability and restlessness at chest, nausea, abdominal pain, or borborygmus with diarrhea; white, slippery tongue coating, and wiry pulse
Ban Xia Xie Xin Tang Qu Gan Jiang, Gan Cao, Jia Zhi Shi Xing Ren Tang [Pinellia Decoction to Drain the Epigastrium without Dry Ginger, Licorice, and with the Addition of Immature Orange and Apricot Seed]	As mentioned in formula name	Treats summer dampness at Yang Ming, slippery rapid pulse, loss of appetite, absence of hunger and absence of bowel movement, turbid phlegm accumulation, and focal distention at the epigastrium
Jia Jian Ban Xia Xie Xin Tang [Modified Pinellia Decoction to Drain the Epigastrium]	Jiang Ban Xia 9 g Huang Lian 3 g Huang Qin 6 g Hua Shi 12 g Tong Cao 4 g Zhu Li 1 Tbs Ginger juice 4 drops	Treats damp–heat at Qi level, internal the collaterals and clears obstruction of orifices, delirium and slurred speech, and greasy tongue coating
Ren Shen Xie Xin Tang [Ginseng Decoction to Drain the Epigastrium]	Core Formula – Ban Xia Gan Cao Da Zao + Zhi Shi Bai Shao Yao	Treats damp–heat, unclear Upper Jiao, internal deficiency and sinking, clouded spirit, slippery tongue coating, and moderate pulse
Jia Jian Xie Xin Tang [Modified Drain the Epigastrium Decoction]	Huang Lian Huang Qin Gan Jiang Jin Yin Hua Shan Zha (charred) Bai Shao Yao Mu Xiang juice	Treats lock-jawed dysentery, dry heaves, abdominal pain, and tenesmus

Comparison of Unregulated Ascending and Descending Function Formulas

All of the above formulas mainly treat vomiting and diarrhea due to ascending and descending dysfunction of Spleen and Stomach. All can ascend the clear and descend turbid. However, the following should be considered:

1. Huo Xiang Zheng Qi San [Agastache Powder to Rectify the Qi] and Liu He Tang [Harmonize the Six Decoction] emphasize the function of aromatic damp-transforming and drying-the-dampness herbs. These herbs recover the transportation and transformation function of Spleen and Stomach, ascend clear, and descend turbid.
2. Can Shi Tang [Silk Worm Excrement Decoction] and Lian Po Yin [Coptis and Magnolia Bark Decoction] emphasize the function of clearing heat and relieving toxin herbs. They eliminate the factor that causes the condition to recover the ascending and descending function of the Middle Jiao. Can Shi Tang [Silk Worm Excrement Decoction] treats cholera with twisted tendons and Lian Po Yin [Coptis and Magnolia Bark Decoction] focuses on harmonizing the Stomach and stopping nausea.
3. Three Xie Xin Tang [Drain the Epigastrium Decoction] emphasize the combination of Huang Qin and Huang Lian for clearing heat and relieving toxins and ginger and Ban Xia to warm and heat. This combination can eliminate the pathogenic factor and also recover the function of Middle Jiao.
4. Huo Xiang Zheng Qi San [Agastache Powder to Rectify the Qi] treats syndromes of both exterior and interior condition. Can Shi Tang [Silk Worm Excrement Decoction] and Lian Po Yin [Coptis and Magnolia Bark Decoction] treats acute heat condition. Three Xie Xin Tang [Drain the Epigastrium Decoction] treat syndromes of mixed heat and cold.

INTERNAL ACCUMULATION OF DAMP–HEAT

This indicates damp–heat blazing in the Middle Jiao leading to fever, spontaneous sweating, cold feet, dizziness, heaviness and pain at head, focal distention at chest, abdominal distention, loose stool and unsmooth bowel movement, thirst but not wanting to drink or drinking a little, and greasy, white or greasy, yellow tongue coating. When damp and heat pathogenic factors mix together, just like mixing oil and flour, they become dough that is hard to separate from each other. Thus, under this section, the focus is on heat clearing and damp draining which is assisted by acrid and aromatic herbs to unblock and transform dampness to separate and eliminate damp and heat. The commonly used bitter cold herbs that clear heat are Shan Zhi Zi, Huang Lian, and Huang Qin; acrid warm herbs to dry dampness herbs are Ban Xia, Hou Po, Chen Pi, Cang Zhu, and Cao Guo; promote urination and leach out dampness herbs are Fu Ling, Yi Yi Ren, Hua Shi, Tong Cao, Zhu Ling, Ze Xie, Fang Ji, Yin Chen Hao, and Lu Gen. These herbs can be combined with aromatic damp transforming herbs, such as Huo Xiang, Bai Dou Kou, Sha Ren, Shi Chang Pu, and Pei Lan; and acrid herbs to unblock Lung Qi such as Pi Pa Ye, Jie Geng, and Xing Ren. Among all the herbs, Fu Ling, Yi Yi Ren, and Tong Cao are sweet bland herbs that leach out dampness without damaging the Spleen. Thus, these are especially suitable for chronic damp–heat syndromes.

Furthermore, if damp is stagnated at the superficial part of the body leading to muscle ache and restlessness, then one should combine with Fang Ji, Yi Yi Ren, Can Sha, Jiang Huang, Qin Jiu, Hai Tong Pi, Wei Ling Xian, Si Gua Luo, or Hai Feng Teng. These herbs are effective in treating the damp at the muscle layer and to unblock the channel and stop pain. If blood movers are added, such as Chuan Xiong and Gui Zhi, then the analgesic function is even greater. If symptoms include distending pain of the head, headache, throat pain, soreness and painful limbs and joints, slight chills and fever and other Upper Jiao syndromes, then one can combine acrid herbs to unblock Lung Qi. If one also sees Lower Jiao syndrome, such as difficulty urinating, frequent, urgent and painful urination, and loose stool, then drain-damp herbs to promote urination should be increased.

Gan Lu Xiao Du Dan [Sweet Dew Special Pill to Eliminate Toxin] *Wen Re Jing Wei*

Huo Xiang	10 g
Bai Dou Kou	10 g
Shi Chang Pu	6 g
Bo He (add near end)	9 g
Chuan Bei Mu	9 g
Lian Qiao	15 g
She Gan	10 g
Huang Qin	15 g
Yin Chen Hao	30 g
Hua Shi	20 g
Mu Tong	10 g

Main Indication: Damp–heat in San Jiao, evil pathogens at the Qi level. Symptoms include fever, sore limbs, stifling sensation at chest, abdominal distention, dark urine, constipation, or unsmooth diarrhea, or with foul smell, thirst, swollen throat, absence of sweating, restlessness, or fever with sweating, or jaundice, white or thick greasy or dry greasy coating, and soggy and rapid pulse.

Preparation: Decoction.

Treatment Principles: Eliminate dampness, transform turbidity, clear heat, and relieve toxins.

This formula is the main formula in treating damp warm febrile disease or seasonal epidemic disorders with heat excess. It is often used in summer season when the weather is hot and humid. Clinically, the key points in syndrome differentiation are fever, fatigue, heavy sensation, thirst, dark urine; thick, greasy white or dry, yellow tongue coating. Today, it can be used in treating enteritis, Yin jaundice, leptospirosis, and encephalitis B.

Modification:

Addition	Indication
Di Yu Ku Shen Huang Lian Ma Chi Xian	Treats enteritis, typhoid, and paratyphoid These herbs can clear heat and relieve toxicity
Qu Mai Bian Xu Bai Mao Gen Da Ji Xiao Ji Shi Wei	Treats pyelonephritis with the addition of these promoting urination herbs
Yin Chen Hao (increase dosage) Shan Zhi Zi Huang Bai Da Huang	To treat infectious hepatitis with jaundice, these drain damp and alleviating jaundice herbs are added
Qing Hao Jin Yin Hua Ban Lan Gen Chuan Xin Lian Bai Mao Gen Da Ji Xiao Ji Huang Qin (large dose)	Treat leptospirosis with the addition of these clearing heat and relieving toxins, cooling the blood and stop bleeding herbs
Da Qing Ye Ban Lan Gen	Treat encephalitis B

San Ren Tang [Three-Nut Decoction] *Wen Bing Tiao Bian*

Xing Ren	10 g
Bai Dou Kou	10 g
Yi Yi Ren	24 g
Hou Po	12 g
Ban Xia	12 g
Tong Cao	6 g
Hua Shi	18 g
Dan Zhu Ye	6 g

Main Indication: Beginning stage of damp warm febrile disease, evil pathogens at the Qi level with stronger damp pathogen than heat. Symptoms include headache, chills, heavy and painful sensation at the body, pale and sallow complexion, stifling sensation at chest, loss of appetite, fever in the afternoon, absence of thirst, pale tongue, and wiry, thin, and soggy pulse.

Preparation: Decoction.

Treatment Principles: Clear and drain damp–heat, aromatically transform, and blandly leach out dampness.

This formula is used for syndromes with excess dampness and slight heat in the Middle Jiao, which includes warm febrile diseases, summer dampness, Bi syndromes, edema, and Lin syndrome. It can also be used for hepatitis, gastroenteritis, nephritis, and scleritis.

There are three common situations that cause this condition to be misdiagnosed. First is mistaking the headache, chills, heavy sensation of the body, and painful body as exterior syndrome and then use the sweating method. When this method is used, it can lead to confusion, diminished visual and aural acuity, and aversion to speaking. Second, if abdominal fullness with loss of appetite is mistaken as internal accumulation and stagnation and then the downward draining method is applied, severe diarrhea will result due to overdraining. Third, if afternoon fever is mistaken as Yin deficiency with Yin nourishing method applied, then the condition will progress to a deeper level and cannot be relieved due to the cloying property of the herbs.

Thus, this formula is appropriate in treating mixed damp and heat stagnation because it is acrid to open the Lung Qi and transforming to open the upper source. It is also aromatic, distinguishing the turbid by waking up the Spleen, then eliminating the turbid. This is also the same as aromatic, transforming and leaching out dampness to expel dampness in the Middle Jiao and sweet–bland, leaching out dampness from the Lower Jiao.

Hua Shi Huo Xiang Tang [Talcum and Agastache Decoction] *Wen Bing Tiao Bian*

Huo Xiang	6 g
Hou Po	6 g
Bai Dou Kou	3 g
Chen Pi	3 g
Fu Ling Pi	9 g
Zhu Ling Pi	6 g
Bai Tong Cao	3 g
(Fei) Hua Shi	9 g

Main Indication: Damp obstructing the Middle Jiao with excess damp, mild heat. Symptoms include stifling sensation, loss of appetite, vomiting, diarrhea, difficult urination, thirst but drinking little, flabby slippery tongue, gray yellow tongue coating, and soggy or slippery pulse.

Preparation: Decoction.

Treatment Principles: Aromatic herbs to transform and leach out dampness, and disperse and unblock the Middle Jiao.

Modification:

Minus	Addition	Indication
	Huang Qin Huang Lian Mu Xiang Bing Lang	The original formula treats "pouring down dampness with red and white dysentery." However, the strength is not enough, thus these four herbs are added
Hua Shi Tong Cao	Cang Zhu Gan Jiang Cao Guo	Cool herbs are eliminated and warm herbs are added to treat cold–damp type of diarrhea

Xing Ren Hua Shi Tang [Apricot Kernel and Talcum Decoction] *Wen Bing Tiao Bian*

Xing Ren	10 g
Huang Qin	6 g
Huang Lian	3 g
Ju Hong	5 g
Ban Xia	10 g
Hou Po	6 g
Yu Jin	6 g
Hua Shi	10 g
Tong Cao	3 g

Main Indication: Damp–heat obstructing San Jiao. Symptoms include sweating with thirst, focal distention, and stifling sensation at the chest, nausea, diarrhea, tidal fever, scanty urine; gray-white tongue coating, and soggy or slippery pulse.

Preparation: Decoction. Taken in three doses.

Treatment Principles: Clear heat, eliminate dampness, disperse, transform, and leach out dampness.

This formula can transform dampness in the Three Jiao. It can also clear heat and relieve toxicity, treating a mixed and equal amount of damp and heat in the body.

Xuan Qing Dao Zhuo Tang [Disperse, Clear, and Guide the Turbid Decoction] *Wen Bing Tiao Bian*

Zhu Ling	15 g
Fu Ling	15 g
Han Shui Shi	18 g
Can Sha	12 g
Zao Jia	9 g

Main Indication: Chronic damp turbid obstruction of the San Jiao. Symptoms include confusion, obstructed orifices, lower abdominal hardness and fullness, difficult urination and constipation, thick greasy or dry, thick yellow tongue coating, and wiry and slippery pulse.

Preparation: Decoction. Taken in two doses and monitored by the bowel movement (discontinue once bowels start moving).

Treatment Principles: Disperse, clear, and guide the turbid.

This syndrome is caused by damp turbid stagnation and obstruction at Qi level. Thus, ascending the clear and descending the turbid are the main treatment principles to open and clear the orifices of the Upper Jiao and unblock the urination and bowel movement in the Lower Jiao. Today, this formula can be used in food poisoning, and acute and chronic toxic bacillary dysentery.

Huang Qin Hua Shi Tang [Scutellaria and Talcum Decoction] *Wen Bing Tiao Bian*

Huang Qin	10 g
Hua Shi	10 g
Fu Ling Pi	10 g
Da Fu Pi	6 g
Bai Dou Kou	3 g
Tong Cao	3 g
Zhu Ling	10 g

Main Indication: Damp–warm febrile pathogens at the Middle Jiao, affecting both channel and collaterals. Symptoms include fever, body ache, relief of fever due to sweating that turns into heat, thirst with drinking just a little, or absence of thirst; thin, yellow, and slippery tongue coating, and slippery and soft pulse.

Preparation: Decoction.

Treatment Principles: Clear heat and drain dampness.

This formula is suitable for damp–heat obstructing the Middle Jiao and also spreading into channels and collaterals, which leads to heavy sensation and body ache.

Related Formula:

Formula	Herbs	Indication
Fu Ling Pi Tang [Poria Skin Decoction]	Fu Ling Pi 15 g Yi Yi Ren 15 g Zhu Ling 9 g Da Fu Pi 9 g Bai Tong Cao 9 g Dan Zhu Ye 9 g	Decoction. This formula treats absorbed turbidity which spreads to San Jiao leading to steaming heat, distending headache, body ache, nausea, difficulty urinating, confused or unconscious mind, thirst without drinking much water, white tongue, and slippery wiry pulse. One should take An Gong Niu Huang Wan [Calm the Palace Pill with Cattle Gallstone] to unblock the spirit and benefit the orifices, leach out dampness, and separate and clear the damp and turbidity

Yi Yi Zhu Ye San [Coicis and Lophatherus Powder] *Wen Bing Tiao Bian*

Yi Yi Ren	15 g
Dan Zhu Ye	9 g
Hua Shi	15 g
Bai Dou Kou	5 g
Lian Qiao	9 g
Fu Ling (chunks)	15 g
Bai Tong Cao	5 g

Main Indication: Damp–heat stagnation at the collaterals. Symptoms include fever, body ache, profuse sweat, diarrhea, pale skin rash at chest and abdomen, white or thin gray, yellow coating, and floating, tight or soggy rapid pulse.

Preparation: Make the above ingredients into powder. Take 15 g each time, three times a day. It can also be taken as a decoction.

Treatment Principles: Acrid cool herbs to clear heat, sweet and bland herbs to drain damp.

This syndrome is a combination of external and internal evil. Pure acrid herbs reach the exterior, and pure bitter herbs clear heat. Both are not appropriate in treating this syndrome. Only by using acrid cool herbs, one can release the heat at muscle and exterior, and by using sweet and bland herbs to leach out the interior dampness. Thus, the exterior evil is released and scattered from the surface, and the interior evil is eliminated from the urine. These are the main treatment principle of this formula.

Xuan Bi Tang [Disband Painful Obstruction Decoction] *Wen Bing Tiao Bian*

Xing Ren	10 g
Lian Qiao	12 g
Shan Zhi Zi	10 g
Ban Xia	12 g
Hua Shi	18 g
Chi Xiao Dou	15 g
Yi Yi Ren	30 g
Fang Ji	12 g
Can Sha (wrapped)	24 g

Main Indication: Damp–heat obstructing the channels and collaterals. Symptoms include soreness and pain at joints, shaking chills, fever, sallow complexion, yellow sclera; thick, greasy tongue coating, and soggy, rapid or wiry pulse.

Preparation: Decoction.

Treatment Principles: Clear heat, drain damp, disband the painful obstruction, and stop pain.

This formula treats damp–heat Bi syndrome. The greasy tongue coating and yellow sclera indicates damp and heat in the body. Fever, shaking chills, soreness, and pain at the joints indicate damp–heat obstructing the channels and collaterals. Furthermore, invigorate-blood herbs are added in the formula based on the classic treatment principle that "treating wind needs to treat the blood first; when the blood is nourished, then the wind will be extinguished."

Modification:

Minus	Addition	Indication
	Jiang Huang 10 g Hai Tong Pi 15 g	Severe pain
	Cang Zhu 10 g Huo Xiang 10 g Pei Lan 10 g	Severe damp turbid
Ban Xia		Tongue coating that is not thick and greasy

Jia Jian Mu Fang Ji Tang [Modified Cocculus Decoction] *Wen Bing Tiao Bian*

Fang Ji	18 g
Gui Zhi	10 g
Shi Gao	20 g
Xing Ren	10 g
Hua Shi	12 g
Bai Tong Cao	6 g
Yi Yi Ren	15 g

Main Indication: Damp–heat obstructing the channels and collaterals. Symptoms include soreness and pain in muscles, red swollen joints, fever with slight chills, thick greasy or yellow tongue coating, and wiry, rapid or soggy rapid pulse.

Preparation: Decoction. Taken in three doses.

Treatment Principles: Clear heat, drain damp, disband painful obstruction, and stop pain.

This formula is used to treat summer damp Bi syndrome, thus Guang Fang Ji is used, which is effective in eliminating damp from the muscles as the chief herb. Today, this formula can also treat wind damp–heat such as rheumatic fever, similar to the function of Cang Zhu Bai Hu Tang [White Tiger Decoction with Atractylodes] and Gui Zhi Bai Hu Tang [Cinnamon Twig and White Tiger Decoction].

Modification:

Addition	Indication
Gui Zhi Sang Ye	For excess wind leading to spasm and wandering pain
Hua Shi Bei Xie Cang Zhu	For excess damp causing swelling
Fang Ji (increase dosage) Gui Zhi Jiang Huang Hai Tong Pi	For pain due to excess cold
Shi Gao 30 to 50 g Zhi Mu 15 to 20 g	For red face and drooling
Qiang Huo Cang Zhu	For absence of sweating
Huang Qi Zhi Gan Cao	For profuse sweating
Hou Po Ban Xia Chen Pi	For concomitant Tan Yin syndrome

Gou Duan Luo Xuan Ti Bing Fang [Leptospirosis Formula] (Cheng Du Zhong Yi Xue Yuan Fang)

Jin Yin Hua	30–60 g
Lian Qiao	30–60 g
Huang Qin	18 g
Shan Zhi Zi	18 g
Tong Cao	9 g
Huo Xiang	12 g
Bai Mao Gen	30 g
Lu Gen	30 g
Chuan Xin Lian	19 g
Ban Lan Gen	19 g

Main Indication: Epidemic toxins transform into heat, with severe heat and mild damp type of leptospirosis. Symptoms include fever, headache, red eyes, swollen inguinal or axillary lymph nodes, severe sharp pain of the gastrocnemius, red tongue, yellow tongue coating, and rapid big pulse.

Preparation: Decoction, one bag per day. In severe condition, two bags per day, taken every 4 hours.

Treatment Principles: Clear heat, relieve toxins, transform turbid, and drain damp.

This formula treats the beginning stage or mild condition of leptospirosis.

Comparison of Internal Accumulation of Damp–Heat Formulas

The above described 10 formulas are all composed of heat clearing herbs and damp draining herbs. These formulas also not only can clear heat but can also drain damp:

1. Gan Lu Xiao Du Yin [Sweet Dew Decoction to Eliminate Toxin] treats damp–heat syndrome with the evil pathogens at the Qi level and more heat manifestations. Thus, the formula focuses on heat clearing. San Ren Tang [Three-Nut Decoction] treats damp–heat syndrome with the evil pathogens at the Qi level and more damp manifestations. Thus it focuses on transforming turbidity and draining damp.
2. Hua Shi Huo Xiang Tang [Talcum and Agastache Decoction] and Xing Ren Hua Shi Tang [Apricot Kernel and Talcum Decoction] both treat vomiting and diarrhea as the main symptoms. The focus of these formulas is the Middle Jiao. However, Hua Shi Huo Xiang Tang [Talcum and Agastache Decoction] is indicated for more dampness symptoms with mild heat symptoms. And Xing Ren Hua Shi Tang [Apricot Kernel and Talcum Decoction] is indicated for both damp and heat symptoms.
3. Xuan Qing Dao Zhuo Tang [Disperse, Clear, and Guide the Turbid Decoction] treats damp–heat syndrome, but its clinical manifestation is different from other damp–heat syndrome. The main indication for this formula is confusion, obstructed orifices, hardness and fullness of the lower abdomen, and blocked urination and bowel movement. Thus, in this formula, Fu Ling, Zhu Ling, and Han Shui Shi transform the formless Qi, and Can Sha and Zao Jia expel the dampness.
4. The heat-clearing function of the Huang Qin Hua Shi Tang [Scutellaria and Talcum Decoction] is stronger than San Ren Tang [Three-Nut Decoction], but weaker than Gan Lu Xiao Du Dan [Sweet Dew Special Pill to Eliminate Toxin]. The damp draining power of Huang Qin Hua Shi Tang [Scutellaria and Talcum Decoction] is similar to San Ren Tang [Three-Nut Decoction]. However, the herbal combinations are different in these two formulas because of the location of the accumulated dampness. Huang Qin Hua Shi Tang [Scutellaria and Talcum Decoction] uses stronger damp draining and urine promoting herbs to regulate the waterways to eliminate dampness in the muscles. San Ren Tang [Three-Nut Decoction] uses more aromatic herbs to wake up the Spleen and expel phlegm. The purpose of the herbs is to awaken the Spleen Yang that is overwhelmed by the dampness and to recover the function of Spleen and Stomach.
5. Yi Yi Zhu Ye San [Coicis and Lophatherus Powder] treats the faint skin rash at chest and abdomen. Thus, Yi Yi Ren is used to eliminate dampness from the muscle and Lian Qiao is used to release stagnant heat. Thus, when dampness is eliminated from the channels, then the skin rash will disappear. When the heat is cleared from the muscle, then fever can be relieved.
6. Xuan Bi Tang [Disband Painful Obstruction Decoction] and Jia Jian Mu Fang Ji Tang [Modified Cocculus Decoction] treat damp–heat type of painful obstruction. The strength of clearing-heat, draining damp and, stopping pain function of Xuan Bi Tang [Disband Painful Obstruction Decoction] is stronger than Jia Jian Mu Fang Ji Tang [Modified Cocculus Decoction].

DAMP–COLD STRANDED SPLEEN

Damp–cold strands the Spleen and weakens the Middle Jiao Yang function. Symptoms often include stifling sensation at the epigastrium, abdominal distention, loose watery stool, and white greasy tongue coating or soreness and muscle ache. Cang Zhu, Bai Zhu, Hou Po, Chen Pi, Ban Xia, and other warm herbs that transform cold–damp are often used with Fu Ling and Ze Xie that are bland and leaching out dampness herbs to form Spleen transporting and damp draining formula.

If damp–cold is severe, then one should combine warm the Spleen Yang herbs, such as Gan Jiang and Cao Guo, or warm Yang Qi, transforming Gui Zhi to transform Qi and move water to alleviate the cold–damp in Spleen.

Ping Wei San [Calm the Stomach Powder] *He Ji Ju Fang*

Cang Zhu	10 g
Hou Po	10 g
Chen Pi	10 g
Gan Cao	3 g

Main Indication: Damp–cold obstructing the Middle Jiao. Symptoms include fullness and distention at epigastrium and abdomen, belching, sour regurgitation, poor appetite, no taste in mouth, nausea, vomiting, loose watery stool, fatigue, somnolence, heavy sensation and soreness of the body; white, greasy, and thick tongue coating, and soggy or slight wiry pulse.

Preparation: Decoction. Or grind into powder and taken 6 g each time, with ginger and Da Zao decoction. Better taken before meals.

Treatment Principles: Dry dampness, transform Spleen, move Qi, and harmonize Stomach.

This is a basic formula to treat cold–damp stranded Spleen. This can be used to treat chronic gastritis, enteritis, and gastroptosis with damp–cold stranded the Spleen type.

Modification:

Addition	Indication
Gan Jiang Rou Gui	Severe cold and damp
Huang Qin Huang Lian	With damp and heat

Related Formulas:

Formula	Herbs	Indication
Zha Qu Ping Wei San [Calm the Stomach Powder with Hawthorne and Medicated Leaven]	Core Formula + Shan Zha Shen Qu Mai Ya	Treats food stagnation, focal distention, distention and sour regurgitation, poor appetite, fatigue, and somnolence This is a formula for draining damp and guiding the stagnation
Kai Wei Jian Pi Wan [Stimulating Appetite and Strengthen the Spleen Pill]	Core Formula – Gan Cao + Zhi Shi	Treats Stomach and Spleen disharmony; no taste in mouth, fullness and distention at epigastrium and abdomen, nausea, vomiting, and sour regurgitation This is a stronger formula to eliminate focal distention and distended Qi than the original formula
Zhi Zhu Pi Wei San [Calm the Stomach Powder with Immature Bitter Orange and Atractylodes Macrocephala]	Core Formula + Zhi Zhu Tang [Immature Bitter Orange and Atractylodes Macrocephala Decoction]	Stronger in the function to guide the Qi downward, eliminate focal distention, strengthen the Spleen, and drain dampness. This formula treats Spleen deficiency with excess dampness
Bu Huan Jin Zheng Qi San [Rectify the Qi Powder Worth More than Gold]	Core Formula + Huo Xiang Ban Xia	Stronger in the function of aromatic transforming damp, descending rebellion, and stopping nausea and vomiting. This can also treat damp stagnation in the Middle Jiao with concomitant exterior syndrome
Xiang Sha Ping Wei San [Calm the Stomach Powder with Aucklandia and Amomum]	Core Formula + Mu Xiang Sha Ren	Treats Spleen deficiency, injured by diet, focal distention, poor appetite, and nausea and vomiting. It has stronger function to move Qi, strengthen Spleen, and stop vomiting

(continued)

Formula	Herbs	Indication
Wei Ling Tang [Calm the Stomach and Poria Decoction]	Core Formula + Wu Ling San [Five-Ingredient Powder with Poria]	Treats damp–cold stranded the Spleen leading to diarrhea. This formula presents the principle of promoting urination to solidify the stools. This is a combination of drying and draining dampness. It is more effective if Gan Jiang is used instead of Gui Zhi
Yin Chen Wei Ling Tang [Calm the Stomach and Poria Decoction with Artemisia Yinchenhao]	Wei Ling Tang [Calm the Stomach and Poria Decoction] + Yin Chen Hao	Treats Yin jaundice with the presentation of smoky yellow, or gradual transformation into heat, yellow slippery tongue coating, dry mouth, and not wanting to drink. This formula treats Liver and Stomach at the same time
Hou Po Yin [Magnolia Bark Decoction]	Core Formula + Gan Jiang	Treats severe cold type
Gan Dan Guan Jie Shi Fang [Formula for Stones in the Hepatic Duct and Biliary Duct]	Core Formula + Yin Chen Hao 30 g Shan Zhi Zi 10 g Yu Jin 10 g Jin Qian Cao 30–60 g	Treats sandy accumulation in the Liver and bile duct. It is effective when the patient takes this formula for a long time
Xia Si Tai Fang [Purge the Dead Fetus Formula]	Core Formula + Mang Xiao	This can promote the expulsion of the dead fetus

Huo Po Xia Ling Tang [Agastache, Magnolia Bark, Pinellia, and Poria Decoction] *Yi Yuan*

Huo Xiang	10 g
Dan Dou Chi	10 g
Bai Dou Kou	3 g
Hou Po	6 g
Ban Xia	10 g
Xing Ren	6 g
Fu Ling	10 g
Zhu Ling	6 g
Ze Xie	9 g
Yi Yi Ren (raw)	20 g

Main Indication: Beginning stage of damp warm febrile conditions. Symptoms include fever, chills, fatigue in extremities, stifling sensation at the chest, greasy taste in mouth, white slippery tongue coating, and soggy and moderate pulse.

Preparation: Decoction.

Treatment Principles: Release exterior, transform dampness, disperse the Upper Jiao, and leach out dampness from the Lower Jiao.

This formula treats damp excess with mild heat syndrome of the damp warm febrile conditions. The focus is on using aromatic herbs to transform dampness and drying dampness herbs. Then it is assisted by opening and dispersing Lung Qi, leaching out dampness, and promoting urination. The structure of this formula is similar to San Ren Tang [Three-Nut Decoction] that treats damp turbidity syndrome of the Upper, Middle, and Lower Jiao. Even though this is a formula to treat damp warm febrile condition, its heat clearing strength is very weak. Thus, it is listed under this category.

Ling Gui Zhu Gan Tang [Poria, Cinnamon Twig, Atractylodes Macrocephalae, and Licorice Decoction] *Shang Han Lun*

Fu Ling	20 g
Gui Zhi	15 g
Bai Zhu	15 g
Gan Cao	6 g

Main Indication: Insufficient Yang in Middle Jiao, internal accumulation of fluid (Shui Yin). Symptoms include fullness sensation at the chest and hypochondrium, vertigo, or palpitation, shortness of breath, cough with watery thin phlegm, and white slippery tongue coating with wiry and slippery pulse.

Preparation: Decoction. Taken warm in three doses.

Treatment Principles: Warm and transform Tan Yin, strengthen Spleen, and drain damp.

This formula is effective in treating syndromes of Middle Jiao Yang deficiency with internal water and fluid congestion (Shui Yin). The manifestations are palpitation at the epigastrium, focal distention and hardness at the epigastrium, gushing Qi sensation up to the chest, and difficult urination. This formula can also treat internal water and fluid congestion leading to deafness, vertigo, abnormal growth or film in the eye, and damp stagnation becoming Wei Syndrome.

Modification:

Addition	Indication
Che Qian Zi	For abnormal growth or film in the eye, dizziness, eye pain, swollen eyelids, and profuse eye discharge
Xiong Huang San (Chuan Xiong Da Huang)	Treats deafness, dizziness, and vertigo, as in the presentation of Menier's syndrome

Gan Cao Gan Jiang Fu Ling Bai Zhu Tang [Licorice, Ginger, Poria, and Atractylodes Macrocephalae Decoction] (also known as Shen Zhuo Tang [Fixed at Kidney Decoction] *Jin Gui Yao Lue*

Gan Cao	6 g
Gan Jiang	15 g
Fu Ling	15 g
Bai Zhu	12 g

Main Indication: Spleen Qi deficiency with excess damp accumulated in the muscle. Symptoms include heavy sensation at body, cold pain below the waist as if sitting in water, absence of thirst, copious urine, normal appetite, heavy sensation at the lower back as if carrying a heavy weight, white slippery tongue coating, and deep, slow pulse.

Preparation: Decoction. Taken warm in three doses.

Treatment Principles: Warm the Spleen, drain dampness, and stop pain.

This formula originally treats cold and heavy sensation of the lower back, which is referred to as Shen Zhuo (Fixed at Kidney). It can also treat cold–damp Bi syndrome.

Modification:

Addition	Indication
Ze Xie Che Qian Zi	Swollen labia and profuse urination
Fu Zi	Enuresis in kids, incontinence in elderly, and heavy cold pain of the lower back region and legs

Comparison of Cold–Damp Stranded Spleen Formulas

Ping Wei San [Calm the Stomach Powder] treats the syndromes with the manifestations of fullness and distention in epigastrium and abdomen and greasy thick white coating. Huo Po Xia Ling Tang [Agastache, Magnolia Bark, Pinellia, and Poria Decoction] treats damp turbid stranded the Spleen earth. The sticky greasy damp turbid is hard to resolve, thus, aromatic herbs that transform turbid dampness are used to disperse to the surface and leach out the dampness from the interior. When the upper Jiao is open and urine is flowing in Lower Jiao, then damp turbid pathogens can be eliminated from the three Jiao. Ling Gui Zhu Gan Tang [Poria, Cinnamon Twig, Atractylodes Macrocephala, and Licorice Decoction] treats Tan Yin syndrome due to Middle Jiao Yang deficiency and other syndromes due to Tan Yin pathogens attacking the Upper and Lower Jiao. Gan Cao Gan Jiang Fu Ling Bai Zhu Tang [Licorice, Ginger, Poria, and Atractylodes Macrocephala Decoction] treats syndromes due to cold–damp accumulation in and below the lower back leading to cold pain, swelling, and heavy sensation.

DAMP ACCUMULATION FORMING PHLEGM

Spleen is the source of phlegm formation. When phlegm is formed, it can cause different types of pathological conditions. There is a saying that strange conditions often involve phlegm. For example, when phlegm accumulates in the chest and hypochondriac region, there is focal distention and congestion at the epigastrium, or pain in the chest or hypochondrium, or hunger with rumbling noise, or frequent cold sensation in the middle back. When phlegm obstructs the communication between fire and water, then it causes disharmony of Heart and Kidney leading to insomnia with excess dreaming. When phlegm ascends with the Stomach Qi, then there is nausea and vomiting. If phlegm attacks the Lung then there is cough and wheezing. If wind evil pathogens ascend with phlegm, then there is dizziness, vertigo, and headache. If phlegm stagnates in the Heart, then there is palpitation with fright and severe palpitation. If phlegm obstructs the pericardium, the wind stroke or epilepsy can be the result of obstructed clear orifices. When phlegm descends down to the intestine, then there is diarrhea. If it pours into the vaginal area, then there is leucorrhoea. When phlegm obstructs the uterine area, then there can be irregular menstruation or infertility. When phlegm traps between the muscles and four limbs, then there can be numbness, loss of sensation, and pain and heavy sensation at the joints. Phlegm can also accumulate and form lumps, swelling, and nodules.

Phlegm can be divided into different types. They are wind, cold, damp, and heat phlegm. Damp phlegm has the main symptoms of stifling sensation at the chest and nausea. Often, Ban Xia, Cang Zhu, Bai Zhu, and Tian Nan Xing are used to dry the dampness and transform the phlegm. The main symptoms of wind phlegm are dizziness, vertigo, and heavy and painful sensation at the head. Often, Dan Nan Xing, Bai Fu Zi, Tian Ma, Tian Zhu Huang, Ju Hua, Gou Teng, and Bai Ji Li are used to expel wind and scour phlegm. If there is hypertension, then one can also add Xia Ku Cao, Huang Qin, Di Long, Niu Xi, Qing Mu Xiang, and Luo Bu Ma to clear Liver and lower blood pressure. Cold–damp often manifests as copious clear phlegm with slippery tongue coating. Often, Gan Jiang, Rou Gui, Gui Zhi, and Fu Zi are used to warm and transform the cold phlegm. Heat phlegm is often seen in Gallbladder heat with insomnia or thick yellow phlegm and yellow greasy tongue coating. Often, Zhu Ru, Huang Qin, Huang Lian, Sang Bai Pi, Zhe Bei Mu, and Dan Nan Xing are added to clear heat and transform phlegm. When phlegm flows into chest and hypochondrium, it causes fullness and discomfort. Often Mu Xiang, Xiang Fu, Bai Jie Zi, Xuan Fu Hua, and Ju Luo are added. When phlegm is at the upper part of the body, then Chen Xiang, Da Huang, and Mang Xiao are used to guide the phlegm downward. When phlegm obstructs the clear orifices, Meng Shi, Shi Chang Pu, and Niu Huang can expel phlegm and open the orifices. When phlegm attacks the Lung leading to cough and wheezing, Jie Geng, Bai Qian, Qian Hu, Kuan Dong Hua Xing Ren, and Ma Huang are added to expel phlegm, stop cough, disperse Lung Qi, and stop wheezing. When phlegm traps between muscles and at limbs, then there is numbness, decreased function, and pain. One should use ginger juice, Zhu Li, Bai Jie Zi, Zao Jia, and Di Long to unblock the channel and collaterals. When phlegm damp pouring downward causes leukorrhea, Che Qian Zi, Ze Xie, Fu Ling, Bai Guo, Qian Shi, Chun Gen Pi, Mu Jin Pi, and Lian Xu should be added. When phlegm obstructs uterine region leading to irregular menstruation or infertility, Chuan Xiong, Dang Gui, Dan Shen, Ze Lan, Xiang Fu, and Yi Mu Cao can be added.

Furthermore, if Qi is not treated, then phlegm cannot be treated. When Qi flow is smooth, then in the whole body Jin and Ye fluid can flow smoothly. Thus, one should remember to combine Chen Pi, Ju Hong, Xiang Fu, Zhi Ke, and other Qi regulating herbs in the formula.

Er Chen Tang [Two-Cured Decoction] *He Ji Ju Fang*

Chen Pi	10 g
Ban Xia	15 g
Fu Ling	15 g
Gan Cao	3 g

Main Indication: Syndromes of damp phlegm. Symptoms include cough with copious white phlegm, easy expectoration, focal distention, and stifling sensation at the chest and epigastrium, nausea, vomiting, fatigue at the extremities, or dizziness, palpitation; and white, moist tongue coating, and slippery pulse.

Preparation: Decoct with seven slices of ginger and one piece of Wu Mei.

Treatment Principles: Regulate Qi and Middle Jiao, dry damp, and expel phlegm.

This formula is the main formula to treat damp phlegm. Phlegm comes from dampness. When dampness accumulates, it becomes phlegm. This formula can be modified according to the symptoms to treat various other phlegm syndromes.

Modification:

Addition	Indication
Shan Zhi Zi Huang Qin Xing Ren Pi Pa Ye Jie Geng	Treat phlegm mostly located at the Upper Jiao
Huang Lian Dou Juan Huo Xiang Hou Po Yi Yi Ren Sha Ren	Treat phlegm mostly located at the Middle Jiao
Ku Shen Huang Bai Hua Shi Mu Tong	Treat phlegm mostly located at the Lower Jiao

(continued)

Addition	Indication
Nan Xing Bai Fu Zi Zao Jiao	Treat wind phlegm
Zhi Ban Xia Ginger juice	Treat cold phlegm
Hai Feng Teng Luo Shi Teng Wei Ling Xian Qin Jiao Fang Ji Gui Zhi Cang Er Zi	Treat wind damp Bi pain
Shi Gao Qing Dai	Treat fire phlegm
Cang Zhu Bai Zhu	Treat damp phlegm
Gua Lou Ren Xing Ren	Treat dry phlegm
Shan Zha Mai Ya Shen Qu	Treat food phlegm
Zhi Shi Hai Fu Shi Mang Xiao	Treat old phlegm
Xiang Fu Zhi Ke	Treat Qi phlegm
Zhu Li	Treat phlegm at the four limbs
Bai Jie Zi	Treat hypochondriac phlegm located between the skin and the membrane

Related Formulas:

Formula	Herbs	Indication
Su Xing Er Chen Tang [Two-Cured Decoction with Perilla and Apricot Kernel]	Core Formula + Zi Su Ye Xing Ren	Treat damp phlegm with exterior syndrome
Ma Xing Er Chen Tang [Two-Cured Decoction with Ephedra and Apricot Kernel]	Core Formula + Ma Huang Xing Ren	Treat cold evil attacking the Lung leading to cough, wheezing, and copious phlegm. This modified formula is a combination of the core formula with San Ao Tang [Three-Unbinding Decoction]
He Wei Er Chen Tang [Two-Cured Decoction to Harmonize the Stomach]	Core Formula + Gan Jiang Sha Ren	Treat cough and vomit of thin phlegm, nausea, vomiting, fullness and stifling sensation at chest and diaphragm This is the result of Yang deficiency in the Middle Jiao
Gui Fu Er Chen Tang [Two-Cured Decoction with Cinnamon and Aconite]	Core Formula + Rou Gui Fu Zi	Treats Yang deficiency in the Lower Jiao, cold water flowing upward, and cough with watery thin phlegm, difficulty urinating, and deep pulse
Lian Ru Er Chen Tang [Two-Cured Decoction with Coptis and Bamboo Shaving]	Core Formula + Huang Lian Zhu Ru	Treats severe vomiting due to heat in Gallbladder. In severe cases, add Huang Qin, Xuan Fu Hua, and Dai Zhe Shi
Lou Bei Er Chen Tang [Two-Cured Decoction with Trichosanthis and Fritillaria]	Core Formula + Gua Lou Bei Mu	Treat cough with scanty phlegm, sticky thick that is hard to expectorate
Hai Ge Er Chen Tang [Two-Cured Decoction with Pumice and Clam Shell]	Core Formula + Hai Ge Ke Hai Fu Shi	Treat old phlegm, manifesting as hard fullness, and focal distention at chest, and lumpy nodules at the abdomen

(continued)

Formula	Herbs	Indication
Er Zhu Er Chen Tang [Two-Cured Decoction with Two Atractylodes]	Core Formula + Cang Zhu Bai Zhu	Treats Spleen deficiency failing to transport, nausea, and vomiting clear fluid
Jiu Zhi Er Chen Tang [Two-Cured Decoction with Chinese Chive Juice]	Core Formula + Jiu Cai juice Lai Fu Zi Xiang Fu	Treat sharp pain at hypochondrium
Zao Li Er Chen Tang [Two-Cured Decoction with Gleditsia and Bamboo Sap]	Core Formula + Zao Jiao Bai Jie Zi Ginger juice Zhu Li	Treat wind phlegm obstructing the channels and collaterals leading to numbness or pain at the limbs
Zha Qu Er Chen Tang [Two-Cured Decoction with Hawthorn and Medicated Leaven]	Core Formula + Shan Zha Shen Qu Mai Ya	Also treat sour regurgitation
Ju Ban Gui Ling Zhi Jiang Tang [Tangerine, Pinellia, Cinnamon Twig, Immature Bitter Orange, and Ginger Decoction]	Ban Xia Ju Pi Fu Ling Gui Zhi Zhi Shi Sheng Jiang	Treat passing out gas from the vagina in people who drink a lot, are wiry, and have slow pulse
Jia Wei Er Chen Tang [Augmented Two-Cured Decoction]	Core Formula + Zhi Nan Xing Chun Gen Pi Che Qian Zi Bai Guo Huang Bai	Treat phlegm damp pouring downward causing sticky thick leucorrhoea, stifling sensation at chest, nausea, excess phlegm, greasy tongue coating, and soggy pulse
Xiong Gui Er Chen Tang [Two–Cured Decoction with Ligusticum Chuanxiong and Tangkuei]	Core Formula + Chuan Xiong Dang Gui	Treat phlegm damp obstruction leading to irregular menstruation and copious vaginal discharge
Ban Xia Tang [Pinellia Decoction]	Ban Xia Gan Cao Gui Zhi	Treat cold type of painful throat

Ban Xia Bai Zhu Tian Ma Tang [Pinellia, Atractylodes Macrocephala, and Gastrodia Decoction] *Yi Xue Xin Wu*

Zhi Ban Xia	12 g
Chen Pi	10 g
Fu Ling	15 g
Gan Cao	3 g
Bai Zhu	12 g
Tian Ma	10 g

Main Indication: Upward disturbance of wind phlegm. Symptoms include dizziness, vertigo, headache or headache with cloudy sensation, stifling sensation at the chest, nausea, copious phlegm, white greasy tongue coating, and wiry slippery pulse.

Preparation: Decoct with one slice of ginger and two pieces of Da Zao.

Treatment Principles: Strengthen the Spleen, dry dampness, transform dampness and extinguish wind.

This formula treats dizziness, vertigo, and headache caused by upward disturbance of wind phlegm. The wind phlegm is a result of Spleen deficiency generating phlegm and Spleen deficiency failing to nourish Liver.

Modification:

Addition	Indication
Man Jing Zi 15 g	Same as the main formula
Bai Zhu reduce to 6 g	This one comes from *Yi Xue Xin Wu* (Chapter: Headache)

Dao Tan Tang [Guide Out Phlegm Decoction] *Ji Sheng Fang*

Chen Pi	10 g
Ban Xia	15 g
Fu Ling	12 g
Gan Cao	3 g
Zhi Ke	10 g
Dan Nan Xing	12 g
Sheng Jiang	10 g

Main Indication: Wind phlegm leading to dizziness, vertigo, headache, nausea, vomiting, cough, wheezing, stifling sensation and fullness at the chest, poor appetite, copious phlegm or even syncope due to phlegm, greasy tongue coating, and wiry slippery pulse.

Preparation: Decoction.

Treatment Principles: Dry dampness and expel phlegm.

This formula is a strong formula to expel phlegm. It can treat various syndromes of wind phlegm.

Related Formula:

Formula	Herbs	Indication
Jia Wei Dao Tan Tang [Augmented Guide Out Phlegm Decoction]	Core Formula + Bai Zhu Huang Qin Huang Lian Sheng Jiang juice Zhu Li	Treat phlegm misting the Heart orifice, drooling, continuous laughing and smiling, dark tongue with yellow greasy coating, and flooding and big pulse

Wen Dan Tang [Warm the Gallbladder Decoction] *Qian Jin Fang*

Chen Pi	10 g
Ban Xia	12 g
Fu Ling	12 g
Gan Cao	3 g
Zhi Shi	6 g
Zhu Ru	10 g

Main Indication: Disharmony of Gallbladder and Stomach, internal disturbance of phlegm heat. Symptoms include restlessness, insomnia, nausea, vomiting, hiccup, hunger with rumbling sounds, fright and palpitation, even epilepsy in severe cases. It can also be damp–heat syndromes with evil pathogens accumulating in the Qi level, cold and heat sensations, focal distention at the epigastrium, abdominal distention, yellow scanty frequent urination, yellow greasy tongue coating, and slippery rapid pulse or wiry rapid pulse.

Preparation: Decoct with five slices of Sheng Jiang and one piece of Da Zao.

Treatment Principles: Clear heat, expel phlegm, harmonize Gallbladder and Stomach. This formula is effective in treating various phlegm-heat syndromes when it is modified appropriately according to the presentation. This formula can also treat sty that belongs to phlegm-heat type.

Modification:

Addition	Indication
Bai Shao Yao Dai Zhe Shi Huang Qin	With severe dizziness and vertigo
Huang Lian Zi Su Ye	With severe vomiting
Suan Zao Ren Hu Po Mu Li	With severe insomnia
Mu Li Ze Xie	With palpitation
Huang Lian (fried with Ginger juice)	Hunger with rumbling sounds
Dan Nan Xing Sheng Jiang juice Zhu Li Zao Jiao Yu Jin Shi Chang Pu	Epilepsy
Shi Chang Pu Ma Huang Xi Xin	With deafness or ear congestion

Related Formulas:

Formula	Herbs	Indication
Huang Lian Wen Dan Tang [Coptis Decoction to Warm the Gallbladder]	Core Formula + Huang Lian	Treat severe cases of phlegm heat
Jia Wei Wen Dan Tang [Augmented Warm the Gallbladder Decoction]	Core Formula + Chai Hu Gou Teng Shi Chang Pu Tong Cao Ju Hua	Treat damp–heat syndrome. After the body has cooled off, the residual evil pathogen still remains in the Gallbladder channel causing tinnitus or deafness
Shi Wei Wen Dan Tang [Ten-Ingredients Decoction to Warm the Gallbladder]	Core Formula + Suan Zao Ren Yuan Zhi Wu Wei Zi Shu Di Huang Ren Shen – Zhu Ru	Treat timidity due to heat and Gallbladder deficiency, fright in contact with the outside world, restlessness, palpitation, stifling sensation, puffy and swollen four limbs, and no taste in mouth

Xiao Xian Xiong Tang [Minor Sinking into the Chest Decoction] *Shang Han Lun*

Ban Xia	12 g
Huang Lian	6 g
Gua Lou	15 g

Main Indication: Phlegm-heat knotted in chest and diaphragm. Symptoms include focal distention, stifling sensation at the chest and epigastrium is worse with palpation, cough with yellow sticky phlegm, yellow greasy tongue coating with slippery rapid pulse.

Preparation: Decoction.

Treatment Principles: Clear heat, transform phlegm, open the chest and scatter the knotted.

This formula treats pathological condition of minor congestion in chest of syndromes with phlegm-heat knotted below the Heart and is painful upon palpation. Main symptoms are focal distention, pain at chest, and greasy and yellow tongue coating. In clinical practice, as long as these two symptoms are present, no matter whether there is damp–heat or phlegm–heat cough, one can use this formula.

Furthermore, it can also be used to treat phlegm-heat above the diaphragm with cough, red face, heat at chest, abdomen, and hypochondrium with cool hands and feet and flooding pulse. It can also treat phlegm obstructing the chest leading to spasm and stiffness of the limbs.

Li Zhong Hua Tan Wan [Regulate the Middle and Transform Phlegm Pill] *Ming Yi Za Zhu*

Ren Shen	10 g
Bai Zhu	10 g
Gan Jiang	12 g
Zhi Gan Cao	3 g
Zhi Ban Xia	15 g
Fu Ling	15 g

Main Indication: Deficient cold in Spleen and Stomach, phlegm and fluid congested internally. Symptoms include scanty food intake, loose stool, nausea and vomiting clear water, difficulty digesting food, cough with phlegm and saliva, white slippery tongue coating, and deep, slow pulse.

Preparation: Decoction or pill form.

Treatment Principles: Warm the Middle Jiao and transform phlegm.

Modification:

Addition	Indication
Chen Pi Gui Zhi	Severe phlegm and saliva discharge and severe obstruction in the Middle Jiao

Related Formula:

Formula	Herbs	Indication
Li Zhong Jiang Tan Wan [Regulate the Middle and Direct Phlegm Decoction]	Core Formula + Zi Su Zi	With concomitant wheezing presentation

San Zi Yang Qin Tang [Three-Seed Decoction to Nourish One's Parents] *Han Shi Yi Tong*

Zi Su Zi	9 g
Bai Jie Zi	6 g
Lai Fu Zi	9 g

Main Indication: Phlegm congestion with Qi stagnation. Symptoms include cough, wheezing, copious phlegm, focal distention at chest, scanty food intake, difficulty in digestion, sticky greasy tongue coating, and slippery pulse.

Preparation: Crush after slightly fried, wrap in cheesecloth, and then make into decoction. Take the decoction frequently. For person with chronic constipation, add small amount of honey. During winter cold weather, add three slices of Sheng Jiang.

Treatment Principles: Transform phlegm, reduce food stagnation, descend Qi, and relax the diaphragm.

In this formula, Zi Su Zi descends the Qi and moves the phlegm, Lai Fu Zi reduces food stagnation and moves phlegm, and Bai Jie Zi relaxes the diaphragm and moves phlegm. All three seeds can move the phlegm, however, Zi Su Zi is stronger in the action to descend Qi, which is the main herb in treating Qi rebellion. Bai Jie Zi is effective in relaxing the diaphragm. Thus, it is the main herb when hypochondriac pain and copious phlegm are the main symptoms. Lai Fu Zi is effective in reducing food stagnation, which is the chief herb in treating scanty food intake with distending and fullness sensation. When three herbs are used together, the combined functions are smoothing Qi flow, descending the rebellion, reducing food stagnation, and transforming phlegm. For senior people with Middle Jiao deficiency, even though there is Qi and phlegm obstruction, one should still treat it with moderated movement to unblock the stagnation. Strong purging formulas are not suitable in this case. This is the main idea of this formula.

Fu Ling Wan [Poria Pill] *Bai Yi Xuan Fang* cited from *Zhi Mi Fang* also known as Zhi Mi Fu Ling Wan [Poria Pill from *Zhi Mi Fang*]

Jiang Ban Xia	60 g
Fu Ling	30 g
Zhi Ke	15 g
Po Xiao	8 g

Main Indication: Spleen not transforming dampness, phlegm turbid obstructing the Middle Jiao. Symptoms include sore and painful arms, puffiness, fatigue and weakness of the four limbs, white greasy tongue coating, and wiry slippery or deep pulse. Sometimes, one can also manifest as cough with copious phlegm, nausea, and stifling sensation at the chest.

Preparation: Grind into powder, mix with ginger juice, and make into pills of 5 to 10 mm in diameter. Take 6 g each time with ginger decoction.

Treatment Principles: Dry dampness, move Qi, soften the hardness, and reduce phlegm.

This formula treats syndromes of phlegm accumulating in the Middle Jiao and overflowing into the four limbs. One should not treat this condition as wind–damp Bi syndrome.

Xiang Fu Xuan Fu Hua Tang [Cyperus and Inula Decoction] *Wen Bing Tiao Bian*

Sheng Xiang Fu	10 g
Xuan Fu Hua (wrapped)	10 g
Zi Su Zi Shuang	10 g
Chen Pi	10 g
Fu Ling	10 g
Yi Yi Ren	20 g
Ban Xia	15 g

Main Indication: Damp stagnation in Liver collaterals. Symptoms include hypochondriac pain, presence or absence of cough, absence of chills, but with tidal fever, or alternating chills and fever like malaria, thin yellow or white coating, and wiry pulse.

Preparation: Decoct in 8 cups of water until 3 cups left, divided the decoction into three doses, and take warm.

Treatment Principles: Dry dampness, soothe the Liver, unblock the collaterals, and stop pain.

This formula treats damp stagnation in Liver collaterals, accumulating at hypochondriac region causing hypochondriac pain and spasm. In the formula, Xuan Fu Hua and Xiang Fu are effective in unblocking Liver collaterals and chasing the damp congestion downward from the hypochondriac region.

Modification:

Addition	Indication
Hou Po	Abdominal fullness
Jiang Xiang	Severe pain

Related Formula:

Formula	Herbs	Indication
Xuan Fu Hua Tang [Inula Decoction]	Xuan Fu Hua 10 g Qian Cao 6 g Cong Bai 15 g	Expel stasis and unblock collaterals Treat hypochondriac pain due to blood stasis in Liver meridian. This can also treat Gan Zhuo (see Glossary)

Comparison of Damp Accumulation Forming Phlegm Formulas

Er Chen Tang [Two-Cured Decoction] dries dampness and transforms phlegm. It is not only the main formula for expelling phlegm but also a fundamental formula for modification. When Bai Zhu and Tian Ma are added, it becomes Ban Xia Bai Zhu Tian Ma Tang [Pinellia, Atractylodes Macrocephala, and Gastrodia Decoction], which treats dizziness, vertigo, and headache due to wind phlegm upward disturbance. When Zhi Ke and Dan Nan Xing are added, Dao Tan Tang [Guide Out the Phlegm Decoction] is formed which treats various syndromes caused by wind phlegm. This formula has stronger function to dry dampness and expel phlegm. When Zhi Shi and Zhu Ru are added, Wen Dan Tang [Warm the Gallbladder Decoction] is formed which can regulate Qi, transform phlegm, clear Gallbladder, harmonize Stomach. It is effective in treating disharmony of Gallbladder and Stomach and internal disturbance of phlegm-heat leading to deficient type of restlessness, insomnia, nausea, vomiting, fright and palpitation, or even epilepsy. When Ren Shen, Bai Zhu, and Gan Jiang are added with elimination of Chen Pi, then it becomes Li Zhong Hua Tan Wan [Regulate the Middle and Transform Phlegm Pill]. It can warm the Middle Jiao, transform phlegm and is effective in treating various digestive syndromes caused by internal congestion of phlegm.

Xiao Xian Xiong Tang [Minor Sinking into the Chest Decoction] treats focal distention and pain at the chest due to knotted phlegm and heat. San Zi Yang Qin Tang [Three-Seed Decoction to Nourish One's Parents] treats phlegm and Qi obstructing the Middle Jiao leading to various Qi rebellion syndromes in Lung and digestive system. Both formulas are moderate and gentle in their function.

Fu Ling Wan [Poria Pill] is effective in moving Qi, softening the hardness, promoting urination, and unblocking the bowels to create an outlet for dampness. It is suitable for phlegm turbidity obstructing the Qi mechanism in the Middle Jiao leading to soreness, weakness, and pain at the four limbs. Xiang Fu Xuan Fu Hua Tang [Cyperus and Inula Decoction] treats damp obstructing the Liver collaterals leading to pain and spasm at the hypochondrium.

SPLEEN DEFICIENCY WITH WATER ACCUMULATION

Spleen deficiency failing to transport and transform water and dampness, in mild cases, leads to damp obstructing the Middle Jiao and in severe cases, water and dampness accumulating, causing swelling and distention. Spleen Deficiency with water accumulation refers to the latter type. The treatment method needs to tonify Spleen and also promote urination to reduce swelling. Often, Fu Ling, Bai Zhu, Gan Cao, Huang Qi, and Fang Ji are used in combination with Shang Jiang and Da Fu Pi.

Wu Pi Yin [Five-Peel Decoction] *Zhong Zang Jing* also known as Wu Pi San [Five-Peel Powder]

Sang Bai Pi	9 g or equal dosage
Chen Ju Pi	9 g
Sheng Jiang Pi	9 g
Da Fu Pi	9 g
Fu Ling Pi	9 g

Main Indication: Spleen deficiency with excess dampness which is referred to as Pi Shui (superficial edema). Symptoms include edema at face and limbs, heaviness at body and limbs, distention and fullness at the epigastrium and abdomen, rapid breathing and wheezing, difficult urination or edema during pregnancy, white greasy tongue coating with slow and moderate pulse.

Preparation: Decoction or grind into coarse powder and take 9 g each time as draft.

Treatment Principles: Drain damp, reduce edema, regulate Qi, and strengthen the Spleen.

This is a common formula in treating Pi Shui (superficial edema) that focuses on moving water and reducing edema.

Modification:

Addition	Indication
Bai Zhu Huang Qi	Severe Spleen deficiency
Fang Feng Qiang Huo	Swelling that is more severe above the waist with wind pathogens
Wu Ling San [Five-Ingredient Powder with Poria]	Swelling that is more severe below the waist with scanty urination

Related Formulas:

Formula	Herbs	Indication
Wu Pi Yin [Five-Peel Decoction] *He Ji Ju Fang*	Sheng Jiang Pi Da Fu Pi Fu Ling Pi Wu Jia Pi Di Gu Pi	Same indications This formula is weaker in moving Qi
Wu Pi Yin [Five-Peel Decoction] *Ma Ke Huo Ren Quan Shu*	Wu Jia Pi Chen Pi Sheng Jiang Pi Da Fu Pi Fu Ling Pi	Same indications. This formula can also unblock the collaterals and expel wind damp
Qi Pi Yin [Seven-Peel Decoction]	Core Formula – Sang Bai Pi + Qing Pi Di Gu Pi Gan Cao Pi	Taken as warm decoction Same indications
Quan Sheng Bai Zhu San [Atractylodes Macrocephala Powder from *Quan Sheng Ji*]	Core Formula – Sang Bai Pi + Bai zhu	Treat edema during pregnancy. Symptoms include puffiness at face, edema at the four limbs, sallow complexion, poor appetite, loose stool, bland taste in mouth, scanty urination, weak, big and slippery pulse which is caused by Spleen deficiency
Jia Jian Wu Pi Yin [Modified Five-Peel Decoction]	Fu Ling Pi Da Fu Pi Wu Jia Pi Sang Zhi Fang Ji Cang Zhu Shi Chang Pu Yin Chen Hao	Taken as warm decoction Treats edema during pregnancy, fullness at chest, palpitation, edema at limbs, weak knees and lower back pain, white greasy tongue coating, deep and slippery pulse which is caused by Spleen deficiency with water accumulation and affecting the Heart

Fang Ji Huang Qi Tang [Stephania and Astragalus Decoction] *Jin Gui Yao Lue*

Fang Ji	15 g
Huang Qi	30 g
Bai Zhu	10 g
Gan Cao	6 g

Main Indication: Wind–water (Feng shui) and wind–damp due to Wei Qi not consolidating the exterior. Symptoms include sweating, aversion to wind, heaviness of the body, difficult urination, pale tongue with white tongue with coating, and floating pulse.

Preparation: Decoct with ginger and Da Zao. Mild sweating occurs after taking the decoction.

Treatment Principles: Strengthen the Spleen, promote urination, tonify Qi, and expel wind.

Today, this formula is often used for edema due to exterior deficiency as seen in rheumatoid arthritis and glomerulonephritis.

Related Formula:

Formula	Herbs	Indication
Fang Ji Fu Ling Tang [Stephania and Poria Decoction]	Fang Ji 9 g Huang Qi 9 g Gui Zhi 9 g Fu ling 18 g Gan Cao 6 g	Taken as warm decoction This formula can tonify Qi, unblock Yang, and promote urination. It treats Pi Shui (skin water), swelling of the four limbs, superficial edema where the water is accumulated between the skin and muscle leading to twitching

Shi Pi Yin [Bolster the Spleen Decoction] *Ji Sheng Fang* also known as Shi Pi San [Bolster the Spleen Powder]

Hou Po	12 g
Bai Zhu	10 g
Mu Gua	12 g
Mu Xiang	6 g
Cao Guo	10 g
Gan Cao	3 g
Gan Jiang	10 g
Fu Ling	15 g
Zhi Fu Zi	15 g
Bing Lang Ke	18 g

Main Indication: Edema due to Spleen and Kidney Yang deficiency. Symptoms include edema that is worse below the waist, cold hands and feet, absence of thirst, distention and fullness at the chest and abdomen, loose stool, clear urine, heaviness of the body, too lazy to eat, thick greasy and moist tongue coating, and deep and slow pulse.

Preparation: Decoct with five slices of fresh ginger and one piece of Da Zao.

Treatment Principles: Warm Yang, strengthen the Spleen, move Qi, and promote urination.

This formula treats Yin edema due to Spleen and Kidney Yang deficiency with internal water accumulation. In modern times, it is being used to treat chronic glomerulonephritis, early stage of Liver cirrhosis with slight ascites, swelling of the lower limb, and slight edema and poor appetite in mild congested Heart failure.

Li Yu Tang [Carp Decoction] *Qian Jin Yao Fang*

Li Yu (remove scales and gut)	about 500 g
Bai Zhu	10 g
Fu Ling	15 g
Chen Pi	10 g
Sheng Jiang	10 g
Dang Gui	10 g
Bai Shao Yao	12 g

Main Indication: Edema during pregnancy due to Spleen deficiency. Symptoms include water congested in the uterus, abnormal enlargement of the abdomen, generalized swelling, fullness and stifling sensation at the chest, wheezing and restlessness, poor appetite, scanty urination, pale tongue with thin and moist coating, and deep and slippery pulse. The presentation is similar to excess amniotic fluid.

Preparation: Boil the fish in water. Once the fish is cooked, remove from soup. Decoct the herbs in soup. Take the warm decoction before meal on an empty stomach.

Treatment Principles: Strengthen the Spleen, promotes urination, nourish blood, and calm the fetus.

Modification:

Addition	Indication
Rou Gui	Yang deficiency. Rou Gui can warm the Yang and transform Qi

Comparison of Spleen Deficiency with Water Accumulation Formulas

1. Wu Pi Yin [Five-Peel Decoction] and Fang Ji Huang Qi Tang [Stephania and Astragalus Decoction] treat Yang edema. Shi Pi Yin [Bolster the Spleen Decoction], Quan Sheng Bai Zhu San [Atractylodes Macrocephala Powder from *Quan Sheng Ji*], and Li Yu Tang [Carp Decoction] treat Yin edema.
2. According to the pathologies, Wu Pi Yin [Five-Peel Decoction] treats depressed Spleen and Qi stagnation. Fang Ji Huang Qi Tang [Stephania and Astragalus Decoction] treats both Spleen and Lung deficieny. Shi Pi Yin [Bolster the Spleen Decoction] treats both Spleen and Kidney deficieny. Quan Sheng Bai Zhu San [Atractylodes Macrocephala Powder *Quan Sheng Ji*] and Li Yu Tang [Carp Decoction] treat Spleen deficiency with edema.
3. According to the location of the disease and the characteristics, Wu Pi Yin [Five-Peel Decoction] treats water accumulation between the skin and muscle giving a shiny complexion. Fang Ji Huang Qi Tang [Stephania and Astragalus Decoction] treats water accumulation in the muscle accompanied with sweating and chills. Shi Pi Yin [Bolster the Spleen Decoction] treats internal accumulation of water. Patient usually does not look heavy from the outside, but feels abdominal distention and swelling at the lower limbs. Quan Sheng Bai Zhu San [Atractylodes Macrocephala Powder *Quan Sheng Ji*] and Li Yu Tang [Carp Decoction] treat water accumulation in the uterus during pregnancy.
4. According to the treatment focus, the first two formulas emphasize on promoting urination and are assisted by tonifying the Spleen. The latter three formulas focus on tonifying the Spleen, promoting urination, and leaching out dampness.

YANG EDEMA EXCESS SYNDROME (YANG SHUI)

Edema that belongs to the Yang edema excess type with the symptoms of a big, hard, full abdomen, constipation, and difficult and hesitant urination should be treated by promoting urination. Herbs which harshly drive out water pathogens such as Gan Sui, Da Ji, Yuan Hua, Ting Li Zi, and Fang Ji are used in combination with downward draining herbs like Da Huang, Bing Lang, and Qi moving herbs such as Hou Po and Mu Xiang to form a formula. The formula drains downward and drives out water to force the internally accumulated water out from urine and stool. If there is excess evil pathogen with deficient Zheng Qi, then one should not take such formula for a long time, but should alternate with Qi tonifying and Spleen strengthening formula.

This method only treats the acute manifestation and not the root condition. When the accumulated water is driven out from the body and the distention is reduced, then regulating and tonifying the body would be the next step.

Zhou Che Wan [Vessel and Vehicle Pill] *Jing Yue Quan Shu*

Black Qian Niu Zi (ground)	120 g
Gan Sui (dip in flour paste, then bake in ember)	30 g
Yuan Hua (fried with vinegar)	30 g
Da Ji (fried with vinegar)	30 g
Da Huang	60 g
Qing Pi	15 g
Chen Pi	15 g
Mu Xiang	15 g
Bing Lang	15 g
Qing Fen	3 g

Main Indication: Internal congestion of water and heat leading to edema and water distention. Symptoms include edema in robust patient, thirst, coarse breathing, abdominal hardness, constipation, difficulty urinating, yellow thick greasy tongue coating, and deep, rapid, and forceful pulse.

Preparation: Grind the above ingredients into powder and form into pills with water. Take 6 to 10 g each time, once per day in the morning on an empty Stomach.

Treatment Principles: Drain downward and drive out water.

This formula is modified from Shi Zao Tang [Ten-Jujube Decoction]. It is a harsh formula to drive out water and drain downward from the body. If the edema, fullness, and distention of the patient who is robust and strong in his or her constitution are not relieved after taking the formula, then the same or decreased dose should be taken again the next day or the day after. However, the formula contains Qing Fen, Yuan Hua, Da Ji, and Gan Sui, which are toxic, thus one should take minimal amounts and should not be taken for a long time.

This formula is contraindicated for pregnant women and a person with weak constitution.

This can also treat patient who has a strong constitution and suffering from ascites due to Liver cirrhosis with symptoms of big and hard abdomen, and deep, rapid, and forceful pulse.

Related Formulas:

Formula	Herbs	Indication
Xiao Shui San [Reduce the Water Powder]	Gan Sui 10 g Hu Po 5 g Qian Niu Zi (black & white) 3 g each Chen Xiang 2 g	Grind into powder, take 3 to 8 g each time, one to two times a day Treats early stage or mild cases of ascites due to Liver cirrhosis
San Hua Shen You Wan [Protected by Deities Pill with Three Flowers]	Gan Sui 15 g Da Ji 15 g Yuan Hua 15 g Qian Niu 60 g Da Huang 30 g Qing Fen 3 g	Grind into powder and make pills with water to 5 to 10 mm in diameter Take 5 pills for the initial dose and increase by 5 pills each time, one to two times per day Treat fullness in the center, abdominal distention, cough, blocked urination, and all other swelling and fullness due to edema
Shu Zao Yin Zi [Unblock and Dig Formula]	Qiang Huo Qin Jiao Shang Lu Bing Lang Da Fu Pi Fu Ling Pi Sheng Jiang Pi Jiao Mu (seed of Hua Jiao) Mu Tong Ze Xie Chi Xiao Dou (equal amounts)	Decoction Treats generalized edema, wheezing, thirst, difficulty urinating, and constipation

Ji Jiao Li Huang Wan [Stephania, Zanthoxylum, Descurainia, and Rhubarb Pill] *Jin Gui Yao Lue*

Fang Ji	30 g
Jiao Mu	30 g
Ting Li Zi	30 g
Da Huang	30 g

Main Indication: Spleen not transporting fluid, water and fluid accumulation. Symptoms include abdominal fullness, dry mouth, dry tongue, thick dry coating, and deep excess pulse.

Preparation: Grind into powder, make into pills with honey the size of 5 to 10 mm in diameter. Take 1 pill after meal, three times per day.

Treatment Principles: Drive out water, scour out congested fluids, and eliminate dampness through both urine and stool.

The main symptom for this formula is frequent abdominal fullness at a fixed location. This indicates accumulation of water in the abdomen and is different from Qi stagnation leading to distention. This formula can be used in mild cases of ascites due to Liver cirrhosis and mild congestive Heart failure due to rheumatic Heart conditions with edema and shortness of breath.

Modification:

Addition	Indication
Mang Xiao 15 g	Jin Ye in the mouth and presence of thirst
Da Fu Pi Bing Lang Jin Qian Cao Xia Ku Cao	For severe cases

Comparison of Formulas Treating Yang Edema Excess Syndromes

1. Zhou Che Wan [Vessel and Vehicle Pill] and San Hua Shen You Wan [Protected by Deities Pills with Three Flowers] are formulas that drive out water and drain downward. They are used in robust patients with severe conditions. The toxicity and side effect of these two formulas are also more severe, thus one should use these formulas with care and limit the use and follow the principle that the formula should be stopped once the majority of the evil is weakened.
2. Xiao Shui San [Reduce the Water Powder] and Shu Zao Yin Zi [Unblock and Dig Decoction] and Ji Jiao Li Huang Wan [Stephania, Zanthoxylum, Descurainia, and Rhubarb Pill] are less harsh than the formulas in 1. They are used to treat early stage or mild cases of Yang edema excess syndromes.
3. Shu Zao Yin Zi [Unblock and Dig Decoction] treats water damp overflowing in both superficial and deep areas where the body has generalized edema. Thus the formula reduces the swelling both in exterior and interior with herbs that drain downward, drive out water, disperse wind, and release exterior. Ji Jiao Li Huang Wan [Stephania, Zanthoxylum, Desurainia, and Rhubarb Pill] treats damp-stranded-the-water accumulating in the Middle Jiao. The water and fluid cannot be disperse upward or descend downward. Thus, the formula reduces water accumulation from the front and the back. The clear water is out of the body through urination and turbid water from stools to get rid of water accumulation, distention, and fullness.

STOMACH YIN DEFICIENCY

This syndrome often manifests as dry mouth, thirst, dry throat, dry tongue or with dry, incomplete stool, scanty urination, dry red tongue with dry coating or scanty coating in severe cases, and thin pulse. This syndrome often is often caused by Stomach Yin deficiency or late stage of warm-febrile disease with damaged Stomach fluid. Herbs like Bei Sha Shen, Yu Zhu, Shi Hu, Mai Men Dong, Sheng Di Huang, Xuan Shen, Xi Gua Cui Yi, Zhi Mu, Li Zhi, Zhe Jiang (sugar cane juice), and Xi Yang Shen are used.

Yi Wei Tang [Benefit the Stomach Decoction] *Wen Bing Tiao Bian*

Bei Sha Shen	10 g
Mai Men Dong	15 g
Yu Zhu	10 g
Sheng Di Huang	15 g
Rock sugar	3 g

Main Indication: Injured Stomach Yin. Symptoms include dry mouth, dry tongue, thirst, and thin or thin and rapid pulse due to warm febrile disease that damages Stomach and exhausts Stomach Yin.

Preparation: Decoct the first four herbs, strain decoction, and then dissolve rock sugar into decoction.

Treatment Principles: Nourish Stomach Yin.

All five herbs are sweet, cold and fluid generating. After treating heat at the Yang Ming Qi level with sweating and draining method, if the fever is reduced, yet the Stomach Yin is not recovered, then this formula is indicated.

Furthermore, this formula can be used for hiccup due to Stomach Yin deficiency, Qi and Yin deficient type of medication induced dermatitis, Spleen, and Stomach deficient type of burned patients.

Wu Zhi Yin [Five-Juice Decoction] *Wen Bing Tiao Bian*

Pear Juice
Bo Qi Juice (water chestnut juice)
Fresh Lu Gen Juice
Mai Men Dong Juice
Ou Zhi or Zhe Jiang (sugar cane sap)

Main Indication: Injured Stomach and Lung Yin. Symptoms include dry mouth, thirst, white foamy sticky saliva, thin or thin rapid pulse, and thin even tongue coating.

Preparation: One can modify the dosage according the clinical manifestations. Mix well and take as cold juice. If one has aversion to cold, then take as warm decoction.

Treatment Principles: Sweet and cold to rescue Yin.

This formula is designed for exhausted Stomach fluid caused by warm febrile disease damages, fluid or heat not relieved after the appearing of rashes in warm febrile disease. It can also be used for thirst due to blood deficiency and exhausted fluid to rescue the Yin fluid in order to increase blood volume.

Zeng Ye Tang [Increase the Fluids Decoction] *Wen Bing Tiao Bian*

Xuan Shen	30 g
Mai Men Dong	24 g
Sheng Di Huang	24 g

Main Indication: Yang Ming warm febrile disease. Symptoms include insufficient Jin and Ye (fluids), constipation, thirst, dry red tongue, and thin, slight, and rapid pulse.

Preparation: Decoction. If bowel movement does not happen after taking the formula, one should take the formula again.

Treatment Principles: Nourish Yin, clear heat, and replenish water to move the stool.

This formula is effective in treating constipation due to fluid deficiency: "insufficient water to move the boat." The dosage should be larger.

It can also treat insufficient Stomach Yin manifesting as mirrored, deep red tongue, dry tongue, and dry lips. It is also a basic formula for all other Yin deficient with fluid deficient conditions.

Modification:

Addition	Indication
Da Huang Mang Xiao	For severe constipation It is also called Tiao Wei Cheng Qi Tang [Order of Qi Decoction to Regulate Stomach]

Qing Shu Yi Qi Tang [Clear Summer Heat and Augment Qi Decoction] *Wen Re Jing Wei*

Xi Yang Shen	6 g
Shi Hu	12 g
Mai Men Dong	12 g
Huang Lian	3 g
Dan Zhu Ye	10 g
He Geng	20 g
Gan Cao	2 g
Zhi Mu	10 g
Xi Gua Cui Yi	30 g
Geng Mi	15 g

Main Indication: Summer stroke with injured Qi and fluid. Symptoms include fever, excess sweating, restlessness, thirst, scanty dark urine, fatigue, shortness of breath, apathy, big but deficient pulse or deficient and rapid pulse, and dry tongue.

Preparation: Decoction.

Treatment Principles: Clear heat, scour summer heat, tonify Qi, and generate fluid.

Modification:

Minus	Addition	Indication
Huang Lian		Summer heat not severe, with severely injured fluid
Mai Men Dong		Accompanied by dampness
	Bai Hu Jia Ren Shen Tang [White Tiger Decoction with Ginseng]	Severe heat
	Zeng Ye Tang [Increase the Fluids Decoction] or Sheng Mai San [Generate the Pulse Powder]	Severe damage of Qi and fluid

Related Formula:

Formula	Herbs	Indication
Qing Shu Yi Qi Tang [Clear Summer Heat and Augment Qi Decoction] *Pi Wei Lun*	Huang Qi 6 g Cang Zhu 3 g Sheng Ma 3 g Ren Shen 1.5 g Shen Qu 3 g Chen Pi 15 g Bai Zhu 3 g Mai Men Dong 3 g Dang Gui 1 g Zhi Gan Cao 1 g Qing Pi 1 g Huang Bai 3 g Ge Gen 3 g Ze Xie 3 g Wu Wei Zi 3 g	Decoction Clear summer heat, tonify Qi, eliminate dampness, and strengthen the Spleen. Treat constitutional Spleen deficiency with external invasion of summer dampness

Mai Men Dong Tang [Ophiopogonis Decoction]
Jin Gui Yao Lue

Mai Men Dong	30 g
Ban Xia	10 g
Ren Shen	6 g
Gan Cao	6 g
Da Zao	10 g
Geng Mi	10 g

Main Indication: Lung and Stomach Yin deficiency with fluid deficiency, with deficient fire blazing upward. Symptoms include wheezing, cough, shortness of breath, hard to expectorate phlegm, dry mouth and throat or nausea and vomiting due to Qi rebellion, thirst, red tongue with scanty coating, and deficient rapid pulse.

Preparation: Take as warm decoction.

Treatment Principles: Nourish Lung and Stomach Yin and descend rebellious Qi.

This formula can also treat dry cough due to tuberculosis in Lung.

Related Formula:

Formula	Herbs	Indication
Jia Wei Mai Men Dong Tang [Augmented Ophiopogonis Decoction]	Core Formula – Geng Mi + Shan Yao Bai Shao Yao Dan Shen Tao Ren	Treats inverted menstruation manifesting as absence of menstrual flow, but with epistaxis

Gan Lu Yin [Sweet Dew Decoction] *He Ji Ju Fang*

Sheng Di Huang	10 g
Shu Di Huang	10 g
Shi Hu	10 g
Tian Men Dong	12 g
Mai Men Dong	12 g
Huang Qin	10 g
Yin Chen Hao	10 g
Zhi Ke	10 g
Gan Cao	6 g
Pi Pa Ye	24 g

Main Indication: Yin deficiency with dampness. Symptoms include swollen gums with pus, mouth sores, and sore throat. Other symptoms can include eye pathological conditions with redness or distended blood vessels at conjunctiva. The blood vessels remain at the same location while pushing the eye lids with the hand, and there is also pain and aversion to light. Tongue is red with scanty or thin yellow coating, and the pulse is thin and soggy.

Preparation: Decoction.

Treatment Principles: Nourish Yin, clear heat, disperse Lung, and drain damp.

This formula is often used for electric ophthalmia, keratitis, or early stage of cataract.

Comparison of Formulas Treating Yang Edema Excess Syndromes

1. Yi Wei Tang [Benefit the Stomach Decoction] and Wu Zhi Yin [Five-Juice Decoction] treats dry mouth, dry throat, and dry tongue due to fluid damage at the late stage of a heat-febrile disease. Zeng Ye Tang [Increase the Fluids Decoction] also treats fluid damage at the late stage of a heat febrile disease. However, constipation is the main symptom.
2. Qing Shu Yi Qi Tang [Clear Summer Heat and Augment Qi Decoction] not only treats fluid deficiency with Qi exhaustion, but can also clear heat and expel summer heat. Qing Shu Yi Qi Tang [Clear Summer Heat and Augment Qi Decoction] from *Pi Wei Lun* focuses on strengthening the Spleen. Expelling damp, clearing summer heat, and tonifying Qi are achieved through tonifying the Spleen.
3. The main indications for Mai Men Dong Tang [Ophiopogonis Decoction] are Qi rebellion and throat discomfort. Even though the manifestation is at the Lung, the root is at the Stomach. Thus, the formula consists of herbs to benefit the Stomach, generate fluid, tonify the Spleen, and benefit the Lung.
4. The etiology for using Gan Lu Yin [Sweet Dew Decoction] is Lung, Kidney, and Stomach Yin deficiency with damp–heat. Though the herbs that nourish Yin and drain damp when used together seem to have opposing effect, they work together in this formula.

SPLEEN FAILING TO CONTROL BLOOD

Bleeding due to Middle Jiao Yang deficiency is referred to as Spleen failing to control blood. The common manifestations are hematemesis, hemoptysis, and blood in stool, and Beng Lou. The concomitant symptoms include dark color of blood or pale watery thin blood, sallow complexion, pale tongue with white coating, cold hands and feet, aversion to cold and preference to heat, deep thin weak pulse, and other symptoms of Yang deficiency. Warm or neutral stop bleeding herbs, such as Fu Long Gan (Zao Xin Tu), E Jiao, Ai Ye, San Qi, and Pao Jiang are often used and combined with warm Yang herbs and scatter cold herbs such as Fu Zi and Gan Jiang to form a formula. These formulas recover the Yang in the Middle Jiao and return the Spleen's and Stomach's ability to control the blood.

Furthermore, warming Yang to control blood method is sometimes used together with tonifying Qi to control blood method, such as Li Zhong Tang [Regulate the Middle Decoction] and Wen Jing She Xue Tang [Warm the Menses and Control the Blood Decoction].

Gan Cao Gan Jiang Tang [Licorice and Ginger Decoction] *Jin Gui Yao Lue*

Pao Jiang	10 g
Gan Cao	6 g

Main Indication: Yang deficiency, Spleen and Lung deficient cold. Symptoms include hemoptysis, consumptive Lung disease, vomiting of saliva but no cough, absence of thirst, enuresis, frequent urination, pale tongue with white coating, and slow pulse.

Preparation: Decoction.

Treatment Principles: Warm Yang and stop bleeding.

This formula can also be used for abdominal pain, vomiting and nausea due to deficient cold and frequent urination, vomiting saliva, shortness of breath, dizziness, and difficulty moving in seniors.

Bai Ye Tang [Biota Leaves Decoction] *Jin Gui Yao Lue*

Ce Bai Ye	20 g
Pao Jiang	6 g
Ai Ye	10 g

Main Indication: Spleen-deficient cold failing to control blood. Symptoms include unceasing vomiting of blood, sallow face, pale tongue, and deficient rapid pulse that disappears with increase in palpation pressure.

Preparation: Decoction, taken with one cup of urine of a young boy.

Treatment Principles: Warm Yang and stop bleeding.

This formula is composed of clearing Liver and warming the Spleen herbs. Bai Ye clears and descends. A young boy's urine is salty and descending. Pao Jiang and Ai Ye warm the middle. Thus the formula can be used for any type of bleeding syndrome due to Middle Jiao deficiency with rebellion.

Modification:

Addition	Indication
Ren Shen San Qi	Severe vomiting of blood

Huang Tu Tang [Yellow Earth Decoction] *Jin Gui Yao Lue*

Zao Zhong Huang Tu	30 g
Dried Sheng Di Huang	19 g
E Jiao (dissolved in decoction)	12 g
Zhi Fu Zi	10 g
Huang Qin	12 g
Bai Zhu	10 g
Gan Cao	3 g

Main Indication: Middle Jiao deficient cold, failing to control blood. Symptoms include blood in the stool, vomiting blood, epistaxis, Beng Lou in female, dark blood color, cold extremities, sallow face, pale tongue with white coating, and deep, thin, and weak pulse.

Preparation: Decoct Zao Xin Tu first, strain decoction, and then add remainder of herbs into decoction.

Treatment Principles: Warm the Middle Jiao, strengthen the Spleen, nourish blood, and stop bleeding.

This formula is effective in treating various types of bleeding syndromes due to Yang Qi deficiency.

Wen Jing She Xue Tang [Warm the Menses and Control the Blood Decoction] *Zhong Yi Fu Ke Zhi Liao Xue*

Hong Ren Shen	10 g
Bai Zhu	10 g
Pao Jiang	10 g
Zhi Gan Cao	3 g
Wu Zhu Yu	3 g
Jiao Ai Ye	10 g

Main Indication: Middle Jiao deficient cold, loss in the ability to control blood. Symptoms include Beng Lou, pale thin watery blood, lower abdominal distending pain which is better with heat, preference of heat, aversion of cold, poor appetite, loose stool, pale tongue with white coating, and deficient and slow pulse.

Preparation: Decoction.

Treatment Principles: Warm Yang, control the blood.

This formula is a modification of Li Zhong Tang [Regulate the Center Decoction]. It treats Spleen deficiency failing to govern blood and Beng Lou due to Qi deficiency failing to control blood.

Modification:

Addition	Indication
Wu Zei Gu 15 g	Severe bleeding
Yan Hu Suo Tan 6 g	For Beng Lou (spotting)

Comparison of Formulas Treating Yang Edema Excess Syndromes

Gan Cao Gan Jiang Tang [Licorice and Ginger Decoction] treats vomiting blood and coughing up blood due to Spleen and Stomach deficient cold. Bai Ye Tang [Biota Leave Decoction] treats Spleen deficient cold type of vomiting blood due to bleeding in the Stomach. Huang Tu Tang [Yellow Earth Decoction] treats deficient cold in the Middle Jiao causing blood in the stool. Wen Jing She Xue Tang [Warm the Menses and Control the Blood Decoction] treats deficient cold type of abnormal uterine bleeding.

KNOTTED HEAT

Yang Ming Fu Syndrome discussed in Shang Han Lun or heat in the Middle Jiao as in warm febrile disease both are excess heat syndrome with dry bounded stool, focal distention at the chest, and fullness of the abdomen. This is treated with bitter cold, downward draining herbs. Commonly,

Da Huang, Mang Xiao, and Qian Niu Zi are chief herbs to drain heat and purge the accumulation. Zhi Shi and Hou Po are often the assisting herbs to move Qi and alleviate the fullness. Thus, focal distention, fullness, dryness, and excess all are eliminated.

Bitter cold downward draining method is used to treat not only constipation, but also other syndromes such as food accumulation in the Middle Jiao affecting the transporting and transforming function leading to diarrhea. Furthermore, excess heat in the Upper Jiao, such as pathological conditions of the mouth and throat, reckless moving of the blood causing vomiting of blood, epistaxis, and mania due to excess heat, can also be treated with this method. This method guides the heat down and out through the stool. This treatment principle shows that by treating the lower part of the body for the pathological condition of the upper body, one can "eliminate the fire by removing the firewood."

This type of formula sometimes is added with Jie Geng, Xing Ren, and Gua Lou, herbs that open and diffuse Lung Qi. Thus, the excess syndrome of Lung heat with wheezing, coughing, phlegm, and rebellious Lung Qi is treated with bitter cold downward draining formula, such as Tiao Wei Cheng Qi Tang [Regulate the Stomach and Order the Qi Decoction] or Liang Ge San [Cool the Diaphragm Powder]. This formula unblocks the Qi of the Fu organs and helps to descend the Lung Qi, guide the Lung excess heat downward. This is an example of treating the Fu organ for Zang organ pathology.

Da Cheng Qi Tang [Major Order the Qi Decoction] *Shang Han Lun*

Da Huang (add near end)	12 g
Mang Xiao (add to decoction)	12 g
Zhi Shi	10 g
Hou Po	15 g

Main Indication:

1. Yang Ming Fu excess syndrome. Symptoms include constipation, abdominal pain that refuses pressure or abdominal hardness with palpation, focal distention, and fullness of the epigastrium and abdomen, tidal fever or delirious speech in severe cases, sweating at the hands and feet; yellow, dry, thorny tongue coating or burned, black, dry, cracked tongue coating, and excess, deep, and forceful pulse.
2. Constipation with occasional watery discharge due to accumulation of heat evil. Symptoms include watery diarrhea, green in color, pain at umbilicus and abdomen, hard or lumpy sensation with palpation, dry tongue and mouth, and slippery and excess pulse.
3. Internal heat excess syndrome that manifests as heat syncope and mania.

Preparation: Decoction. Add Da Huang near the end and then dissolve Mang Xiao in the decoction. Once the bowel movement happens, stop taking the formula.

Treatment Principles: Purge the heat and accumulation downward.

This is the main formula of bitter cold downward draining formulas. There are four points that indicate the use of this formula: *pi* (focal distention, stifling and congested sensation, and hardness at epigastrium), *man* (distention and fullness at the chest, hypochondrium, epigastrium, and abdomen), *zao* (dry stool in the intestine that cannot move out), *shi* (hardness and fullness of the abdomen, pain with palpation, refuse palpation, and constipation or watery diarrhea that does not alleviate the hardness and fullness at the abdomen, yellow tongue coating, and excess pulse).

This formula can also be used to treat acute simple bowel obstruction, acute cholecystitis, and acute appendicitis, with constipation and yellow tongue coating. It is also used to treat excess Yang Ming Fu syndromes with mania, delirium, and delirious speech.

Related Formulas:

Formula	Herbs	Indication
Lai Po Tong Jie Tang [Open the Blockage Decoction with Radish Seeds and Magnolia Bark]	Lai Fu Zi (fried) 30 g Hou Po 30 g Qian Niu Zi 15 g Gan Sui powder (add to decoction) 2 g Da Huang 15 g	This formula is indicated for bowel obstruction with excess mucous and water accumulation in the bowel
Gan Sui Tong Jie Tang [Open the Blockage Decoction with Euphorbia]	Gan Sui powder (add to decoction) 1 g Tao Ren 10 g Chi Shao Yao 15 g Niu Xi 10 g Chuan Hou Po 15–30 g Da Huang (add near end) 18–24 g Mu Xiang 10 g	Treat severe cases of bowel obstruction
Shi Quan Ku Han Jiu Bu Tang [All-Inclusive Bitter and Cold Decoction to Rescue and Tonify]	Core Formula + Huang Qin Huang Lian Huang Bai Shi Gao Zhi Mu Xi Jiao (Shui Niu Jiao)	Treats severe cases of rashes due to pestivirus or pestilence toxins. The manifestations include unconsciousness, coarse breathing, and black tongue coating. This is a strong forceful formula to drain fire and relieve toxins. It combines downward draining, heat clearing, relieving toxicity, clearing Qi, and blood methods

Xiao Cheng Qi Tang [Minor Order of Qi Decoction] *Shang Han Lun*

Da Huang	12 g
Hou Po	6 g
Zhi Shi	9 g

Main Indication: Yang Ming Fu excess syndrome. Symptoms include constipation, focal distention and fullness at chest and abdomen, tidal fever, delirious speech; old, yellow tongue coating, and slippery and swift pulse.

Preparation: Decoction.

Treatment Principles: Bitter cold to purge down the heat accumulation.

This formula can also be used in beginning stage of dysentery with unbearable abdominal pain or distending pain and tenesmus.

Related Formula:

Formula	Herbs	Indication
Shun Qi San [Smooth the Qi Powder] *Ba Cui Fang*	Same as the core formula	Treat Middle Jiao type of Xiao Ke, which has heat in the Stomach, hunger, and dark yellow urine. This formula is downward draining, causing decreased appetite to treat the syndrome. This can also be used for food accumulation

Tiao Wei Cheng Qi Tang [Regulate the Stomach and Order the Qi Decoction] *Shang Han Lun*

Da Huang	12 g
Mang Xiao	12 g
Gan Cao	9 g

Main Indication: Excess heat in the Stomach, dry and hard stools. Symptoms include constipation, thirst, restlessness, fever, abdominal fullness which refuses pressure, or with delirious speech, yellow tongue coating, and slippery rapid pulse. Other symptoms include rashes, epistaxis, vomiting blood, and swollen and painful throat and teeth due to heat excess in Stomach and intestine.

Preparation: Decoct Da Huang and Gan Cao and drain the decoction. Dissolve Mang Xiao into the decoction and boil for few seconds. Take the warm decoction little by little.

Treatment Principles: Clear heat and drain the accumulation.

This formula is most effective in treating dry hard stool due to heat excess in Stomach and intestine without focal distention and fullness.

In *Wai Ke Shu Yao*, this formula is ground into powder, mixed with honey, and formed into pills, which is called Po Guan Dan [Break the Coffin Pills]. This treats extreme heat causing ulceration, copious sweating, big thirst, constipation, delirious speech, and mania.

In *Yu Ji Wei Yi*, this formula is named Tiao Wei Wan [Regulate the Stomach Pill], which treats toothache with continuous bleeding. In *Kou Chi Lei Yao*, this formula treats excess heat in the Middle Jiao, constipation, and swollen, painful throat, and ulceration of the tongue and mouth.

This formula can also treat blocked excess Yang Ming Fu syndrome causing wheezing due to Lung Qi failing to descend. This is an example of treating a Fu organ for a Zang organ's pathological condition.

Related Formulas:

Formula	Herbs	Indication
Da Huang Gan Cao Tang [Rhubarb and Licorice Decoction]	Da Huang Gan Cao	Treats vomiting after meal
Tao Ren Cheng Qi Tang [Order of Qi Decoction with Peach Kernel]	Core Formula + Tao Ren Wu Ling Zhi Sheng Pu Huang Fresh Sheng Di Huang Xi Jiao juice (Shui Niu Jiao)	Treats mania due to Xu Xue (accumulation of blood stasis), profuse urination, and black stool. This quickly purges stagnant heat from the intestine
Dang Gui Cheng Qi Tang [Order of Qi Decoction with Tangkuei]	Core Formula + Dang Gui Sheng Jiang Da Zao	Treats warm febrile disease, heat entering the blood chamber, mania, and yelling at family due to Yang excess and Yin deficiency. This formula purges heat accumulation, and nourishes blood and Yin
Wu Ji Wan [Infinity Pill]	Da Huang	Treats postpartum internal blood stagnation, abnormal mental state. Ground wine-fried Da Huang into powder form and take 9 g with tea, two to three times a day, to treat dizziness and vertigo due to damp–heat

Fu Fang Da Cheng Qi Tang [Revised Major Order of Qi Decoction] *Zhong Xi Yi Jie He Zhi Liao Ji Fu Zheng*

Lai Fu Zi (fried)	30 g
Hou Po	15 g
Zhi Shi	9 g
Tao Ren	9 g
Chi Shao Yao	15 g
Da Huang (add near end)	15 g
Mang Xiao (dissolve in decoction)	9–15 g

Main Indication: Acute simple bowel obstruction. Symptoms include sudden distending pain at epigastrium and abdomen, accompanied by episodic spasm pain, vomiting, inability to eat, obstructed bowel, wiry slippery pulse, thin yellow or greasy tongue coating, and dark tongue body.

Preparation: Make 200 mL of decoction and divide into two doses taken with meal; 2 to 3 hours after taking the formula, one can use the same formula as enema.

Treatment Principles: Unblock the interior and purge downward, move Qi, and activate blood.

This formula is only suitable for simple bowel obstruction with more severe Qi distention. Other more severe cases of bowel obstruction should be treated with surgery.

Related Formula:

Formula	Herbs	Indication
Xiao Fu Tang	Fresh Turnip 1000 g Mang Xiao 60 g	Make 200 mL of decoction and take two to three doses per day Treat mild and early stages of bowel obstruction

Liang Ge San [Cool the Diaphragm Powder] *He Ji Ju Fang*

Da Huang	600 g
Mang Xiao	600 g
Gan Cao	600 g
Shan Zhi Zi	300 g
Bo He Ye	300 g
Huang Qin	300 g
Liang Qiao	1250 g

Main Indication: Heat accumulation in the Upper and Middle Jiao. Symptoms include fever, thirst, red face, dry burned lips, agitation and heat at the chest and diaphragm, ulcerations in mouth and tongue, sore throat, swollen tongue, red eyes, epistaxis, constipation, dark urine, red tongue, dry yellow coating, and slippery and rapid pulse.

Preparation: Grind the ingredients into coarse powder. Decoct 9 g of the powder with seven pieces of Dan Zhu Ye and add some honey. Take the warm decoction after meals.

Treatment Principles: Drain fire and unblock the bowel.

This pattern has unformed heat accumulation in the Upper Jiao and formed heat accumulation in the Middle Jiao. Thus, in this formula, Tiao Wei Cheng Qi Tang [Regulate the Stomach and Order the Qi Decoction] is used to unblock the bowel to guide the heat downward which is like removing the wood from the fire. The rest of the herbs clear, drain, and open the chest and diaphragm, and clear the stagnant heat. Thus, heat is cleared from Upper, Middle, and Lower Jiao.

Recently, this formula has been used to treat lobar pneumonia.

Related Formula:

Formula	Herbs	Indication
Qing Xin Liang Ge San [Clear the Heart and Cool the Diaphragm Powder] also called Jie Geng Tang [Platycodon Decoction]	Shan Zhi Zi Huang Qin Bo He Ye Lian Qiao Jie Geng or also add Shi Gao	Treats beginning stage of eruptive epidemic disease

Xuan Bai Cheng Qi Tang [Order of Qi Decoction to Diffuse the White] *Wen Bing Tiao Bian*

Sheng Shi Gao	15 g
Sheng Da Huang	10 g
Xing Ren	6 g
Gua Lou Pi	6 g

Main Indication: Heat in the Lung and Large Intestine. Symptoms include rapid wheezing, agitation and heat, cough with yellow thick phlegm, constipation, red tongue/yellow tongue coating, and excess and big pulse at the right cun position.

Preparation: Decoction. Divide into two doses.

Treatment Principles: Diffuse Lung Qi and move the stools.

This formula drains heat and unblocks the bowel to treat the Zang and Fu organs together. It is effective in treating heat in the Lung with constipation.

Xian Xiong Cheng Qi Tang [Sinking into the Chest and Order of Qi Decoction] *Tong Su Shang Han Lun*

Gua Lou Ren (crushed slightly)	20 g
Zhi Shi	5 g
Ban Xia	10 g
Huang Lian	3 g
Sheng Da Huang	10 g
Mang Xiao	5 g

Main Indication: Yang Ming excess heat with phlegm fire accumulating and binding in the chest. Symptoms include focal distention and pain at the chest and diaphragm, coma or delirious speech in severe cases, abdominal fullness, constipation, and yellow burned tongue coating with slippery and rapid pulse.

Preparation: Decoction.

Treatment Principles: Scour phlegm, open the chest, drain heat, and unblock the bowel.

This formula treats phlegm accumulation and obstruction above and excess fu below.

Yi Nian Jin [One Pinch of Gold Decoction] *Yi Zong Jin Jian*

Sheng Da Huang	Equal amount
Qian Niu Zi (black)	
Qian Niu Zi (white)	
Bing Lang	
Ren Shen	

Main Indication: Heat toxin accumulation and Fu excess in newborn. Symptoms include constipation, distention and fullness of the abdomen, restlessness and agitation, frequent crying or aversion to drinking milk, red face, red lips, dry tongue and mouth, and green purple vein.

Preparation: Grind into powder and mix a small amount of powder with honey water.

Treatment Principles: Drain heat and open the bowel.

This formula treats newborn with heat accumulated in the Stomach and intestine causing constipation or vomiting of milk. The heat was passed from the toxin in the uterus of the mother. Four purgative herbs are used to drain heat toxin and Ren Shen to protect the young Yin and young Yang.

Related Formula:

Formula	Herbs	Indication
Hei Bai San [Black and White Powder]	Core Formula – Ren Shen + Xuan Ming Fen Chen Pi Gan Cao	Treat the same syndrome as the original formula. It is stronger in moving Qi and unblocking bowels

Da Xian Xiong Tang [Major Sinking into the Chest Decoction] *Shang Han Lun*

Da Huang	12 g
Mang Xiao	10 g
Gan Sui powder	1–2 g

Main Indication: Excess syndrome due to evil accumulation in the chest. Symptoms include hardness, fullness, and pain that it is unbearable with any touch on the epigastrium or at any place on the entire abdomen; severe constipation, tidal fever at dusk, or shortness of breath, restlessness and agitation, dry tongue and thirst, and tight and forceful pulse.

Preparation: Decoct Da Huang first, strain, then dissolve Mang Xiao in the decoction. Take the decoction with Gan Sui powder. This formula should be discontinued once results are achieved.

Treatment Principles: Drains heat and drives out water.

This formula is designed for knotted heat and water accumulating in Stomach and epigastrium causing the clumping in the chest syndrome (Jie Xiong Zheng). The main symptoms are epigastric pain and rock hardness with palpation. Recently, this formula is effective in treating acute pancreatitis with edema.

Related Formula:

Formula	Herbs	Indication
Da Xian Xiong Wan [Major Sinking into the Chest Pill]	Core Formula + Ting Li Zi Xing Ren Bai Mi (Honey)	Grind the ingredients into powder and form pills with honey. Take 5–10 g each time with warm water Treats clumping-in-the-chest syndrome, hardness, fullness, and pain at the chest, and stiff neck as if in mild spasm

Zeng Ye Cheng Qi Tang [Increase the Fluids and Order the Qi Decoction] *Wen Bing Tiao Bian*

Sheng Di Huang	24 g
Xuan Shen	30 g
Mai Men Dong	12 g
Da Huang	9 g
Mang Xiao	5 g

Main Indication: Fluid exhaustion and dry accumulation. Symptoms are caused by Yang Ming warm febrile disease that damages the Jin fluids to cause dry stools that are difficult to pass, yellow tongue coating, and dry red tongue with wiry and thin pulse.

Preparation: Decoct the herbs until two cups are left. Take one cup first; if there are no changes then take the second cup.

Treatment Principles: Nourish Yin, increase fluids, drain heat, and move stools.

This formula is a combination of Zeng Ye Tang [Increase the Fluid Decoction] and Tiao Wei Cheng Qi Tang [Regulate the Stomach and Major Order of Qi Decoction] minus Gan Cao. It represents the method of "increase the water to move the boat." From the symptom of the inability to move the stool, one knows that there is "not enough water to move the boat."

Xin Jia Huang Long Tang [Newly-Augmented Yellow Dragon Decoction] *Wen Bing Tiao Bian*

Sheng Di Huang	15 g
Xuan Shen	15 g
Mai Men Dong	15 g
Hai Shen (sea cucumber)	2 pieces
Sheng Da Huang	10 g
Mang Xiao	3 g
Sheng Gan Cao	6 g
Ren Shen (decoct separately)	5 g
Dang Gui	5 g
Sheng Jiang juice	6 tbs

Main Indication: Heat accumulation, internal excess with Qi and Yin deficiency. Symptoms include constipation, distention, fullness and hardness of the abdomen, apathy, shortness of breath, dry mouth and throat, cracked lips, burned tongue, dry yellow or burned black, dry cracked tongue coating, and deep and excess pulse.

Preparation: Decoction. Mix with Ren Shen decoction and ginger juice. Take 1 cup first and if no bowel movement, then take 1 more cup. Once bowel movement begins, then patient can take one dose of Yi Wei Tang [Benefit the Stomach Decoction].

Treatment Principles: Nourish Yin, tonify Qi, open the Fu organ, and drain heat.

This formula is a combination of Zeng Ye Tang [Increase the Fluid Decoction], Tiao Wei Cheng Qi Tang [Regulate the Stomach and Order of Qi Decoction], Ren Shen, Dang Gui, and ginger juice. It drains heat, opens the bowel, nourishes Yin, and tonifies Qi which is a formula with both draining and tonifying methods.

Related Formula:

Formula	Herbs	Indication
Huang Long Tang [Yellow Dragon Decoction]	Da Huang 12 g Mang Xiao 9 g Zhi Shi 9 g Hou Po 12 g Gan Cao 3 g Ren Shen 6 g Dang Gui 9 g Jie Geng 3 g Sheng Jiang Da Zao	Take the warm decoction Treat interior heat excess syndrome with Qi and blood deficiency. This formula is stronger in its purging downward and draining heat functions

Comparison of Cold Bitter Downward Draining Formulas for Knotted Heat

All of the above formulas treat constipation due to excess heat syndromes. However, there are different concomitant symptoms with the constipation, thus there are differences among all the formulas.

Tiao Wei Cheng Qi Tang [Regulate the Stomach and Order of Qi Decoction] simply drains downward with bitter cold herbs. With the accompanying symptoms of focal distention at the chest and abdominal distention, Zhi Shi and Hou Po are added to the formula, as in Da Cheng Qi Tang [Major Order of Qi Decoction] and Xiao Cheng Qi Tang [Minor Order of Qi Decoction]. When the accompanying symptoms include dry intestine with fluid exhaustion, Sheng Di Huang, Xuan Shen, and Mai Men Dong are added, as in Zeng Ye Cheng Qi Tang [Increase the Fluid and Order of Qi Deccotion]. If the body Zheng Qi is deficient, with excess evil pathogens; then Ren Shen, Dang Gui, and Sheng Di Huang are added, as in Huang Long Tang [Yellow Dragon Decoction] and Xin Jai Huang Long Tang [Newly-Augmented Yellow Dragon Decoction]. When there is heat toxin accumulating in the Lungs, then Shan Zhi Zi, Huang Lian, Huang Qin, Lian Qiao, and Bo He are added as in Liang Ge San [Cool the Diaphragm Powder] and Xian Xiong Cheng Qi Tang [Sinking into the Chest and Order of the Qi Decoction]. Or Shi Gao and Gua Lou can be added to diffuse above and drain below as in Xuan Bai Cheng Qi Tang [Order of Qi Decoction to Diffuse the White] and Bai Hu Cheng Qi Tang [White Tiger and Order of Qi Decoction]. If there is also knotted water and heat accumulating in the intestine, then Gan Sui is added as in Da Xian Xiong Tang [Major Sinking into the Chest Decoction] and Lai Po Tong Jie Tang [Open the Blockage Decoction with Radish Seed and Magnolia Bark]. If there is also high fever, restlessness and thirst, delirium, and rash due to blazing heat in Qi and blood, Shi Gao, Zhi Mu, Xi Jiao (Shui Niu Jiao), and Sheng Di Huang are added as in She Quan Gu Han Jiu Bu Tang [All-Inclusive Bitter and Cold Decoction to Rescue and Tonify]. For early stage bowel obstruction, blood- and Qi-moving herbs are added such as in Fu Fang Da Cheng Qi Tang [Revised Major Order of Qi Decoction].

COLD ACCUMULATION

Cold accumulation refers to cold causing constipation. The treatment should warm the Yang and guide the stagnation; often Da Huang and Mang Xiao are combined with Fu Zi and Gan Jiang. With this combination, the cold bitter property of Da Huang and Mang Xiao are eliminated and only the downward draining function remains. This is referred to as "eliminating the property and utilizing the function." However, the dosage of acrid warm herbs should be greater than the cold cool herbs in order to "eliminate the property and utilize the function."

Da Huang Fu Zi Tang [Rhubarb and Prepared Aconite Decoction] *Jin Gui Yao Lue*

Da Huang	10 g
Zhi Fu Zi	15 g
Xi Xin	3 g

Main Indication: Cold accumulation and internal excess. Symptoms include abdominal pain, constipation, pain at the hypochondrium, cold hands and feet or fever, deep, wiry and tight pulse, and white greasy tongue coating.

Preparation: Decoction. Take the warm decoction three times a day.

Treatment Principles: Warm, unblock, and break the accumulation.

In this formula Fu Zi is combined with Xi Xin. This is Zhang Zhong Jing's method to treat deep, hidden, and cold evils. This combination can warm and scatter. The dosage of Fu Zi is much greater than Da Huang to eliminate the bitter and cold property of Da Huang and maintain the moving stagnation and breaking the accumulation function.

This formula can treat constipation due to cold accumulation. It can also treat cold congealed in the channel at one side of the body leading to hypochondriac pain, chest, and abdominal pain.

Wen Pi Tang [Warm the Spleen Decoction] *Qian Jin Yao Fang*

Da Huang (decoct later)	12 g
Fu Zi	12 g
Gan Jiang	6 g
Ren Shen	9 g
Gan Cao	3 g

Main Indication: Cold accumulation due to Spleen Yang deficiency. Symptoms include constipation due to cold accumulation, abdominal pain that is better with warmth, cold hands and feet, or chronic diarrhea with blood and mucous, white greasy tongue coating, and deep and wiry pulse.

Preparation: Decoct herbs and add Da Huang later.

Treatment Principles: Warm Spleen Yang and guide stagnation.

This formula is modified from Da Huang Fu Zi Tang [Rhubarb and Prepared Aconite Decoction] with Ren Shen, Gan Jiang, Gan Cao, and remove Xi Xin. It is also Si Ni Tang [Frigid Extremities Decoction] with Ren Shen and Da Huang. Since the formula warms the Lower Jiao and also tonifies Qi, it is used for chronic diarrhea with Qi deficient syndromes.

Related Formulas:

Formula	Herbs	Indication
Wen Pi Tang [Warm the Spleen Decoction] *Qian Jin Fang* (Chapter 15)	Core Formula – Gan Cao + Gui Xin	Treat more severe cold type of cold dysentery
Wen Pi Tang [Warm the Spleen Decoction] *Qian Jin Fang* (Chapter 13)	Core Formula + Mang Xiao Dang Gui	Treat constipation due to cold accumulation with spasm pain below the navel
Wen Pi Tang [Warm the Spleen Decoction] *Ben Shi Fang*	Core Formula – Ren Shen + Gui Xin Hou Po	Treat severe coldness in Middle Jiao leading to abdominal pain and diarrhea

San Wu Bei Ji Wan [Three-Substance Pill Prepared for Emergency] *Jin Gui Yao Lue*

Da Huang	30 g
Gan Jiang	30 g
Ba Dou	30 g

Main Indication: Cold excess accumulation. Symptoms include sudden sharp distending pain at the chest and abdomen region, rapid breathing, mouth tightly closed, constipation, greenish complexion, or loss of consciousness. Pulse is deep, tight, and forceful, and the tongue coating is white and greasy.

Preparation: Grind into powder. Take 0.6 to 1.5 g for adults with soup or warm water. Dosage is decreased for children. If mouth is tightly closed, one can feed the formula through nose.

Treatment Principles: Break and expel cold accumulation.

This formula is specifically designed to treat sudden onset of cold food stagnation and obstructed Stomach and intestines, focal distention, and obstructed Qi mechanism. The focus is on attacking and eliminating the cold accumulation. Once vomiting or diarrhea occurs after taking the herbs, evil pathogens are gone and the Zheng Qi is in peace.

In the formula, Ba Dou is very acrid and very hot with high toxicity, thus it is contraindicated for seniors, weak constitution people, and during pregnancy.

If diarrhea cannot be stopped after taking the herbs, then one can take cold congee to stop the diarrhea.

Comparison of Formulas Treating Cold Accumulations

Da Huang Fu Zi Tang [Rhubarb and Prepared Aconite Decoction] is suitable for more severe cases of cold accumulation without Qi deficiency. Wen Pi Tang [Warm the Spleen Decoction] is good for Qi deficiency with cold accumulation or chronic diarrhea with Qi deficiency syndrome. San Wu Bei Ji Wan [Three-Substance Pill Prepared for Emergency] treats acute type of cold accumulating in Stomach and intestines, causing obstructed Qi mechanism, severe abdominal pain, and severe constipation.

FLUID-DEFICIENT CONSTIPATION

This type of constipation is caused by insufficient fluid in the Large Intestine which inhibits the normal transportation and movement of the stools. Seeds that have high oil content are commonly used to moisten the intestine and move stools, such as Ma Zi Ren, Bai Zi Ren, Tao Ren, Xing Ren, Rou Cong Rong, and He Shou Wu. If there is also dry heat in the intestine, one should moisten the intestine and at the same time drain heat, as in Ma Zi Ren Wan [Hemp Seed Pill].

Run Chang Wan [Moisten the Intestines Pill] *Shen Shi Zun Sheng Shu*

Dang Gui	10 g
Sheng Di Huang	30 g
Tao Ren	10 g
Huo Ma Ren	15 g
Zhi Ke	10 g

Main Indication: Blood deficiency, Yin exhaustion, and dry accumulation in the intestine. Symptom of fluid exhausted constipation seen in seniors, deficient persons, postpartum females, and postsurgery patients. Tongue is red with scanty coating and pulse is thin.

Preparation: Make honey pills, take 15 g each time with warm water on an empty stomach.

Treatment Principles: Moisten intestines and unblock bowels.

Wu Ren Wan [Five-Seed Pill] *Shi Yi De Xiao Fang*

Tao Ren	equal amount
Xing Ren	
Bai Zi Ren	
Song Zi Ren (pine seed)	
Yu Li Ren	
Chen Pi	

Main Indication: Exhausted fluid and dry intestine. Symptoms include difficult bowel movement, constipation in seniors, and postpartum females. Dry tongue and thin pulse.

Preparation: Make into honey pill in the size of 5 to 10 mm in diameter, and take 15 g each time on an empty Stomach with warm water.

Treatment Principles: Moisten the intestine and unblock the bowel.

This is a classical formula in moistening the intestine and moving the stool. It is composed of five seeds that are high in oil with Chen Pi to move Qi.

Ma Zi Ren Wan [Hemp Seed Pill] *Shang Han Lun* also known as Pi Yue Ma Ren Wan [Constraint Spleen Hemp Seed Pill]

Ma Zi Ren	500 g
Shao Yao	250 g
Zhi Zhi Shi (prepared Zhi Shi)	250 g
Da Huang	500 g
Zhi Hou Po	250 g
Xing Ren	250 g

Main Indication: Yin and fluid deficiency in Tai Yin, and excess dry heat in Yang Ming. Symptoms include constipation, frequent urination, or habitual constipation or constipation with hemorrhoids, dry tongue, yellow tongue coating, and minute slippery pulse.

Preparation: Grind the ingredients into powder and make into honey pill. Take 10 g, three times a day. Decrease the dosage when taking this formula as decoction.

Treatment Principles: Moisten intestine, drain heat, move Qi, and unblock bowel.

The syndrome that is indicated by this formula is called Pi Yue, which refers to constipation due to Spleen dysfunction in *Shang Han Lun*. It is caused by dry heat in the Stomach and fluid deficiency in the Spleen.

This formula is a modification of Xiao Cheng Qi Tang [Minor Order of Qi Decoction] with Ma Ren, Xing Ren, Shao Yao, and Bai Mi. It is most effective in treating constipation due to dryness in the intestine with stagnation. It can be used in treating dry intestine type of constipation in seniors and postpartum females. However, if the constipation is due to simple blood deficiency and fluid exhaustion or postpartum constipation, then this formula is contraindicated.

Ji Chuan Jian [Benefit the River (Flow) Decoction] *Jing Yue Quan Shu*

Dang Gui	15 g
Niu Xi	6 g
Rou Cong Rong	9 g
Ze Xie	6 g
Sheng Ma	3 g
Zhi Ke	3 g

Main Indication: Constipation due to Kidney deficiency. Symptoms include constipation, frequent profuse urination, dizziness, vertigo, weakness of the lower back and knees, flabby pale tongue, and deep pulse, especially Chi pulse.

Preparation: Decoction. Taken before meal.

Treatment Principles: Warm the Kidney, moisten the intestine, and move stool.

Modification:

Minus	Addition	Indication
	Ren Shen	Qi Deficiency
	Huang Qin	With deficient heat
	Shu Di Huang	Kidney deficiency
Zhi Ke		Severe Kidney Qi deficiency

Comparison of Formulas Treating Fluid-Deficient Constipation

Run Chang Wan [Moisten the Intestine Pill] treats constipation in seniors due to Yin and blood deficiency and in postpartum women due to lost of blood and damaged Yin. Wu Ren Wan [Five-Seeds Pill] moistens the dryness that is indicated in dry withered intestinal tract leading to constipation. Ma Zi Ren Wan [Hemp Seed Pill] moistens the deficient Tai Yin and drains dry heat from the Stomach, thus treating constipation due to Spleen dysfunction. Ji Chuan Jian [Benefit the River (Flow) Decoction] treats constipation due to Kidney deficiency and insufficient circulating Qi function.

DAMP–HEAT IN THE LOWER JIAO

Damp–heat in the Lower Jiao often manifests as urgent frequent diarrhea, even watery or it can manifest as dysentery, abdominal pain, tenesmus, pus and blood in the stool, with burning at the anus, restlessness, thirst, yellow greasy tongue coating, and rapid and soggy pulse. Herbs that clear heat, dry damp, relieve toxins, and stop dysentery are used, such as Huang Lian, Huang Qin, Qin Pi, Bai Tou Weng, Huang Bai, and Ma Chi Xian. The main herbs are combined with regulating Qi and blood herbs according to the syndromes, such as Mu Xiang, Hou Po, Dang Gui, Shao Yao, and Mu Dan Pi. This follows the principle that "when Qi is regulated, then the heaviness at the anus is alleviated, and when the blood is regulated then the pus and blood in the stool will be healed." If it is accompanied by food stagnation, then Shan Zha, Da Huang, Bing Lang, and other reducing food stagnation and unblocking stagnation herbs can be added. If damp–heat excess damages Spleen Yang, then Gan Jiang and Huang Lian should be added.

The above formulary combination has good effect in treating bacillary dysentery. If the diarrhea is due to amebic dysentery, then Chun Gen Pi and Shi Liu Pi should be added, or one should take Ya Dan Zi at the same time.

The focus and application of the treatment principles to clear heat, drain damp, relieve toxin, and stop diarrhea or dysentery all depend on the strength of evil pathogens in the body Zheng Qi, as well as on the severity and duration of the disease.

Ge Gen Huang Qin Huang Lian Tang [Kudzu, Scutellaria, and Coptis Decoction] *Shang Han Lun*

Ge Gen	24 g
Huang Qin	15 g
Huang Lian	10 g
Gan Cao	6 g

Main Indication: Damp–heat in the Lower Jiao. Symptoms include fever, sweating, thirst, diarrhea with foul smell, burning pain at the anus, yellow tongue coating, and rapid pulse.

Preparation: Decoction.

Treatment Principles: Clear heat and stop diarrhea.

Clinically, this formula can be used to treat food poisoning, acute enteritis, bacillary dysentery, and so on. If the diarrhea is not accompanied by fever, and the pulse is deep weak and slow which belongs to deficient cold type, one should not take this formula.

Modification:

Addition	Indication
Mu Xiang Bai Shao Yao	Abdominal pain

Related Formula:

Formula	Herbs	Indication
Jia Wei Ge Gen Huang Qin Huang Lian Tang [Augmented Kudzu, Coptis, and Scutellaria Decoction]	Ge Gen Huang Lian Huang Qin Jin Yin Hua Mu Dan Pi Bai Shao Yao Mang Xiao Ma Chi Xian	Treats toxic type of bacillary dysentery in children. Symptoms include high fever, agitation, vomiting, nausea, abdominal distention, frequent bowel movement with blood and mucous, fright and syncope in severe cases, delirium, scanty urination, purple lips, red face, deep red tongue, dry tongue coating, and swift pulse. Add half pill of An Gong Niu Huang Wan [Calm the Palace Pill with Cattle Gallstone] for delirium, fright, and syncope

Huang Qin Tang [Scutellaria Decoction] *Shang Han Lun*

Huang Qin	12 g
Bai Shao Yao	12 g
Zhi Gan Cao	6 g
Da Zao	4 pieces

Main Indication: Disharmony of the Liver and Spleen, damp–heat in the Lower Jiao. Symptoms include fever, bitter mouth, diarrhea, abdominal pain, or burning diarrhea with abdominal pain, red tongue, yellow tongue coating, and rapid pulse.

Preparation: Decoction.

Treatment Principles: Clear heat, stop diarrhea, and regulate Liver and Spleen.

In this formula, Huang Qin and Bai Shao Yao are used together to drain heat from the blood. Thus this formula is appropriate in treating heat type of diarrhea with abdominal pain. To treat dysentery, Mu Xiang, Bing Lang, Huang Lian, Ku Shen, Guan Zhong, Di Yu, and Mu Dan Pi should be added to increase the effectiveness.

Related Formulas:

Formula	Herbs	Indication
Huang Qin Jia Ban Xia Sheng Jiang Tang [Scutellaria Decoction with Pinellia and Ginger]	Core Formula + Ban Xia Sheng Jiang	Treats the same syndrome with nausea and rebellious Qi. It is used in acute gastroenteritis
Huang Qin Shao Yao Tang [Scutellaria and Peony Decoction]	Core Formula – Da Zao	Treats heat type of dysentery, abdominal pain, or fire rising leading to epistaxis. To treat epistaxis, one can add Qing Hao, Qing Dai, Shan Zhi Zi, and Mu Dan Pi
Ku Shen Zhi Li Jian [Sophorae to Stop Diarrhea Decoction]	Ku Shen Huang Qin Bai Shao Yao Shan Zha (burned) Bing Lang	Take the decoction on empty Stomach. Treat dysentery

Bai Tou Weng Tang [Pulsatilla Decoction] *Shang Han Lun*

Bai Tou Weng	15 g
Huang Bai	12 g
Huang Lian	6 g
Qin Pi	12 g

Main Indication: Heat type of dysenteric disorder. Symptoms include fever, abdominal pain, tenesmus, diarrhea with pus and blood, more blood than mucous, burning pain at the anus, restlessness, thirst, red tongue, yellow tongue coating, and wiry and rapid pulse.

Preparation: Decoction. One bag per day. In severe cases take two bags per day.

Treatment Principles: Clear heat, relieve toxicity, cool blood, and stop dysentery.

This formula treats damp–heat type of fulminant dysentery, such as excess heat type of amoebic dysentery or acute bacillary dysentery.

Modification:

Addition	Indication
Shi Liu Pi	Amoebic dysentery
Mu Xiang Bai Shao Yao Di Yu	Acute bacillary dysentery

Related Formulas:

Formula	Herbs	Indication
Bai Tou Weng Jia Gan Cao E Jiao Tang [Pulsatilla Decoction plus Licorice and Ass-Hide Gelatin]	Core Formula + Gan Cao E Jiao	Treat postpartum blood deficient type of hot dysenteric disorder
Jia Wei Bai Tou Weng Tang [Augmented Pulsatilla Decoction]	Core Formula + Huang Qin Bai Shao Yao	Treat the same syndrome, but this formula is slightly stronger
Jia Jian Bai Tou Weng Tang [Modified Pulsatilla Decoction]	Core Formula – Huang Bai + Jin Yin Hua Di Yu Bai Shao Yao Mu Xiang Gan Cao	Treat acute bacillary dysentery. Bowel movement with pus and blood, tenesmus, yellow tongue coating, and rapid pulse. This formula is stronger to clear heat, relieve toxicity, and stop dysentery
Bian Tong Bai Tou Weng Tang [Adapted Pulsatilla Decoction]	Shan Yao Bai Tou Weng Qin Pi Sheng Di Yu Bai Shao Yao Gan Cao San Qi (powder) Ya Dan Zi	Grind San Qi and Ya Dan Zi. Take half of the powder with sugar water first and take the remaining powder with the decoction Treat severe type of heat dysenteric disorder with abdominal pain, and diarrhea with severe pus and blood

Shao Yao Tang [Peony Decoction] *Bao Ming Ji*

Shao Yao	15 g
Huang Qin	12 g
Huang Lian	10 g
Dang Gui	6 g
Rou Gui	6 g
Gan Cao	6 g
Mu Xiang	10 g
Bing Lang	10 g
Da Huang	10 g

Main Indication: Damp–heat dysentery. Symptoms include abdominal pain, frequent diarrhea with pus and blood more than ten times a day, tenesmus, burning pain at the anus, scanty dark urination, fever, yellow greasy tongue coating, and slippery and rapid pulse.

Preparation: Take the warm decoction after meals.

Treatment Principles: Drain heat, relieve toxins, and stop dysentery.

This formula is designed for syndromes of damp–heat, epidemic toxins in the Large Intestine with obstruction of Qi function and circulation. This formula can clear heat, dry dampness, relieve toxins, move Qi and blood which is an effective formula in treating dysentery with blood and mucous. From the result of antibacterial experiment, it shows that this formula has inhibitory function to the bacteria causing dysentery.

Related Formulas:

Formula	Herbs	Indication
Xiang Lian Wan [Aucklandia and Coptis Pill]	Mu Xiang Huang Lian	Treat mild cases of dysentery with blood and mucous, or the chronic type
Jia Jian Shao Yao Tang [Modified Peony Decoction]	Bai Shao Yao Huang Qin Huang Lian Hou Po Mu Xiang Chen Pi	Treat dysentery with abdominal distention and pain

Zhu Che Wan [Stationed Vehicle Pill] *Qian Jin Fang*

Huang Lian	180 g
Gan Jiang	60 g
Dang Gui	90 g
E Jiao	9 g

Main Indication: Chronic dysentery which damages Yin. Symptoms include dysentery with blood and mucous, bowel incontinence, chronic abdominal pain, red tongue, scanty coating, and thin or thin and wiry pulse.

Preparation: Grind the first three herbs into fine powder, dissolve E Jiao, and mix everything together with vinegar to form pills. Take 6 g, twice a day.

Treatment Principles: Clear heat, nourish Yin.

This formula uses both heat and cold herbs. It supports the body Zheng Qi and expels evil pathogens. It is effective in treating chronic dysenteric disorders with remaining evil pathogens, damaged Yin and blood, and cold in the Zang organs.

Di Yu Wan [Sanguisorba Pill] *Pu Ji Fang*

Di Yu	15 g
Huang Lian	15 g
Dang Gui	15 g
E Jiao	15 g
Mu Xiang	15 g
He Zi Rou	15 g
Wu Mei	15 g

Main Indication: Chronic dysenteric disorders damaging Ying and blood. Symptoms include chronic dysentery with blood or defecating bloody water, damaging Ying and blood, pale tongue, thin, yellow greasy tongue coating, and thin pulse.

Preparation: Taken as pills, 6 g each on an empty stomach.

Treatment Principles: Clear heat, relieve toxins, regulate Ying, consolidate, and astringe.

This formula can relieve the residual epidemic toxins of dysentery. It clears the damp–heat in order to get rid of the pathogenic factors. It can also support the body Zheng Qi by regulating blood, nourishing Yin, and consolidating Qi to prevent prolapse. It is suitable in treating chronic dysentery with blood, with damaged Ying and blood, and signs of rectal prolapse.

Comparison of Formulas Treating Damp–Heat in the Lower Jiao

Ge Gen Huang Qin Huang Lian Tang [Kudzu, Scutellaria, and Coptis Decoction] and Huang Qin Tang [Scutellaria Decoction] are the main formulas in treating damp–heat diarrhea. They can be used in treating diarrhea or dysentery, as long as it is a damp–heat in the Lower Jiao pattern. However, Ge Gen Huang Qin Huang Lian Tang [Kudzu, Scutellaria, and Coptis Decoction] is stronger in clearing heat and relieving toxicity. Huang Qin Tang [Scutellaria Decoction] is stronger in moderating the acute pain.

Bai Tou Weng Tang [Pulsatilla Decoction] and Shao Yao Tang [Peony Decoction] are main formulas in treating dysenteric disorders. However, the former focuses on eliminating the pathogenic factors that cause dysentery. The latter also reduces and moves stagnation, harmonizes Qi and blood.

Zhu Che Wan [Stationed Vehicle Pill] and Di Yu Wan [Sanguisorba Pill] treat chronic dysenteric disorder with remaining pathogenic factors and weakened body Zheng Qi. The former formula uses both heat and cold, and the latter formula clears and astringes.

PROLAPSE AND FECAL INCONTINENCE

The manifestations of deficient cold type of chronic diarrhea or dysenteric disorders are chronic diarrhea, poor appetite, apathy, cool four limbs, or chronic diarrhea with dark old blood and mucous, abdominal pain that is better with heat and pressure, or Spleen and Kidney Yang deficiency and 5 A.M. diarrhea. The main symptoms indicating prolapse or fecal incontinence, thus the treatment principles are to warm the center and astringe. The commonly used herbs are Gan Jiang, Rou Gui with He Zi, Rou Dou Kou, Chi Shi Zhi, Wu Mei, Wu Bei Zi, Ying Su Ke, and Wu Wei Zi. However, if this method is used with remaining pathogenic factors in the body, then the pathogenic factors will be trapped inside the body and will not be cleared.

Spleen and Stomach prefer warmth and dislike cold; Yang Qi is easily damaged, thus warming the Middle Jiao is often the treatment principle, especially with chronic patterns. Li Zhong Tang [Regulate the Middle Decoction] is an example of warming the Middle Jiao and tonifying Qi. Fu Zi Li Zhong Wan [Prepared Aconite Pill to Regulate the Middle], Da, and Xiao Jiang Zhong Tang [Major] and [Minor Construct the Middle Decoction] are examples of warming the Middle Jiao and stopping pain. An example of warm the middle and reduce accumulation is Zha Qu Li Zhong Tang [Regulate the Middle Decoction with Hawthorn and Fermented Leaven]. Examples of warm the middle and transform dampness are Ling Gui Zhu Gan Tang [Poria, Cinnamon Twig, Atractylodes Macrocephala, and Licorice Decoction] and Li Zhong Hua Tan Tang [Regulate the Middle and Transform Phlegm Decoction]. Examples of warm the middle and control the blood formulas are Gan Cao Gan Jiang Tang [Licorice and Dried Ginger Decoction], Huang Tu Tang [Yellow Earth Decoction], Bai Ye Tang [Biota Decoction], and Wen Jing She Xue Tang [Warm the Menses and Control the Blood Decoction]. Examples of warm the middle and move the stagnation formulas are Da Huang Fu Zi Tang [Rhubarb and Prepared Aconite Decoction] and Wen Pi Tang [Warm the Spleen Decoction]. Examples of warm the middle and astringe formulas are Yang Zang Tang [Nourishing the Organs Decoction] and Si Sheng Wan [Four-Miracle Pill], which is the combination of the two treatment principles.

Yang Zang Tang [Nourishing the Organs Decoction] *He Ji Ju Fang*

Ren Shen	15 g
Bai Zhu	10 g
Gan Cao	6 g
Bai Shao Yao	12 g
Dang Gui	10 g
Rou Gui	3 g
Rou Dou Kou	12 g
He Zi	15 g
Ying Su Ke	15 g
Mu Xiang	6 g

Main Indication: Spleen and Kidney deficient cold, prolapse and fecal incontinence. Symptoms include fecal incontinence and rectal prolapse, abdominal pain that prefers warmth and pressure, flaccidity of the anus, fatigue, little food intake or stool with blood and pus day and night, tenesmus, "hollow" pain around umbilicus and abdomen, pale tongue, and slow pulse.

Preparation: Take the warm decoction. Avoid cold, greasy foods and alcohol.

Treatment Principles: Astringe the intestine, treat prolapse, warm and tonify Spleen and Kidney.

In modern times, this formula can be used to treat long-term diarrhea or rectal prolapse due to chronic enteritis and chronic dysentery.

Modification:

Addition	Indication
Fu Zi	Severe diarrhea with rectal prolapse
Sheng Ma Chai Hu	Frequent rectal prolapse

Si Shen Wan [Four-Miracle Pill] *Zheng Zhi Zhun Sheng*

Bu Gu Zhi	120 g
Wu Wei Zi	60 g
Rou Dou Kou	60 g
Wu Zhu Yu	30 g

Main Indication: Deficient cold in Spleen and Kidney. Symptoms include 5:00 A.M. diarrhea, poor appetite, indigestion, or chronic diarrhea, abdominal pain, cold limbs, sore lower back, apathy, fatigue, pale tongue with white coating, and deep, slow, and weak pulse.

Preparation: Grind into powder form and take 6 to 9 g each, one to two times a day on an empty stomach with warm water. If form into pills, one can add Sheng Jiang and Da Zao.

Treatment Principles: Warm and tonify Spleen and Kidney, astringe the intestine, and stop diarrhea.

This formula treats Spleen and Kidney Yang deficiency leading to 5:00 A.M. diarrhea. It is effective in treating diarrhea due to chronic enteritis, intestinal tuberculosis, and irritable bowel syndrome.

Si Shen Wan [Four-Miracle Pill] is a combination of Er Shen Wan [Two-Miracle Pill] from *Ben Shi Fang* (Bu Gu Zhi and Rou Dou Kou) and Wu Wei Zi San [Schisandra Powder] (Wu Wei Zi and Wu Zhu Yu), which specializes in treating 5:00 A.M. diarrhea with undigested food in the stool.

Tao Hua Tang [Peach Blossom Decoction] *Shang Han Lun*

Chi Shi Zhi	30 g
Gan Jiang	10 g
Geng Mi	30 g

Main Indication: Middle Jiao deficient cold. Symptoms include chronic diarrhea, pus and dark old blood in the stool, difficult urination, abdominal pain which prefers warmth and pressure, pale tongue, flabby moist tongue, and deep pulse especially at guan position.

Preparation: Take the decoction three times a day.

Treatment Principles: Warm the Middle Jiao and astringe the intestine.

The formula name comes from the color of Chi Shi Zhi, which is like peach blossom.

Modification:

Addition	Indication
Ren Shen Fu Zi	Severe deficient cold

Related Formula:

Formula	Herbs	Indication
Chi Shi Zhi Yu Yu Liang Tang [Halloysitum and Limonite Decoction]	Chi Shi Zhi 30 g Yu Yu Liang 30 g	Chronic diarrhea, prolapse, and fecal incontinence This formula treats the manifestation

INTESTINAL PARASITES

Common intestinal parasites include roundworms, tapeworms, and hookworms.

The main manifestation include rumbling hunger sound from the Stomach, episodic pain around the navel, or with color blotches, craving but sallow complexion, or grinding teeth during sleep, round small red spots on the tongue, white spots on face, and frequently green spots on sclera. Commonly used herbs include Ku Lian Gen Pi, Lei Wan, He Shi, Chuan Jiao, Shi Jun Zi, and Bing Lang in combination with downward-draining herbs such as Da Huang, Mang Xiao, and Qian Niu Zi.

Intestinal hookworm often causes epigastric discomfort, dull ache, sallow complexion, weakness of the four limbs, palpitation, shortness of breath, tinnitus, blurry vision, craving for raw rice or dirt, face, palm or whole body appearing as waxy yellow, or mild to medium edema. Often herbs like Bai Fan, Qing Fan, Hou Po, and Cang Zhu are used. Patients with tapeworms often see white flakes in their stool which is treated with Nan Gua Zi and Bing Lang.

Gou Chong Wan [Hookworm Pill] *Chen Zhang Wei Fang*

Shan Zhi Zi	10 g
Huang Bai	10 g
Gan Song	10 g
Bai Fan	10 g
Gan Cao	10 g
Yin Chen Hao	20 g
Cang Zhu	20 g
Qing Fan	20 g

Main Indication: Hookworm. Symptoms include palpitation, shortness of breath, sallow complexion and skin, or yellow edema, or craving of raw rice and dirt.

Preparation: Grind the above ingredients into powder and form into pills with brown sugar. Taken once in the morning and once at night with rice soup and finish the herbs in 7 days. In severe cases, one can take four to five bags of this formula.

Treatment Principles: Dry dampness, kill parasites, promote the function of Gallbladder, and alleviate jaundice.

This formula specifically treats hookworm (ancylostomiasis). It is also effective in treating jaundice due to chronic hepatitis and chronic cholecystitis.

Related Formulas:

Formula	Herbs	Indication
Xiao Shi Fan Shi San [Niter and Alum Powder]	Xiao Shi Fan Shi (calcined)	Equal amount of the powdered ingredient. Take 3 g, three times a day with Da Mai congee This treats chronic hepatitis, Liver cirrhosis with ascites
Zao Fan Wan [Red Date and Alum Pill]	Da Zao 50 pieces Bai Fan 15 g	Calcined Bai Fan, cook with pitted red date and form into pills. Take 3 g with ginger tea for 1 month Treat chronic hepatitis, hypochondriac pain with mass, and accumulation at the hypochondrium and umbilical region

Hua Chong Wan [Dissolve Parasites Pill] *He Ji Ju Fang*

He Shi	10 g
Hu Fen (lead powder) (fried)	10 g
Ku Lian Gen Pi	18 g
Bing Lang	15 g
Wu Yi	9 g
Ku Fan	6 g
Shi Jun Zi	15 g

Main Indication: Roundworm, tapeworm, pinworm, fasciolopsis infestations, and other intestinal parasites.

Preparation: Fry He Shi with Hu Fen, then remove Hu Fen. Grind herbs into powder and mix with flour to form pills. Take 9 g, twice a day with warm rice soup. Take 2 g per day for kids.

Treatment Principles: Resolve and kill various parasites in the intestine.

This formula is often used to kill roundworms, tapeworms, and pinworms.

Nan Gua Zi Fen Bing Lang Jian [Pumpkin Seed Powder and Betel Nut Decoction] *Jing Yan Fang*

Nan Gua Zi (powder)	60–120 g
Bing Lang	30–60 g

Main Indication: Tapeworm. Ancient medical classics referred to it as white worm or white inch worm. The main symptom is white flakes in the stool. This formula is neutral and even, thus the dosage should be large. Taking of the draining herbal formula after taking this formula facilitates the expelling of parasites out of the body.

Preparation: Decoct Bing Lang, take with Nan Gua Zi powder with meal. Take a draining formula half an hour after taking this formula.

Treatment Principles: Expel and kill intestinal tapeworms. In old classic texts, tapeworms are referred to as white worm or cun white worm that is characterized by white flakes in stools. This formula is gentle, thus dosage should be big. After taking this formula, one should continue taking draining herbs to expel the parasites from the body.

Fei Er Wan [Strengthen Baby Pill] *He Ji Ju Fang*

Shen Qu	300 g
Huang Lian	300 g
Rou Dou Kou	150 g
Shi Jun Zi	150 g
Mai Ya	150 g
Bing Lang	120 g
Mu Xiang	60 g

Main Indication: Parasites accumulation causing malnutrition. Symptoms include poor digestion, sallow complexion, thin build, distention and big Stomach, foul breath, fever, and loose stool.

Preparation: Grind herbs into thin powder, mix with pig bile, and form into small pills. Each pill weighs 1 g. Take one pill each time on an empty Stomach. Decrease dosage for children.

Treatment Principles: Kill parasites, reduce accumulation, tonify Spleen, and clear heat.

Bu Dai Wan [Cloth Sack Pill] *Bu Yao Xiu Zhen Xiao Er Fang Lun*

Ye Ming Sha	60 g
Wu Yi	60 g
Shi Zhun Zi	60 g
Bai Fu Ling	15 g
Bai Zhu	15 g
Ren Shen	15 g
Gan Cao	15 g
Lu Hui	15 g

Main Indication: Parasites and malnutrition in children. Symptoms include fever, sallow complexion, thin four limbs, big abdomen, dry hair, and dark sclera.

Preparation: Grind into fine powder and take 3 g each time with pork soup in the morning on an empty Stomach.

Treatment Principles: Expel parasites, reduce malnutrition, and strengthen Spleen.

This syndrome belongs to Spleen malnutrition. Long-term parasites cause deficient body Zheng Qi and excess evil. Thus Si Jun Zi Tang [Four-Gentleman Decoction] is used to support the root, and the rest of the herbs expel parasites.

Comparison of Formulas Treating Intestinal Parasites Gou Chong Wan [Hookworm Pill], Hua Chong Wan [Reduce Parasites Pill], Nan Gua Zi Fen Bing Lang Jian [Pumpkin Seed Powder and Betel Nut Decoction] are better in treating hookworm, roundworm, and tapeworm, focusing on expelling and killing the parasites. Fei Er Wan [Strengthen Baby Pill] and Bu Dai Wan [Cloth Sack Pill] not only expel parasites but also tonify Spleen. The strength of these formulas is milder which is suitable for chronic parasite with malnutrition and for pediatrics.

CHAPTER 4

Spleen and Stomach with Heart, Liver, or Gallbladder Syndromes

SPLEEN AND HEART DEFICIENCY

This method refers to Spleen Qi and Heart blood deficiency. The manifestations are fright, palpitation which can be severe, forgetfulness, excess dreaming, fatigue, poor appetite, watery stool, and irregular menstruation, Beng Lou in females. The formulas are made with Ren Shen, Bai Zhu, Fu Ling, Gan Cao, Shan Yao, Huang Qi, Da Zao, and Hua Sheng Yi and Suan Zao Ren, Bai Zi Ren, Yuan Zhi, Dang Gui, and Shu Di Huang, two groups of herbs to tonify Qi and nourish blood.

Gui Pi Tang [Restore the Spleen Decoction] *Ji Sheng Fang*

Bai Zhu	10 g
Huang Qi	24 g
Fu Shen	12 g
Ren Shen	12 g
Gan Cao	6 g
Mu Xiang	6 g
Yuan Zhi	6 g
Suan Zao Ren	15 g
Long Yan Rou	15 g
Dang Gui	12 g

Main Indication: Spleen and Heart deficiency and Spleen not controlling the blood. Symptoms include palpitation, fright, forgetfulness, insomnia, fatigue, small amount of food intake, sallow complexion, pale tongue, thin, white tongue coating, and thin, moderate pulse. Symptoms can also include early menstruation in women with heavy pale flow, continuous spotting, leukorrhea, Beng Lou, or blood in the stool.

Preparation: Decoct with Sheng Jiang 6 g and Da Zao three to five pieces. One can also double the dosage and form into honey pills. Take 10 g each time, two to three times a day.

Treatment Principles: Tonify Qi, nourish blood, strengthen Spleen, and nourish Heart.

Today, this formula is used to treat Spleen and Heart deficient type of neurasthenia, anemia, heart diseases, functional uterine bleeding, and thrombocytopenia.

Modification:

Addition	Indication
Korean Ren Shen Hua Sheng Yi Da Zao (increase dose)	Heavy bleeding type

Related Formula:

Formula	Herbs	Indication
Xin Pi Shuang Bu Wan [Tonify Both Heart and Spleen Pill]	Xi Yang Shen Bai Zhu Fu Ling Gan Cao Sheng Di Huang Dan Shen Suan Zao Ren Yuan Zhi Wu Wei Zi Mai Men Dong Xuan Shen Bai Zi Ren Huang Lian Zhi Xiang Fu Chuan Bei Mu Jie Geng Long Yan Rou	Treat severe cases of Spleen and Heart deficiency

DISHARMONY OF SPLEEN AND LIVER

The main manifestations of this syndrome are abdominal distention and pain, nausea, vomiting, bitter taste in mouth, watery stool, and wiry pulse. The commonly used herbs include Chai Hu, Mu Xiang, Yu Jin, Zhi Ke, Yan Hu Suo, and Chuan Lian Zi to smooth Liver and regulate Qi. Huang Qin and Huang Lian can clear Liver, and Shao Yao and Gan Cao can soften Liver. Fu Ling, Bai Zhu, Ban Xia, and Gan Cao can strengthen Spleen and harmonize the Middle Jiao.

Disharmony of Liver and Spleen should also be differentiated into cold, heat, excess, or deficient types. Cold type of Liver Qi invading Stomach which causes nausea and vomiting should be treated with Wu Zhu Yu Tang [Evodia Decoction]. Heat type is treated by Xiao Chai Hu Tang [Minor Bupleurum Decoction]. For deficient type of Liver overacting on Spleen causing abdominal pain, Xiao Jian Zhong Tang [Minor Construct the Middle Decoction], Chai Shao Liu Jun Zi Tang [Six-Gentleman Decoction with Bupleurum and Peony], and Xiao Yao San [Relax Wanderer Powder] can be used. On the other hand, Da Chai Hu Tang [Major Bupleurum Decoction], Qing Yi Tang [Clear the Pancreas Decoction], and Lan Wei Hua Yu Tang [Transform the Blood Stasis Decoction for Appendicitis] are used to excess type. For severe abdominal pain, Gui Zhi Jia Shao Yao Tang is used [Cinnamon Twig Decoction with Peony]. If severe Liver Qi stagnation causes abdominal and epigastric pain, or even cold limbs in severe cases, one can use Si Ni San [Frigid Extremities Powder]. If there is also stifling sensation in the chest and poor appetite, then Yue Ju Wan [Escape Restraint Pill] is used. These are examples of treating Liver and Spleen at the same time.

In terms of syndrome differentiation, if the main focus of the formula is for Spleen deficiency, Xiao Jian Zhong Tang [Minor Construct the Middle Decoction], Chai Shao Liu Jun Zi Tang [Six-Gentleman Decoction with Bupleurum and Peony], and Xiao Yao San [Relax Wanderer Powder] are "tonifying earth to nourish wood." If the main focus of the formula is for Liver Qi stagnation, Shao Yao Gan Cao Tang [Peony and Licorice Decoction] is "control wood to support earth. " If Liver Qi stagnation and Spleen deficiency are both severe, then one can use Dang Gui Shao Yao San [Tangkuei and Peony Powder] and Xiao Yao San [Relax Wanderer Powder] which nourishes wood to support earth. The modification of the core formula usually consists of two groups of herbs, smoothing Liver Qi herbs and strengthening Spleen herbs such as Bai Shao Yao, Dang Gui, and Chai Hu, and Bai Zhu, Gan Cao, and Yi Tang. Bai Shao Yao and Bai Zhu are often used together especially to treat pathological conditions of both Liver and Spleen.

Furthermore, if Liver Qi stagnation overacts on Stomach leading to Stomach Qi rebellion, the symptoms are nausea and vomiting; at the same time, if there are also Gallbladder symptoms such as bitter taste in mouth, hypochondriac distention, and fullness and pain in abdominal area, and Spleen symptoms such as nausea, vomiting, stifling sensation at the epigastrium, and poor appetite, then the focus of the treatment should be on Gallbladder and modified with descending and stop vomiting formulas such as Hao Qin Qing Dan Tang [Artemisia Annua and Scutellaria Decoction to Clear the Gallbladder].

Hao Qin Qing Dan Tang [Artemisia Annua and Scutellaria Decoction to Clear the Gallbladder] *Tong Su Shang Han Lun*

Qing Hao	12 g
Huang Qin	10 g
Zhi Shi	10 g
Zhu Ru	10 g
Chen Pi	10 g
Ban Xia	12 g
Fu Ling	12 g
*Bi Yu San	10 g

***Bi Yu San [Jasper Powder]:** Hua Shi, Gan Cao, and Qing Dai

Main Indication: Damp–heat in Shao Yang. Symptoms include chills and fever like malaria, mild chills and severe fever, fullness and pain at chest and hypochondriac region, bitter taste in mouth, vomiting sour and bitter water or vomiting yellow sticky saliva, dry heaves and hiccups in severe cases, red tongue, white or yellow greasy tongue coating, and rapid pulse, slippery at right and wiry at left.

Preparation: Decoction.

Treatment Principles: Clear Gallbladder, drain damp, harmonize Stomach, and transform phlegm.

This formula is formed by adding Qing Hao, Huang Qin, Hua Shi, and Qing Dai to Wen Dan Tang [Warm the Gallbladder Decoction]. It is effective in treating damp–heat and phlegm heat in Shao Yang and San Jiao. When modified appropriately, it can treat damp–heat in Upper, Middle, or Lower Jiao.

Modification:

Addition	Indication
Dai Zhe Shi	Severe nausea
Yu Jin Yin Chen Hao Shan Zhi Zi Da Huang	Acute yellow jaundice
Bai Shao Yao Man Jing Zi Dai Zhe Shi	Dizziness and vertigo
Mu Dan Pi Sang Ye Mu Li	Damp–heat in Liver and Gallbladder, profuse sweating
Shi Chang Pu Gou Teng Ju Hua Ze Xie Tong Cao	Damp–heat obstructing Shao Yang Channel leading to tinnitus or deafness
Shi Chang Pu Lian Qiao	Late stage of encephalitis B after high fever has decreased but damp–heat obstructing the clear orifices causing unconsciousness
Hu Po Gua Lou Pi Huang Lian Ze Xie	Palpitation, insomnia due to phlegm heat
Yu Xing Cao Lu Gen Dong Gua Ren	Cough and chest pain due to phlegm heat accumulating in Lung If with coarse breathing, then add Ma Huang, Xing Ren, and Shi Gao
Shan Zhi Zi Gua Lou Pi	Liver counteracts on Lung, causing hemoptysis
Mu Tong Shan Zhi Zi Chai Hu	Lin syndrome due to damp–heat in Shao Yang and San Jiao with symptoms of lower back pain, frequent urination, urgency, and pain

Shao Yao Gan Cao Tang [Peony and Licorice Decoction]
Shang Han Lun

Bai Shao Yao	30–60 g
Zhi Gan Cao	10–20 g

Main Indication: Insufficient Liver Yin, tendons not nourished. Symptoms include abdominal spasm and pain or contraction of hands and feet, red tongue, scanty coating, and thin pulse.

Preparation: Decoction.

Treatment Principles: Soften Liver and relieve spasm.

This formula can treat spasm pain at any part of the body such as intestinal spasms or spasms of the gastrocnemius muscle.

Modification:

Addition	Indication
Huo Luo Xiao Ling Dan [Fantastically Effective Pill to Invigorate the Collaterals]	Sciatic pain
Long Dan Cao	Trigeminal neuralgia

Related Formulas:

Formula	Herbs	Indication
Shao Yao Gan Cao Fu Zi Tang [Peony, Licorice, and Prepared Aconite Decoction]	Core Formula + Fu Zi	Aversion to cold, especially coldness in the lower limbs
Shao Gan Zhi Jing Tang [Peony and Licorice Decoction to Stop Spasms]	Core Formula + Quan Xie Wu Gong	Treats encephalitis B after fever has decreased, cool body, tightness and contraction of the hands and feet, back and neck stiffness
Jia Wei Shao Gan Tang [Augmented Peony and Licorice Decoction]	Core Formula + Zi Shi Ying Da Zao Fried Xiao Mai	Treats opisthotonos, during attack. Back cannot touch the bed; after attack everything returns to normal

Dang Gui Shao Yao San [Tangkuei and Peony Powder] *Shang Han Lun*

Dang Gui	10 g
Chuan Xiong	10 g
Shao Yao	30 g
Bai Zhu	12 g
Fu Ling	15 g
Ze Xie	10 g

Main Indication:

1. Spasmic abdominal pain during pregnancy.
2. Diarrhea, abdominal pain, difficulty urinating, and numbness and weakness of the lower back and feet during pregnancy or postpartum.
3. Swelling and enlargement of the Liver and pain at right hypochondriac area.
4. Tongue is pale with thin white coating; pulse is wiry, slippery, and forceless.

Preparation: Grind into fine powder; take 3 g each time, mixed with alcohol, twice a day. It can also be taken as decoction.

Treatment Principles: Nourish blood, soften Liver, alleviate spasm, and stop pain.

This formula treats abdominal pain due to Liver blood deficiency and disharmony of Liver and Spleen.

Modification:

Addition	Indication
Zhi Shi Yu Jin Mu Xiang	For swelling and enlargement of liver
Gui Zhi Wu Zhu Yu	With more cold
Mu Dan Pi Shan Zhi Zi Yin Chen Hao	With more heat
Da Huang Shan Zha	With blood stasis

Note: To be effective, formula should be taken continuously until several dozen bags have been used.

Related Formulas:

Formula	Herbs	Indication
Jia Wei Dang Gui Shao Yao San [Augmented Tangkuei and Peony Powder]	Core Formula + Xu Duan 10 g Tu Si Zi 12 g	Treats abnormal fetal position at 7 to 9 months of pregnancy
Bao Sheng Wu You San [Protect the Fetus without Any Worries Powder]	Dang Gui Bai Shao Yao Chuan Xiong Huang Qi Gan Cao Hou Po Zhi Ke Tu Si Zi Ai Ye Jing Jie Sui Qiang Huo Chuan Bei Mu	Treats abnormal fetal position which belongs to blood deficiency, Qi stagnation, and cold damp in the Lower Jiao

Si Ni San [Frigid Extremities Powder] *Shang Han Lun*

Chai Hu	12 g
Zhi Ke	12 g
Shao Yao	30–100 g
Gan Cao	6 g

Main Indication: Frigid extremities syndrome of Shao Yin pathological condition which is coldness of hands and feet. It can also treat fullness and pain of the chest and hypochondriac region, focal distention of the epigastrium, abdominal pain or diarrhea with heavy downward sensation, and dark tongue with deep thin pulse.

Preparation: Taken as powder or decoction.

Treatment Principles: Pacify Liver and regulate Spleen, vent evils, and relieve stagnation.

This syndrome is caused by Liver Qi stagnation leading to disharmony of Liver and Spleen, and pain at epigastrium, abdomen, and hypochondriac region. It can also treat chronic hepatitis, pancreatitis, and fever with cold limbs in children.

Modification:

Addition	Indication
Wu Wei Zi Gan Jiang	Presence of cough
Gui Zhi	With palpitation
Fu Ling	Difficulty urinating
Zhi Fu Zi	Abdominal pain
Xie Bai	Severe diarrhea

Related Formulas:

Formula	Herbs	Indication
Zhi Shi Shao Yao San [Immature Bitter Orange and Peony Powder]	Zhi Shi Shao Yao	Taken as powder Treats postpartum abdominal pain, restlessness, and fullness that does not permit lying down
Chai Hu Shu Gan San [Bupleurum Powder to Spread the Liver]	Core Formula + Chen Pi Chuan Xiong Xiang Fu	Treats hypochondriac pain and spasm, alternating chills and fever which belong to Liver Qi stagnation and blood deficiency
Si Mo Tang [Four Milled-Herb Decoction]	Bing Lang 9 g Wu Yao 9 g Ren Shen 3 g Chen Xiang 3 g	Liver Qi stagnation attacking Upper Jiao causing rapid breathing and wheezing and attacking Middle Jiao causing irritability; stifling sensation and fullness at epigastrium; lack of appetite
Wu Mo Tang [Five Milled-Herb Decoction]	Mu Xiang 6 g Chen Xiang 6 g Bing Lang 9 g Zhi Shi 9 g Wu Yao 9 g	Extreme anger damaging Liver, causing heart and abdominal distention and pain or shooting pain
Liu Mo Tang [Six Milled-Herb Decoction]	Bing Lang 12 g Wu Yao 12 g Chen Xiang 6 g Zhi Shi 12 g Mu Xiang 9 g Da Huang 9 g	Same indication as the Wu Mo Tang, but with constipation or phlegm obstruction

Dan Bai Si Ni San [Moutan and Phellodendron Powder for Frigid Extremities] (Lu Zhou Yi Xue Yuan Fang)

Chai Hu	12 g
Bai Shao Yao	30 g
Zhi Shi	12 g
Gan Cao	12 g
Mu Dan Pi	12 g
Huang Bai	12 g

Main Indication: Beginning stage of appendicitis. Symptoms include lower right quadrant pain, fever or nausea, diarrhea, yellow tongue coating, red tongue, and rapid pulse.

Preparation: Decoction. One dose per day and the formula should be taken for a few days continuously.

Treatment Principles: Clear heat, smooth Liver, alleviate spasm, and transform stasis.

This formula treats beginning stage of appendicitis which belongs to Qi stagnation in the Liver channel and blood stasis. Thus, the formula uses higher doses of Bai Shao Yao to soften the Liver and stop pain. It is combined with heat clearing and blood cooling herbs, such as Mu Dan Pi and Huang Bai, to resolve stasis, scatter accumulation, reduce inflammation, and stop pain.

Modification:

Addition	Indication
Hong Teng Bai Jiang Cao	Severe cases

Tong Xie Yao Fang [Important Formula for Painful Diarrhea] *Jing Jue Quan Shu* taken from *Liu Cao Chuang Fang*

Bai Zhu	20 g
Bai Shao Yao	20 g
Chen Pi (fried)	9 g
Fang Feng	6 g

Main Indication: Excess Liver and deficient Spleen causing painful diarrhea. Symptoms include borborygmus, abdominal pain, continuous diarrhea with abdominal pain, thin white tongue coating, and wiry and moderate pulse at both guan positions.

Preparation: Decoction or can be taken as powder.

Treatment Principles: Tonify Spleen and drain Liver.

When Spleen is deficient, then there is diarrhea. When Liver is excess, then there is pain. Bai Shao Yao and Bai Zhu control wood and support earth, thus they can treat painful diarrhea.

Modification:

Addition	Indication
Sheng Ma	Chronic continuous diarrhea
Fu Ling Che Qian Zi Gan Jiang	Watery stool
Bai Tou Weng Huang Qin Huang Lian Huang Bai	Blood and pus in stool
Bing Lang Mu Xiang	Tenesmus
Bai Shao Yao (double the dosage) Qing Pi Xiang Fu	Severe abdominal pain
Wu Zhu Yu Gan Jiang	Chronic diarrhea leading to deficient cold

Yue Ju Wan [Escape Restraint Pill], also known as Xiong Zhu Wan [Ligusticum and Atractylodes Macrocephala Pill] *Dan Xi Xin Fa*

Chuan Xiong	equal amount
Cang Zhu	
Xiang Fu	
Shen Qu	
Shan Zhi Zi	

Main Indication: Mild Qi, blood, phlegm, fire, damp, and food stagnation syndrome. Symptoms include focal distention and stifling sensation at the chest and diaphragm, distending pain of the epigastrium and abdomen, sour regurgitation, nausea, vomiting, belching, indigestion, white or yellow greasy tongue coating, and wiry or slippery pulse.

Preparation: Grind into powder and form pills. Take 6 to 9 g each time. It can also be taken as decoction.

Treatment Principles: Smooth Liver and regulate Spleen.

This formula focuses on Qi stagnation, thus the main treatment principle is to guide Qi and relieve stagnation. In clinical practice, it can be modified to treat different types of stagnation by increasing the dosage of the corresponding herb. It can be used to treat chronic hepatitis, gastritis, indigestion due to neurasthenia, abdominal distention, and sour regurgitation or apathy.

Modification:

Addition	Indication
Mu Xiang Qing Pi Wu Yao Fo Shou Chai Hu	Severe Qi stagnation
Tao Ren Hong Hua Dan Shen Ru Xiang Mo Yao Chuan Shan Jia	Severe blood stasis
Huang Qin Huang Lian	Severe fire accumulation
Chen Pi Ban Xia Tian Nan Xing	Severe phlegm stagnation
Fu Ling Yi Yi Ren	Severe damp stagnation
Wu Zhu Yu Gui Zhi	With cold symptoms
Zhi Ke Hou Po	Focal distention and fullness

Related Formula:

Formula	Herbs	Indication
Hai Zao Yu Hu Tang [Sargassum Decoction for the Jade Flask]	Hai Zao Kun Bu Ban Xia Chen Pi Qing Pi Lian Qiao Zhe Bei Mu Dang Gui Chuan Xiong Du Huo Hai Dai Gan Cao	Transforms phlegm, softens hardness, reduces and scatters accumulation Treats goiters due to disorders of Liver and Spleen, Qi stagnation, and congealed phlegm Symptoms include swollen nodules at neck, no skin color change, absence of pain and ulceration, dark tongue or flabby tongue with scalloped edges, thick white tongue coating, and wiry, slippery pulse

Comparison of Formulas Treating Liver and Spleen Disharmony

Shao Yao Gan Cao Tang [Peony and Licorice Decoction] and Dang Gui Shao Yao San [Tangkuei and Peony Powder] both are analgesic formulas which focus on tonification. Both formulas soften and tonify Liver, alleviate spasms, and stop pain.

Si Ni San [Frigid Extremities Powder], Tong Xie Yao Fang [Important Formula for Painful Diarrhea], and Yue Ju Wan [Escape Restraints Pill] all treat syndrome of Liver Qi stagnation, overact on Spleen, and unregulated Qi mechanism. However, Si Ni San [Frigid Extremities Powder] focuses on cold hands and feet. Tong Xie Yao Fang [Important Formula for Painful Diarrhea] treats abdominal pain followed by diarrhea. Yue Ju Wan [Escape Restraints Pill] focuses on treating Liver Qi stagnation leading to all other types of stagnations.

Dan Bai Si Ni San [Moutan and Phellodendron Powder] is designed for the early stage of appendicitis. Hao Qin Qing Dan Tang [Artemisia Annua and Scutellaria Decoction to Clear the Gallbladder] focuses on treating Stomach and Gallbladder disharmony, damp–heat in Shao Yang syndrome with main symptoms of bitter taste in the mouth, alternating chills and fever, and wiry and rapid pulse.

QI AND BLOOD DEFICIENCY

If Liver blood and Spleen Qi are both deficient with symptoms of dizziness, vertigo, palpitation, shortness of breath, fatigue, sallow complexion, pale tongue with thin white coating, and weak thin pulse, treatment should tonify Qi and nourish blood at the same time. Often blood nourishing herbs such as Dang Gui, Bai Shao Yao, Shu Di Huang, and He Shou Wu are often used together with Qi tonics such as Ren Shen, Huang Qi, Bai Zhu, and Gan Cao.

Ba Zhen Tang [Eight-Treasure Decoction] *Zheng Ti Lei Yao*

Ren Shen	3 g
Bai Zhu	10 g
Fu Ling	12 g
Gan Cao	6 g
Shu Di Huang	15 g
Dang Gui	12 g
Bai Shao Yao	12 g
Chuan Xiong	6 g

Main Indication: Qi and blood deficiency. Symptoms include sallow or pale face, shortness of breath, too lazy to talk, fatigue in four limbs, dizziness, blurry vision, palpitation maybe with fright, pale tongue with thin white coating, and thin weak or deficient and big but forceless pulse.

Preparation: Decoct with three slices of Sheng Jiang and two pieces of Da Zao. Take before meal.

Treatment Principles: Tonify Qi and blood together.

This formula is a combination of Si Jun Zi Tang [Four-Gentleman Decoction] and Si Wu Tang [Four-Substance Decoction]. It is a common formula for tonifying both Qi and blood at the same time. It is often used in treating excess bleeding, postpartum or postsurgery Qi, blood exhaustion and leukorrhea, and chronic fatigue syndrome.

Related Formulas:

Formula	Herbs	Indication
Shi Quan Da Bu Tang [All-Inclusive Great Tonifying Decoction]	Core Formula + Rou Gui 9 g Huang Qi 15 g	Qi and blood deficiency with cold
Ren Shen Yang Ying Tang [Ginseng Decoction to Nourish the Nutritive Qi]	Core Formula – Chuan Xiong + Rou Gui Huang Qi Yuan Zhi Chen Pi Wu Wei Zi	Treat Qi and blood deficiency, with restless feeling With presence of diarrhea or nocturnal emission, add Long Gu If with wheezing and coughing, add E Jiao

Dang Gui Bu Xue Tang [Tangkuei Decoction to Tonify Blood] *Nei Wai Shang Bian Huo Lun*

Huang Qi	30 g
Dang Gui	6 g

Main Indication: Blood deficiency with floating Yang. Symptoms include overwork, fatigue, internal injury, Qi and blood deficiency, floating Yang surfacing, red face, muscle heat, irritability and thirst, pale tongue with thin white coating, flooding big and deficient pulse, minute upon heavy palpation. It can also treat postpartum blood deficiency leading to fever and headache or chronic ulceration with difficulty in healing.

Preparation: Decoction. Take before meals, three times a day.

Treatment Principles: Tonifies Qi to generate blood.

In this formula, the dosage of Huang Qi is five times that of Dang Gui which is based on the fact that "blood substance cannot be generated rapidly, so formless Qi should be consolidated quickly."

Therefore, this formula can be used to treat false heat syndrome due to blood deficiency with floating Yang or bleeding syndromes due to Qi deficiency failing to control blood.

The false heat symptoms of floating Yang which is indicated for using this formula are easily confused with the "Big Four" that is indicated for Bai Hu Tang [White Tiger Decoction]. If one differentiates the symptoms carefully, then one would notice that the pulse of this syndrome is flooding and big but soft and deficient; there is thirst but a preference for warm drinks, and fever but not very high with an absence of profuse sweat.

Related Formulas:

Formula	Herbs	Indication
Jia Wei Dang Gui Bu Xue Tang I [Augmented Tangkuei Decoction to Tonify Blood I]	Core Formula + Dang Shen Bai Zhu Tong Cao Yu Zhu Gan Cao, Bie Jia Cook with pig's feet	Treats insufficient lactation due to Qi and blood deficiency
Jia Wei Dang Gui Bu Xue Tang II [Augmented Tangkuei Decoction to Tonify Blood II]	Core Formula + San Qi 10 g Sang Ye 6 g	Treats seniors' excessive uterine bleeding which belongs to Spleen failing to control blood
Gu Ben Zhi Beng San [Stabilize the Root and Stop Excessive Uterine Bleeding Powder]	Core Formula + Ren Shen Bai Zhu Sheng Jiang Tan (charred) Shu Di Huang	Treats excessive uterine bleeding (Beng) and spotting (Lou), continuous blood flow, blurry vision, dizziness, vertigo, and thin weak pulse
Si Miao Tang [Four-Marvel Decoction]	Core Formula + Jin Yin Hua Gan Cao	Treats ulcerations of boils with residual toxins accumulation and slow healing process

Comparison of Formulas Treating Qi and Blood Deficiency

Ba Zhen Tang [Eight-Treasure Decoction] is a basic formula to tonify Qi and nourish blood which is gentle and all around. Shi Quan Da Bu Tang [All-Inclusive Great Tonifying Decoction] is based on Ba Zhen Tang [Eight-Treasure Decoction] and is slightly warmer. Ren Shen Yang Ying Tang [Ginseng Decoction to Nourish the Nutritive Qi] is based on Shi Quan Da Bu Tang [All-Inclusive Great Tonifying Decoction], modified to be calmer and more nourishing to Heart Shen and to benefit Heart Qi. Dang Gui Bu Xue Tang [Tangkuei Decoction to Tonify Blood], is a small formula, yet versatile. It can quickly tonify Qi and blood, and also can stop bleeding due to Qi failing to control blood. It can also treat false heat due to blood deficiency leading to floating Yang. Furthermore, this formula can be added to other formulas to strengthen the function of tonifying Qi and nourishing blood.

CHAPTER 5

Liver Syndromes and Formulas (Gallbladder Syndromes)

Liver stores blood and regulates the blood volume for whole body functions. It dominates free flow and circulation, controls bile, assists digestion, unites tendons and membranes, and connects and nourishes joints and muscle to promote normal functions. Thus, to maintain the normal functions of Liver and Gallbladder, one must maintain the free flow of Qi, blood, fluid, and emotions.

COLD IN THE LIVER AND LIVER CHANNEL

Cold stagnation in the Liver often manifests as cold extremities, abdominal pain, green or purple nails, or flaccid constriction of the penis, spasm of gastrocnemius muscle, and wiry or thin deep pulse. This is a direct attack of cold which is acute. Furthermore, weakness of Liver function, and Yang deficiency with excess Yin can manifest as fatigue, lost of endurance, fright, cold limbs and a deep slow pulse which is more chronic. To treat these conditions, Wu Zhu Yu, Chuan Jiao, Gui Zhi, Fu Zi, and Xi Xin are combined with Wu Yao and Xiang Fu to smooth the Liver, then Ren Shen and Dang Gui are added to warm and scatter the cold evil pathogens. Wu Zhu Yu Tang [Evodia Decoction] and Nuan Gan Jian [Warm the Liver Decoction] focus on warming and nourishing Liver. Dang Gui Si Ni Tang [Tangkuei Decoction for Frigid Extremities] and Wu Yu Mu Gua Tang [Evodia and Chinese Quince Fruit Decoction] focus on scattering cold and alleviating spasms.

Dang Gui Si Ni Tang [Tangkuei Decoction for Frigid Extremities] *Shang Han Lun*

Dang Gui	9 g
Bai Shao Yao	12 g
Gui Zhi	9 g
Xi Xin	3 g
Gan Cao	6 g
Tong Cao	6 g
Da Zao	9 g

Main Indication: Cold injured Jue Yin and congealed and stagnant blood in channels. Symptoms include coldness in the four limbs or cold pain at lower back, inguinal area, legs, and feet; pale tongue with white coating, and thin, minute or deep thin pulse.

Preparation: Decoction.

Treatment Principles: Warm channel, scatter cold, nourish blood, and unblock channel.

In clinical practice, this formula can be used for congealed coldness and blood stasis type of wind–damp Bi pain, frostbite, and menstrual pain. It can also be used for cold type of one-sided abdominal pain or limb pain.

Related Formulas:

Formula	Herbs	Indication
Dang Gui Si Ni Jia Wu Zhu Yu Sheng Jiang Tang [Tangkuei Decoction for Frigid Extremities plus Evodia and Fresh Ginger]	Core Formula + Wu Zhu Yu 9 g Sheng Jiang 9 g	Treats flaccid contraction of penis, abdominal pain, cold extremities, or cold abdominal pain before menstruation in a person with cold constitution. This formula is stronger in its warming channel and scattering cold function
Ji Sheng Tong Mai Si Ni Tang [Unblock the Channel for Frigid Extremities Decoction from Formulas that Aid the Living]	Core Formula + Fu Zi	Treats cholera with more cold symptoms, cold flesh, and disappearing pulse

Shang Han Lun discussed Si Ni Tang [Frigid Extremities Decoction], Si Ni San [Frigid Extremities Powder], and Dang Gui Si Ni Tang [Tangkuei Decoction for Frigid Extremities]. However, the compositions and indications of these three formulas are different. Si Ni Tang [Frigid Extremities Decoction] focuses on rescuing Yang. Si Ni San [Frigid Extremities Powder] focuses on harmonizing Spleen and Liver. Dang Gui Si Ni Tang [Tangkuei Decoction for Frigid Extremities] focuses on nourishing blood and unblocking the channels.

Wu Zhu Yu Tang [Evodia Decoction] *Shang Han Lun*

Wu Zhu Yu	9 g
Ren Shen	9 g
Da Zao	9 g
Sheng Jiang	9 g

Main Indication: Deficient cold in Liver and Stomach with rebellious turbid Yin. Symptoms include dry heaves, vomiting saliva, vertex headache, epigastric and abdominal pain, sour regurgitation, pale tongue with white slippery coating, and wiry slow pulse. Symptoms of Shao Yin are vomiting, diarrhea, cold extremities, severe restlessness, and agitation to the point that the patient wants to die.

Preparation: Decoction. Taken as cold decoction.

Treatment Principles: Warm the Middle Jiao and descend the rebellion.

This formula is effective in treating deficient cold-type of vertex Jue Yin headache and Yang Ming nausea and vomiting.

Modification:

Minus	Addition	Indication
Ren Shen	Dang Shen	Nausea and vomiting without cold limbs
	Chen Xiang	Abdominal pain with vomiting food
	Mu Gua	Spasm of gastrocnemius muscle after vomiting and diarrhea

Related Formula:

Formula	Herbs	Indication
Wu Zhu Yu Jia Fu Zi Tang [Evodia Decoction with Prepared Aconite]	Core Formula + Fu Zi	Treats cold type of hernia with lower back pain that radiates to testicles and deep slow pulse at Chi position

Wu Yu Mu Gua Tang [Evodia and Chinese Quince Fruit Decoction] *Yan Fang Xin Bian*

Wu Zhu Yu	9 g
Mu Gua	12 g
Salt	2 g

Main Indication: Cold evil invasion, and spasms of gastrocnemius muscle. Symptoms include intermittent spasm of gastrocnemius muscle during sleep, pale tongue with thin white coating, and deep and wiry pulse.

Preparation: Decoction.

Treatment Principles: Warm the Liver and relax tendons.

This formula is effective in treating spasms of gastrocnemius muscle due to cold injuries.

Nuan Gan Jian [Warm the Liver Decoction] *Jing Yue Quan Shu*

Xiao Hui Xiang	6 g
Rou Gui	6 g
Wu Yao	6 g
Chen Xiang (or Mu Xiang)	3 g
Fu Ling	9 g
Dang Gui	9 g
Gou Qi Zi	12 g
Sheng Jiang	9 g

Main Indication: Cold stagnation in Liver and Kidney. Symptoms include lower abdominal pain, hernia, pale dusky tongue with thin white coating, and deep slow or deep wiry pulse.

Preparation: Grind Chen Xiang into powder, decoct other herbs, and take the Chen Xiang powder after drinking the decoction.

Treatment Principles: Warm Kidney and Liver, move Qi, and stop pain.

This formula mainly treats cold type of hernia pain from external genitalia to lower abdomen. Patient often has internal excessive cold and Yin with external cold evil invasion.

Modification:

Addition	Indication
Wu Zhu Yu	Severe cold with sharp pain
Gan Jiang	If the pain is even more severe, then add Fu Zi

This formula is contraindicated for damp–heat downward pouring, and swelling of the external genitalia with heat pain.

Related Formula:

Formula	Herbs	Indication
Bu Gan Tang [Tonify the Liver Decoction]	Fu Ling 45 g Zhi Wu Tou 4 g Yi Yi Ren 37 g Du Huo 37 g Zhi Fu Zi 2 g Bai Zi Ren 75 g Fang Feng 75 g Xi Xin 15 g Shan Zhu Yu 5 g Rou Gui 5 g Zhi Gan Cao 18 g	Grind all ingredients into powder. Take 5 g of the powder with Da Zao decoction (2 pieces), two to three times a day Treat severe type of cold hernia pain

Comparison of Formulas Treating Cold in Liver and Liver Channel

Dang Gui Si Ni Tang [Tangkuei Decoction for Frigid Extremities] and Wu Yu Mu Gua Tang [Evodia and Chinese Quince Fruit Decoction] treat cold evil damaging Jue Yin with tight tendons and acute pain. The focus of these formulas is to warm the Liver, move Qi, and moderate and stop the pain. Wu Zhu Yu Tang [Evodia Decoction] and Nuan Gan Jian [Warm the Liver Decoction] treat internal excess of coldness in Liver or damage to Stomach or Kidney. The focus of this formula is to warm Liver, Stomach, and Kidney, move Qi, and stop pain.

HEAT IN THE LIVER AND LIVER CHANNEL

Liver heat syndrome often manifests as irritability, stifling sensation, dry mouth, dark yellow urine, swollen pain at testicles, or vaginal burning pain and itching. Liver fire often refers to headache, distending sensation, red face, red eyes, tinnitus, deafness, and bitter taste in mouth. The difference between Liver fire and Liver heat is that Liver fire has signs of upward rushing rebellion. Both syndromes belong to excess, thus cold bitter herbs such as Long Dan Cao, Huang Qin, Qing Dai, Shan Zhi Zi, and Xia Ku Cao should be used as main herbs. The main herbs are then combined with Yin nourishing and blood nourishing herbs such as Sheng Di Huang, Bai Shao Yao, and Dang Gui. Ze Xie, Che Qian Zi, and Mu Tong which promote urination and expel dampness are added to form the formula that clears and drains Liver fire.

Long Dan Xie Gan Tang [Gentiana Longdancao Decoction to Drain the Liver] *Yi Fang Ji Jie*

Long Dan Cao	9 g
Sheng Di Huang	12 g
Dang Gui	6 g
Chai Hu	9 g
Mu Tong	9 g
Ze Xie	9 g
Che Qian Zi	12 g
Shan Zhi Zi	9 g
Huang Qin	12 g
Gan Cao	6 g

Main Indication: Excess (damp) heat in Liver and Gallbladder. Symptoms include headache, red face, hypochondriac pain, bitter taste, dizziness, vertigo, tinnitus, swollen ear, vaginal itch, swollen external genitalia, painful and turbid urination, foul fetid leukorrhea, red tongue with yellow tongue coating, and wiry, rapid, forceful pulse.

Preparation: Decoction. Also can be taken as pills. Take 6 to 9 g each time, twice a day.

Treatment Principles: Clear Liver and Gallbladder of excess fire, and clear damp–heat from Liver channel.

The indicated syndrome for this formula is excess Liver and Gallbladder fire blazing upward and damp–heat pouring downward to Liver channel. Clinically this formula can treat Liver and Gallbladder (damp) heat type of stubborn migraine headache, hypertension, acute keratitis, sinusitis, acute hepatitis with jaundice, acute cholecystitis, eczema at the external genitalia, and inflammation of the urinary tract.

Related Formulas:

Formula	Herbs	Indication
Dong Yuan Long Dan Xie Gan Tang [Gentiana Longdancao Decoction to Drain the Liver by Li Dong Yuan]	Core Formula – Huang Qin Shan Zhi Zi Gan Cao	Treats damp–heat type of vaginal burning and itch with foul smell
Jia Wei Long Dan Xie Gan Tang [Augmented Gentiana Longdancao Decoction to Drain the Liver]	Core Formula + Lian Xu Chi Shao Yao	Treats damp–heat in the Liver channel, pink red leukorrhea which looks like blood but is not blood, hypochondriac distention or pain, bitter taste, yellow urine, red tongue, yellow tongue coating, and wiry rapid pulse
Jia Jian Long Dan Xie Gan Tang [Modified Gentiana Longdancao Decoction to Drain the Liver]	Long Dan Cao Huang Qin Shan Zhi Zi Bai Shao Yao Ze Lan Mu Dan Pi Bie Jia Niu Xi Mao Gen Qing Hao (in the presence of tidal fever)	Treats reverse menstruation (absence of menstrual flow but epistaxis), scanty early menstruation or amenorrhea, with dizziness, tinnitus, intermittent tidal fever, irritability, dry mouth, red lips, yellow tongue coating, and wiry, rapid pulse
Xie Qing Wan [Drain the Green Pill]	Dang Gui Long Dan Cao Chuan xiong Shan Zhi Zi Da Huang Qiang Huo Fang Feng	Grind the herbs into fine powder, taken with Dan Zhu Ye Decoction. Treats stagnant fire in Liver. Symptoms include swollen painful red eyes, irritability and agitation, easily angered, inability to sleep, yellow urine, constipation, and flooding excess pulse. It can also treat fright convulsion in children or spasms due to heat excess

Dang Gui Long Hui Wan [Tangkuei, Gentiana Longdancao, and Aloe Pill] *Dan Xi Xin Fa*

Long Dan Cao	30 g
Dang Gui	30 g
Lu Hui	15 g
Da Huang	15 g
Shan Zhi Zi	30 g
Huang Qin	30 g
Huang Lian	30 g
Huang Bai	30 g
Mu Xiang	6 g
Qing Dai	15 g
She Xiang	1.5 g

Main Indication: Excess fire in Liver and Gallbladder; symptoms include dizziness, vertigo, tinnitus, deafness, unsettled mind, radiating pain from hypochondrium to lower abdomen, constipation, dark yellow urine, delirium or mania in severe cases, rash, hematemesis, epistaxis, red tongue with yellow coating, and wiry forceful pulse.

Preparation: Form into honey pills and take 6 g each time, twice a day.

Treatment Principles: Clear and drain excess fire of Liver and wih Gallbladder.

This formula treats severe excess heat syndrome of blazing excess fire in Liver channel which disturbs the Heart and mind, causing reckless blood flow. Thus, extreme bitter and cold herbs are used to drain the excess fire through both urination and bowel movement.

Shi Jue Ming San [Haliotidis Shell Powder] *Yan Ke Liu Jing Fa Yao*

Shi Jue Ming	30 g
Cao Jue Ming	30 g
Qing Xiang Zi	15 g
Mu Zei	15 g
Shan Zhi Zi	15 g
Chi Shao Yao	15 g
Da Huang	6 g
Qiang Huo	3 g
Jing Jie	6 g

Main Indication: Punctate keratitis. Symptoms include black spots appearing at cornea, swollen eyelids, headache, red eyes, tearing, red tongue tip and sides, thin yellow tongue coating, and wiry, floating, rapid pulse.

Different parts of the eye represent different organs: eyelids belong to Spleen, inner corner of the eye belongs to Heart, sclera belongs to Lung, iris belongs to Liver, and the pupil belongs to Kidney. The pathology of this syndrome is located at Liver. Internal blazing of Liver fire combines wind–heat and moves upward to attack the eyes.

Preparation: Grind into powder, and take 6 g each time with Mai Men Dong (12 g) decoction.

Treatment Principles: Clear Liver, expel wind; eye is brightened and nebula cleared.

This formula is effective for nebula due to keratitis if taken long term.

Chun Gen Pi Wan [Ailanhti Bark Pill] *She Sheng Zhong Miao Fang*

Gao Liang Jiang (burnt ash)	9 g
Huang Bai (ash)	6 g
Shao Yao	6 g
Chun Gen Pi	45 g

Main Indication: Damp–heat downward pouring in Liver channel. Symptoms include vaginal discharge with blood and mucous, foul smell, dark yellow urine or urination with burning pain, red tongue, yellow, greasy tongue coating, and wiry or slippery pulse.

Preparation: Grind into powder and form pills with flour paste into size of seeds from Chinese parasol tree. Take 30 g each time on empty stomach with rice water. Or take with water.

Treatment Principles: Clear heat, dry dampness.

Mucous type of leukorrhea is due to excessive dampness. Bloody leukorrhea is due to excessive heat. Thus, vaginal discharge with blood and mucous is caused by damp–heat accumulation pouring downward. This formula is designed for such a syndrome.

Related Formula:

Formula	Herbs	Indication	Notes
Yu Dai Wan [Cure Discharge Pill]	Core formula + Shu Di Huang Dang Gui Chuan Xiong	Treat syndromes of Ren and Chong channels not consolidated, and excess leukorrhea with blood and mucous	Yu Dai Wan [Cure Vaginal Discharge Pill] compared to Chun Gen Pi Wan [Ailanthi Bark Pill] has both draining and tonifying function. It consolidates and nourishes Ren channel, consolidates and astringes Chong channel, and also clears and transform damp–heat

Qing Re Zhi Dai Tang [Clear Heat and Stop Discharge Decoction] (Jing Yan Fang)

Chai Hu	9 g
Xiang Fu	9 g
Chuan Lian Zi (charcoaled)	9 g
Tu Fu Ling	30 g
Yin Hua Teng	30 g
Pu Gong Ying	30 g
Yu Xing Cao	30 g
Guan Zhong	15 g
Ye Ju Hua	15 g
Long Dan Cao	9 g
Cang Zhu	9 g
Hong Teng	30 g

Main Indication: Liver stagnation with damp–heat in the Lower Jiao. Symptoms include fever, lower abdominal pain which refuses pressure and palpation, excess vaginal discharge with foul fishy smell, yellow urine, yellow greasy tongue coating, and wiry, rapid pulse.

Preparation: Decoction.

Treatment Principles: Clear heat, relieve toxins, disperse Liver, move Qi, stop vaginal discharge.

This formula treats Lower Jiao damp–heat type of acute pelvic inflammatory disease, vaginitis, and urinary tract infection.

Modification:

Addition	Indication
Che Qian Zi Jin Qian Cao Ze Xie	Severe dampness

Yin Jia Wan [Silver Armour Pill] *Cheng Du Zhong Yi Da Xue Fang*

Chuan Lian Zi (charcoaled)	9 g
Xiang Fu	12 g
Wu Yao	15 g
Dang Gui	12 g
Chuan Xiong	6 g
Chi Shao Yao	15 g
Hu Po	9 g
Bie Jia	30 g
Chuan Shan Jia	12 g
Xia Ku Cao	15 g
Si Gua Luo	15 g
Zi Hua Di Ding	30 g
Pu Gong Ying	30 g
Lian Qiao	30 g
Jin Yin Hua	30 g
Hong Teng	30 g

Main Indication: Damp–heat accumulation, Qi stagnation, and blood stasis. Symptoms include pain and distention of the lower abdomen, lower back, and sacrum; excess vaginal discharge, or lumps and nodules at the sides of the lower abdomen, irregular menstruation, excess vaginal discharge, or with blood and mucous, yellow urine, yellow tongue coating, and wiry or slippery pulse.

Preparation: Grind into fine powder, form into pills with honey. Each pill is 9 g; take one pill each time, three times a day.

Treatment Principles: Clear heat, relieve toxins, disperse Liver, unblock collaterals, soften hardness, and scatter the accumulation.

Clinically, this formula is used to treat knotted damp–heat with Qi stagnation and blood stasis type of chronic pelvic inflammatory disease, endometritis, uterine polyps, uterine myoma, vaginitis, ovarian cysts, and infertility. It is effective when taken long term.

Zuo Jin Wan [Left Metal Pill] *Dan Xi Xin Fa also known as Si Ling Wan [Four Order Pill]*

Huang Lian (fried with ginger juice)	180 g
Wu Zhu Yu	30 g

The ratio of Huang Lian and Wu Zhu Yu is 6:1.

Main Indication: Liver fire invading Stomach. Symptoms include distending pain at hypochondrium, rumbling noise in Stomach, sour regurgitation, nausea, vomiting, bitter taste in mouth, focal distention at epigastrium, belching, red tongue with yellow coating, and wiry rapid pulse.

Preparation: Grind into fine powder, mix with water to form pills, and take 2 to 3 g each time, two to three times a day. This formula can also be taken as decoction; the dosage should be modified.

Treatment Principles: Clear Liver, drain fire, descend rebellion, and stop vomiting.

This is a small formula which is a fundamental one in treating Liver and Stomach pathology. When using this formula in clinical practice, one should modify the dosage according to the presence of heat or cold. For cold type, Wu Zhu Yu should be the main herb.

Related Formulas:

Formula	Herbs	Indication	Notes
Wu Ji Wan [Fifth and Sixth Heavenly Stem Pill]	Core formula + Bai Shao Yao	Take 150 g of each herb, grind into fine powder, and form pills Treat damp–heat in the Lower Jiao and abdominal pain	Numbering of the formula name is according to Heavenly Stem
Zhi Yu Wan [Gardenia and Evodia Pill]	Wu Zhu Yu Shan Zhi Zi Xiang Fu	Treat fire stagnation in Liver channel, hypochondriac pain, abdominal pain, and sour regurgitation	

Comparison of Formulas Treating Liver Heat and Liver Fire

Long Dan Xie Gan Tang [Gentiana Longdancao Decoction to Drain Liver] has a wide application. It can be modified to treat any syndromes of upward blazing Liver fire, and damp–heat in the Lower Jiao. Dang Gui Long Hui Wan [Tangkuei, Gentiana Longdancao, and Aloe Pill] is effective in treating Liver fire, upward blazing. Shi Jue Ming San [Haliotidis Shell Powder] treats eye conditions due to Liver fire and wind–heat. Qing Re Zhi Dai Tang [Clear Heat and Stop Vaginal Discharge Decoction], Chun Gen Pi Wan [Ailanthi Bark Pill], and Yin Jia Wan [Silver Armour Pill] treat vaginal discharge with blood and mucous due to damp–heat in the Lower Jiao. Zuo Jin Wan [Left Metal Pill] often used to treat Liver fire invading Stomach causing hypochondriac, stomach, and epigastric pain. Among the seven formulas, the formula that has the strongest heat clearing and relieving toxins function is Qing Re Zhi Dai Tang [Clear Heat and Stop Discharge Decoction]. Long Dan Xie Gan Tang [Gentiana Longdancao Decoction to Drain Liver] is the strongest to drain dampness. Dang Gui Long Hui Wan [Tangkuei, Gentiana Longdancao, and Aloe Pill] is the best to drain upward attacking fire toxins. Yin Jia Wan [Silver Armour Pill] is better in dispersing Liver, regulating Qi, moving blood, and unblocking channels. Zuo Jin Wan [Left Metal Pill] is effective in dispersing Liver, harmonizing Stomach, and stopping pain.

LIVER BLOOD DEFICIENCY

The main manifestations of Liver blood deficiency are dizziness, blurry vision, lustrerless facial complexion and nails, tinnitus, palpitation, scanty menstrual flow, unceasing spotting, or Beng Lou, thin pulse, and pale tongue. The formulas are often formed with Shu Di Huang, Dang Gui, Bai Shao Yao, E Jiao, He Shou Wu, and Ai Ye.

Furthermore, blood nourishing formulas are often combined with Qi tonic herbs to strengthen Yang in order to generate Yin. Blood nourishing formulas are often combined with blood moving herbs to prevent blood stasis. However, the dosage should be small.

Jiao Ai Tang [Ass-Hide Gelatin and Mugwort Decoction] *Jin Gui Yao Lue*

Chuan Xiong	6 g
E Jiao	9 g
Ai Ye	9 g
Gan Cao	6 g
Dang Gui	9 g
Shao Yao	12 g
Shu Di Huang	18 g

Main Indication: Deficient and damaged Ren and Chong channels. Symptoms include Beng Lou, menorrhagia, unceasing bleeding or unceasing bleeding after miscarriage, or bleeding during pregnancy, abdominal pain, pale tongue with white coating, and thin or slightly wiry pulse.

Preparation: Decoction. Strain or add moderate amount of alcohol, melt E Jiao in the decoction, and take warm.

Treatment Principles: Nourish blood, stop bleeding, regulate menstruation, and calm fetus.

This formula is formed by adding E Jiao and Ai Ye to Si Wu Tang [Four-Substance Decoction] and Shao Yao Gan Cao Tang [Peony and Licorice Decoction]. The purpose is to nourish blood, consolidate Chong and Ren channels, stop bleeding, calm fetus, and stop pain. It is a common nourishing formula for miscarriage, Beng Lou, and bleeding during pregnancy and postpartum.

Blood deficient manifestation from unceasing bleeding due to hemorrhoids and ulcerative colitis can also be treated with this formula.

Related Formulas:

Formula	Herbs	Indication
Da Jiao Ai Tang [Major Ass-Hide Gelatin and Mugwort Decoction]	Core Formula + Gan Jiang	Treats injury to the five Zang organs due to falling from high places. In mild cases, patient spits out blood; in severe cases, patient has hematemesis
Dang Gui Tang [Tangkuei Decoction]	Core formula + Ren Shen	Treat restless fetus caused by fright to the pregnant woman
Ding Xiang Jiao Ai Tang [Clove, Ass-Hide Gelatin, and Mugwort Decoction]	Core Formula – Gan Cao + Ding Xiang	Treat unceasing Beng Lou with icy cold sensation below the umbilicus; preference for thick clothing, and excess vaginal discharge with intermittent dribbling sensation
Fu Bao Dan [Women's Treasure Pill]	Core formula – Gan Cao + Fu Zi	Treat deficient cold uterus, irregular menstruation, and infertility due to cold uterus
Qi Xiao Si Wu Tang [Miraculous Effective Four-Substance Decoction]	Core formula – Gan Cao + Huang Qin	Treat insufficient Yin and blood, excess Yang evil pathogen, and Beng syndrome due to Yin deficiency and Yang excess

Si Wu Tang [Four-Substance Decoction] *He Ji Ju Fang*

Shu Di Huang	12 g
Dang Gui	10 g
Bai Shao Yao	12 g
Chuan Xiong	3 g

Main Indication: Deficient and damaged Chong and Ren channels, and Liver blood deficiency. Symptoms include irregular menstruation, abdominal pain, Beng Lou, or menstruation with clots, intermittent abdominal pain, or restless fetus, bleeding during pregnancy, or postpartum retention of lochia, mass and nodule formation, lower abdomen hardness and pain, and intermittent coldness and heat, pale tongue, white tongue coating, thin or wiry or slippery pulse.

Preparation: Decoction. Taken on empty stomach, three times a day.

Treatment Principles: Nourish blood, regulate blood, and consolidate Chong and Ren channels.

This is a fundamental formula for building blood. It is also a basic formula to regulate menstruation. One can modify this formula according to the syndrome and manifestations to treat any type of blood deficiency syndromes and menstrual disorders.

Modification:

Addition	Indication
Pao Jiang Gui Zhi Xiang Fu Zhi Ke	Blood stasis with cold causing lower back and abdominal pain during menstruation
Huang Qin Mu Dan Pi	Blood deficiency with stagnant heat
Dang Shen Huang Qi Bai Zhu	Qi deficiency not controlling blood
E Jaio Ai Ye Zhi Gan Cao	Spotting during pregnancy
Dan Shen Tao Ren Hong Hua	Blood deficiency with blood stasis

Related Formulas:

Formula	Herbs	Indication
Sheng Yu Tang [Sage-Like Healing Decoction]	Core Formula + Ren Shen Huang Qi	Treat early menstruation, excess pale blood, and fatigue
Tao Hong Si Wu Tang [Four-Substance Decoction with Peach Pit and Safflower]	Core Formula + Tao Ren Hong Hua	Treat irregular menstruation, excess menstrual flow with clots, purple and sticky blood, abdominal pain, and distention It can also treat hyphema and sudden loss of vision in eye disorders
Gui Fu Si Wu Tang [Four-Substance Decoction with Cinnamon and Prepared Aconite]	Core Formula + Rou Gui Fu Zi	Treat delayed menstruation with pale menstrual blood
Guo Qi Yin [Delayed Menstruation Decoction]	Core Formula + Tao Ren Hong Hua Xiang Fu E Zhu Mu Tong Rou Gui Gan Cao	Treat blood stasis type of delayed and scanty menstruation. The manifestations are scanty menstrual flow, dark purple in color with clots, and abdominal distending pain which refuses pressure, slightly alleviated with passing clots
Si Wu Tang [Four-Substance Decoction with Lindera Decoction]	Core Formula + Xiang FU Wu Yao	Treat Liver blood deficiency and Liver Qi stagnation type of irregular menstruation and dysmenorrhea
Di Gu Pi Yin [Lycium Root Bark Decoction]	Core Formula + Di Gu Pi Mu Dan Pi	Treat deficient heat type of early menstruation, steaming heat at bone, and tidal fever which is worse at night
Xiao Ying Jian [Minor Nourishing Decoction]	Core Formula + Shan Yao Gou Qi Zi Gan Cao – Chuan Xiong	Treat Yin deficiency, blood exhaustion, and scanty watery menstruation

(continued)

Formula	Herbs	Indication
Qin Lian Si Wu Tang [Four-Substance Decoction with Scutellaria and Coptis]	Core Formula + Huang Qin Huang Lian	Treat excess heat syndrome of early menstruation. Excess menstrual flow with dark red blood or sticky red leukorrhea, intermittent tidal fever, palpitation, insomnia, dry mouth, and irritability
San Huang Si Wu Tang [Three-Yellow and Four-Substance Decoction]	Core Formula + Huang Qin Huang Lian Da Huang	Treat excess heat type of vomiting and epistaxis before menstruation
Yu Zhu San [Jade Candle Decoction]	Core Formula + Da Huang Ma Xiao Gan Cao	Treat excess heat type of blood stasis amenorrhea
Ai Fu Nuan Gong Wan [Mugwort and Cyperi Rhizome Pill for Warming the Womb]	Core Formula + Ai Ye Xiang Fu Wu Zhu Yu Rou Gui Xu Duan Huang Qi	Treat lower abdominal coldness and pain at 3–4 months of pregnancy, restless fetus, and pale tongue with white coating
Dang Gui San [Tangkuei Powder]	Core Formula – Shu Di Huang + Huang Qin Bai Zhu	This formula can calm and protect the fetus, and it should be taken frequently during pregnancy
Wu Fu Yin [Five-Fortune Decoction]	Ren Shen Shu Di Huang Dang Gui Chen Pi Pao Jiang	Decoction Tonify Qi, nourish blood, warm Spleen, and expel evil pathogens Treat taxation malaria due to Qi and Blood deficiency
Jia Wei Si Wu Er Chen Tang [Augmented Four-Substance and Two-Cured Decoction]	Core Formula + Chen Pi Ban Xia Fu Ling Hong Teng Mu Dan Pi Xiang Fu Hai Zao	Treat knotted damp, turbid, and blood causing amenorrhea. The stasis obstructs the uterus. Lower abdomen is distended, hard and painful
He Ren Yin [Polygonum Multiflorum Root and Ginseng Decoction] *Jing Yue Quan Shu*	He Shou Wu Ren Shen Dang Gui Chen Pi Sheng Jiang	Tonify Qi, nourish blood, regulate Liver, and expel evil pathogens Treat taxation malaria due to Qi and blood deficiency. It can also treat anemia due to Spleen and Liver deficiency

Dang Gui Sheng Jiang Yang Rou Tang [Mutton Stew with Tangkuei and Fresh Ginger] *Jin Gui Yao Lue*

Dang Gui	60 g
Sheng Jiang	120 g
Yang Rou (lamb)	250 g

Main Indication: Postpartum blood deficiency with cold. Symptoms include postpartum hollow pain at the abdomen, pale tongue with white coating, wiry or thin pulse.

Preparation: Decoction. Taken as a meal.

Treatment Principles: Warm the channel, nourish blood, scatter cold, and stop pain.

This formula tonifies but does not cause stagnation. It is warm but not drying. Furthermore, it is taken as food which is convenient during long-term treatment. This is an example of food therapy in Chinese medicine. It is also effective in treating dysmenorrhea due to deficient cold.

This formula can also treat flank pain due to Liver blood deficiency or hernial pain due to cold stagnation in the Liver channel.

Related Formulas:

Formula	Herbs	Indication
Qian Jin Dang Gui Tang [Tangkuei Decoction from Thousand Ducat Formula]	Core Formula + Chi Shao Yao 30–60 g	Treat cold type hernia in female and postpartum spasmodic abdomen pain
Ji Sheng Dang Gui Yang Rou Tang [Tangkuei and Mutton Decoction from the Formulas that Aid the Living]	Core Formula + Ren Shen Huang Qi	Treat postpartum fever, spontaneous sweating, and body and extremity pain. If there is only postpartum fever and spontaneous sweating without body and extremity pain, then one should use Dang Gui Bu Xue Tang [Tangkuei Decoction to Nourish Blood] and Sheng Mai San [Generate the Pulse Powder]

Comparison of Formulas Treating Liver Blood Deficiency

Si Wu Tang [Four-Substance Decoction] is a fundamental formula to nourish Liver blood and consolidate Chong and Ren channels. Because the formula is small and nourishes and invigorates blood at the same time, it can be taken for a long time. It can also be modified with other herbs to treat presenting syndromes. E Jiao and Ai Ye in the Jiao Ai Tang [Ass-Hide Gelatin and Mugwort Decoction] warm the uterus and nourish Chong and Ren channels. Thus Jiao Ai Tang [Ass-Hide Gelatin and Mugwort Decoction] is stronger than Si Wu Tang [Four-Substance Decoction] in terms of its functions of blood nourishing and consolidating Chong and Ren channels. Furthermore, this formula also contains Shao Yao Gan Cao Tang [Peony and Licorice Decoction] which moderates and stops pain. Thus Jiao Ai Tang [Ass-Hide Gelatin and Mugwort Decoction] is even more appropriate in treating gynecological complaints due to blood deficiency and damaged Chong and Ren channels with presence of pain. Dang Gui Sheng Jiang Yang Rou Tang [Mutton Stew with Tangkuei and Fresh Ginger] is a basic food recipe for treating deficient cold type of lower abdominal pain. Dull pain of the lower abdomen with a cold and empty sensation will be relieved from this formula.

LIVER BLOOD STASIS

Liver stores blood and also regulates the volume of blood in the body. However, the transportation, generation, and distribution of blood need to rely on the participation of Heart, Spleen, Lung, and other Zang organs. The manifestations of blood stasis often include irregular menstruation, dysmenorrhea, amenorrhea, abdominal mass, profuse lochia, and retention of lochia. It can also be seen in traumatic injuries with bruises, purple spots, stagnant nodules, fixed pain, purple dusky tongue or with purple spots on tongue, and choppy or wiry pulse. Herbs such as Tao Ren, Hong Hua, Dang Gui, Chuan Xiong, Chi Shao Yao, Mu Dan Pi, Ru Xiang, Mo Yao, Ze Lan, Niu Xi, Yan Hu Suo, Su Mu, Dan Shen, Zhe Chong, and Gan Xi which can invigorate the blood and dispel stasis are used.

The cold type of blood stasis is often seen with delayed menstruation, dark menstrual flow with clots, cold pain at the lower abdomen, aversion to cold and preference for heat, and slow and choppy pulse. Herbs such as Gui Zhi, Gan Jiang, and Wu Zhu Yu are often combined with blood moving and stasis dispelling herbs.

Heat type of blood often seen with early menstruation, bright excess menstrual flow or unceasing bleeding, a preference for cold and aversion to heat, and rapid or slippery pulse. Da Huang, Mu Dan Pi, Shui Zhi, and Mang Chong are often added.

Blood stasis in the collaterals is often seen in chronic conditions such as Chest Bi with obstruction of the collaterals, stifling sensation, distention, and pain at the chest area. Then, herbs that can unblock the collaterals and dispel Bi pain, such as Xie Bai, Bai Jiu (wine), Gui Zhi, Gua Lou, Xi Xin, and Fu Zi, are added to blood moving herbs. If blood stasis obstructs the sea of marrow causing hemiplegia and mouth and eye deviations, then acrid aromatic herbs such as She Xiang, Di Long, Chuan Shan Jia, and Bing Pian should be used with blood moving herbs.

For traumatic injuries which damage blood vessels and channels leading to unbearable pain, reducing-swelling and stopping pain herbs such as Ru Xiang, Mo Yao, Tao Ren, Hong Hua, Su Mu, Tu Bie Chong, Ma Qian Zi, Chuan Wu, and Cao Wu should be added. They should be combined with moving unblocking channel herbs such as She Xiang and Bing Pian. For severe traumatic injury with broken tendons and bone fractures, one should also combine Xu Duan, Zi Ran Tong, Du Zhong, and bobcat bone.

Moving blood and dispelling stasis can also be used in bleeding syndromes caused by blood stasis. In these syndromes, if the stasis is not cleared, then bleeding cannot be stopped. For example, in some postpartum unceasing bleeding, unceasing bleeding after abortion, or some bleeding due to traumatic injuries, moving blood and stopping bleeding are the main treatment principles. Formulas, such as Shi Xiao San [Sudden Smile Powder], Qi Li San [Seven-Thousandths of a Tael Powder], and Yun Nan Bai Yao [White Powder of Yun Nan] represent these treatment principles.

Furthermore, blood following Qi movement also becomes stasis with Qi stagnation which is most often seen with Liver Qi stagnation. Thus, formulas of blood moving and stasis dispelling often have one or two herbs that can disperse Liver and regulate Qi flow. Even though these are not main herbs in the formula, they cannot be omitted.

In ancient times, blood moving formulas were often decocted with alcohol, taken with alcohol, or soaked with alcohol; the purpose being to increase blood circulation and strengthen the blood moving function.

Xue Fu Zhu Yu Tang [Drive Out Stasis in the Mansion of Blood Decoction] *Yi Lin Gai Cuo*

Tao Ren	12 g
Hong Hua	9 g
Dang Gui	9 g
Sheng Di Huang	9 g
Chuan xiong	6 g
Chi Shao Yao	6 g
Chuan Niu Xi	9 g
Jie Geng	6 g
Chai Hu	3 g
Zhi Ke	6 g
Gan Cao	3 g

Main Indication: Blood stasis obstruction in the chest. Symptoms include chest and hypochondriac pain, chronic stifling sensation and tightness at the chest area, sharp and fixed pain, or insomnia with excessive dreaming, palpitation or with fright, irritability, easily angered, or tidal fever in the evening, dark red tongue with purple spots at the tongue or the sides, dark lips or dark circles around eyes, and choppy or tight, wiry pulse.

Preparation: Decoction.

Treatment Principles: Invigorate blood, expel stasis, move Qi, and stop pain.

This formula is a combination of Tao Hong Si Wu Tang [Four-Substance Decoction with Peach Pit and Safflower], Si Ni San [Frigid Extremities Powder], Jie Geng, and Chuan Niu Xi. It invigorates blood without damaging it, and relaxes Liver but does not exhaust Qi. With modification, it is often used to treat angina pectoris in coronary heart disease, rheumatic Heart disease, chest and hypochondriac pain, and intercostals neuralgia due to bruise or costochondritis, headache, forgetfulness, insomnia, depression, and retinal bleeding in postconcussion syndrome. This formula can be used in treating any type of blood stasis with internal obstruction.

Related Formula:

Formula	Herbs	Indication
Guan Xin Er Hao Fang [No. 2 Formula for Coronary Artery Disease]	Chuang Xiong 15 g Chi Shao Yao 15 g Hong Hua 15 g Jiang Xiang 15 g Dan Shen 30 g	Treat blood stasis type of angina pectoris in coronary heart disease

Ge Xia Zhu Yu Tang [Drive Out Blood Stasis Below the Diaphragm Decoction] *Yi Lin Gai Cuo*

Tao Ren	9 g
Hong Hua	6 g
Wu Ling Zhi	9 g
Dang Gui	9 g
Chuan Xiong	6 g
Mu Dan Pi	6 g
Wu Yao	6 g
Yan Hu Suo	6 g
Xiang Fu	6 g
Zhi Ke	6 g
Gan Cao	9 g
Chi Shao Yao	9 g

Main Indication: Blood stasis and obstruction below the diaphragm. Symptoms include epigastric and hypochondriac pain, or mass or nodules below the diaphragm, or abdominal pain with fixed location, or sinking sensation while lying down as if something were in the abdomen, dark red tongue with purple spots, and wiry, tight or choppy pulse.

Preparation: Decoction.

Treatment Principles: Invigorate blood, move Qi, and stop pain.

Sheng Hua Tong Jing Tang [Generation and Transformation Decoction to Unblock the Menses] *Zhong Yi Fu Ke Zhi Liao Xue*

Dan Shen (prepared with wine)	12 g
Chuan Niu Xi	9 g
Tao Ren	6 g
Hong Hua	3 g
Ze Lan	12 g
Dang Gui Wei	9 g
Xiang Fu	9 g

Main Indication: Irregular menstruation due to blood stasis. Symptoms include early or delayed menstruation, sometimes excessive menstrual flow, sometimes scanty, purple color with clots, intermittent lower abdominal pain which dislikes pressure or palpation, dry mouth with no desire to drink water, dusky complexion, dark red tongue or with purple spots, and deep wiry forceful pulse.

Preparation: Decoction.

Treatment Principles: Invigorate blood, unblock menstruation.

This formula also treats amenorrhea due to blood stasis.

Shi Xiao San [Sudden Smile Powder] *He Ji Ju Fang*

Wu Ling Zhi	equal amount
Pu Huang	

Main Indication: Postpartum blood stasis with internal obstruction. Symptoms include retention of lochia, lower abdominal pain which refuses pressure, or profuse lochia, postpartum blood dizziness which refers to dizziness, vertigo, inability to sit, stifling sensation and fullness at the epigastrium; nausea, vomiting, or phlegm congestion, rapid breathing, or coma and lockjaw in severe cases. The tongue is dark, and pulse is wiry, choppy.

Preparation: Grind into fine powder, take 9 g each time, two to three times a day.

Treatment Principles: Invigorate blood and expel stasis.

Postpartum abdominal pain is the main indication for using this formula with accompanying blood stasis and bleeding symptoms. Thus, Wu Ling Zhi is used to scatter stasis, open the channel, and stop pain. Pu Huang is used to invigorate blood and stop bleeding.

This formula can also treat acute lower abdominal pain right before the onset of menstruation due to blood stasis and internal obstruction.

Related Formula:

Formula	Herbs	Indication
Jia Wei Shi Xiao San [Augmented Sudden Smile Powder]	Core Formula + Yan Hu Suo Mu Dan Pi Tao Ren Xiang Fu Wu Yao	Treat dysmenorrhea due to Qi stagnation and blood stasis

Jia Jian Huo Luo Xiao Ling Dan [Modified Fantastically Effective Special Pill to Invigorate the Collaterals] (Jing Yan Fang)

Dan Shen	15 g
Chi Shao Yao	9 g
Ru Xiang	6 g
Mo Yao	9 g
Tao Ren	9 g

Main Indication: Blood stasis and accumulation type of ectopic pregnancy. Symptoms include sudden one-sided lower abdominal pain as if cut by knife, vaginal bleeding which begins with scanty purple dark blood, followed by excessive bleeding, purple red tongue, and wiry slippery pulse.

Preparation: Decoction.

Treatment Principles: Warm the channel and expel stasis.

This formula is indicated with the presence of lower abdominal spasm pain, vaginal bleeding, and mass with palpation upon gynecological exam. The treatment principles are moving blood and expelling stasis to remove the cause of bleeding.

Modification:

Addition	Indication
Du Shen Tang [Single Ginseng Decoction] or Shen Fu Tang [Ginseng and Prepared Aconite Decoction]	Excessive bleeding
San Leng E Zhu	Miscarriage with mass and swelling

Related Formulas:

Formula	Herbs	Indication
Huo Luo Xiao Ling Dan [Fantasticaly Effective Special Pill to Invigorate the Collaterals]	Dang Gui 15 g Dan Shen 15 g Ru Xiang 15 g Mo Yao 15 g	Treat any type of mass, nodules, or accumulations
Gong Wai Yun Yi Hao Fang [No. 1 Extrauterine Pregnancy Formula]	Dan Shen 15 g Chi Shao Yao 15 g Tao Ren 9 g	Treat eruption of extrauterine pregnancy However, if the patient is in shock due to excessive bleeding, she should be sent to the emergency room
Gong Wai Yun Er Hao Fang [No. 2 Extrauterine Pregnancy Formula]	Above Formula + San Leng 6 g E Zhu 6 g	

Wen Jing Tang [Warm the Menses Decoction] *Jin Gui Yao Lue*

Wu Zhu Yu	9 g
Gui Zhi	9 g
Dang Gui	12 g
Bai Shao Yao	12 g
Chuan Xiong	6 g
Ren Shen	6 g
Gan Cao	6 g
E Jiao	9 g
Mai Men Dong	9 g
Ban Xia	9 g
Chuan Xiong	6 g
Sheng Jiang	9 g

Main Indication: Deficient cold of Chong and Ren channels, blood stasis, and internal obstruction. Symptoms of irregular menstruation either early or delayed, or prolonged menstruation, spotting, lower abdominal cold pain, dry lips, dry mouth, five center heat, irritability, tidal fever, or infertility, dark red tongue, white tongue coating, and thin deep pulse.

Preparation: Decoction.

Treatment Principles: Warm the menses, scatter cold, expel stasis, and nourish blood.

This formula has all cold, heat, reducing, and tonification methods. However, the focus is warming and nourishing Chong and Ren channels. It is a common gynecological formula for regulating menstruation. It is mainly used for irregular menstruation, dysmenorrhea, Beng Lou, and infertility due to deficient cold Chong and Ren channels with blood stasis.

Related Formulas:

Formula	Herbs	Indication
Ju Fang Wen Jing Tang [Warm the Menses Decoction from Imperial Grace Formula]	Gui Zhi Dang Gui Chuan Xiong Bai Shao Yao Mu Dan Pi Ren Shen Gan Cao E Zhu Niu Xi	Treat the same indications as the core formula, but weaker in warming and nourishing function and stronger in moving blood and expelling stasis. Thus, it is suitable for severe cases of stasis and stagnation
Nuan Gong Wan [Mugwort and Cyperi Rhizome Pill for Warming the Womb]	Ai Ye Xiang Fu Wu Zhu Yu Chuan Xiong Bai Shao Yao Huang Qi Xu Duan Sheng Di Huang Rou Gui Dang Gui	Grind into powder, mix with rice vinegar to form pills. Take 6 g with mild vinegar water each time, two to three times a day Treat deficient cold uterus, white vaginal discharge, sallow complexion, extremity pain, fatigue, poor appetite, irregular menstruation, intermittent abdominal pain, infertility. This formula is stronger in its warming the menstruation and nourishing blood function than Wen Jing Tang [Warm the Menses Decoction]. However, it is weaker in its expelling stasis function. It is often used in deficient cold with blood deficiency of the uterus

Shen Tong Zhu Yu Tang [Drive Out Blood Stasis from a Painful Body Decoction] *Yi Lin Gai Cuo*

Qin Jiao	3 g
Chuan Xiong	6 g
Tao Ren	9 g
Hong Hua	9 g
Gan Cao	6 g
Qiang Huo	3 g
Mo Yao	6 g
Dang Gui	9 g
Wu Ling Zhi	6 g
Xiang Fu	3 g
Chuan Niu Xi	9 g
Di Long	6 g

Main Indication: Stasis and blockage of the channels and collaterals. Symptoms include chronic shoulder, arm, lower back, leg or whole body pain, dark tongue, and wiry or thin pulse.

Preparation: Decoction.

Treatment Principles: Expel stasis, unblock the collaterals, unblock painful obstruction, and stop pain.

This formula treats severe and difficult to treat problems of the entire body, extremities, and joint pain due to chronic blood stasis entering collaterals.

Gui Zhi Fu Ling Wan [Cinnamon Twigs and Poria Pill] *Jin Gui Yao Lue*

Gui Zhi	equal amount
Fu Ling	
Shao Yao	
Mu Dan Pi	
Tao Ren	

Main Indication: Blood stasis in the uterus. Symptoms include female lower abdominal mass, abdominal pain which refuses pressure or palpation, or amenorrhea with abdominal distention and pain, or postpartum profuse lochia, or unceasing menstrual flow with dark purple menstrual blood, dark red tongue or tongue with purple spots, and slippery or choppy pulse.

Preparation: Form into honey pills. Take 10 g each time. It can also be taken as decoction, in which case the dosage for each herb is 9 g.

Treatment Principles: Invigorate blood, expel stasis, and reduce mass and accumulation.

This formula is a mild formula for moving blood and expelling stasis, and needs to be taken for a long time to be effective. Clinically, it is used to treat uterine myoma, uterine polyps, ovarian cysts, chronic salpingitis, and chronic pelvic inflammatory disease with the manifestation of abdominal distention and pain, or with a moderate size of mass or one that is hard to palpate.

In *Ji Yin Gang Mu*, this formula is prepared as decoction and is called Cui Sheng Tang [Induce Labor Decoction]. It is used during prelabor period when the water breaks with abdominal and lower back pain. It has the function of inducing labor.

Modification:

Addition	Indication
Da Huang	Severe blood stasis

Shao Fu Zhu Yu Tang [Drive Out Blood Stasis in the Lower Abdomen Decoction] *Yi Lin Gai Cuo*

Xiao Hui Xiang (fried)	9 g
Gan Jiang (fried)	9 g
Rou Gui	6 g
Dang Gui	9 g
Chuan Xiong	6 g
Chi Shao Yao	9 g
Wu Ling Zhi (fried)	9 g
Pu Huang	9 g
Mo Yao	6 g
Yan Hu Suo	6 g

Main Indication: Cold in the Liver, Qi stagnation, and blood stasis. Symptoms are mass in the lower abdomen, presence or absence of pain, or pain without mass, or abdominal distention and fullness, or lower back soreness and lower abdominal distention during menstruation, or three to five times of menstrual flow in one month with purple or black menstrual blood, or with clots, or Beng Lou with lower abdominal pain, dark red tongue or with purple spots at the sides of the tongue, and choppy or wiry pulse.

Preparation: Decoction.

Treatment Principles: Warm the menses and expel stasis.

This formula is effective in treating lower abdominal mass, irregular menstruation, and dysmenorrhea due to cold congelation and blood stasis. Clinically, it is used to treat uterine myoma, polyps, or endometriosis.

Modification:

Minus	Addition	Indication
	Dang Shen E Jiao	Qi and blood deficiency
Rou Gui Gan Jiang	Sheng Jiang Huang Qin (fried)	Yellow tongue coating, dry mouth, constipation
	Chai Hu Yu Jin Qing Pi	Severe Liver Qi stagnation
	Fu Ling Ze Xie	Excessive vaginal discharge

Related Formulas:

Formula	Herbs	Indication
Qian Jin Dang Bao Tang [Cleanse the Uterus Decoction from Thousand Ducat Formula]	Mang Xiao 9 g Mu Dan Pi 9 g Dang Gui 9 g Da huang 9 g Tao Ren 9 g Hou Po 6 g Jie Geng 6 g Ren Shen 6 g Chi Shao Yao 6 g Fu Ling 6 g Gui Xin 6 g Gan Cao 6 g Chuan Niu Xi 6 g Chen Pi 6 g Fu Zi 18 g Meng Chong 10 pieces Shui Zhi 10 pieces	Decoct with water and alcohol, taken in four doses, three during the day and one at night. Cover with blanket after taking the decoction to induce mild sweating Treat blood stasis obstructing uterus causing infertility

Tuo Hua Jian [Release from Flower Decoction] *Jing Yue Quan Shu*

Dang Gui	3 g
Rou Gui	6 g
Chuan Xiong	6 g
Chuan Niu Xi	6 g
Hong Hua	3 g
Che Qian Zi	9 g

Main Indication: Difficult labor due to cold congelation and blood stasis. Symptoms include long time in labor due to retention of fetus, purple and blue green complexion of the mother, distending, severe pain at lower back and abdomen, pale tongue, and deep tight pulse.

Preparation: Take the decoction warm. If one can drink alcohol, drink one glass of alcohol after taking the decoction for increased effectiveness.

Treatment Principles: Warm the channels, invigorate blood, and purge the fetus.

This formula also has a function in abortion. In severe cases of long time difficult labor, one should be sent to the hospital.

Related Formulas:

Formula	Herbs	Indication
Jia Wei Cui Sheng Xiong Gui Tang [Augmented Inducing Labor Decoction with Ligusticum and Tangkuei]	Dang Gui 6 g Chuan Xiong 9 g Yi Mu Cao 30 g	Treat difficult labor without coldness; dark red tongue and deep excess pulse
Cui Sheng Yin [Induce Labor Decoction]	Dang Gui equal amount Chuan Xiong Da Fu Pi Zhi Ke Bai Zhi	Treat difficult labor with Qi stagnation, depression, thin white greasy tongue coating, and deep, wiry pulse
Tong Yu Jian [Unblock the Stasis Decoction]	Dang Gui Wei 12 g Shan Zha 6 g Xiang Fu 6 g Hong Hua 6 g Wu Yao 5 g Qing Pi 5 g Mu Xiang 2 g Ze Xie 5 g	
Wu Tou Tang [Aconite Decoction]	Chuan Wu 6 g Ma Huang 6 g Bai Shao Yao 9 g Huang Qi 9 g Zhi Gan Cao 9 g Feng Mi (Honey) 60 g	Warm the meridian, scatter cold, alleviate spasms, and stop pain Treat wind–cold obstructing meridians leading to pain, numbness, tingling, and painful obstruction

Sheng Hua Tang [Generation and Transformation Decoction] *Fu Qing Zhu Nu Ke*

Quan Dang Gui	25 g
Chuan Xiong	9 g
Tao Ren	9 g
Pao Jiang	2 g
Zhi Gan Cao	2 g

Main Indication: Postpartum blood deficiency and coldness, blood stasis, and obstruction. Symptoms include retention of lochia or hesitant flow of lochia, dark color, lower abdominal cold pain, pale or purple tongue, and wiry or thin pulse.

Preparation: Decoction or decoct with moderate amount of yellow wine.

Treatment Principles: Invigorate blood, transform stasis, warm channels, and stop pain.

This formula is a common postpartum formula, especially effective in treating postpartum cold congelation and blood stasis with lower abdominal pain.

Modification:

Addition	Indication
Rou Gui	Eating cold foods during postpartum, causing severe abdominal pain or mass in severe cases

Tao He Cheng Qi Tang [Peach Pit Decoction to Order the Qi] *Shang Han Lun*

Tao Ren	12 g
Da Huang	15 g
Mang Xiao (add near end)	6 g
Gan Cao	6 g
Gui Zhi	6 g

Main Indication: Accumulation of blood in the Lower Jiao. Symptoms include acute accumulation at the Lower Jiao, black stool, urinary incontinence, delirious speech, irritability, thirst, fever at night, or mania in severe cases. Other symptoms include retention of lochia, hardness and pain in the lower abdomen, or amenorrhea due to blood stasis, dysmenorrhea, or internal accumulation of stagnant blood due to traumatic injury and pain with inability to turn the body; purple red tongue, and slippery and excess pulse.

Preparation: Decoction.

Treatment Principles: Drain heat, crack blood, and expel stasis.

This formula treats syndrome of knotted blood stasis and heat, and blood accumulation in the Lower Jiao. This formula is a combination of Tiao Wei Cheng Qi Tang [Regulate the Stomach and Order the Qi Decoction] with Tao Ren and Gui Zhi. Thus it can drain heat, break accumulation, and guide stagnant heat out through bowel movement.

Related Formula:

Formula	Herbs	Indication
Xia Yu Xue Tang [Purge the Stagnant Blood Decoction]	Da Huang 9 g Tao Ren 9 g Zhe Chong 9 g	Amenorrhea due to blood stasis or postpartum abdominal pain due to internal obstruction of stagnant blood

Fu Yuan Huo Xue Tang [Revive Health by Invigorating the Blood Decoction] *Yi Xue Fa Ming*

Chai Hu	15 g
Tian Hua Fen	12 g
Dang Gui	9 g
Hong Hua	6 g
Gan Cao	6 g
Pao Chuan Shan Jia	9 g
Da Huang (soaked in alcohol)	30 g
Tao Ren	9 g

Main Indication: Blood stasis obstructing collaterals. Symptoms include blood stasis obstruction at hypochondrium after traumatic injury, severe unbearable pain, purple spots on tongue, and wiry or choppy pulse.

Preparation: Decoction.

Treatment Principles: This formula treats rib or flank pain due to internal obstruction of blood stasis or chronic rib and flank pain. It can also be used in pain due to internal obstruction of blood stasis at the channels caused by any type of traumatic injury or soft tissue injury. It can also be combined with Xian Fang Huo Ming Yin [Sublime Formula for Sustaining Life] to treat Liver abscess.

Bie Jia Jian Wan [Decocted Turtle Shell Pill] *Jin Gui Yao Lue*

Bie Jia (zhi)	90 g
She Gan (zhi)	22.5 g
Huang Qin	22.5 g
Di Shi (zhi)	22.5 g
Gan Jiang	22.5 g
Da Huang	22.5 g
Gui Zhi	22.5 g
Shi Wei (remove hair)	22.5 g
Hou Po	22.5 g
Qu Mai	22.5 g
Zi Wei	22.5 g
E Jiao	22.5 g
Ren Shen	7.5 g
Ban Xia	7.5 g
Ting Li Zi	7.5 g
Qiang Lang	45 g
Chai Hu	45 g
Chi Shao Yao	37 g
Mu Dan Pi	37 g
Zhe Chong	37 g
Lu Feng Fang	30 g
Chi Xiao	90 g
Tao Ren	15 g

Main Indication: Malaria with splenomegaly. Symptoms include chronic malarial symptoms, focal distention, hardness and unmovable mass below the ribs (enlargement and cirrhosis of Liver and Spleen), abdominal pain, emaciation, decreased food intake, intermittent chills and fever, amenorrhea in female, dark purple tongue with purple stagnant spots, and deep, thin choppy pulse.

Preparation: Soak 1500 g of burned ash in 5000 g of yellow alcohol. Take the filtered solution to decoct Bie Jia until it becomes gelatinous. Grind the rest of the 22 herbs into a fine powder and mix with Bie Jia decoction and honey to form into small pills. Take 3 to 6 g each time, three times a day.

Treatment Principles: Move Qi, invigorate blood, reduce mass, and resolve accumulation.

The reducing mass and resolving accumulation function of this formula is accomplished through four aspects: moving Qi, moving blood, expelling stasis, and promoting urination. It can be used in mild enlargement of Liver and Spleen, Liver cirrhosis, and ascites which is the main indication of this formula.

Da Huang Zhe Chong Wan [Rhubarb and Eupolyphaga Pill] *Jin Gui Yao Lue*

Da Huang (steamed)	300 g
Huang Qin	60 g
Gan Cao	90 g
Tao Ren	60 g
Xing Ren	60 g
Bai Shao Yao	120 g
Sheng Di Huang	300 g
Gan Qi	30 g
Mang Chong	60 g
Shui Zhi	60 g
Qi Cao	60 g
Zhe Chong	30 g

Main Indication: Blood stasis forming mass. Symptoms include internal accumulation of stagnant blood, dry blood leading to consumption, emaciation and weakness, abdominal fullness with inability to eat, scaly skin, dark eyes, red purple tongue with purple spots at the sides, and wiry or choppy pulse, especially at left guan position.

Preparation: Grind into fine powder, mix with honey to form pills of 3 g each. Take one pill three times a day. It can also be taken as decoction with one tenth of the above dosage.

Treatment Principles: Break stasis and reduce mass.

This formula treats internal accumulation of stagnant blood syndrome due to consumptive disease which damages Ying and Wei channels and collaterals. Thus, the formula utilizes blood cracking, expelling stasis, and tonification methods at the same time. Pill form can slowly scatter and slowly reduce the accumulation. It can be used in Liver cirrhosis or fibrosis of Liver due to overconsumption of alcohol.

Bu Yang Huan Wu Tang [Tonify the Yang to Restore Five-Tenths Decoction] *Yi Lin Gai Cuo*

Huang Qi	120 g
Dang Gui Wei	9 g
Chi Shao Yao	9 g
Di Long	6 g
Chuan Xiong	6 g
Tao Ren	6 g
Hong Hua	6 g

Main Indication: Sequelae of stroke syndrome. Symptoms include hemiplegia, mouth and eye deviation, difficult and slurred speech, drooling, dry stool, frequent urination or urinary incontinence, white greasy tongue coating or same symptoms with only one half of the tongue white and greasy and shiny on the other half, deviated tongue, and moderate pulse.

Preparation: Decoction.

Treatment Principles: Tonify Qi, invigorate blood, and unblock collaterals.

This formula treats stasis syndromes due to deficiency which is caused by deficient Zheng Qi, obstruction of stagnant blood in channels and collaterals, and loss of nourishment of the muscle, channels, and tendons. It should be taken for a long time to be effective.

Modification:

Addition	Indication
Shi Chang Pu Yuan Zhi	Severe difficulty and slurred speech
Bai Fu Zi Jiang Can Quan Xie	Severe mouth and eye deviation
Ma Zi Ren Xing Ren Jiu Da Huang	Severe constipation

(continued)

Addition	Indication
Shu Di Huang Shan Zhu Yu Wu Wei Zi	Urinary incontinence
Di Long (double the dosage) Chuan Niu Xi 30 g Bai Shao Yao 60 g	Wiry, slippery, forceful pulse which indicates excessive evil pathogens
Mang Chong Shui Zhi	Chronic hemiplegia without significant effect
Huai Niu Xi Du Zhong Sang Ji Sheng	Weakness and atrophy of the lower limbs

Xiao Huo Luo Dan [Minor Invigorate the Collaterals Special Pill] *Sheng Ji Zong Lu* also called Huo Luo Dan [Invigorate the Collaterals Special Pill]

Zhi Chuan Wu	180 g
Zhi Cao Wu	180 g
Zhi Nan Xing	180 g
Di Long	180 g
Ru Xiang	60 g
Mo Yao	60 g

Main Indication: 1. Sequelae of stroke. Stagnation of damp–phlegm and dead blood in the channel and collaterals. Symptoms of chronic numbness of hands and feet, heavy sensation, dullness or pain in lower back and leg, tongue deviation, and uneven pulse of the left from right. 2. Wind–cold–damp evil obstructing the channels and collaterals leads to tightness and contraction of tendon and channels, stiffness of the joints, pain at extremities, fixed or wandering pain, pale or dark red tongue, and wiry or choppy pulse.

Preparation: Grind into fine powder and mix with honey to form pills of 3 g each. Take one, twice a day with water or wine.

Treatment Principles: Invigorate blood, unblock collaterals, expel wind, eliminate dampness, and stop pain.

This formula treats wind–cold–damp or phlegm–damp and blood obstructing channels and collaterals. Qi and blood flow are obstructed leading to contracting pain, stiffness of the joints, and other symptoms. Recently, this formula has been used in treating postcerebral vascular accidents, and pain and numbness due to wind–damp or rheumatoid arthritis. It is effective when taken for a longer period. However, this formula is only appropriate for persons with strong constitutions; Yin-deficient persons and pregnant women should use with caution.

Related Formula:

Formula	Herbs	Indication
Da Huo Luo Dan [Major Invigorate the Collaterals Special Pill]	Bai Hua She 60 g Wu Shao She 60 g Wei Ling Xian 60 g Liang Tou Jian 60 g Cao Wu 60 g Tian Ma 60 g Quan Xie 60 g He Shou Wu 60 g Gui Ban 60 g Ma Huang 60 g Guan Zhong 60 g Zhi Gan Cao 60 g Qiang Huo 60 g Rou Gui 60 g Huo Xiang 60 g Wu Yao 60 g Huang Lian 60 g Shu Di Huang 60 g Da Huang (steamed) 60 g Mu Xiang 60 g Chen Xiang 60 g Xi Xin 30 g Chi Shao Yao 30 g Ding Xiang 30 g Ru Xiang 30 g Jiang Can 30 g Zhi Nan Xing 30 g Qing Pi 30 g Gu Sui Bu 30 g Bai Dou Kou 30 g An Xi Xiang 30 g Zhi Fu Zi 30 g Huang Qin 30 g Fu Ling 30 g Zhi Xiang Fu 30 g Xuan Shen 30 g Bai Zhu 30 g Fang Feng 45 g Ge Gen 45 g Hu Gu (substitute with bobcat bone) 45 g Dang Gui 45 g Xue Jie 21 g Zhi Di Long 15 g Xi Jiao (with Shui Niu Jiao) 15 g She Xiang 15 g Song Zhi 15 g (pine seed) Niu Huang 4.5 g Bing Pian 4.5 g Ren Shen 90 g	Grind the above 49 herbs into fine powder and form into honey pill the size of a Longan pit. Take one pill, one to two times a day with alcohol Treat poststroke, paralysis, or hemiplegia, atrophy, painful obstruction, coma, or coldness of extremities due to phlegm obstruction, difficult-to-treat deep rooted boils, fistula, and traumatic injuries.

Tong Qiao Huo Xue Tang [Unblock the Orifices and Invigorate the Blood Decoction] *Yi Lin Gai Cuo*

Chi Shao Yao	9 g
Chuan Xiong	9 g
Tao Ren	9 g
Hong Hua	9 g
Hong Zao	9 g
Sheng Jiang	9 g
Lao Cong (old Cong Bai)	6 g
She Xiang (wrapped separately)	0.3 g

Main Indication: Blood stasis obstructing the collaterals of head and face. Symptoms include headache, dizziness, deafness, loss of hair, blocked nasal passages, inability to smell, purple green complexion, or rosacea, purple green tongue tip or with purple spots, weak pulse especially at cun position.

Preparation: Decoction with water and 250 g of yellow alcohol, then mix with She Xiang. Take one to two times per day.

Treatment Principles: Invigorate blood, unblock orifices.

This formula treats headache caused by blood stasis obstructing collaterals. Patient presents with history of traumatic injury and chronic, fixed sharp pain. Furthermore, this formula can also be used for sequelae of epilepsy, meningitis, encephalitis B, and postconcussion syndrome (forgetfulness, decreased concentration, headache, dizziness, and slow response).

Related Formula:

Formula	Herbs	Indication
Tong Jing Zhu Yu Tang [Drive Out Blood Stasis and Unblock the Channel Decoction]	Tao Ren 9 g Hong Hua 9 g Chi Shao Yao 9 g Zao Jiao Ci 9 g Di Long 9 g Chuan Shan Jia 6 g Chai Hu 12 g Lian Qiao 12 g She Xiang 0.3 g	This formula has a stronger function to invigorate blood and unblock collaterals than Tong Qiao Huo Xue Tang [Unblock Orifices and Invigorate Blood Decoction] which is suitable for severe cases of blood stasis and stagnation

Jia Jian Gua Lou Xie Bai Tang [Modified Trichosanthes Fruit and Chinese Chive Decoction] (Zhong Guo Zhong Yi Yan Jiu Yuan Fang)

Gua Lou Ke	24 g
Xie Bai	15 g
Gui Zhi	9 g
Yu Jin	9 g
Zhi Xiang Fu	12 g
Hong Hua	9 g
Tao Ren	9 g

Main Indication: Qi stagnation, blood stasis, phlegm obstruction, and blocked chest Yang. Symptoms include sharp pain at the chest area or under the sternum, stifling sensation and pain, palpitation and restlessness; in severe cases, greenish complexion, purple lips and nails, cold limbs, purple dark tongue, and thin or choppy pulse.

Preparation: Decoction with moderate amount of wine or water.

Treatment Principles: Unblock Yang, disperse Bi, transform stasis, and unblock collaterals.

This formula is modified from Gua Lou Xie Bai Bai Jiu Tang [Trichosanthes Fruit, Chinese Chive, and Wine Decoction] *Jin Gui Yao Lue*. The main indication is pain due to chest Bi syndrome or mild case of angina pectoris. However, this is a basic formula which needs to be modified according to patient's presentation of Qi stagnation, blood stasis, and phlegm obstruction.

Related Formulas:

Formula	Herbs	Indication
Gua Lou Xie Bai Bai Jiu Tang [Trichosanthes Fruit, Chinese Chive, and Wine Decoction]	Gua Lou Shi 24 g Xie Bai 30 g Bai Jiu moderate amount	Take the decoction warm. Treat chest Bi with inability to lie down, wheezing, coughing with copious sputum, chest and back pain, shortness of breath
Gua Lou Xie Bai Ban Xia Tang [Trichosanthes Fruit, Chinese Chive, and Pinellia Decoction]	Gua Lou Shi 24 g Xie Bai 20 g Ban Xia 15 g Bai Jiu moderate amount	Take the decoction warm. Treat chest Bi with severe phlegm obstruction
Zhi Shi Xie Bai Gui Zhi Tang [Immature Bitter Orange, Chinese Chive, and Cinnamon Twig Decoction]	Zhi Shi 12 g Hou Po 12 g Xie Bai 30 g Gui Zhi 10 g Gua Lou 24 g	Take the warm decoction Treat chest Bi due to Qi stagnation, blood stasis, and phlegm obstruction

Qi Li San [Seven-Thousandths of a Tael Powder] *Liang Fang Ji Ye*

Xue Jie	30 g
She Xiang	0.4 g
Bing Pian	0.4 g
Ru Xiang	5 g
Mo Yao	5 g
Zhu Sha	4 g
Er Cha	7.5 g

Main Indication: Blood stasis, swelling, and pain due to traumatic injury, broken tendon, bone fracture or bleeding, and pain due to cuts. It can also treat any type of unknown swelling, toxins, and burns.

Preparation: Grind into super fine powder and store in a sealed container. Take 0.22 to 2 g each time with warm water or wine. It can also be used externally by mixing with wine before placing on injury site.

Treatment Principles: Invigorate blood, scatter stasis, stop pain, reduce swelling.

This formula cannot be taken in a quantity more than seven-thousandths of a tael (one tael equals 38 g or 1.3 oz), thus it is called Qi Li San [Seven-Thousandths of a Tael Powder].

This is a famous formula in traumatology. It is especially effective if used internally and externally together for trauma, trip and fall, blood stasis, swelling, and pain. It is also effective for pain due to blood stasis, hematemesis, and blood in urine due to internal injuries.

Modification:

Addition	Indication
San Qi Tu Bie Chong Zi Ran Tong Zhi Ma Qian Zi	Severe swelling and pain

Related Formulas:

Formula	Herbs	Indication
Ba Li San [Eight-Thousandths of a Tael Powder]	Xue Jie 9 g Ru Xiang 9 g Mo Yao 9 g Ban Liang Qian 3 g Su Mu 3 g Zi Ran Tong (calcine and dip in vinegar seven times) 9 g Hong Hua 6 g Fan Mu Bie 3 g Ding Xiang 2 g She Xiang 0.3 g	Grind into fine powder. Take 0.24 to 2.2 g each time with warm wine or urine of a boy. This formula contains herbs to set the bone, Ban Liang Qian, Zi Ran Tong, and Fan Mu Bie which stops pain in comparison to Qi Li San [Seven-Thousandths of a Tael Powder]. It is better for severe pain due to broken tendon and bones
Duo Ming San [Seize the Life Powder]	Xue Jie 3 g Mo Yao 6 g	Grind into fine powder, mix with sugar, take in two doses with wine Treat postpartum retention or hesitant discharge of stagnant blood leading to vertigo and dizziness

Die Da Wan [Trauma Pill] *Quan Guo Zhong Cheng Yao Chu Fang Ji*

Dang Gui	30 g
Chuan Xiong	30 g
Xue Jie	30 g
Tu Bie Chong	30 g
Zi Ran Tong	30 g
Ru Xiang	90 g
Mo Yao	30 g
Ma Huang	60 g

Main Indication: Swelling, distending fixed pain after sprain, strain, and other traumatic injuries.

Preparation: Grind into powder, mix with honey, and form pills of 3 g each. Take one to two pills, one to two times a day.

Treatment Principles: Invigorate blood, stop pain.

Modification:

Addition	Indication
Ma Qian Zi She Xiang	Severe swelling and pain

Jie Gu Wan [Set the Bone Pill] (Cheng Du Ti Yu Da Xue Fang)

Ding Xiang	30 g
Mu Xiang	30 g
Xue Jie	30 g
Er Cha	30 g
Zhi Da Huang	30 g
Hong Hua	30 g
Dang Gui	60 g
Lian Zi (without the pit)	60 g
Fu Ling	60 g
Bai Shao Yao	60 g
Mu Dan Pi	15 g
Gan Cao	9 g
Zi Ran Tong	30 g
Tu Bie Chong	30 g

Main Indication: Injury of tendon and bone. Presence of pain, stasis, and swelling after bone and tendon have been corrected to the right place.

Preparation: Grind into fine powder and mix with honey to form pills. Take 2 to 3 g, two to three times a day.

Treatment Principles: Reduce swelling, stop pain, set the bone, and connect the tendon.

This formula can regulate Qi, invigorate blood, stop pain, and also tonify Qi and blood to speed the recovery time.

Related Formula:

Formula	Herbs	Indication
Zheng Gu Zi Jin Dan [Purple Gold Pill for Correcting the Bone]	Core formula – Zi Ran Tong Tu Bie Chong	Slightly weaker in setting bones and connecting tendons in comparison to the core formula

Huo Xue Jiu [Invigorate the Blood Wine] (Cheng Du Zhong Yi Yao Da Xue Fu Yuan Fang)

Ru Xiang	9 g
Mo Yao	9 g
Dang Gui	9 g
Chuan Xiong	9 g
Zi Jing Pi	9 g
Gui Zhi	9 g
Du Huo	9 g
Qiang Huo	9 g
Hu Gu (substitute with bobcat bone)	9 g
Mu Gua	9 g
Zhe Bei Mu	9 g
Zi Ran Tong	9 g
Xu Duan	9 g
Mu Xiang	9 g
Hou Po	9 g
Xiang Fu	9 g

(continued)

Xiao Hui Xiang (fried)	9 g
Xue Jie	6 g
Zhi Chuan Shan Jia	6 g
Bai Zhi	3 g
Zhi Chuan Wu	3 g
Zhi Cao Wu	3 g
She Xiang	1.5 g

Main Indication: Wind–cold–damp invasion after traumatic injury and sports injury. Symptoms include soreness and weakness of the whole body, joint pain, worse with rainy cold weather, and better with heat, purple tongue or with purple spots, and wiry or soggy pulse.

Preparation: Grind the above 23 herbs into powder.

1. Internal use: 15 g of powder soaked in 500 g of wine for 7 to 10 days. Take 2 to 3 g, twice a day.
2. External use: Mix with hot water into a paste for external use.

Treatment Principles: Invigorate blood, move Qi, dispel wind, expel dampness, invigorate collaterals, and stop pain.

This formula can also treat general type of wind–damp with generation of dampness and stasis in a chronic situation.

Shu Jin Huo Luo Tang [Relax the Tendon and Invigorate Collaterals Decoction] *Wai Ke Bu Yao*

Dang Gui	12 g
Niu Xi	12 g
Wu Jia Pi	12 g
Zhi Ke	12 g
Xu Duan	15 g
Du Zhong	15 g
Hong Hua	6 g
Qing Pi	9 g
Jing Jie	9 g
Fang Feng	9 g
Qiang Huo	9 g
Du Huo	9 g

Main Indication: Blood stasis and obstruction, adhesion of tendon and membrane after external or internal traumatic injuries. Locally, the symptoms include stiffness, spasm, and contraction of tendon and muscle, inability to extend or flex the joint, purple spots on tongue, and choppy and uneven pulse.

Preparation: Decoction.

Treatment Principles: Relax tendon and invigorate collaterals.

This formula can be used in the presence of stagnant blood or bruises for a long time after a traumatic injury. It can also treat chronic Bi leading to stasis or chronic diseases leading to stasis with Kidney deficiency and atrophy of the bones. Clinically, it is used to treat postoperational adhesions, spasms, and contractions after injury to tendons and broken bones, chronic rheumatic arthritis, joint deformation in rheumatoid arthritis, and osteoporosis.

Related Formula:

Formula	Herbs	Indication
Huo Xue Shu Jin Tang [Invigorate the Blood and Relax the Tendon Decoction]	Dang Gui Wei Xu Duan Chi Shao Yao Jiang Huang Shen Jin Cao Song Jie Gui Zhi Chuan Niu Xi Hai Tong Pi Fang Feng Qiang Huo Du Huo Gan Cao	Decoction. Treats the same syndrome as the core formula

Dan Shen Yin [Salvia Decoction] *Shi Fang Ge Kuo*

Dan Shen	30 g
Tan Xiang	5 g
Sha Ren	5 g

Main Indication: Chest, Heart, stomach pain, deep red tongue, and wiry pulse.

Preparation: Decoction.

Treatment Principles: Invigorate blood, move Qi, and stop pain.

This formula treats Heart pain, chest pain, and stomach pain due to Qi stagnation and blood stasis.

Comparison of Formulas Treating Liver Blood Stasis

Xue Fu Zhu Yu Tang [Drive Out Stasis in the Mansion of Blood Decoction] treats the syndrome of stasis and obstruction in the chest. Ge Xia Zhu Yu Tang [Drive Out Blood Stasis Below the Diaphragm Decoction] treats stasis and accumulation below the diaphragm with mass and distending pain at flank area and abdomen. Shao Fu Zhu Yu Tang [Drive Out Blood Stasis in the Lower Abdomen Decoction] treats blood stasis in the lower abdomen with focal distention, mass, and cold. It also treats irregular menstruation and dysmenorrhea. Shen Tong Zhu Yu Tang [Drive Out Blood Stasis from the Painful Body Decoction] often treats chronic stagnant blood obstructing channels and collaterals leading to pain in the body, extremities, and joints. Tong Qiao Huo Xue Tang [Unblock Orifices, Invigorate Blood Decoction] treats any syndromes due to stagnant obstruction at the head and face. Wen Jing Tang [Warm the Menses Decoction] treats deficient cold type of irregular menstruation and dysmenorrhea. Sheng Hua Tong Jing Tang [Generation and Transformation Decoction to Unblock the Menses] treats early or delayed menstruation and amenorrhea due to blood stasis. Tuo Hua Jian [Release from Flower Decoction] can induce labor. Shi Xiao San [Sudden Smile Powder] and Sheng Hua Tang [Generation and Transformation Decoction] treat postpartum retention of lochia and lower abdominal pain. Gui Zhi Fu Ling Wan [Cinnamon Twig and Poria Pill], Bie Jia Jian Wan [Decoct Turtle Shell Pill], and Da Huang Zhe Chong Wan [Rhubarb and Eupolyphaga Pill] are ancient formulas to break stasis and reduce mass and accumulations. Jia Jian Huo Luo Xiao Ling Dan [Modified Fantastically Effective Special Pill to Invigorate the Collaterals] is an empirical formula to treat extrauterine pregnancy. Fu Yuan Huo Xue Tang [Revive Health by Invigorating the Blood Decoction], Tao He Cheng Qi Tang [Peach Pit Decoction to Order the Qi], Qi Li San [Seven-Thousandths of a Tael Powder], Die Da Wan [Trauma Pill], Jie Gu Wan [Set the Bone Pill], Huo Xue Jiu [Invigorate Blood Wine], and Shu Jin Huo Xue Tang [Relax

Tendon and Invigorate Blood Decoction] all treat swelling and pain due to traumatic injury, with injured tendons and bones and internal accumulation of blood stasis. Jia Jian Gua Lou Xie Bai Tang [Modified Trichosanthes Fruit and Chinese Chive Decoction] treats pain due to chest Bi syndrome. Dan Shen Yin treats Heart, chest, and Stomach pain. Bu Yang Huan Wu Tang [Tonify the Yang to Restore Five-Tenths Decoction] and Xiao Huo Luo Dan [Minor Invigorate the Collaterals Pill] treat sequalae of stroke.

LIVER FAILING TO STORE BLOOD

Liver fails to store blood is often caused by Liver heat. Blood–heat in the Liver channel can cause menorrhalgia or even Beng Lou in severe cases. Liver fire scorches Stomach can cause hematemesis and blood in the stool. Liver fire counteracting on Lung can cause hemoptysis and epistaxis. Often herbs that clear heat, cool blood, and stop bleeding are used, such as Sheng Di Huang, Huang Qin, Di Yu, Xi Jiao (Shui Niu Jiao), Da Ji, Xiao Ji, Ce Bai Ye, Bai Mao Gen, and Huai Jiao. Other heat clearing herbs, such as Huang Lian, Shan Zhi Zi, Qing Dai, and stop bleeding herbs, such as San Qi, E Jiao, Ai Ye, Ou Jie, Hua Rui Shi, Xue Yu Tan, Xian He Cao, Zong Lu Tan, Qian Cao Gen, and Hua Sheng Yi, are used together as formulas.

The selection of which of the stop bleeding herbs to use while making the formula depends on the location of bleeding. Sheng Di Huang, Huang Qin, San Qi, Zong Lu Tan, Wu Zei Gu, Bai Ji, and Hua Rui Shi are often used in hematemesis. Shan Zhi Zi, Huang Qin, Bai Mao Gen, Da Ji, and Xiao Ji are used in epistaxis. E Jiao, Bai Ji, Xue Yu Tan, and Xiao Ji are often used in hemoptysis. For vaginal bleeding, E Jiao, Ai Ye, Sheng Di Huang, Huang Qin, and Ce Bai Ye are suitable. For blood in the urine, Da Ji, Xiao Ji, and Bai Mao Gen should be used. For blood in the stool, Huai Jiao and Di Yu are used.

When Liver heat forces blood flow recklessly, it always leads to Yang hyperactivity, thus Sheng Di Huang, Bai Shao Yao, Long Gu, and Mu Li which nourish Yin and subdue Yang are often added to heat formulas that clear and stop bleeding.

Furthermore, from the point of view of the ascending descending function of Qi mechanism, when bleeding is located in the upper part of the body, one should also combine herbs that disperse and descend Qi, such as Gua Lou Ke, Jiang Xiang, Chen Xiang, and Da Huang. When bleeding is located in the lower part of the body, ascending lifting herbs should be added, such as Ren Shen, Huang Qi, Sheng Ma, and Chai Hu.

Shi Hui San [Ten Partially Charred Substances Powder]
Shi Yao Shen Shu

Da Ji	equal amount
Xiao Ji	
He Ye	
Ce Bai Ye	
Bai Mao Gen	
Qian Cao Gen	
Zong Lu Pi	
Mu Dan Pi	
Shan Zhi Zi	
Da Huang	

Main Indication: Reckless blood flow due to blood–heat. Symptoms include hematemesis, hemoptysis, acute sudden onset, red tongue, and slippery and rapid pulse.

Preparation: Partially char all herbs, grind into super fine powder, mix 9 to 15 g of powder with lotus juice, radish juice, or warm water, and take. This formula also can be taken as decoction.

Treatment Principles: Cool blood, and stop bleeding.

This formula treats syndromes of excess blood–heat causing bleeding in the digestive tract and respiratory ducts. It can also treat uterine bleeding. Because all ten herbs in the formula are charred and ground into powder, the formula is called Shi Hui San [Ten Partially Charred Powder].

Zhi Xue San [Stop Bleeding Powder] (Jing Yan Fang)

Hua Rui Shi (heated, then dipped in vinegar)	30 g
E Jiao	30 g
Da Ji	18 g
Xiao Ji	18 g
Ce Bai Ye Tan	9 g
Jiao Shan Zhi Zi	15 g
Mu Li	24 g
Long Gu	24 g
Dai Zhe Shi	24 g

Main Indication: Bleeding of the digestive tract. It treats blood–heat causing hematemesis, copious dark red blood or with undigested food, red tongue, and rapid pulse.

Preparation: Grind into fine powders, 3 to 6 g, take twice a day with water.

Treatment Principles: Clear heat and stop bleeding.

This formula treats heat type of bleeding due to ulcerations in digestive tract.

Huai Hua San [Sophora Japonica Flower Powder] *Ben Shi Fang*

Hua Hua (fried)	30 g
Ce Bai Ye (toasted)	15 g
Jing Jie Tan	9 g
Zhi Ke (fried)	9 g

Main Indication: Bleeding due to intestinal wind or toxins in Zang organs. Symptoms include bleeding before or after bowel movement, or blood in the stool, or acute severe onset, splashing to every direction, or bleeding due to hemorrhoids with bright red blood, red or purple tongue, and slippery or rapid pulse. (Note: Fresh clear red blood indicates intestinal wind, and turbid dark blood indicates toxins in the Zang organs.)

Preparation: Grind into fine powder, take 6 g, two to three times a day with clear rice soup, or it can be taken as decoction.

Treatment Principles: Clear heat, stop bleeding, and disperse wind.

This formula treats acute excess heat type of bleeding.

Modification:

Addition	Indication
Di Yu Wu Zei Gu	Excessive bleeding
Ku Shen Huang Qin Shan Zhi Zi Qing Dai	Excess heat

Related Formula:

Formula	Herbs	Indication
Huai Jiao Wan [Sophora Japonica Fruit Pill]	Huai Jiao (fried) 500 g Fang Feng 250 g Di Yu 250 g Dang Gui (toasted) 250 g Huang Qin 250 g Zhi Ke 250 g	Grind into powder, mix with wine to form pill the size of seeds from Chinese parasol tree. Take 30 pills with rice water. Treats wind evil heat toxins or damp–heat type of bleeding in hemorrhoids, anal prolapse

Si Sheng Wan [Four-Fresh (Herbs/Ingredients) Pill] *Fu Ren Liang Fang*

Sheng He Ye	9 g
Sheng Ai Ye	9 g
Sheng Ce Bai Ye	12 g
Sheng Di Huang	15 g

Main Indication: Blood–heat and reckless blood flow. Symptoms include hemoptysis, epistaxis, bright red blood, dry mouth, dry throat, red tongue or with deep red color, and wiry and rapid pulse.

Preparation: Decoction.

Treatment Principles: Cool blood and stop bleeding.

This formula treats hemoptysis, hematemesis, and epistaxis due to heat in the blood level forcing the upward movement of blood.

Qing Jing Zhi Xue Tang [Cool the Menses and Stop Bleeding Decoction] *Zhong Yi Fu Ke Zhi Liao Xue*

Sheng Di Huang	30 g
Mu Dan Pi	9 g
Huang Qin	15 g
Huang Bai	12 g
Bai Mao Gen	30 g
Di Yu	30 g
Chao Pu Huang	15 g
Yi Mu Cao	15 g
Zong Lu Pi Tan	24 g

Main Indication: Excess condition, blood–heat. Symptoms include sudden heavy or gushing menstrual flow, irritability, feeling hot, thirst, red tongue, and slippery, rapid, and forceful pulse.

Preparation: Decoction.

Treatment Principles: Clear heat and stop bleeding.

This formula treats blood–heat in the Liver channel causing reckless menstrual blood flow which manifests as heavy menstruation or sudden gushing bleeding.

Qing Re Zhi Beng Tang [Clear Heat and Stop Excessive Uterine Bleeding Decoction] *Zhong Yi Fu Ke Zhi Liao Xue*

Duan Gui Ban	15 g
Bai Shao Yao	30 g
Sheng Di Huang	24 g
Mu Dan Pi	9 g
Shan Zhi Zi	9 g
Chao Huang Qin	15 g
Huang Bai	9 g
Chun Gen Pi	30 g
Ce Bai Ye Tan	30 g
Di Yu	24 g

Main Indication: Excessive uterine bleeding (Beng) due to blood heat. Symptoms include sudden gushing bleeding with bright red blood, dry mouth, dry lips, yellow tongue coating, and rapid pulse.

Preparation: Decoction. One dose per day.

Treatment Principles: Clear heat and stop bleeding.

This formula treats excessive uterine bleeding (Beng) syndromes due to heat in the blood level and heat entering the blood chamber. It is especially effective in treating sudden excessive uterine bleeding due to unconsolidated Chong and Ren channels.

Related Formula:

Formula	Herbs	Indication
Xiao Pin Sheng Di Huang Tang [Small Rehmannia Decoction]	Sheng Di Huang 30 g Ce Bai Ye 15 g Huang Qin 9 g E Jiao 15 g Gan Cao 9 g	Treat Yin deficiency with blood heat. Symptoms include sudden gushing menstruation with bright red blood, red tongue without coating, and thin rapid pulse

Zhi Beng Zheng Ji Yan Fang [Extreme Effective Formula in Treating Beng Syndromes] *Nu Ke Ji Yao*

Jiao Shan Zhi Zi	9 g
Huang Qin	12 g
Chuan Lian Zi	9 g
Mu Dan Pi	9 g
Chao Di Yu	24 g
Lian Xu	9 g
Mu Li	15 g
Bai Shao Yao	24 g
Gan Cao	6 g
Sheng Di Huang	30 g

Main Indication: Yin deficiency and Liver heat with reckless blood flow. Symptoms include excessive uterine bleeding (Beng), bright red blood, dry mouth, dry lips, yellow tongue coating, red tongue, and rapid and forceful pulse.

Preparation: Decoction. Continue taking a few doses of this formula.

Treatment Principles: Clear heat, nourish Yin, and stop bleeding.

This formula focuses on clearing heat. Nourishing Yin and stopping bleeding are secondary. This formula is used to treat Beng syndromes due to extreme Liver heat and Yin deficiency.

Modification:

Addition	Indication
Ren Shen Huang Qi	Qi deficiency, excessive bleeding
Wu Mei Han Lian Cao Ce Bai Ye	Unceasing bleeding

Gu Jing Wan [Stabilize the Menses Pill] *Yi Xue Ru Men*

Gui Ban	30 g
Bai Shao Yao	30 g
Huang Qin	30 g
Huang Bai	9 g
Chun Gen Pi	30 g
Xiang Fu	6 g

Main Indication: Yin deficiency and blood–heat. Symptoms include unceasing menstruation or uterine bleeding alternating between spotting and gushing, dark red blood, restlessness with heat at chest, abdominal pain, dark urine, red tongue, and wiry and rapid pulse.

Preparation: Decoction.

Treatment Principles: Nourish Yin, clear heat, stop bleeding, and stabilize menses.

Sheng Pu Huang Tang [Fresh Cattail Pollen Decoction] *Yan Ke Liu Jing Fa Yao*

Sheng Pu Huang	24 g
Han Lian Cao	24 g
Dan Shen	15 g
Mu Dan Pi	12 g
Jing Jie Tan	12 g
Yu Jin	15 g
Sheng Di Huang	12 g
Chuan Xiong	6 g

Main Indication: Blood–heat and retinal bleeding. Symptoms include seeing red or dark red color and decreased vision or even blindness. Other signs include a red tongue tip and rapid pulse. The patient is diagnosed with retinal bleeding after an eye exam. If patient's vision gradually decreases and becomes blurry, then it belongs to Kidney essence deficiency and not to the above pattern.

Preparation: Decoction.

Treatment Principles: Invigorate blood, stop bleeding, cool blood, and scatter stasis.

In this formula, Sheng Pu Huang, Han Lian Cao, Sheng Di Huang, and Jing Jie Tan cool blood and stop bleeding. Dan Shen, Mu Dan Pi, Yu Jin, and Chuan Xiong invigorate blood and scatter stasis. Thus, this is an example of using both invigorating blood and stop bleeding treatment methods.

Comparison of Formulas That Stop Bleeding

This type of formula is formed by two groups of herbs: clearing heat and stop bleeding. It demonstrates treatment of both the root and the branch at the same time by clearing the source and removing the blockage.

Shi Hui San [Ten Partially Charred Substances Powder] has a wide application. It can be used for any type of bleeding and is especially effective in treating bleeding of the Lung and Stomach. Zhi Xue San [Stop Bleeding Powder] is effective in treating hematemesis. Ke Xue Fang (please refer to Chapter 6 for more information) [Coughing of Blood Formula] treats hemoptysis. Si Sheng Wan [Four-Fresh (Herbs/Ingredients) Pill] can treat hemoptysis, hematemesis, and epistaxis, but the formula is weaker. Huai Hua San [Sophora Japonica Flower Powder] focuses on treating blood in the stool. Qing Jing Zhi Xue Tang [Clear the Menses and Stop Bleeding Decoction], Qing Re Zhi Beng Tang [Clear Heat and Stop Excessive Uterine Bleeding Decoction], Zhi Beng Zheng Ji Yan Fang [Extremely Effective Formula in Treating Beng Syndrome], and Gu Jing Wan [Stabilize the Menses Pill] treat abnormal uterine bleeding. However, Qing Re Zhi Beng Tang [Clear Heat and Stop Excessive Uterine Bleeding Decoction] and Gu Jing Wan [Stabilize the Menses Pill] can also nourish Yin and stabilize the Chong and Ren channels. Sheng Pu Huang Tang [Fresh Cattail Pollen Decoction] is a specialized formula for treating retinal bleeding.

STAGNANT LIVER QI

Symptoms of Liver Qi stagnation often include hypochondriac distention and pain, distending pain at epigastrium, or irregular menstruation, dysmenorrhea, and distending pain at lumbar or sacrum. When Liver pathology affects Spleen, other symptoms such as poor appetite, abdominal distention, diarrhea, and abdominal pain are present. The herbs that are often used to smooth Liver and regulate Qi are Xiang Fu, Xiang Yuan, Fo Shou, Qing Pi, Zhi Ke, Mu Xiang, and Wu Yao. Blood nourishing and Liver regulating herbs such as Dang Gui, Bai Shao Yao, Chuan Xiong, and Dan Shen are combined in formulas.

Liver Qi stagnation with cold is often seen in inguinal hernia, cold pain in the lower abdomen or delayed menstruation with dark menstrual blood, and distending pain in lower abdomen and lower back. Soothing Liver and regulating Qi herbs that are warm in nature are often used, such as Wu Zhu Yu, Wu Yao, Mu Xiang, Xiao Hui Xiang, and Chuan Jiao. Gan Jiang, Rou Gui, and Fu Zi can be added to smooth Liver formulas to reinforce the warming of Yang and scattering cold functions.

Liver Qi stagnation with heat often manifests as bitter taste in the mouth, sour regurgitation, red sides of the tongue, yellow tongue coating, irritability, early menstruation, and wiry rapid pulse. Soothing Liver and regulating Qi herbs that are cold in nature, such as Chuan Lian Zi and Qing Hao, are used. Others, such as Shan Zhi Zi, Huang Qin, and Huang Lian, are added to strengthen the function of clearing Liver heat.

There are also formulas that combine both hot and cold herbs such as Huang Lian and Wu Zhu Yu in Zuo Jin Wan [Left Metal Pill] (please refer to this chapter, on Liver heat).

Syndromes of both Liver Qi stagnation and blood stasis are often manifest as distention and fullness at hypochondriac, chest, and abdominal region, irregular menstruation due to hesitant blood flow, scanty menstruation with clots, and abdominal pain worsening with pressure. Treatment should include both regulating Qi and invigorating blood with such formulas as Shu Gan Jie Yu Tang [Spread the Liver and Relieve Stagnation Decoction]. The ratio between the regulating Qi herbs and invigorating blood herbs should be determined according to the severity of Qi stagnation and blood stasis.

Xiao Yao San [Relax Wanderer Powder] *He Ji Ju Fang*

Chai Hu	9 g
Dang Gui	9 g
Bai Shao Yao	15 g
Bai Zhu	9 g
Fu Ling	15 g
Gan Cao	6 g

Main Indication: Liver Qi stagnation and Spleen deficiency. Symptoms include breast distention, hypochondriac pain, alternating chills and fever, headache, blurry vision, poor appetite, apathy, irregular menstruation, wiry big and deficient pulse, and distending sublingual vein or purple dark vein.

Preparation: Decoct with small amount of Sheng Jiang and Bo He. It can also be taken as powder or pills, 6 to 9 g, two to three times a day.

Treatment Principles: Spread Liver, alleviate Qi stagnation, strengthen Spleen, and harmonize Ying.

This formula has a wide application. Any type of Liver Qi stagnation and Spleen deficiency with the manifestation of hypochondriac and abdominal pain, dysmenorrhea, early or delayed menstruation, and depressed emotions can be treated with this formula. With modification, it can also treat lumps and nodules of the breasts or abdomen which change with emotions.

Related Formulas:

Formula	Herbs	Indication
Dan Zhi Xiao Yao San [Relax Wanderer Powder with Moutan and Gardenia]	Core formula + Mu Dan Pi Shan Zhi Zi	Treat syndromes of Spleen and Liver blood deficiency transforming into heat
Hei Xiao Yao San [Black Relax Wanderer Powder]	Core Formula + Sheng Di Huang or Shu Di Huang	Treat Liver and Spleen blood deficiency with manifestations of premenstrual abdominal pain, and wiry, deficient pulse
Jia Jian Xiao Yao San [Modified Relax Wanderer Powder] *Nei Ke Zhai Yao*	Core Formula + Mu Dan Pi Shan Zhi Zi Xiang Fu Ze Lan	Treat early menstruation with heavy bleeding, bright red blood, hypochondriac pain, restlessness, and excess dreaming With the presence of dizziness and fever, Dang Gui should be eliminated and Yi Mu Cao should be added
Qing Gan Da Yu Tang [Clear the Liver to Reach Stagnation Decoction]	Core Formula – Bai Zhu + Mu Dan Pi Shan Zhi Zi Ju Ye Ju Bai Ju Hua	Treat Liver Qi stagnation, blood heat with the manifestations of early menstruation, scanty red blood, tidal fever, spontaneous sweating, dizziness, restlessness, red tongue, and wiry, thin pulse
Shu Yu Qing Gan Tang [Spread the Depression and Clear the Liver Decoction]	Core Formula – Fu Ling + Xiang Fu Yu Jin Huang Qin Shan Zhi Zi Mu Dan Pi	Treat Liver stagnation with slight heat Symptoms include premenstrual distending pain at hypochondrium and abdomen, impatience, easily angered, dizziness, bitter taste and dry mouth, heavy red menstruation, or with clots, red tongue with yellow coating, and wiry and rapid pulse

(continued)

Formula	Herbs	Indication
Fu Pi Shu Gan Tang [Assist Spleen and Spread the Liver Decoction]	Core formula – Dang Gui Gan Cao + Pao Shen Chao Pu Huang Xue Yu Tan Jiao Ai Ye	Treat anger damaged Liver causing sudden heavy menstruation or unceasing spotting, purple blood with blood clots, distention and fullness in the lower abdomen, chest and hypochondrium, shortness of breath, apathy, and poor appetite and indigestion
Shu Yu Qing Gan Yin [Spread the Depression and Clear the Liver Beverage]	Core Formula – Dang Gui Gan Cao + Sheng Di Huang Shan Zhi Zi Yi Mu Cao	Treat bleeding or spotting during pregnancy, bitter taste, dry throat, hypochondriac distention, restlessness, insomnia, five center heat, red tongue, slight yellow coating, and wiry, rapid, and slippery pulse
Jia Wei Xiao Yao San [Augmented Relax Wanderer Powder]	Core Formula + Mu Dan Pi Shan Zhi Zi Di Yu Tan Ce Bai Ye Chun Gen Pi	Treat black or red vaginal discharge due to stagnant heat in Liver channel
Jia Wei Dan Zhi Xiao Yao San [Augmented Relax Wanderer Powder with Moutan and Gardenia]	Core formula + Mu Dan Pi Shan Zhi Zi Che Qian Zi	Treat Liver and Spleen blood deficiency with heat and syndromes of painful hesitant urination
Jia Jian Xiao Yao San [Modified Relax Wanderer Powder]	Chai Hu Chi Shao Yao Dang Gui Gan Cao Chao Chuan Shan Jia Qing Pi Chen Pi Chuan Xiong	Treat breast lumps that are fixed but change in size
Ru Fang Zhang Tong Jing Yan Fang [Empirical Formula for Breast Distention and Pain]	Core Formula – Gan Cao + Wang Bu Liu Xing Ji Xue Teng Dan Shen Xiang Fu	Treat breast distention and pain due to cystic enlargement of mammary gland
Ru Xian Zeng Sheng Fang [Mastoplasia Formula]	Chai Hu Chi Shao Yao Dang Gui Wei Gan Cao Zhi Ke Qing Pi Shi Chang Pu Xiang Fu Yu Jin Chuan Xiong Sheng Di Huang Bing Lang Ru Xiang Mu Li Pu Gong Ying	Treat mastoplasia

Chai Hu Shu Gan San [Bupleurum Powder to Spread the Liver] *Jing Yue Quan Shu*

Chai Hu	0 g
Chen Pi	9 g
Zhi Ke	9 g
Xiang Fu	9 g
Chuan Xiong	6 g
Bai Shao Yao	9 g
Zhi Gan Cao	3 g

Main Indication: Hypochondriac pain due to Liver Qi stagnation. Symptoms include distending pain at hypochondrium, dysmenorrhea, alternating chills and fever, depression, no desire to eat, bitter taste, pale or dark tongue, sublingual vein is distented or purple and wiry pulse.

Preparation: Decoction, taken before meals.

Treatment Principles: Spread Liver and stop pain.

This is a classic formula to spread Liver and relieve depression. It also treats epigastric pain due to Liver Qi overacting on Stomach. It can also treat a type of lower back pain that occurs just before dawn.

Modification:

Addition	Indication
Mu Xiang Wu Yao	Severe Qi stagnation
Dang Gui Dan Shen	Severe blood stasis

Related Formula:

Formula	Herbs	Indication
Jia Wei Chai Hu Shu Gan San [Augmented Bupleurum Powder to Spread the Liver]	Core formula + Shan Zhi Zi Pao Jiang	Decoction. Take the warm decoction before meal Treat anger and fire damaged Liver causing hypochondriac pain and hematemesis If hematemesis occurs, then add small amount of Tong Bian (urine of a young boy) to decoction

Jia Wei Wu Yao Tang [Augmented Lindera Decoction] *Ji Yin Gang Mu*

Wu Yao	12 g
Xiang Fu	12 g
Mu Xiang	9 g
Sha Ren	6 g
Yan Hu Suo	9 g
Gan Cao	3 g

Main Indication: Liver Qi stagnation. Symptoms include premenstrual distention and pain at lower abdomen, hypochondrium, and breast; more distention than pain; hesitant menstrual flow; depression; stifling sensation at chest; nausea, belching; lower back soreness and distention; pale tongue with thin white coating; and wiry, choppy pulse.

Preparation: Decoction.

Treatment Principles: Spread Liver and regulate Qi.

This formula treats abdominal distention and pain, and more distention than pain before menstruation. It can also treat syndromes with chest and hypochondriac distention, and pain due to Liver Qi stagnation.

Modification:

Addition	Indication
Wu Zhu Yu Xiao Hui Xiang	Cold type of hernia

Related Formula:

Formula	Herbs	Indication
Wu Yao Tang [Lindera Decoction]	Core Formula – Sha Ren Yan Hu Suo + Dang Gui	Treat pain in uterus

Jia Jian Cang Sha Yin [Modified Atractylodes and Cyperi Decoction] *Zhong Yi Fu Ke Zhi Liao Xue*

Cang Zhu	6 g
Fu Ling	9 g
Xiang Fu	9 g
Wu Yao	6 g
Pao Jiang	3 g
Hong Ze Lan	12 g
Dang Gui	9 g
Chuan Xiong	9 g
Xue Mu Tong	9 g

Main Indication: Liver cold and Qi stagnation. Symptoms include delayed menstruation, dark scanty menstrual blood, lower abdominal pain, distention at lower back, slight aversion to cold, pale tongue, white tongue coating, and slow pulse.

Preparation: Take the decoction warm.

Treatment Principles: Warm Liver and move Qi.

This formula is suitable for treating Liver Qi stagnation with cold.

Tian Tai Wu Yao San [Top-Quality Lindera Powder] *Yi Xue Fa Ming*

Wu Yao	12 g
Mu Xiang	6 g
Xiao Hui Xiang	6 g
Qing Pi	6 g
Chao Gao Liang Jiang	9 g
Bing Lang	6 g
Chuan Lian Zi	9 g
Ba Dou	70 pieces

Main Indication: Cold type of inguinal hernia. Symptoms include lower abdominal pain radiating to the testicles, swelling, distention and sinking sensation, pale tongue with thin white coating, and deep slow or wiry pulse.

Preparation: Fry Ba Dou and Chuan Lian Zi until black. Remove Ba Dou, then decoct with the rest of the herbs. Taken with moderate amount of yellow wine. One can also grind the processed Chuan Lian Zi and the rest of the herbs into fine powder. Take 3 g with yellow wine.

Treatment Principles: Spread Liver, scatter cold, and stop pain.

This formula treats inguinal hernia due to cold stagnation in Liver channel. It can also treat dysmenorrhea due to Liver cold and Qi stagnation.

In the formula, Chuan Lian Zi is fried with Ba Dou so that the cold, bitter property of Chuan Lian Zi is removed and strengthens the functions of spreading Liver Qi, stopping pain and scattering accumulation.

Related Formulas:

Formula	Herbs	Indication
Hui Xiang San [Fennel Powder]	Core formula – Wu Yao Gao Liang Jiang (Chuan Lian Zi does not fry with Ba Dou)	Treats appendicitis with lower abdominal pain and genital pain
San Ceng Hui Xiang Wan [Three Layers Fennel Pill]	Hui Xiang Bei Sha Shen Chuan Lian Zi Mu Xiang Grind into fine powder, form into pills with rice powder. If it is not effective, then add Bi Ba and Bing Lang powder to form pill If condition is still not recovered, add Fu Ling and Fu Zi powder and form into pills	Treats cold type of hernia Symptoms include pain around navel and abdomen, enlarged testicles, swelling, distention, and sinking sensation of the scrotum which affects walking, or testicles that are cold, hard like stones, and gradually enlarged
Dao Qi Tang [Guide the Qi Decoction]	Chuan Lian Zi Mu Xiang Xiao Hui Xiang Wu Zhu Yu	Treats pain due to cold type of hernia

Mu Xiang Wu Gong San [Aucklandia and Centipede Powder]
Chen Ji Rong Fang

Mu Xiang	10 g
Wu Gong	3 pieces, remove head and feet

Main Indication: Distention at the testicles due to cold congealed in the Liver channel. Symptoms include one sided or bilateral swelling and pain in the testicles, green purple sublingual vein, and wiry pulse.

Preparation: Grind into fine powder, divide into three doses. For adults, take with yellow wine. For children, take with glutinous rice wine. This formula can be continuously taken for several days.

Treatment Principles: Warm the channel, move Qi, stop pain, and reduce swelling.

This is a simple yet effective formula. If there is severe cold in Liver channel with radiating pain from testicles to lower abdomen or lower limbs, then one can also take Si Ni San [Frigid Extremities Powder].

Ju He Wan [Tangerine Seed Pill] *Ji Sheng Fang*

Ju He (fried)	30 g
Hai Zao	30 g
Hai Dai	30 g
Kun Bu	30 g
Chuan Lian Zi (fried)	30 g
Tao Ren (fried)	30 g
Hou Po (fried with ginger juice)	30 g
Mu Tong	30 g
Zhi Shi (fried)	30 g
Yan Hu Suo (fried)	30 g
Gui Xin	30 g
Mu Xiang	15 g

Main Indication: Cold–damp in Liver channel causing herniation with Qi stagnation, blood stasis, phlegm congelation. Symptoms include swelling and distention of testicles, or testicles hard like stone, or radiating spasm pain at navel and abdomen, dark red tongue, sublingual vein distention, and wiry or choppy pulse.

Preparation: Grind into fine powder, form with wine into pills the size of seeds from Chinese parasol tree. Take 70 pills on empty stomach with warm wine or salt water. It can also be taken as decoction.

Treatment Principles: Move Qi, invigorate blood, soften hardness, and scatter accumulation.

This formula is effective in treating cold–damp type of inguinal hernia with Qi stagnation and blood stasis. The manifestations are swelling and pain in the testicles.

Modification:

Addition	Indication
San Leng E Zhu	Severe blood stasis
Cang Zhu Fu Ling	Severe damp

Jin Ling Zi San [Melia Toosendan Powder] *Su Wen Bing Ji Qi Yi Bao Ming Ji*

Jin Ling Zi (Chuan Lian Zi)	30 g
Yan Hu Suo	30 g

Main Indication: Liver Qi stagnation turning into fire leading to intermittent pain at chest, abdomen, hypochondrium, agitation, restlessness, bitter taste in mouth, red tongue with yellow coating, and wiry, rapid pulse.

Preparation: Grind into fine powder. Take 9 g with wine or warm water.

Treatment Principles: Clear heat, soothe Liver, and stop pain.

This is a basic formula for any type of pain due to Qi stagnation and blood stasis. The formula is small, which is easy to add other herbs. It can also be added to other Qi regulating and blood invigorating formulas.

Related Formulas:

Formula	Herbs	Indication
Yan Hu Suo San [Corydalis Powder]	Dang Gui Yan Hu Suo Pu Huang Chi Shao Yao Rou Gui Jiang Huang Ru Xiang Mo Yao Mu Xiang Gan Cao Sheng Jiang	Take the warm decoction before meals Treat pain syndromes due to Qi stagnation and blood stasis, such as dysmenorrhea and irregular menstruation (This formula is slightly warm)
Jin Zhu Wan [Melia Toosendan and Evodia Pill]	Jin Ling Zi Wu Zhu Yu	Treat cold type hernia, pain due to Qi stagnation, and slight swelling of the testicles
Nian Tou San [Twist the Head Powder]	Grind core formula into fine powder and take 2 to 3 g with rice soup, adding a few drops of vegetable oil	Treat difficult urination in children
Qing Zhong Tang [Clear the Middle Decoction]	Core formula + Huang Lian Shan Zhi Zi Xiang Fu Gan Cao Chen Pi	Treat epigastric and abdominal pain, dry tongue, dry lips, and rapid forceful pulse due to heat in the Stomach and Intestine
Chen Xiang San [Aquilaria Powder]	Core formula – Yan Hu Suo Chen Xiang Fu Zi	Treat intermittent cold type of hernia, cold weak Zang organ, and pain at navel
Xuan Hu Tang [Corydalis Decoction]	Core formula – Cuan Lian Zi + Fu Zi Mu Xiang	Treat any type of hernia with cold pain at epigstrium and abdomen, borborygmus, cold body, spontaneous sweating, and diarrhea or loose stool which indicates deficient cold in Middle Jiao
Shen Ying San [Magical Response Powder]	Core formula – Chuan Lian Zi + Pepper powder	Treat cold type of hernia and unbearable spasm pain at chest and abdominal area
Jin Ling San [Melia Toosendan Powder]	Chuan Lian Zi San Leng E Zhu Qing Pi Chen Pi Mu Xiang Xiao Hui Xiang Zhi Ke Bing Lang Gou Teng Gan Cao Chi Fu Ling	Treat stubborn hernia with abdominal pain and severe Qi stagnation and blood stasis
Chen Xiang Jiang Qi San [Aquilaria to Direct Qi Downward Powder]	Core formula + Chen Xiang Sha Ren Xiang Fu Gan Cao	Treat Qi stagnation with pain in the Stomach and Intestines

(continued)

Formula	Herbs	Indication
Xuan Hu Ku Liang Tang [Corydalis and Bitter Melia Toosendan Decoction]	Core Formula + Rou Gui Fu Zi Gan Cao Shu Di Huang Huang Bai	Treat cold, shrinking pain below the navel, and cold at the genitals which belongs to interior cold and exterior heat or "true cold, false heat"

Tong Ru San Jie Tang [Unblock Breast Milk and Scatter Accumulations Decoction] *Zhong Yi Fu Ke Zhi Liao Xue*

Quan Gua Lou	9 g
Qing Pi	9 g
Si Gua Luo	9 g
Ju Luo	6 g
Tong Cao	6 g
Ju Ye	6 g
Yu Jin	6 g
Ci Ji Li	12 g
Pu Gong Ying	15 g

Main Indication: Liver Qi stagnation and obstructed breast milk. Symptoms include galactostasis or hesitant flow of breast milk, hardness, fullness, and distending pain of the breasts, or even red and swollen (sometimes with chills and fever) pale tongue with white coating and wiry and rapid pulse.

Preparation: Decoction.

Treatment Principles: Soothe Liver, unblock collaterals, and scatter accumulation.

Related Formulas:

Formula	Herbs	Indication
Tong Jing Huo Luo Tang [Unblock the Channel and Invigorate the Collateral Decoction]	Gua Lou Qing Pi Si Gua Luo Ju Luo Xiang Fu Tong Cao Bian Dou Dang Gui	Treat postpartum Liver Qi stagnation leading to galactostasis, distending pain of the breasts, hypochondriac pain, and deep and choppy pulse
Gong Xiao He Jie Ruan Jian Tang [Attack, Reduce, Harmonize, Resolve, and Soften Hardness Decoction]	Chuan Shan Jia (processed) 9 g Dang Gui 9 g Chi Shao Yao 9 g Qing Pi 9 g Chen Pi 9 g Ru Xiang 9 g Mo Yao 9 g Lian Qiao 9 g Jiang Can 9 g Gua Lou 18 g Tian Hua Fen 12 g Mu Li 15 g Xia Ku Cao 15 g Jin Yin Hua 15 g Pu Gong Ying 30 g Gan Cao 3 g Ju Ye 12 g	Continue taking five to ten bags; if not effective add 1.5 g processed Ma Qian Treat breast nodule This formula is stronger in clearing Liver heat, unblocking collaterals, and scattering accumulations

(continued)

Formula	Herbs	Indication
Zhi Ru Yong Yan Fang [Empirical Formula in Treating Mastitis]	Gua Lou Zhe Bei Mu Jin Yin Hua Dang Gui Ru Xiang Mo Yao Gan Cao Pu Gong Ying Zao Jiao Ci Qing Pi Lou Lu Hong Teng Cong Bai	Treat beginning stage of mastitis before the formation of pus
Ruan Jian San Jie Tang [Soften the Hardness and Scatter the Accumulation Decoction]	Chai Hu Zhi Ke Qing Pi Chi Shao Yao Chuan Xiong Hong Hua Chuan Shan Jia Tong Cao Zhe Bei Mu Mu Li Xia Ku Cao Gua Lou Ke Tian Kui Zi Zao Xiu Lian Qiao Gan Cao	Continue taking 20 to 30 doses. Treats hormone imbalance, chronic cystic galactocele, or lumps and nodules of the breasts

Jia Jian Niu Xi Tang [Modified Achyranthis Decoction] *Zhong Yi Fu Ke Zhi Liao Xue*

Chuan Niu Xi	9 g
Dang Gui Wei	9 g
Dan Shen	9 g
Tao Ren	9 g
Xiang Fu	9 g
Yan Hu Suo	9 g
Tan Xiang	9 g
Wu Yao	6 g

Main Indication: Qi stagnation, blood stasis. Symptoms include delayed menstruation, scanty dark blood with clots, or prolonged menstruation, distending pain of the lower abdomen which is worse with pressure, and better with passing clots; sallow complexion; dark purple tongue; thin white tongue coating; deep, tight or wiry, choppy pulse.

Preparation: Decoction, taken warm.

Treatment Principles: Move Qi, invigorate blood.

This formula treats syndromes of delayed menstruation, spotting, and dysmenorrhea due to Qi stagnation, blood stasis, and slight cold.

Shu Gan Jie Yu Tang [Spread the Liver and Resolve the Stagnation Decoction] *Zhong Yi Fu Ke Zhi Liao Xue*

Xiang Fu	9 g
Qing Pi	9 g
Chai Hu	9 g
Yu Jin	9 g
Chuan Xiong	9 g
Dan Shen	12 g
Hong Ze Lan	12 g
Yan Hu Suo	6 g
Chuan Lian Zi Tan	6 g

Main Indication: Qi stagnation and blood stasis. Symptoms include hesitant menstruation, scanty with clots, distending pain at chest, hypochondriac region, lower back, and abdomen, occasional belching, and wiry pulse.

Preparation: Decoction.

Treatment Principles: Soothe Liver, resolve stagnation, invigorate blood, and remove stasis.

This formula is effective in treating dysmenorrhea or delayed menstruation which has more Qi stagnation than blood stasis, or blood stasis due to Qi stagnation.

Comparison of Formulas That Treat Liver Qi Stagnation

1. Xiao Yao San [Relax Wanderer Powder] and Chai Hu Shu Gan San [Bupleurum Powder to Spread the Liver] are mild and neutral and can be used for many syndromes. However, the former formula treats slight deficient syndromes, and the later treats slight excess syndromes. Xiao Yao San [Relax Wanderer Powder] also treats both Liver and Spleen.
2. Jia Wei Wu Yao Tang [Augmented Lindera Decoction] and Jia Jian Cang Sha Yin [Modified Atractylodis Cyperi Decoction] mainly treat irregular menstruation and abdominal distending pain during menstruation. However, the moving Qi and stopping pain function of the former formula is stronger than the latter. The latter formula is stronger in invigorating blood and expelling dampness.
3. Tian Tai Wu Yao San [Top-Quality Lindera Powder], Mu Xiang Wu Gong San [Aucklandia and Centipede Powder], and Ju He Wan [Tangerine Seed Pill] all treat inguinal hernial pain and swollen testicles. However, the first one is better in scattering cold, cracking Qi stagnation, and treating pain due to cold congelation and Qi stagnation. The middle formula is better in moving Qi and relieving toxins which is the specific formula in treating swollen painful testicles. The last formula moves Qi, invigorates blood, expels phlegm, and softens hardness which is effective in treating swollen hard testicles due to Qi stagnation, blood stasis, and phlegm accumulation.
4. Jin Ling Zi San [Melia Toosendan Powder] and Tong Ru San Jie Tang [Unblock the Breast, Scatter Accumulation Decoction] clear heat and soothe Liver. The former formula is better in stopping pain and the latter is specific for pain due to blocked breast milk and distending fullness of the breasts.
5. Jia Jian Niu Xi Tang [Modified Achyranthis Decoction] and Shu Gan Jie Yu Tang [Spread the Liver and Resolve Stagnation Decoction] treat Qi stagnation and blood stasis. The former formula focuses on invigorating blood and transforming stasis. The latter focuses on moving Qi and resolving stagnation.

YIN DEFICIENCY AND LIVER QI STAGNATION

This syndrome is a combination of Liver Yin deficiency and Liver Qi stagnation. The symptoms include hypochondriac pain, distention, and fullness of the chest and abdomen, dry throat, dry mouth, red tongue, scanty fluids, wiry or thin pulse. Often, Bai Shao Yao, Shu Di Huang, Dang Gui, and Gou Qi Zi are combined with Chuan Lian Zi and Ci Ji Li to form formulas that soften the Liver and spread the stagnation.

In syndromes of Yin deficiency with Liver Qi stagnation, it is contraindicated to use acrid, warm, aromatic, and dry herbs. If these herbs are misused, then blood and Yin are further damaged, leading to more severe Liver Qi stagnation.

Yi Guan Jian [Linking Decoction] *Liu Zhou Yi Hua*

Bei Sha Shen	9 g
Mai Men Dong	9 g
Dang Gui	9 g
Sheng Di Huang	18 g
Gou Qi Zi	12 g
Chuan Lian Zi	5 g

Main Indication: Yin deficiency with Liver Qi stagnation. Symptoms include pain in chest and epigastrium, sour regurgitation, bitter vomit, dry throat, dry mouth, red tongue, scanty coat, scanty fluid, and thin, weak, or deficient, rapid, wiry pulse.

Preparation: Decoction, taken warm.

Treatment Principles: Nourish Yin and spread Liver.

This formula treats Liver and Kidney Yin deficiency, and Liver Qi rebellion, flowing horizontally, causing hypochondriac pain. It can also treat irregular menstruation, amenorrhea, and dysmenorrhea due to Yin deficiency and Liver Qi stagnation.

Modification:

Addition	Indication
Xuan Shen Bai Shao Yao	Severe cases of Yin deficiency, Liver Qi stagnation, and severe pain
Gua Lou Ren	Constipation
Di Gu Pi	Deficient heat or excess sweating
Shi Hu	Red dry tongue
Chuan Bei Mu	Copious phlegm
Bie Jia	Hypochondriac pain and hardness upon palpation
Zhi Mu Shi Gao	Irritability, heat sensation, and thirst
Bai Shao Yao Gan Cao	Abdominal pain
Suan Zao Ren	Insomnia
Huang Lian 1.5 g	Bitter taste and dry mouth

Zi Shui Qing Gan Yin [Nourish Water and Clear the Liver Decoction] (Gao Gu Feng Fang)

Sheng Di Huang	24 g
Shan Yao	18 g
Shan Zhu Yu	12 g
Mu Dan Pi	9 g
Fu Ling	12 g
Ze Xie	6 g
Dang Gui	9 g
Bai Shao Yao	30 g
Shan Zhi Zi	9 g
Da Zao	9 g
Chai Hu	9 g

Main Indication: Yin deficiency and Liver Qi stagnation. Symptoms include intermittent pain at hypochondrium, stomach, and epigastrium; red tongue; scanty or no coating; and wiry or thin pulse.

Preparation: Decoction.

Treatment Principles: Nourish Yin and blood, clear heat, and spread Liver.

This formula is basically a combination of Dan Zhi Xiao Yao San [Rambling Powder with Moutan and Gardenia] and Liu Wei Di Huang Wan [Six-Ingredient Pill with Rehmannia]. Since the Yin is deficient, Bai Zhu, which is warm in nature, is removed.

Comparison of the Two Formulas Treating Yin Deficiency and Liver Qi Stagnation

Yi Guan Jian [Linking Decoction] focuses on nourishing Yin and blood and secondarily treats Liver Qi stagnation which is caused by Yin deficiency. Zi Shui Qing Gan Yin [Nourish Water and Clear the Liver Decoction] nourishes Yin and spreads Liver at the same time and also treats syndromes of heat accumulation in the Liver channel due to chronic Liver Qi stagnation.

LIVER WIND INTERNAL STIRRING

This syndrome is caused by hyperactive Liver Yang, excess heat generating wind, and Yin deficiency producing wind. To treat hyperactive Liver Yang and upward stirring of Liver wind, one should sedate the Liver to extinguish wind. To treat excess heat generating wind, cool the Liver to extinguish wind. To treat Liver Yin deficiency generating wind, one should nourish Yin to extinguish the wind.

Hyperactive Liver Yang often manifests as dizziness, distending sensation at the eyes, tinnitus, intermittent headaches, loss of sense of direction or ability to recognize others, drunken appearance and complexion, and wiry, long, and forceful pulse. The commonly used chief herbs to anchor and subdue Yang are Long Gu, Mu Li, Shi Jue Ming, Dai Zhe Shi, and Gui Ban. Then, Yin nourishing herbs such as Sheng Di Huang, Xuan Shen, and Bai Shao Yao are added with heat clearing and spreading Liver herbs such as Chuan Lian Zi and Qing Hao. Early stage of hypertension has similar manifestation as this syndrome of hyperactive Liver Yang. Thus, herbs that can decrease blood pressure by calming Liver and clearing heat can be used, such as Xia Ku Cao, Qing Mu Xiang, Huang Qin, Chuan Niu Xi, Du Zhong, Gou Teng, Ju Hua, Cao Jue Ming, Di Long, and Yu Mi Xu.

Excess heat generating wind is often caused by warm heat evil pathogens entering Jue Yin. Symptoms include high fever, delirium, irritability, agitation, stifling sensation, disturbances, dizziness, spasms of hands and feet or convulsion, burned dry tongue with thorny prickles, and wiry, rapid, and forceful pulse. These are signs of extreme heat generating wind. Herbs, such as Gou Teng, Ling Yang Jiao (Shan Yang Jiao), Ju Hua, Xuan Shen, Sheng Di Huang, or Bai Shao Yao are combined with Da Qing Ye, Ban Lan Gen, Lian Qiao, and Jin Yin Hua. If delirium is present with extreme heat, one can also take An Gong Niu Huang Wan [Calm the Palace Pill with Cattle Gallstone] or Zhi Bao Dan [Greatest Treasure Special Pill] at the same time to clear heat and open the orifices.

Yin deficiency generating wind is often seen in late stage of febrile diseases where Yin is damaged, fluid is exhausted, Liver fails to be nourished by blood, and Liver wind is stirring internally. The manifestations are trembling hands and feet, spasms and convulsions of the tendons and channels, dry mouth, dry lips, deep red tongue with scanty coating, and thin rapid pulse. Commonly used herbs are Ji Zi Huang (yolk), E Jiao, Sheng Di Huang, Bai Shao Yao, Gui Ban, Bie Jia, and Mu Li.

Zhen Gan Xi Feng Tang [Sedate the Liver and Extinguish the Wind Decoction] *Yi Xue Zhong Zhong Can Xi Lu*

Huai Niu Xi	30 g
Sheng Zhe Shi (grind into fine powder)	30 g
Sheng Long Gu (crushed)	15 g
Sheng Mu Li (crushed)	15 g
Sheng Gui Ban (crushed)	15 g
Bai Shao Yao	15 g
Xuan Shen	15 g
Tian Men Dong	15 g
Chuan Lian Zi (crushed)	6 g
Yin Chen Hao	6 g
Sheng Mai Ya	6 g
Gan Cao	4.5 g

Main Indication: Wind stroke syndrome due to internal stirring of Liver wind. Symptoms include dizziness, vertigo, distending sensations at eyes, tinnitus, heat pain inside the head, irritability and heat sensation in Heart and chest, drunken appearance and complexion, frequent belching, or gradual loss of sensation in the limbs, or gradual deviation of mouth and eyes. In severe cases there might be dizziness, sudden loss of consciousness, and awakening by moving the person, but paralysis or deviation persisting, atrophy of the limbs, tongue deviation, yellow greasy tongue coating, wiry, long pulse bilaterally or on one side while the other is weak and small.

Preparation: Decoction.

Treatment Principles: Sedate the Liver and extinguish wind, nourish Yin, and anchor Yang.

This syndrome is caused by hyperactive Liver Yang, upward stirring of wind and Yang, and disturbance of the Qi and blood flow. Thus, the treatment principles include nourishing Liver and Kidney Yin to control Yang, at the same time sedating Liver and extinguishing wind. Herbs that harmonize the Middle Jiao and protect stomach are added to prevent the minerals from damaging the stomach. This formula can be used in acute stage of paralysis due to wind stroke which is similar to stroke due to hypertension.

Modification:

Addition	Indication
Shi Gao	Severe heat at the chest
Dan Nan Xing	Copious phlegm
Shu Di Huang Shan Zhu Yu	Chi pulse is deficient upon heavy palpation

Related Formulas:

Formula	Herbs	Indication
Zhen Gan Xi Feng Tang Jia Jian [Modified Sedate the Liver Extinguish Wind Decoction]	Sheng Long Gu 15 g Shen Mu Li 15 g Sheng Zhe Shi 15 g Bai Shao Yao 9 g Xuan Shen 9 g Tian Men Dong 9 g Chuan Lian Zi 3 g Mu Gua 9 g Gou Teng 9 g	This formula is specifically for postpartum spasms and convulsions. Treats postpartum occasional fever, dizziness, vertigo, tendon tightness, spasms and convulsion of the four limbs, red tongue, and thin, rapid pulse If spasms and convulsions are severe, increase Bai Shao Yao to 30 to 90 g and add Gan Cao
Jian Ling Tang [Construct Roof Tiles Decoction]	Huai Shan Yao 30 g Huai Niu Xi 30 g Sheng Zhe Shi 24 g Sheng Long Gu 18 g Sheng Mu Li 18 g Sheng Di Huang 18 g Bai Shao Yao 12 g Bai Zi Ren 12 g	Decoct with the water from washing rusted iron Treat hyperactive Liver Yang, dizziness, tinnitus, distending sensation at the ear, palpitation, forgetfulness, irritability, agitation, insomnia, excess dreaming, and wiry, hard, and long pulse

Ling Yang Gou Teng Yin [Antelope Horn and Uncaria Decoction] *Tong Su Shang Han Lun*

Ling Yang Jiao Powder (wrapped) or replaced by Shan Yang Jiao	9 g (27–45 g)
Gou Teng (add near end)	12 g
Sang Ye	9 g
Ju Hua	12 g
Fresh Sheng Di Huang	15 g
Bai Shao Yao	9 g
Gan Cao	3 g
Chuan Bei Mu	6 g
Zhu Ru	12 g
Fu Shen	15 g

Main Indication: Heat enters Jue Yin, extreme heat-generating wind. Symptoms include high persistent fever, dizziness, agitation, irritability, stifling sensation and disturbed feeling, spasm and convulsion of the hands and feet, syncope or delirium, deep red tongue, burned dry tongue with thorny prickles, and wiry, rapid, forceful pulse.

Preparation: Decoct all herbs except Ling Yang Jiao (Shan Yang Jiao) Powder. Mix the powder into the decoction, take three times a day.

Treatment Principles: Cool the Liver, extinguish wind, supplement fluid, and relax the tendons.

This formula is a representative formula for treating extreme heat-generating wind. It can also be used for warm febrile disease with manifestations of high fever, irritability, agitation, and spasm and convulsion of the limbs. It can also be used for epilepsy during pregnancy with high fever and delirium.

Modification:

Addition	Indication
Da Qing Ye Ban Lan Gen Jin Yin Hua Lian Qiao	Excess internal heat toxins
An Gong Niu Huang Wan [Calm the Palace Pill with Cattle Gallstone] Zi Xue Dan [Purple Snow Special Pill]	Internal closed obstruction due to heat evil pathogens

Qing Re Xi Feng Tang [Clear Heat and Extinguish Wind Decoction] (Jing Yan Fang)

Shi Gao (decoct first)	60 g
Jin Yin Hua	30 g
Lian Qiao	30 g
Da Qing Ye	30 g
Lian Xin	6 g
Tian Zhu Huang	6 g
Chao Shan Zhi Zi	6 g
Quan Xie	6 g
Wu Gong	6 g
Jiang Can	6 g
Chan Tui	6 g
Shi Chang Pu	6 g
Gou Teng (add near end)	9 g
Di Long	9 g

Main Indication: Heat enters Jue Yin, extreme heat generating wind, evil attacks pericardium, and Shen disturbance. Symptoms include high fever, unconsciousness, delirious speech, convulsion, deep red tongue, and wiry, rapid pulse.

Preparation: Decoction.

Treatment Principles: Clear heat, resolve toxins, extinguish wind, relieve convulsion, cool blood, and open orifices.

This formula includes a group of heat clearing, resolving toxicity herbs that are not in Ling Yang Gou Teng Yin [Antelope Horn and Uncaria Decoction]. This formula has one more group of extinguishing wind, relieving convulsion herbs than does Qing Wen Bai Du Yin [Clear Epidemics and Overcome Toxin Decoction]. Thus, it is effective in treating extreme heat generating wind, and evil toxins obstructing the Heart orifice. If Sheng Di Huang, Mu Dan Pi, Bai Shao Yao, and Xuan Shen are added, which can cool the blood and nourish Yin, then the formula is even more effective.

Gou Teng Yin [Uncaria Decoction] *You Ke Xin Fa*

Gou Teng (add near end)	9 g
Ling Yang Jiao (grind, take with the decoction) or replace with Shan Yang Jiao	3 g (9–15 g)
Quan Xie (remove toxicity)	6 g
Tian Ma	6 g
Ren Shen	3 g
Gan Cao	3 g

Main Indication: Infantile convulsion. Symptoms include clenched teeth, spasms, and convulsion of limbs, fright, palpitation, high fever, eyes looking upward, head tilting back, red dry tongue, and the vein on the index finger is "reaching life gate."

Preparation: Mix Ling Yang Jiao (Shan Yang Jiao) powder with decoction, three to four times a day.

Treatment Principles: Clear heat, extinguish wind, relieve convulsion, and tonify Qi.

Excess heat easily causes wind and disturbs Shen in children which can also damage Zheng Qi. In this formula, other than treating the manifestations by clearing heat, relieving convulsion, and opening orifices, Qi tonics are also added to support the Zheng Qi and treat the root. This prevents the damage of body Zheng Qi which is hard to recover once the evil pathogens are eliminated.

Modification:

Addition	Indication
Da Qing Ye Ban Lan Gen	Severe heat toxins

E Jiao Ji Zi Huang Tang [Ass-Hide Gelatin and Egg Yolk Decoction] *Tong Su Shang Han Lun*

Ji Zi Huang (yolk) (cook and prepare first)	2
E Jiao (melt in decoction)	9 g
Sheng Di Huang	12 g
Bai Shao Yao	12 g
Shi Jue Ming (decoct first)	18 g
Sheng Mu Li (decoct first)	18 g
Fu Shen	12 g
Gou Teng (add near end)	9 g
Luo Shi Teng	9 g
Gan Cao	6 g

Main Indication: "Heat evil damages Yin, deficient Yin generates wind." Symptoms include spasms and convulsion of tendons and meridians, worm like twitching of the limbs, or dizziness, vertigo, dry mouth, dry lips, deep red tongue with scanty coating, and thin, rapid pulse.

Preparation: Decoction. Add Ji Zi Huang (yolk) and melt E Jiao in the decoction. Take warm.

Treatment Principles: Nourish Yin and extinguish wind.

This formula treats Yin deficiency internally generating wind. This syndrome belongs to deficient type with heat.

Related Formula:

Formula	Herbs	Indication
E Jiao Ji Zi Huang Tang [Ass-Hide Gelatin and Chicken Egg Yolk Decoction] *Guang Wen Re Lun*	E Jiao 4.5 g Mu Li 15 g Sheng Di Huang 12 g Bai Shao Yao 6 g Nu Zhen Zi 9 g Ju Hua 6 g Ji Zi Huang 1 Boy's urine 1 cup	Treat upward disturbing of Liver wind, vertigo, Heart palpitation, tinnitus, agitation, or manic and sycope syndromes. This formula is effective in sedating and extinguishing Liver wind

Da Ding Feng Zhu [Major Arrest Wind Pearl] *Wen Bing Tiao Bian*

Bai Shao Yao	18 g
E Jiao (dissolve in decoction)	9 g
Gui Ban	12 g
Sheng Di Huang	18 g
Huo Ma Ren	6 g
Wu Wei Zi	6 g
Mu Li	12 g
Mai Men Dong	18 g
Zhi Gan Cao	12 g
Bie Jia	12 g
Ji Zi Huang (yolk)	2

Main Indication: Extreme exhaustion of Yin, internal disturbance of deficient wind. Other symptoms include apathy, deficient tremor with flaccidity in the four limbs, deep red tongue, scanty coat, and deficient weak pulse.

Preparation: Decoct all herbs except Ji Zi Huang and E Jiao. Add E Jiao and Ji Zi Huang to strained decoction and take warm.

Treatment Principles: Nourish Yin and extinguish wind.

This formula treats chronic warm febrile condition where evil heat burns and damages true Yin or by improper treatment of sweating or purging that damages Yin and fluid. This is a severe critical syndrome with signs of possible collapse. Thus, in this formula, rich, flavor nourishing herbs are used to nourish Yin and provide fluid, supplementing the near-exhausted true Yin, and calming and extinguishing the internal deficient wind.

Modification:

Addition	Indication
Ren Shen	Wheezing and shortness of breath
Long Gu Ren Shen Fu Xiao Mai	Spontaneous sweating
Fu Shen Ren Shen Fu Xiao Mai	Palpitation

Related Formula:

Formula	Herbs	Indication
Xiao Ding Feng Zhu [Minor Arrest the Wind Pearl]	Ji Zi Huang 1 raw E Jiao (melt in decoction) 6 g Gui Ban 18 g Boy's urine 1 cup Dan Cai (mussel) 9 g	Treat long term accumulation of warm evil pathogen at the Lower Jiao which burns the Liver fluid leading to spasms. This also disturbs the Chong channel, causing dry heaves. Pulse is thin, wiry

Comparison of Formulas Treating Liver Wind Internal Stirring

Zhen Gan Xi Feng Tang [Sedate the Liver and Extinguish Wind Decoction] treats hyperactive Liver Yang, causing upward stirring Liver wind. It can also be used for hyperactive Liver Yang type of hypertension at the beginning stage (better result with Ju Hua, Huang Qin, Xia Ku Cao and Di Long). Ling Yang Gou Teng Tang [Antelope Horn and Uncaria Decoction],

Qing Re Xi Feng Tang [Clear Heat and Extinguish Wind Decoction], and Gou Teng Yin [Uncaria Decoction] treat syndromes of excess heat generating wind. However, the first formula focuses on cooling the blood, softening the Liver, transforming phlegm, and calming the Heart. The middle formula has one extra group of herbs to clear heat and relieve toxicity. It also uses insects to strengthen the extinguishing wind, relieving spasm function. This is effective in treating extreme heat, severe toxicity, and severe spasms and convulsions. The last formula is a rescue formula for infantile convulsion.

E Jiao Ji Zi Huang Tang [Ass-Hide Gelatin and Chicken Egg Yolk Decoction] and Da Ding Feng Zhu [Major Arrest the Wind Pearl] treat Yin deficiency generating wind. This is a deficient wind. The focus of the treatment is nourishing Yin and fluid to rescue the collapsing true Yin. However, the former formula is used in milder conditions and the latter is used in more severe conditions.

SPLEEN NOT NOURISHING LIVER, DEFICIENT WIND INTERNAL STIRRING

This syndrome is often seen after vomiting and diarrhea where Jin and Ye fluids are damaged, and tendons failed to be nourished. The symptoms include spasm and twitching of the limbs, upward starring of the eyes, cold limbs and body, undigested food in stools, cool breath through nose and mouth, weak respiration, pale sallow complexion, and other symptoms of Yang Qi deficiency. Thus, Ren Shen, Bai Zhu, Fu Ling, Gan Cao, Gan Jiang, and other tonifying Qi and strengthening Spleen herbs are used to treat the root. Jiang Can, Quan Xie, Wu Gong, and other relieving spasms herbs are used to treat the manifestations. Representing formulas are Xing Pi San [Awaken the Spleen Powder] and Wu Xie Si Jun Zi Tang [Four-Gentleman Decoction with Aconite and Scorpion] (please see Chapter 3, Si Jun Zi Tang [Four-Gentleman Decoction]).

Xing Pi San [Awaken the Spleen Powder] *Gu Jin Yi Tong*

Ren Shen	equal amount
Bai Zhu	
Fu Ling	
Mu Xiang	
Quan Xie	
Jiang Can	
Tian Ma	

Main Indication: Infantile convulsion. Chronic pediatric condition, or Spleen is not nourishing Liver with internal stirring of deficient wind after vomiting and diarrhea. Symptoms include cold limbs and body, undigested food in stools, cool breath at mouth and nose, pale and sallow complexion, minute respiration, twitching and spasms of limbs, upward starring of eyes, somnolence, pale tongue or flabby with teeth marking, thin white coating, and pale purple vein at the wind or Qi gate of the index finger.

Preparation: Decoction, taken warm. Or grind into fine powder, add to food for children.

Treatment Principles: Tonify Spleen and relieve spasms.

Modification:

Addition	Indication
Gan Jiang	Abdominal cold pain

Related Formula:

Formula	Herbs	Indication
Huan Gan Li Pi Tang [Moderate the Liver, Regulate the Spleen Decoction]	Ren Shen Bai Zhu Fu Ling Zhi Gan Cao Chen Pi Shan Yao Bian Dou Gou Teng Jiang Can Sheng Jiang Da Zao	Treat infantile convulsion More severe Spleen deficiency than internal wind stirring

EXTERNAL WIND CAUSING THE STIRRING OF INTERNAL WIND

This syndrome is often seen with clenched teeth, tight lips, body stiffness, opisthotonos, and mouth and eye deviations. Herbs that are strong in relieving spasm such as Quan Xie, Wu Gong, Jiang Can, Chan Tui, and Bai Fu Zi are used as chief herbs in combination with expelling wind, scattering evil herbs such as Qiang Huo, Fang Feng, Jing Jie, and Bai Zhi. This syndrome is quite similar to modern day tetanus.

The treatments for tetanus from ancient times until today all emphasize sweating after taking the formula or covering the patient with thick blanket to induce sweating after taking the formula. The effectiveness of the formula is determined by whether sweating has occurred or not. Thus, expelling wind herbs in the formula are added to disperse wind, scatter evil pathogens, and eliminate toxins.

Yu Zhen San [True Jade Powder] *Wai Ke Zheng Zong*

Tian Nan Xing	equal amount
Fang Feng	
Bai Zhi	
Tian Ma	
Qiang Huo	
Bai Fu Zi	

Main Indication: Tetanus. Symptoms include clenched teeth, tight lips, body stiffness, opisthotonos, and wiry or rapid pulse. Other symptoms can be spasms and twitching of the four limbs.

Preparation: Grind into fine powder, mix 3 g with hot wine or boy's urine each time. Can also be used topically.

Treatment Principles: Expel wind and relieve spasms.

In this formula, Bai Fu Zi and Tian Nan Xing are used raw, which can become toxic; thus one should not overdose.

This formula is effective if sweating occurs after taking the formula. It can also treat headache due to wind phlegm attacking the head.

Modification:

Addition	Indication
Quan Xie Wu Gong Jiang Can	Severe cases of clenched teeth and opisthotonos

Wu Hu Zhui Feng Tang [Five-Tiger Decoction to Pursue Wind] *Shang Han Lun*

Chan Tui	30 g
Zhi Nan Xing	6 g
Tian Ma	6 g
Quan Xie	7–9 pieces
Jiang Can (fried)	7–9 pieces
Zhu Sha (add to decoction)	1.5 g

Main Indication: Tetanus. Spasms, twitching, severe case of opisthotonos, and wiry and forceful pulse.

Preparation: Decoct the herbs, drain, then add 60 mL of yellow wine followed by fine Zhu Sha powder. Take one dose per day for 3 days. The effectiveness of the formula is indicated by sweating at the five centers after taking the herbs.

Treatment Principles: Expel wind, relieve spasms, and stop pain.

Zhi Jing San [Stop Spasms Powder] (Jing Yan Fang)

Quan Xie (bake until dry)	equal amount
Wu Gong (bake until dry)	

Main Indication: Spasms and syncope. Symptoms include continuous spasms of the four limbs, or body stiffness, opisthotonos, red tongue, yellow or greasy tongue coating, and rapid pulse.

Preparation: Grind into fine powder. For children, take 1 to 1.5 g each time. For adults, 6 g, two to four times a day.

Treatment Principles: Expel wind, relieve spasms.

This formula treats spasms and sycope similar to the spasms caused by warm febrile conditions such as encephalitis B or opisthotonos seen after warm febrile diseases. It is also effective in treating stubborn headache and joint pain.

Modification:

Addition	Indication
Da Qing Ye Ban Lan Gen Jin Yin Hua Lian Qiao	Severe heat toxins in warm febrile diseases

Related Formulas:

Formula	Herbs	Indication
Jin She Zhi Jing San [Stop Spasms Powder with Viper]	Core formula + Jin Qian She 6 g	Same indications. Stronger expelling wind, unblocking collaterals functions
Jia Wei Zhi Jing San [Augmented Stop Spasms Powder]	Core formula + Tian Ma Jiang Can	Stronger in relieving spasms function
Zheng She San [Rectify Tongue Deviation]	Quan Xie Wei (soak in vinegar, fried) 9 g Fu Ling 30 g Mix with ginger juice, sun dry, grind into fine powder, divide into six doses, take with warm water	Treat Liver heat generating wind, phlegm obstructing orifices, and tongue deviation and stiffness

Qian Zheng San [Lead to Symmetry Powder] *Yang Shi Jia Cang Fang*

Bai Fu Zi	equal amount
Jiang Can	
Quan Xie (remove toxin)	

Main Indication: Wind phlegm obstruction and stagnation, and cold type of mouth and eye deviation syndrome. Tongue is deviated, pale with white coating and floating tight or wiry pulse.

Preparation: Grind into fine powder, take 3 g of raw powder with warm water or warm wine, one to two times per day. It can also be taken as decoction or topically.

Treatment Principles: Expel wind, transform phlegm, and stop spasms.

This formula is suitable for mouth and eye deviation due to wind evil pathogens attacking the channels and tendons of the face. It should not be used if the cause is internal Liver wind.

Related Formula:

Formula	Herbs	Indication
Jia Wei Qian Zheng San [Augmented Lead to Symmetry Powder]	Core formula + Wu Gong Gou Teng Bai Zhi Grind into powder and take with Fang Feng Decoction	It is indicated for more severe syndromes. It is stronger in relieving spasms and expelling the wind function than the core formula

Jia Wei Tian Ma Tang [Augmented Gastrodia Decoction] *Zhong Yi Fu Ke Zhi Liao Xue*

Tian Ma	12 g
Bai Fu Zi (roasted)	9 g
Tian Nan Xing (roasted)	9 g
Ban Xia	9 g
Quan Xie (fried)	9 g
Gou Teng	9 g
Chen Pi	9 g

Main Indication: Wind phlegm upward disturbing (wind stroke). Symptoms include obese constitution, difficulty in speech, or locked jaw without speech, copious phlegm and saliva, phlegm sounds at throat, focal distention and stifling sensation at chest and epigastrium, paralysis of the four limbs, white, greasy tongue coating, and wiry and slippery pulse.

Preparation: Decoction.

Treatment Principles: Expel phlegm and extinguish wind.

This formula originally is designed for postpartum wind stroke due to wind phlegm upward disturbing the clear orifices. Clinically, it can also be used to treat syndromes of vertigo, dizziness, headache, numbness, and radiating pain in the four limbs due to upward disturbing wind phlegm.

Comparison of Formulas Treating External Wind Causing the Stirring of Internal Wind

Yu Zhen San [True Jade Powder] and Wu Hu Zhui Feng Tang [Five-Tiger Decoction to Pursue Wind] both are common formulas in treating tetanus. However, the former formula is stronger in its expelling wind and transforming phlegm function. It is weak in relieving spasms. The latter formula is good at expelling wind and relieving spasms. Zhi Jing San [Stop Spasms Powder] and Qian Zheng San [Lead to Symmetry Powder] both can expel wind and relieve spasms. However, Zhi Jing San [Stop Spasms Powder] is often used in body stiffness, spasms, and twitching in the later stage of warm febrile diseases. On the other hand, Qian Zheng San [Lead to Symmetry Powder] treats cold type of mouth and eye deviations due to wind phlegm obstructing the collaterals. Jia Wei Tian Ma Tang [Augmented Gastrodia Decoction] treats wind stroke in obese postpartum woman due to wind phlegm upward disturbing the clear orifices. It can also treat syndromes of vertigo and headache as in Meniere's disease, or syndromes of numbness and radiating pain in people with phlegm damp constitution.

SHAO YANG SYNDROME

When evil pathogens attack Shao Yang, the manifesting symptoms include alternating chills and fever, fullness and discomfort at chest and hypochondrium, bitter taste, dry throat, vertigo, restlessness, frequent nausea, no desire for food, and wiry pulse. The commonly used herbs are Chai Hu, Huang Qin, Qing Hao, Shan Zhi Zi, Ban Xia, and Dang Shen. Cold and hot herbs are used to expel evil pathogens, support the Zheng Qi, and harmonize Shao Yang.

The membrane that connects and lies between the internal organs and the body is referred to as the membrane source. It is half interior and half exterior. When epidemic and turbid pathogens enter the body through the mouth and nose, lodging in the membrane source, symptoms manifest as aversion to cold, high fever, malaria-like complaints, headache, stifling sensation at chest, nausea and vomiting, and greasy and dirty tongue coating. Commonly used herbs include Bing Lang, Cao Guo, Hou Po, Huang Qin, and Zhi Mu. It is similar to Shao Yang syndromes.

Xiao Chai Hu Tang [Minor Bupleurum Decoction] *Shang Han Lun*

Chai Hu	12 g
Huang Qin	9 g
Ren Shen	6 g
Ban Xia	9 g
Zhi Gan Cao	3 g
Sheng Jiang	9 g
Da Zao	4 pieces

Main Indication: Shao Yang syndrome. Symptoms include alternating chills and fever, distending pain of the chest and hypochondrium, no desire for food, restlessness, frequent vomiting, bitter taste, dry throat, vertigo, thin white tongue coating, and wiry pulse.

Preparation: Decoction, taken warm.

Treatment Principles: Harmonize Shao Yang.

This formula is the chief formula for harmonizing Shao Yang. It has wide applications in clinical practice for symptoms of alternating chills and fever, distending pain at chest and hypochondrium, vertigo, bitter taste, no desire for food, restlessness, and frequent vomiting. As long as one of the symptoms appear, one can use this formula. It can also be used in Shao Yang syndromes of malarial disorders, jaundice, and Liver/Gallbladder pathological conditions.

Modification:

Minus	Addition	Indication
Huang Qin	Bai Shao Yao	Abdominal pain
Huang Qin	Fu Ling	Palpitation, difficult urination
Ren Shen Da Zao Sheng Jiang	Wu Wei Zi Gan Jiang	Cough

Related Formulas:

Formula	Herbs	Indication
Chai Hu Zhi Jie Tang [Bupleurum, Bitter Orange, and Platycodon Decoction]	Chai Hu Zhi Ke Jiang Ban Xia Sheng Jiang Huang Qin Jie Geng Chen Pi Lu Cha (green tea)	Treat alternating chills and fever, headache at the corners of the head, deafness, vertigo, fullness and pain at chest and hypochondrium, white slippery tongue coating, wiry slippery pulse at right, and wiry floating big pulse at left
Chai Hu Gui Zhi Tang [Bupleurum and Cinnamon Twig Decoction]	Core forumula + Gui Zhi Bai Shao Yao	Treat the core syndrome with annoying pain at the limbs and unresolved exterior syndrome
Chai Hu Gui Jiang Tang [Bupleurum, Cinnamon, and Ginger Decoction]	Chai Hu Huang Qin Gan Jiang Mu Li Gua Lou Gen Rou Gui Da Zao	Treat malarial disorders with more cold and less heat or just cold without heat
Chai Hu Jia Long Gu Mu Li Tang [Bupleurum plus Dragon Bone and Oyster Shell Decoction]	Core formula + Gui Zhi Fu Ling Qian Dan Da Huang Long Gu Mu Li Da Huang (add near end)	Treat Xiao Chai Hu Tang [Minor Bupleurum Decoction] syndrome with irritability, fright, delirious speech, and seminal emission
Chai Hu Jia Mang Xiao Tang [Bupleurum Decoction plus Mirabilite]	Core formula + Mang Xiao	Treat the core syndrome with constipation
Chai Hu Jia Lu Gen Tang [Bupleurum Decoction plus Reed]	Core formula + Lu Gen 60 g	Treat cough and vomiting bile
Jia Jian Xiao Chai Hu Tang [Modified Minor Bupleurum Decoction]	Core formula – Ren Shen Da Zao + Tao Ren Sheng Di Huang Xi Jiao (Shui Niu Jiao) Mu Dan Pi Shan Zha	Treat heat entering uterus during menstruation, battling and knotted heat and stasis in lower back, hypochondrium, and pulling pain in lower abdomen; refuses palpation and pressure If with delirium, remove Xi Jiao (Shui Niu Jiao) and add Qing Dai, Mang Xiao, and Da Huang
Chai Hu Si Wu Tang [Bupleurum and Four-Substance Decoction]	Core formula – Ren Shen Da Zao + Dang Gui Sheng Di Huang Bai Shao Yao Chuan Xiong	Treat evil attacking Jue Yin. Symptoms include alternating chills and fever as in malarial disorders; and radiating pain at hypochondrium, worsening at night

(continued)

Formula	Herbs	Indication
Chai Ping Tang [Bupleurum and Calm the Stomach Decoction]	Core formula – Ren Shen Da Zao + Hou Po Chi Fu Ling Cang Zhu Chen Pi	Treat severe dampness type of malarial disorder. It is more effective if Qing Hao is added
Qing Pi Yin [Clear the Spleen Decoction]	Chai Hu Huang Qin Ban Xia Qing Pi Hou Po Bai Zhu Cao Guo Fu Ling Gan Cao	Treat malarial disorder with severe cold and mild heat, fullness and stifling sensation at chest and diaphragm, no desire for food, white and slippery or white greasy tongue coating, and soggy and moderate pulse

Da Chai Hu Tang [Major Bupleurum Decoction] *Jin Gui Yao Lue*

Chai Hu	15 g
Huang Qin	9 g
Bai Shao Yao	9 g
Ban Xia	9 g
Zhi Shi	9 g
Da Huang (add near end)	9 g
Sheng Jiang	9 g
Da Zao	4 pieces

Main Indication: Concomitant Shao Yang and Yang Ming syndrome. Symptoms include alternating chills and fever, fullness and pain at chest and hypochondrium, unceasing vomiting, focal distention and hardness at epigastrium, or distending pain below the Heart; constipation, or heat type diarrhea; slight irritability; yellow tongue coating; and wiry and forceful pulse.

Preparation: Decoction, taken warm.

Treatment Principles: Harmonize Shao Yang and drain the internal accumulated heat.

This formula is formed by removing Ren Shen and Gan Cao, and adding Da Huang, Zhi Shi, and Bai Shao Yao to Xiao Chai Hu Tang [Minor Bupleurum Decoction]. In clinical practice, the indicative symptoms for this syndrome are alternating chills and fever, fullness and pain at the chest, hypochondrium and epigastrium, yellow tongue coating, and constipation. It can be used in treating acute pancreatitis, acute cholecystitis, and cholelithiasis which belongs to excess heat in Gallbladder and stomach. It can also treat upward blazing of Liver and Gallbladder fire which manifests as headache, tinnitus, deafness, red eyes, and mania.

Modification:

Addition	Indication
Mu Xiang Yu Jin Yin Chen Hao Mang Xiao	Severe condition
Qing Dai Shan Zhi Zi Mu Dan Pi Mang Xiao	Upward blazing of Liver fire causing mania. If there is focal distention and obstruction at epigstrium and severe movement at Ren 17, then add Tie Fen (iron powder)

Related Formulas:

Formula	Herbs	Indication
Jia Wei Da Chai Hu Tang [Augmented Major Bupleurum Decoction]	Core formula + Tao Ren Chi Shao Yao	Treat pathogenic heat entering the uterus right at the end of menstruation with symptoms of fullness and pain at lower back, hypochondrium, and lower abdomen
Jia Jian Da Chai Hu Tang [Modified Major Bupleurum Decoction]	Core formula – Ban Xia Sheng Jiang Da Zao + Chuan Lian Zi Yan Hu Suo Mu Xiang Pu Gong Ying Gan Cao	Treat acute pelvic syndromes. In severe case of infection in pelvic cavity, add Jin Yin Hua and Lian Qiao With constipation, add Mang Xiao With severe case of blood stasis, add Tao Ren, Hong Hua, Sheng Pu Huang, and Chuan Xiong
Chai Hu Xian Xiong Tang [Bupleurum Decoction for Sinking into the Chest]	Chai Hu Ban Xia Huang Qin Mu Xiang Yu Jin Zhi Shi (fried) Huang Lian Da Huang Mang Xiao Bai Shao Yao	Treat acute cholecystitis or acute attack of chronic cholecystitis with alternating chills and fever; pain at upper right quadrant of the abdomen; refuses pressure, constipation, and yellow, greasy tongue coating
Dan Nang Yan Da Chai Hu Tang [Major Bupleurum Decoction for Cholecystitis]	Chai Hu Mu Xiang Zhi Ke Chuan Lian Zi Yan Hu Suo Sheng Da Huang (add near end) Huang Qin	Treat Qi stagnation type of cholecystitis with distending pain of the right hypochondrium which is better with belching; stifling sensation at chest; poor appetite, thin tongue coating; and wiry pulse
Jia Jian Yin Chen Da Chai Hu Tang [Modified Major Bupleurum Decoction with Artemesia Yinchenhao]	Yin Chen Hao 30 g Chai Hu 9 g Shan Zhi Zi 12 g Huang Qin 9 g Da Huang (add near end) 9 g Mang Xiao (add to decoction) 9 g Zhi Ke 6 g Qing Pi 6 g Mu Xiang 9 g	Treat damp–heat type of cholecystitis, cholelithiasis, acute pancreatitis, with bitter taste, red tongue, yellow tongue coating, and wiry, rapid pulse
Hou Po Qi Wu Tang [Seven-Substance Decoction with Magnolia Bark]	Hou Po 15 g Gan Cao 6 g Da Huang 9 g Zhi Shi 9 g Gui Zhi 6 g Da Zao 4 pieces Sheng Jiang 12 g	Treat interior syndrome with unresolved exterior syndrome. Symptoms include abdominal fullness, fever, floating rapid pulse, constipation, and bitter taste or dry mouth

Da Yuan Yin [Reach the Membrane Source Decoction] *Wen Yi Lun*

Bing Lang	12 g
Hou Po	9 g
Cao Guo	3 g
Zhi Mu	9 g
Bai Shao Yao	6 g
Huang Qin	9 g
Gan Cao	3 g

Main Indication: Warm epidemics or malarial disorder with evil pathogens at the membrane source. Symptoms include aversion to cold, high fever once a day or two to three times a day at random times, stifling sensation at chest, nausea and vomiting, headache, irritability and agitation, deep red tongue sides, thick powdery tongue coating, and wiry rapid pulse.

Preparation: Decoction, taken warm after lunch.

Treatment Principles: Disperse and vent the membrane source, separate clear from turbid, and transform turbid.

This formula can treat flu and malarial disorder.

Modification:

Minus	Addition	Indication
Bai Shao Yao Zhi Mu	Pei Lan Yin Chen Hao	For more coldness and slight fever, more dampness than heat in flu
Bing Lang	Bai Wei Shan Zhi Zi	For flu with more heat than damp
	Qing Hao 60–90 g	Remove decoction for a few seconds after boiling (Qing Hao is not good for cooking long) Severe type of malarial disorders

Related Formulas:

Formula	Herbs	Indication
Chai Hu Da Yuan Yin [Bupleurum Decoction to Reach the Membrane Source]	Core formula – Zhi Mu Bai Shao Yao + Chai Hu Zhi Ke Qing Pi Jie Geng He Geng (lotus stem)	Treat phlegm damp obstructing the membrane source. Symptoms include fullness and focal distention of the chest and diaphragm, restlessness, irritability, vertigo, greasy taste in mouth, unproductive cough, tertian malarial complaints; thick tongue coating like powder, rough and resistant to being scraped off; and wiry, slippery pulse
Xuan Tou Mo Yuan Fa [Disperse and Vent the Membrane Source Formula]	Core formula – Zhi Mu Bai Shao Yao + Huo Xiang Ye Ban Xia Sheng Jiang	Treat warm malarial disorders with severe chills, slight fever, body ache with sweating, heavy limbs, and epigastric fullness
Xin Ding Da Yuan Yin [Newly-Settled Reach the Membrane Source Decoction]	Hou Po Bing Lang Cao Guo Zhi Ke Jian Shan Zhi Zi Dan Dou Chi Huang Qin Jie Geng Xian He Ye wrap Liu Yi San [Six-to-One Powder] Zhi Mu Lu Gen Xi Xin	Treat warm malarial disorder, severe heat, mild cold, even coma, and offending foul smell
Chang Shan Cao Guo Yin [Dichroa and Amomi Tsao-ko Decoction]	Chang Shan Cao Guo Ban Xia Chen Pi Hou Po Cang Zhu Gna Cao	Harmonize Shao Yang; warm, transform, and reach the evil pathogens, transform phlegm, eliminate food stagnation, dry dampness, and stop malaria

San Xiang Tang [Three-Fragrant Decoction] *Wen Bing Tiao Bian*

Dan Dou Chi	6 g
Jiang Xiang	10 g
Yu Jin	6 g
Gua Lou Pi	10 g
Jie Geng	10 g
Shan Zhi Zi (burned)	10 g
Zhi Ke	6 g

Main Indication: Damp–heat turbid, obstructing the membrane source. Symptoms include absence of hunger, no appetite, misted orifices, delirious speech in milder cases, mania in severe cases or delirium and coma in extreme cases; or tidal fever, stifling sensation at chest, white greasy tongue coating, or thick and soggy or soggy, rapid pulse.

Preparation: Decoction, divided into two doses.

Treatment Principles: Clear heat, transform phlegm, relieve stagnation, and open orifices.

This formula mainly treats abnormal emotional or psychological syndromes due to damp–heat obstructing the clear orifices.

Related Formula:

Formula	Herbs	Indication
Qing Zhang Tang [Clear Miasma Decoction]	Qing Hao Chai Hu Fu Ling Zhi Mu Chen Pi Ban Xia Huang Qin Huang Lian Zhi Shi Chang Shan Zhu Ru Hua Shi Gan Cao	Relieve toxicity, clear miasma, disperse Liver, promote Gallbladder, clear heat, and reserve body fluid Treat Shao Yang summer damp, miasma, and miasmic malaria

Qing Dan Xuan Bi Tang [Clear the Gallbladder and Disband Painful Obstruction Decoction] (Jing Yan Fang)

Pi Pa Ye	10 g
She Gan	12 g
Yu Jin	10 g
Zhi Ke	12 g
Qing Hao (add near end)	30 g
Huang Qin	10 g
Ban Lan Gen	30 g
Yin Chen Hao	30 g
Ban Xia	12 g
Bing Lang	15 g
Lu Gen	30 g
Hua Shi	10 g
Shi Chang Pu	10 g

Main Indication: Damp–heat in San Jiao. Symptoms include intermittent chills and feverish sensation, dizziness, distending headache, distention and fullness at chest and hypochondrium, bitter taste in mouth, nausea and vomiting, stifling sensation at epigastrium, absence of hunger, thirst without desire to drink, thick greasy tongue coating, and soggy, rapid pulse.

Preparation: Decoction.

Treatment Principles: Clear heat, dry dampness, transform dampness, and promote urination.

Modification:

Addition	Indication
Bai Dou Kou Huo Xiang	Very thick and greasy tongue coating
Cao Guo Hou Po	Severe stifling sensation at chest, nausea, and vomiting

Related Formula:

Formula	Herbs	Indication
Yu Shu Dan [Jade Pivot Special Pill]	Shan Ci Gu Xu Sui Zi Da Ji She Xiang Niu Huang Zhu Sha Wu Bei Zi	Expel phlegm, open orifices, move the turbid downward, and transform it Treat dry cholera due to turbid phlegm obstructions, convulsions due to phlegm in children, mumps in children, and nausea and vomiting

Sheng Jiang San [Ascending Descending Powder] *Han Wen Tiao Bian*

Jiang Can (fried with wine)	6 g
Chan Tui	3 g
Jiang Huang	9 g
Sheng Da Huang	12 g

Main Indication: Big chills and big fever of the exterior, interior, and San Jiao. Syndromes due to dysfunction of the ascending and descending Qi mechanism and abnormal flow of blood.

Preparation: Grind into fine powder, mix with one bottle of yellow wine, 15 g of honey, and divide into three to four doses. After recovery, discontinue use. If the formula is taken and made as honey pills, it is called Tai Ji Wan [Tai Ji Pill].

Treatment Principles: Ascend the clear, descend the turbid, and treat exterior and interior at the same time.

In this formula, Jiang Can and Chan Tui ascend the clear Yang within the Yang. Jiang Huang and Da Huang descend the turbid Yin of the Yin. The ascending and descending actions and the harmonizing of the interior and exterior eliminate the toxic evil pathogens and epidemic factors. Thus, this formula has wide applications.

Comparison of Formulas Treating Shao Yang Syndromes

Xiao Chai Hu Tang [Minor Bupleurum Decoction] is the representative formula in treating Shao Yang syndrome. It is elegantly formulated with cold and hot herbs. It is used to treat a variety of conditions and has wide applications. Often this formula can be used and modified as long as one of the indicated symptoms is presented. Da Chai Hu Tang [Major Bupleurum Decoction] treats concomitant Shao Yang and Yang Ming syndrome. Thus, it not only harmonizes but also unblocks the Fu organs. Da Yuan Yin [Reach the Membrane Source Decoction] and San Xiang Tang [Three-Fragrance Decoction] treat evil pathogens at the membrane source. The former mainly manifests as chills and malaria-like fever and the latter manifests as an abnormal emotional state. Qing Dan Xuan Bi Tang [Clear the Gallbladder and Disband Painful Obstruction Decoction] can dry and transform dampness and promote urination. Thus, it is used to treat various syndromes of damp–heat obstructing San Jiao. Sheng Jiang San [Ascending and Descending Powder] is indicated for severe syndromes of both exterior and interior where cold and heat symptoms are seen at the same time, with the dysfunction of ascending and descending and abnormal flow of Qi and blood.

STAGNANT HEAT IN GALLBLADDER

This syndrome often manifests as gallstones and jaundice. Symptoms include yellow appearance at the body, eyes, yellow urine, spasmodic or distending pain at Gallbladder, bitter taste in mouth, dry throat, constipation, and wiry or wiry slippery pulse. The commonly used herbs are Shan Zhi Zi, Huang Qin, Huang Lian, Long Dan Cao, Hu Zhang, Pu Gong Ying, Ban Lan Gen, Da Qing Ye, Qing Hao, Qing Pi, Mu Xiang, Zhi Ke, Chuan Lian Zi, Yu Jin, Yin Chen Hao, Hua Shi, Mu Tong, Jin Qian Cao, Da Huang, and Mang Xiao. These herbs clear Liver and Gallbladder, and unblock and descend Gallbladder Qi, thus the stagnation can be eliminated.

Yin Chen Hao Tang [Artemisia Yinchenhao Decoction] *Shang Han Lun*

Yin Chen Hao	30 g
Shan Zhi Zi	15 g
Da Huang	9 g

Main Indication: Damp–heat jaundice. Symptoms include whole body and face jaundice with a color like fresh tangerine, slight abdominal fullness, thirst, difficulty urinating, yellow greasy tongue coating, and deep excessive or slippery rapid pulse.

Preparation: Decoction.

Treatment Principles: Clear heat, drain damp, and reduce jaundice.

This formula treats Yang Huang–Yang type of jaundice where heat pathogen is more than damp. It can be used in jaundice due to acute infectious hepatitis, cholelithiasis, and cholecystitis.

Related Formulas:

Formula	Herbs	Indication
Zhi Zi Bai Pi Tang [Gardenia, Phellodendron Decoction]	Shan Zhi Zi 15 g Gan Cao 6 g Huang Bai 9 g	Decoction, taken warm Treats fever and jaundice due to cold-induced disorder
Zhi Zi Da Huang Tang [Gardenia and Rhubarb Decoction]	Shan Zhi Zi 12 g Da Huang 12 g Zhi Shi 9 g Dan Dou Chi 9 g	Decoction, taken warm. Treats wine jaundice, irritability, heat and pain, and difficulty urinating
Da Huang Xiao Shi Tang [Rhubarb and Mirabilite Decoction]	Da Huang 12 g Huang Bai 12 g Xiao Shi (add near end) 12 g Shan Zhi Zi 9 g	Treats jaundice, abdominal fullness, difficult and dark urination, and spontaneous sweating

Qing Dan Xing Qi Tang [Clear the Gallbladder and Move Qi Decoction] (Tian Jin Nan Kai Yi Yuan Fang)

Chai Hu	9 g
Huang Qin	9 g
Ban Xia	9 g
Mu Xiang	9 g
Yu Jin	9 g
Da Huang (add near end)	9 g
Zhi Ke	9 g
Xiang Fu	9 g
Yan Hu Suo	9 g
Bai Shao Yao	15 g

Main Indication: Qi stagnation type of cholecystitis. Symptoms include distending, radiating, or spasmodic pain at hypochondrium, impatience and easily angered, bitter taste in mouth, dry throat, dizziness, no desire to eat, slight red tongue tip, thin white or slightly yellow tongue coating, and wiry pulse.

Preparation: Decoction.

Treatment Principles: Spread Liver Qi and promote Gallbladder; stop pain.

Qing Dan Li Shi Tang [Clear the Gallbladder and Drain Damp Decoction] (Tian Jin Nan Kai Yi Yuan Fang)

Chai Hu	9 g
Huang Qin	9 g
Ban Xia	9 g
Mu Xiang	9 g
Yu Jin	9 g
Da Huang (add near end)	9 g
Che Qian Zi	9 g
Mu Tong	9 g
Shan Zhi Zi	9 g
Yin Chen Hao	15 g

Main Indication: Damp–heat type of cholecystitis and cholelithiasis. Symptoms include distending or spasmodic pain at the hypochondrium, bitter taste in mouth, dry throat, dizziness, no desire to eat, alternating chills and fever; yellow sclera, yellow body and yellow urine the color of a tangerine; constipation; red tongue yellow with greasy coating, and wiry, slippery, or slippery, rapid pulse.

Preparation: Decoction.

Treatment Principles: Clear Gallbladder, promote urination, stop pain, and reduce jaundice.

Qing Dan Xie Huo Tang [Clear the Gallbladder and Drain Fire Decoction] (Tian Jin Nan Kai Yi Yuan Fang)

Chai Hu	15 g
Huang Qin	15 g
Ban Xia	15 g

(continued)

Mu Xiang	15 g
Yu Jin	15 g
Da Huang (add near end)	15 g
Mang Xiao (add to decoction)	15 g
Shan Zhi Zi	9 g
Long Dan Cao	9 g
Yin Chen Hao	30 g

Main Indication: Excess fire type of cholecystitis. Symptoms include continuous distending pain at hypochondrium, bitter taste in mouth, dry throat, dizziness, no desire to eat, alternating chills and fever, whole body jaundice with a tangerine-like color, yellow turbid urine, constipation, abdominal distention and fullness, deep red tongue, prickled or yellow dry tongue coating, and wiry, slippery, and rapid pulse.

Preparation: Decoction.

Treatment Principles: Clear Gallbladder, unblock Fu organs, drain fire, and stop pain.

Modification:

Addition	Indication
Ban Lan Gen Jin Yin Hua Lian Qiao	Severe heat
Da Huang (increase dosage) Mang Xiao (increase dosage) Hou Po	Severe constipation
Chuan Lian Zi Yan Hu Suo	Severe pain
Zhu Ru	Vomiting
Huo Xiang Pei Lan Shan Zha	Poor appetite
Tao Ren Dang Gui Chi Shao Yao Hong Hua	Blood stasis

Dan Dao Pai Shi Tang [Expelling Stones from Biliary Tract Decoction] (Jing Yan Fang)

Jin Yin Hua	15 g
Huang Qin	9 g
Zhi Ke	9 g
Mu Xiang	9 g
Bai Shao Yao	30 g
Jin Qian Cao	30 g
Yin Chen Hao	30 g
Da Huang (add near end)	9 g
Mang Xiao (add to decoction)	15 g
Chai Hu	15 g

Main Indication: Acute attack of cholelithiasis or cholecystitis. Symptoms include severe right upper quadrant pain which refuses pressure, nausea, vomiting, thirst with preference for cold drinks, high fever, aversion to cold, jaundice, constipation, red tongue, yellow greasy tongue coating, and wiry and rapid pulse.

Preparation: Decoction.

Treatment Principles: Clear heat, promote Gallbladder, unblock Fu organs, expel stones, and stop pain.

After taking this formula, some patients report unblocking sensation from upper part of the Liver area to the lower part. One can also combine acupuncture, needling points on Liver and Gallbladder channel or the corresponding back Shu points to promote expelling stones. During expelling of stones, intermittent spasmodic pain can occur at Gallbladder. Patient should be monitored carefully and observed for stones in stools.

Modification:

Minus	Addition	Indication
Bai Shao Yao	Zhi Shi Hou Po	Cholelithiasis
	Bai Shao Yao to 60 g	Spasmodic pain during acute attack of cholecystitis

Related Formulas:

Formula	Herbs	Indication
Jia Jian Dan Dao Pai Shi Tang [Modified Expelling Stone from Biliary Tract Decoction]	Core formula – Bai Shao Yao Jin Qian Cao + Zhi Shi	Treat cholelithiasis
Pai Shi Tang [Expelling Stone Decoction]	Core formula – Jin Yin Hua Huang Qin Mang Xiao + Yu Jin	Cholelithiasis without obvious constipation
Li Dan Wan [Promote Gallbladder Pill]	Yin Chen Hao 120 g Long Dan Cao 90 g Yu Jin 90 g Mu Xiang 90 g Zhi Ke 90 g	Grind into fine powder, add 500 g of pig or goat bile, and then add moderate amount of honey to form pills. Each pill is 9 g. Take one in the morning and at night Treat chronic cholelithiasis and, especially, sandy type of stones

Comparison of Formulas Treating Stagnant Heat in Gallbladder

Yin Chen Hao Tang [Artemisia Yinchenhao Decoction] is a famous and effective formula in treating jaundice. It is also a basic formula. In the formula, Da Huang unblocks the bowel which can spread the Liver and promote Gallbladder. Thus, it became the model for the later dynasties to use Da Huang to purge downward and invigorate blood to treat Liver and Gallbladder disorders.

Qing Dan Xing Qi Tang [Clear the Gallbladder and Move Qi Decoction], Qing Dan Li Shi Tang [Clear the Gallbladder and Drain Damp Decoction], and Qing Dan Xie Hua Tang [Clear the Gallbladder and Drain Fire Decoction] all can clear and drain heat from Gallbladder, spread Liver, and regulate Qi. They all can treat hypochondriac pain and jaundice due to cholecystitis and cholelithiasis. However, the first formula focuses on moving Qi and stopping pain which is suitable for Qi stagnation type. The middle formula clears Liver and promotes urination at the same time, thus it is good for damp–heat type. The latter formula focuses on clearing Liver, promoting Gallbladder, unblocking the bowels, and draining heat. Thus, it is good for excess heat fire type.

Dan Dao Pai Shi Tang [Expelling Stone from Biliary Tract Decoction] is used in acute attack of cholecystitis or cholelithiasis which focuses on quickly draining Gallbladder heat, unblocking bowels, and expelling stones. At the same time, the patient should be carefully monitored while taking the formula. Once the stones are out, the formula should be discontinued.

PARASITES IN BILE DUCT/BILIARY ASCARRIASIS

Intestinal parasites can be stimulated by heat, cold, or hunger that move upward into the bile duct. The symptoms include intermittent spasmodic pain or heavy distending pain below the xiphoid process, or at the right hypochondriac region. In severe cases, strong pain can cause screaming, with shortness of breath, bitter taste in mouth, nausea, vomiting sour bitter bile and gastric acid, and wiry or rapid pulse. All symptoms can be absent between attacks.

When roundworms encounter sourness, they are calmed; when they encounter bitterness, they retreat back to the intestines; and when they encounter acridity, they hide. Thus, Wu Mei, Huang Lian, Huang Bai, Chuan Jiao, Xi Xin, and other sour, bitter, and acrid herbs are used in combination with Zhi Ke, Mu Xiang, Yu Jin, and Bing Lang which move Qi, promote the Gallbladder function to numb the parasites, relieve spasms of the bile duct, and promote bile flow. Then, the parasites can return to the intestines from the bile duct and be expelled out of the body.

Wu Mei Wan [Mume Pill] *Shang Han Lun*

Wu Mei	480 g
Xi Xin	180 g
Gan Jiang	300 g
Huang Lian	480 g
Dang Gui	120 g
Fu Zi	180 g
Chuan Jiao	120 g
Gui Zhi	180 g
Ren Shen	180 g
Huang Bai	180 g

Main Indication: Spells of attacks by roundworms. Symptoms include sudden intermittent spasmodic pain at epigastrium or heavy distending pain; irritability; vomiting or even vomiting roundworms after eating; greenish complexion with sweating after severe pain attack; cold hands and feet; and hidden pulse. The patient is normal between attacks.

Preparation: Soak Wu Mei in 50% vinegar overnight, remove the pits, then roast with all other herbs until dry. Grind into powder and mix with honey to form pill. Take 9 g, one to three times a day on an empty stomach. It can also be taken as decoction; dosage should be reduced proportionally.

Treatment Principles: Warm the Zang organs and calm roundworms.

This formula is an effective formula in treating roundworms in the bile duct. It can also be used to treat chronic diarrhea and dysentery when Zheng Qi is deficient and the evil pathogens are not cleared. Since Wu Mei, Chuan Jiao, Huang Lian, and Huang Bai have strong antibacterial function, this formula is also effective in treating skin conditions due to fungal infections.

Related Formulas:

Formula	Herbs	Indication
Jiao Mei Tang [Pepper and Mume Decoction]	Huang Lian 6 g Huang Qin 6 g Ban Xia 6 g Gan Jiang 6 g Ren Shen 6 g Zhi Shi 6 g Bai Shao Yao 9 g Chuan Jiao 9 g Wu Mei 9 g	Treat summer warm remains at Jue Yin, Xiao Ke, hardness and focal distention at epigstrium, nausea, vomiting roundworms, mixed chills and fever, and diarrhea with blood
Li Zhong An Hui Wan [Regulate the Middle and Calm Roundworms Pill]	Ren Shen Bai Zhu Fu Ling Chuan Jiao Wu Mei Gan Jiang	Taken as decoction or pill Treat deficient cold of Spleen and Stomach, loose stool; clear urine, abdominal pain, borborygmus, cold limbs, deficient and moderate pulse, and vomiting roundworms or passing out the roundworms through stools
Lian Mei An Hui Tang [Picrorhizae and Mume Decoction to Calm Roundworms]	Hu Huang Lian Chuan Jiao Lei Wan Wu Mei Huang Bai Bing Lang	Decoction Treat worm accumulation and abdominal pain due to Liver and stomach excess heat. Other symptoms include no desire to eat and vomiting worms after eating; severe cases include irritability, collapse, red face, dry mouth, red tongue, and rapid pulse and fever

Qu Hui Tang [Expel Roundworm Decoction] (Jing Yan Fang)

Bing Lang	30 g
Shi Jun Zi	30 g
Ku Lian Gen Pi	20 g
Wu Mei	15 g
Mu Xiang	12 g
Zhi Ke	12 g
Mang Xiao (taken with decoction)	9 g
Chuan Jiao	6 g
Xi Xin	5 g
Gan Jiang	6 g

Main Indication: Roundworm in bile duct or intestine.

Preparation: Decoction.

Treatment Principles: Purge, drain, and expel roundworms.

This formula calms and expels roundworms, and treats both root and branch symptoms.

Modification:

Addition	Indication
Dang Shen Dang Gui Bai Shao Yao	Deficiency symptoms
Da Huang	Severe excess symptoms
Rou Gui Fu Zi	Severe coldness
Huang Qin Huang Lian Shan Zhi Zi Yin Chen Hao Lian Qiao	Severe heat symptoms

Gan Cao Fen Mi Tang [Licorice, Lead, and Honey Decoction] *Jin Gui Yao Lue*

Gan Cao	6 g
Qian Fen	3 g
Bai Mi (honey)	12 g

Main Indication: Roundworms in bile duct. Symptoms are as indicated in Wu Mei Tang [Mume Pill].

Preparation: Decoct Gan Cao and add Qian Fen and Bai Mi into the decoction. Take with meal.

Treatment Principles: Kill parasites and stop pain.

After taking this formula, pain will stop in 1 to 2 hours.

Comparison of Formulas Treating Parasites in Bile Duct Formulas

Wu Mei Wan [Mume Pill], Qu Hui Tang [Expel Roundworms Decoction], and Gan Cao Fen Mi Tang [Licorice, Lead, and Honey Decoction] not only treat roundworms in the bile duct, but are also effective in treating intestinal parasites. The first formula focuses on calming the roundworms and making them retreat into the intestines. The middle formula focuses on calming and killing roundworms and expelling them out of the body. The latter formula kills the roundworms by using honey to lure them into eating lead but does not have the function of promoting expulsion of roundworms.

CHAPTER 6

Liver and Lung, Kidney Syndromes and Formulas

LIVER AND KIDNEY SYNDROMES AND FORMULAS

Liver and Kidney Deficiency

This syndrome manifests as emaciation, dizziness, tinnitus, sore and weak lower back and legs, nocturnal emission, night sweats, poor appetite, apathy, red tongue with scanty coating, and thin pulse. The commonly used herbs to nourish Kidney and Liver are Shu Di Huang, Shan Zhu Yu, Nu Zhen Zi, Han Lian Cao, Gou Qi Zi, Tu Si Zi, Bai Shao Yao, and Tian Men Dong.

Gu Yin Jian [Stabilize the Yin Decoction] *Jin Yue Quan Shu*

Dang Shen	12 g
Shu Di Huang	12 g
Shan Yao	12 g
Shan Zhu Yu	9 g
Yuan Zhi	3 g
Tu Si Zi	9 g
Xu Duan	9 g
Wu Wei Zi	3 g
Zhi Gan Cao	6 g

Main Indication: Liver and Kidney deficiency. Symptoms include dizziness, tinnitus, deafness, sore lower back and legs, increased volume and frequency of nocturia, apathy, poor appetite, loose stool, and poor memory; or irregular menstruation, sometimes scanty and sometimes heavy flow, pale thin menstrual blood; difficulty falling asleep, excessive dreaming; pale tongue with thin white coating, and deep or thin, weak pulse.

Preparation: Decoction. Taken warm on empty stomach.

Treatment Principles: Nourish Liver and Kidney.

This formula is designed for perimenopause or menopausal syndrome due to Liver blood and Kidney essence deficiency. It can also treat lower energy and essence after middle age in males.

Comparison of the Two Formulas Treating Liver and Kidney Deficiency

Gu Yin Jian [Stabilize the Yin Decoction] and Zuo Gui Wan (please refer to Chapter 9 for more information) [Restore the Left (Kidney) Pill] both treat Liver and Kidney deficiency. The former formula treats the beginning stage of this syndrome where there is no heat manifestation. The latter formula treats later stage of this syndrome which is more severe with heat manifestations. Liver blood and Kidney essence are greatly supplemented with Gui Ban Jiao and Lu Jiao Jiao to unblock Ren and Du channels and to promote the formation of essence and blood.

YIN DEFICIENCY WITH DEFICIENT HEAT

This syndrome manifests as dry mouth, dry throat, malar flush, afternoon fever, night sweats, sore lower back and spine, nocturnal emission, yellow urine, red tongue, and thin, rapid pulse. For treatment, herbs that clear deficient heat and steaming bone symptoms are added to nourishing Liver and Kidney herbs, such as Huang Bai, Zhi Mu, Mu Dan Pi, Di Gu Pi, Yin Chai Hu, and Hu Huang Lian. In addition, Long Gu, Mu Li, Gui Ban and Bie Jia are added to subdue Yang to form comprehensive formulas.

For weak and atrophic tendons and bones, herbs that nourish and strengthen tendon and bone are added to nourishing Yin and clearing deficient heat formulas, such as Huai Niu Xi, Suo Yang, Xu Duan, Du Zhong, and bobcat bone.

At later stages of heat febrile disease, when Yin is damaged and heat pathogens are hidden at the Lower Jiao, the syndrome manifests as coolness in the morning, heat in late afternoon, decreased fever without sweating, red tongue, and thin, rapid pulse. To treat this condition, one should nourish Yin, clear heat, and vent the evil pathogens at the same time.

Liang Di Tang [Rehmannia and Lycium Root Bark Decoction]
Fu Qing Zhu Nu Ke

Sheng Di Huang	30 g
Xuan Shen	15 g
Mai Men Dong	9 g
Bai Shao Yao	12 g
Di Gu Pi	9 g
E Jiao (melt in decoction)	12 g

Main Indication: Yin deficiency and blood heat. Symptoms include early menstruation, bright red blood, scanty volume, tidal fever, dizziness, restlessness, red tongue, and deficient pulse.

Preparation: Decoction.

Treatment Principles: Nourish Yin and clear heat.

This formula focuses on nourishing Yin and clearing heat as a support to treating Yin deficiency with mild heat syndrome.

Related Formulas:

Formula	Herbs	Indication
Jia Jian Liang Di Tang [Modified Rehmannia and Lycium Root Bark Decoction]	Core formula – Mai Men Dong + Yi Mu Cao Jiao Ai Ye	Treat Yin deficiency, blood heat, which manifests as heavy menstrual flow, persisting for a few days, dark red blood or purple with clots, distending pain at lumbar and abdomen, red face, sometimes tidal fever, red dry lips, red tongue, yellow tongue coating, and thin, rapid pulse.
Jia Jian Yi Yin Jian [Modified One Yin Decoction]	Sheng Di Huang Shu Di Huang Bai Shao Yao Mai Men Dong Di Gu Pi Zhi Mu Gan Cao	Treat Yin deficiency and blood heat, early menstruation, heavy purple flow, abdominal distention, lumbar pain, restlessness, tidal fever, pale red tongue, thin yellow coating, and deficient, rapid pulse.

Bie Jia Yang Yin Jian [Nourish Yin Decoction with Soft-Shelled Turtle Shell] *Zhong Yi Fu Ke Zhi Liao Xue*

Bie Jia	15 g
Gui Ban	15 g
Sheng Di Huang	24 g
Bai Shao Yao	12 g
Gou Qi Zi	12 g
Mu Dan Pi	9 g
Di Gu Pi	12 g
Ye Jiao Teng	15 g
Fu Shen	12 g

Main Indication: Yin deficiency, blood deficiency, and Yang hyperactivity. Symptoms include amenorrhea, malar flush, tidal fever, night sweats, restlessness, insomnia, five center heat, dry mouth, red lips, red tongue with scanty coating, and thin, rapid pulse.

Preparation: Decoct Bie Jia and Gui Ban for 2 hours, then decoct with rest of the herbs.

Treatment Principles: Nourish Yin, clear heat, and subdue Yang.

This formula can treat not only gynecological complaints due to Yin and blood deficiency, but also chronic insomnia, steaming heat, and restlessness and agitation due to Yin and blood deficiency with hyperactive Yang.

Jia Jian Qing Jing Tang [Modified Clear the Menses Decoction] *Zhong Yi Fu Ke Zhi Liao Xue*

Dan Shen	9 g
Di Gu Pi	15 g
Bai Shao Yao	9 g
Sheng Di Huang	9 g
Xuan Shen	12 g
Zhi Mu	12 g
Huang Bai	6 g

Main Indication: Yin deficiency, blood heat. Symptoms include early menstruation, deep red and heavy menstrual blood, sometimes with clots, red face, red lips, thirst and preference for cool drinks, deep red tongue, and wiry rapid pulse.

Preparation: Decoction.

Treatment Principles: Nourish Yin and descend fire.

This formula focuses on clearing the fire and supports by nourishing Yin. It can also treat epistaxis and gum bleeding due to blood heat.

Related Formula:

Formula	Herbs	Indication
Qing Jing Tang [Clear the Menses Decoction]	Mu Dan Pi 9 g Di Gu Pi 12 g Bai Shao Yao 12 g Shu Di Huang 15 g Qing Hao 12 g Fu Ling 9 g Huang Bai 9 g	Treat the same indications as the core formula. The clearing-heat function is slightly weaker. Thus, it is effective in treating mild heat and menstruation with little clots. If Shu Di Huang is replaced by Sheng Di Huang, then the nourishing Yin and clearing heat functions are stronger

Da Bu Yin Wan [Great Tonify the Yin Pill] *Dan Xi Xin Fa originally known as Da Bu Wan [Great Tonify Pill]*

Huang Bai (fried until brown)	120 g
Zhi Mu (fried)	120 g
Shu Di Huang (steamed with wine)	180 g
Gui Ban (zhi)	180 g

Main Indication: Liver and Kidney Yin deficiency with deficient fire upward blazing. Symptoms include steaming bones, tidal fever, night sweats, nocturnal emission, or cough with blood, irritability, easily angered, quick to get hungry, weakness, pain or heat at knee and feet, red tongue, scanty tongue coating, and rapid and forceful Chi pulse.

Preparation: Grind into fine powder, steam with pig spine marrow until done, and form into small honey pills. Take 6 to 10 g in the morning and at night with light salt water. This can also be taken as decoction with one tenth of the dosage.

Treatment Principles: Nourish Yin and descend fire.

This formula is designed for Yin deficiency with deficient fire. It nourishes the true Yin to control the fire which is stronger and faster than Liu Wei Di Huang Wan [Six-Ingredient Pill with Rehmannia].

Modification:

Addition	Indication
Mu Li Fu Xiao Mai	Night sweats
Qing Dai Huang Qin Ce Bai Ye E Jiao	Hematemesis

Hu Qian Wan [Hidden Tiger Pill] *Dan Xi Xin Fa*

Huang Bai (fried with wine)	150 g
Gui Ban (honey fried with wine)	120 g
Shu Di Huang	60 g
Zhi Mu (fried with wine)	60 g
Chen Pi	60 g
Bai Shao Yao	60 g
Suo Yang	45 g
Hu Gu (bobcat bone) (zhi)	30 g
Huai Niu Xi	30 g
Dang Gui	60 g

Main Indication: Kidney and Liver deficiency, Yin deficiency with internal heat, and soreness and atrophy of the tendons and bones. Symptoms include sore and weak lumbar and knees, atrophy, and weak tendons and bones, emaciation of the legs, difficulty walking, steaming bones, tidal fever, red tongue, scanty coating, and thin weak pulse.

Preparation: Grind into fine powder, form honey pills. Take 15 g, twice a day with light salt water.

Treatment Principles: Nourish Yin, bring down fire, and strengthen the tendons and bones.

Modification:

Minus	Addition	Indication
	Shan Yao	Too weak to talk
Shu Di Huang	Sheng Di Huang	Severe heat

Related Formulas:

Formula	Herbs	Indication
Bu Yi Wan [Tonify and Augment Pill]	Core formula + Gan Jiang Bai Zhu Fu Ling Gan Cao Wu Wei Zi Tu Si Zi Zi He Che	Treat the same condition Stronger tonifying spleen and Kidney function
Long Hu Ji Yin Dan [Dragon and Tiger to Benefit Yin Special Pill]	Core formula + Long Gu	Treat nocturnal emission
Long Gu Wan [Dragon Bone Pill]	Core formula – Niu Xi	Treat the same indications

Qin Jiao Bie Jia San [Gentiana Qinjiao and Soft-Shelled Turtle Shell Powder] *Wei Sheng Bao Jian*

Bie Jia (honey fried until crispy)	30 g
Di Gu Pi	30 g
Chai Hu	30 g
Qin Jiao	15 g
Zhi Mu	15 g
Dang Gui	15 g

Main Indication: Wind consumptive disease which is Yin deficiency with deficient fire. Symptoms include steaming bone, night sweats, soreness and pain at limbs, emaciation, red tongue, red cheeks, fatigue, red tongue with scanty coating, and minute rapid pulse.

Preparation: Grind into coarse powder. Take 15 g of powder decoct with Qing Hao 9 g, Wu Mei one piece in 1 cup of water until 70% is left. Take once in the morning and once at night on empty stomach.

Treatment Principles: Nourish Yin and clear heat.

Wind consumptive disease is caused by external pathogenic factors that move internally, cause deficiency, and damage Yin. The manifestations are consumptive heat and steaming bones. Thus, the treatment not only nourishes Yin and clears heat but also vents evil pathogens outward.

Qing Gu San [Cool the Bones Powder] *Zheng Zhi Zhun Sheng*

Yin Chai Hu	9 g
Hu Huang Lian	6 g
Qin Jiao	6 g
Bie Jia (fried with vinegar)	6 g
Di Gu Pi	6 g
Qing Hao	9 g
Zhi Mu	6 g
Gan Cao	3 g

Main Indication: Yin deficiency with internal heat, deficient consumption, and steaming bones. Symptoms include fever in the afternoon or at night, thirst, irritability, dry throat, night sweats, malar flush, red tongue, scanty coating, and thin, rapid pulse.

Preparation: Decoction.

Treatment Principles: Clear deficient heat and steaming bone symptoms.

This formula focuses on clearing heat, decreasing steaming bone symptoms, and nourishing Yin. It is indicated for exuberant internal deficient fire.

Modification:

Addition	Indication
Sheng Di Huang Mai Men Dong Bai Shao Yao Tian Men Dong	Severe Yin deficiency and blood deficiency
Wu Wei Zi E Jiao	Dry cough

Qing Hao Bie Jia Tang [Artemisia Annua and Soft-Shelled Turtle Shell Decoction] *Wen Bing Tiao Bian*

Qing Hao	9 g
Bie Jia	15 g
Sheng Di Huang	12 g
Zhi Mu	9 g
Mu Dan Pi	9 g

Main Indication: Late stage of warm febrile disease with residual heat pathogens in the body and damaged Yin and fluid. Symptoms include fever at night and cool temperature in the morning, absence of sweating when the fever is gone, red tongue, scanty coating, and thin, rapid pulse.

Preparation: Decoction.

Treatment Principles: Nourish Yin, vent heat.

This formula treats chronic fever with unknown cause, and Yin deficient type of consumptive fever in chronic conditions, as well as fever due to heat hidden in Ying and blood level. Zhu Ye Shi Gao Tang [Lophatherus and Gypsum Decoction] treats fever due to heat in the Qi level. The indications for these two formulas are different.

Related Formulas:

Formula	Herbs	Indication
Jia Jian Qing Hao Bie Jia Tang [Modified Artemisia Annua and Soft-Shelled Turtle Shell Decoction]	Core formula – Sheng Di Huang + Sang Ye Tian Hua Fen	Treat severe heat and hidden heat in Shao Yang
Di Gu Yang Yin Jian [Lycium Root Bark to Nourish Yin Decoction]	Core formula + Di Gu Pi Hu Huang Lian Bai Shao Yao Ze Lan	Treat fever during menstruation, scanty maroon-colored blood, irritability, heat sensation, five center heat, red tongue, and thin, rapid pulse
Bie Jia Qing Hao Yin [Soft-Shelled Turtle Shell and Artemisa Annua Decoction]	Core formula – Mu Dan Pi + Yin Chai Hu Gan Cao Chi Shao Yao Hu Huang Lian Di Gu Pi Deng Xin Cao	Treat beginning stage of fever in infant malnutrition (Gan Ji) due to improper feeding

Dang Gui Liu Huang Tang [Tangkuei and Six-Yellow Decoction] *Lan Shi Mi Cang*

Dang Gui	equal amount
Sheng Di Huang	
Shu Di Huang	
Huang Qin	
Huang Bai	
Huang Lian	
Huang Qi	Double the dosage

Main Indication: Yin deficiency with excess fire. Symptoms include fever, night sweats, red face, irritability, dry mouth and lips, dry stools, yellow urine, red tongue, and rapid pulse.

Preparation: Grind into coarse powder. Decoct 15 g with 2 cups of water until 1 cup left. Take before meal.

Treatment Principles: Nourish Yin, drain fire, consolidate exterior, and stop sweating.

This formula treats both exterior and interior at the same time. It not only nourishes Yin and drains fire, but also tonifies Qi to consolidate the exterior.

Comparison of the Formulas Treating Liver and Kidney Deficiency and Deficient Heat

Liang Di Tang [Rehmannia and Lycium Root Bark Decoction], Bie Jia Yang Yin Jian [Nourish Yin Decoction with Soft-Shelled Turtle Shell], and Jia Jian Qing Jing Tang [Modified Clear the Menses Decoction] all treat irregular menstruation due to Yin deficiency. However, Liang Di Tang [Rehmannia and Lycium Root Bark Decoction] treats early menstruation due to Yin deficiency and blood heat. Bie Jia Yang Yin Jian [Nourish Yin Decoction with Soft-Shelled Turtle Shell] treats amenorrhea due to Yin and blood deficiency. Jia Jian Qing Jing Tang [Modified Clear the Menses Decoction] treats syndromes of Yin deficiency and blood heat with blood stasis.

Qin Jiao Bie Jia San [Gentiana Qinjiao and Soft-Shelled Turtle Shell Powder], Qing Gu San [Cool the Bones Powder], and Qing Hao Bie Jia Tang [Artemisia Annua and Soft-Shelled Turtle Shell Decoction] all treat Yin deficiency with fever. However, the main indication for the first formula is wind consumption disease which is external pathogens moving internally and damaging Yin. Thus, this formula nourishes Yin, clears heat, expels wind, and harmonizes. The main indication for the middle formula is a deficiency from steaming bones due to overstrain; thus, this formula mainly clears deficient heat. The latter formula mainly treats heat conditions damaging Yin with pathogens hidden at Yin level; thus, this formula nourishes Yin and at the same time vents heat.

Da Bu Yin Wan [Great Tonify the Yin Pill] and Dang Gui Liu Huang Tang [Tangkuei and Six-Yellows Decoction] treat both night sweats and tidal fever due to Yin deficiency with deficient heat. However, the former mainly supplements Yin to control the deficient fire. The latter formula mainly clears internal heat, and at the same time, nourishes Yin and consolidates the exterior.

Hu Qian Wan [Hidden Tiger Pill] treats weakness and atrophy of tendons and bones, emaciation, unsteady walking due to Liver and Kidney deficiency, and Yin deficiency with internal heat. Also can be used for the promotion of healing of bone fracture.

LIVER AND KIDNEY DEFICIENCY WITH WIND–DAMP BI SYNDROME

When Liver and Kidney are both deficient, Ying and Wei, both empty, wind–damp–cold pathogens, can enter the body causing heavy pain at lumbar and knee, weakness of the legs, and difficulty moving. The treatment should tonify Liver and Kidney and at the same time stop pain. The herbs that are commonly used are Du Zhong, Sang Ji Sheng, Niu Xi, Dang Gui, Bai Shao Yao, Sheng Di Huang, Chuan Xiong, Dang Shen, Huang Qi and Du Huo, Fang Feng, Xi Xin, Qin Jiao, and Qiang Huo.

Du Huo Ji Sheng Tang [*Angelica Pubescens and Sangjisheng Decoction*] *Qian Jin Fang*

Du Huo	12 g
Fang Feng	9 g
Xi Xin	9 g
Qin Jiao	9 g
Sang Ji Sheng	9 g
Du Zhong	9 g
Chuan Niu Xi	9 g
Rou Gui Xin	9 g
Dang Gui	9 g
Chuan Xiong	9 g
Bai Shao Yao	9 g
Sheng Di Huang	9 g
Dang Shen	9 g
Fu Ling	9 g
Gan Cao	9 g

Main Indication: Liver and Kidney deficiency and Bi syndrome. Symptoms include heavy pain at lumbar and knee, weakness of leg, difficulty moving or numbness and tingling, aversion to cold, preference for heat, palpitation, shortness of breath, pale tongue with white coating, and thin, weak, or slow pulse.

Preparation: Decoction.

Treatment Principles: Expel wind–damp, stop Bi pain, tonify Kidney and Liver, and tonify Qi and blood.

This is a common formula to treat chronic Bi syndrome with Qi, blood, and Liver and Kidney deficiency. It is especially indicated for seniors. In modern times, it is also used to treat chronic arthritis, rheumatoid arthritis, and soreness, weakness, and atrophy of lumbar and leg.

Modification:

Minus	Addition	Indication
	Chuan Wu Di Long Bai Hua She	Severe pain
	Fu Zi Gan Jiang	Severe dampness and cold
	Sang Zhi Fang Ji	Severe dampness all over the body
	Hong Hua Ru Xiang Mo Yao	Severe blood stasis
Sheng Di Huang Dang Shen		Less obvious Zheng Qi deficiency

Related Formula:

Formula	Herbs	Indication
Jia Jian San Bi Tang [Modified Three Painful-Obstruction Decoction]	Core formula – Sang Ji Sheng + Xu Duan Huang Qi Sheng Jiang Da Zao	Treat wind–cold–damp Bi, spasms, and convulsion of the extremities due to Qi deficiency and Qi stagnation

Da Qin Jiao Tang [Major Gentiana Qinjiao Decoction] *Su Wen Bing Ji Qi Yi Bao Ming Ji*

Qin Jiao	90 g
Gan Cao	60 g
Chuan Xiong	60 g
Dang Gui	60 g
Bai Shao Yao	60 g
Xi Xin	15 g
Qiang Huo	30 g
Fang Feng	30 g
Huang Qin	30 g
Bai Zhi	30 g
Bai Zhu	30 g
Sheng Di Huang	30 g
Shu Di Huang	30 g
Fu Ling	30 g
Shi Gao	60 g
Du Huo	60 g

Main Indication: Empty channel and collaterals and wind evil invasions. Symptoms include sudden deviation of mouth and eyes, tongue stiffness, inability to speak, inability to move hands and feet, numbness and tingling at skin, or hemiplegia, with exterior syndrome, thin, white tongue coating, and floating pulse.

Preparation: Grind into powder, decoct 30 g each time. The dosage can also be decreased to make decoction directly.

Treatment Principles: Expel wind, unblock collaterals, nourish blood, and invigorate blood.

This formula is designed for beginning stage of wind evil attacking the meridians and collaterals. It is effective in treating annoying pain and stubborn numbness and tingling of the limbs due to wind Bi syndrome. However, the formula contains many expel wind herbs that are very acrid and dry, thus it can damage Yin and blood.

Modification:

Minus	Addition	Indication
Qiang Huo Fang Feng Xi Xin Bai Zhi Du Huo	Chi Shao Yao Mu Dan Pi Dan Shen Tian Men Dong	Treat Yin deficiency with severe heat
	Wu Gong	Speech difficulty
Shi Gao Huang Qin		Absence of heat signs
	Zhi Shi	Focal distention below the heart at the epigastrium

LIVER AND LUNG SYNDROMES AND FORMULAS

Liver Fire Insulting Lung

This syndrome manifests as dry cough or hemoptysis, or phlegm with blood, red cheeks, irritability, pain at chest and hypochondriac region, agitation, easily angered, red tongue, and wiry pulse. The commonly used herbs are Shan Zhi Zi, Qing Dai, Huang Qin and Gua Lou, Hai Fu Shi, and He Zi. Calming and extinguishing Liver fire is the treatment for the root cause. Opening the chest, and stopping cough and bleeding are branch treatments.

Ke Xue Fang [Coughing of Blood Formula] *Dan Xi Xin Fa*

Qing Dai	9 g
Shan Zhi Zi	9 g
Gua Lou Ren (remove oil)	9 g
He Zi	9 g
Hai Fu Shi	9 g

Main Indication: Liver fire insulting Lung. Symptoms include cough with sticky phlegm and blood, unproductive cough; or irritability, easily angered, sharp pain at chest and hypochondriac region, red cheeks, constipation, red tongue, yellow tongue coating, and wiry, rapid pulse.

Preparation: Grind into fine powder and mix with honey and ginger juice to form pills. It can also be taken as decoction.

Treatment Principles: Clear Liver fire, calm Lung.

This formula can clear Liver fire, transform phlegm, stop cough. it is effective in treating cough or phlegm with blood due to Liver fire burning and damaging Lung collaterals. In modern times, it has also been used to treat hemoptysis due to bronchiectasis or Lung tuberculosis.

Modification:

Addition	Indication
Xing Ren	Severe cough
Huang Qin Qing Hao Mu Dan Pi	Epistaxis

CHAPTER 7

Heart Syndromes and Formulas

HEAT IN YING AND BLOOD LEVEL

When evil pathogenic heat enters Ying level, it often manifests as delirium or delirious speech worse at night, presence or absence of thirst, dark red tongue, scanty tongue coating, and weak and rapid pulse. When pathogenic heat enters blood level, it often manifests as irritability, insomnia, dark red tongue, and rapid pulse. When excess heat causes reckless flow of blood, it manifests as hematemesis, epistaxis, blood in stool or urine, or rash. Sweet cold or salty cold herbs are often used, such as Xi Jiao (Shui Niu Jiao), Xuan Shen, Sheng Di Huang, Mu Dan Pi, Da Qing Ye, Ban Lan Gen, Zi Cao, Jin Yin Hua, and Lian Qiao. These herbs can cool blood, relieve toxicity, and nourish Yin, so "when the heat is cleared, then blood is calm."

Qing Ying Tang [Clear the Nutritive Level Decoction]
Wen Bing Tiao Bian

Xi Jiao (or replaced by Shui Niu Jiao)	3 g (15–30 g)
Sheng Di Huang	15 g
Xuan Shen	9 g
Zhu Ye Xin	3 g
Mai Men Dong	9 g
Dan Shen	6 g
Huang Lian	5 g
Jin Yin Hua	9 g
Lian Qiao	6 g

Main Indication: Heat enters Ying (nutritive) level. Symptoms include fever worse at night, irritability, restless sleep, sometimes delirious speech, presence or absence of thirst, or faint skin rash, dark red tongue, rapid pulse, or big and deficient pulse.

Preparation: Decoction. Decoct one to two bags per day, divided into three doses.

Treatment Principles: Clear Ying, vent heat, nourish Yin, and invigorate blood.

This is a representative formula in treating evil heat pathogens entering Ying (nutritive) level. It can cool blood, nourish Yin, relieve toxicity, and can also vent heat to Qi level. When the heat is cleared from the interior, it is transferred to Qi level, then out of the body.

Modification:

Addition	Indication
Shi Gao Zhi Mu	Excess heat at Qi level
Increase dosage Xi Jiao (Shui Niu Jiao) Sheng Di Huang Xuan Shen + Chi Shao Yao Mu Dan Pi Da Qing Ye Ban Lan Gen	Qi level heat already decreased, but Ying-level heat still excess
Huang Qin Di Long Chuan Niu Xi	Cerebral vascular accident caused by hypertension, manifesting high fever and irritability
Ban Lan Gen Da Qing Ye Huang Lian	Meningitis, encephalitis B, and septicemia, presenting high fever and dark red and dry tongue
An Gong Niu Huang Wan [Calm the Palace Pill with Cattle Gallstone] Zhi Bao Dan [Greatest Treasure Special Pill]	Delirium, delirious speech, tongue stiffness, and spasms of the limbs
Ling Yang Jiao (Shan Yang Jiao) Di Long Gou Teng or Zi Xue Dan [Purple Snow Special Pill]	Convulsion

Related Formula:

Formula	Herbs	Indication
Qing Gong Tang [Clear the Palace Decoction]	Xuan Shen Xin Lian Zi Xin Zhu Ye Xin Lian Qiao Xin Mai Men Dong Xi Jiao (substitute with Shui Niu Jiao)	Mistreating warm febrile disease with sweating method, causing evil pathogens to enter pericardium, with delirium and delirious speech

Xiao Ban Qing Dai Yin [Indigo Decoction to Clear Rash] (Tao Jie An Fang)

Qing Dai	15 g
Huang Lian	15 g
Xi Jiao (or Shui Niu Jiao)	3 g (or 15–30 g)
Shi Gao	30 g
Zhi Mu	9 g
Xuan Shen	9 g
Shan Zhi Zi	9 g
Sheng Di Huang	9 g
Chai Hu	9 g
Ren Shen	9 g
Gan Cao	3 g

Main Indication: Heat in both Qi and blood level. Symptoms include big fever, extreme irritability, Yang toxins, rash, red tongue, yellow dry tongue coating, flooding, and big, slippery, rapid pulse.

Preparation: Decoct with Sheng Jiang and Da Zao. Add 1 tablespoon of vinegar to decoction.

Treatment Principles: Drain fire, relieve toxins, cool blood, and clear rash.

This formula specifically treats heat entering Ying (nutritive) level forcing blood outside the vessels leading to rashes and excess heat at Qi level.

Modification:

Minus	Addition	Indication
Ren Shen	Da Huang	Constipation
Chai Hu	Qing Hao	Excess Yang

Shen Xi Dan [Magical Rhinoceros Special Pill] *Wen Re Jing Wei*

Xi Jiao (or substitute with Shui Niu Jiao)	180 g (or 1000–1500 g)
Shi Chang Pu	180 g
Huang Qin	180 g
Fresh Sheng Di Huang juice	500 g
Jin Yin Hua	500 g
Lian Qiao	300 g
Jin Zhi *	300 g
Ban Lan Gen	270 g
Dan Dou Chi	240 g
Xuan Shen	210 g
Tian Hua Fen	120 g
Zi Cao	120 g
* Purified feces	

Main Indication: Warm, heat, summer epidemics, pathogens enter Ying blood, severe heat with severe toxicity, fluid, and Yin damage. Symptoms include high fever, delirium, delirious speech, purple rash, ulcerations in mouth and throat, irritability, agitation, red eye, dark red purple tongue or dry tongue, and thin, rapid pulse.

Preparation: Sun dry the above herbs (other than the juices), then grind into fine powder. Mix with Sheng Di Huang juice and Jin Zhi to form pill. Each pill is 3 g. Take one with cool water, twice a day. Reduce dosage for children.

Treatment Principles: Clears heat, opens orifice, cools blood, and relieves toxicity.

This formula is strong in cooling blood and relieving toxicity function, thus it is indicated for severe heat and toxicity conditions.

Xi Jiao Di Huang Tang [Rhinoceros Horn and Rehmannia Decoction] *Qian Jin Yao Fang*

Xi Jiao (or substitute with Shui Niu Jiao)	3 g (or 15–30 g)
Sheng Di Huang	30 g
Chi Shao Yao	12 g
Mu Dan Pi	9 g

Main Indication:

1. Heat entering blood level causing reckless blood flow. Symptoms include hematemesis, epistaxis, and blood in stool and urine.
2. Heat disturbing Shen. Symptoms include delirium, delirious speech, dark purple rash, dark red tongue with thorn-like prickles on tongue, and rapid pulse.
3. Syndromes due to blood stasis and heat evil pathogens. Symptoms include easily forgetful as if manic, prefer to rinse the mouth but not drink the water, irritability and pain at chest, abdominal fullness, and black and easily passed stools.

Preparation: Decoction.

Treatment Principles: Clear heat, cool blood, and scatter stasis.

This formula is indicated for any bleeding syndromes due to reckless blood flow. It can also treat hepatic coma, hepatic atrophy, uremia, purpura, and acute leukemia with manifestations of high fever and bleeding; and eye disorders, such as pus accumulation (empyema) of anterior chamber, blood accumulation (hematocele) of anterior chamber, inflammation of the iris and ciliary body (iridocyclitis), glaucoma, and other bleeding conditions with blood stasis.

Modification:

Addition	Indication
Da Huang Huang Qin	Easily forgetful, as in mania
Ce Bai Ye Bai Mao Gen San Qi Hua Rui Shi	Hemoptysis
Bai Mao Gen Huang Qin Qing Hao	Epistaxis
Huai Hua Di Yu	Blood in stool
Bai Mao Gen Xiao Ji	Blood in urine
Qing Dai Zi Cao	Rash and purpura
Ren Shen	Qi deficiency not controlling blood with excess bleeding
An Gong Niu Huang Wan [Calm the Palace Pill with Cattle Gallstone] Zhi Bao Dan [Greatest Treasure Special Pill]	Extreme heat and toxins, severe case of delirium

Related Formulas:

Formula	Herbs	Indication
Ju Fang Xi Jiao Di Huang Wan [Rhinoceros Horn and Rehmannia Pill]	Core formula + Huang Qin Huang Lian	Treat excess heat with hemoptysis and epistaxis. The clearing heat and relieving toxicity functions are stronger than in the core formula
Xi Jiao Di Huang He Xie Xin Tang [Rhinoceros Horn and Rehmannia, and Drain the Epigastrium Decoction]	Core formula + Huang Qin Huang Lian Da Huang	This formula is especially effective in treating excess heat causing bleeding
Xi Di Qing Luo Yin [Clear the Collaterals Decoction with Rhinoceros Horn and Rehmannia]	Core formula + Tao Ren Bai Mao Gen Lian Qiao Deng Xin Cao Ginger juice	Treat knotted blood stasis and heat at blood level, irritability, agitation, delirium; sometimes mentally clear, sometimes delirious; dry mouth; will rinse mouth but does not like to drink; dark purple tongue; and thin, rapid, or choppy pulse

Comparison of the Four Formulas Treating Heat Entering Blood and Ying Level

Qing Ying Tang [Clear the Nutritive Level Decoction] is indicated for beginning stage of heat entering Ying level, where blood flow is not yet affected. Xiao Ban Qing Dai Yin [Indigo Decoction to Clear Rash] clears both Qi and blood, but focuses on clearing rash and purpura, invigorating blood and cooling blood. Shen Xi Dan [Magical Rhinoceros Special Pill] is indicated for severe heat and toxins with delirium, delirious speech and rash manifesting at the same time. Xi Jiao Di Huang Tang [Rhinoceros Horn and Rehmannia Decoction] is indicated for any bleeding conditions due to heat causing reckless blood flow. However, the heat clearing and relieving toxicity functions are weaker.

HEAT ENTERING PERICARDIUM, DELIRIUM WITH BLOCKED ORIFICES

Heat toxins entering Pericardium cause this syndrome. Symptoms include fever, irritability, agitation, delirium, delirious speech, dark red tongue, and rapid pulse. In severe cases, there can be spasms, stroke, and blocked orifices. The herbs that are commonly used are Niu Huang, Xi Jiao (Shui Niu Jiao), Ling Yang Jiao (Shan Yang Jiao), Huang Qin, Huang Lian, Shan Zhi Zi, with She Xiang, Bing Pian, An Xi Xiang, Shi Chang Pu, Yu Jin, Zhu Sha, Zhen Zhu, Ci Shi, Hu Po, Dai Mao, and Jin Bo to form formulas that clear heat, relieve toxicity, open orifices, and calm Shen.

If there is also presence of excess heat in Qi level, add Shi Gao, Hua Shi, Han Shui Shi, and Zhi Mu. With the presence of turbid phlegm, add Tian Zhu Huang, Xiong Huang, and Dan Nan Xing.

An Gong Niu Huang Wan [Calm the Palace Pill with Cattle Gallstone] *Wen Bing Tiao Bian*

Niu Huang	30 g
Xi Jiao (substitute with Shui Niu Jiao)	30 g (150–300 g)
Huang Qin	30 g
Huang Lian	30 g
Shan Zhi Zi	30 g
Zhu Sha	30 g
Xiong Huang	30 g
Yu Jin	30 g
Bing Pian	7.5 g
She Xiang	7.5 g
Zhen Zhu	15 g
Jin Bo	appropriate amount

Main Indication: Warm heat attacks Pericardium, delirium, and obstruction of the orifices. Symptoms include high fever, irritability, agitation, delirium, delirious speech, stroke, coma, and infantile convulsion which all belongs to internal obstruction due to evil heat. Tongue is dark red and dry with rapid pulse.

Preparation: Grind all ingredients expect Jin Bo into extremely fine powder. Mix with honey to form pills of 3 g each, coated with Jin Bo (gold foil), and protect with wax. Take one pill, one to three times a day. Reduce dosage for children.

Treatment Principles: Clear heat, relieve toxicity, open orifices, and calm the Shen.

This formula can treat epidemic meningitis, encephalitis B, uremia, hepatic coma, or cerebral vascular accidents with symptoms of fever, irritability, agitation, delirium, and delirious speech.

Modification:

Addition	Indication
Steep Da Huang 9 g in water and take with two pills.	Excess in Fu organs. Symptoms include delirium, short tongue, constipation, and thirst that cannot be alleviated by drinking. This is Niu Huang Cheng Qi Tang [Cattle Gallstone Decoction to Order the Qi]
Decoction of Jin Yin Hua and Bo He. To be taken with the pill.	Rapid and forceful pulse
Ren Shen Decoction	Rapid and deficient pulse

Related Formula:

Formula	Herbs	Indication
Niu Huang Qing Xin Wan [Cattle Gallstone Decoction to Clear the Heart]	Niu Huang Zhu Sha Huang Lian Huang Qin Shan Zhi Zi Yu Jin Grind the above herbs into fine powder and mix with honey to form pills of 3 g each Take 1 pill, two to three times a day. Reduce to half for children	Strong in clearing heat and relieving toxicity function. Weaker in opening the orifices and calming the Shen function. It is suitable for mild syndromes of heat obstruction with manifestations of fever, irritability, agitation, delirium, infantile convulsion, restlessness, excess phlegm, and saliva

Zi Xue Dan [Purple Snow Special Pill] *He Ji Ju Fang*

Han Shui Shi	1500 g
Duan Ci Shi	1500 g
Hua Shi	1500 g
Shi Gao	1500 g
Mang Xiao	1000 g
Huo Xiao	1000 g
Xuan Shen	500 g
Sheng Ma	500 g
Gan Cao	240 g
Ling Yang Jiao (substitute with Shan Yang Jiao)	150 g (500–1000 g)
Xi Jiao (substitute with Shui Niu Jiao)	150 g (500–1000 g)
Qing Mu Xiang	150 g
Chen Xiang	150 g
Zhu Sha	60 g
Ding Xiang	30 g
She Xiang	38 g
Huang Jin (gold)	300 g

Main Indication: Heat attacks the Pericardium, blazing heat in both Qi and blood. Symptoms include high fever, irritability, agitation, delirium, delirious speech, spasms and syncope, thirst, burned lips, dark urine, constipation, or convulsion of the four limbs, dark red tongue, no coating, and rapid and forceful pulse. It is also indicated for heat type of infantile convulsion.

Preparation: Decoct Shi Gao, Hua Shi, Ci Shi, Xuan Shen, Mu Xiang, Chen Xiang, Sheng Ma, Gan Cao, and Ding Xiang. Filter the above decoction and then concentrate it to form syrup. Grind the rest of the herbs into fine powder, add to the syrup. Take 1.5 to 3 g, twice a day. Reduce the dosage appropriately for children.

Treatment Principles: Clear heat, open orifices, stop spasms, calm the Shen, and extinguish wind.

This formula is stronger in clearing heat, opening orifices, calming the Shen and extinguishing wind. It treats both internal and external heat, high fever, delirium, infantile high fever and convulsion, measles in children, internal excess heat toxins, purplish red rash, high fever with rapid wheezing, and coma.

It can treat epidemic meningitis, encephalitis B, typhus, scarlet fever with symptoms of high fever, irritability, agitation, delirium, spasms and syncope, and bleeding or rash.

Zhi Bao Dan [Greatest Treasure Special Pill] *He Ji Ju Fang*

Xi Jiao (substitute with Shui Niu Jiao)	30 g (150–300 g)
Zhu Sha	30 g
Hu Po	30 g
Xiong Huang	30 g
Dai Mao	30 g
Bing Pian	7.5 g
She Xiang	7.5 g
Niu Huang	15 g
An Xi Xiang	45 g
Yin Bo (silver foil)	50 pieces
Jin Bo (gold foil)	50 pieces

Main Indication: Summer heat stroke, wind stroke, and internal obstruction due to phlegm heat in warm febrile disease. Symptoms include delirium, delirious speech, fever, irritability, agitation, excess phlegm, coarse breathing, red tongue, yellow greasy coating, and slippery and rapid pulse. It is also indicated for any pediatric epilepsy, fright, and convulsions which belong to internal obstruction due to phlegm heat.

Preparation: Grind into fine powder, after sieving; mix with honey to form pills of 3 g each. Take 1 pill each time and decrease dosage for kids. For patient with Zheng Qi deficiency, take with Ren Shen decoction. For copious phlegm causing nausea and vomiting, take the pill with ginger juice and boy urine.

Treatment Principles: Clear heat, unblock orifices, transform turbidity, and relieve toxins.

This formula treats all severe and excess syndromes of excess evil heat, with phlegm misting and blocking the orifices. This formula can also unblock the orifices and wake the mind in cases of summer heat stroke and wind stroke. However, this formula is not indicated for Yin deficiency with Yang excess type of delirium and delirious speech. Pregnant women should take this formula with caution due to its having more aromatic, acrid, and drying herbs.

Bao Long Wan [Hold the Dragon Pill] *Xiao Er Yao Zheng Zhi Jue*

Tian Zhu Huang	30 g
Xiong Huang	3 g
Zhu Sha	15 g
She Xiang	15 g
Dan Nan Xing	120 g

Main Indication: Infantile convulsion, and internal congestion of phlegm heat. Symptoms include fever, delirium, coarse breathing, fright, convulsion, spasms of the four limbs, red tongue with yellow greasy coating, and red vein reaching the life gate on index finger.

Preparation: Grind into fine powder; mix with Gan Cao decoction to form pills. Take 1 to 2 g, two to three times a day.

Treatment Principles: Open the orifices, relieve spasm, clear heat, and transform phlegm.

Related Formulas:

Formula	Herbs	Indication
Bao Long Wan [Hold the Dragon Pill] *He Ji Ju Fang*	Core formula + Quan Xie	Stronger at relieving spasms function
Niu Huang Bao Long Wan [Hold the Dragon Pill with Cattle Gallstone] *Ming Yi Za Zhu*	Core formula + Niu Huang	Stronger at clearing heat, anchoring fright function
Jin San Bao Long Wan [Hold the Dragon Pill] *Jin San*	Core formula + Hu Po Jiang Can Quan Xie Niu Huang Chi Fu Ling Jin Bo Deng Xin Cao Bo He	This is indicated for severe cases of heat excess causing fright convulsion
Hu Po Bao Long Wan [Hold the Dragon Pill with Amber]	Core formula – She Xiang + Hu Po Shan Yao Ren Shen Fu Ling Gan Cao Zhi Ke Zhi Shi Tan Xiang	Treat infantile convulsion with weak constitution
Niu Huang Bao Long Wan [Hold the Dragon Pill with Cattle Gallstone]	Core formula + Niu Huang Zhen Zhu Jiang Can Quan Xie Tian Ma Fang Feng Zhe Bei Mu Ju Hong Chen Xiang Shi Chang Pu	Stronger heat clearing, opening orifices, and relieving spasms and scouring out phlegm function

Xiao Er Hui Chun Dan [Return of Spring Special Pill] *Jing Xiu Tang Yao Shuo*

Chuan Bei Mu	37.5 g
Chen Pi	37.5 g
Mu Xiang	37.5 g
Bai Dou Kou	37.5 g
Zhi Ke	37.5 g
Ban Xia	37.5 g
Chen Xiang	37.5 g
Tian Zhu Huang	37.5 g
Jiang Can	37.5 g
Quan Xie	37.5 g
Tan Xiang	37.5 g
Niu Huang	12 g
She Xiang	12 g
Dan Nan Xing	60 g
Gou Teng	240 g
Da Huang	60 g
Tian Ma	37.5 g
Gan Cao	26 g
Zhu Sha	small amount

Main Indication: Acute infantile convulsion, phlegm heat obstructing the orifices. Symptoms include fever, irritability, agitation, delirium, fright and convulsion, or nausea and vomiting, vomiting milk, crying at night, cough with phlegm, wheezing, breathlessness, abdominal pain, and diarrhea. Red tongue, dry yellow tongue coating, and red vein reaching Qi or life gate on the index finger.

Preparation: Grind into fine powder; form into small pills of 0.09 g each. Give one for baby less than 1 year of age. Take 2 pills, two to three times a day for children 1 to 2 years old.

Treatment Principles: Open orifices, relieving fright, clear heat, and transform phlegm.

This formula is an effective formula in treating acute infantile convulsion. The pathological factors of acute infantile convulsions are heat, phlegm, wind, and fright. This formula has the functions of clearing heat, transforming phlegm, opening orifices, extinguishing wind, and relieving fright.

Xing Jun San [Marching Powder] *Huo Luan Lun*, Also Called Zhu Ge Xing Jun San or Wu Hou Xing Jun San

Niu Huang	3 g
She Xiang	3 g
Zhen Zhu	3 g
Peng Sha	3 g
Xiong Huang	24 g
Xiao Shi	0.9 g
Jin Bo (gold foil)	20 pieces

Main Indication: Summer toxic and poisonous pathogens causing distention. Symptoms include vomiting, diarrhea, abdominal pain, stifling sensation, irritability as if dying, dizziness, vertigo, unconsciousness, red tongue, yellow greasy dirty tongue coating, and slippery or wiry pulse. This formula also treats mouth ulcerations and throat pain. When it is used as eye drops, it can clear wind heat and growth in eyes. When it is used as nasal spray, it can prevent epidemic pathogenic factors from invading the body.

Preparation: Grind the above ingredients into extremely fine powder; take 0.3 to 0.9 g each time, two to three times a day.

Treatment Principles: Open orifices, separate the turbid, and relieve toxins.

This formula is acrid, aromatic, moving, and scurrying. The functions are strong, so use caution with pregnant women.

Related Formula:

Formula	Herbs	Indication
Ren Ma Ping An San [Safe Powder for People and Horses]	Core formula – Niu Huang Zhen Zhu + Zhu Sha	Same indications as the core formula, but weaker

Appendix:

Su He Xiang Wan [Liquid Styrax Pill] *He Ji Ju Fang*

Bai Zhu	60 g
Mu Xiang	60 g
Xi Jiau (substitute with Shui Niu Jiao)	60 g (600 g)
Xiang Fu	60 g
Zhu Sha	60 g
He Zi	60 g
Tan Xiang	60 g
An Xi Xiang	60 g
Chen Xiang	60 g
She Xiang	60 g
Ding Xiang	60 g
Bi Bo	60 g
Bing Pian	30 g
Su He Xiang You	30 g
Ru Xiang	30 g

Main Indication: Cold, closed disorder. Symptoms include sudden collapse, loss of consciousness, clenched jaw, or Middle Jiao coldness with Qi obstruction, sudden spasm pain at the abdomen and chest, nauseous but cannot throw up, urges to have diarrhea but does not occur, or falling even into a coma and collapsing, or phlegm obstructing the Qi flow leading to sudden loss of consciousness. The tongue is pale with thick or greasy white coating. The pulse is deep, forceful, or slow.

Preparation: Grind into fine powder; mix well and form pill with An Xi Xiang paste and honey. Each pill is 3 g. Take one pill one to two times a day with warm water.

Treatment Principles: Aromatic to transform turbidity, open the orifices, and awaken the Shen.

This is a representative formula for warming and opening the Shen. Aromatic herbs are moving and scurrying, so the formula is contraindicated for pregnant women.

This formula can also be used in angina pectoris and psychological disorders of cold syndromes, and with greasy, thick tongue coating, and deep, slow pulse.

Related Formula:

Formula	Herbs	Indication
Guan Xin Su He Xiang Wan [Liquid Styrax Pill for Coronary Heart Conditions]	Su He Xiang 50 g Bing Pian 105 g Zhi Ru Xiang 105 g Tan Xiang 210 g Qing Mu Xiang 210 g	Treats Qi stagnation and phlegm obstruction type of angina pectoris. Symptoms include stifling sensation of the chest and holding of breath

Comparison of Formulas Treating Heat Entering Pericardium, Delirium with Blocked Orifices

1. The strength of the formulas to clear heat and relieve toxicity, from strongest to weakest, are: An Gong Niu Huang Wan [Calm the Palace Pill with Cattle Gallstone], Zi Xue Dan [Purple Snow Special Pill], and Zhi Bao Dan [Greatest Treasure Special Pill].
2. The strength of the formulas to open orifices and calm the Shen, from strongest to weakest are: Zhi Bao Dan [Greatest Treasure Special Pill], Zi Xue Dan [Purple Snow Pill], and An Gong Niu Huang Wan [Calm the Palace Pill with Cattle Gallstone].
3. Zi Xue Dan [Purple Snow Special Pill] treats heat in both Qi and blood. It is effective in relieving spasms and settling the mind.
4. Bao Long Wan [Hold the Dragon Pill] and Xiao Er Hui Chun Dan [Return of Spring Special Pill] both treat acute infantile convulsion, but the latter formula is stronger.
5. Xing Jun San [Marching Powder] treats nausea, vomiting, and phlegm obstructing the pericardium due to phlegm turbid accumulation during summer months.
6. Su He Xiang Wan [Liquid Styrax Pill] is suitable for cold type of obstructed orifices because it warms and opens the orifices.

PHLEGM OBSTRUCTING THE CLEAR ORIFICES

This syndrome often manifests with sudden fall due to wind stroke, confused mind, unconsciousness, phlegm-like sound in the throat, greasy tongue coating, and slippery pulse. Commonly used herbs include Tian Nan Xing, Ban Xia, Bei Mu, Tian Zhu Huang, Zhu Li, Shi Chang Pu, Yu Jin, and Yuan Zhi.

Phlegm fire raging upward will cause symptoms like mania, agitation, restlessness, sometimes crying, sometimes laughing, nonsense speech, suddenly taking off clothes and walking outside, climbing up to high places to sing, frequent changes of emotions, yellow greasy tongue coating, and slippery rapid pulse. One should clear fire and scour out phlegm with Meng Shi Gun Tan Wan [Vaporize Phlegm Pill with Chlorite–Mica].

Di Tan Tang [Scour Phlegm Decoction] *Ji Sheng Fang*

Chen Pi	9 g
Ban Xia	15 g
Fu Ling	12 g
Gan Cao	3 g
Zhi Shi	9 g
Zhu Ru	15 g
Dan Nan Xing	12 g
Shi Chang Pu	6 g
Ren Shen	3 g

Main Indication: Wind stroke, phlegm obstructing the Heart orifices. Symptoms include tongue stiffness, inability to talk, or absent-mindedness; white, greasy, thick tongue coating or slight deviation of the tongue, and wiry or slippery, forceful pulse.

Preparation: Decoct with Sheng Jiang and Da Zao.

Treatment Principles: Scour phlegm and open orifices.

This formula is modified from Er Chen Tang [Two-Cured Decoction] by adding more herbs to eliminate dampness, expel and scour phlegm, open orifices, and also clear heat. It can be used to treat difficult speech with internal phlegm damp obstruction.

Modification:

Minus	Addition	Indication
	Huang Qin Huang Lian Jiang Zhi (ginger juice) Zhu Li	Obvious heat symptoms
Zhu Li Dan Nan Xing	Tian Nan Xing	Without heat symptoms

Related Formula:

Formula	Herbs	Indication
Bian Tong Shi Wei Wen Dan Tang [Adapted Ten Flavored Warm the Gallbladder Decoction]	Ju Hong Fu Shen Ban Xia Gan Cao Zhi Shi Sheng Di Huang Suan Zao Ren Yuan Zhi Shi Chang Pu Zhu Li	Treat dementia with the manifestations of absent-mindedness, frequent crying or laughing without reason, fright, palpitation, insomnia, and dull reaction

Juan Yin Liu Shen Tang [Six-Miracle Decoction to Expel Congested Fluid] *Nu Ke Ji Yao*

Ban Xia Qu	15 g
Ju Hong	9 g
Fu Shen	9 g
Dan Nan Xing	6 g
Xuan Fu Hua	9 g
Shi Chang Pu	6 g

Main Indication: Postpartum phlegm turbid obstructing orifices. Symptoms include delirium, delirious speech, manic, lochial discharge, or even deviation of mouth and tongue, hemiplegia, dark tongue with thick greasy tongue coating, and wiry, slippery, or choppy pulse.

Preparation: Decoction.

Treatment Principles: Scour phlegm and open orifices.

This formula is designed for postpartum phlegm obstructing clear orifices. It can also be used for increased vaginal discharge for patient with a phlegm damp constitution.

Modification:

Addition	Indication
Da Huang Taken with yellow wine	Mania with lochial retention

Gun Tan Wan [Vaporize Phlegm Pill] (Wang Yin Jun Fang), Also Called Meng Shi Gun Tan Pill [Vaporize Phlegm Pill with Chlorite–Mica]

Da Huang	240 g
Huang Qin	240 g
Meng Shi (calcined with Huo Xiao)	30 g
Chen Xiang	15 g

Main Indication: Excess heat with old phlegm. Symptoms include mania, fright palpitation, coma, or cough wheezing with sticky phlegm, or stifling and focal distention of the chest and epigastrium, or vertigo and dizziness, constipation, or lymphadenoid around the neck, or insomnia, or twitching of the mouth and eyes, or dreaming about strange objects, or difficulty describing pain at joints, yellow thick greasy tongue coating, and slippery, rapid, and forceful pulse.

Preparation: Grind into fine powder, and make into small pills with water. Take 5 to 10 g, one to two times a with warm water.

Treatment Principles: Drain fire and expel phlegm.

This formula is strong, thus it is used to treat excess heat type of stubborn phlegm. It should be used with caution in pregnant women and weak constitution people.

Comparison of Formulas Treating Phlegm Obstructing Clear Orifices

Di Tan Tang [Scour Phlegm Decoction] treats phlegm obstructing the Heart collaterals and the Heart orifice, causing tongue stiffness and inability to talk. Bian Tong Shi Wei Wen Dan Tang [Adapted Ten Flavored Warm the Gallbladder Decoction] treats dementia syndrome due to phlegm obstructing Heart and mind. Juan Yin Liu Shen Tang [Six-Miracle Decoction to Expel Congested Fluid *He Ji Ju Fang*] treats postpartum phlegm obstructing clear orifices or phlegm obstructing the channels and collaterals leading to coma or hemiplegia. Gun Tan Tang [Vaporize Phlegm Decoction] treats various syndromes transformed from chronic excess heat and stubborn old phlegm.

HEART YIN DEFICIENCY

Palpitation, forgetfulness, insomnia, or excess dreaming are symptoms occuring with dry mouth, dry throat, restlessness, red tongue, scanty coating, and rapid pulse, and indicate Heart Yin deficiency. Commonly used herbs are Dang Gui, E Jiao, Sheng Di Huang, Xuan Shen, Mai Men Dong, Suan Zao Ren, and Bai Zi Ren to nourish Heart Yin. These herbs are combined with tonifying Heart Qi and calming the Heart and Shen herbs such as Ren Shen, Fu Shen, Bai Zi Ren, and Lian Xin. They can combine with anchoring the mind and calming the Shen herbs such as Long Gu and Mu Li, or clearing Heart and calming the Shen herbs such as Hu Po, Zhu Sha, Zhen Zhu Mu, and Shi Jue Ming.

Furthermore, insufficient Yin can lead to hyperactive Yang, thus this group of formula often has clear heat herbs such as Zhi Mu in the formula.

Tian Wang Bu Xin Dan [Emperor of Heaven's Special Pill to Tonify the Heart] *She Sheng Mi Po*

Ren Shen	9 g
Wu Wei Zi	9 g
Yuan Zhi	9 g
Jie Geng	9 g
Tian Men Dong	9 g
Mai Men Dong	9 g
Xuan Shen	12 g
Dan Shen	15 g
Fu ling	12 g
Dang Gui	20 g
Bai Zi Ren	30 g
Suan Zao Ren	30 g
Sheng Di Huang	30 g

Main Indication: Heart Yin and blood deficiency. Symptoms include insomnia, palpitation, forgetfulness, nocturnal emission with dreams, sluggish thinking, ulcerations in mouth and on tongue, dry stool, red tongue, scanty tongue coating, and thin and rapid pulse.

Preparation: Grind into fine powder; mix with honey to form pills. Coat the pill with Zhu Sha. Take 9 to 15 g, twice a day.

Treatment Principles: Nourish Heart and calm Shen.

This formula treats syndrome of Heart Yin deficiency, blood deficiency, and hyperactive Heart Yang. In recent years, it has been found effective in treating neurasthenia and neurotic tachycardia.

Related Formula:

Formula	Herbs	Indication
Bai Zi Yang Xin Wan [Biota Seed Pill to Nourish the Heart]	Bai Zi Ren Gou Qi Zi Mai Men Dong Dang Gui Shi Chang Pu Fu Shen Xuan Shen Shu Di Huang Gan Cao	Nourish Heart and calm Shen, and tonify Heart blood and Kidney Yin Treat palpitation with or without fright, excessive dreaming, night sweats, and forgetfulness or absent-mindedness due to Heart and Kidney disharmony

Jia Jian Fu Mai Tang [Modified Restore the Pulse Decoction] *Wen Bing Tiao Bian*

Zhi Gan Cao	18 g
Sheng Di Huang	18 g
Bai Shao Yao	18 g
Mai Men Dong	15 g
E Jiao	9 g
Huo Ma Ren	9 g

Main Indication: Late stage of warm febrile disorders, and Yin and blood exhaustion. Symptoms include fever, red face, dry mouth, dry tongue, or even black teeth, cracked lips, heat in palms and soles, or even at the back of hands and feet, deficient big pulse. Symptoms can also include deafness after 6 or 7 days of warm febrile disorders. This formula is also indicated for mistreating the warm febrile disorder with ascending and scattering method, which leads to knotted or intermittent pulse, or other irregular pulse.

Preparation: Decoction. Divide the decoction into three doses and melt the E Jiao in the decoction.

Treatment Principles: Nourish blood, astringe Yin, generate fluid, and moisten dryness.

This formula is modified from Zhi Gan Cao Tang [Honey-Fried Licorice Decoction] written by Zhang Zhong Jing. The acrid warm herbs are removed from the formula, and Bai Shao Yao is added. Thus, nourishing both Yin and Yang function of the original formula is changed to nourishing Yin and blood.

Modification:

Addition	Indication
Increase Gan Cao to 30 g Sheng Di Huang 24 g Bai Shao Yao 24 g Mai Men Dong 21 g	Severe Yin exhaustion

Related Formulas:

Formula	Herbs	Indication
Jiu Ni Tang [Rescue the Critical Condition Decoction]	Core formula – Huo Ma Ren + Sheng Long Gu 12 g Sheng Mu Li 12 g Ren Shen 6 g for pulse that is empty, big, and almost scattered	This formula is indicated for mistreatment of warm febrile disease with sweating method. This method damages fluid, causes spontaneous sweating, empties the Middle Jiao, and soon collapses Yang. The treatment principle is to strengthen the Heart, consolidate the collapsing Yang, and tonify both Qi and Yin
Yi Jia Fu Mai Tang [One-Shell Decoction to Restore the Pulse]	Core formula – Huo Ma Ren + Mu Li 30 g	Treats warm febrile disease at the Lower Jiao with loose stool. So Mu Li is added to consolidate and astringe the Lower Jiao to prevent draining of Yin
Er Jia Fu Mai Tang [Two-Shell Decoction to Restore the Pulse]	Core formula + Sheng Mu Li 15 g Sheng Bie Jia 24 g	Treat heat evil pathogens located deep in the Lower Jiao; deep rapid pulse, dry tongue, dark or black teeth, and worm-like movement of the hands and fingers which should be treated by nourishing Yin and subduing Yang to prevent spasms and cold extremities
San Jia Fu Mai Tang [Three-Shell Decoction to Restore the Pulse]	Core formula + Sheng Mu Li 15 g Sheng Bie Jia 24 g Sheng Gui Ban 30 g	Treat warm febrile disease of the Lower Jiao; heat lodged deep in the Lower Jiao and syncope; thin, hurried pulse, hard palpitation, or even chest pain. This should be treated by nourishing Yin, anchoring Yang, strengthening the Heart, and restoring the pulse

Yang Xin Tang [Nourish the Heart Decoction] *Gu Jin Yi Tong*

Dang Gui	9 g
Sheng Di Huang	9 g
Shu Di Huang	9 g
Fu Shen	9 g
Ren Shen	12 g
Mai Men Dong	12 g
Bai Zi Ren	6 g
Zhi Gan Cao	3 g
Wu Wei Zi	3 g

Main Indication: Heart Qi and Yin deficiency. Symptoms include weak constitution, or overthinking and worrying after recovering from diseases, insomnia, or light restless sleep, fright, and palpitation, Heart deficiency, restlessness, flabby tongue with white coating, or red tongue tip, and thin, rapid, or empty big pulse.

Preparation: Decoct with Deng Xin Cao and Lian Zi Xin.

Treatment Principles: Nourish Heart blood, and calm Shen.

This formula is designed for people with weak constitution or for overthinking and worrying. It nourishes Heart and calms Shen. This can also treat irregular heart rhythm and palpitation due to Heart blood and Heart Qi deficiency.

Suan Zao Ren Tang [Sour Jujube Decoction] *Jin Gui Yao Lue*

Fu Ling	9 g
Chuan Xiong	3 g
Suan Zao Ren (fried)	18 g
Gan Cao	3 g
Zhi Mu	9 g

Main Indication: Ying and blood deficiency fail to nourish Heart, and deficient fire disturbs the mind. Symptoms include restlessness due to deficient heat, insomnia, Heart palpitation, night sweats, dizziness, vertigo, dry throat, dry mouth, red tongue with a scanty coating, and thin rapid pulse.

Preparation: Decoction. Divided in three doses and take the decoction warm.

Treatment Principles: Nourish Heart blood, calm Shen, clear deficient heat, eliminate restlessness.

This formula treats any type of insomnia and deficient restlessness due to syndromes of Yin deficiency and floating Yang. This formula tonifies and drains mildly.

Related Formula:

Formula	Herbs	Indication
Ding Zhi Wan [Settle the Emotion Pill]	Ren Shen Fu Ling Fu Shen Shi Chang Pu Jiang Yuan Zhi Form into honey pill and coat with Zhu Sha Take this pill before sleep	Treats apprehensiveness, easily angered, and restless sleep. This formula focuses on tonifying the Heart, calming Shen, and benefiting the mind

Zhen Zhu Mu Wan [True Mother of Pearl Pill] *Ben Shi Fang* same as Zhen Zhu Mu Wan [Mother of Pearl Pill]

Zhen Zhu Mu (powdered)	22.5 g
Dang Gui	45 g
Shu Di Huang	45 g
Ren Shen	30 g
Suan Zao Ren	30 g
Bai Zi Ren	30 g
Xi Jiao (substitute with Shui Niu Jiao) (powdered)	15 g
Fu Shen	15 g
Chen Xiang	15 g
Long Gu	15 g

Main Indication: Heart and Liver blood deficiency, hyperactive Yang. Symptoms include insomnia, restless sleep, irritability, sometimes fright with palpitation, dizziness, vertigo, red tongue with scanty coating; or thin, rapid, and slight, wiry pulse.

Preparation: Grind into fine powder, mix with honey to form pill the size of seeds from a Chinese parasol tree. Coat the pill with Zhu Sha. Take 40 to 50 pills with gold, silver, and Bo He decoction at noon and bedtime.

Treatment Principles: Nourish Yin, nourish blood, anchor the Heart, and calm the Shen.

This formula nourishes Yin and blood, and at the same time calms Heart and smoothes Liver. It is indicated for syndromes of Yin blood deficiency with hyperactive Yang. However, it is not indicated for types like phlegm heat or phlegm fire insomnia, and fright and palpitation, which may mistakenly retain the evil pathogens by tonifying.

Huang Lian E Jiao Tang [Coptis and Ass-Hide Gelatin Decoction] *Shan Han Lun*

Huang Lian	6 g
Huang Qin	12 g
Bai Shao Yao	15 g
E Jiao (melt in decoction)	12 g
Ji Zi Huang (yolk)	1 piece

Main Indication: Heat entering Shao Yin, and exhausted true Yin with deficient fire. Symptoms include restlessness, insomnia, red tongue, scanty tongue coating, very red tongue tip, and thin, rapid pulse.

Preparation: Decoct the first three herbs, then melt E Jiao and mix into Ji Zi Huang.

Treatment Principles: Nourish Yin and clear heat.

Related Formula:

Formula	Herbs	Indication
E Jiao Huang Lian Tang [Ass-Hide Gelatin and Coptis Decoction]	Core formula + Sheng Di Huang	Same indication as the core formula. This formula has stronger Yin nourishing and clearing heat functions

Zhu Sha An Shen Wan [Cinnabar Pill to Calm the Spirit] *Lan Shi Mi Cang*, Also Called An Shen Wan [Calm the Spirit Pill]

Huang Lian	6 g
Zhu Sha	15 g
Sheng Di Huang	24 g
Dang Gui	12 g
Zhi Gan Cao	15 g

Main Indication: Heart Yin deficiency with hyperactive Heart Yang. Symptoms include irritability and restlessness of Heart and Shen, insomnia, excessive dreaming, fright and palpitation; in severe cases, urge to vomit but not able to do so, heat and restless disturbance at the Heart and chest, agitation, red tongue, and thin, rapid pulse.

Preparation: Decoction. Grind Zhu Sha into powder, divide into three doses, and take with the decoction. This formula can be formed into pills, and then coated with Zhu Sha. Take 6 to 9 g each time.

Treatment Principles: Anchor Heart, calm the Shen, nourish Yin, and drain fire.

This formula nourishes Heart Yin and clears Heart fire, balancing Yin and Yang. So insomnia, agitation, restlessness, and other symptoms can be cleared. In modern times, this formula is used to treat insomnia due to deficient type of excitation in neurasthenia.

Related Formula:

Formula	Herbs	Indication
Sheng Tie Luo Yin [Iron Filings Decoction]	Sheng Tie Luo (decoct for 3 hours, and use the decoction to decoct other herbs) Tian Men Dong Mai Men Dong Chuan Bei Mu Dan Nan Xing Ju Hong Yuan Zhi Shi Chang Pu Lian Qiao Fu Ling Fu Shen Xuan Shen Gou Teng Dan Shen Zhu Sha	Treat manic syndrome due to phlegm fire disturbing Heart

Comparison of Formulas Treating Heart Yin Deficiency

Tian Wang Bu Xin Dan [Emperor of Heaven's Special Pill to Tonify the Heart], Yang Xin Tang [Nourish the Heart Decoction], Jia Jian Fu Mai Tang [Modified Restore the Pulse Decoction], and Suan Zao Ren Tang [Sour Jujube Decoction] all nourish Yin and blood, calm the Shen, and eliminate restlessness. However, the first two formulas nourish Heart Yin and tonify Heart Qi. Tian Wang Bu Xin Dan [Emperor of Heaven's Special Pill to Tonify the Heart] is stronger between the two. Jia Jian Fu Mai Tang [Modified Restore the Pulse Decoction] treats late stage of warm febrile disease when Yin and blood are damaged and exhausted, and there is the presence of dryness and Heat, with an empty, knotted, or intermittent pulse. Suan Zao Ren Tang [Sour Jujube Decoction] is a mild formula to nourish Heart, calm the Shen, and clear deficient heat.

Zhen Zhu Mu Wan [Mother of Pearl Pill], Zhu Sha An Shen Wan [Cinnabar Pill to Calm the Spirit], and Huang Lian E Jiao Tang [Coptis and Ass-Hide Gelatin Decoction] all treat fright, palpitation, insomnia, and excessive dreaming syndromes due to Yin deficiency and hyperactive Yang. However, the strength of Zhen Zhu Mu Wan [Mother of Pearl Pill] is to nourish Yin and blood. The strength of Zhu Sha An Shen Wan [Cinnabar Pill to Calm the Spirit] is draining fire and clearing Heart. For Huang Lian E Jiao Tang [Coptis and Ass-Hide Gelatin Decoction], the strength is to supplement the true Yin "to calm the minister fire."

HEART YIN AND YANG DEFICIENCY

This syndrome presents with both Yin and Yang deficiency. So both Yin and Yang are tonified. Commonly used herbs are Ren Shen, Gui Zhi, and Zhi Gan Cao, in combination with Sheng Di Huang, Shu Di Huang, Tian Men Dong, and E Jiao. If nocturnal emission in men and dreaming of sexual intercourse in women occur, it indicates Heart and spirit dissipation. One should then add Long Gu and Mu Li to astringe and anchor the Yang.

Zhi Gan Cao Tang [Honey-Fried Licorice Decoction] *Shang Han Lun*, Also Called Fu Mai Tang [Restore the Pulse Decoction]

Zhi Gan Cao	15 g
Ren Shen	6 g
Sheng Di Huang	30 g
Gui Zhi	9 g
E Jiao (melt in decoction)	6 g
Mai Men Dong	9 g
Huo Ma Ren	9 g
Da Zao	9 g
Sheng Jiang	9 g

Main Indication:

1. Heart Yin and Yang deficiency. Symptoms include knotted or intermittent pulse, palpitation, emaciation, and shortness of breath, pale tongue with scanty coating, or mirrored tongue, with scanty fluid.
2. Consumptive disease with Lung atrophy. Symptoms include dry cough without phlegm, or cough with scanty phlegm, or phlegm with blood tinge, emaciation, shortness of breath, restlessness, insomnia, spontaneous sweating, or night sweat, dry throat, dry tongue, or occasional deficient fever, constipation, red tongue, scanty tongue coating, and deficient, rapid pulse.

Preparation: Decoct the above herbs. Add 10 ml of rice wine to the decoction and melt the E Jiao in the decoction. Mix well, divided into three doses, and take in 1 day.

Treatment Principles: Tonify Qi, nourish Yin, and restore pulse.

This formula treats deficient Yang Qi that cannot disperse and opens the pulse and syndromes of Yin deficiency that cannot nourish Heart blood. Today, this formula is used to treat irregular Heart rhythm due to various causes. It can also treat Qi and Yin deficiency type of consumptive disease and dry cough as in the late stage of Lung tuberculosis.

Modification:

Minus	Indication
Sheng Jiang Gui Zhi Rice Wine (or use above ingredients in small amount)	Severe Yin exhaustion with heat symptoms

Gui Zhi Jia Long Gu Mu Li Tang [Cinnamon Twig Decoction plus Dragon Bone and Oyster Shell] *Jin Gui Yao Lue*

Gui Zhi	9 g
Bai Shao Yao	9 g
Sheng Jiang	9 g
Gan Cao	6 g
Da Zao	9 g
Long Gu	24 g
Mu li	24 g

Main Indication: Yin and Yang deficiency. Symptoms include nocturnal emission in men, dreaming of sexual intercourse in women, lower abdominal contraction and pain, cold sensation at external genitalia, blurry vision, loss of hair, red tongue tip with thin white coating, and thin or deep, thin, slightly rapid pulse.

Preparation: Decoction. Take warm in three doses.

Treatment Principles: Regulate and harmonize Yin and Yang, anchor the rebellion, and astringe the essence.

In this formula Gui Zhi, Sheng Jiang, and Gan Cao are acrid and sweet which transforms Yang. Bai Shao Yao, Gan Cao, and Da Zao are sour and sweet which transforms Yin. Long Gu anchors Yang and astringes. So this formula is effective in treating nocturnal emission and dreaming of sexual intercourse due to both Yin and Yang being deficient.

Comparison of Formulas Treating Both Yin and Yang Deficiency

Zhi Gan Cao Tang [Honey-Fried Licorice Decoction] nourishes Heart Yin to support the Heart, and warms Heart Yang to restore pulse in order to recover the main motor of the body. So this formula not only can treat knotted or intermittent pulse, but also can assist the recovery at late-stage Lung tuberculosis with Qi and Yin both deficient.

Gui Zhi Jia Long Gu Mu Li Tang is an empirical formula. It is composed of Gui Zhi Tang [Cinnamon Twig Decoction] to regulate and harmonize Yin and Yang, with Long Gu and Mu Li to anchor Yang. It is mainly used for nocturnal emission and dreaming of sexual intercourse due to Yin and Yang imbalance. It is not a nourishing or tonifying formula.

HEART QI DEFICIENCY AND EXHAUSTION

With conditions of Heart Qi deficiency and exhaustion, danger of collapse often manifests as pale sallow complexion, apathy, dull expression, weak feeble breathing, bead-like spontaneous sweating, and minute disappearing pulse. If it occurs at late stage of warm febrile disease, symptoms also include dry mouth, dry tongue, deficient rapid pulse or fever, irritability, agitation, symptoms worse during the night, palpitation, anxiousness, dark red tongue, and scanty tongue coating. Commonly used herbs include Ren Shen, Xi Jiao (Shui Niu Jiao), Mai Men Dong, and Wu Wei Zi.

Du Shen Tang [Unaccompanied Ginseng Decoction] *Shang Han Da Quan*

Ren Shen	10–30 g

Main Indication:

1. Sudden hemorrhage (severe Beng) in women or postpartum dizziness and severe hemoptysis.
2. Any severe syndrome with the manifestations of sallow, pale complexion, apathy, dull expression, spontaneous sweating, feeble breathing, and minute thin disappearing pulse.

Preparation: Decoct the herb under low heat until concentrated. Take as one dose.

Treatment Principles: Tonify Qi, consolidate the collapsing Qi.

The dosage of this formula is better when bigger. It can also greatly tonify Yuan Qi, benefit Qi, consolidate the collapsing Qi, and save a life.

Ren Shen Xi Jiao Tang [Ginseng and Rhinoceros Horn Decoction] (Cao Mian Wei Fang)

Ren Shen	10–15 g
Xi Jiao (or substitute with Shui Niu Jiao)	3–9 g (5–10 times that of the Xi Jiao dosage)

Main Indication: Heart disturbance at Ying level, Heart Qi exhaustion. Symptoms include fever, irritability, restlessness, sometimes delirious speech, worse at night, palpitation, shortness of breath, thirst with desire to drink, dark red tongue with scanty coating, deficient rapid pulse with more than seven beats per breath.

Preparation: Decoction.

Treatment Principles: Cool Ying, tonify and rescue collapsed Qi.

Comparison of Formulas Treating Heart Qi Deficiency and Exhaustion

Du Shen Tang [Unaccompanied Ginseng Decoction] treats severe loss of blood when Qi has no support, and is thus in danger of collapsing. It can also treat a patient in critical condition when Yuan Qi is deficient and in danger of collapsing.

Ren Shen Xi Jiao Tang [Ginseng and Rhinoceros Horn Decoction] treats syndromes of heat disturbing Heart energy with Heart exhaustion due to heat entering Ying level, damaging Yin and exhausting Qi.

HEART YANG DEFICIENCY AND EXHAUSTION

Heart Yang deficiency and exhaustion have different severity, thus the following three types of syndromes:

1. Yang deficiency with cold extremities. Symptoms include cold extremities, diarrhea with undigested food, and minute thin pulse. Commonly used herbs include Fu Zi, Gan Jiang, and Rou Gui to rescue the Yang.
2. Yin excess expelling Yang. Symptoms include diarrhea with undigested food, cold limbs, and minute pulse. At the same time, fever or red face can be observed. This is true cold false heat syndrome. Gan Jiang and Fu Zi are used to benefit the fire and reduce Yin. One can also add a small amount of heat healing herbs as in Tong Mai Si Ni Jia Zhu Dan Zhi Tang [Unblock the Pulse Decoction with Pig Bile for Frigid Extremities].
3. Collapsing exhausted Yang. Symptoms include unconsciousness, coma, wheezing, rapid breathing, bead-like sweat, cold limbs, and minute disappearing pulse. One should use Ren Shen and Fu Zi to rescue the collapsing Yang.

In the past, people categorized Si Ni Tang [Frigid Extremities Decoction] and Shen Fu Tang [Ginseng and Prepared Aconite Decoction] in formulas that warm Kidney Yang. However, these formulas can strengthen the Heart and prevent collapse, thus they are discussed under this section. However, this does not change the fact that they have warming Kidney function.

Si Ni Tang [Frigid Extremities Decoction] *Shang Han Lun*

Fu Zi	15–30 g
Gan Jiang	6–9 g
Zhi Gan Cao	6–12 g

Main Indication:

1. Shao Yin syndrome. Symptoms include cold extremities, diarrhea with undigested food, or nausea, vomiting, abdominal pain, absence of thirst, apathy, somnolence, white slippery tongue coating, and minute, thin, and weak pulse.
2. Tai Yang syndrome that is mistakenly treated with sweating method, which leads to collapsing Yang.

Preparation: Decoct Fu Zi first for 1 hour, then add other herbs. Take the decoction warm.

Treatment Principles: Rescue collapsing Yang.

This is the representing formula to rescue collapsing Yang. It is specially indicated for cold extremities and minute pulse as seen in Yang deficiency with excess cold. From clinical research, this formula shows stimulatory function of Heart and digestive function, and of blood circulation, thus it can treat collapse due to decreased or weakened metabolism. In clinical practice, it is often used in deficient or collapse condition after vomiting and diarrhea due to acute gastroenteritis or after big sweating in febrile diseases. It can also treat deficient cold type of gastroptosis.

Related Formulas:

Formula	Herbs	Indication
Tong Mai Si Ni Tang [Unblock the Pulse Decoction for Frigid Extremities]	Core formula with double dosage of Gan Jiang	Treat Shao Yin syndrome with diarrhea with undigested food, cold extremities with absence of aversion to cold; red face, or abdominal pain, dry heaves, sore throat, and minute and disappearing pulse
Tong Mai Si Ni Jia Zhu Dan Zhi Tang [Unblock the Pulse Decoction with Pig Bile for Frigid Extremities]	Tong Mai Si Ni Tang [Unblock the Pulse Decoction for Frigid Extremities] Pig Bile 10 ml	This formula is indicated after vomiting has stopped; sweating leads to cold and contraction of the extremities; minute and disappearing pulse
Si Ni Jia Ren Shen Tang [Ginseng Decoction for Frigid Extremities]	Core formula + Ren Shen 3 g	Treats cold extremities, aversion to cold, sleeping in curled up position, minute pulse, even though diarrhea has stopped but other symptoms still remain, or unceasing diarrhea. This is Yang deficiency and collapse syndrome
Yin Chen Si Ni Tang [Artemisia Yinchenhao Decoction for Frigid Extremities]	Core formula + Yin Chen Hao	Treat cold damp in the Middle Jiao with gray slippery coat, cold extremities, dusky, yellow complexion, and sclera
Yin Chen Zhu Fu Tang [Artemisia Yinchenhao, Atractylodes, and Prepared Aconite Decoction]	Core formula + Yin Chen Hao Bai Zhu	Treat the same indication as Yin Chen Si Ni Tang [Artemisia Yinchenhao Decoction for Cold Extremities] This formula is stronger in tonifying Spleen and draining damp

(continued)

Formula	Herbs	Indication
Hui Yang Ji Jiu Tang [Restore and Revive the Yang Decoction]	Core formula + Rou Gui Ren Shen Bai Zhu Fu Ling Chen Pi Ban Xia Wu Wei Zi She Xiang	Treat cold evil directly, attacks three Yin, and true Yang becomes weak and minute. Symptoms include aversion to cold, sleeping in fetal position, cold extremities, shivering, abdominal pain, vomiting, nausea, absence of thirst; or purple toe nails and lips; or spitting thin and thick saliva; and deep, slow, forceless pulse or absence of pulse

Shen Fu Tang [Ginseng and Prepared Aconite Decoction] *Zheng Ti Lei Yao*

Ren Shen	15 g
Zhi Fu Zi	15–60 g

Main Indication: Yang Qi suddenly collapses. Symptoms include cold extremities, bead-like sweat, wheezing, rapid breathing, shortness of breath, dizziness, and minute pulse.

Preparation: Cook Zhi Fu Zi for 90 to 120 minutes. Steam ginseng to collect decoction. Mix the two together before taking them.

Treatment Principles: Rescue collapsing Yang.

In this formula, the number of different herbs is small, but the dosages are high to make the formula effective. In critical conditions, this formula can be used as a rescue remedy. Ren Shen in the formula cannot be replaced by Dang Shen.

Related Formulas:

Formula	Herbs	Indication
Qi Fu Tang [Astragalus and Prepared Aconite Decoction]	Huang Qi Zhi Fu Zi	Treats spontaneous sweating due to Yang deficiency
Zhu Fu Tang [Atractylodes and Prepared Aconite Decoction]	Bai Zhu Zhi Fu Zi	Treat spontaneous sweating, low back cold pain and heaviness sensation, and frequent urination, which belongs to the cold–damp pattern
Bai Zhu Fu Zi Tang [Atractylodes and Prepared Aconite Decoction]	Bai Zhu Zhi Fu Zi Gan Cao	Treat dizziness, vertigo, heaviness of the head, and discomfort from deficient wind rising upward
Shen Fu Long Mu Tang [Ginseng, Prepared Aconite, Dragon Bone, and Oyster Shell Decoction]	Core formula + Long Gu Mu Li	Treat same indications as Shen Fu Tang [Ginseng and Prepared Aconite Decoction] but can also anchor Yang

Comparison of Formulas Treating Heart Yang Deficiency and Exhaustion

Si Ni Tang [Frigid Extremities Decoction] can rescue collapsed Yang to treat cold extremities due to Yang deficiency and excess cold.

Shen Fu Tang [Ginseng and Prepared Aconite Decoction] can rescue the collapsing Yang. It is indicated for critical conditions with cold extremities, minute pulse due to Yang deficiency, and collapsing Yang.

HEAT TOXIN TYPE OF PYOGENIC INFECTIONS AND ULCEROUS DISEASE OF THE SKIN

This syndrome is due to heat toxins congested and stagnated at the skin and muscle layers, leading to carbuncles, swellings, deep-rooted boils, and furuncles. In *Su Wen*, it is discussed that any types of carbuncles, pururitis, and skin infections belong to Heart, so one should treat these conditions from the Heart. This syndrome often manifests as redness, swelling, heat, pain, and other Yang symptoms. Herbs that clear heat and relieve toxicity, such as Shan Zhi Zi, Huang Qin, Huang Lian, Jin Yin Hua, Lian Qiao, Ye Ju Hua, Pu Gong Ying, Zi Hua Di Ding, and Zao Xiu, are used to make the formula.

At the beginning stage of the deep-rooted boils and carbuncles, and before the formation of pus, one should combine herbs that move blood, open collaterals, disperse wind, and scatter evil pathogens, such as Fang Feng and Jing Jie. If fire toxins are excessive internally, which damages Ying and blood, then one should combine cooling blood herbs with clearing Ying, such as Xi Jiao (Shui Niu Jiao) and Mu Dan Pi to clear and eliminate blood heat toxins and relieve symptoms of fever and toxicity.

In TCM dermatology, Chan Su, Qing Fen, Xiong Huang, Ku Fan, and Niu Huang are strong herbs to relieve toxins and are often used in combination with blood and Qi moving herbs, such as Ru Xiang and Mo Yao with She Xiang to unblock channels.

Xian Fang Huo Ming Yin [Sublime Formula For Sustaining Life] *Wai Ke Fa Hui*, Also Called Zhen Ren Huo Ming Yin [True Man's Formula for Sustaining Life]

Zhi Chuan Shan Jia	3 g
Tian Hua Fen	3 g
Gan Cao	3 g
Ru Xiang	3 g
Mo Yao	3 g
Chi Shao Yao	3 g
Dang Gui Wei	3 g
Zao Jiao Ci (fried)	3 g
Bai Zhi	3 g
Zhe Bei Mu	3 g
Fang Feng	3 g
Jin Yin Hua	9 g
Chen Pi	9 g

Main Indication: Beginning stage of swelling, toxic ulcerations and carbuncles, accumulation of heat toxins, Qi stagnation, and blood stasis, which belong to Yang syndromes. Symptoms include local redness, swelling, heat, and pain, or fever and chills, red tongue, and thin white tongue coating or yellow coating, with rapid forceful pulse.

Preparation: Decoction. Or decoct with half water and half wine.

Treatment Principles: Clear heat, relieve toxicity, move blood, reduce swelling, and stop pain.

This formula treats swollen toxic ulcerations, swelling, and carbuncles. Before the formation of pus, it can eliminate and scatter the skin condition. After the formation of pus, the formula can speed up the ulceration. This formula should not be taken after the ulceration and is contraindicated for Yin syndromes. In modern practice, this formula can also be used in treating Yang heat types of mastitis, osteomyelitis, sty, and acute dacryadenitis.

Modification:

Addition	Indication
Da Huang	Constipation
Wei Ling Xian	Swelling of the neck
Jing Jie Bo He	Red and swollen face

Related Formula:

Formula	Herbs	Indication
Si Miao Yong An Tang [Four-Valiant Decoction for Well-Being]	Jin Yin Hua 90 g Xuan Shen 90 g Dang Gui 30 g Gan Cao 15 g Decoction Continuous taking of 10 bags	Treat sloughing ulcer (Tuo Ju) caused by excess heat toxins. Symptoms include ulcerations on a limb that is dark red, slightly swollen, burning, foul smell, severe pain or accompanied with fever, thirst, red tongue rapid pulse. Today, it is used in treating heat toxin type of thromboangitis obliterans

Wu Wei Xiao Du Yin [Five-Ingredient Decoction to Eliminate Toxin] *Yi Zong Jin Jian*

Jin Yin Hua	20 g
Ye Ju Hua	30 g
Zi Hua Di Ding	30 g
Pu Gong Ying	30 g
Zi Bei Tian Kui	15 g

Main Indication: Fire toxin accumulation type of skin ulcerations, swelling, and furuncle or toxic carbuncles. Symptoms include local swelling, redness, heat and pain, generalized fever, chills, any type of deep-rooted boil toxins with manifestations like a chestnut or nail; hard and deep-rooted lesions, and red tongue with yellow tongue coating and rapid pulse.

Preparation: Decoction. Taken with 1 to 2 tablespoon of wine. The dregs can be mashed and used externally.

Treatment Principles: Clear heat, relieve toxicity, and eliminate deep-rooted boils, carbuncles, and toxins.

This formula is effective in treating any type of suppurative infections, such as deep-rooted boils, carbuncles, furuncles, pus and swelling with the manifestation of redness, swelling, heat, pain, red tongue, and rapid pulse. It can also treat heat toxic type of mastitis.

Modification:

Addition	Indication
Lian Qiao 15 g Mu Dan Pi 6 g Huang Qin 12 g	Severe heat toxins
Xia Ku Cao 30 g Qi Ye Yi Zhi Hua 9 g Chan Su Wan [Toad Cake Pill]	Deep-rooted boils and sores
Gua Lou 15 g Zhe Bei Mu 9 g Qing Pi 9 g Ju Ye 9 g	Breast carbuncles

(continued)

Addition	Indication
Bai Xian Pi 24 g Di Fu Zi 9 g Wu Shao She 9 g	Damp–heat rash
Huang Lian 9 g Huang Bai 9 g	Other types of carbuncles or furuncles

Yin Hua Jie Du Tang [Honeysuckle Decoction to Relieve Toxicity] *Yang Ke Xin De Ji*

Jin Yin Hua	10–20 g
Zi Hua Di Ding	10–20 g
Lian Qiao	10–20 g
Chuan Lian Zi	10–20 g
Xia Ku Cao	10–20 g
Chi Fu Ling	10–20 g
Mu Dan Pi	10–20 g
Xi Jiao (or substitute with 5–10 times of Shui Niu Jiao)	10–20 g

Main Indication: Damp–heat fire toxins invade Ying and blood. Symptoms include carbuncles, swelling, deep-rooted boils, furuncles, sores, red, swollen, heat, pain, or ulceration with discharge of pus and blood, or fever, chills, red tongue, yellow dry coating, and rapid, forceful pulse.

Preparation: Decoction.

Treatment Principles: Clear heat, relieve toxicity, drain fire, and cool blood.

This formula treats carbuncles, swelling, boils, and furuncles which belong to syndromes of excess toxins, severe swelling, and severe heat. It is effective in treating severe redness and swelling during the high fever stage of acute suppurative infection.

Modification:

Minus	Indication
Xi Jiao (Shui Niu Jiao)	Mild blood heat condition

Xi Huang Wan [Rhinoceros Gallstone Pill] *Wai Ke Quan Sheng Ji*

Xi Huang (substitute with Niu Huang)	15 g
She Xiang	75 g
Ru Xiang	500 g
Mo Yao	500 g

Main Indication: Breast cancer, scrofula, subcutaneous nodule, multiple abscess, bubo, pulmonary abscess, and intestinal abscess with a moderate constitution.

Preparation: Grind into extremely fine powder. Mix with 500 g of cooked old rice to form pills. Sun dry. Take 9 g with wine each time. If the location of disease is in the upper part of the body, take the pills before bedtime. If the location is at the lower part of the body, take the pill on empty stomach.

Treatment Principles: Move blood, expel stasis, relieve toxin, eliminate abscess, transform phlegm, and scatter the nodule.

In recent times, this formula has been used in treating lymphadenitis, mastitis, cystic hyperplasia of the breast, mammary cancer, multiple abscess, osteomyelitis with red tongue, and slippery rapid pulse. This formula should be used carefully when there is profuse pus and water discharge after ulceration, and deficiency of Qi and blood.

Related Formula:

Formula	Herbs	Indication
Xing Xiao Wan [Awaken and Eliminate Pill]	Ru Xiang 30 g Mo Yao 30 g She Xiang 6 g Xiong Huang 15 g Cooked yellow rice 30 g	Mix and smash to form pill in the size of Lai Fu Zi. Sun dry, and take 9 g with wine. Cover patient with blanket to induce sweating. Treats swollen red toxic abscesses

Chan Su Wan [Toad Cake Pill] *Wai Ke Zheng Zong*

Chan Su (mix with wine)	6 g
Qing Fen	1.5 g
Ku Fan	3 g
Han Shui Shi (rusted)	3 g
Tong Lu	3 g
Ru Xiang	3 g
Mo Yao	3 g
Dan Fan	3 g
She Xiang	3 g
Xiong Huang	6 g
Wo Niu (snail)	21 pieces
Zhu Sha	9 g

Main Indication: Deep-rooted boils, furuncles and carbuncles, boils at nape area, brain abscess, mastitis, bone tuberculosis, and other severe ulcerations due to fire toxin accumulation, Qi, and blood stagnation.

Preparation: Grind the above herbs into fine powder. Mash Wo Niu (snails) and mix well with Chan Su. Then mix with all other powder to form pills the size of mung beans. Chew three pills with 5 inches of white scallion and one cup of warm wine. Cover with blanket to induce sweating. In severe cases, take one more dose.

Treatment Principles: Relieve toxin, eliminate swelling, move blood, and stop pain.

This formula resolves toxicity through sweating. If the formula is used externally, it can transform the rotten, ulcerated tissue and generate muscle. So it can eliminate unformed skin conditions and speed up the ulceration of formed skin conditions. It is contraindicated for pregnant women or if the carbuncles and furuncles are already ulcerated. It should be used with caution for Qi and blood deficient constitution.

Comparison of Formulas Treating Heat Toxins Type of Pyogenic Infections and Ulcerous Disease of the Skin

1. Xian Fang Huo Ming Yin [Sublime Formula for Sustaining Life], Si Miao Yong An Tang [Four-Valiant Decoction for Well-Being], Wu Wei Xiao Du Yin [Five-Ingredient Decoction to Eliminate Toxin], and Yin Hua Jie Du Tang [Honeysuckle Decoction to Relieve Toxicity] can all clear heat and relieve toxicity to treat carbuncles and furuncles on the body surface. However, the first formula focuses on moving blood and Qi, reducing swelling, softening the hardness. It is used for swelling and toxins due to excess heat toxins by assisting and resolving the ulceration. Si Miao Yong An Tang [Four-Valiant Decoction for Well-Being]

is used in sloughing ulcers due to heat toxins transforming into fire, especially seen more often at the lower extremities. It is similar to thromboangitis due to heat toxins. Wu Wei Xiao Du Yin [Five-Ingredient Decoction to Eliminate Toxin] focuses on clearing heat, relieving toxin, and eliminating and removing deep-rooted boils and sores. The last formula focuses on draining fire, and cooling blood. It is used in severe heat toxins type of abscess, ulcerations, carbuncles, and furuncles.

2. Xi Huang Wan [Rhinoceros Gallstone Pill], Xing Xiao Wan [Awaken and Eliminate Pill], and Chan Su Wan [Toad Cake Pill] all can move blood and resolve the accumulation. However, Xi Huang Wan [Rhinoceros Gallstone Pill] is strongest in clearing heat and resolving toxicity among the three, followed by Xing Xiao Wan [Awaken and Eliminate Pill] and the Chan Su Wan [Toad Cake Pill]. It can also warm and scatter. Xi Huang Wan [Rhinoceros Gallstone Pill] and Xing Xiao Wan [Awaken and Eliminate Pill] can treat Yang type of abscess and swelling. Chan Su Wan [Toad Cake Pill] treats both abscesses and ulcerations.

CHAPTER 8

Heart and Kidney Syndromes and Formulas

HEART AND KIDNEY DISHARMONY

This syndrome is caused by Heart Yang not warming the Kidney Yin, and Kidney Yin not able to support the Heart Yang, leading to disharmony of the Yin and Yang. Symptoms include palpitation, insomnia, excessive dreaming, being easily frightened, and nocturnal emission. The formula uses Heart and Kidney tonic herbs with Yuan Zhi, Shi Chang Pu, Lian Zi, Fu Shen, Huang Lian, and Rou Gui to harmonize the center and aromatic herbs to transform turbidity, and awaken the Shen. When the Middle Jiao is harmonized, then the communication between Heart and Kidney can be established.

An Shen Ding Zhi Wan [Calm the Spirit and Settle the Will Pill]
Yi Xue Xin Wu

Ren Shen	10 g
Fu Ling	12 g
Fu Shen	10 g
Long Chi	15 g
Yuan Zhi	10 g
Shi Chang Pu	10 g

Main Indication: Heart Qi deficiency, Heart and Kidney disharmony, restlessness, and anxiousness. Symptoms include excessive dreaming, being easily frightened, palpitation, insomnia, or nocturnal emission, or epilepsy attacks, pale tongue, and thin, weak pulse.

Preparation: Grind into fine powder, and mix with honey to form pill. Take 9 g, two to three times a day. It can also be taken as decoction.

Treatment Principles: Nourish Heart, calm Shen, and harmonize Kidney and Heart.

This formula mainly treats anxiousness and restlessness with fright, palpitation, and insomnia. It focuses on treating the Heart and supported by harmonizing the Heart and Kidney.

Related Formulas:

Formula	Herbs	Indication
Yuan Zhi Wan [Polygalae Pill]	Core formula + Zhu Sha	Treat insomnia and restless sleep after being frightened or scared
Qian Jin Fu Shen Tang [Poria Cocos Pararadicis Decoction from Thousand Ducat Formula]	Fu Shen 30 g Fu Ling 30 g Ren Shen 30 g Shi Chang Pu 15 g Chi Xiao Dou 40 pieces	Treat Heart Qi deficiency leading to restlessness, agitation, fright, and palpitation

Jiao Tai Wan [Grand Communication Pill] *Han Shi Yi Tong*

Rou Gui Xin	3 g
Huang Lian	18 g

Main Indication: Hyperactive Heart fire, and Heart and Kidney disharmony. Symptoms include fright and palpitation, restless sleep, agitation, excessive dreaming, inability to concentrate; or nocturnal emission; dreaming of sexual intercourse; red tongue tip; thin rapid pulse; both Cun positions floating and rapid, and both Chi positions thin and deep.

Preparation: Grind into fine powder and mix with honey to form pill. It can also be taken as decoction, but dosage should be decreased.

Treatment Principles: Harmonize Heart and Kidney.

This formula uses Huang Lian to clear Heart fire to control the hyperactive Heart Yang and assists Rou Gui to guide the Heart fire down to the Kidney. This reestablishes the communication between Heart and Kidney and thus symptoms can be cleared. This formula is small and can be used by itself or added to other formulas.

Modification:

Addition	Indication
Ban Xia Geng Mi	Disharmony of the Middle Jiao

Related Formula:

Formula	Herbs	Indication
Xin Shen Jiao Tai Tang [Grand Communication Decoction for Heart and Kidney]	Bei Sha Shen 9 g Sheng Di Huang 9 g Mai Men Dong 5 g Dang Gui 5 g Yuan Zhi 2 g Bai Shao Yao 5 g Fu Shen 9 g Zhi Gan Cao 1.5 g Huang Lian 2 g Rou Gui 1 g Geng Mi 15 g	Treat inability to sleep during day or night, fear or fright once asleep, sinking or collapsing sensation, screaming or yelling, and being sometimes sad and sometimes crying

Sang Piao Xiao San [Mantis Egg Case Powder] *Ben Cao Yan Yi*

Sang Piao Xiao	30 g
Yuan Zhi	30 g
Shi Chang Pu	30 g
Long Gu	30 g
Ren Shen	30 g
Fu Shen	30 g
Dang Gui	30 g
Gui Ban (vinegar and honey fried)	30 g

Main Indication: Heart and Kidney deficiency and disharmony. Symptoms include frequent urination, or urine-like cloudy rice water, absent-mindedness, forgetfulness, no desire to eat, emaciation, enuresis, nocturnal emission without dreams, pale tongue, thin white tongue coating, and deficient thin pulse.

Preparation: Grind into fine powder and take 6 g with Ren Shen decoction before going to sleep. Dosage can be reduced when prepared as decoction.

Treatment Principles: Regulate and tonify Heart and Kidney, astringe, consolidate, and stop discharges.

This formula is designed for various syndromes due to Heart deficiency failing to nourish Shen (spirit), and Kidney deficiency failing to consolidate and astringe. In modern times it is used to treat pediatric enuresis and apathy with weak constitution and adult incontinence due to Kidney deficiency.

Comparison of the Formulas Treating Heart and Kidney Disharmony

An Shen Ding Zhi Wan [Calm the Spirit and Settle the Will Pill] focuses on tonifying Heart Qi to calm Heart and Shen. It also assists the communication between Heart and Kidney. It is mainly used to treat fright, palpitation, and insomnia. Jiao Tai Wan [Grand Communication Pill] focuses on clearing the Heart, calming the Shen, and guiding the fire back to its source. It is mainly used in treating palpitation and insomnia. Sang Piao Xiao San [Mantis Egg Case Powder] focuses on tonifying Heart and consolidating the Kidney. It is mainly used in treating absent-mindedness, nocturnal emissions, and is better for children and seniors.

CHAPTER 9

Kidney Syndromes and Formulas (Urinary Bladder)

Kidney is the congenital foundation and an important Zang organ of the body. It stores the essence that is the root of the source of Yin and Yang. It controls bone, generates marrow, dominates the reproductive system, controls development of the body and water metabolism, and opens to the ear. Once it is out of balance, then, there can be Yin and Yang deficiency, failure to consolidate and astringe, water accumulation, Kidney deficient type of leukorrhea, damp–heat in the Lower Jiao, and other syndromes.

KIDNEY YIN DEFICIENCY AND EXHAUSTION

This syndrome often manifests as nocturnal emission, night sweats, dizziness, vertigo, lumbar soreness, tinnitus, and thin Chi pulse. It should be treated with Shu Di Huang, Zhi He Shou Wu, Gou Qi Zi, Han Lian Cao, and Gui Ban Jiao. If there is Yin deficiency with hyperactive Yang, then Mu Dan Pi, Zhi Mu, and Huang Bai should be added to drain deficient fire. If Yin deficiency leads to floating Yang and Kidney Yang fails to grasp the Qi, then Wu Wei Zi and Shan Zhu Yu should be added.

Liu Wei Di Huang Wan [Six-Ingredient Pill with Rehmannia] *Xiao Er Yao Zheng Zhi Jue, Originally Called Di Huang Wan [Rehmannia Pill]*

Shu Di Huang	240 g
Shan Zhu Yu	120 g
Shan Yao	120 g
Ze Xie	90 g
Fu Ling	90 g
Mu Dan Pi	90 g

Main Indication: Kidney Yin deficiency and exhaustion, upward blazing of deficient fire. Symptoms include lower back weakness and knee soreness, dizziness, vertigo, tinnitus, deafness, night sweats, nocturnal emission, loose teeth, or toothache due to deficient fire, dry mouth, dry throat, heel pain, or steaming bone, tidal fever, five center heat, or Xiao Ke, frequent erection, or unclosed fontanel in children, delayed or slow development, red tongue, scanty tongue coating, thin and rapid pulse.

Preparation: Grind into powder and mix with honey to form pills 5 to 10 mm in diameter. Take three pills (9 g) with warm water on empty stomach, two to three times a day. The dosage can be reduced to make a decoction.

Treatment Principles: Nourish Yin and tonify Kidney.

This formula is modified from Jin Gui Shen Qi Wan [Kidney Qi Pill from the Golden Cabinet] *Jin Gui Yao Lue* by removing Gui Zhi and Fu Zi. The original formula is indicated for syndromes of pediatric Liver and Kidney Yin deficiency. Later, it became the base formula for nourishing Kidney Yin and many formulas have evolved from this one. Today, it is used in treating chronic pyelonephritis, chronic nephritis, diabetes, hypertension, hyperthyroidism, neurasthenia, functional uterine bleeding and other syndromes that belong to Liver and Kidney Yin type deficiency.

Related Formulas:

Formula	Herbs	Indication
Du Qi Wan [Capital Qi Pill]	Core formula + Wu Wei Zi	Treat Kidney deficient type of wheezing, with red face and hiccups. It is used in bronchial asthma, chronic bronchitis, wheezing, and asthma due to the Kidney failing to grasp the Qi
Qi Ju Di Huang Wan [Lycium Fruit, Chrysanthemum, and Rehmannia Pill]	Core formula + Gou Qi Ju Hua	Treat dizziness, vertigo, blurry vision, tearing with wind, dry, and painful eyes that belong to Liver and Kidney deficient syndrome
Mai Wei Di Huang Wan [Ophiopogonis, Schisandra, and Rehmannia Pill] also called Ba Xian Chang Shou Wan [Eight-Immortal Pill for Longevity]	Core formula + Mai Men Dong Wu Wei Zi	Treat cough, wheezing, tidal fever, and night sweats and nocturnal emission, which belongs to Lung and Kidney Yin deficiency
Er Long Zuo Ci Wan [Pill for Deafness that is Kind to the Left (Kidney)]	Core formula + Duan Ci Shi Shi Chang Pu Wu Wei Zi	It treats Kidney deficiency causing tinnitus, deafness, or late stage febrile disease; after fever is gone, body returns to coolness, but tinnitus and deafness remain due to Kidney deficiency and exhaustion
Zhi Bai Di Huang Wan [Anemarrhena, Phellodendron, and Rehmannia Pill] also called Zhi Bai Ba Wei Wan [Eight-Ingredient Pill with Anemarrhena and Phellodendron]	Core formula + Zhi Mu Huang Bai	Treat steaming bones, tidal fever, deficient restlessness, night sweats, lower back and spine soreness and pain; and nocturnal emission that belongs to Yin deficiency with hyperdeficient fire.
Zuo Gui Yin [Restore the Left (Kidney) Decoction]	Core formula – Mu Dan Pi Ze Xie + Gou Qi Gan Cao	Treat the same indication as the core formula. Slightly stronger in the functions

Er Zhi Wan [Two-Ultimate Pill] *Yi Fang Ji Jie*

Nu Zhen Zi	equal amount
Han Lian Cao	

Main Indication: Kidney (Liver) Yin deficiency. Symptoms include blurry vision, insomnia, excessive dreaming, bitter taste in mouth, dry throat; weak, sore, lower back and knees, atrophy or weakness of the lower legs, nocturnal emission, early graying of hair, thin white tongue coating, and thin or deep, thin pulse.

Preparation: Grind Nu Zhen Zi into fine powder. Make Han Lian Cao Extract or decoct with equal amount of Sang Shen to form syrup. Mix honey and Nu Zhen Zi powder to form pill. Each pill weighing 15 g. Take one in the morning and one at night with warm water.

Treatment Principles: Tonify Kidney and nourish Liver.

This is a mild formula to nourish Kidney and Liver. It is indicated for mild chronic Kidney Yin deficiency, leakage of Kidney essence, and Liver Yin deficiency without heat manifestations. It needs to be taken for a long period of time.

Related Formula:

Formula	Herbs	Indication
Sang Ma Wan [Mulberry Leaf and Sesame Seed Pill] also called Fu Sang Wan [Assist with Mulberry Leaf Pill]	Sang Ye 300 g Hei Zhi Ma 120 g Bai Mi (honey) 300 g Cook Sang Ye to make concentrated decoction. Mix with powdered Zhi Ma and honey to form pills. Take 10 g in the morning with salt water and with wine at night.	Treat Yin deficiency with blood dryness type of dizziness, blurry vision, chronic cough, constipation, wind–damp numbness, Bi syndrome, and dry skin.

Shi Hu Ye Guang Wan [Dendrobium Pill for Night Vision] *Yuan Ji Qi Wei*

Tian Men Dong	60 g
Ren Shen	60 g
Fu Ling	60 g
Mai Men Dong	30 g
Shu Di Huang	30 g
Sheng Di Huang	30 g
Tu Si Zi	23 g
Ju Hua	23 g
Cao Jue Ming	23 g
Xing Ren	23 g
Shan Yao	23 g
Gou Qi	23 g
Huai Niu Xi	23 g
Wu Wei Zi	23 g
Ci Ji Li	15 g
Shi Hu	15 g
Rou Cong Rong	15 g
Chuan Xiong	15 g
Zhi Gan Cao	15 g
Zhi Ke	15 g
Qing Xiang Zi	15 g
Fang Feng	15 g
Huang Lian	15 g
Shui Niu Jiao (to replace Xi Jiao)	60 g
Ling Yang Jiao (substitute with Shan Yang Jiao)	15 g (60 g)

Main Indication: Kidney and Liver deficiency with deficient fire. Symptoms include dilated pupil, blurry vision, light sensitivity, tearing, dizziness, vertigo, and cataract or macular degeneration. It can also treat overuse of the eyes on computers and video games, causing dry eyes and or tearing, dizziness, vertigo, and headache. The tongue is red with scanty tongue coating and the pulse is thin and rapid.

Preparation: Grind into fine powder and mix with honey to form pills 5 to 10 mm in diameter. Take 30 to 50 pills each in the morning and at night with warm wine or salt water.

Treatment Principles: Nourish Kidney Yin, calm Liver, and brighten the eyes.

Zuo Gui Wan [Restore the Left (Kidney) Pill] *Jing Yue Quan Shu*

Shu Di Huang	240 g
Shan Yao (fried)	120 g
Gou Qi Zi	120 g
Shan Zhu Yu	120 g
Chuan Niu Xi (steamed)	90 g
Tu Si Zi (zhi)	120 g
Lu Jiao Jiao	120 g
Gui Ban Jiao	120 g

Main Indication: Liver blood and Kidney essence deficiency. Symptoms include dizziness, vertigo, sore lower back and legs, nocturnal emission, spontaneous sweating, night sweats, dry mouth, dry throat, mirror tongue with scanty coating, and thin or rapid pulse.

Preparation: Grind into fine powder and form honey pills of 15 g each. Take one in the morning and one at night on empty stomach with bland soup.

Treatment Principles: Tonify and supplement Kidney Yin, and nourish Liver blood.

This formula is indicated for syndromes of exhaustion of Liver blood and Kidney essence, and insufficient Qi and blood in Ren and Du channels. This syndrome is often seen after severe conditions, chronic conditions, or surgery.

Modification:

Addition	Indication
Chen Pi Sha Ren	Middle Jiao weakness

Related Formula:

Formula	Herbs	Indication
Qi Bao Mei Ran Dan [Seven-Treasure Special Pill for Beautiful Whiskers]	He Shou Wu 300 g Fu Ling 150 g Huai Niu Xi 150 g Dang Gui 150 g Gou Qi Zi 150 g Tu Si Zi 150 g Bu Gu Zhi 120 g	Nourish Kidney water and benefit Liver blood Treat deficient Liver and Kidney, early graying of hair, loose teeth, nocturnal emission with or without dreaming, and lumbar and knee soreness and weakness

Da Zao Wan [Great Creation Pill] *Fu Shou Jing Fang*

Zi He Che	1 piece
Zhi Gui Ban	120 g
Huang Bai	30 g
Du Zhong	60 g
Niu Xi	60 g
Mai Men Dong	60 g
Tian Men Dong	60 g
Sheng Di Huang	180 g
Ren Shen	30 g

Main Indication: Kidney Yin deficiency with Yang hyperactivity. Symptoms include steaming bone, exhausted heat, cough, emaciation, shortness of breath, wheezing with mouth open, night sweats, red tongue, scanty tongue coating, and thin rapid pulse; empty big pulse at Cun, and rapid and thin at Chi.

Preparation: Grind into fine powder, and mix with honey to form pills. Take 9 g, two to three times a day.

Treatment Principles: Nourish Kidney Yin, drain heat, and anchor Yang.

This formula can treat consumptive cough, and emaciation due to Yin deficiency with hyperactive Yang. It can also treat scanty menstruation or amenorrhea, emaciation, and infertility that belongs to Chong and Ren deficiency. It can also treat fever, night sweats, and cough due to Lung tuberculosis.

Related Formula:

Formula	Herbs	Indication
Da Zao Wan [Great Creation Pill]	Ren Shen 3–30 g Shan Yao (fried) 6 g Shu Di Huang 9 g Du Zhong 6 g Gou Qi Zi 9 g Dang Gui 9 g Shan Zhu Yu 3 g Zhi Gan Cao 6 g	Nourish and tonify Lung and Kidney Treat lung TB due to Lung and Kidney deficiency, nocturnal emission, epilepsy syndrome (Kidney and Heart deficient type), delayed menstruation due to blood deficiency, and impotence due to Kidney deficiency

Zhu Jing Wan [Preserve Vistas Pill] *Zheng Zhi Zhun Sheng*

Tu Si Zi	150 g
Shu Di Huang	120 g
Che Qian Zi	60 g

Main Indication: Liver and Kidney Yin deficiency. Symptoms include decreased eyesight, blurry vision, thin white tongue coating, and thin pulse.

Preparation: Grind into fine powder and mix with honey to form pills. Take 6 to 12 g with Fu Ling and Shi Chang Pu decoction on empty stomach two to three times a day.

Treatment Principles: Nourish Yin and brighten the eyes.

This formula nourishes Yin and brightens the eyes with the function of Che Qian Zi which drains Liver and Kidney evil heat and does not cause stagnation. It can be used to treat decreased vision in seniors due to cataract and glaucoma.

Comparison of Formulas Treating Kidney Yin Deficiency and Exhaustion

Liu Wei Di Huang Wan [Six-Ingredient Pill with Rehmannia] is the representing formula in nourishing Kidney Yin. It can be used in a variety of syndromes. The herbal combination of this formula is balanced and comprehensive. It is suitable for taking long term. Er Zhi Wan [Two-Ultimate Pill] is a mild formula for nourishing Liver and Kidney. It is indicated for the beginning stage or mild conditions of Yin deficiency. The formula is small, thus it is easy to add this formula to other formulas. Shi Hu Ye Guang Wan [Dendrobium Pill for Night Vision] and Zhu Jing Wan [Preserve Vistas Pill] both treat eye conditions due to Liver and Kidney deficiency. However, the former formula treats photophobia, tearing, dilated pupils, or even red, dry, painful eyes due to Yin deficiency with deficient fire. The latter treats blurry vision due to Yin deficiency. Zuo Gui Wan

[Restore the Right (Kidney) Pill] nourishes Kidney Yin and essence. It is stronger than Liu Wei Di Huang Wan [Six-Ingredient Pill with Rehmannia] in its Yin nourishing function. At the same time, there are herbs in Zuo Gui Wan [Restore the Right (Kidney) Pill] to tonify Kidney Yang in order to generate Yin. Da Zao Wan [Great Creation Pill] treats steaming bone, tidal fever, emaciation, cough due to Yin deficiency, and Yang hyperactivity. It can also treat gynecological disorders due to damages and deficiency of Chong and Ren channels.

KIDNEY YANG DEFICIENCY AND EXHAUSTION

This syndrome is also called deficiency of Ming Men fire. Symptoms include soreness and weakness of lower back and knees, dizziness, tinnitus, aversion to cold, copious clear urination, or frequent urination or incontinence, impotence, pale or flabby moist tongue, and weak pulse, especially at right Chi position. Commonly used herbs are Rou Gui, Fu Zi, Ba Ji Tian, Suo Yang, Rou Cong Rong, Bu Gu Zhi, Du Zhong, Tu Si Zi, Yin Yang Huo, Lu Rong, Qiang Lang, and Jiu Zi. These herbs then are combined with Shan Yao, Shu Di, and Gui Ban Jiao to nourish Yin. Thus, the formula tonifies Yang and nourishes Yin at the same time. When the Yang is generated, Yin is also supported which demonstrates the idea of seeking Yang within Yin.

If Yang is deficient and Yin is excess, and there is an internal accumulation of water and dampness, it can lead to difficult urination, edema, damp transforming into phlegm, and cough with thin, watery phlegm. For those conditions, use Rou Gui and Fu Zi as the main herbs and combine them with Fu Ling, Ze Xie, Che Qian Zi, Zhu Ling, Sheng Jiang Pi, Gui Zhi, and Ban Xia.

For frequent urination, enuresis, incontinence, or dribbling urination due to Kidney deficiency and loss of bladder control, add Yi Zhi Ren, Wu Wei Zi, Sang Piao Xiao, and Fu Pen Zi to the tonifying Kidney Yang herbs. If Kidney deficiency failing to store essence leads to nocturnal emission, use Kidney Yang tonics with Sha Yuan Ji Li, Qian Shi, Lian Xu, Long Gu, Mu Li, Shan Zhu Yu, Wu Wei Zi, and Jin Ying Zi. For Kidney deficient impotence, deficient cold, and infertility, add Jiu Zi, She Chuang Zi, Yin Yang Huo, Xiong Can E, Hai Long, and Hai Ma to warm the Yang. For sore and weak lower back and knees due to Kidney deficiency, use Sang Ji Sheng, Hu Tao Rou, Du Zhong, Xu Duan, Huai Niu Xi, Chuan Niu Xi, Gou Qi, Gou Ji, and Yin Yang Huo. If cold congeals at the tendons and bones with Kidney Yang deficiency causing Yin type of deep-rooted carbuncles and multiple abscesses, add Ma Huang, Xi Xin, and Bai Jie Zi to warm the Kidneys formulas.

Jin Gui Shen Qi Wan [Kidney Qi Pill from the Golden Cabinet] *Jin Gui Yao Lue*

Sheng Di Huang	240 g
Shan Yao	120 g
Shan Zhu Yu	120 g
Ze Xie	90 g
Mu Dan Pi	90 g
Fu Ling	90 g
Gui Zhi	30 g
Pao Fu Zi	30 g

Main Indication: Kidney Yang deficiency. Symptoms include sore and weak lower back and knees, frequent cold sensation in the lower part of the body, difficult urination, or frequent profuse urination, pale flabby tongue; thin, white, moist tongue coating, and deep, thin, Chi pulse.

Preparation: Grind into fine powder, and then mix with honey to form pills 5 to 10 mm in diameter. Take 15 to 25 pills, once in the morning and once at night with light salt water. If it is taken as a decoction, dosage should be decreased.

Treatment Principles: Warm and tonify Kidney Yang.

This formula supports Yang from Yin source. The main idea is not focusing on tonifying the fire, but on slowly and gently generating the fire, which is the Kidney Qi. In clinical practice, this formula is indicated when the symptoms are present. Thus, it can treat senior and pediatric incontinence, enuresis in children, edema with cold Bi of the lumbar, leg, or foot, decreased sensation at lower abdomen, difficult urination; Lin syndrome, frequent nocturia in children with slight pain after urination, impotence, turbid urination, and leukorrhea. It can also treat phlegm Yin syndrome (congested fluid), dysuria due to pressure of the fetus, beriberi, and Xiao Ke syndrome.

Modification:

Minus	Addition	Indication
	Lu Rong Chen Xiang	Dribbling, turbid, white urine
	Wu Wei Zi	Copious and frequent nocturia
	Bu Gu Zhi Lu Rong	Frequent copious urination, sallow pale face, and emaciation
Ze Xie	Wu Wei Zi	Nocturnal emission, itching at the inside of the penis, and incontinence, which indicates Prime Minister Fire upsetting the body
	Ren Shen Lu Rong Ba Ji Tian Rou Cong Rong Suo Yang Gou Qi Zi	Impotence and decline in semen
	Sheng Mai San [Generate the Pulse Powder]	Vomit blood, cold feet, red face, irritability, and thirst

Related Formulas:

Formula	Herbs	Indication
Ji Sheng Shen Qi Wan [Kidney Qi Pill from Formulas that Aid the Living] also called Jia Wei Shen Qi Wan [Augmented Kidney Qi Pill]	Core formula + Chuan Niu Xi Che Qian Zi	Kidney Yang deficiency, heaviness at lumbar, foot edema, and difficulty urinating
Shi Bu Wan [Ten-Tonic Pill]	Core formula + Lu Rong Wu Wei Zi	Treat Kidney Yang deficiency and exhaustion, dark complexion, cold feet, swollen feet, tinnitus, deafness, soreness and weakness of feet and knees, difficult urination, and lower back pain
Xin Jia Ba Wei Di Huang Wan [Newly-Augmented Eight-Ingredient Pill with Rehmannia]	Core formula + Zi Shi Ying Sheng Tie Luo Hei Xi Dan [Lead Special Pill] – Mu Dan Pi	This formula treats excess above and deficiency below

(continued)

Formula	Herbs	Indication
You Gui Yin [Restore the Right (Kidney) Decoction]	Core formula – Fu Ling Ze Xie Mu Dan Pi + Gan Cao Gou Qi Du Zhong	Warm Kidney Yang function is stronger than the core formula. It is better for impotence and nocturnal emission
He Che Ba Wei Wan [Eight-Flavor Pill with Placenta]	Zi He Che Shu Di Huang Mu Dan Pi Da Zao Fu Ling Ze Xie Shan Yao Mai Men Dong Wu Wei Zi Rou Gui Fu Zi Shan Zhu Yu Lu Rong	Make herbs into pill. Take 10 to 15 g, two to three times a day. Warm and tonify Kidney Qi, supplement Kidney marrow, and tonify Ren and Du channels. Treats Kidney Qi deficiency, essence exhaustion, deficient and empty Du and Ren channels leading to slow development in children and those with weak constitutions. It can also treat female infertility due to essence deficiency, coldness in uterus, delayed and scanty menstruation, low libido, and impotence or infertility due to low sperm count in men.

You Gui Wan [Restore the Right (Kidney) Pill] *Jing Yue Quan Shu*

Shu Di Huang	240 g
Shan Yao	120 g
Shan Zhu Yu (fried)	90 g
Gou Qi (fried)	120 g
Lu Jiao Jiao	120 g
Tu Si Zi (fried)	120 g
Du Zhong (fried)	120 g
Dang Gui	90 g
Rou Gui	60–120 g
Zhi Fu Zi	60–180 g

Main Indication: Kidney Yang deficiency, and weak Ming Men fire. Symptoms include exhausted Qi, apathy, aversion to cold, and cold limbs in chronic conditions; or impotence, nocturnal emission, or infertility due to weak Ming Men fire; or loose stool with undigested food; or incontinence; or sore and weak lower back and knees, edema of the lower extremities, pale and flabby moist tongue, and deep thin pulse, especially at Chi position.

Preparation: Grind into fine powders and mix with honey to form pills. Each pill weighs 15 g. Take one in the morning and one at night. If taken as a decoction, reduce the dosage.

Treatment Principles: Warm the Kidney Yang, supplement essence, and nourish blood.

This formula is modified from Jin Gui Shen Qi Wan [Kidney Qi Pill from the Golden Cabinet] by removing three draining herbs (Fu Ling, Ze Xie, and Mu Dan Pi) and then adding Lu Jiao Jiao, Tu Si Zi, Du Zhong, and Gou Qi. It has a stronger function for nourishing and tonifying Kidney Yin and Yang, and a weaker draining function than Jin Gui Shen Qi Wan [Kidney Qi Pill from the Golden Cabinet]. It also shows the idea of seeking Yang within Yin source.

Modification:

Addition	Indication
Ren Shen	Qi deficiency and Yang exhaustion
Bu Gu Zhi (fried with wine)	Yang deficiency, nocturnal emission without dreaming, leukorrhea, and loose stool
Gan Jiang	Poor appetite and digestion, nausea, vomiting, and sour regurgitation
Wu Zhu Yu	Unceasing abdominal pain
Hu Tao Rou	Soreness and painful lower back and knees
Ba Ji Tian Rou Cong Rong Huang Gou Shen	Kidney Yin and Yang both deficient, and impotence

Nei Bu Wan [Internal Tonify Pill] *Nu Ke Qie Yao*

Lu Rong	equal amount
Tu Si Zi	
Sha Yuan Ji Li	
Huang Qi	
Rou Gui	
Sang Piao Xiao	
Rou Cong Rong	
Zhi Fu Zi	
Fu Ling	
Bai Ji Li	
Zi Wan Rong	

Main Indication: Exhausted minute Kidney Yang, and damaged and deficient Chong and Ren channels. Symptoms include excessive, unceasing, watery leukorrhea, lower back pain as if the back is broken, frequent urination, pale face, apathy, aversion to cold; dizziness, vertigo, palpitation, shortness of breath, 5 A.M. diarrhea, pale tongue, thin white tongue coating, and deep slow pulse.

Preparation: Grind into powder and mix with honey to form pill as in size of Mung bean. Take 20 pills on empty stomach, two to three times a day.

Treatment Principles: Warm and tonify Kidney Yang. Stabilize Chong and Ren channels.

Er Xian Tang [Two-Immortal Decoction] *Zhong Yi Fang Ji Lin Chuang Shou Ce*

Xian Mao	10 g
Yin Yang Huo	10 g
Ba Ji Tian	10 g
Huang Bai	10 g
Zhi Mu	10 g
Dang Gui	10 g

Main Indication: Kidney Yin and Yang deficiency leading to menopause with hypertension developing around menopause; irregular menstruation, amenorrhea, tidal fever, sweating, apathy, depression, irritability, agitation, insomnia, palpitation, pale tongue with white coating or red tongue with scanty coating, and thin, weak or thin, rapid pulse.

Preparation: Decoction.

Treatment Principles: Balance Yin and Yang, warm Kidney, nourish Yin, and drain heat.

It is indicated for menopause, nodules, or hyperplasia of mammary gland due to Chong and Ren channel disorders.

Tu Si Zi Wan [Cuscuta Seed Pill] *Ji Sheng Fang*

Tu Si Zi	60 g
Wu Wei Zi	30 g
Duan Mu Li	60 g
Rou Cong Rong	60 g
Zhi Fu Zi	30 g
Ji Nei Jin	15 g
Lu Rong	30 g
Sang Piao Xiao	30 g
Yi Zhi Ren	30 g
Wu Yao	30 g
Shan Yao	30 g

Main Indication: Kidney Yang deficiency and leakage of fluids. Symptoms include apathy, aversion to cold, emaciation, dizziness, soreness in the lower back, weakness of the feet, frequent dribbling urination, pale tongue, thin white tongue coating, and deep, thin pulse especially at Chi position.

Preparation: Grind into powder and mix with wine to form pill 5 to 10 mm in diameter. Take 70 pills with salt water, twice a day before meals.

Treatment Principles: Warm Kidney and astringe fluids.

In the formula, Wu Yao is good at scattering the cold between Kidney and Urinary Bladder. Ji Nei Jin treats frequent dribbling urination. In combination, Wu Wei Zi astringes Kidney Qi. There is also a group of herbs to warm the Kidney. Thus, this formula is stronger in the function of warming Kidney, astringing the leakage. It is indicated for recovery of severe conditions or chronic conditions; seniors' aversion to cold and apathy due to Kidney Yang and Kidney Qi deficiency; and dribbling urination.

Hei Xi Dan [Lead Special Pill] *He Ji Ju Fang*

Jin Ling Zi (steamed)	30 g
Hu Lu Ba (fried)	30 g
Mu Xiang	30 g
Pao Fu Zi	30 g
Rou Dou Kou (bake after coated with flour paste)	30 g
Bu Gu Zhi (fried)	30 g
Chen Xiang	30 g
Hui Xiang (fried)	30 g
Yang Qi Shi	30 g
Rou Gui	15 g
Hei Xi	60 g
Liu Huang	60 g

Main Indication:

1. Exhaustion and deficiency of true Yang, Kidney failing to grasp the Qi, excess above and deficient below, turbid Yin upward flowing. Symptoms include phlegm congested in the chest, Qi rebellions, rapid wheezing, cold extremities, unceasing cold sweats, pale tongue with white coating, and deep, minute pulse.

2. Ben Teng Qi. Symptoms include a rushing sensation starting from lower abdomen, rushing to the chest; distending pain at the chest, hypochondrium, epigastrium, and abdomen, or cold hernial pain, borborygmus, uncontrolled bowel movements; or impotence and cold essence in male; in female, deficient and coldness of uterus, irregular menstruation, watery leukorrhea, and infertility; pale tongue, white tongue coating, and deep, slow pulse.

Preparation: Grind into fine powder, and then mix with wine to form pills 5 to 10 mm in diameter. Take 3 to 9 g with warm water.

Treatment Principles: Warm the source, descend the hyperactivity, and anchor the floating Yang.

This formula can be used in syndromes of excess above and deficiency below where the excess above is a manifestation and the deficiency below is the root of the condition, and syndromes of floating Yang, nearing collapse. It can also treat deficient cold of the lower source, which can be seen in syndromes of cold hernial disorder, cold essence, Ben Teng Qi, and infertility.

Suo Quan Wan [Shut the Sluice Pill] *Fu Ren Liang Fang*

Wu Yao	equal amount
Yi Zhi Ren	

Main Indication: Deficient cold of Kidney and Urinary Bladder. Symptoms include frequent urination, pediatric enuresis, pale tongue, thin white tongue coating, and deep, thin or slow pulse.

Preparation: Grind into fine powder, mix with wine and cooked Shan Yao powder to form pill 5 to 10 mm in diameter. Take 70 pills with salt water or rice water each time. It can also be taken as a decoction.

Treatment Principles: Warm Kidney and astringe the urination. It is good for chronic and mild syndromes of enuresis and frequent urination that belongs to deficient cold of lower source.

Gu Jing Wan [Stabilize the Essence Pill] *Ji Sheng Fang*

Rou Cong Rong	equal amount
Yang Qi Shi	
Lu Rong	
Chi Shi Zhi	
Ba Ji Tian	
Jiu Zi	
Fu Ling	
Lu Jiao Shuang	
Long Gu	
Zhi Fu Zi	

Main Indication: Exhaustion and deficiency of the lower Yuan Qi, and leakage of Kidney essence. Symptoms include frequent nocturnal emission, pale complexion, dispiritedness, pale tongue, white tongue coating, and deep, weak pulse.

Preparation: Grind into fine powder and mix with wine to form pills. Take 10 g each time on an empty Stomach with salt water.

Treatment Principles: Tonify Kidney and astringe Kidney essence.

Modification:

Addition	Indication
Wu Wei Zi Mu Li Sang Piao Xiao	Severe frequent nocturnal emission without dreams

Zan Yu Dan [Special Pill to Aid Fertility] *Jing Yue Quan Shu*

Shu Di Huang	240 g
Bai Zhu	240 g
Dang Gui	180 g
Gou Qi	180 g
Du Zhong	120 g
Xian Mao	120 g
Ba Ji Tian	120 g
Shan Zhu Yu	120 g
Yin Yang Huo	120 g
Rou Cong Rong	120 g
Jiu Zi	120 g
She Chuang Zi	60 g
Zhi Fu Zi	60 g
Rou Gui	60 g

Main Indication: Waning Ming Men fire and deficient cold of essence Qi. Symptoms include impotence, scanty or exhausted semen, lower abdominal coldness and empty sensation, infertility due to deficiency; cold, pale tongue; thin, white, or moist tongue coating; and deep, weak pulse, especially at right Chi position. One can also see pale complexion, dispiritedness, and soreness and weakness of the lower back and feet.

Preparation: Grind into fine powder, mix with honey to form pills, and take 6 to 9 g twice a day.

Treatment Principles: Warm and tonify the lower source.

This formula is designed for impotence, infertility, and cold in the uterus due to waning of the Ming Men fire. When impotence occurs, there is about 70 to 80% chance of waning fire. It can also treat impotence due to cold–damp downward pouring in Yang Ming, causing flaccidity of the tendons. (In the formula, Bai Zhu warms the Spleen and transforms phlegm.)

Modification:

Addition	Indication
Ren Shen	Qi deficiency and apathy
Lu Rong	Severe Yang deficiency and coldness

Related Formulas:

Formula	Herbs	Indication
Gui Lu Er Xian Jiao [Tortoise Shell and Deer Antler Syrup]	Lu Jiao 50 g Gui Ban 25 g Gou Qi Zi 15 g Ren Shen 5 g	Decoct the herbs until a thick consistency is formed. Mix 3 g in clear wine and take with lightly salted water every morning Benefit Qi, strengthen Yang, and supplement essence and blood in Ren and Du channel Treat weak constitution, nocturnal emission, impotence, blurry vision, and lumbar soreness and pain

(continued)

Formula	Herbs	Indication
Wu Zi Yan Zong Wan [Five-Seeds Pill to Continue the Lineage] *Shi Yi De Xiao Fang*	Gou Qi Zi 24 g Che Qian Zi 6 g Wu Wei Zi 6 g Fu Pen Zi 12 g Tu Si Zi 24 g	Make into honey pills and take 6–9 g, two to three times a day. It tonifies Liver, benefits Kidney, warms Yang, and supplements essence It treats Liver and Kidney deficiency, essence deficiency, and infertility. It can also treat impotence due to weak Ming Meng fire

Qing E Wan [Young Maiden Pill] *He Ji Ju Fang*

Bu Gu Zhi	120 g
Du Zhong (fried)	240 g
Hu Tao Rou (fried with ginger juice)	300 g

Main Indication: Kidney deficiency, lower back pain. Symptoms include unceasing chronic back pain, worse with exertion, better with lying down, weakness of leg and knee, tightness at the lower abdomen, pale complexion, cold extremities, pale tongue, no taste in mouth, and deep, thin pulse.

Preparation: Grind into fine powder and mix with 120 g of mashed garlic to form pills. Take 3 g, two to three times a day with warm wine.

Treatment Principles: Tonify Kidney and strengthen the lower back.

This formula treats Kidney deficient lower back pain with some Yang deficiency. It can also treat mild lower back pain in Kidney stones, and Kidney tuberculosis.

Yang He Tang [Yang-Heartening Decoction] *Wai Ke Quan Sheng Ji*

Shu Di Huang	30 g
Rou Gui	3 g
Lu Jiao Jiao	9 g
Bai Jie Zi	6 g
Ma Huang	2 g
Jiang Tan	2 g
Gan Cao	3 g

Main Indication: Yin deep-rooted carbuncles due to Yang deficiency and cold congelation. Symptoms include deep-rooted bone carbuncles, multiple abscesses, phlegm nodules, and He Xi Feng (enlarged, deformed, and pronounced knee joint). Symptoms include local swelling, soreness and pain without heat, normal skin color, absence of thirst, pale tongue and thin, white tongue coating, and deep thin pulse.

Preparation: Decoction.

Treatment Principles: Warm Yang, nourish blood, scatter cold, and unblock stagnation.

This formula treats Yin deep-rooted carbuncles. It is formed in patient with deficient Ying and blood constitution, cold congelation, phlegm stagnation, and Bi obstruction in the muscle, tendon, blood vessels, and joints. With Yang type of sores and ulcerations, with redness, swelling, pain or Yin deficiency with heat manifestation, or ulcerated Yin deep-rooted carbuncles, this formula cannot be used. Caution: Ma Huang can be used before the ulcerations. Once the carbuncles are ulcerated, then Ma Huang is contraindicated.

In modern times, this formula has been used to treat deficient cold, Yin type of bone tuberculosis, tuberculosis of peritoneum, tuberculosis of lymphadenitis, thromboangitis obliterans, chronic deep abscess, protrusion of intervertebral disc, and hypertrophy of lumber vertebrae.

Related Formula:

Formula	Herbs	Indication
Zhong He Tang [Middle-Heartening Decoction]	Ren Shen 6 g Chen Pi 6 g Huang Qi 5 g Bai Zhu 5 g Dang Gui 5 g Bai Zhi 5 g Fu Ling 3 g Chuan Xiong 3 g Zao Jiao Ci (fried) 3 g Ru Xiang 3 g Mo Yao 3 g Jin Yin Hua 3 g Gan Cao 3 g Decoct with half wine and half water	Treat half-Yang half-Yin type of carbuncles and sores where the sores seem to be ulcerated but are not; swelling with slight pain, and pale red without heat. This formula can tonify Qi, vent the carbuncles, harmonize blood, and eliminate and scatter boils

Comparison of Formulas Treating Kidney Yang Deficiency and Exhaustion

1. Jin Gui Shen Qi Wan [Kidney Qi Pill from the Golden Cabinet] is a formula that tonifies Kidney Qi to gently generate Kidney Yang. It is used in mild syndromes or chronic conditions of Kidney Yang deficiency.
2. You Gui Wan [Restore the Right (Kidney) Pill] and Nei Bu Wan [Internal Tonify Pill] tonify Kidney, strengthen Yang, open Du channel, and consolidate Chong and Ren channel, igniting Ming Men fire. It is used in severe conditions of Kidney Yang deficiency, and essence and blood exhaustion. It can also be used in gynecological disorders due to Kidney deficiency and Chong and Ren channels damage. However, Nei Bu Wan [Internal Tonify Pill] focuses on warming the Yang and You Gui Wan [Restore the Right (Kidney) Pill] focuses on both warming the Yang and nourishing essence and blood.
3. Er Xian Tang [Two-Immortal Decoction] regulate the Yin and Yang within the Kidney. It nourishes Kidney Yin and drains deficient heat. It is mainly used in gynecology to treat women with mammary disorders and menopause.
4. Tu Si Zi Wan [Cuscuta Seed Pill], Suo Quan Wan [Shut the Sluice Pill], and Gu Jing Wan [Stabilize the Essence Pill] all warm Kidney and astringe leakage. However, the first formula is strong in both warming Yang and astringing essence, which can be used in severe cases of Yang deficiency with cold and leakage. The second formula is a mild small formula, which is used in mild cases of frequent urination and enuresis in children. The last formula focuses on astringing Kidney essence to treat syndromes of nocturnal emission and dispiritedness.
5. Hei Xi Dan [Lead Special Pill] treats syndromes of waning Ming Men fire, almost collapsing true Yang, and deficient cold of the lower source. It can warm Ming Men, consolidate the lower source, calm wheezing, relieve cold extremities, and stop cold sweat. Thus, the excess of the upper body is decreased and the lower deficiency is nourished, leading to the balance between the upper and lower parts of the body.
6. Zan Yu Dan [Special Pill to Aid Fertility] warms and tonifies the lower source, which is designed to treat infertility due to cold in the uterus and impotence with scanty semen.
7. Qing E Wan [Young Maiden Pill] warms Yang and stops pain, which treats unceasing chronic lower back pain due to Kidney Yang deficiency.
8. Yang He Tang [Yang-Heartening Decoction] warms Yang, harmonizes Ying, scatters cold, and expels stasis. It is indicated for Yin syndromes of carbuncles and sores with internal swelling causing obstruction.

KIDNEY DEFICIENCY FAILING TO HOLD

This syndrome is due to Kidney deficiency failing to hold Kidney essence, leading to nocturnal emission with or without dreams. On the basis of tonify Kidney Yin and Yang, astringent herbs are added such as Sha Yuan Ji Li, Qian Shi, Lian Xu, Long Gu, Mu Li, Shan Zhu Yu, Wu Wei Zi, Sang Piao Xiao, Jin Ying Zi, and Fu Pen Zi to form formulas that stabilize the Kidney and astringe essence.

Jin Suo Gu Jin Wan [Metal Lock Pills to Stabilize the Essence] *Yi Fang Ji Jie*

Sha Yuan Ji Li (fried)	60 g
Qian Shi (steamed)	60 g
Lian Xu	60 g
Long Gu (calcined)	30 g
Mu Li (calcined)	30 g

Main Indication: Kidney deficiency failing to hold essence. Symptoms include chronic nocturnal emission with or without dreams, apathy, lack of strength, night sweat, restlessness, soreness and weakness of the four limbs, soreness of the lower back, tinnitus, pale or red tongue with thin white coating, and deficient or thin pulse especially at Chi position.

Preparation: Grind into fine powder; mix with Lian Zi powder paste to form pills. Take 9 g each, twice a day with salt water on empty stomach. It can also be taken as decoction with 60 g of Lian Zi Rou and reduced dosages of all herbs.

Treatment Principles: Tonify Kidney and astringe essence.

Kidney Yin deficiency with Prime Minister Fire upsetting the body causes the leakage of the essence, which manifests as nocturnal emission, with or without dreams. This is a common formula to treat such a syndrome. However, the formula focuses on stabilizing and astringing. Thus, Kidney tonic formulas should be used when the leakage stops in order to treat the root. It can also be used in pediatric enuresis. In modern times, this formula is often used to treat nocturnal emission with or without dreams, and insomnia and excessive dreaming due to neurasthenia. If the nocturnal emission is due to Heart and Liver fire or Lower Jiao damp–heat, then one cannot use this formula.

Modification:

Addition	Indication
Wu Wei Zi Jin Ying Zi Tu Si Zi	Severe nocturnal emission and weakness
Sheng Di Huang Mu Dan Pi Zhi Mu Huang Bai	Yin deficiency and hyperactive Yang with heat
Lu Jiao Jiao Bu Gu Zhi Shan Yu Rou	Kidney Yang deficiency

Related Formula:

Formula	Herbs	Indication
Shui Lu Er Xian Dan [Water and Earth Immortals Special Pill]	Qian Shi Jin Ying Zi Equal amount Decoct Jin Ying Zi into paste then mix with powdered Qian Shi to form pills. Take 9 g, twice a day on empty stomach with salt water. It can also be taken as decoction	Treat turbid nocturnal emission and leukorrhea due to Kidney deficiency failing to hold

Comparison of Formulas Treating Kidney Deficiency Failing to Hold

Jin Suo Gu Jin Wan [Metal Lock Pill to Stabilize the Essence], Shui Lu Er Xian Dan [Water and Earth Immortals Special Pill], and Gu Jin Wan [Stabilize the Metal Pill] consolidate the Kidney and astringe the essence. However, the first formula is stronger in both tonifying the Kidney and astringing the essence than the middle formula. The latter formula focuses on warming and tonifying Kidney Yang, and at the same time astringing essence. It is used in Kidney Yang deficiency, causing cold type of nocturnal emission. Jin Suo Gu Jin Wan [Metal Lock Pill to Stabilize the Essence] focuses on nourishing Kidney Yin by astringing and stabilizing. It is used in treating frequent nocturnal emission without obvious heat or cold.

LEUKORRHEA DUE TO KIDNEY DEFICIENCY

This syndrome is due to Kidney deficiency failing to hold Kidney fluids leading to leukorrhea and copious watering discharge. Commonly used herbs are Rou Gui, Fu Zi, Tu Si Zi, Lu Rong, and Ai Ye which warm Kidney Yang and, combining with Qian Shi, Lian Zi, Lian Xu, Bai Guo, Wu Zei Gu, and Chun Gen Pi, stabilize, astringe, and stop leukorrhea in the formulas that tonify Kidney and stop leukorrhea.

Lu Jiao Tu Si Wan [Deer Antler and Cuscuta Seed Pill] (Jing Yan Fang)

Lu Jiao Shuang	30 g
Tu Si Zi	15 g
Mu Li	15 g
Bai Zhu	9 g
Du Zhong	18 g
Lian Xu	15 g
Yin Xing	15 g
Qian Shi	15 g

Main Indication: Kidney deficiency and Dai channel not stabilized, leading to leukorrhea. Symptoms include unceasing, clear, watery leukorrhea, pale face, apathy, fatigue, aversion to cold, cold limbs, dizziness, vertigo, palpitation, shortness of breath, lumbar pain as if there were a broken bone or heavy cold sensation below the waist, frequent urination, pale or flabby tongue, moist white tongue coating, and deep slow pulse.

Preparation: Decoction.

Treatment Principles: Tonify Kidney, stabilize, astringe, and stop leukorrhea.

This formula is designed for cold pain in the lower back and leukorrhea due to Kidney deficiency and Dai channel failing to hold. The purpose of the formula is to tonify Kidney and stabilize the Dai. It can also treat threatened miscarriage due to Kidney deficiency.

Shou Se Zhi Dai Tang [Astringe and Stop Discharge Decoction] (Fu Jian Min Jian Fang)

Shan Yao	15 g
Qian Shi	15 g
Ji Guan Hua	15 g
Tu Si Zi	12 g
Du Zhong	12 g
Xu Duan	12 g
Bai Zhu	12 g
Chun Gen Pi	30 g

Main Indication: Kidney Qi failing to hold, Chong and Ren channel damage and deficiency. Symptoms include leukorrhea that is chronic and unceasing, lower back pain, or soreness and weakness, or lower abdominal pain, unceasing menstrual spotting; clear, watery, pale blood; pale tongue, thin white tongue coating, and deep, thin pulse.

Preparation: Decoction.

Treatment Principles: Tonify Kidney and stabilize Chong and Ren channels.

This formula treats leukorrhea due to damages in Chong and Ren channels, or due to multiple births or miscarriages, or due to postpartum exhaustion, or excessive bleeding in postpartum or due to damages from operation. The formula tonifies Kidney, strengthens Chong and Ren channels, stops leukorrhea, and stops bleeding.

In modern times, this formula is indicated for leukorrhea, unceasing menstrual bleeding due to any type of damage, infertility due to damage from operations, or habitual miscarriage.

Modification:

Addition	Indication
Gan Jiang Xiao Hui Xiang	With cold
Huang Qin Haung Bai Long Dan Cao Yu Xing Cao	With heat
Fu Ling Yi Yi Ren Bei Xie Cang Zhu	More dampness
Long Gu Mu Li Jin Ying Zi	Chronic and hard to treat leukorrhea
Xiao Yao San [Relax Wanderer Powder]	Irregular menstruation

Comparison of Formulas Treating Leukorrhea Due to Kidney Deficiency

Lu Jiao Tu Si Wan [Deer Antler and Cuscuta Seed Pill] focuses on tonifying Kidney Yang and stabilizing Chong and Ren channels. It is used to treat lower back pain and cold manifestations in leukorrhea due to Kidney deficiency. Shou Se Zhi Dai Tang [Astringe and Stop Discharge Decoction] tonifies Kidney Qi and stabilizes Chong and Ren channels. It is used in leukorrhea due to Kidney Qi deficiency and damaged Chong and Ren channels without obvious cold or heat manifestations.

EDEMA DUE TO KIDNEY DEFICIENCY

This syndrome is caused by Kidney Yang deficiency failing to transform Qi and move water. Thus, water accumulates and manifests as edema. The symptoms include difficult urination, heaviness of the limbs, or edema or pain, cough with clear, watery phlegm, cold hands and feet, pale flabby tongue with teeth marks, and deep, slow, weak pulse. Commonly used herbs are Rou Gui and Fu Zi to warm Yang and transform Qi, combined with Fu Ling and Ze Xie to drain fluid by promoting urination.

Zhen Wu Tang [True Warrior Decoction] *Shang Han Lun*

Fu Ling	9–15 g
Bai Shao Yao	9–15 g
Bai Zhu	6–12 g
Sheng Jiang	9–18 g
Zhi Fu Zi	9–15 g

Main Indication:

1. Spleen and Kidney Yang deficiency and internal accumulation of water. Symptoms include difficult urination and deep, heavy, painful limbs; abdominal pain, loose stools, or cough, nausea, or edema of the extremities, flabby tongue with teeth marks, white slippery tongue coating, absence of thirst, and deep, thin pulse.
2. Tai Yang syndromes. Symptoms include excessive sweating, persistent fever, severe Yang Qi deficiency, palpitation, dizziness, vertigo, shaking, shivering, near fainting episodes, flabby tongue with moist tongue coating, and thin, deep pulse.

Preparation: Decoction Fu Zi first for 30 minutes, then add all other herbs. Divide the decoction into three doses.

Treatment Principles: Warm Yang and drain water by promoting urination.

This is an important formula to treat Spleen and Kidney Yang deficiency with internal accumulation of water. The key diagnostic points are flabby pale tongue with teeth marks, and deep, thin weak pulse. It can also treat syndromes of Yang deficiency with water accumulation under the following conditions: difficult urination in cases of prostate enlargement (one should take 10 bags continuously), edema and scanty urination in chronic nephritis, joint pain without redness and swelling in rheumatic arthritis, cough, wheezing, facial swelling, inability to lie down in rheumatic heart conditions, vertigo, dizziness, headache, frequent nocturia in hypertension, chest pain, shortness of breath, cold extremities and body in coronary heart disease, cough, wheezing, vomiting watery clear phlegm in Lung and heart disorders, tachycardia, bradycardia, nausea, vomiting, diarrhea, common cold due to Yang deficiency, and postpartum spontaneous sweating due to Yang deficiency.

Modification:

Minus	Addition	Indication
	Wu Wei Zi Gan Jiang Xi Xin	Cough
Fu Ling		Normal urination
Bai Shao Yao	Gan Jiang	Diarrhea
Fu Zi	Sheng Jiang (increase dosage)	With vomiting

(continued)

Minus	Addition	Indication
	Chuan Niu Xi Ze Xie Sang Ji Sheng	Hypertension
	Fang Ji Huang Qi Ting Li Zi	Wheezing and coughing, edema
	Gui Zhi Ze Xie Da Fu Pi	Chronic nephritis
	Gua Lou Pi Xie Bai Ban Xia	Coronary heart disease
	Ban Xia Chen Pi Gui Zhi or Wu Ling San [Five-Ingredient Powder with Poria]	Lung and heart diseases
	Wu Zhu Yu Ban Xia Gan Jiang	Nausea and vomiting, diarrhea
	Gui Zhi Ma Huang Fang Ji Yi Yi Ren Chuan Xiong Jiang Huang	Cold pain at the joints

Related Formula:

Formula	Herbs	Indication
Fu Zi Tang [Prepared Aconite Decoction]	Core formula – Sheng Jiang + Ren Shen Double the dosage of Fu Zi Bai Zhu	Treat Yang deficiency and cold evil pathogens attacking internally, leading to joint pains. The formula focuses on warming and tonifying in order to expel cold–damp

Wu Ling San [Five-Ingredient Powder with Poria] *Shang Han Lun*

Zhu Ling	9 g
Ze Xie	15 g
Bai Zhu	9 g
Fu Ling	12 g
Gui Zhi	6 g

Main Indication: Spleen deficiency, unable to transport dampness, and water and dampness accumulating. Symptoms include headache, fever, irritability, thirst with desire to drink, or vomiting right after drinking water; difficult urination, white tongue coating, and floating pulse. Other symptoms can include edema, diarrhea, difficult urination, or vomiting and diarrhea in cholera, throbbing pulsation below the navel, vomiting saliva, dizziness, vertigo, shortness of breath, and cough.

Preparation: Decoction or powder. Take 3 to 6 g with warm water until sweating occurs.

Treatment Principles: Promote urination, leach out dampness, and warm Yang to transform Qi. This formula mainly leaches out dampness, promotes urination, and tonifies Spleen to transform Qi. It is used in difficult urination, edema due to internal water and damp accumulation, and also diarrhea due to damp water pouring downward. This formula promotes urination to drain out dampness, thus diarrhea stops. It is also used in phlegm Yin (congested fluid) syndrome with throbbing pulsation below the navel. When phlegm is expelled, dampness is drained, the congested fluid is cleared, then palpitation stops. This formula is also indicated for turbid damp type of cholera. Furthermore, it can also treat turbid, cloudy urine due to withdrawal or holding semen during ejaculation, moving semen into the bladder.

Related Formulas:

Formula	Herbs	Indication
Jia Wei Wu Ling San [Augmented Five-Ingredient Powder with Poria]	Core formula + Qiang Huo	Cold–damp body aches, thirst, and difficult urination
Yin Chen Wu Ling San [Artemisia Yinchenhao and Five-Ingridient Powder with Poria]	Core formula + Yin Chen Hao	Damp–heat yellow jaundice and difficult urination
Chun Ze Tang [Spring Pond Decoction]	Core formula + Ren Shen	Treat Middle Jiao Qi deficiency with difficult urination and stress incontinence with coughing
Si Ling San [Four-Ingredient Powder with Poria]	Core formula – Gui Zhi	Treat internal damages by food, damp accumulation, scanty dark urination, and loose stool
Si Ling San Jia Mu Gua Cao Guo Ban Xia Tang [Four-Ingredient Powder with Poria plus Chaenomelis Fruit, Amomi Tsaoko, and Pinellia]	Core formula – Gui Zhi + Mu Gua Cao Guo Ban Xia	Treat cold–damp stranded Spleen symptoms including cold extremities, abdominal diarrhea, yellow sclera, white slippery tongue coating, apathy, and no desire to talk; tongue stiffness and difficulty speaking.
Wei Ling Tang [Calm the Stomach and Five-Ingredient Powder with Poria Decoction]	Wu Ling San [Five-Ingredient Powder with Poria] + Ping Wei San [Calm the Stomach Powder]	Decoct with Sheng Jiang and Da Zao, taken on empty Stomach Treat damages to Spleen and Stomach due to cold when summer changes to fall. Symptoms include undigested food, unceasing diarrhea, or edema, abdominal distention and difficult urination

Comparison of Formulas Treating Edema Due to Kidney Deficiency

Zhen Wu Tang [True Warrior Decoction] warms Yang, drains water, which is indicated for Spleen and Kidney Yang deficiency, and cold water congealed at limbs, leading to deep heavy edema. Fu Zi Tang [Prepared Aconite Decoction] warms, tonifies, and expels cold damp. The main indications are Yang Qi deficiency; cold evil attacks internally, leading to joint pain. The former formula focuses on warming, scattering, and expelling water. The latter focuses on warming and tonifying to eliminate congealed coldness deep into the joints.

Wu Ling San [Five-Ingredient Powder with Poria] is indicated for Spleen deficiency failing to transform water and dampness, leading to difficult urination, edema, and diarrhea. The focus is to transform, transport, and separate the water and dampness. At the same time, it tonifies Spleen and unblocks Yang. In the formula, Gui Zhi can warm and unblock Kidney Yang, "as if adding wood to fire." Si Ling San [Four-Ingredient Powder with Poria] does not have Gui Zhi, thus it leaches out dampness without supporting Spleen Yang function. Wei Ling Tang [Calm the Stomach and Five-Ingredient Powder with Poria Decoction] is indicated for water dampness obstructing internally, causing abdominal distention, diarrhea, difficult urination, and edema. This formula drains damp and harmonizes the Stomach at the same time.

LOWER JIAO DAMP–HEAT

The manifestations of this syndrome are Lin syndrome or urinary retention. The symptoms include urgent, frequent, hesitant, painful urination or urinary retention, or with fever, shivering, lower abdominal spasmic pain, distention, and sinking sensation (which is seen in cystitis, urinary tract infection, and pyelonephritis). The treatment should drain fire and unblock the Lin. Commonly used herbs are Bian Xu, Qu Mai, Mu Tong, Hua Shi, Che Qian Cao, Bai Mao Gen, and Shi Wei in combination with Shan Zhi Zi, Huang Qin, Da Huang, Jin Yin Hua, Zi Hua Di Ding, Ye Ju Hua, and Tian Kui Zi. These symptoms indicate stones in Kidney, ureter, or urinary bladder: dull pain in the lower back with occasional spasmic pain radiating from Kidney area to ureters, inguinal, and genital area or even to the medial aspect of the thigh, or suddenly stop during urination with severe sharp pain radiating to genital area, with frequent, urgent urination and blood in the urine. One should add Jin Qian Cao, Hai Jin Sha, Ji Nei Jin, Hu Po, Huo Xiao, Da Huang, and other herbs to transform and eliminate stones.

If there is turbid cloudy urine, dribbling urination as seen in prostatitis, UTI, epididymitis, and glacturia, one can add Dan Zhu Ye, Huang Lian, Jin Yin Hua, Lian Qiao, Mu Tong, Deng Xi Cao, Lu Gen, Bei Xie, Huang Bai, and Zhi Mu to clear heat, drain damp, clear heat, and relieve toxins.

Er Miao San [Two-Marvel Powder] *Dan Xi Xin Fa*

Huang Bai (fried)	15 g
Cang Zhu (fried)	15 g

Main Indication: Damp–heat pouring downward. Symptoms include soreness and weakness of the legs, or red swollen painful knees, or damp–heat leukorrhea, or damp sores at genital area, or pain at tendons and bones, scanty yellow urine, yellow greasy tongue coating, and slippery rapid pulse.

Preparation: Grind into powder and take 3 to 5 g each time. It can also be prepared as pills, or decoction.

Treatment Principles: Clear heat and dry dampness.

This formula treats any syndromes of damp–heat downward pouring.

Modification:

Addition	Indication
Wu Jia Pi Xi Xian Cao	Treat damp–heat Wei syndrome
Yi Yi Ren Mu Gua Bing Lang	Damp–heat beriberi
Qian Shi Chun Gen Pi Fu Ling	Damp–heat leukorrhea
Yi Yi Ren Chi Xiao Dou Long Dan Cao	Genital oozing sores

Related Formulas:

Formula	Herbs	Indication
San Miao Wan [Three-Marvel Pill] *Yi Xue Zheng Zhuan*	Huang Bai 120 g Cang Zhu 180 g Chuan Niu Xi 60 g	Damp–heat pouring downward, numbness of the legs as if on fire
Si Miao Wan [Four-Marvel Pill] *Cheng Fang Bian Du*	Huang Bai 200 g Yi Yi Ren 200 g Cang Zhu 120 g Huai Niu Xi 120 g	Damp–heat pouring downward, numbness, atrophy, and swollen, painful feet

Ba Zheng San [Eight-Herb Powder for Rectification] *He Ji Ju Fang*

Che Qian Zi	500 g
Qu Mai	500 g
Bian Xu	500 g
Hua Shi	500 g
Shan Zhi Zi	500 g
Gan Cao Shao	500 g
Mu Tong	500 g
Da Huang	500 g

Main Indication: Heat and bloody Lin syndrome due to damp–heat downward pouring. Symptoms include turbid, dark urine, fullness of the lower abdomen, hesitant pain during urination; dribbling and hesitant urine or urinary retention, dry throat, dry mouth, yellow greasy tongue coating, and slippery and rapid pulse.

Preparation: Grind into powder. Decoct 6 to 10 g with 300 to 500 g of water and Deng Xin Cao. Take the decoction warm. It can also be prepared as decoction with reduced dosage.

Treatment Principles: Clear heat, drain fire, promote urination, and unblock Lin syndrome.

This formula treats damp–heat downward pouring and accumulating in the Lower Jiao. It can be used in damp–heat type of cystitis, urinary tract infection, acute prostatitis, stones in urinary system, and pyelonephritis.

Modification:

Addition	Indication
Jin Yin Hua Zi Hua Di Ding Ye Ju Hua Chai Hu	High fever and chills
Da Ji Xiao Ji Bai Mao Gen Han Lian Cao	Blood in the urine with hesitant pain
Jin Qian Cao Hai Jin Sha Hu Po	Stone Lin and Sandy Lin syndrome
Bei Xie Shi Chang Pu	Cloudy Lin syndrome

Related Formulas:

Formula	Herbs	Indication
Liu Yi San [Six-to-One Powder]	Hua Shi 6 portions Gan Cao 1 portion	Take 6 to 9 g of the powder with Deng Xin Cao decoction, two to three times a day Clear summer heat and drain damp Treat internal obstruction of summer damp, dark scanty urine, or summer heat, and damp type common cold
Bi Yu San [Jasper Powder]	Hua Shi 180 g Gan Cao 30 g Qing Dai 15 g	Take 6 to 9 g of powder, two to three times a day Expel summer heat Treat summer heat and damp with stagnant heat in Liver and Gallbladder
Ba Zheng Jie Du Tang [Eight-Herb Decoction for Rectification and Reliving Toxicity]	Core formula + Xiong Huang 3–6 g Take the decoction with 1.5 g of Zi Jin Ding * [Purple Gold Tablet]	Cloudy urination due to filariasis

* Zi Jin Ding [Purple Gold Tablet] is a patent formula. It reduces swelling, stops pain, and separates clear from turbid. It treats abdominal pain, vomiting, diarrhea, sores, carbuncles, deep-rooted boils, and swelling.

Jia Jian Chai Ling Tang [Modified Bupleurum and Poria Decoction] (Chong Qing Shi Zhong Yi Yuan Fang)

Chai Hu	15 g
Huang Qin	15 g
Ban Xia	9 g
Zhu Ling	9 g
Fu Ling	12 g
Ze Xie	9 g
Hua Shi	18 g
Gan Cao	6 g
Yin Hua Teng	30 g
Jin Qian Cao	30 g

Main Indication: Shao Yang syndrome and damp–heat pouring downward, becoming Lin. Symptoms include beginning stage of pyelonephritis, alternating chills and fever, restlessness, urge to vomit, bitter and dry mouth, no desire for food, lower back pain, lower abdominal pain, frequent burning urination, thin yellow tongue coating, and soggy, rapid pulse.

Preparation: Decoction. One bag per day.

Treatment Principles: Harmonize Shao Yang, drain heat, and unblock Lin.

This formula is a modification of Xiao Chai Hu Tang [Minor Bupleurum Decoction] and Wu Ling San [Five-Ingredient Powder with Poria]. In the formula, Yin Hua Teng and Jin Qian Cao are frequently used to clear heat, unblock Lin, control inflammation, and maintain the smooth flow of urine.

Modification:

Addition	Indication
Xiao Ji Bai Mao Gen Han Lian Cao	Bloody urine
Qing Hao Che Qian Cao Bai Jiang Cao Huang Lian Huang Bai Da Huang	Severe syndrome

Zhu Ling Tang [Polyporus Decoction] *Shang Han Lun*

Zhu Ling (remove skin)	12 g
Fu Ling	12 g
Ze Xie	12 g
E Jiao (melt)	9 g
Hua Shi	18 g

Main Indication: Yin deficiency, knotted heat, and water. Symptoms include hesitant and painful urination, blood in urine, fullness and pain in the lower abdomen, or difficult urination, fever, thirst with desire to drink, restlessness, insomnia, thin yellow tongue coating, and thin rapid pulse or soggy rapid pulse.

Preparation: Decoct all herbs except E Jiao. Divide E Jiao into two portions and melt in the decoction.

Treatment Principles: Drain water, clear heat, and nourish Yin.

This formula focuses on draining and leaching out dampness. Clearing heat and nourishing Yin are its secondary functions. This formula treats damp–heat in the Lower Jiao.

Wu Lin San [Five-Ingredient Powder for Painful Urinary Dysfunction] *He Ji Ju Fang*

Chi Fu Ling	18 g
Dang Gui	9 g
Gan Cao	6 g
Chi Shao Yao	15 g
Shan Zhi Zi	15 g

Main Indication: Heat accumulation in the Lower Jiao and difficult urination. Symptoms include frequent, painful urination, lower abdominal pain, or cloudy, turbid urination, or with sands or stone, or bloody urination, red tongue, yellow tongue coating, and rapid pulse.

Preparation: Grind into powder. Take 6 g on empty stomach. It can also be prepared as decoction.

Treatment Principles: Clear heat, drain fire, drain dampness, and unblock Lin.

This is a common formula to treat heat Lin. With appropriate modification, this formula can treat any type of Lin syndromes.

Modification:

Addition	Indication
Jin Qian Cao Hai Jin Sha Hu Po	Stone Lin
Xiao Ji Bai Mao Gen Han Lian Cao	Blood Lin
Wu Yao Chen Xiang	Qi Lin
Bei Xie Fen Qing Yin [Dioscorea Hypoglauca Decoction to Separate the Clear]	Cloudy Lin

Xiao Ji Yin Zi [Cephalanoplos Decoction] *Ji Sheng Fang*

Sheng Di Huang	30 g
Xiao Ji	15 g
Hua Shi	15 g
Mu Tong	9 g
Pu Huang (fried)	9 g
Ou Jie	9 g
Dan Zhu Ye	9 g
Dang Gui	6 g
Shan Zhi Zi	9 g
Gan Cao	6 g

Main Indication: Heat accumulation in Lower Jiao, and damaged channels and collaterals. Symptoms include blood in urine, frequent dark urination, burning painful and hesitant; or blood in urine with thirst, irritability, red tongue tip, thin yellow tongue coating, and rapid, forceful pulse.

Preparation: Decoction.

Treatment Principles: Drain fire, unblock Lin, cool blood, and stop bleeding.

This formula is modified from Dao Chi San [Guide Out the Red Powder]. It focuses on cooling the blood and stopping bleeding. Draining dampness and unblocking Lin functions are secondary. It is a common formula in treating bloody Lin syndrome and blood in urine. In modern times, it is also used in pyelonephritis with obvious blood in urine. With severe heat manifestations, Ba Zheng San [Eight-Herb Powder for Rectification] is added.

Shi Wei San [Pyrrosiae Powder] *Pu Ji Fang*

Che Qian Zi	12 g
Qu Mai	12 g
Shi Wei	30 g
Dong Kui Zi	15 g
Hua Shi	30 g
Yu Bai Pi	15 g
Mu Tong	15 g
Chi Fu Ling	12 g

Main Indication: Stone Lin syndrome. Symptoms include sand or stone found in urine, difficult urination; yellow, dark, and turbid urine, or interrupted urination with burning pain, difficult to hold or with lower back pain, lower abdominal pain; or blood in the urine, normal tongue, greasy or greasy yellow tongue coating, and wiry or rapid pulse.

Preparation: Decoction.

Treatment Principles: Unblock Lin syndrome and expel stones.

This is a common formula to treat Shi Lin (Stone Lin) or Sha Lin (Sandy Lin). The effectiveness is better if the formula is taken continuously for a period of time.

Modification:

Addition	Indication
Jin Qian Cao Ji Nei Jin	Severe syndrome

Related Formulas:

Formula	Herbs	Indication
Shi Wei San [Pyrrosiae Powder] *Zheng Zhi Hui Bu*	Shi Wei Dong Kui Zi Qu Mai Hua Shi Che Qian Zi	Same indication as the core formula, but the function is weaker
Shi Wei San [Pyrrosiae Powder] *Ben Shi Fang*	Core formula + Chi Shao Yao Gan Cao	Treat the same indications as the core formula. It is stronger in cooling blood and stopping pain function

(continued)

Formula	Herbs	Indication
Chen Xiang San [Aquilariae Powder]	Chen Xiang Shi Wei Hua Shi Dang Gui Ju Pi Bai Shao Yao Dong Kui Zi Gan Cao Wang Bu Liu Xing	Treat Qi Lin syndrome

Sha Lin Wan [Pill for Sandy Lin Syndrome] *Yi Xue Zhong Zhong Can Xi Lu*

Ji Nei Jin	30 g
Sheng Huang Qi	24 g
Zhi Mu	24 g
Chi Shao Yao	18 g
Pu Xiao	15 g
Xiao Shi	15 g
Peng Sha	18 g

Main Indication: Sandy Lin syndrome, Stone Lin syndrome that belong to Qi, and Yin deficiency.

Preparation: Grind into fine powder and mix with honey to form pills 5 to 10 mm in diameter. Take 9 g on empty stomach twice a day.

Treatment Principles: Unblock Lin syndrome, transform stones, tonify Qi, and nourish Yin.

This formula is indicated for chronic Sandy Lin or Stone Lin syndrome where Qi and Yin are damaged.

Related Formula:

Formula	Herbs	Indication
Hua Shi San [Transform the Stone Powder]	Hua Shi 30 g Gan Cao 9 g Huo Xiao 18 g Ji Nei Jin 15 g Take 20 to 30 bags continuously	Treat stones in kidney and ureters

Zao Shi Wan [Chisel the Stone Pill] (Hu Nan Zhong Yi Xue Yuan Fang)

Huo Xiao	15 g
Hu Po	9 g
Hai Jin Sha	15 g
Fu Ling	9 g
Ze Xie	9 g
Di Long	9 g
Chi Shao Yao	18 g
Gan Cao Shao	6 g
Hua Shi	30 g
Dong Kui Zi	15 g
Chen Xiang	6 g
Chuan Niu Xi	15 g

Main Indication: Stone in urinary tract and especially for stones in ureters. Symptoms for stones in upper portion of the ureters are dull pain at lumbar, occasional wrenching pain, and pain radiating down from Kidney area. Stones in lower portion of the ureters give pain located near the inguinal groove or wrenching pain at the sides of the lower abdomen. Pulse is wiry.

Preparation: Grind into fine powder and make into small pills. Take 15 g, two to three times a day before meal.

Treatment Principles: Unblock Lin syndrome and transform stones.

This formula can make the stones smaller, relax the ureters, increase urine volume, and encourage Qi and blood to move downward. The combined above functions expel the stones out the body.

Related Formula:

Formula	Herbs	Indication
Mi Niao Dao Pai Shi Fang [Expel Stone from the Urinary Tract Formula]	Jin Qian Cao 60 to 180 g Hai Jin Sha 24 g Hua Shi Fen 24 g Chuan Shan Jia 6 g Chuan Niu Xi 6 g Chuan Lian Zi 6 g	Treat stones in ureters

Dao Chi San [Guide Out the Red Powder]
Xiao Er Yao Zheng Zhi Jue

Sheng Di Haung	24 g
Mu Tong	12 g
Gan Cao Shao	6 g
Dan Zhu Ye	6 g

Main Indication: Yin deficiency and heat excess in heart meridian transferring to small intestine. Symptoms include restlessness and heat at chest and heart, thirst with desire for cold drinks, red face, or ulcerations on tongue and mouth, dark painful urination, red tongue especially at tip, and rapid pulse.

Preparation: Decoct with 12 g of Deng Xin Cao.

Treatment Principles: Clear heart heat, nourish Yin, drain water, and unblock Lin.

Modification:

Addition	Indication
Huang Qin	Severe heart heat
Huang Lian Shi Chang Pu	Mouth and tongue ulcers

Related Formulas:

Formula	Herbs	Indication
Jia Wei Dao Chi San [Augmented Guide Out the Red Powder]	Core formula + Mai Men Dong Lian Zi Xin	Treat residual heat in warm febrile diseases; patient likes to talk but makes many mistakes
Dao Chi San He Jia Wei Hu Zhang San [Guide Out the Red and Augmented Polygoni Cuspidati Powder]	Core formula + Chuan Niu Xi 30 g Hu Zhang 30 g Chong Wei Zi 6 g Hu Po 2 g She Xiang 0.3 g	Treat uremia
Xie Xin Dao Chi Tang [Drain the Epigastrium and Guide Out the Red Decoction]	Core formula – Dan Zhu Ye + Huang Lian	Treat heart heat, protruding of tongue in children
Qing Xin Lian Zi Yin [Lotus Seed Decoction to Clear the Heart]	Huang Qin Mai Men Dong Di Gu Pi Che Qian Zi Gan Cao Lian Zi Bai Fu Ling Huang Qi Ren Shen	Treat Qi and Yin deficiency with excess heart fire and damp–heat downward pouring. Symptoms include nocturnal emission, turbid urination, heavy menstrual bleeding, leukorrhea, worse with over exertion, or Kidney Yin deficiency with dry mouth, dry tongue, irritability, agitation, and fever

Bei Xie Fen Qing Yin [Dioscorea Hypoglauca Decoction to Separate the Clear] *Yi Xue Xin Wu*

Bei Xie	24 g
Huang Bai	6 g
Shi Chang Pu	3 g
Fu Ling	6 g
Bai Zhu	6 g
Lian Zi Xin	6 g
Dan Shen	6 g
Che Qian Zi	12 g

Main Indication: Damp–heat downward pouring. Symptoms include turbid urination, burning painful hesitant urination, red tongue tip, and thin yellow tongue coating or greasy at the tongue root; and rapid pulse or empty big pulse at the Chi position.

Preparation: Decoction.

Treatment Principles: Clear heat, drain damp, separate the clear, and guide the turbidity.

This formula treats cloudy turbid urine due to damp–heat. In modern practice, it is used to treat Lower Jiao heat type prostatitis.

Related Formulas:

Formula	Herbs	Indication
Dan Xi Bei Xie Fen Qing Yin [Dioscorea Hypoglauca Decoction to Separate the Clear from *Dan Xi Xin Fa*]	Bei Xie Wu Yao Yi Zhi Ren Shi Chang Pu	Kidney Yang deficiency failing to transform damp, turbid, frequent urination with cloudy greasy thick urine, cold damp leukorrhea in female, and Yang deficient type of prostatitis
Zhi Qian Lie Xian Yan Fang [Formula that Treats Prostatitis]	Bei Xie 15 g Yi Zhi Ren 6 g Huang Bai 6 g Shi Chang Pu 6 g Sheng Di Huang 15 g Shan Yao 6 g Shan Zhu Yu Pi 6 g Fu Ling 6 g Ze Xie 5 g Mu Dan Pi 5 g	Decoction Treat acute or chronic prostatitis. If patient presents with more cold, then add Rou Gui and Fu Zi

Zi Shen Tong Guan Wan [Enrich the Kidneys and Open the Gates Pill] *Lan Shi Mi Cang*

Huang Bai	30 g
Zhi Mu	30 g
Rou Gui	2 g

Main Indication: Damp–heat accumulation, dysfunction of Qi transformation. Symptoms include difficult urination, or even retention, distention and fullness of the lower abdomen, red tongue, greasy tongue coating at the root, rapid thin pulse worse at Chi position.

Preparation: Grind into fine powder; mix with honey to form pills. Take 9 g, two to three times a day.

Treatment Principles: Drain heat and transform water.

This formula drains Prime Minister Fire to unblock urination. The formulation is unique. Small dosage of Rou Gui is used to assist the Qi transformation function.

Comparison of Formulas Treating Lower Jiao Damp–Heat

1. Er Miao San [Two-Marvel Powder] specially treats various Lower Jiao syndromes, painful tendons and bone, or stubborn hard to heal conditions due to damp–heat downward pouring or an accumulation of damp–heat.
2. Ba Zheng San [Eight-Herb Powder for Rectification] is a bitter cold formula to unblock urination. Any damp–heat type of Lin syndrome can use this formula. It is a common formula for Lower Jiao damp–heat.
3. Zhu Ling Tang [Polyporus Decoction], Wu Lin San [Five-Ingredient Powder for Painful Urinary Dysfunction], and Xiao Ji Yin Zi [Cephalanoplos Decoction] all treat bloody Lin syndrome. However, the first one is weaker in its function of stopping bleeding. It focuses on nourishing Yin and clearing heat. The middle formula clears heat and cools blood. Its strength is in the middle among the three formulas. The latter formula clears heat, nourishes Yin, cools blood, and transforms stasis. Its functions are strong and comprehensive.

4. Shi Wei San [Pyrrosiae Powder], Sha Lin Wan [Pill for Sandy Lin Syndrome], and Zao Shi Wan [Chisel the Stone Pill] all treat Sha Lin and Shi Lin syndromes. However, the first formula is stronger in draining fire, unblocking Lin, and expelling stones. It is suitable for acute attacks of stone in kidney, urinary bladder, or ureters. The middle formula is indicated for chronic conditions with Qi and Yin deficiency. The latter formula is specifically designed for stones in ureters. It can transform stone, relieve spasms, unblock Lin, and expel stone. The functions are strong and the formula is effective. It is indicated for a short term during the expelling stone period.
5. Dao Chi San [Guide the Red Out Powder] treats Yin deficiency with heart fire in children. The manifestations are tongue and mouth ulcers and burning, painful urination. If it is combined with Zi Shen Tong Guan Wan [Enrich the Kidneys and Open the Gates Pill], which drains Prime Minister Fire and strengthens the Qi transformation function, it can be used in heat and Kidney disharmony, insomnia, restlessness, excessive dreaming, and nocturnal emissions in adults.
6. Jia Jian Chai Ling Tang [Modified Bupleurum and Poria Decoction] is a modern formula in treating pyelonephritis or beginning stage of urinary tract infection with concomitant Shao Yang syndrome. Bei Xie Fen Qing Yin [Dioscorea Hypoglauca Decoction to Separate the Clear] focuses on separating the clear from turbid, which treats damp–heat downward pouring, leading to turbid cloudy urination. In modern times, it can be used for prostatitis or leukorrhea.

CHAPTER 10

Kidney and Lung, Spleen Syndromes and Formulas

KIDNEY AND LUNG SYNDROMES

Lung and Kidney Deficiency

This syndrome often seen as Kidney failing to grasp the Qi, and as Lung Qi rebellion, causing wheezing, breathlessness, and inability to lie flat, or as Kidney Yin deficiency with deficient fire upward blazing, causing Lung Yin deficient cough or hemoptysis. Commonly used herbs to tonify Kidney are Hu Tao, Ge Jie, Sheng Di Huang, and Shu Di Huang, and herbs to tonify Lung are Bei Sha Shen, Tian Men Dong, Mai Men Dong, and Ren Shen.

Bai He Gu Jin Tang [Lily Bulb Decoction to Preserve the Metal] *Zhao Ji An Fang*

Sheng Di Huang	12 g
Shu Di Huang	9 g
Xuan Shen	9 g
Dang Gui	9 g
Bai Shao Yao	9 g
Bai He	15 g
Mai Men Dong	6 g
Chuan Bei Mu	6 g
Sheng Gan Cao	3 g
Jie Geng	6 g

Main Indication: Lung and Kidney Yin deficiency with deficient fire upward blazing. Symptoms include cough and phlegm with a little blood, dry and painful throat, cough, wheezing, warm hands and feet, steaming bone, night sweats, red tongue, scanty tongue coating, and thin, rapid pulse.

Preparation: Take the decoction on empty stomach, twice a day.

Treatment Principles: Moisten Lung, nourish Kidney, transform phlegm, and stop cough.

This formula treats both the root and manifestations. It is used for dry cough, unproductive cough, or phlegm with blood in Lung tuberculosis, bronchitis, and bronchiectasis.

Modification:

Addition	Indication
Zhi Mu Yu Xing Cao	Severe heat
Ce Bai Ye Xian He Cao	Hemoptysis
Bai Ji Bai Bu Xia Ku Cao	Lung tuberculosis

Ren Shen Gu Ben Wan [Ginseng Pill to Reserve the Source] *Gu Fang Ba Zhen*

Ren Shen	60 g
Mai Men Dong	120 g
Tian Men Dong	120 g
Sheng Di Huang	120 g
Shu Di Huang	120 g

Main Indication: Lung and Kidney Yin deficiency, deficient fire scorching Lung. Symptoms include cough without phlegm, hemoptysis, red tongue, scanty tongue coating, dry throat or sore throat, and thin, rapid pulse.

Preparation: Grind into fine powder and mix with honey to form pills. Take 6 g, twice a day.

Treatment Principles: Sweet and cold to nourish Yin, and stop cough.

This is a classical formula to nourish Lung and Kidney.

Ren Shen Hu Tao Tang [Ginseng and Walnut Decoction] *Ji Sheng Fang*

Ren Shen	6 g
Hu Tao Ren	30 g

Main Indication: Lung and Kidney Qi deficiency, and Kidney not grasping the Qi. Symptoms include wheezing, rebellious Qi, more exhalation and less inhalation, or with dry cough, shortness of breath, too deficient to talk, spontaneous sweating worse with wheezing, or with poor appetite, urinary incontinence, pale or flabby tongue, and deficient or big pulse.

Preparation: Decoct with Sheng Jiang and Da Zao. Take the warm decoction after meal.

Treatment Principles: Tonify deficiency and stop wheezing.

This is a representing formula which treats Kidney failing to grasp Qi, and Lung and Kidney Qi deficiency.

Related Formulas:

Formula	Herbs	Indication
Ren Shen Ge Jie San [Ginseng and Gecko Powder]	Ge Jie 1 pair Ren Shen 45 g Grind into fine powder. Cook one bowl of sticky rice congee and mix with 20 g of the herbs. Eat the congee slowly while it is still hot	Treats wheezing, and coughing with swollen face and extremities
Ren Shen Ge Jie San [Ginseng and Gecko Powder] *Wei Sheng Bao Jian*	Ge Jie 1 pair Xing Ren 150 g Gan Cao 150 g Ren Shen 60 g Fu Ling 60 g Zhe Bei Mu 60 g Sang Bai Pi 60 g Zhi Mu 60 g Grind into powder, and take 6 g, once in the morning, once at night	Treat chronic cough, chronic wheezing, sticky yellow phlegm, cough or vomiting of pus and blood, restlessness and heat sensation at the chest, emaciation, or swollen face and eyelids, or Lung atrophy in chronic condition, and floating deficient pulse

Comparison of Formulas Treating Lung and Kidney Deficiency

Bai He Gu Jin Tang [Lily Bulb Decoction to Preserve the Metal] is a comprehensive nourishing Yin and clearing deficient heat formula. It treats dry cough due to Lung and Kidney Yin deficiency or phlegm with blood. Ren Shen Gu Ben Wan [Ginseng Pill to Preserve the Source] is a formula to treat Yin deficiency with deficient fire scorching lungs and hemoptysis. It nourishes Lung and Kidney Yin, also tonifies Qi and controls the blood. Ren Shen Hu Tao Tang [Ginseng and Walnut Decoction] is a specialized formula to treat Lung and Kidney Qi deficiency and Kidney failing to grasp the Qi. Ren Shen Ge Jie San [Ginseng and Gecko Powder] is a mild formula treating edema due to Lung and Kidney Qi deficiency, and inability to distribute fluid and transform Qi. In *Wei Sheng Bao Jian*, Ren Shen Ge Jie San [Ginseng and Gecko Powder] is indicated for chronic cough and wheezing where Lung abscess or Lung atrophy has already occurred. In this case, Lung and Kidney are both already deficient, but the evil pathogens still remain, thus one needs to clear Lung, transform phlegm, stop bleeding, calm coughing and wheezing, and at the same time tonify Lung and Kidney Qi to preserve the source.

KIDNEY AND SPLEEN SYNDROMES

Spleen and Kidney Yang Deficiency

Spleen and Kidney Yang deficiency often causes internal accumulation of water and dampness. It manifests as lower body edema or diarrhea due to damp accumulation and pouring downward. Commonly used herbs are Fu Zi, Rou Gui, Fu Ling, and Bai Zhu to warm Yang and drain damp, and to strengthen the spleen to eliminate dampness, as in Zhen Wu Tang [True Warrior Decoction] and Si Shen Wan [Four-Miracle Pill].

Si Shen Wan [Four-Miracle Pill] *Zheng Zhi Zhun Sheng*

Rou Dou Kou	60 g
Bu Gu Zhi	120 g
Wu Wei Zi	60 g
Wu Zhu Yu (fried)	30 g

Main Indication: Spleen and Kidney deficient cold. Symptoms include 5 A.M. diarrhea, no desire to eat, or chronic diarrhea, abdominal pain, lower back soreness, cold limbs, apathy, fatigue, pale, flabby, moist tongue, and deep weak pulse especially at the Chi position.

Preparation: Grind into powder, and take 9 g, one to two times a day with Sheng Jiang and Da Zao decoction before meals. It can also be taken as decoction with reduced dosages.

Treatment Principles: Warm and tonify Spleen and Kidney, and astringe Intestines to stop diarrhea.

"Four Miracles" refers to the effectiveness of the four herbs in the formula. This formula is modified from Er Shen Wan [Two-Miracle Pill] of *Ben Shi Fang* and Wu Wei Zi San [Schisandra Powder]. Er Shen Wan [Two-Miracle Pill] is composed of Rou Dou Kou and Bu Gu Zhi, and Wu Wei Zi San [Schisandra Powder] is composed of Wu Wei Zi and Wu Zhu Yu. It is effective in treating 5 A.M. diarrhea, and chronic diarrhea due to Spleen and Kidney deficient cold. However, the formula should be taken for 1 to 2 months.

PART II

Formulas According to Allopathic Medical Systems

CHAPTER 11

Blood-Related Disorders

ALLERGIC PURPURA

This condition is discussed under bleeding at muscle (Ji Nu) and Zi Dian Feng. It is caused by internal excess of heat toxins or external wind–heat toxins steaming and accumulating at Ying and blood level that manifest as purpura of the skin. In Western medicine, this is also called hemorrhagic capillary blood poisoning syndrome. It is an allergic reaction of the body to certain conditions leading to increased permeability and fragility of the capillaries. Thus there is bleeding or edema at skin, mucous membrane, and internal organs. The manifestations are frequent, symmetrical, purple spots on the skin and medial sides of the arms, hip, and legs, especially the latter. Some patients also present with swollen painful joints, abdominal pain; and hemafecia, edema, hematuria, and proteinuria. TCM treatment focuses on expelling dampness, dispersing wind, invigorating blood, cooling blood, relieving toxicity, and strengthening Spleen.

During Attack Period

Stagnant Heat Obstructing Collaterals Syndrome

Treatment Principle: Cool blood, invigorate blood, relieve toxins, and disperse wind.

Use Xiao Feng San [Eliminate Wind Powder] plus Sheng Di Huang, Chi Shao Yao, Zi Cao, Mu Dan Pi, Pu Gong Ying, Tu Fu Ling, and Huang Qin.

Related conditions and modifications are

- Dark purpura, deep red tongue tip: Add Shui Niu Jiao and Huang Lian. Or change to Yin Hua Jie Du Tang [Honeysuckle Decoction to Relieve Toxicity].
- Constipation: Minus Fang Feng, Cang Zhu and add Da Huang, Zhi Shi, and Hou Po.
- Swollen painful joints and inability to extend and flex: Add Qin Jiao and Wei Ling Xian.
- Severe painful upper extremity: Add Sang Zhi and Qiang Huo.
- Severe lower extremity pain: Add Chuan Niu Xi, Han Fang Ji, and Yi Yi Ren.
- Abdominal pain, nausea, and vomiting: Add Mu Xiang, Bai Shao Yao, Ban Xia, and Hou Po.
- Vomiting blood and blood in stool: Add Xian He Cao and San Qi powder.
- Swollen distention around the purpura, swollen face, and itchy skin: Add Qiang Huo, Ma Huang, Cang Zhu, Yin Chen Hao, and Che Qian Zi.
- Hemoptysis, chest pain, cough, and wheezing: Remove Fang Feng, Chan Tui, Tu Fu Ling, and add Sang Bai Pi, Qing Dai, Xing Ren, and Ce Bai Ye.

Yin Deficiency and Blood Heat Syndrome

Treatment Principle: Nourish Yin and clear blood heat.

Use Er Zhi Wan [Two-Ultimate Pill] plus Qing Ying Tang [Clear the Nutritive Level Decoction] plus Bie Jia, Di Gu Pi, Zhi Mu, Qian Cao, Hong Hua, and Mu Dan Pi.

Related conditions and modifications are

- Spleen deficiency, loose stool, and abdominal distention: Add Bai Zhu, Sha Ren, Chen Pi, and Shan Yao.
- Severe blood heat: Add Zi Cao and Huang Lian.

Spleen Not Controlling Blood Syndrome

Treatment Principle: Tonify Heart, Spleen, and dissipate purple spots.

Use Gui Pi Tang [Restore the Spleen Decoction] plus Bai Shao Yao, Sheng Di Huang, Gou Qi Zi, E Jiao, and San Qi powder.

Related conditions and modifications are

- Sinking Middle Jiao Qi: Add Chai Hu and Sheng Ma.
- Spleen Yang deficiency: Add Gan Jiang and Shu Di Huang.

Remission Period

1. Decoct 40 g of Bai Mao Gen and 60 g of Da Zao; drink as tea.
2. Continuously take Liu Wei Di Huang Wan [Six-Ingredient Pill with Rehmannia] for Kidney Yin deficiency or Jin Gui Shen Qi Wan [Kidney Qi Pill from the Golden Cabinet] for Kidney Yang deficiency or Gui Pi Wan [Restore the Spleen Pill] for Qi and blood deficiency.
3. Patients who use steroids often present with Yin deficiency and hyperactive Yang. To help reduce the side effects of steroids, use herbs that nourish Yin and clear heat such as Sheng Di Huang, Mu Dan Pi, Zhi Mu, Xuan Shen, Gui Ban, Shan Zhu Yu, Mai Men Dong, and Nu Zhen Zi. This can be given to patients who have used steroids for a long time.

Furthermore, during the active stage, one should disperse wind, relieve toxicity, cool blood, and invigorate blood. One should select Fang Feng, Ma Huang, Qiang Huo, Chan Tui, Jing Jie, Bo He, Bai Xian Pi, Lian Qiao, Jin Yin Hua, Ban Lan Gen, Sheng Di Huang, Zi Cao, Mu Dan Pi, and Han Lian Cao, Dan Shen, Hong Hua, Chi Shao Yao, Tao Ren, Chuan Xiong, Ze Lan, San Qi, and Di Long, according to syndrome differentiation to increase the effectiveness.

CHRONIC GRANULOCYTIC LEUKEMIA

This condition is discussed under abdominal mass and consumptive disease. It is caused by internal injuries leading to unregulated Qi and blood flow, stagnation and stasis of the channels and collaterals, and insufficient formation of Ying and Wei. This condition often occurs in middle-aged and young people. The main cause is excessive formation of granulocytes. The clinical manifestations are dizziness, fatigue, palpitation, emaciation, excessive sweating, decreased food intake, low-grade fever; and bleeding from nose, eyes, gums, and skin or internal organs. In females, there can be amenorrhea and in males there can be excessive erections or enlarged Liver and Spleen (especially the Spleen) and pain upon palpation of the Sternum. Qing Dai and Xiong Huang are effective in treating this condition and can be added to the formula.

Qi Stagnation and Blood Stasis Syndrome

Treatment Principle: Regulate Qi and invigorate blood.

Use Ge Xia Zhu Yu Tang [Drive Out Blood Stasis Below the Diaphragm Decoction] plus Qing Dai and Xiong Huang.

Zheng Qi Deficiency and Blood Stasis Syndrome

Treatment Principle: Tonify Qi, nourish blood, invigorate blood, and transform stasis.

Use Ba Zhen Tang [Eight-Treasure Decoction] plus Ge Xia Zhu Yu Tang [Drive Out Blood Stasis Below the Diaphragm Decoction] plus Yan Hu Suo, Qing Dai, and Xiong Huang.

Related conditions and modifications are

- Qi and Yin deficiency: Add Ren Shen, Mai Men Dong, Wu Wei Zi, and Shan Zhu Yu.
- Bleeding: Add Da Zao, E Jiao, San Qi powder, and Han Lian Cao.
- Tidal fever and night sweats: Add Hu Huang Lian, Yin Chai Hu, and Mu Dan Pi.

Furthermore, one can also combine patent medicine to increase the effectiveness, such as Liu Shen Wan [Six-Miracle Pill], Niu Huang Jie Du Wan [Cattle Gallstone Pill to Relieve Toxicity], and Dang Gui Long Hui Wan [Tangkuei, Gentiana Longdancao, and Aloe Pill].

CHRONIC LYMPHOCYTIC LEUKEMIA

This condition is similar to consumptive disease, scrofula, and abdominal mass in TCM. It is caused by invasion of evil toxins while the body constitution is weak. This damages blood and marrow, leading to deficiency and exhaustion of Qi and blood, accumulation of phlegm–damp, exhaustion of Yin and blood, and knotted accumulated phlegm fire. In Western medicine, it is described as being caused by a clonal proliferation of a type of small lymphocytes. The accumulations of these lymphocytes infiltrate the marrow, blood, lymph nodes, and other organs and lead to severe condition of hemopoietic failure. The manifestations are slow onset, often without obvious symptoms. Some patients can experience swollen lymph nodes and anemia. The condition is diagnosed when they have blood tests done. There can also be dizziness, palpitation, shortness of breath, weight loss, fever, bleeding of the gums, hemafecia, skin purpura, and pain in the bones. In 80% of the patients there are swollen lymph nodes, especially cervical lymph nodes. Chinese herbs are effective in treating the early stages and stable stage of this condition by supporting Zheng Qi, softening hardness, and scattering accumulations.

Early Stage: Insidious Evil Toxins

Treatment Principle: Support Zheng Qi and expel toxins.

Use Si Jun Zi Tang [Four-Gentleman Decoction], Er Chen Tang [Two-Cured Decoction] plus Bai Hua She She Cao, Long Kui, Ban Zhi Lian, Chi Shao Yao, Huang Yao Zi, E Zhu, and Shan Ci Gu.

Middle and Late Stage

Knotted Phlegm and Stasis Syndrome

Treatment Principle: Transform phlegm, scatter accumulation, invigorate blood, and transform stasis.

Use Di Tan Tang [Scour Phlegm Decoction] remove Ren Shen, and add Dan Shen, Chi Shao Yao, E Zhu, Xia Ku Cao, Zhe Bei Mu, Kun Bu, Huang Yao Zi, Chuan Xiong, and Yu Jin.

Related conditions and modifications are

- Severe tidal fever: Add Di Gu Pi and Yin Chai Hu.
- Severe night sweats: Add Sheng Mu Li and Ma Huang Gen.

Both Qi and Yin Deficiency with Blood Stasis Syndrome

Treatment Principle: Tonify Qi, nourish Yin, and invigorate blood.

Use Yang Yin Qing Fei Tang [Nourish the Yin and Clear the Lungs Decoction] and add Dan Shen, Tai Zi Shen, Dang Shen, Shan Ci Gu, Wu Wei Zi, and Yu Zhu.

Related conditions and modifications are

- Severe Qi deficiency: Add Huang Qi, Fu Ling, and Bai Zhu.
- Severe Yin deficiency: Add Shi Hu and Tian Men Dong.

Phlegm Stasis and Damp–Heat Syndrome

Treatment Principle: Invigorate blood, transform phlegm, clear heat, and drain damp.

Use San Xiang Tang [Three-Fragrance Decoction] plus Yin Chen Hao, San Leng, E Zhu, Huang Yao Zi, Xia Ku Cao, Zhu Ling, Zhe Bei Mu, and Bei Xie.

Related conditions and modifications are

- Heat excess: Add Long Dan Cao and Jin Qian Cao.
- Skin itch and rash: Add Di Fu Zi, Bai Xian Pi, Fang Feng, Ku Shen, and Chan Tui.

IRON DEFICIENCY ANEMIA

This condition is discussed under deficiency, sallow complexion, and general edema with yellowish skin. It is often caused by deficient constitution, Spleen and Stomach deficiency, insufficient transformation, emotional stagnation; and Yin and blood exhaustion or chronic diseases leading to Zang Fu organs damages. It can also be caused by unbalanced diet or parasites accumulation. In Western medicine, it refers to insufficient iron storage in the body, thus affecting the formation of hemoglobin and leading to hypochromic microcytic anemia. Treatment in TCM focuses on nourishing blood by first tonifying the Qi and combining with food therapy.

Treatment According to Syndrome Differentiation

Qi and Blood Deficiency Syndrome

Treatment Principle: Tonify Qi and nourish blood.

Use Ba Zhen Tang [Eight-Treasure Decoction] plus Gou Qi Zi, Wu Wei Zi, E Jiao, Zhi Gan Cao, Zhi He Shou Wu, and Da Zao.

Related conditions and modifications are

- Abdominal distention, loose stool, nausea, vomiting, and no taste in mouth: Add Huo Xiang, Hou Po, Chen Pi, Ban Xia, Gu Ya, Shen Qu, and Sheng Jiang.
- Intestinal dryness due to blood deficiency, constipation, dry mouth and dry throat: Add 30 to 40 g of Ji Xue Teng, Huo Ma Ren, or Rou Cong Rong.

- Insomnia, excessive dreaming, forgetfulness, and tongue pain: Add Suan Zao Ren, Bai Zi Ren, and Mai Men Dong.
- Delayed menstruation, scanty flow with pale blood, and lower abdominal pain with empty sensation: Add Yi Mu Cao, Ji Xue Teng, and increase Chuan Xiong to 15 g.
- Subcutaneous purple spots and tarry dark stool: Add Xian He Cao and Han Lian Cao.

Spleen and Stomach Deficiency Syndrome

Treatment Principle: Strengthen Spleen and nourish blood.

Use Shen Ling Bai Zhu San [Ginseng, Poria, and Atractylodes Macrocephala Powder] plus Ji Xue Teng, and Huang Qi.

Related conditions and modifications are

- Loose stool: Add burned Shan Zha, Shen Qu, Gu Ya, and Mai Ya.
- Facial swelling: Add Yi Yi Ren, Bai Bian Dou, and Ze Xie.
- Undigested food and cool limbs: Add Wu Zhu Yu, Gan Jiang, and Gui Zhi.

Herbal Food Therapy

1. Mu Er Da Zao Tang [Wood Ear and Red Date Soup]: Black wood ear fungus 50 g and Da Zao 30 pieces. Cook together and take frequently.
2. Tian Qi Ji Tang [Chicken Soup with Pseudoginseng]: Chicken 250 g, Tian Qi 15 g, sesame oil, and salt. Cook for 4 hours and drink the soup for 7 days.

Patent Medicine

One can also take Gui Pi Wan [Restore the Spleen Pill], Ba Zhen Wan [Eight-Treasure Pill], and Ren Shen Yang Ying Wan [Ginseng Pill to Nourish the Nutritive Qi] to strengthen the effect and also during traveling.

LEUKOPENIA AND AGRANULOCYTOSIS

Leukopenia is a condition with reduced white blood cells. Symptoms are more deficient, especially Qi and blood deficiencies. Agranulocytosis is caused by heat toxins and often progresses according to Wei, Qi, Ying, and blood. In Western medicine, white blood cells count in blood is lower than 4.0×10^9/l for leukopenia. For agranulocytosis, white blood cell count is lower than 2.0×10^9/l and neutrophils are disappearing or are lower than 0.5×10^9/l. Treatment focuses on removing the cause, increasing white blood cells, granulocytes, and controlling infections.

Leukopenia

Ji Xue Teng, Zhi Chuan Shan Jia, E Jiao, Bu Gu Zhi, Ren Shen, Huang Qi, Da Zao, and Nu Zhen Zi can stimulate the formation of white blood cells, and can be selected according to syndrome differentiation. If there is a decrease in red blood cells, then one can add Dang Shen, Shu Di Huang, Long Yan Rou, Gou Qi Zi, and Da Zao. Huang Jing, Ren Shen, Mai Men Dong, and Wu Wei Zi can be used for hypotension. Bai Bian Dou, Gu Ya, Bai Zhu, and Ji Nei Jin can be added for poor appetite. Mu Dan Pi and Zhi Bie Jia can be added for enlarged Spleen.

Furthermore, from clinical results the following formulas can protect and stimulate the formation of white blood cells for patients with decreased white blood cells due to radiation therapy or chemotherapy.

White blood cell protection formula: Ren Shen 10 g, Huang Qi 30 g, Zhi He Shou Wu 30 g, Ji Xue Teng 30 g, Da Zao 30 g, Shu Di Huang 15 g, Bu Gu Zhi 15 g, Nu Zhen Zi 15 g, Shan Zhu Yu 15 g, E Jiao 15 g, and Hu Zhang 15 g.

Heart and Spleen Deficiency Syndrome

Treatment Principle: Nourish Heart blood and tonify Spleen Qi.

Use Gui Pi Tang [Restore the Spleen Decoction] plus E Jiao.

Heart Blood and Heart Yin Deficiency Syndrome

Treatment Principle: Tonify blood and nourish Yin.

Use Zhi Gan Cao Tang [Honey-Fried Licorice Decoction] plus Suan Zao Ren, Ji Xue Teng, Huang Qi, and Gou Qi Zi.

Spleen Deficiency with Internal Heat Syndrome

Treatment Principle: Benefit Qi, tonify Spleen, ascend Yang, and scatter fire.

Use Bu Zhong Yi Qi Tang [Tonify the Middle Jiao and Augment the Qi Decoction] plus Qing Hao, Bie Jia, and Di Gu Pi.

Kidney Essence Deficiency Not Transforming Blood Syndrome

- Mild Kidney Yang deficiency.
  Warm Kidney and replenish essence: You Gui Wan [Restore the Right (Kidney) Pill] plus Bu Gu Zhi, Ren Shen, Huang Qi, and Ji Xue Teng.
- Mild Kidney Yin deficiency.
  Nourish Kidney and replenish essence: Zuo Gui Wan [Restore the Left (Kidney) Pill] plus E Jiao, Er Zhi Wan [Two-Ultimate Pill], Di Gu Pi, Ren Shen, Da Zao, and Huang Qi.

Agranulocytosis

This condition has an acute onset high fever and chills with ulceration; or necrosis of the mucous membrane in oral cavity, throat, and rectum. There can also be severe infections of the Lung and urinary system. In acute severe cases, the patient should be referred to his or her medical doctor to control the infection, then herbs can be taken to increase granulocytes.

Residual Heat in Upper Jiao Syndrome

Treatment Principle: Clear Upper Jiao heat and relieve toxicity.

Use Yin Qiao San [Honeysuckle and Forsythia Powder] plus Ban Lan Gen, Shan Dou Gen, Huang Lian, Huang Qin, and Qing Dai.

Residual Heat Toxins in San Jiao Syndrome

Treatment Principle: Drain heat from San Jiao and relieve toxicity.

Use Zhu Ye Shi Gao Tang [Lophatherus and Gypsum Decoction] plus Tiao Wei Cheng Qi Tang [Regulate the Stomach and Order the Qi Decoction].

Residual Heat in Ying and Blood Syndrome

Treatment Principle: Cool blood, nourish Yin, and relieve toxicity.

Use Qing Ying Tang [Clear the Nutritive Level Decoction] plus She Gan, Ban Lan Gen, and Jie Geng.

Related condition and modification are

- Constipation: Add Da Huang.

CHAPTER 12

Cancer Disorders

BRIEF INTRODUCTION

The earliest record of the character for "tumor" appeared in *The Oracle-Bone Scripture*. In ancient classics, cancer and tumor is discussed under the categories of Ying Liu (goiter), Yi Ge (dysphagia), Fan Wei (regurgitation), Zheng Jia (abdominal mass), Ji Ju (accumulated lumps), Tan Pi (clumped phlegm), Fu Liang (cancer), E Chuang (bad ulcerations), Du Ju (toxic deep rooted phlegm), Shi Ying (cervical carcinoma with cachexia), Yan (cancer), and Ai (cancer). In general, it is caused by deficient Zheng Qi, disorder of Yin and Yang, disharmony of Zang Qi, invasions of evil toxins or self-generated toxins leading to Qi stagnation; blood stasis, congealed phlegm, heat toxins; and knotted accumulation and formation of swellings, and lumps over a period of time.

TCM treatments combine herbs, acupuncture, diet, external application, and Qi Gong therapy. These can stimulate Zheng Qi and improve the body condition, alleviate symptoms, and stabilize the condition. Furthermore, these techniques have minimal side effects and can be use widely. Some cancer patients' conditions improve after chemotherapy and radiation therapy, and they see a shrinking in the size of tumors, but their life is not prolonged. However, some patients who incorporate TCM treatments as part of therapy do not see significant decrease in tumor size, but they are able to maintain their quality of living, to sustain their ability to work, and to prolong life. At the same time, TCM treatments can moderate and reduce the toxic side effects of Western pharmaceuticals, assist in clearing and eliminating residual cancer cells after surgery, strengthen the effectiveness of pharmaceuticals, and prevent recurrence and spreading of cancer.

However, large cancers or tumors, high malignancy, and fast-developing cancers are not indicated for TCM treatments.

SELECTION OF CHINESE HERBS

Strong Anticancer Herbs

Dosage of these herbs should be carefully selected.

There are two groups of strong anticancer herbs:

- Toxin-attack toxin herbs: Shan Ci Gu, Ya Dan Zi, Lang Du, Nao Sha, Lu Feng Fang, Wu Gong, Bi Hu, Ban Mao, Chan Su, and Pi Shi.
- Crack blood stasis and soften hardness herbs: E Zhu, San Leng, Shui Zhi, Ban Xia, Tian Nan Xing, and Qing Dai.

Nontoxic Anticancer Herbs

Dosage of this category herbs can be increased.

Often use Bai Hua She She Cao, Bai Jiang Cao, Xian He Cao, Zhong Jie Feng, Zao Xiu, Ban Bian Lian, Ban Zhi Lian, Sheng Yi Yi Ren, Teng Li Gen, Tian Men Dong, and He Tao Shu Zhi.

Selection of Herbs According to Herbal Functions

Often choose

- Clear heat and anticancer herbs: Bai Hua She She Cao, Zao Xiu, Bai Jiang Cao, and She Mei.
- Transform stasis and anticancer herbs: Shi Jian Chuan, Tie Shu Ye, Long Kui, and E Zhu.
- Anti-inflammatory and anticancer herbs: Shan Ci Gu, Ban Xia, and Tian Nan Xing.
- Regulate Qi and anticancer herbs: Ba Yue Zha.
- Support Zheng Qi and anticancer herbs: Sheng Yi Yi Ren, Tian Men Dong, and He Tao Shu Zhi.

Selection of Herbs According to Disease

- Throat and vocal cord cancer: Ma Bo and Shan Dou Gen.
- Nasopharynx cancer: Bai Ying and Shi Shang Bai.
- Liver and lung cancer: Long Kui, Ban Zhi Lian, Zi Cao, and Zao Xiu.
- Leukemia: Qing Dai, Zi Cao Gen, Niu Huang, and Zhu Yang Yang.
- Esophageal and cardiac sphincter cancer: Dong Ling Cao.
- Digestive tract and urinary bladder cancer: Da Huang, Bai Hua She She Cao, Zhong Jie Feng, and Feng Wei Cao.
- Cervical cancer: Bai Jiang Cao, Bai Hua She She Cao, Chun Gen Pi, and Mu Hui Tou.
- Mammary and thyroid cancer: Shan Ci Gu and Huang Yao Zi.
- Malignant lymphoma: Bai Hua She She Cao.

COMMON TCM TREATMENTS

Regulate Qi and Descend Rebellion Method

It is indicated for distending pain at chest and hypochondriac region, focal distention and fullness of epigastrium and abdomen, belching, nausea, poor appetite, difficulty swallowing, cough, radiating pain at abdomen, breast distention, and thin and wiry or wiry pulse.

Conditions and treatments are

- Liver, Gallbladder, pancreas, and mammary cancer with Liver Qi stagnation: Use Chai Hu Shu Gan San [Bupleurum Powder to Spread the Liver] and Xiao Yao San [Relax Wanderer Powder].
- Gastrointestinal and esophageal cancer with Qi stagnation at Middle Jiao and Stomach Qi rebellion: Use Mu Xiang Bing Lang Wan [Aucklandia and Betel Nut Pill] and Xuan Fu Dai Zhe Tang [Inula and Haematite Decoction].
- Lung cancer with phlegm, Qi congestion, and Lung Qi rebellion: Use Ting Li Da Zao Xie Fei Tang [Descurainia and Jujube Decoction to Drain the Lungs], Qing Qi Hua Tan Wan [Clear the Qi and Transform Phlegm Pill], and Su Zi Jiang Qi Tang [Perilla Fruit Decoction for Directing Qi Downward].

Invigorate Blood and Transform Stasis Method

This is for immovable masses or tumors, sharp stabbing pain, fixed and dark facial complexion, scaly skin or tarry stool; purple tongue or purple-spotted tongue; and thin and choppy or choppy and wiry pulse.

Conditions and treatments are

- General pain: Use Huo Luo Xiao Ling Dan [Fantastically Effective Pill to Invigorate the Collaterals] and Shi Xiao San [Sudden Smile Powder].
- Bleeding and stasis obstruction leading to pain: Use Tao Hong Si Wu Tang [Four-Substance Decoction with Peach Pit and Safflower] and Xue Fu Zhu Yu Tang [Drive Out Stasis in the Mansion of Blood Decoction].
- Swelling pain: Use Bie Jia Jian Wan [Decocted Soft-Shelled Turtle Shell Pill].

Clear Heat and Relieve Toxicity Method

This is for fever, aversion to heat, thirst, copious sweating, irritability, agitation, skin rash, bleeding, red face, red eyes, tongue and mouth ulcerations, foul breath, coarse breathing; dry stool, yellow scanty urine, or with tenesmus, or urgent and painful urination; or local burning, redness, swelling and pain, or bright yellow sclera; and yellow body, thick, yellow, greasy or rough tongue coating, red tongue, and rapid pulse.

Conditions and treatments are

- Acute leukemia: Bai Hu Tang [White Tiger Decoction] plus Xi Jiao Di Huang Tang [Rhinoceros Horn and Rehmannia Decoction].
- Chronic granulocytic leukemia: Dang Gui Long Hui Wan [Tangkuei Gentiana Longdancao and Aloe Pill].
- Lung cancer due to heat and stasis obstructing the Lung: Wei Jing Tang [Reed Decoction] plus Sang Bai Pi and Huang Qin.
- Nasopharynx cancer due to upward blazing of wind–fire: Ling Yang Gou Teng Tang [Antelope Horn and Uncaria Decoction].
- Pharyngeal cancer due to Yin deficiency and deficient fire: Zhi Bai Di Huang Wan [Anemarrhena, Phellodendron, and Rehmannia Pill] plus Yin Qiao Ma Bo San [Honeysuckle, Forsythia, and Puffball Powder].
- Urinary bladder cancer and rectal cancer due to Lower Jiao damp–heat: Ba Zheng San [Eight-Herb Powder for Rectification] and Bai Tou Weng Tang [Pulsatilla Decoction].

Eliminate Phlegm and Scatter Accumulation Method

This is indicated for swelling, mass accumulation, nodules under the skin, abdominal mass, mild pain, and masses soft in quality on palpation. There is also nausea, poor appetite, focal distention, fullness, discomfort, vomiting phlegm, and thin saliva, stifling sensation at chest, cough and wheezing, and unproductive cough. There can also be headache, dizziness, nausea and vomiting, tongue stiffness, speech difficulty, and confusion. The tongue is flabby and is big with greasy tongue coating. The pulse is wiry and slippery or thin and slippery.

Conditions and treatments are

- Stomach cancer due to congealed phlegm and Qi obstruction: Gun Tan Wan [Vaporize Phlegm Pill].
- Thyroid cancer and malignant lymphoma: Hai Zao Yu Hu Tang [Sargassum Decoction for the Jade Flask], Si Hai Shu Yu Wan [Spread the Stagnation Pill for Four Oceans], and Zhi Mi Fu Ling Wan [Poria Pill to Point Out Confusion].
- Mammary cancer: Xi Huang Wan [Rhinoceros Gallstone Decoction].

- Lung cancer: Er Chen Tang [Two-Cured Decoction], Dao Tan Tang [Guide Out Phlegm Decoction], and Gun Tan Wan [Vaporize Phlegm Pill].

Toxin-Attack-Toxin Method

This is indicated for localized and limited pathological change and bigger cancerous tumors, which are not effectively treated by the above methods; strong Zheng Qi and severe pain or unbearable pain. This can also treat "shallow" cancer that can be touched and to which medicine can be applied topically.

Conditions and treatments are

- Skin cancer, cervical cancer: Local topical herbs are often used, such as Qing Fen, Bai Fan, Nao Sha, Peng Sha, Xiong Huang, Shui Yin, Ban Xia, Tian Nan Xing, Ya Dan Zi, Ba Dou, Wu Bei Zi, and Ma Qian Zi.
- Leukemia, metastasized cancer: Xi Shu, Shi Suan, Chang Chun Hua, and Qiu Shui Xian.
- Digestive tract, pancreatic cancer: Zhong Jie Feng.
- Liver cancer: Ban Mao.
- Skin cancer: Teng Huang.
- Mammary cancer: Shan Ci Gu.
- Thyroid cancer, chronic granulocytic leukemia: Mao Zhua Cao.
- Leukemia: Xiong Huang and Pi Shi.
- Lung cancer: Ya Dan Zi.

(Note: Dosage should be determined according to the toxicity level.)

- Mild toxicity: Zhang Lang and Qiang Lang.
- Moderate toxicity: Quan Xie, Wu Gong, and Lu Feng Fang.
- Highly toxic: Ya Dan Zi, Tian Nan Xing, and Chuan Wu.
- Extreme toxic: Ma Qian Zi, Pi Shi, Ba Dou, Ban Mao, and She Du.

Nourish Yin and Moisten Dryness Method

This is indicated for dry lips, dry mouth, dry nose, thirst, dry cough with scanty phlegm, phlegm with blood tinge, dry skin, difficulty swallowing, dry stool, hesitant urination, dry rough tongue coating, scanty body fluid, scanty coating or peeled coating, and thin, rapid pulse. Or emaciation, dizziness, vertigo, tinnitus, forgetfulness, blurry vision, tidal fever, night sweats, hair loss, decreased menstruation and infertility, deep red tongue with peeled coating, and thin or thin, rapid pulse.

Conditions and treatments are

- Cancers of lung, esophagus, larynx, nasopharynx, and leukemia which belong to Lung and Stomach fluid exhaustion: Sha Shen Mai Dong Tang [Glehnia and Ophiopogonis Decoction], Yi Wei Tang [Benefit the Stomach Decoction], and Wu Zhi Yin [Five-Juice Decoction].
- Cancers of Liver, cervix, bone, and brain tumor which belong to Liver and Kidney Yin deficiency: Yi Guan Jian [Linking Decoction], Use Zuo Gui Yin [Restore the Left (Kidney) Decoction], Liu Wei Di Huang Wan [Six-Ingredient Pill with Rehmannia], and Da Bu Yin Wan [Great Tonify the Yin Pill].

Tonify Kidney and Warm Yang Method

This is for apathy, cold limbs, soreness and weakness of the lumbar and knees, dizziness, tinnitus, clear frequent urination and nocturia, impotence, premature ejaculation, pale tongue, and deep, thin pulse. Or rapid wheezing, worsening with movement; or swelling, scanty urine; or diarrhea with undigested food; or 5 A.M. diarrhea.

Conditions and treatments are

- Lung cancer: Du Qi Wan [Capital Qi Pill], Ren Shen Ge Jie San [Ginseng and Gecko Powder], Zhen Wu Tang [True Warrior Decoction], and Jin Gui Shen Qi Wan [Kidney Qi Pill from Golden Cabinet].
- Postsurgery, postchemotherapy, and postradiation therapy for digestive tract cancer: Si Shen Wan [Four-Miracle Pill] and Zhen Ren Yang Zang Tang [Pure Man's Decoction for Nourishing the Organs].
- Late stage cancer, anemia, aversion to cold and deficient constitution: Tu Si Zi Wan [Cuscuta Seed Pill], You Gui Wan [Restore the Right (Kidney) Pill], and Nei Bu Wan [Tonify the Internal Pill].

Tonify Qi and Nourish Blood Method

This is indicated for shortness of breath, too lazy to talk, apathy, weakness, dizziness, vertigo, palpitation, spontaneous sweating, poor appetite, loose stool, pale or sallow complexion, pale tongue, white tongue coating, and deficient, big, and soggy pulse. There can also be prolapse of internal organs, purpura, bleeding with scanty pale blood and they might be recurrent.

Related conditions and modifications are

- Anemia, decreased WBC and platelet after chemotherapy: Ba Zhen Tang [Eight-Treasure Decoction] and Shi Quan Da Bu Tang [All-Inclusive Great Tonifying Decoction].
- Nausea, vomiting, and loose stool due to chemotherapy: Xiang Sha Liu Jun Zi Tang [Six-Gentleman Decoction with Aucklandia and Amomum] and Shen Ling Bai Zhu San [Ginseng, Poria, and Atractylodes Macrocephala Powder].
- Copious urination, chronic diarrhea, and sinking of internal organs: Bu Zhong Yi Qi Tang [Tonify the Middle and Augment the Qi Decoction].
- Decreased platelet count, purpura, and heavy menstruation or unceasing spotting: Gui Pi Tang [Restore the Spleen Decoction] and Dang Gui Bu Xue Tang Jia Wei [Augmented Tangkuei Decoction to Tonify the Blood].
- Exterior deficiency, spontaneous sweating, and catches cold easily: Yu Ping Feng San Jia Wei [Augmented Jade Windscreen Powder].

BRAIN TUMOR

This is discussed under headache, deep-rooted brain lumps, and consumption disease. It is caused by wind, cold, damp, heat pathogens invasion, turbid phlegm, and blood stasis obstructing the clear orifices leading to formation of tumor. In Western medicine, this is a general term referring to growth of tissue inside the skull. There are two types, primary and secondary. Secondary brain tumor is often seen in mammary and lung cancer. Beginning clinical symptoms include headache, nausea, vomiting, visual disturbance, somnolence, fright, syncope, and paralysis. After confirming the diagnosis, surgery is the treatment. As for TCM treatment, it improves symptoms before the surgery to prepare patients for surgery physically and mentally. After surgery, TCM treatments strengthen body constitution, stimulate recovery, and prevent recurrence. For bigger brain tumors, it metastasizes widely at later stages. Chinese herbal treatment is combined with chemotherapy to maintain the whole body function, strengthen anticancer ability, reduce pain such as treating headache, nausea, vomiting, and epilepsy. For blood stasis type of high intracranial pressure, treatments of invigorating blood, transforming stasis, promoting urination, and leaching out dampness are used. For recurrent epilepsy after surgery, treatments of clearing heat, expelling wind, and stopping spasms are used.

Treatments According to Syndrome Differentiation

Internal Obstruction of Phlegm–Damp Syndrome

Treatment Principle: Dry damp, transform phlegm, open orifices, and stop pain.

Use Dao Tan Tang [Guide Out Phlegm Decoction] plus Shi Chang Pu, Wei Ling Xian, Cang Zhu, Zhu Ru, Yu Jin, Zhu Ling, Gua Lou Pi, and Bai Zhu.

Related conditions and modifications are

- Palpitation with or without fright: Add Yuan Zhi, Fu Shen, and Bai Zi Ren.
- Greasy, yellow tongue coating, and wiry, rapid, and slippery pulse: Add Huang Qin and Tian Zhu Huang.

Qi Stagnation and Blood Stasis Syndrome

Treatment Principle: Invigorate blood, transform stasis, unblock collaterals, and scatter accumulation.

Use Tong Qiao Huo Xue Tang [Unblock the Orifices and Invigorate the Blood Decoction] plus Dan Shen, Sheng Di Huang, Zhi Ke, Jie Geng, Chuan Niu Xi, Di Long, Chuan Shan Jia, and Gou Teng.

Related conditions and modifications are

- Severe headache: Add Quan Xie, Wu Gong, and Di Bie Chong.
- Dizziness, forgetfulness, and insomnia: Remove She Xiang, and add He Shou Wu, Gou Qi Zi, and Suan Zao Ren.

Upward Disturbance of Liver Fire Syndrome

Treatment Principle: Clear Liver, drain fire, relieve toxicity, and transform phlegm.

Use Long Dan Xie Gan Tang [Gentiana Longdancao Decoction to Drain the Liver] plus Xia Ku Cao, Chuan Niu Xi, and Che Qian Cao.

Related conditions and modifications are

- Delirium or delirious speech: Add Shi Chang Pu, Tian Zhu Huang, and Shui Niu Jiao.
- Tongue stiffness and inability to talk: Add Qing Dai, Niu Huang, and Bo He.
- Headache and distending pain, blurry vision: Add Ju Hua and Man Jing Zi.

Liver and Kidney Yin Deficiency Syndrome

Treatment Principle: Nourish and tonify Liver and Kidney, and clear and benefit the head and eyes.

Use Qi Ju Di Huang Wan [Lycium Fruit, Chrysanthemum, and Rehmannia Pill] plus Huai Niu Xi, Gou Teng, Bai Shao Yao, Xuan Shen, Nu Zhen Zi, and Dai Zhe Shi.

Related conditions and modifications are

- Tidal fever: Add Xia Ku Cao and Shan Zhi Zi.
- Severe distending pain in the head: Add Bai Ji li, Long Gu, and Mu Li.
- For deviation of mouth and eyes: Add Jiang Can and Quan Xie.

Spleen and Kidney Yang Deficiency Syndrome

Treatment Principle: Warm and tonify Spleen and Kidney, tonify brain, and supplement marrow.

Use You Gui Wan [Restore the Right (Kidney) Pill] plus Fu Ling, Yi Zhi Ren, Ze Xie, Chuan Niu Xi, Che Qian Zi, Bu Gu Zhi, and Shan Ci Gu.

Related conditions and modifications are

- Sharp pain in the head: Add Chuan Xiong, Chi Shao Yao, and Shui Zhi.
- Low-grade fever and irritability: Add Mu Dan Pi and Zhi Mu.

Treatments According to Presenting Symptoms

Headache

This is often caused by phlegm stasis obstructing the clear orifices.

This formula consists of Quan Xie, Chuan Xiong, Jiang Can, Che Qian Zi, Zhu Ling, Ze Xie, and Mu Tong.

Increased Intracranial Pressure

This formula combines with Western medicine to decrease intracranial pressure by dehydrating agents.

The formula consists of Chi Xiao Dou, Qian Niu Zi, Da Fu Pi, Shang Lu, Fang Ji, Chuan Jiao Mu, and Ji Xue Teng.

CERVICAL CANCER

This is caused by accumulation of Qi stagnation, congealed phlegm, and damp–heat stasis obstructing the uterine collaterals over time. In Western medicine, this is a malignant tumor growing at the cervix which often occurs in middle-aged women. There are no significant signs at the beginning stage or there may be scanty white leukorrhea, bleeding after sexual intercourse or gynecological exam, and purulent or watery leukorrhea, sometimes with blood and foul smell. At the late stage, there can be abdominal pain, abdominal mass; frequent, painful, and hesitant urination; tenesmus, blood or pus in stool, fever, anemia, and emaciation.

At the early stage, surgery and radiation therapy are performed. TCM treatments begin after surgery to regulate and support Zheng Qi and anticancer therapy. At the late stage, the treatments support Zheng Qi, consolidate the root, relieve toxicity; strengthen the body's immune function to fight cancer, reduce recurrence and metastasis, and reduce the side effects of radiation therapy.

Liver Qi Stagnation Syndrome

Treatment Principle: Disperse Liver, regulate Qi, relieve toxicity, and anticancer.

Use Xiao Yao San [Relax Wanderer Powder] plus Yu Jin, Xiang Fu, E Zhu, Qing Pi, Sha Shen, Bai Hua She She Cao, Ban Zhi Lian, and Bai Jiang Cao.

Related conditions and modifications are

- Lower abdominal pain: Add Chuan Lian Zi and Wu Ling Zhi.
- Copious leukorrhea: Add Chun Gen Pi.
- Vaginal bleeding: Add Han Lian Cao and Jing Jie Tan.

- Spleen deficiency: Add Tai Zi Shen, Lian Zi, and Shan Yao.

Damp–Heat Stasis Toxins Syndrome

Treatment Principle: Clear heat, drain damp, relieve toxicity, and transform stasis.

Formula consists of Ye Ju Hua, Lian Qiao, Shan Ci Gu, Tu Bei Mu, Dan Shen, Zi Cao, Lou Lu, Bie Jia, Bai Hua She She Cao, Bai Jiang Cao, Chun Gen Pi, Ju Ye, Dang Gui, Shan Yao, and Yi Yi Ren.

Related conditions and modifications are

- Constipation with yellow thick coating: Add Da Huang, Zhi Ke, and Bing Lang.
- Loose stool with copious yellow leukorrhea: Add Cang Zhu, Tu Fu Ling, Hou Po, Huang Bai, and Qian Shi.
- Lumbar soreness, tidal fever, dizziness, and tinnitus: Add Sang Ji Sheng, Shu Di Huang, Nu Zhen Zi, and Shan Zhu Yu.
- Vaginal bleeding: Add Han Lian Cao, Xian He Cao, Di Yu Tan, Sheng Di Huang, and E Jiao.
- Pus and blood in stool: Add Qin Pi, Huai Hua, Bai Shao Yao, Mu Xiang, and Bai Tou Weng.

Liver and Kidney Yin Deficiency Syndrome

Treatment Principle: Nourish Liver and Kidney, clear heat, and relieve toxicity.

Use Zhi Bai Di Huang Wan [Anemarrhena, Phellodendron, and Rehmannia Pill] plus Er Zhi Wan [Two-Ultimate Pill] and Tai Zi Shen, Sheng Di Huang, Bie Jia, Mu Dan Pi, Ban Zhi Lian, Bai Hua She She Cao, Wu Gong, and Lu Feng Fang.

Related conditions and modifications are

- Fatigue, apathy, shortness of breath, and heart palpitation: Remove Zhi Mu and Huang Bai and add Xi Yang Shen and Huang Qi.
- Copious vaginal bleeding: Add E Jiao, fried Pu Huang, and Ce Bai Tan.
- Lumbar soreness and pain: Add Xu Duan, Sang Ji Sheng, and Du Zhong.
- Severe lumbar pain: Add Xue Jie, Ru Xiang, and Mo Yao.

Spleen and Kidney Yang Deficiency Syndrome

Treatment Principle: Warm and tonify Spleen and Kidney, relieve toxicity, and expel damp.

Use You Gui Wan [Restore the Right (Kidney) Pill] plus Bu Gu Zhi, Dang Shen, Fu Ling, Ba Ji Tian, Huang Qi, Chi Xiao Dou, and Yi Yi Ren.

Related conditions and modifications are

- Foul smelling leukorrhea: Add Bai Jiang Cao, Bai Hua She She Cao, and Hong Teng.
- Dry stool: Add Rou Cong Rong, Dang Gui, and Huo Ma Ren.
- Nausea and vomiting due to chemotherapy: Add Sha Ren and Ban Xia.

COLORECTAL CANCER

This is often caused by improper diet generating damp and heat, or evil toxins accumulating at colon, and accumulation of phlegm–damp–heat toxins over time forming mass. In Western medicine, cancer of the large intestine includes colon cancer and rectal cancer. They occur more often in people aged 30 to 60 years and more so in males. It is believed to relate to genetics, high-fat, low-fiber diets, and chronic Large Intestinal pathologies. At the early stage, there are no significant symptoms. After certain pathological changes of the large intestine, then there are changes in shape of the stool, blood in stool, and abdominal pain or obstructed symptoms. Other symptoms include fatigue, anemia, and weight loss. At the late stage there can be metastasis to Liver, Lung, and brain; or spreading to urinary bladder, uterus, and peritoneum.

Colorectal cancer usually progresses slowly and metastasis occurs at the later stage. Therefore, once the diagnosis is confirmed, surgery should be done to remove the cancer and followed by chemotherapy or radiation therapy. TCM treatments can prevent reoccurrence or metastasis of the cancer, and reduce toxic side effects of chemotherapy and radiation therapy. The treatment principles are to support Zheng Qi, fight cancer, and relieve toxicity.

Steaming Damp–Heat

Treatment Principle: Clear heat, drain damp, and relieve toxicity and anticancer treatment.

Use Ge Gen Huang Qin Huang Lian Tang [Kudzu, Coptis, and Scutellaria Decoction] plus Bai Jiang Cao, Hong Teng, Ban Zhi Lian, Bai Hua She She Cao, Yi Yi Ren, Qin Pi, Bai Shao Yao, Shan Ci Gu, and Mu Xiang.

Related conditions and modifications are

- Abdominal distention: Bing Lang, Zhi Ke, and Hou Po.
- Blood in stool: Add Huai Hua, Shan Zhi Zi, Xian He Cao, and San Qi powder.
- Metastasis to lymph nodes: Add Tie Shu Ye and E Zhu.
- Lung metastasis: Add Quan Gua Lou, Long Kui, and Yi Yi Ren.

Spleen Deficiency with Excess Damp Syndrome

Treatment Principle: Strengthen Spleen, transform damp; anticancer treatment and relieve toxicity.

Use Shen Ling Bai Zhu San [Ginseng, Poria, and Atractylodes Macrocephala Powder] plus Zhu Ling, Bai Shao Yao, Gou Qi Zi, E Zhu, Dan Shen, Ji Nei Jin, and Xian He Cao.

Related condition and modification are

- Severe diarrhea: Add Zhi He Zi and Rou Dou Kou.

Liver and Kidney Yin Deficiency

Treatment Principle: Nourish Kidney, soften Liver, clear heat, and relieve toxicity.

Use Yi Guan Jian [Linking Decoction] plus Nu Zhen Zi, Shan Zhu Yu, Bai Hua She She Cao, Ban Zhi Lian, and Xian He Cao.

Related conditions and modifications are

- Abdominal distention: Remove Shu Di Huang and add Zhi Ke and Bing Lang.
- Dry stool: Add Zhi Da Huang and Huo Ma Ren.
- Loose stool: Remove Shu Di Huang and add Bai Zhu and Fu Ling.

Qi Stagnation and Blood Stasis Syndrome

Treatment Principle: Regulate Qi, transform stasis, and unblock Fu organs.

Use Tiao Wei Cheng Qi Tang [Regulate the Stomach and Order the Qi Decoction] plus Shi Xiao San [Sudden Smile Powder] and Tao Ren, Hong Hua, Hong Teng, Bai Jiang Cao, Ban Zhi Lian, Bai Hua She She Cao, and Bing Lang.

Related conditions and modifications are

- Distending pain at sides of the lower abdomen: Add Yan Hu Suo and Chuan Lian Zi.
- Fever and thirst: Add Chai Hu and Huang Lian.
- Dry mouth and red tongue: Add Sheng Di Huang and Xuan Shen.

ESOPHAGEAL CANCER

This is discussed under dysphagia in TCM. It is often caused by accumulated Qi stagnation, blood stasis, and phlegm–damp at the esophagus. In Western medicine, esophageal cancer is a common malignant tumor of the digestive tract which is seen less in females than in males between the age of 40 to 70 years. At the middle and late stages of this disease, the main clinical manifestations are progressive swallowing difficulties, chest and back soreness and pain, and vomiting saliva or food. In severe cases, there is anemia, dehydration, and emaciation due to difficulty swallowing.

Surgery should be done as early as possible. After surgery, TCM treatments can support Zheng Qi, relieving toxicity, and stabilize the effectiveness of anticancer therapies. For esophageal cancer at the upper or middle part of the esophagus or when surgery is not suitable, then radiation therapy is the main treatment. TCM treatment mainly prevents the side effects of radiation therapy. Thus, taking herbal decoctions after surgery can support Zheng Qi and other therapies.

Treatments According to Syndrome Differentiation

Liver Qi Stagnation Syndrome

Treatment Principle: Regulate Qi, spread Liver, transform stasis, and unblock accumulation.

Chai Hu Shu Gan San [Bupleurum Powder to Spread the Liver] plus Xiang Fu, Zhe Bei Mu, Shan Dou Gen, Xie Bai, Dan Shen, Gua Lou Pi, Bai Hua She She Cao, Yi Yi Ren, and Ban Zhi Lian.

Related conditions and modifications are

- Dry mouth and scanty fluid: Add Tian Hua Fen, Mai Men Dong, and Lu Gen.
- Dry stool: Add Huo Ma Ren and Sheng Da Huang.

Spleen Deficiency and Phlegm–Damp Syndrome

Treatment Principle: Strengthen Spleen and expel phlegm.

Use Shen Ling Bai Zhu San [Ginseng, Poria, and Actractylodes Macrocephala Powder] plus Huang Yao Zi, Hou Po, Ban Xia, Yi Yi Ren, and He Tao Shu Zhi.

Internal Obstruction of Stasis and Toxins Syndrome

Treatment Principle: Invigorate blood, transform stasis, eliminate accumulation, and unblock the knots.

This formula consists of Sheng Di Huang, Bai Shao Yao, Dan Shen, Yu Jin, Wei Ling Xian, Tian Hua Fen, E Zhu, Wu Ling Zhi, Pu Huang, Zhi Ke, Tao Ren, Long Kui, Huang Yao Zi, Bai Hua She She Cao, and Sha Shen.

Qi and Yin Deficiency Syndrome

Treatment Principle: Benefit Qi, nourish Yin, support Zheng Qi, and relieve toxicity.

Use Sha Shen Mai Dong Tang [Glehnia and Ophiopogonis Decoction] plus Er Zhi Wan [Two-Ultimate Pill] plus Tai Zi Shen, Huang Jing, Gou Qi Zi, Huang Qi, Ji Nei Jin, Bai Hua She She Cao, and He Tao Shu Zhi.

Related conditions and modifications are

- Five-center heat: Add Qing Hao, Zhi Mu, Mu Dan Pi, and Sheng Di Huang.
- Tidal fever and night sweats: Add Zhi Mu, Gui Ban, and Sheng Mu Li.
- Edema with scanty urine: Add Fu Ling, Che Qian Zi, and Ze Xie.

Treatments for Complications After Surgery for Esophageal Cancer

Diarrhea

Use Tong Xie Yao Fang [Important Formula for Painful Diarrhea] plus Ge Gen, Fu Ling, and Che Qian Zi.

Related condition and modification are

- Chronic diarrhea: Add Bu Gu Zhi, Wu Wei Zi, and Zhi Fu Zi.

Cough

Use Er Chen Tang [Two-Cured Decoction] plus Kuan Dong Hua, Bai Bu, Bai Jie Zi, and Zi Su Zi.

Reflux Esophagitis

Use Zuo Jin Wan [Left Metal Pill] plus Gua Lou Ren, Ban Xia, Hou Po, Xie Bai, and Fo Shou.

KIDNEY CANCER

This is often caused by damp–heat toxin accumulation, Qi stagnation, and blood stasis. In Western medicine, it often occurs in people who are above 50 and more so in males than females (2:1). The cause is still not clear. At the early stage, there is no significant sign of kidney cancer. At middle and late stages, the manifestations are blood in urine, lumbar pain, abdominal mass, and discomfort throughout the body. At late stage, there is also metastasis to Lung, bone, and brain.

TCM treatments are given before and after the surgery, during or after chemotherapy and radiation therapy. Moving Qi, invigorating blood, and promoting urination are the methods to stimulate the expelling of cancerous toxins and toxins from drugs. It can also support Zheng Qi and strengthen immunity.

There are several key points in using herbs. Herbs should be mild and moderate. Avoid using harsh and strong herbs. Commonly used herbs Bai Hua She She Cao, Yi Yi Ren, Zhong Jie Feng, Bai Jiang Cao, Fu Ling, Zhu Ling, and Shi Wei have anticancer properties, can promote urination functions to clear and eliminate cancerous toxins. These herbs can be used in larger dosages over the long term. Late stage treatments focus on regulating Spleen and Stomach to prolong life by using sweet and bland herbs to bolster Spleen, acrid and fragrant herbs to stimulate Stomach, and aromatic herbs to awaken Spleen.

Damp–Heat Accumulating in Lower Jiao

Treatment Principle: Clear heat, drain damp, benefit Kidney, and unblock Lin.

Use Ba Zheng San [Eight-Herb Powder for Rectification] plus Hai Jin Sha, Huang Bai, Sheng Di Huang, Chi Shao Yao, Huang Qi, Tu Fu Ling, and Bai Mao Gen.

Related conditions and modifications are

- Low-grade fever: Add Zhi Mu and Qing Hao.
- Copious blood in urine: Add Da Ji and Xiao Ji.
- Mass and swelling in abdomen and lumbar region: Add Chuan Shan Jia and Jiang Can.

Internal Obstruction of Blood Stasis

Treatment Principle: Invigorate blood, transform stasis, regulate Qi, and scatter accumulation.

Use Tao Hong Si Wu Tang [Four-Substance Decoction with Peach Pit and Safflower] plus Mu Xiang, Yan Hu Suo, Zhi Ke, Dan Shen, Xiang Fu, Chi Shao Yao, Ban Zhi Lian, and Long Kui.

Related conditions and modifications are

- Incomplete sensation with urination: Add Che Qian Zi, Mu Tong, and Lu Lu Tong.
- Constipation: Add Lai Fu Zi and Da Huang. If bowel movement is still not regular, add Mang Xiao.
- Severe pain: Add Shi Xiao San [Sudden Smile Powder].

Spleen and Kidney Deficiency Syndrome

Treatment Principle: Tonify Spleen and Kidney, soften hardness, and scatter accumulation.

Use Ba Zhen Tang [Eight-Treasure Decoction] plus Huang Qi, Di Gu Pi, Ban Zhi Lian, Jiang Can, Tian Hua Fen, Nu Zhen Zi, and Shi Hu.

Related conditions and modifications are

- More Yang deficiency: Change to You Gui Wan [Restore the Right (Kidney) Pill]; remove Lu Jiao Jiao, Fu Zi, and Rou Gui, and add Bu Gu Zhi, Bai Zhi, Huang Jing, and Huai Niu Xi.
- Small quantity of food intake and abdominal distention: Add Ji Nei Jin, Shen Qu, and Lai Fu Zi.
- Copious blood in urine: Add Da Ji, Xiao Ji, Xian He Cao, and San Qi powder.
- Bitter taste and dry mouth, and persistent low grade fever: Add Huang Qin, Chai Hu, and Qing Hao.

LIVER CANCER

This is often caused by knotted Qi stagnation, blood stasis, damp–heat, stagnant toxins, and phlegm Yin (congested fluid) which obstruct the Liver meridian and lead to accumulation over a period of time. In Western medicine, liver cancer occurs more in young and middle-aged people and more so in males than females. In general, chronic liver diseases such as hepatitis B, hepatitis C, and liver cirrhosis are related to liver cancer. At the early stage, patient may not have significant symptoms. At the middle and late stages, the manifestations are pain at right hypochondriac area, distention and fullness at epigastrium, decreased appetite, progressive swelling and hardening of liver, fever, fatigue, and emaciation. At late stage, there often are ascites, jaundice, and metastasis to lung and brain.

Surgery is the first choice in early stages. At middle and late stages, when patients lose the opportunity to have surgery, radiation therapy and chemotherapy can be combined with Chinese herbal medicine. The aim of the herbs is to increase the anticancer effect and reduce toxic side effects of radiation therapy and chemotherapy.

Treatments According to Syndrome Differentiation

Liver Qi Stagnation and Spleen Deficiency Syndrome

Treatment Principle: Disperse Liver, strengthen Spleen, support Zheng Qi, and relieve toxicity.

Use Xiao Yao San [Relax Wanderer Powder] plus Dang Shen, Ban Xia, E Zhu, Shan Yao, Dan Shen, Bie Jia, Mu Li, Shi Shang Bai, Ma Bian Cao, and Ba Yue Zha.

Related conditions and modifications are

- Yang deficiency after surgery and chemotherapy: Remove Shi Shang Bai, Ma Bian Cao, and E Zhu and add Huang Qi, Gui Zhi, and Da Zao.
- Abdominal distention, belching, and nausea: Add Da Fu Pi, Bing Lang, Hou Po, and Sha Ren.
- Poor appetite, stifling sensation at chest, and greasy tongue coating: Add Huo Xiang, Cang Zhu, Hou Po, and Yi Yi Ren.

Damp–Heat, Stasis, and Toxin Syndrome

Treatment Principle: Clear heat, drain damp, cool blood, and transform stasis.

Use Long Dan Xie Gan Tang [Gentiana Longdancao Decoction to Drain the Liver] plus Ban Zhi Lian, Shi Shang Bai, Bai Hua She She Cao, Bai Jiang Cao, Bie Jia, Dan Shen, and Sheng Di Huang.

Related conditions and modifications are

- Constipation and abdominal distention: Add Da Huang. If symptoms still persist, add Mang Xiao and Zhi Shi.
- Tongue center dryness: Add Tian Hua Fen and Lu Gen.
- Burning heat sensation at Liver area: Minus Dang Gui and Chai Hu, add Shi Gao and Zhi Mu.
- Jaundice: Add Yin Chen Hao and Da Huang.
- Foul smell of stool, burning sensation at anus, and frequent bowel movement: Remove Chai Hu, Ban Zhi Lian, Shan Zhi Zi, Dang Gui, and Sheng Di Huang, and add Ge Gen, Bai Tou Weng, Qin Pi, Bai Shao Yao, Mu Xiang, He Zi, and Huang Lian.

Qi Stagnation and Blood Stasis Syndrome (Commonly Seen with Liver Cirrhosis)

Treatment Principle: Regulate Qi, transform stasis, relieve toxicity, and scatter accumulation.

Use Tao He Cheng Qi Tang [Peach Pit Decoction to Order the Qi] plus Si Wu Tang [Four-Substance Decoction] Chai Hu, Zhi Ke, San Leng, E Zhu, Zhe Chong, Bie Jia, Dan Shen, Lou Lu, Ma Bian Cao, Wa Leng Zi, and Shi Jian Chuan.

Related conditions and modifications are

- Fatigue and apathy: Add Dang Shen, Tai Zi Shen, and Bai Zhu.
- Aversion to cold and cold limbs: Add Gui Zhi.

Yin Deficiency with Internal Heat Syndrome (Seen in Patients Who Take Diuretics after Chemotherapy and Radiation Therapy)

Treatment Principle: Nourish Liver and Kidney and clear heat.

Use Qing Hao Bie Jia Tang [Artemisia Annua and Soft-Shelled Turtle Shell Decoction] plus Sha Shen, Mai Men Dong, Xuan Shen, Gou Qi Zi, Nu Zhen Zi, Bai Hua She She Cao, and Bai Ying.

Related conditions and modifications are

- Tidal fever: Add Huang Bai and Hu Huang Lian.
- Mirror tongue with mild restlessness: Remove Qing Hao and add Wu Wei Zi and Shan Yao.
- Apathy, shortness of breath, palpitation, and spontaneous sweating: Remove Qing Hao and add Xi Yang Shen, Huang Qi, Wu Wei Zi, and Shan Zhu Yu.
- Epistaxis, bleeding gum, subcutaneous bleeding, and dark red tongue: Add Shui Niu Jiao, Zi Cao, Han Lian Cao, Bai Mao Gen, and E Jiao.

Treatments According to Symptoms

Pain

Use 30 g each of Zhi Chuan Wu, Huang Yao Zi, Qi Ye Yi Zhi Hua, Yan Hu Suo, Lu Gen, and Shan Ci Gu, and 6 g of Bing Pian. Grind all ingredients into powder. Take 1 to 2 g each time, one to two times per day with Xian He Cao and Gan Cao decoction.

Ascites

Strengthen Spleen, regulate Qi, leach out damp, and promote urination.

Use Che Qian Zi, Fu Ling, Ze Xie, Mu Tong, Chuan Jiao Mu, Zhu Ling, Chi Xiao Dou, Bing Lang, Da Fu Pi, Bai Zhu, Ting Li Zi, Sang Bai Pi, Chuan Niu Xi, Yi Mu Cao, Ban Bian Lian, Hua Shi, Gui Zhi, and Long Kui.

LUNG CANCER

It is caused by stasis and heat damaging Yin, and Qi deficiency with damp–phlegm. In Western medicine, this is the most commonly seen primary malignant tumor in lung. It often begins at the bronchial wall of people who are above 40. The ratio of occurrences in males and females is 5 to 1. The main manifestations are irritating cough, hemoptysis, chest pain, fever, shortness of breath or even hoarseness, edema at upper limbs and face, and fluid accumulation in chest. For early stage (limited) in the lung cancer patient, surgery is the first choice. Then, TCM treatment with chemotherapy is used after surgery. If patient has heart, lung, liver, and kidney pathological conditions that are not suitable for surgery, then radiation therapy with Chinese herbs can be used. When the pathological changes are diffused with metastasis, chemotherapy is combined with Chinese herbs. TCM treatments strengthen the anticancer effect of the pharmaceutical drugs, reduce the toxic side effects of chemotherapy and radiation therapy, and assist patients in completing the therapeutic course.

Yin Deficiency with Internal Heat Syndrome

Treatment Principle: Nourish Yin, moisten Lung, clear heat, and stop bleeding.

Use Sha Shen Mai Dong Tang [Glehnia and Ophiopogonis Decoction] plus Qing Hao, Bie Jia, Bai Hua She She Cao, Ban Zhi Lian, Shi Da Gong Lao Ye, Xing Ren, Chuan Bei Mu, and Pi Pa Ye.

Related conditions and modifications are

- Hemoptysis: Add Xian He Cao and San Qi powder.
- Copious sticky phlegm and pus: Remove Chuan Bei Mu and add Yu Xing Cao, Dong Gua Zi, Lian Qiao, Huang Qin, and Wei Jing.
- Chest pain: Add Tao Ren, Yu Jin, Si Gua Luo, and San Qi powder.
- Night sweats: Add Mu Li, Fu Xiao Mai, and Wu Wei Zi.

Spleen Deficiency with Phlegm Damp Syndrome

Treatment Principle: Benefit Qi, strengthen Spleen, transform phlegm, and relieve toxicity.

Use Shen Ling Bai Zhu San [Ginseng, Poria, and Atractylodes Macrocephala Powder] plus Kuan Dong Hua, Zhu Ling, Ting Li Zi, Xie Bai, Gua Lou Pi, Bai Hua She She Cao, and Tian Nan Xing.

Related condition and modification are

- Stifling sensation in chest and copious phlegm: Add Su Zi, Bai Jie Zi, and Huang Qin.

Qi Stagnation and Blood Stasis Syndrome

Treatment Principle: Regulate Qi, transform stasis, eliminate accumulation, and soften hardness.

Use Tao Hong Si Wu Tang [Four-Substance Decoction with Peach Pit and Safflower]; remove Shu Di Huang; and add Hai Zao, Kun Bu, Gua Lou, Tie Shu Ye, Dan Shen, Mu Li, Bie Jia, Yi Yi Ren, Bai Hua She She Cao, E Zhu, and San Leng.

Related conditions and modifications are

- Hemoptysis: Remove Dang Gui and add Qing Dai and San Qi powder. If blood is bright red and copious, then remove San Leng and E Zhu.
- Chest and hypochondriac pain: Add Yan Hu Suo, Ru Xiang, Mo Yao, and Si Gua Luo.
- Swollen lymph nodes above the clavicle and under armpit: Add Huang Yao Zi and Shan Ci Gu.

Qi and Yin Deficiency Syndrome

Benefit Qi, nourish Yin, and support Zheng Qi and anticancer therapies.

Use Sha Shen Mai Dong Tang [Glehnia and Ophiopogonis Decoction] plus Si Jun Zi Tang [Four-Gentleman Decoction] with Tai Zi Shen, Bai He, Tian Men Dong, Huang Qi, Ban Zhi Lian, Long Kui, Bai Hua She She Cao, Wu Wei Zi, and Xing Ren.

Related condition and modification are

- Deficient heat damaging collaterals and hemoptysis: Add Bai Ji powder, Mu Dan Pi, Bai Mao Gen, and Han Lian Cao.

Yin and Yang Deficiency Syndrome

Treatment Principle: Warm Kidney, nourish Yin, support Zheng Qi, and anticancer therapies.

Use You Gui Yin [Restore the Right (Kidney) Decoction] plus Rou Cong Rong, Yin Yang Huo, Suo Yang, Ba Ji Tian, Bu Gu Zhi, Tai Zi Shen, Hong Shen, Sha Shen, Xian He Cao, and Dong Chong Xia Cao.

Related conditions and modifications are

- Wheezing with movements: Add Wu Wei Zi, Zi Su Zi, and Ge Jie powder.
- Edema: Add Che Qian Zi, Fu Ling, Ze Xie, Chi Xiao Dou, and Huang Qi.

MALIGNANT LYMPHOMA

This disease is caused and formed by accumulation of external wind–heat blood dryness, congealed cold–damp and stagnation, internal Liver Qi stagnation transforming into fire, phlegm stagnation transforming into fire; Qi stagnation and blood stasis. In Western medicine, it is described as being divided into Hodgkin's and non-Hodgkin's syndromes. Clinical manifestations are progressive enlargement of the lymph nodes and symptoms caused by the compression of these lymph nodes. It can almost invade every tissue and organ of the body.

During surgery, chemotherapy, or radiation therapy, TCM treatment primarily supports Zheng Qi and secondarily clears heat, transforms phlegm, expels stasis, and scatters accumulations. On the basis of syndrome differentiation, the following anticancer herbs can be added to the formulas: Bai Hua She She Cao, Xia Ku Cao, Zhe Bei Mu, Shan Ci Gu, Chuan Shan Jia, Xuan Shen, Wu Gong, Huang Yao Zi, Jiang Can, Zao Jiao Ci, Mao Zhua Cao, Kun Bu, and Hai Zao.

Wind–Heat Blood Dryness Syndrome

Treatment Principle: Disperse wind, relieve toxicity, moisten dryness, and scatter accumulation.

Use Fang Feng Tong Shen San [Ledebouriella Powder that Sagely Unblocks] plus Zeng Ye Tang [Increase the Fluids Decoction], Shan Ci Gu, and Bai Hua She She Cao.

Internal Congealed Cold–Phlegm Syndrome

Treatment Principle: Warm and transform cold phlegm, soften hardness, and scatter accumulation.

Use Yang He Tang [Yang-Heartening Decoction] plus Zao Jiao Ci and Tian Nan Xing.

Knotted Phlegm and Stasis Syndrome

Treatment Principle: Transform phlegm, expel stasis, relieve stagnation, and scatter accumulation.

Use Er Chen Tang [Two-Cured Decoction] plus Shi Xiao San [Sudden Smile Powder], Xing Ren, Zhe Bei Mu, San Qi, Bai Shao Yao, Xia Ku Cao, Lu Feng Fang, Huang Yao Zi, and Mao Zhua Cao.

Liver and Kidney Yin Deficiency, Qi and Blood Deficiency Syndrome

Treatment Principle: Tonify Qi and blood and tonify Liver and Kidney.

Use Ba Zhen Tang [Eight-Treasure Decoction] plus Xiang Fu, Chen Pi, Kun Bu, Zhe Bei Mu, Bie Jia, Hong Hua, and Sheng Ma.

Above four syndromes can be modified according to the following:

- Fever: Add Yin Chai Hu, Qing Hao, and Shi Gao.
- Persistent high fever: Add Han Shui Shi, Niu Huang, and Shui Niu Jiao.
- Night sweat: Add Fu Xiao Mai, calcined Long Gu, calcined Mu Li, Huang Qi, and Wu Wei Zi.
- Itchy skin: Add Di Fu Zi, Bai Xian Pi, and Ku Shen.

- Anemia: Add E Jiao, Lu Jiao Jiao, Gu Sui Bu, and Xian He Cao.
- Swollen Liver and Spleen: Add Da Huang Zhe Chong Wan [Rhubarb and Eupolyphaga Pill].

MAMMARY CANCER

This disease is often caused by internal injuries due to seven emotions, which lead to Qi stagnation, congealed phlegm, and blood stasis; and at the last stage, internal accumulation of evil toxins affecting the Zang organs and depleted essence lead to termination of life. In Western medicine, it is described as more commonly seen in females between the ages of 40 to 59. This can be related to estrogen level in the body, especially dysfunction of the ovaries.

The key for treating mammary cancer is early diagnosis and early treatment (surgery). TCM treatments strengthen the patient's constitution, improves quality of life, and prolongs life. It can also reduce the toxic side effects of chemotherapy and radiation therapy.

Treatments According to Stages

Early Stage

For small lumps in the breast, no pain, unsmooth surface, and unclear border. Zheng Qi remains still strong.

Move Qi, eliminate phlegm, invigorate blood, transform stasis, soften hardness, and scatter accumulation.

Use Lou Bei Er Chen Tang [Two-Cured Decoction with Trichosanthes and Fritillaria] plus Chai Hu Shu Gan San [Bupleurum Powder to Spread the Liver] and Qing Pi, E Zhu, Shan Ci Gu, Lu Feng Fang, Xia Ku Cao, Chuan Shan Jia, Yu Jin, and Shi Jian Chuan.

Related conditions and modifications are

- Hard lumps with dull pain: Add San Leng, Chi Shao Yao, and Dan Shen.
- Purple and dark lumps with visible capillaries, easily angered, and irritated, red face, and red eyes: Add Mu Dan Pi, Sheng Di Huang, Xuan Shen, Mu Li, and Bie Jia.
- Irregular menstrual cycle, scanty menstruation, cold sensation, and thin pulse: Add Yin Yang Huo, Ba Ji Tian, and Lu Jiao Jiao.

Late Stage

For big lumps in the breast, raised nipple, swelling, redness, and pain or edema with skin that looks like an orange peel. At this stage, Qi is exhausted and fluids are deficient. Body Zheng Qi is deficient and evil pathogens excessive.

Support Zheng Qi; anticancer therapies expel evil pathogens, tonify Qi and blood, and transform phlegm and, stasis.

Use Si Miao Yong An Tang [Four-Valiant Decoction for Well-Being] plus Huang Qi, Dang Shen, Shu Di Huang, Ji Xue Teng, Chen Pi, Bai Zhu, Shan Yao, Bai Hua She She Cao, and Ban Zhi Lian.

Related conditions and modifications are

- Liver and Kidney Yin deficiency: Add Er Zhi Wan [Two-Ultimate Pill], Wu Wei Zi, and Shan Zhu Yu.
- Severe hardness of the lumps: Add Chuan Shan Jia, Jiang Can, Wang Bu Liu Xing, Ru Xiang, Mo Yao, and San Qi powder.
- Ulcerations: Add Tu Fu Ling, Zao Xiu, Xuan Shen, Pu Gong Ying, and Zi Cao.
- Qi deficiency: Add Ren Shen.
- Blood deficiency: Add E Jiao and Gui Ban Jiao.

Treatments According to Symptoms

Edema at Upper Limbs after Surgery

Formula consists of Huang Qi, Ji Xue Teng, Sang Zhi, Dan Shen, Wei Ling Xian, Zhu Ling, Fu Ling, Lu Lu Tong, Si Gua Luo, and Fang Ji.

Treatment Given with Radiation Therapy

Formula consists of Sha Shen, Huang Qi, Tai Zi Shen, Nu Zhen Zi, Ling Zhi, Tu Si Zi, Zhu Ling, Mai Men Dong, Bai Zhu, Gou Qi Zi, and Xuan Shen.

Related condition and modification are

- Nausea, vomiting, and poor appetite after radiation therapy: Add Dai Zhe Shi, Yi Yi Ren, Yu Zhu, Shi Hu, and Chen Pi.

Treatment Given with Chemotherapy

Formula consists of Ren Shen, Huang Qi, Dang Gui, Bai Zhu, Yin Hua Teng, Ji Xue Teng, Fu Ling, Shan Ci Gu, and Lu Feng Fang.

NASOPHARYNGEAL CANCER

In TCM, this is often caused by phlegm stasis and toxic heat attacking the Upper Jiao with wind obstructing the clear orifices and forming swollen mass. In Western medicine, this is described as referring to cancerous swelling at the mucous membrane of the nose and pharynx. This occurs more in males than females. The clinical manifestations often occurs during middle and late stages with bloody nasal discharge from the nose; or with cough, nasal congestion, migraine headache; swelling and lumps at neck; or with tinnitus, decreased hearing, drooping eyelids, double vision, and hoarseness. Radiation therapy can kill and eliminate cancerous cells, thus it is the best choice. TCM treatments can moderate the symptoms such as heat burning and damaged Yin, and damaged salivary gland that are caused by radiation therapy. After radiation therapy, TCM treatments focus on supporting Zheng Qi, relieving toxins, clearing residual cancerous cells, and preventing recurrence and metastasis. If patient has purple tongue, it indicates poor prognosis as compared to light red tongue, thus invigorating blood, relieving toxicity and anticancer treatments are recommended. Even after the patient feels better, with decreased or disappearance of lumps, they should continue following this treatment for few months.

Treatments According to Syndrome Differentiation

Knotted Phlegm Stasis Syndrome

Treatment Principle: Move Qi, transform stasis, eliminate phlegm, and soften hardness.

Use Qing Qi Hua Tan Wan [Clear the Qi and Transform Phlegm Pill] plus Lian Qiao, Zhe Bei Mu, Jiang Can, Ci Ji Li, Ye Ju Hua, Cang Er Zi, and Shi Shang Bai.

Related conditions and modifications are

- Headache or migraine: Add Chuan Xiong, Wu Gong, Tian Ma, Lu Feng Fang, and Bai Shao Yao.
- Copious epistaxis: Remove Cang Er Zi and add Xian He Cao, Ce Bai Ye, and fried Shan Zhi Zi.
- Swollen lymph nodes in neck: Add Quan Xie, Wu Gong, E Zhu, Zhe Chong, and Shan Ci Gu.

Accumulation of Fire Toxins Syndrome

Treatment Principle: Drain fire, relieve toxicity, and scatter swelling.

Use Long Dan Xie Gan Tang [Gentiana Longdancao Decoction to Drain the Liver] plus Qing Dai, Bai Shao Yao, Shan Dou Gen, Sang Ji Sheng, Gou Teng, Xia Ku Cao, Long Kui, Shi Shang Bai, and Chong Lou.

Related condition and modification are

- Severe headache: Add Shi Jue Ming, Chuan Niu Xi, Shan Yang Jiao, and Yi Mu Cao.

Excess Evil Pathogens and Deficient Zheng Qi Syndrome

Treatment Principle: Nourish Kidney, strengthen Spleen, relieve toxicity, and eliminate accumulation.

Use Zhu Jing Wan [Preserve Vistas Pill] plus Shen Ling Bai Zhu San [Ginseng, Poria, and Atractylodes Macrocephala Powder] plus Dan Shen, Huang Jing, Sha Shen, Tu Si Zi, Gou Qi Zi, Xian He Cao, He Tao Shu Zhi, and Chong Lou.

Related conditions and modifications are

- Dizziness, vertigo, and tinnitus: Add Rou Cong Rong, Ba Ji Tian, and Yin Yang Huo.
- Loose stool and poor appetite: Add Shan Yao, He Zi, Sha Ren, and Bai Dou Kou.

Post-Radiation Treatments

After radiation therapy, some patients present with symptoms of damaged Yin due to heat toxins. Thus, nourishing-Yin and moistening-dryness herbs can be used, such as Mai Men Dong, Sha Shen, Xuan Shen, Shi Hu, Lu Gen, Tian Hua Fen, Tian Men Dong, Sheng Di Huang, Bai Mao Gen, Shan Yao, Bai Shao Yao, and Wu Mei. Liu Wei Di Huang Wan [Six-Ingredient Pill with Rehmannia] or Qi Ju Di Huang Wan [Lycium Fruit, Chrysanthemum, and Rehmannia Pill] can also be added. If Qi is exhausted due to damaged Yin, then Qi tonics such as Tai Zi Shen, Dang Shen, Huang Qi, and Xi Yang Shen can be added. If stasis spots appear at the sides of the tongue after radiation therapy, blood-moving herbs such as Dan Shen, Chi Shao Yao, and Mu Dan Pi can be added. Blood crackers are not appropriate in this case. Clearing heat and relieving toxicity herbs such as Lian Qiao, Jin Yin Hua, Zhi Mu, and Shan Zhi Zi can be used. However, cold, bitter, and transform dryness herbs are not suggested for use.

OVARIAN CANCER

This is often caused by accumulation of internal injuries due to seven emotions, Qi stagnation; blood stasis obstructing the collaterals over time, causing blood stasis at the sides of lower abdomen; and formation of mass. Spleen deficiency can also cause phlegm accumulation and mass formation. In Western medicine, this can occur at any age. The cause is still not clearly known. Clinical manifestations at early stages include only lower abdominal discomfort, followed by bloating, increased waist with increasing tumor size, lower abdominal mass, emaciation, irregular vaginal bleeding, and severe bleeding after menopause.

Main treatment is surgery combined with chemotherapy and radiation therapy. However, the location of the disease is often deep, and it is hard to release all cancerous toxins. TCM treatments focus on clearing and differentiating excess and deficient conditions, supporting Zheng Qi, and expelling evil pathogens.

Excess Evil Pathogen Syndrome

Qi Stagnation and Blood Stasis

Treatment Principle: Move Qi, invigorate blood, soften hardness, and scatter accumulation.
Use Tong Yu Jian [Unblock the Stasis Decoction] plus Tu Fu Ling and Tao Ren.

Congealed Phlegm and Damp and Accumulation

Treatment Principle: Transform phlegm, eliminate damp, move Qi, and scatter accumulation.
Use Hai Zao Yu Hu Tang [Sargassum Decoction for the Jade Flask] plus Wang Bu Liu Xing.

Accumulation and Stagnation of Water and Damp

Treatment Principle: Drain water and guide damp.
Use Shu Zao Yin Zi [Spread and Chisel Decoction] plus Long Kui, Bai Ying, and Ban Bian Lian.

Accumulation and Knotted Damp and Heat

Treatment Principle: Clear heat and transform damp.
Use San Miao Wan [Three-Marvel Pill] plus Huang Lian Jie Du Tang [Coptis Decoction to Relieve Toxicity] and Ku Shen, Bai Hua She She Cao, Ban Bian Lian, and Lou Lu.

Deficient Zheng Qi Syndrome

Yin Deficiency with Internal Heat Treatment Principle:

Nourish Yin and clear heat.
Use Zhi Bai Di Huang Wan [Anemarrhena, Phellodendron, and Rehmannia Pill] plus Sheng Di Huang and Nu Zhen Zi.

Liver and Kidney Yin Deficiency

Treatment Principle: Nourish Liver and Kidney.
Use Qi Bao Mei Ran Dan [Seven-Treasure Special Pill for Beautiful Whiskers].

Qi Deficiency with Spleen Weakness

Treatment Principle: Tonify Qi and strengthen Spleen.
Use Bu Zhong Yi Qi Tang [Tonify the Middle and Augment the Qi Decoction].

Qi and Blood Deficiency

Treatment Principle: Tonify Qi and nourish blood.
Use Ba Zhen Tang [Eight-Treasure Decoction] plus Zhu Ling, Zi He Che, and Wu Wei Zi.

Yin and Yang Deficiency

Treatment Principle: Warm Yang and nourish Yin.
Use You Gui Wan [Restore the Right (Kidney) Pill].

SKIN CANCER

This is often caused by knotted internal toxins and external pathogens at the skin, which prevents the toxins from being expelled outside and manifests as skin cancer. In Western medicine, it is said to be caused by many reasons when the immune surveillance and immune defense are damaged, leading to abnormal hyperplasia of skin cells. For squamous cell carcinoma the clinical characteristic is infiltrated hard mass which can form patches and nodules with ulceration at the surface. For basal cell carcinoma, the ulceration slowly expands with pearl-like eminence at the border.

The key for treatment is to remove cancerous tissue at early stage to prevent spreading and metastasis. If surgery cannot be performed or is refused by patients, then radiation therapy is administered with Chinese medicine. TCM treatments are to tonify Qi, nourish Yin, clear heat, relieve toxicity, transform stasis, and scatter accumulation.

Wind Toxins Accumulation

Treatment Principle: Disperse wind, relieve toxicity, nourish blood, and invigorate blood.

Use Yin Qiao San [Honeysuckle and Forsythia Powder] plus Bai Hu Tang [White Tiger Decoction] and Xing Ren, Sheng Di Huang, Chi Shao Yao, Mu Dan Pi, and Shen Qu.

Prepared as decoction, take four to six times a day. One bag of herbs is used for 2 days and 10 bags constitute one course.

Related conditions and modifications are

- Constipation: Add Huo Ma Ren, Bai Zi Ren, Yu Li Ren, and Da Huang.
- Red tongue with scanty or no coating: Remove Jing Jie and Dan Zhu Ye and add Xuan Shen, Mai Men Dong, Bai He, and Ren Shen.

Damp–Heat Accumulation

Treatment Principle: Eliminate damp, relieve toxicity, transform phlegm, and unblock collaterals.

Use Zhi Mi Fu Ling Wan [Poria Pill from Clear the Confusion Formula] plus Zhi Nan Xing, Tu Fu Ling, Bai Hua She She Cao, Ban Zhi Lian, Bei Xie, Huo Xiang, Yin Chen Hao, Shan Ci Gu, Dong Gua Ren, Yi Yi Ren, Mu Dan Pi, Si Gua Luo, and Chi Shao Yao.

One bag of herbs per day. Make into decoction and divide into four to six doses to be taken in a day. Thirty bags constitute one course.

Related conditions and modifications are

- Poor appetite and focal distention at epigastrium: Add Cang Zhu and Hou Po.
- Reduced tongue coating and persistent pain worsening at night: Add Ren Shen and Ji Xue Teng.

Fire Toxins Accumulation

Treatment Principle: Drain fire, relieve toxicity, cool blood, and anticancer therapies.

Use Long Dan Xie Gan Tang [Gentiana Longdancao Decoction to Drain the Liver] plus Ye Ju Hua, Zao Xiu, Bai Mao Gen, Mu Dan Pi, Bai Hua She She Cao, Huang Yao Zi, Shan Ci Gu, Zhe Bei Mu, and Da Huang.

One bag of herbs per day. Make into decoction and divide into four to six doses to be taken in a day.

Related conditions and modifications are

- Severe pain: Add Bai Shao Yao, Mu Xiang, and Gan Cao.
- Bleeding: Add fried Pu Huang, Di Yu Tan, Guan Zhong, and Xi Yang Shen.

Qi and Blood Deficiency

Treatment Principle: Tonify Qi and blood, vent toxins, and anticancer therapies.

Use Shi Quan Da Bu Wan [All-Inclusive Great Tonifying Decoction] plus Tu Fu Ling, Yi Yi Ren, Ji Xue Teng, Bai Hua She She Cao, and Da Zao.

One bag of herbs per day. Make decoction and divide into three doses to be taken in a day. One course has 30 bags.

Related conditions and modifications are

- Diarrhea: Add Gan Jiang.
- Cold Limbs: Add Gui Zhi.

STOMACH CANCER

The abdominal mass is formed by Qi stagnation, phlegm accumulation, and blood stasis which are caused by many factors. In Western medicine, stomach cancer is the most common cancer of the digestive tract. It occurs more in males than females and more in the age group of 40 to 60. In the early stages, there are no significant symptoms or there is mild epigastric pain and a bloating, fullness sensation, easily mistaken as gastritis. As the condition progresses, there is increasing pain that may radiate to the back, decreased appetite, aversion to animal products, progressive emaciation, anemia or difficulty swallowing, or nausea and vomiting. After surgery, TCM treatments clear residual cancerous cells, decrease the side effects from chemotherapy or radiation therapy, and prevent reoccurrence and metastasis. It can also reduce some symptoms, such as pain, poor appetite, and nausea and vomiting.

Disharmony of Liver and Stomach Syndrome

Treatment Principle: Disperse Liver, harmonize Stomach, descend rebellion, and stop pain.

Use Xiao Yao San [Relax Wanderer Powder] plus Xuan Fu Dai Zhe Tang [Inula and Haematite Decoction] plus Bai Hua She She Cao.

Related conditions and modifications are

- Bitter taste in mouth, dry mouth, and distending burning pain at stomach and epigastrium: Add Huang Qin and Huang Lian.
- Sour regurgitation and foul breath: Add Shan Zha, Shen Qu, Lian Qiao, and Lai Fu Zi.

Spleen and Stomach Deficient Cold Syndrome

Treatment Principle: Warm the middle, harmonize Stomach, and strengthen Spleen.

Use Fu Zi Li Zhong Wan [Prepared Aconite Pill to Regulate the Middle] plus Fu Ling, Sha Ren, Huang Qi, Yi Yi Ren, Chen Pi, and Ban Xia.

Related condition and modification are

- Belching, nausea, and white, greasy, and thick tongue coating: Remove Ren Shen and add Huo Xiang, Cang Zhu, and Cao Guo.

Stomach Heat Damaging Yin Syndrome

Treatment Principle: Clear heat, harmonize Stomach, nourish Yin, and moisten dryness.

Use Yi Wei Tang [Benefit the Stomach Decoction] plus Zhi Mu, Bai Hua She She Cao, Xi Yang Shen, Shan Yao, and Gan Cao.

Related conditions and modifications are

- Nausea and vomiting: Add Ban Xia and Huang Lian.
- Epigastric pain and abdominal distention: Add Mu Xiang, Da Fu Pi, and Yan Hu Suo.

Internal Obstruction of Blood Stasis Syndrome

Treatment Principle: Regulate Qi, invigorate blood, soften hardness, and eliminate accumulation.

Use Shi Xiao San [Sudden Smile Powder] plus Tao Ren, Hong Hua, Wu Yao, Yan Hu Suo, Sheng Di Huang, E Zhu, Zhe Cong, Lu Feng Fang, and Tie Shu Ye.

Related conditions and modifications are

- Apathy and fatigue: Add Huang Qi and Dang Shen.
- Black stool: Add E Jiao, Bai Ji, and Xue Yu Tan or combine with Yun Nan Bai Yao [White Powder from Yun Nan Province].

Phlegm–Damp Obstructing Stomach Syndrome

Treatment Principle: Dry dampness, harmonize Stomach, and strengthen Spleen.

Use Wen Dan Tang [Warm the Gallbladder Decoction] plus Xie Bai, Zhe Bei Mu, Yi Yi Ren, Tai Zi Shen, Bai Hua She She Cao, and Xia Ku Cao.

Related condition and modification are

- Phlegm and Qi rebellion and nausea and vomiting: Add Sheng Jiang and Huo Xiang.

Qi and Blood Deficiency Syndrome

Treatment Principle: Benefit Qi and nourish blood.

Use Ba Zhen Tang [Eight-Treasure Decoction] plus Huang Qi, Nu Zhen Zi, and E Jiao.

Related conditions and modifications are

- Abdominal distention after taking the herbs: Add Ban Xia and Sha Ren.
- Aversion to cold and cold limbs: Add Gui Zhi.
- Facial edema and swelling at limbs: Add Zhu Ling, Ze Xie, increase Huang Qi to 50 g, and Sheng Jiang Pi.
- Decreased WBC, platelet, dizziness, tinnitus, and soreness and weakness at lumbar and knees: Add Rou Cong Rong, Yin Yang Huo, Tu Si Zi, Du Zhong, and Lu Jiao Jiao.

Furthermore, from modern research there are certain correlations between TCM syndromes of stomach cancer and pathologies. Disharmony between Liver and Stomach is often early stage of stomach cancer, and the tumor is smaller. Spleen and Stomach deficient cold is also seen at early stages and occurs more frequently in patients with weak constitution and chronic gastritis. Ulcerative type of stomach cancer often relates to Qi and blood deficiency or Spleen and Stomach deficient cold syndromes. Polypus type of stomach cancer often relates to Stomach heat-damaging Yin syndrome. Infiltrative type of stomach cancer often relates to Spleen and Kidney Yang deficiency and phlegm–damp obstructing Stomach; the large mass type of stomach cancer is often due to internal obstruction of blood stasis.

URINARY BLADDER CANCER

This is often caused by Spleen and Kidney deficiency, damp–heat and turbid evil toxins invading urinary bladder and accumulating over time, causing problems. At early stages, it is accumulated damp–heat at Lower Jiao. At the middle stage, it is caused by blood stasis and obstruction. At the late stage, the cancerous toxins travel in the body with Qi and blood deficiency. In Western medicine, it is the most common cancer of the urinary system with a high reoccurrence rate. It often occurs in people more than 50 years old. Clinical manifestations are intermittent bloody urine without pain at the beginning stage. At middle and late stage, the manifestations are frequent, urgent, painful urination and metastasis especially to bone, liver, and lung.

Urinary bladder cancer is hard to treat due to its severity and difficulty in both TCM and Western medicine. With severe syndromes, Western medicine treats with surgery first, then combines with radiation therapy and chemotherapy. TCM treatments focus on strengthening Zheng Qi, reducing toxic side effects of chemotherapy and radiation therapy, and reducing reoccurrence. Furthermore, TCM treatments before surgery can improve symptoms and increase tolerance to the surgery. Treatment principles are to tonify and benefit Qi and blood, strengthen Spleen, harmonize Stomach, clear heat, and relieve toxicity.

Treatments According to Syndrome Differentiation

Lower Jiao Damp–Heat

Treatment Principle: Clear heat, drain damp, and stop pain.

Use Xiao Ji Yin Zi [Cephalanoplos Decoction] plus Bai Mao Gen, Zhu Ling, Huang Bai, Hai Jin Sha, Mu Dan Pi, and Chi Shao Yao.

Related conditions and modifications are

- Painful urination: Add Bian Xu.
- Difficult urination: Add Ze Xie and Jin Qian Cao.
- Bloody urine: Add Shui Niu Jiao and Xue Jian Chou.

Liver Qi Stagnation

Treatment Principle: Disperse liver, regulate Qi, unblock, and promote urination.

Use Shi Wei San [Pyrrosia Powder] plus Dan Zhu Ye, Yi Yi Ren, Zhu Ling, Wang Bu Liu Xing, Chen Xiang, Ju Pi, and Dang Gui.

Related conditions and modifications are

- Qi stagnation transforming into fire: Add Long Dan Cao and Shan Zhi Zi.
- Epigastric and abdominal distention: Add Wu Yao and Xiao Hui Xiang.

Blood Stasis and Obstruction

Treatment Principle: Invigorate blood, transform stasis, and drain damp.

Use Xue Fu Zhu Yu Tang [Drive Out Stasis in the Mansion of Blood Decoction] plus Wu Ling Zhi, Dan Shen, Zhu Ling, Yi Yi Ren, and Dan Zhu Ye.

Related condition and modification are

- Hard abdominal mass with or without pain: Add Shi Jian Chuan, Bai Hua She She Cao, and Shan Ci Gu.

Yin Deficiency and Deficient Fire

Treatment Principle: Nourish Yin and descend fire.

Use Er Zhi Wan [Two-Ultimate Pill] plus Mu Dan Pi, Shan Yao, Zhi Mu, Qu Mai, Shi Wei, Dan Zhu Ye, Yi Yi Ren, Shan Zhu Yu, and Tu Si Zi.

Related condition and modification are

- Lumbar pain: Add Xu Duan and Sang Ji Sheng.

Spleen and Kidney Qi Deficiency

Treatment Principle: Tonify Qi, consolidate Kidney, soften hardness, and scatter accumulation.

Use Ju Yuan Jian [Lift the Source Decoction] plus Dang Gui, Bu Gu Zhi, Jin Ying Zi, Tu Fu Ling, Bai Ying, Wang Bu Liu Xing, Shi Wei, Qu Mai, and Yi Yi Ren.

Related condition and modification are

- Scanty urine, irritability, agitation, nausea, vomiting, and dizziness: Add Wu Zhu Yu, Sheng Jiang, Gui Zhi, and Dan Zhu Ye.

Internal Accumulation of Stasis and Toxins

Treatment Principle: Tonify Liver and Kidney, transform stasis, and relieve toxicity.

Use Zhi Bai Di Huang Wan [Anemarrhena, Phellodendron, and Rehmannia Pill] plus Chen Xiang San [Aquilaria Powder] and Yi Yi Ren, Long Kui, Chi Xiao Dou, Mai Men Dong, Shi Hu, Nu Zhen Zi, Wang Bu Liu Xing, Chi Shao Yao, Da Huang, and Bai Zhu.

Related conditions and modifications are

- Abdominal distention: Add Lai Fu Zi and Da Fu Pi.
- Yin deficiency with deficient heat: Add Gui Ban, Bie Jia, and Mu Li.

Cancerous Toxins Traveling in Body; Qi and Blood Deficiency

Treatment Principle: Tonify Qi and blood, support Zheng Qi, and inhibit cancer.

Use Yi Guan Jian [Linking Decoction] plus Sheng Mai San [Generate the Pulse Powder] and add Huang Qi, Yi Yi Ren, Nu Zhen Zi, Dan Zhu Ye, Fu Ling, Huang Jing, Huang Qin, and Xi Yang Shen.

Related conditions and modifications are

- Poor appetite and nausea: Add Bai Zhu, Chen Pi, Shen Qu, and Shan Zha.
- Blood in urine and painful urination with blood clots: Add Chi Shao Yao, Tao Ren, and Dan Shen.

Treatments According to Symptoms

Pain

Grind Xue Jie, Er Cha, Yan Hu Suo, Hong Hua, Mo Yao, Ru Xiang, Bai Shao Yao, and She Xiang into powder. This can be taken internally or mixed with alcohol and applied topically.

Urinary Retention

Grind Wu Zhu Yu, Xi Xin, Hui Xiang, Wu Yao, Gan Jiang, and Bi Ba into powder. Mix with alcohol to form paste and apply topically.

APPENDIX 1: TCM TREATMENT FOR SIDE EFFECTS OF CHEMOTHERAPY AND RADIOTHERAPY

Bone Marrow Inhibition

Reduced white blood cell can easily lead to severe infections. Reduced platelet can lead to bleeding. Reduced red blood cell can lead to anemia.

Strengthen Spleen, tonify Kidney, and tonify Qi and blood.

Reduced WBC

Commonly used herbs are Ren Shen, Huang Qi, Bai Zhu, Dang Shen, Shan Yao, Huang Jing, Gan Cao, Ling Zhi, Ci Wu Jia, Dang Gui, Bai Shao Yao, Shu Di Huang, Gou Qi Zi, He Shou Wu, and Ji Xue Teng.

Formulas include Dang Gui Bu Xue Tang [Tangkuei Decoction to Tonify the Blood], Ba Zhen Tang [Eight-Treasure Decoction], Gui Pi Tang [Restore the Spleen Decoction], Yu Ping Feng San [Jade Windscreen Powder], Bu Zhong Yi Qi Tang [Tonify the Middle and Augment the Qi Decoction], and Da Bu Yuan Jian [Great Tonify the Basal Decoction].

Related conditions and modifications are

- Yin deficiency: Add Sheng Di Huang, Nu Zhen Zi, Shan Zhu Yu, Mai Men Dong, Sha Shen, and Sang Shen Zi.
- Yang deficiency: Add Yin Yang Huo, Ba Ji Tian, Rou Cong Rong, Zi He Che, Bu Gu Zhi, Rou Gui, Zhi Fu Zi, and Lu Rong.

Reduced Platelet

Commonly used herbs are Huang Qi, Tai Zi Shen, Nu Zhen Zi, Bai Zhu, Da Zao, Ji Xue Teng, E Jiao, Long Yan Rou, Han Lian Cao, Xian He Cao, and Zi Zhu Cao.

Formulas include Gui Pi Tang [Restore the Spleen Decoction] and Er Zhi Wan [Two-Ultimate Pill].

Related conditions and modifications are

- Yin deficiency and deficient fire: Add Zhi Mu, Huang Bai, Di Gu Pi, and Shui Niu Jiao.
- Stasis spots on skin or bleeding: Add San Qi, Qian Cao, Xue Yu Tan, and Pu Huang Tan.

Reduced RBC

Commonly used herbs are E Jiao, Gui Ban Jiao, Lu Jiao Jiao, Zhu Ji Sui (pig bone marrow), and Zi He Che.

Cautions:

1. The effect is better if the patient begins to take these herbs a few weeks before chemotherapy and radiation therapy, and once again begins on the week of the chemotherapy and radiation therapy.
2. After chemotherapy and radiation therapy, these herbs should be continued for a few more weeks to prevent delayed bone marrow inhibition.
3. Antibone marrow inhibition treatment is not limited only to tonic herbs. Other herbs such as Hu Zhang, Qian Cao, Wu Zei Gu, Ku Shen, Shan Dou Gen, Huang Lian, Shi Da Gong Lao Ye, and Can Sha can also raise WBC.

Immune Inhibition

Most of the anticancer pharmaceutical drugs have a different degree of inhibiting immune system function. This can lead to a decrease in physiological function, disturbance, and proneness to infections.

Support Zheng Qi and the root, invigorate blood, transform stasis, clear heat, and relieve toxicity.

1. Support Zheng Qi and root to strengthen overall immunity
   Use Huang Qi, Yin Yang Huo, Huang Jing, Nu Zhen Zi, Bai Shao Yao, He Shou Wu, Gou Ji, Lu Rong, Ren Shen, Tu Si Zi, Gou Qi Zi, Han Lian Cao, Bu Gu Zhi, Shu Di Huang, Mai Men Dong, Bai He, Xian Mao, Du Zhong, Bai Zhu, Bian Dou, Shan Yao, Xu Duan, Dang Gui, E Jiao, Shan Zhu Yu, Sha Shen, and Shi Hu. These herbs can be used according to symptoms.
2. Clear heat, relieve toxicity, invigorate blood, and transform stasis, which can stimulate the transformation of lymphocytes
   Use Pu Gong Ying, Zi Hua Di Ding, Huang Qin, Huang Lian, Jin Yin Hua, Hong Hua, Dan Shen, Chuan Xiong, Wang Bu Liu Xing, Chai Hu, Xian He Cao, Sang Ji Sheng, and Lou Lu.
3. Tonify Qi, nourish Yin, drain damp to increase function of macrophages, and stimulate the transformation of lymphocytes
   Use Ren Shen, Dan Zhu Ye, Huang Qi, Dang Gui, Ci Wu Jia, Gou Qi Zi, Nu Zhen Zi, Dong Chong Xia Cao, Wu Wei Zi, Fu Ling, Tian Ma, Zhu Ling, and Hei Mu Er (black fungus).

Patent herbs that strengthen immunity: Liu Wei Di Huang Wan [Six-Ingredient Pill with Rehmannia], Gui Pi Wan [Restore the Spleen Pill], Bu Zhong Yi Qi Wan [Tonify the Middle and Augment the Qi Pill], You Gui Wan [Restore the Right (Kidney) Pill], Wu Ji Bai Feng Wan [Black Chicken and White Phoenix Pill], Ren Shen Yang Rong Wan [Ginseng Pill to Nourish the Nutritive Qi], and Da Bu Yin Wan [Great Tonify the Yin Pill].

Nausea and Vomiting

This is due to Spleen and Stomach damages. Symptoms include dysfunction of ascending and descending mechanisms, which cause Stomach Qi rebellion due to drug toxins.

Strengthen Spleen, harmonize Stomach, and descend rebellion.

Spleen Deficiency and Excess Damp

Use Ping Wei San [Calm the Stomach Powder] plus Er Chen Tang [Two-Cured Decoction]; or use Xiang Sha Liu Jun Zi Tang [Six-Gentleman Decoction with Aucklandia and Amomum].

Damp–Heat in Spleen and Stomach

Use Da Huang Gan Cao Tang [Rhubarb and Licorice Decoction] (steep 15 g of Da Huang and 10 g of Gan Cao in hot water; drink frequently), or take Huang Lian Su Ye Tang [Coptis and Perilla Leaf Decoction]; or San Ren Tang [Three-Nut Decoction].

Stomach Yin Deficiency

Use Yi Wei Tang [Benefit the Stomach Decoction]; or Sha Shen Mai Dong Tang [Glehnia and Ophiopogonis Decoction] with Huang Qi and Tai Zi Shen.

Above formulas can be modified with Ban Xia, Ding Xiang, Chen Xiang, Sheng Jiang, Huo Xiang, Sha Ren, and Dai Zhe Shi to descend rebellion and stop nausea according to the presenting symptoms.

Diarrhea and Constipation

Damp–Heat Diarrhea

Herbs include Ge Gen, Huang Qin, Huang Lian, Bai Shao Yao, Bai Tou Weng, Qin Pi, Dang Gui, Mu Dan Pi, Huo Xiang, and Lu Gen.

Formulas include Ge Gen Huang Qin Huang Lian Tang [Kudzu, Coptis, and Scutellaria Decoction] and Bai Tou Weng Tang [Pulsatilla Decoction].

Spleen Deficiency Diarrhea

Herbs include Dang Shen, Bai Zhu, Huang Qi, Fu Ling, Lian Zi, Yi Yi Ren, Sha Ren, Chen Pi, and fried Shan Zha.

Formulas include Xiang Sha Liu Jun Zi Tang [Six-Gentleman Decoction with Aucklandia and Amomum] and Shen Ling Bai Zhu San [Ginseng, Poria, and Actractylodes Macrocephala Powder].

Related conditions and modifications are

- Spleen Yang deficiency: Add Gan Jiang, Xiao Hui Xiang, and Rou Gui.
- Excess damp: Add Cang Zhu, Huo Xiang, and Ban Xia.

Qi, Blood Deficiency, and Intestinal-Dryness Constipation

Herbs include Sheng Di Huang, Mai Men Dong, Xuan Shen, Gua Lou Ren, Huo Ma Ren, Bai Shao Yao, Hou Po, Zhi Shi, Da Huang, and Mang Xiao.

Formulas are Zeng Ye Cheng Qi Tang [Increase the Fluids and Order the Qi Decoction], Ma Zi Ren Wan [Hemp Seed Pill], and Huang Long Tang [Yellow Dragon Decoction].

Intestinal Numbness and Abdominal Distention

Fry Xiao Hui Xiang and apply topically on abdomen. Combine with Ma Zi Ren Wan [Hemp Seed Pill] and Cao Jue Ming Zi powder.

Stomatitis and Mouth Ulcerations

Heart and Spleen Accumulated Heat

Use Dao Chi San [Guide Out the Red Powder]; or Liang Ge San [Cool the Diaphragm Powder]; or Da Huang Huang Lian Xie Xin Tang [Rhubarb and Coptis Decoction to Drain the Epigastrium].

Spleen and Kidney Deficiency

Use Shen Qi Di Huang Tang [Ginseng, Astragalus, and Rehmannia Decoction] and Gui Fu Di Huang Tang [Cinnamon, Prepared Aconite, and Rehmannia Decoction].

Combine with a mouth rinse: Qing Dai, Hu Zhang, Er Cha, Bing Pian, and Wu Mei; or use Xi Gua Shuang [Watermelon Frost].

Liver Toxicity

Chemotherapeutic drugs often can damage liver to different degrees, especially if it is used in large dosages over a long time. If patient has past liver pathologies, then it can manifest as acute, chronic hepatitis, fatty liver, and liver fibrosis. The symptoms include nausea, aversion to food, ascites, enlargement of liver and spleen, and increased SGOT or SGPT.

Damp–Heat Type

Herbs include Yin Chen Hao, Jin Qian Cao, Long Dan Cao, Tian Ji Huang, Huang Qin, Shan Zhi Zi, Huo Xiang, Fu Ling, Ze Xie, Yi Yi Ren, Che Qian Zi, Hou Po, Da Fu Pi, Hua Shi, and Da Huang.

Formulas include Yin Chen Hao Tang [Artemisia Yinchenhao Decoction], Gan Lu Xiao Du Dan [Sweet Dew Special Pill to Eliminate Toxin], and Long Dan Xie Gan Tang Jia Jian [Modified Gentiana Longdancao Decoction to Drain the Liver].

It is important that bowel movements and urination are regular to reduce SGPT, SGOT, and jaundice.

Spleen Deficiency Type

Use Dang Shen, Huang Qi, Bai Zhu, Tai Zi Shen, Fu Ling, Lian Zi, Yi Yi Ren, Da Fu Pi, and Ji Nei Jin.

Also, Gui Shao Liu Jun Zi Tang [Six-Gentleman Decoction with Tangkuei and Peony], Xiang Sha Liu Jun Zi Tang [Six-Gentleman Decoction with Aucklandia and Amomum], and Fang Ji Huang Qi Tang Jia Jian [Modified Stephania and Astragalus Decoction].

Yin Deficiency Type

Herbs include Shan Zhu Yu, Bai Shao Yao, Sheng Di Huang, Mai Men Dong, Shi Hu, Yu Zhu, and Wu Mei.

Formulas are Yi Guan Jian [Linking Decoction], Zi Shui Qing Gan Yin [Nourish the Water and Clear the Liver Decoction], and Liu Wei Di Huang Wan Jia Jian [Modified Six-Ingredient Pill with Rehmannia].

Furthermore, one can also add invigorating-blood and transforming-stasis herbs, such as Shan Zha, Dan Shen, San Qi, Tu Bie Chong, Bie Jia, and E Zhu, which are effective in treating liver fibrosis and swollen liver and spleen.

Kidney Toxicity

Many chemotherapeutic drugs can damage the kidney to different degrees. At early stage, there can be protein, RBC, WBC in urine, or cylindruria; some patients can have edema, sore back, and fatigue, and can develop dizziness, increased nocturia, hypertension, abnormal kidney functions, and scanty urine or even absence of urine.

Damp–Heat at Lower Jiao

Use Ba Zheng San [Eight-Herb Powder for Rectification] plus E Jiao, Pu Huang, and Han Lian Cao.

Yin Deficiency with Blood Heat

Use Xiao Ji Yin Zi [Cephalanoplos Decoction] plus Er Zhi Wan [Two-Ultimate Pill] or combine with Zhi Bai Di Huang Wan [Anemarrhena, Phellodendron, and Rehmannia Pill].

Qi Deficiency, Blood Stasis, and Blood Not Returning to Meridians

Use Gui Pi Tang [Restore the Spleen Decoction] plus Xue Yu Tan, Pu Huang, and E Jiao.

Absence of Significant Chills or Fever (Early Stage of Kidney Damage)

Use Wu Ling San [Five-Ingredient Powder with Poria].
Related condition and modification are

- Kidney deficiency in seniors: Add Gou Qi, Bu Gu Zhi, Shan Yao, Huang Qi, Shan Zhu Yu, and Sheng Di Huang.

For above four types, Dan Shen, Tao Ren, and Yi Mu Cao can be added to invigorate blood, unblock collaterals, transform Qi, and move water.

Acute Kidney Damage

Disperse wind, clear heat, drain damp, unblock collaterals, and consolidate Kidney Qi.
Herbs include Jing Jie, Fang Feng, Chan Tui, Lian Qiao, Wu Mei, Huang Bai, Fu Ling, Ze Xie, Bai Mao Gen, Dan Shen, Yi Mu Cao, Tai Zi Shen, Sheng Huang Qi, and Nu Zhen Zi.

APPENDIX 2: TCM TREATMENT FOR SIDE EFFECTS OF RADIATION THERAPY

The side effects of radiation therapy belong to fire toxin pathogens. The following are common acute radiation therapy side effects and their treatments.

Stomatitis, Pharyngitis, and Damages to Salivary Gland for Radiation Therapy at Head

Treatment Principle: Nourish Yin, generate fluid, clear heat, and relieve toxicity.

Herbs include Xuan Shen, Mai Men Dong, Sheng Di Huang, Shi Hu, Sha Shen, Wu Mei, Yu Zhu, Bai He, Lu Gen, Tian Hua Fen, Shan Dou Gen, Pu Gong Ying, Lian Qiao, Jin Yin Hua, Huang Qin, Zhi Mu, and Sheng Shi Gao.

Also, Xuan Mai Gan Jie Tang [Scrophularia, Ophiopogonis, Licorice, and Platycodi Decoction], Sha Shen Mai Dong Tang [Glehnia and Ophiopogonis Decoction], Wu Zhi Yin [Five-Juice Decoction]

Related conditions and modifications are

- Severe heat: Use Zhu Ye Shi Gao Tang [Lophatherus and Gypsum Decoction], Dao Chi San [Guide Out the Red Powder], Gan Lu Yin [Sweet Dew Decoction], and Qing Wei San [Clear the Stomach Powder].
- Sore throat and edema: Add She Gan, Niu Bang Zi, Jie Geng, Ma Bo, and Sheng He Zi.
- Stimulate the repair of damages: Huang Qi and Tai Zi Shen.

Combine with mouth and throat rinse. Decoct Bo He, Jing Jie, Fang Feng, Lian Qiao, Jin Yin Hua, Sheng Gan Cao, and Peng Sha. After the decoction is cool, it can be used as a rinse or topical wash.

Radiation Therapy for Nasopharyngeal Cancer Often Causing Inflammation of the Nasal Cavity

Treatment Principle: Clear heat, moisten dryness, and unblock orifices.

Use Xin Yi Hua, Bai Zhi, Cang Er Zi, Xi Xin, Lu Feng Fang, Huang Qin, Sang Bai Pi, Yu Xing Cao, Jin Yin Hua, Lian Qiao, Dong Gua Zi, Mai Men Dong, and Sha Shen.

Also, Cang Er Zi San [Xanthium Powder].

Related condition and modification are

- Epistaxis: Add Xian He Cao.

Radiation Therapy for Esophageal Cancer Often Causing Esophagitis

Symptom: Burning pain at breast bone, worse with swallowing.

Treatment Principle: Clear heat, relieve toxicity, harmonize Stomach, and cool blood.

Herbs include Gua Lou Pi, Ban Xia, Xie Bai, Dai Zhe Shi, Xuan Fu Hua, Hou Po, Zhi Shi, Zhu Ru, Huang Lian, Pu Gong Ying, Ba Yue Zha, and Ji Xing Zi.

Formulas include Gua Lou Xie Bai Ban Xia Tang [Trichosanthes Fruit, Chinese Chive, and Pinellia Decoction], Wen Dan Tang [Warm the Gallbladder Decoction], Xiao Xian Xiong Tang [Minor Sinking into the Chest Decoction], and Ban Xia Xie Xin Tang [Pinellia Decoction to Drain the Epigastrium].

Related condition and modification are

- Ulceration at mucousal layer of esophagus: Add Bai Ji, Qing Dai, and Yun Nan Bai Yao.

Radiation Therapy for Lung Cancer and Esophageal Cancer Causing Radiation Pneumonia

The manifestations are cough, tight chest, stifling sensation at chest, and copious phlegm.

Treatment Principle: Clear Lung, transform phlegm, stop cough, calm wheezing, invigorate blood, and transform stasis.

Herbs include Huang Qin, Zhi Mu, Lian Qiao, Jin Yin Hua, Sang Bai Pi, Zhe Bei Mu, Yu Xing Cao, Xing Ren, Gua Lou Ren, Zhi Ke, Ting Li Zi, Ma Huang, Dan Shen, Chi Shao Yao, and Qian Cao.

Also, Qing Jin Hua Tan Tang [Clear the Metal and Transform Phlegm Decoction], Ma Xing Shi Gan Tang [Ephedra, Apricot Kernel, Gypsum, and Licorice Decoction], Bei Mu Gua Lou San [Fritillaria and Trichosanthes Fruit Powder].

Related condition and modification are

- Dry cough, scanty phlegm, dry throat, and chest pain: Use Bai He Gu Jin Tang [Lily Bulb Decoction to Preserve the Metal] and Er Dong Er Mu Tang [Ophiopogonis, Asparagi, Fritillaria, and *Fritillaria Thunbergi* Decoction].

Radiation Therapy (Internal Irradiation) for Colorectal Cancer and Cervical Cancer Causing Inflammation at Rectum, Colon, or Small Intestine

The manifestations include abdominal pain, diarrhea, tenesmus, burning pain at anus, or recurrent blood in stool.

Treatment Principle: Clear intestine, relieve toxicity, regulate Qi, strengthen Spleen, and harmonize blood.

Herbs include Bai Tou Weng, fried Di Yu, Qin Pi, Ma Chi Xian, Huang Lian, Huang Qin, Bai Zhu, Bai Shao Yao, Bing Lang, Mu Xiang, Shan Zha, Shen Qu, Mai Ya, Dang Gui, and Sheng Huang Qi.

Formulas include Shao Yao Tang [Peony Decoction], Ge Gen Huang Qin Huang Lian Tang [Kudzu, Coptis, and Scutellaria Decoction], and Bai Tou Weng Tang [Pulsatilla Decoction].

Related condition and modification are

- Blood in stool: Add Zi Zhu Cao, Xian He Cao, or Yun Nan Bai Yao. The decoction prepared from the above formula can also be used as an enema.

CHAPTER 13

Cardiovascular Disorders

ATRIOVENTRICULAR BLOCK

Atrioventricular block is discussed under palpitation and palpitation with fright. The root causes of the disease are deficiencies in Qi, blood, Yin, Yang, Heart, Kidney, and Spleen. The branch manifestations are blood stasis, turbid phlegm, Qi stagnation, and heat toxins, which lead to stasis and obstruction of heart vessels. TCM treatment focuses on warming Yang, benefiting Qi, invigorating blood, and transforming stasis.

Qi and Yin Deficiency Syndrome

Treatment Principle: Tonify Qi and nourish Yin.

Use Zhi Gan Cao Tang [Honey-Fried Licorice Decoction] plus Sheng Mai San [Generate the Pulse Powder].

Related conditions and modifications are

- For stifling sensation and pain at chest and stasis dots or spots on tongue: Add Chuan Xiong, Hong Hua, Jiang Xiang, and Chi Shao Yao.
- For dizziness, vertigo, spitting saliva and sputum, stifling sensation at chest, and white, greasy, tongue coating: Add Gua Lou Pi, Ban Xia, Zhu Ru, and Dan Nan Xing.

Heart and Kidney Yang Deficiency Syndrome

Treatment Principle: Warm Yang and benefit Qi.

Use Shen Fu Tang [Ginseng and Prepared Aconite Decoction] plus You Gui Wan [Restore the Right (Kidney) Pill].

Yang Deficiency About to Collapse Syndrome

Treatment Principle: Restore Yang and rescue collapse.

Use Shen Fu Tang [Ginseng and Prepared Aconite Decoction] plus Dang Gui Bu Xue Tang [Tangkuei Decoction to Tonify the Blood] plus Rou Gui, Yin Yang Huo, Chen Xiang, Lu Rong, and San Qi.

Related condition and modification are

- For severe blood stasis: Add Chi Shao Yao, E Zhu, Dan Shen, and Hong Hua.

CARDIAC ARRHYTHMIA

Cardiac arrhythmia is a commonly seen symptom of cardiac diseases. It can be caused by many factors. According to the research on herbal properties and their clinical applications, antiarrhythmia herbs are as follows: Ku Shen inhibits ectopic beating of the heart; Yi Mu Cao can dilate coronary artery, increase blood flow, and improve microcirculation; Dan Shen dilates coronary arteries, increases blood flow, strengthens the contraction of the heart, and regulates heart rate; Chan Tui and Jiang Can can decrease heart rate; Chang Shan, Sheng Di Huang, Dang Gui, Ling Zhi, Yu Zhu, Zhi Fu Zi, Yin Yang Huo, and Cha Shu Gen also function to regulate heart rate.

Antiarrhythmia formulas include:

Sheng Mai San [Generate the Pulse Powder] increases the activity of DNA at the heart muscle and inhibits the activity of adenosine triphosphate (ATP) at heart muscle.

Yu Ping Feng San [Jade Windscreen Powder] increases the activity of NK cells in patients with viral myocarditis, which reduces the symptoms visible on an EKG.

Sick Sinus Syndrome (SSS)

Sick sinus syndrome sometimes is referred to as sick sinus. The dysfunction of sinoatrial node and conduction is often caused by coronary heart disease, rheumatic heart disease, myocardiac disease, and myocarditis. The clinical manifestations are bradycardia, which causes insufficient blood supply to brain, Heart, and Kidney, leading to dizziness, vertigo, insomnia, forgetfulness, slow reactions, or quick irritability, or even convulsion and syncope. This condition is discussed under palpitation, chest Bi, knotted intermittent pulse, and Jue syndrome in TCM. The pulse can be slow, moderate, knotted, or intermittent. The pathogenesis are Yang deficiency and Yin exhaustion. The main organs involved are Heart, Spleen, and Kidney. The treatment focuses on tonifying Qi and warming Yang. It is supported by invigorating blood and nourishing Yin. Patients with heart rate slower than 40 beats per minute should be sent to an emergency room. Other cases can be treated with the formula containing Dang Shen, Huang Qi, Zhi Fu Zi, Ma Huang, Xi Xin, Gan Jiang, Zhi Ke, Rou Gui, and Zhi Gan Cao.

Related conditions and modifications are

- Kidney Yang deficiency: Substitute Dang Shen with Hong Shen and add Lu Rong powder and Wu Zhu Yu.
- Spleen Qi deficiency: Add Huang Jing, fried Bai Zhu, Shan Yao, Fu Ling, and Chen Pi.
- Red tongue or red tongue tip or cracks on tongue: Add Shu Di Huang, Mai Men Dong, Yu Zhu, and Wu Wei Zi.
- Purple spots at sides and tip of the tongue, dark lips, and greenish complexion: Add Dan Shen, Yu Jin, and Xie Bai.
- Low blood pressure: Add Chai Hu and Sheng Ma.
- Repeated occurrences of dizziness or fainting: Add Shi Chang Pu, Yuan Zhi, and Yu Jin or take with Su He Xiang Wan [Liquid Styrax Pill].
- Sudden increase of heart rate to 90 to 140 beats per minute: Remove Ma Huang, Xi Xin, Gan Jiang, Rou Gui, and Fu Zi, and add Sheng Di Huang, Ku Shen, and Bai Zi Ren.
- Constipation: Add Dang Gui and Rou Cong Rong.

Syndrome Differentiation

Heart Yang Deficiency, Yang Deficiency, and Yin Excess Syndrome

Treatment Principle: Warm and unblock Heart Yang, assist Yang, and inhibit Yin.

Use Ma Huang Fu Zi Xi Xin Tang [Ephedra, Prepared Aconite, and Asarum Decoction], Gui Zhi Qu Shao Yao Jia Fu Zi Tang [Cinnamon Twig Decoction minus Peony and add Prepared Aconite Decoction] plus Dang Gui Si Ni Jia Wu Zhu Yu Sheng Jiang Tang [Tangkuei Decoction for Frigid Extremities plus Evodia and Fresh Ginger].

Related conditions and modifications are

- For Qi and blood deficiency: Change to Zhi Gan Cao Tang [Honey-Fried Licorice Decoction].
- For Floating Heart Yang, irritability, and agitation: Add Sheng Long Gu and Sheng Mu Li.
- For Qi deficiency and blood stasis: Add Dang Shen, Huang Qi, and Dan Shen.

Spleen, Heart, and Kidney All-Deficient Syndrome

Treatment Principle: Tonify Heart, Kidney, and Spleen, expel phlegm, and transform turbidity.

Use Fu Zi Li Zhong Tang [Prepared Aconite Decoction to Regulate the Middle] plus Huang Qi, Gui Zhi, Dang Gui, Fu Ling, Wu Wei Zi, and Chen Pi.

Kidney Yang Deficiency and Exhaustion Damages of the Above and Exhaustion of the Below Syndrome

Treatment Principle: Warm Kidney Yang, supplement below, and consolidate above.

Use Gui Zhi Fu Zi Tang [Cinnamon Twig and Prepared Aconite Decoction] plus Fu Zi Tang [Prepared Aconite Decoction], You Gui Wan [Restore the Right (Kidney) Pill] plus Lu Rong.

Premature Beat

Premature beat is a commonly seen irregular heart beat. Premature beat can occur randomly or frequently, and sometimes without symptoms. Frequent occurrences lead to decreased cardiac output which causes palpitation, dizziness, fatigue, or induce angina pectoris in patients with past cardiac pathologies, or worsen the heart failure condition. This condition is discussed under palpitation, palpitation with fright, and chest Bi and heart pain in TCM. Hurried, knotted, or intermittent pulses are often observed in the patients. The pathogenesis are Qi deficiency and blood exhaustion, Heart and Kidney disharmony, internal obstruction of turbid phlegm, and Qi stagnation and blood stasis. The treatment focuses on harmonizing Heart and Kidney, tonify Qi and blood, moving Qi, and transforming stasis and phlegm.

Heart Qi Deficiency, Upward Flowing of Turbid Phlegm Syndrome

Treatment Principle: Tonify and benefit Heart Qi, transform phlegm, and calm spirit.

Use Wen Dan Tang [Warm the Gallbladder Decoction] plus Ban Xia Hou Po Tang [Pinellia and Magnolia Bark Decoction] plus Zi Su Geng.

Related conditions and modifications are

- For stifling pain at chest: Add Gua Lou Pi, Xie Bai, Yu Jin, E Zhu, Chuan Xiong, Ru Xiang, Mo Yao, Chi Shao Yao, Hong Hua, and Jiang Xiang.
- For insomnia, excessive dreaming, and being easily frightened: Add He Huan Hua, Zhi Shou Wu, Ye Jiao Teng, and Hu Po.

Heart Yin Deficiency, Heart and Kidney Disharmony Syndrome

Treatment Principle: Harmonize Heart and Kidney, nourish Heart, and calm spirit.

Use Jiao Tai Wan [Grand Communication Pill], Er Zhi Wan [Two-Ultimate Pill] plus Tai Zi Shen, Sheng Di Huang, Dang Gui, Fu Shen, Dan Shen, Mai Men Dong, Tian Men Dong, Xuan Shen, Yuan Zhi, Wu Wei Zi, Bai Zi Ren, and Suan Zao Ren.

Related conditions and modifications are

- Irritability and agitation: Add Lian Xin and Shan Zhi Zi.
- Dizziness and tinnitus: Add Shu Di Huang, Gou Qi Zi, and Shan Zhu Yu.
- Dry mouth and throat: Add Ge Gen, Sha Shen, Yu Zhu, Shi Hu, and Tian Hua Fen.
- Sore lumbar pain: Add Sang Ji Sheng, Tu Si Zi, Xu Duan, Ba Ji Tian, Huai Niu Xi, and Xian Mao.

Qi and Yin Deficiency, Qi Stagnation, and Blood Stasis Syndrome

Treatment Principle: Tonify Qi, nourish Yin, invigorate blood, and restore pulse.

Use Zhi Gan Cao Tang [Honey-Fried Licorice Decoction] plus Dan Shen, Chi Shao Yao.

Related conditions and modifications are

- Severe Qi deficiency: Add Huang Qi and Huang Jing; replace Dang Shen with Hong Shen.
- Severe Yin deficiency: Remove Gui Zhi and Sheng Jiang, and substitute Dang Shen with Xi Yang Shen.
- Severe Qi stagnation and blood stasis: Add Quan Xie, Wu Gong, and Mang Chong.

Spleen and Heart Deficiency, Qi Deficiency and Blood Stasis Syndrome

Treatment Principle: Nourish blood and Heart, tonify Qi, and invigorate blood.

Use Gui Pi Tang [Restore the Spleen Decoction] plus Dan Shen, Hong Hua, and Bai Zi Ren.

Paroxysmal Tachycardia

The heart rate of paroxysmal tachycardia is generally between 150 to 220 beats per minute. It is a fast regular paroxysmal abnormal heart rate. It often occurs on top of a preexisting chronic cardiac disease due to emotional disturbance, sudden exertion, and changing of body position or posture, or is induced by overeating. During the attacks, there are sudden palpitation and increased regular heart rate, lasting a few seconds, few minutes, few hours or a few days, then suddenly returning to normal. It can also be accompanied by fright, agitation, and copious urine. In TCM, it is called fright palpitation. The pulse can be rapid or swift. The pathogenesis are Gallbladder Qi deficiency with timidity, phlegm heat upward, disturbing the heart; insufficient true Yin, Heart blood deficiency, and blood not nourishing Heart. The treatment focuses on clearing and transforming damp and heat, and nourishing Yin and blood.

Gallbladder Qi Deficiency and Timidity, and Phlegm Heat Upward Disturbing Syndrome

Treatment Principle: Clear and transform hot phlegm, anchor the Heart, and calm the spirit.

Use Huang Lian Wen Dan Tang [Coptis Decoction to Warm the Gallbladder] plus Zhu Sha An Shen Wan [Cinnabar Pill the Calm the Spirit]. Remove Zhu Sha, add Bai Zi Ren, Zhen Zhu Mu, Sheng Mu Li, and Hu Po powder.

Related conditions and modifications are

- Irritability and restlessness, rapid and swift pulse: Add Ku Shen and Yu Jin.
- Insomnia and excess dreaming: Add He Huan Hua, Yuan Zhi, and Shi Chang Pu.
- Liver Qi stagnation and hypochondriac pain: Add Xiang Fu and Qing Pi.

Insufficient True Yin, and Heart Yin Exhaustion Syndrome

Treatment Principle: Nourish Yin and blood, make the Heart tranquil, and calm Shen.

Use Er Zhi Wan [Two-Ultimate Pill] plus Tai Zi Shen, Sheng Di Huang, Sha Shen, Suan Zao Ren, Bai Zi Ren, Dan Shen, Mai Men Dong, Xuan Shen, Dang Gui, Yuan Zhi, Wu Wei Zi, and Fu Shen.

Related conditions and modifications are

- Dizziness and fatigue: Add Xi Yang Shen.
- Presence of "Sparrow Pecking Pulse" or "Leaky Roof Pulse": Remove Sha Shen, Sheng Di Huang, Xuan Shen, Er Zhi Wan [Two-Ultimate Pill], Bai Zi Ren, Suan Zao Ren, and Fu Shen; add Hong Shen, Zhi Fu Zi, Rou Gui, and Zhi Gan Cao.

Atrial Fibrillation

Atrial fibrillation is fast irregular heart beats commonly seen in adults. During attacks, there are irregular movements of the atrium for about 350 to 600 times in a minute. It causes uncoordinate atrial contractions. At the same time, there is also fast and irregular ventricular contraction for 120 to 180 beats per minute. Clinically, the manifestations are palpitation, stifling sensation at chest, fright, or even heart failure, shock, coma, or angina pectoris attacks. This condition is discussed under heart palpitation, fright, and palpitation. The pulses can be hurried, intermittent, or knotted. The pathogenesis are Qi, blood, Yin and Yang exhaustion failing to nourish Heart, or phlegm damp disturbing the Heart. In treatment, both Chinese and Western medicine do not have satisfying results.

Qi and Blood Deficiency Syndrome

Treatment Principle: Tonify Qi and nourish blood.

Use Ren Shen Yang Ying Tang [Ginseng Decoction to Nourish the Nutritive Qi] plus Bai Zi Ren.

Related conditions and modifications are

- Blood deficiency: Add E Jiao.
- Insomnia and severe fright palpitation: Add Suan Zao Ren and Long Yan Rou.

Phlegm and Damp Disturbing the Heart Syndrome

Treatment Principle: Strengthen Spleen, transform dampness, and calm spirit.

Use Shi Wei Wen Dan Tang [Ten-Ingredient Decoction to Warm the Gallbladder].

Related condition and modification are

- Poor appetite and loose stool with undigested food: Add burned Shan Zha and fried Shen Qu.

Qi and Yin Deficiency Syndrome

Treatment Principle: Tonify Qi, nourish Yin, and recover pulse.

Use Zhi Gan Cao Tang [Honey-Fried Licorice Decoction].

Heart and Kidney Yang Deficiency Syndrome

Treatment Principle: Strengthen Heart and Kidney, and tonify Yang Qi.

Use Shen Fu Tang [Ginseng and Prepared Aconite Decoction] plus Gan Jiang, Rou Gui, Zhi Gan Cao, and Gou Qi Zi.

Related conditions and modifications are

- Cough and wheezing: Add Sang Bai Pi, Ting Li Zi, and Da Zao.
- Edema: Add Huang Qi, Fu Ling, and Gui Zhi.

CORONARY HEART DISEASE

In this condition, atheromas build up on the wall of coronary arteries, which leads to narrowing of arteries, obstructing blood flow, and decreasing the blood and oxygen supply to the heart muscle. TCM treatments are more effective in treating Latent Coronary Heart Disease which is also called Asymptomatic Coronary Heart Disease. Treatment should be implemented when there are signs of blood stasis, phlegm obstruction, Qi stagnation, or cold congealed at the heart. Other symptoms may include fat constitution, hyperlipidemia, preference for fatty, greasy food, and lack of exercise. Early implementation of herbal treatments can prevent angina pectoris or myocardial infarction. Some specialists in this area found herbs that nourish blood, and Yin, invigorate blood and expel phlegm can lower blood lipids and elevate HDL. These include He Shou Wu, Shan Zha, Ling Zhi, Du Zhong, Sang Ji Sheng, Huang Jing, Gua Lou Pi, Cao Jue Ming, Ze Xie, Hei Zhi Ma, Ge Gen, Di Gu Pi, Hong Hua, Ci Wu Jia, Yu Jin, San Qi, Da Suan, Dan Shen, Hu Zhang, Mo Yao, Da Huang, Sang Shen, and Dong Chong Xia Cao. Other invigorating blood and nourishing Yin herbs can lower the viscosity of the blood, such as Shan Zhu Yu, Sheng Di Huang, Sang Shen, Gou Qi Zi, Mai Men Dong, Wu Wei Zi, Dan Shen, Hong Hua, Shui Zhi, Chi Shao Yao, He Shou Wu, Nu Zhen Zi, Han Lian Cao, Yi Mu Cao, Gui Ban, and Xuan Shen.

One can also use the following formula as a base which mainly tonifies Qi and invigorates blood, modified according the presenting symptoms. The formula consists of Huang Qi, Zhi Shou Wu, Dan Shen, Ge Gen, and Chuan Xiong.

Related conditions and modifications are

- Severe blood stasis: Add Shi Xiao San [Sudden Smile Powder], Chi Shao Yao, Jiang Xiang, and Yan Hu Suo. If blood stasis is even more severe, then add Quan Xie and Gui Wei.
- Qi and Yang deficiency: Add Gui Zhi Gan Cao Tang [Cinnamon Twig and Licorice Decoction] plus Gao Liang Jiang, and Bi Bo.

- Severe damp phlegm: Add Gua Lou Xie Bai Ban Xia Tang [Trichosanthes Fruit, Chinese Chive, and Pinellia Decoction], Yu Jin, Zhi Ke, and Shi Chang Pu.
- Severe phlegm heat: Add Yin Chen Hao, Huang Qin, Mu Dan Pi, Shan Zhi Zi, and Yu Jin.
- Rapid pulse: Add Sheng Mai San [Generate the Pulse Powder], Sheng Di Huang, Suan Zao Ren, and Shan Zhu Yu.
- Slow pulse: Add Hong Shen, Rou Gui, Yin Yang Huo, Dang Gui, Xi Xin, and Ba Ji Tian.
- Knotted and intermittent pulse: Add Zhi Gan Cao and Ku Shen.
- Constipation: Add Gua Lou Ren and Ji Xue Teng. In severe cases, steep Da Huang and drink. Discontinue when bowel movement occurs.
- Insomnia: Add Suan Zao Ren and Ye Jiao Teng.

HYPERTENSION

Hypertension is discussed under dizziness, vertigo, and headache in TCM. The pathogeneses are Liver, Kidney, Heart and Spleen deficiencies, turbid phlegm, blood stasis, and hyperactive Yang. The clinical manifestations at the beginning stage or in mild cases are dizziness, distending sensation of the head, tinnitus, vertigo, and insomnia. In chronic or severe cases, there can also be damage to the heart, brain, and kidney. TCM treatment not only expels wind, fire, phlegm, and stasis, but also regulates liver, kidney, heart, and spleen. From modern research, herbs that can lower the blood pressure include Du Zhong, Huang Qin, Luo Bu Ma, Xiao Ji, Mu Xiang, Ma Dou Ling, Xia Ku Cao, Ye Ju Hua, Cao Jue Ming, Mu Zei, Huang Lian, Huang Bai, Shan Zhi Zi, Ge Gen, and Long Dan Cao. These herbs can be added to the formulas according to the syndrome differentiation.

Liver Fire Blazing Syndrome

Treatment Principle: Calm Liver, drain heat, and descend fire.

Use Long Dan Xie Gan Tang [Gentiana Longdancao Decoction to Drain the Liver] plus Ju Hua, Xia Ku Cao and Shi Jue Ming.

Related conditions and modifications are

- Dry yellow tongue coating and constipation: Add Zhi Shi, Hou Po, Lu Hui, and Sheng Da Huang.
- Severe headache, dizziness, and vertigo: Add Zhen Zhu Mu and Bai Ji Li.
- Numbness of extremities: Add Fa Xia, Zhu Ru, Dan Nan Xing, and Xi Xian Cao.
- Neck stiffness and pain: Add Bai Shao Yao, Di Long, and Ge Gen.

Yin Deficiency and Hyperactivity of Yang Syndrome

Treatment Principle: Nourish Yin and calm Liver.

Use Tian Ma Gou Teng Yin [Gastrodia and Uncaria Decoction] plus Sheng Di Huang and Ju Hua.

Related conditions and modifications are

- Dry mouth and tongue: Add Xuan Shen and Shi Hu.
- Constipation: Add Huo Ma Ren and Bai Zi Ren.
- Severe headache, numbness at limbs, and twitching: Add Dai Zhe Shi, Bai Shao Yao, Tian Men Dong, Bai Ji Li, Chuan Lian Zi, Long Gu, and Mu Li.

Liver and Kidney Yin Deficiency Syndrome

Treatment Principle: Nourish Liver and Kidney.

Use Yi Guan Jian [Linking Decoction] plus Zhi Shou Wu, Huai Niu Xi, Nu Zhen Zi, Tu Si Zi, Du Zhong, Sang Ji Sheng, Han Lian Cao, and Gui Ban Jiao.

Related conditions and modifications are

- Weakness of lumbar and knees: Add Xu Duan.
- Dry eyes: Add Cao Jue Ming and Gou Qi Zi.
- Tinnitus and vertigo: Add Ju Hua and Ci Shi.
- Frequent nocturia: Add Fu Pen Zi and Sang Piao Xiao.
- Heart and Kidney disharmony: Add Gui Zhi, Huang Lian, and Suan Zao Ren.

Yin and Yang Deficiency Syndrome

Treatment Principle: Nourish Yin and assist Yang.

Use Jin Gui Shen Qi Wan [Kidney Qi Pill from the Golden Cabinet] plus Du Zhong and Huai Niu Xi.

Related conditions and modifications are

- Five-center heat, malar flush, and insomnia: Remove Rou Gui, add He Shou Wu, Gui Ban, and Bie Jia.
- Menopause, dizziness, vertigo, tinnitus, palpitation, insomnia, excessive dreaming, irritability, heat, and night sweats: Change to Er Xian Tang [Two-Immortal Decoction].

Phlegm Damp Obstruction and Rebellion Syndrome

Treatment Principle: Expel phlegm and transform dampness.

Use Wen Dan Tang [Warm the Gallbladder Decoction] plus Cang Zhu, Hou Po and Huang Lian.

Related conditions and modifications are

- Severe dizziness, vertigo, and heaviness at head: Add Jiang Can, Dan Nan Xing, and Shi Chang Pu.
- Copious phlegm and stifling sensation at chest: Add Jie Geng, Gua Lou Pi, and Sheng Jiang.
- Yellow phlegm and bitter taste in mouth: Add Ju Hua, Huang Qin, and Zhe Bei Mu.
- Concomitant exterior syndrome, scanty urine, lower abdominal distention, pale tongue, and white moist tongue coating: Change to Wu Ling San [Five-Ingredient Powder with Poria] plus Huo Xiang, Su Zi, and Hou Po.

Qi Deficiency and Blood Stasis Syndrome

Treatment Principle: Benefit Qi, nourish Yin, and invigorate blood.

Use Guan Xin Er Hao Fang [Number 2 Formula for Coronary Heart Disease] plus Huang Qi, Ze Xie, Ge Gen, and Chuan Niu Xi.

Related conditions and modifications are

- Headache, red face, and dry stool: Add Ju Hua, Zhi Ke, Tao Ren, and Gou Teng.
- Qi stagnation: Add Xie Bai, Guo Lou Pi, and Yu Jin.
- Turbid phlegm: Add Ban Xia, Dan Nan Xing, Di Long, and Jiang Can.
- Yin deficiency: Add Sheng Di Huang and Gou Qi Zi.
- Yang deficiency: Add Tu Si Zi and Du Zhong.

HYPOTENSION

A person with a blood pressure lower than 80/60 mm Hg is considered hypotensive. In TCM, it is related to dizziness, vertigo, or syncope. The causes include a weak prenatal constitution and failing to be nourished postnatal, chronic sickness, fatigue, and exhaustion. These result in deficiencies of

Qi, blood, and Zang Fu organs failing to nourish the Heart and brain and to fill the blood vessels. TCM treatments focus on tonifying the deficiency. It is more effective in treating primary hypotension.

Both Qi and Yin Deficiency Syndrome

Treatment Principle: Benefit Qi and nourish Yin.

Use Sheng Mai San [Generate the Pulse Powder] and Zhi Gan Cao Tang [Honey-Fried Licorice Decoction] plus Zhi Shou Wu, Huang Qi, and Wu Wei Zi.

Related conditions and modifications are

- Blood stasis: Add Dan Shen, Gui Zhi, and Tan Xiang.
- Yang deficiency, cold body, and limbs: Add Zhi Fu Zi and Rou Gui.
- Blazing deficient fire, dry mouth, and throat: Add Xuan Shen and Zhi Mu.

Both Heart and Spleen Deficiency Syndrome

Treatment Principle: Benefit Qi, nourish blood, strengthen Spleen, and calm the Heart.

Use Gui Pi Tang [Restore the Spleen Decoction] plus Gui Zhi and Zhi Ke.

Related condition and modification are

- Food stagnation: Add Mai Ya, Ji Nei Jin, and Chen Pi.

Spleen and Kidney Deficiency Syndrome

Treatment Principle: Strengthen Spleen and tonify Kidney.

Use Si Jun Zi Tang [Four-Gentleman Decoction] and Ren Shen Hu Tao Tang [Ginseng and Walnut Decoction] and Gou Qi Zi, Huang Jing, Yuan Zhi, Suan Zao Ren, Huang Qi, and Bu Gu Zhi.

Related condition and modification are

- 5 A.M. diarrhea: Add Rou Dou Kou and Bu Gu Zhi.

Liver and Kidney Yin Deficiency Syndrome

Treatment Principle: Nourish Liver and Kidney and descend fire.

Use Qi Ju Di Huang Wan [Lycium Fruit, Chrysanthemum, and Rehmannia Pill] plus Dang Gui, Mai Men Dong, Dang Shen, and Huang Qi.

Related conditions and modifications are

- Hypochondriac distention and discomfort: Add Chuan Lian Zi, Zhi Ke, and Yu Jin.
- Malar flush, dry painful throat, and red tongue: Add Zhi Mu, Huang Bai, and Di Gu Pi.

Lower Source Deficiency Syndrome

Treatment Principle: Supplement the lower source.

Use Ren Shen Ge Jie San [Ginseng and Gecko Powder].

Related conditions and modifications are

- Slight Yin deficiency: Use Zuo Gui Wan [Restore the Left (Kidney) Pill].
- Slight Yang deficiency: Use You Gui Wan [Restore the Right (Kidney) Pill].
- Leakage of essence: Add Jin Ying Zi and Qian Shi.

Heart and Kidney Yang Deficiency Syndrome

Treatment Principle: Warm and tonify Heart and Kidney, arouse Yang Qi.

Use Yang Xin Tang [Nourish the Heart Decoction] plus Jin Gui Shen Qi Wan [Kidney Qi Pill from the Golden Cabinet].

Related conditions and modifications are

- Copious nocturia: Add Yi Zhi Ren and Tu Si Zi.
- Swelling of the lower limbs and difficult urination: Add Ze Xie and Che Qian Zi.
- Purple lips and dark purple tongue: Add Dan Shen and Hong Hua.

MYOCARDIAL DISEASE

This condition is discussed under palpitation, palpitation with fright, chest Bi syndrome, and syncope. There often is a congenital abnormal constitution making it susceptible to evil pathogen invasion or overexhaustion leading to blood stasis, Bi obstruction, and blocked heart vessels. The treatment focuses are to unblock channels and collaterals, invigorate the blood, and transform stasis. The main symptoms are difficulty breathing, pain in heart area, dizziness, fatigue, palpitation, fainting, or syncope. At late stage, there can be heart failure, irregular heart beat, and ventricular fibrillation.

Qi Stagnation and Blood Stasis Syndrome

Treatment Principle: Invigorate blood, transform stasis, move Qi, and stop pain.

Use Xue Fu Zhu Yu Tang [Drive Out Stasis in the Mansion of Blood Decoction] plus Dan Shen, Yan Hu Suo, and Yu Jin.

Related conditions and modifications are

- Five-center heat: Add Huang Lian and Shan Zhi Zi.
- Severe chest pain: Add Tan Xiang and Jiang Xiang.

Accumulated Phlegm and Blood Stasis Syndrome

Treatment Principle: Transform stasis, expel phlegm, open the chest, and unblock Yang.

Use Tao Hong Si Wu Tang [Four Substances Decoction with Peach Pit and Safflower] plus Gua Lou Xie Bai Ban Xia Tang [Trichosanthes Fruit, Chinese Chive, and Pinellia Decoction] and Dan Nan Xing and Zhu Ru.

Related conditions and modifications are

- Severe damp phlegm: Add Sheng Jiang, Chen Pi, Xing Ren, Fu Ling, and Ban Xia.
- Chest pain radiating to the back: Add Ru Xiang, Mo Yao, Wu Ling Zhi, and Pu Huang.

Both Qi and Yin Deficiency Syndrome

Treatment Principle: Benefit Qi, nourish Yin, invigorate blood, and unblock collaterals.

Use Sheng Mai San [Generate the Pulse Powder] plus Ren Shen Yang Ying Tang [Ginseng Decoction to Nourish the Nutritive Qi].

Related condition and modification are

- Shortness of breath: Add Tai Zi Shen, Xi Yang Shen, and Huang Jing.
- Dizziness, vertigo, and facial heat: Zhi Shou Wu, Nu Zhen Zi, and Bie Jia.
- Dry stool, dry mouth, and tongue: Add Rou Cong Rong and Shi Hu.

Congested Fluid Attacking Heart Syndrome

Treatment Principle: Warm Yang and drain water.

Use Zhen Wu Tang [True Warrior Decoction] plus Zhu Ling, Gui Zhi, and Chen Pi.

Related conditions and modifications are

- Palpitation, cough, wheezing, inability to lie flat, and difficult urination: Add Ting Li Zi and Da Zao.
- Severe edema: Add Da Fu Pi, Che Qian Cao, and Ze Xie.

Yang Qi About to Collapse Syndrome

Treatment Principle: Return Yang and rescue the collapse.

Use Shen Fu Tang [Ginseng and Prepared Aconite Decoction] plus Wu Wei Zi, Long Gu, and Mu Li.

PREEXCITATION SYNDROME

This syndrome refers to the early conduction of the atrial impulse, which causes the preexcitation of the whole or part of the ventricle, or reverse conduction of ventricular impulse, which causes preexcitation of whole or part of the atrium. The abnormality can be observed on EKG with tachycardia for this syndrome. It is discussed under palpitation, palpitation with fright, and syncope in TCM. The pathogeneses are stagnation and stasis of the Qi mechanisms in heart channels. In mild cases, there is no symptom, but in severe cases there can be severe irregular heart beat. The treatment focuses on smoothing Liver, relieving stagnation, tonifying Qi, and nourishing Yin.

Liver Qi Stagnation Syndrome

Treatment Principle: Smooth Liver and relieve stagnation.

Use Dan Zhi Xiao Yao San [Relax Wanderer Powder with Moutan and Gardenia].

Related conditions and modifications are

- For chest stifling sensation: Add Dan Shen and Chi Shao Yao.
- Liver Qi stagnation transforming into fire: Add Long Dan Cao and Ci Shi.

Phlegm Fire Disturbing the Heart Syndrome

Treatment Principle: Drain fire, scour phlegm, clear Heart, and calm the spirit.

Use Huang Lian Wen Dan Tang [Coptis Decoction to Warm the Gallbladder] plus Suan Zao Ren, Shi Chang Pu, Yu Jin, Shan Zhi Zi, Zhe Bei Mu, Huang Qin, and Guan Lou Pi.

Related conditions and modifications are

- Palpitation with uneasy feeling: Add Zhen Zhu Mu, Long Gu, and Mu Li.

Both Qi and Yin Deficiency Syndrome

Treatment Principle: Benefit Qi, nourish Yin, and restore pulse.
Use Zhi Gan Cao Tang [Honey-Fried Licorice Decoction].
Related conditions and modifications are

- Qi deficiency and blood stasis: Add Huang Qi, Chi Shao Yao, and Chuan Xiong.
- Yin deficiency and deficient fire: Add Huang Lian, Shan Zhi Zi, and Di Gu Pi.

REYNAUD'S DISEASE

This condition belongs to cold syncope or channel Bi syndrome. It is caused by cold stimulation, emotional stress, and other reasons, which lead to clonic spasm of peripheral capillaries. Thus, there is intermittent change of skin color in hands and feet. The pathological changes can be divided into three stages: Spastic ischemic period, dilation period, and congestion period. It is often seen in females 20 to 30 years of age. Pathogenesis of this condition is blood stasis obstructing the distal collaterals and channels of the limbs and is often seen as cold congeal and blood stasis. The treatment focuses on invigorating blood, transforming stasis, relieving spasms of the blood vessels, and improving blood circulations. One can use the formula which consists of Dang Gui, Ji Xue Teng, Wu Shao She, Di Long, Tao Ren, Hong Hua, Shui Zhi, Jiang Can, and Wu Gong.

Related conditions and modifications are

- Upper extremities conditions: Add Sang Zhi and Jiang Huang.
- Lower extremities conditions: Add Chuan Niu Xi.
- Cold congeal and blood stasis: Add Gui Zhi, Gan Jiang, and Zhi Fu Zi.
- Severe case of blood stasis obstructing channels: Add San Leng, E Zhu, and San Qi.
- Emotional stress leading to attacks: Add Chai Hu, Yan Hu Suo, Bai Shao Yao, Yu Jin, and Xiang Fu.
- Swelling stasis at end of extremities: Add Quan Xie.
- Severe pain: Add Zhi Ru Xiang, Mo Yao, Yan Hu Suo, and Ying Su Ke.
- Chronic condition, rough skin, or ulcerations: Add Huang Qi, Tai Zi Shen, and Shu Di Huang.

Furthermore, one can also add external wash, using one of the following:

1. Wen Jing Zhi Tong Xi Fang [Warm the Channel and Stop Pain Wash Formula]: Chuan Wu, Cao Wu, Xi Xin, Su Mu, Chuan Jiao, Bai Zhi, Ma Huang, and Hong Hua.
2. Huo Xue Tong Jing Xi Fang [Invigorate Blood and Unblock Channel Wash Formula]: Tou Gu Cao, Shen Jin Cao, Su Mu, Chuan Xiong, Bai Zhi, Hong Hua, and Ru Xiang.
3. Jie Du Xiao Zhong Xi Fang [Relieve Toxicity and Reduce Swelling Wash Formula]: Jin Yin Hua, Huang Bai, Pu Gong Ying, Bai Jiang Cao, Dang Gui, Ru Xiang, Bai Wei, and Da Huang.

RHEUMATIC VALVULAR HEART DISEASE

This condition belongs to heart Bi, palpitation, and edema in TCM. It is caused by a combination of wind, cold, damp, heat, and evil Qi. The pathogenesis are Heart Qi insufficiency, stasis and obstruction of the heart blood, which chronically lead to various related syndromes. In Western medicine, it is said that chronic heart valvulopathy occurs after rheumatic carditis. It is most often seen in females between the age of 20 to 40. The clinical manifestations at the beginning stage are palpitation, rapid breathing, stifling sensation, red face, purple lips, and increased contraction of the heart after exercise. At the late stage, there can be left or right heart failure. TCM treatment focuses on benefiting Qi, nourishing Heart, invigorating blood, and transforming stasis, especially for the symptoms at beginning stage.

Heart Qi Deficiency Syndrome

Treatment Principle: Benefit Qi and nourish Heart.

Use Yang Xin Tang [Nourish the Heart Decoction] plus Huang Qi, Gui Zhi, Chuan Xiong, and Suan Zao Ren.

Related conditions and modifications are

- Palpitation, scanty food intake, loose stool, and fatigue: Add Bai Zhu, Long Yan Rou, and Mu Xiang.
- Knotted or intermittent pulse, shortness of breath, and scanty tongue coating: Add Sheng Di Huang, E Jiao, and Mai Men Dong.

Bi Obstruction of the Heart Channel Syndrome

Treatment Principle: Invigorate blood, transform stasis, and unblock channel.

Use Tao Hong Si Wu Tang [Four-Substance Decoction with Peach Pit and Safflower] plus Dan Shen, Dang Shen, Huang Qi, Bai Zhu, Fu Ling, Zhi Gan Cao, Xiang Fu, and Jiang Huang.

Related conditions and modifications are

- Severe blood stasis: Add Shui Zhi, Mang Chong, and Tu Bie Chong.
- Severe palpitation: Add Suan Zao Ren and Bai Zi Ren.

Heart and Spleen Yang Deficiency Syndrome

Treatment Principle: Strengthen Spleen, leach out dampness, and restore Yang.

Use Ling Gui Zhu Gan Tang [Poria, Cinnamon Twig, Atractylodes Macrocephala, and Licorice Decoction] plus Hou Po, Ze Xie, Dan Shen, Chi Shao Yao, Fang Ji, and Huang Qi.

Related conditions and modifications are

- Severe Yang deficiency: Add Zhi Fu Zi.
- Abdominal distention: Add Da Fu Pi and Mu Xiang.
- Qi deficiency and spontaneous sweating: Add Ren Shen, Wu Wei Zi, calcined Long Gu, and calcined Mu Li.

VIRAL MYOCARDITIS

This disease includes acute or chronic inflammation at the myocardium due to various viruses. It belongs to Wind Warm, palpitation, Heart Bi, or chest Bi syndrome. Patient has a deficient Zheng Qi and then is attacked by warm or heat evil toxins. Thus, the root cause is the Heart Qi and blood deficiency and the manifestations are heat toxins, phlegm damp, and blood stasis. The treatment principles at the acute stage are clearing heat, relieving toxicity, and assisting by benefiting Qi and nourishing Yin. At chronic stage, treatment is to tonify Yang and nourish Yin, and is assisted by benefiting Qi and unblocking collaterals.

Acute Stage

The main cause is heat toxin invasion. In severe cases, patient can present with Heart Yang deficiency.

Wind–Heat Attacking Lung, Heat-Toxins Attacking Heart Syndrome

Treatment Principle: Disperse wind, clear heat, relieve toxicity, benefit Qi, and nourish Yin.

Use Yin Qiao San [Honeysuckle and Forsythia Powder] plus Da Qing Ye, Chan Tui, Tai Zi Shen, Xuan Shen, and Mai Men Dong.

Related conditions and modifications are

- Severe cough: Add She Gan, Tu Niu Xi, and Shan Dou Gen.
- Frequent premature beats: Add Ku Shen, Dan Shen, and Yu Jin.
- Severe heat: Add Huang Qin, Shan Zhi Zi, Zhi Mu, and Shi Gao.

Damp–Heat Obstructing the Middle Jiao, Internal Damage of Heart Ying Syndrome

Treatment Principle: Clear heat, relieve toxicity, transform dampness, and transport Spleen.

Use Gan Lu Xiao Du Dan [Sweet Dew Special Pill to Eliminate Toxin] plus Huang Qin and Ban Lan Gen.

Related conditions and modifications are

- Palpitation with fright: Add Bai Zi Ren and Suan Zao Ren.
- Stifling sensation at chest: Add Gua Lou Pi and Ban Xia.

Residual Heat, Damaged Heat Yin Syndrome

Treatment Principle: Clear residual heat and nourish Heart Yin.

Use Yu Nu Jian [Jade Woman Decoction] plus Ban Lan Gen, Dan Shen, Tu Niu Xi, and Mu Dan Pi.

Related condition and modification are

- Tidal fever and night sweats: Add Qing Hao and Bai Wei.

Heart Yang Deficient Collapse Syndrome

Treatment Principle: Warm Yang and rescue collapse.

Use Shen Fu Tang [Ginseng and Prepared Aconite Decoction] plus Gan Jiang, Zhi Gan Cao, Long Gu, and Mu Li.

Related condition and modification are

- Lowering blood pressure: Add Sheng Mai San [Generate the Pulse Powder].

Chronic Stage or Recovery Stage

Mainly deficient Zheng Qi and residual evil pathogens.

Both Qi and Yin Deficiency Syndrome

Treatment Principle: Tonify and benefit Qi and Yin, nourish Heart, and calm spirit.

Use Zhi Gan Cao Tang [Honey-Fried Licorice Decoction] plus Dan Shen, Xuan Shen, Bai Shao Yao, and Wu Wei Zi.

Related conditions and modifications are

- Spontaneous sweating and fatigue: Add Huang Qi and calcined Mu Li.
- Severe Yin deficiency: Add Zhu Ye, Yu Zhu, Nu Zhen Zi, and Han Lian Cao.

Yin Deficiency with Deficient Fire Syndrome

Treatment Principle: Nourish Yin, descend fire, nourish Heart, and calm spirit.

Use Tian Wang Bu Xin Dan [Emperor of Heaven's Special Pill to Tonify the Heart] plus Qing Hao, Di Gu Pi, and Zhi Mu.

Related condition and modification are

- Night sweats: Add Fu Xiao Mai and Mu Li.

Heart and Spleen Qi Deficiency Syndrome

Treatment Principle: Benefit Qi, nourish Heart, and calm Shen.

Use Gui Pi Tang [Restore the Spleen Decoction] plus Bai Shao Yao, Wu Wei Zi, and Mu Xiang.

Heart and Spleen Qi Deficiency Syndrome

Treatment Principle: Benefit Qi, unblock Yang, invigorate blood, and transform stasis.

Use Tao Hong Si Wu Tang [Four-Substance Decoction with Peach Pit and Safflower] plus Huang Qi, Dan Shen, Gui Zhi, and Xie Bai.

Related conditions and modifications are

- Stifling sensation at chest: Add Gua Lou Pi and Yu Jin.
- Severe chest pain: Add Yan Hu Suo and San Qi.

Heart and Spleen Yang Deficiency Syndrome

Treatment Principle: Strengthen Spleen, drain water, warm Yang, and arrest palpitation.

Use Ling Gui Zhu Gan Tang [Poria, Cinnamon Twig, Atractylodes Macrocephala, and Licorice Decoction] plus Dan Shen, Da Zao, Sheng Jiang, and Chuan Xiong.

Related conditions and modifications are

- Nausea and Qi rebellion: Add Chen Pi and Ban Xia.
- Severe lower limb edema: Add Huang Qi, Fang Ji, and Fu Zi.
- Slow moderate pulse: Add Fu Zi and Xi Xin.

Heart and Kidney Yang Deficiency Syndrome

Treatment Principle: Warm Yang, drain water, unblock collaterals, and calm Heart.

Use Zhen Wu Tang [True Warrior Decoction] plus Ze Xie, Gui Zhi, Dan Shen, and Zhi Gan Cao, and substitute Bai Shao Yao with Chi Shao Yao.

Related conditions and modifications are

- Severe edema: Add Sheng Jiang, Che Qian Zi, and Fang Ji.
- Rapid wheezing: Add Xi Xin and Wu Wei Zi.

CHAPTER 14

Dermatological Disorders

ACNE

This condition is caused by constitution heat excess in Qi and blood, or accumulated heat in Lung and Stomach which move upward along the meridians, congesting at chest and face.

In Western medicine, it is described as a chronic inflammation of the hair follicle and sebaceous glands. It is also possible to relate to internal increased secretion of androgen. The characteristics are manifestations on face, chest, and back which can form black heads, papule, pustule, nodules, and cysts. It often occurs during teenage years and menopausal period. TCM treatments focus on clearing and dispersing accumulated heat from Lung and Stomach, relieving toxicity, scattering accumulations. They also focus on unblocking the six Fu organs to assist tonification.

Accumulated Heat in Lung and Stomach Syndrome

Treatment Principle: Clear and drain Lung and Stomach, unblock stool, and flatten the skin lesion.

Use Pi Pa Qing Fei Yin [Loquat Leaf Decoction to Clear the Lung] remove Ren Shen, and add Lian Qiao, Huang Qin, Chi Xiao Dou, Sheng Di Huang, Mu Dan Pi, Hong Hua, Zhi Ke, Jin Yin Hua, Zhi Da Huang, and Bai Hua She She Cao.

Qi and Blood Stagnation Syndrome

Treatment Principle: Clear Lung, cool blood, regulate Qi, and transform stasis.

Use Liang Ying Qing Qi Tang [Cool the Nutritive Level and Clear the Qi Decoction] plus Dan Shen and Yi Mu Cao.

Accumulated Knotted Phlegm and Stasis Syndrome

Treatment Principle: Eliminate phlegm, soften hardness, invigorate blood, and scatter accumulation.

Use Qing Qi Hua Tan Wan [Clear the Qi and Transform Phlegm Pill] plus Jin Yin Hua, Dan Shen, Xia Ku Cao, Pu Gong Ying, Sheng Mu Li, Chi Shao Yao, Dang Gui, and Jiang Can.

Related condition and modification are

- Dark excess scars: Add Da Huang Zhe Cong Wan [Rhubarb and Eupolyphaga Pill].

The above formulas can be modified according to presenting symptoms.

Related conditions and modifications are

- Red face: Add Mei Gui Hua (rose flower), fried Huai Hua, and Han Shui Shi.
- Copious pustule: Add Zi Hua Di Ding, Ye Ju Hua, and Zao Xiu.
- Copious swelling nodules: Add Huang Yao Zi, Zao Jiao Ci, and Kun Bu.
- Greasy oily skin: Add Hu Zhang, Chi Fu Ling, and Yin Chen Hao.
- Worsening before periods or during menopause period: Add Yi Mu Cao, Bai Shao Yao, Xian Mao, Yin Yang Huo, and Zhi Mu.

ALOPECIA

Alopecia is often caused by seven emotional disturbances, blood–heat, accumulated heat in Stomach and Intestines, and Liver and Kidney Qi blood deficiencies. In Western medicine, this is a sudden loss of hair in patches. The cause is not fully understood. Clinical characteristics are sudden round or oval patches of hair loss, no fixed amount of patches, absence of local inflammation; smooth, flat, skin surface, and no other symptoms. TCM treatment focuses on clearing and unblocking for excess syndromes. For deficient syndromes, the treatment is to tonify and astringe.

Blood–Heat Generating Wind Syndrome (Rapid Excess Hair Loss)

Treatment Principle: Cool blood, extinguish wind, and nourish Yin.

Use Si Wu Tang [Four-Substance Decoction] and Liu Wei Di Huang Wan [Six-Ingredient Pill with Rehmannia Pill], Er Zhi Wan [Two-Ultimate Pill] plus Xuan Shen and Bai Fu Zi.

Blood Stasis Obstructing Hair Orifices Syndrome (Patch Hair Loss)

Treatment Principle: Open orifices and invigorate blood.

Use Tong Jing Zhu Yu Tang [Unblock Channels and Drive Out Stasis Decoction] plus He Shou Wu.

Accumulated Heat in Lung and Stomach Syndrome (Expanding Type of Hair Loss from the Vertex)

Treatment Principle: Clear and drain Lung and Stomach.

Use Xie Bai San [Drain the White Powder] and Qing Wei San [Clear the Stomach Powder] with Fan Xie Ye 15 g, steep, and drink as tea.

Qi and Blood Deficiency Syndrome (Chronic Hair Loss)

Treatment Principle: Tonify Qi and nourish blood.

Use Shi Quan Da Bu Wan [All-Inclusive Great Tonifying Decoction] plus He Shou Wu, Nu Zhen Zi, and Yuan Zhi.

Liver and Kidney Deficiency Syndrome (Large Area of Even Hair Loss with White Hair)

Treatment Principle: Nourish Liver and Kidney.

Use Qi Bao Mei Ran Dan [Seven-Treasure Special Pill for Beautiful Whiskers] plus Er Zhi Wan [Two-Ultimate Pill] and Hei Zhi Ma, Shan Zhu Yu, Shi Chang Pu, and Rou Cong Rong.

Above five syndromes can be modified according to the following:

- Severe wind and heat, itchy scalp: Add Sang Bai Pi, Tian Ma, and Chong Wei Zi.
- Excessive nightmares or difficulty falling asleep: Add Ye Jiao Teng, Suan Zao Ren, He Huan Pi, Long Gu, and Mu Li.
- Depression and excessive dreaming: Add Xiang Fu and Yu Jin.
- Abdominal distention after meal, selective in eating: Add Gu Ya, Ji Nei Jin, Hou Po, and Mei Gui Hua.
- Greasy scalp, itching like bugs crawling: Add Ku Shen, Wei Ling Xian, Bai Xian Pi, and Di Fu Zi.

ECZEMA

This is often caused by damp–heat obstruction with Ying and Wei disharmony. In Western medicine, it is also called allergic contact dermatitis. It is a common type of allergic inflammation of the skin. The clinical manifestations are irregularly shaped rashes. In the acute stage, there are papules, blisters, or exudates, and in the chronic stage, there are skin damages, local infiltration, thickening, and severe itching; symptoms easily reoccur. TCM treatment focuses on strengthening Spleen and eliminating dampness. In chronic cases, it should combine with nourish Yin and moisten dryness. In general, the treatment is effective.

Acute Eczema

Damp–Heat Excess Syndrome

Treatment Principle: Clear heat, cool blood, eliminate damp, and stop itching.

Use Long Dan Xie Gan Tang [Gentiana Longdancao Decoction to Drain the Liver] plus Da Qing Ye and Hua Shi.

Related conditions and modifications are

- Red burning eczema: Add Mu Dan Pi and Chi Shao Yao.
- Severe itching: Add Di Fu Zi and Bai Xian Pi.
- Dry stool: Steep Fan Xie Ye and drink as tea.

Concomitant Wind and Damp Syndrome

Treatment Principle: Expel wind, eliminate damp, clear heat, and stop itching.

Use Xiao Feng San [Eliminate Wind Powder].

Related conditions and modifications are

- Greasy tongue coating: Add Huo Xiang, Pei Lan, and Hua Shi.
- Eczema at lower extremities: Add Bei Xie, Huang Bai, and Yi Yi Ren.
- Eczema at upper extremities: Add Qiang Huo and Chan Tui.

Wind–Damp–Heat Syndrome

Treatment Principle: Expel wind, eliminate damp, and clear heat.

Use Jing Jie, Fang Feng, Qiang Huo, Ze Xie, Cang Zhu, Huang Qin, Yi Yi Ren, Chuan Xiong, and Lian Qiao.

Furthermore, one can also cook and eat 60 g of Yi Yi Ren, 30 g of Bai Bian Dou, and 30 g of Lu Dou (mung bean) as congee.

Chronic Eczema

Damaged Yin and Residual Damp Syndrome

Treatment Principle: Strengthen Spleen, eliminate damp, and nourish Yin.

Use Sha Shen Mai Dong Tang [Glehnia and Ophiopogonis Decoction] plus Sheng Di Huang, Dang Gui, Fu Ling, Bai Zhu, Shan Yao, Ze Xie, and Bai Xian Pi.

Related conditions and modifications are

- Dry, itchy skin rash: Increase Sha Shen to 20 g, add Shi Hu and Huo Ma Ren.
- Damp, itchy skin rash: Add Di Fu Zi and Bai Mao Gen.
- Blisters and moist skin rash: Add Hua Shi and Dong Gua Pi.

Furthermore, one can also use topical herbs:

Zhi Yang Run Fu Gao [Stop Itch and Moisten Skin Ointment]: Dang Gui 50 g, Bai Ji 30 g, Zi Cao 20 g, Bai Lian 30 g, Bing Pian 2 g, Bai La (wax) 15 g, and Ma You (sesame oil) 400 mL.

Boil sesame oil, fry the first four herbs, and then remove the dregs. Remove the oil from stove and add wax and Bing Pian to form ointment. Apply topically two to three times a day. This is effective in treating dry flaky, itching, pain, and thickening of the skin.

HERPES SIMPLEX

This is caused by invasion of wind–heat and damp–toxin to a person with weak Zang Fu organs. These pathogens follow the meridians and express outwardly on skin. In Western medicine, it is an acute inflammatory skin condition caused by simple herpes virus. The virus enters the body via respiratory tract, mouth, genital organs, and broken skin surface which manifests while the immune system becomes weak. It often occurs on the lip, mouth, or external genital organs. It manifests as single or clusters of small blisters. In general, the blisters heal in 1 to 2 weeks, but often recur. TCM treatments focus on expelling toxins and evil pathogens, and eliminating and scattering herpes simplex. During the attack period, treatment focuses on clearing heat, eliminating dampness, and relieving toxicity. During normal days, one should strengthen Spleen and tonify Qi.

Wind–Heat and Damp–Toxin Syndrome

Treatment Principle: Disperse wind, clear heat, transform damp, and relieve toxicity.

Use Qing Fei Yin [Clear the Lung Decoction] plus Da Qing Ye, Shi Gao, Pi Pa Ye, Sheng Ma, Yi Yi Ren, and Xuan Shen.

Qi and Yin Deficiency Syndrome

Treatment Principle: Tonify Qi and nourish Yin.

Use Ren Shen Gu Ben Wan [Ginseng Pill to Stabilize the Root] plus Sha Shen, Yi Yi Ren, Shan Yao, Huang Qi, Bai Shao Yao, Sheng Ma, Gan Cao, and Ban Lan Gen.

The above two syndromes can be modified according to the following:

- Rash around the orbits: Add Qing Xiang Zi, Ju Hua, and Sang Ye.
- Rash at external genital area: Add Long Dan Cao, Bai Mao Gen, and Che Qian Zi.
- Frequent occurrence of rash that is hard to heal: Add Xi Yang Shen, Bai Lian, and Lu Dou Yi (mung bean skin).
- Local sharp, itching, burning pain: Add Zi Cao, Shi Jue Ming, and Gou Teng.

Furthermore, one should also pay attention to daily treatment to improve immunity.

1. Tonify Spleen, Kidney, and benefit Qi: Sheng Di Huang 8000 g, Fu Ling 1500 g, Xi Yang Shen 750 g, and Feng Mi 5000 g. Decoct the herbs and make into syrup. Take 20 g twice a day.

HERPES ZOSTER

This condition is often caused by internal accumulation of damp–heat, transforming into fire and toxin, obstructing Qi mechanism and flowing in the channels. In Western medicine, this is called an acute inflammatory skin condition caused by viral infections. The clinical characteristics are clusters of small fluid filled blisters surrounded by a small red area. The blisters follow the affected nerve, thus it is often on one side of the body with belt distribution. It is accompanied by persistent nerve pain; or local, swollen, and painful lymph nodes, and blisters that often form scabs in 1 week. This most often occurs at intercostals nerves and the course of the condition lasts about 3 weeks. TCM treatments focuses on relieving toxicity, stopping pain, eliminating and scattering blisters and rashes, and draining damp and clearing heat.

Stop Pain

1. Use Shao Yao Gan Cao Tang [Peony and Licorice Decoction] plus Quan Xie, Ying Su Ke, Suan Zao Ren, and Yan Hu Suo. Take the decoction three times a day.
2. Acupuncture the corresponding section of Hua Tuo Jia Ji points and Ashi points.

Treatments According to Syndrome Differentiation

Damp–Heat Accumulation Syndrome

Treatment Principle: Clear heat and eliminate damp.

Use Long Dan Xie Gan Tang [Gentiana Longdancao Decoction to Drain the Liver] plus Ma Chi Xian. Related condition and modification are

- Excess heat: Add Huang Lian and Che Qian Cao.

Damp–Heat Transforming into Fire Syndrome

Treatment Principle: Clear heat, eliminate damp, and relieve toxicity.

Use Long Dan Xie Gan Tang [Gentiana Longdancao Decoction to Drain the Liver] plus Wu Wei Xiao Du Yin [Five-Ingredient Decoction to Eliminate Toxin] and Ban Lan Gen, Ju Hua, Jin Yin Hua, Zi Hua Di Ding, Ma Chi Xian, and Zao Xiu.

Heat–Toxin Entering Ying Syndrome

Treatment Principle: Drain fire, relieve toxicity, clear heat, and cool blood.

Use Wu Wei Xiao Du Yin [Five–Ingredient Decoction to Eliminate Toxin] plus Chi Shao Yao, Lian Qiao, Mu Dan Pi, Ban Lan Gen, Bai Jiang Cao, and Ma Chi Xian.

Damaged and Exhausted Yin and Blood Syndrome

Treatment Principle: Nourish blood, soften Liver, invigorate blood, and unblock collaterals.

Use Si Wu Tang [Four-Substance Decoction] plus Tian Men Dong, Bai He, Di Long, Zhi He Shou Wu, Quan Xie, and Gan Cao.

Related condition and modification are

- Local burning pain: Remove Chuan Xiong, add Huang Bai and Ren Dong Teng.

Furthermore, one can also use topical application: Grind Qing Dai 50 g, Huang Bai 50 g, Hua Shi 60 g, calcined Shi Gao 60 g, and Da Huang 20 g into extremely fine powder. Mix with sesame oil and apply topically two to four times a day. This is indicated before the blisters rupture. If blisters are already ruptured, then use Huang Lian Ruan Gao [Coptis Soft Gel].

NEURODERMATITIS

This is often caused by blood–heat generating wind, or blood deficiency damaging Yin, or wind evil pathogens accumulating at skin. In Western medicine, it is also called chronic simple lichen. This skin condition has main symptoms of paroxysmal itching and lichenification of the skin. The clinical characteristics are local itching and formation of flat papules after frequent scratching, which gradually develops into bigger patches over a long period of time with a shape similar to lichen. It often occurs on the back of the neck; or on the elbow, chest, and coccyx; or behind the ear. It often relates to nerves or psycho-emotions. TCM treatments focus on clearing heat, cooling blood, expelling wind, and stopping itching.

Blood–Heat Generating Wind, Failing to Nourish Skin Syndrome

Treatment Principle: Clear heat, cool blood, expel wind, and stop itching.

Use Chi Shao Yao, Gan Cao, Ku Shen, Huang Qin, Cang Er Zi, Mu Dan Pi, Di Fu Zi, Sheng Di Huang, and Dan Shen.

Blood Deficiency, Wind Dryness Failing to Nourish Skin Syndrome

Treatment Principle: Nourish blood, moisten dryness, extinguish wind, and stop itching.

Use Dang Gui, Shu Di Huang, Bai Shao Yao, Mai Men Dong, He Shou Wu, Bai Xian Pi, Bai Ji Li, Dan Shen, and Sang Shen Zi.

Related conditions and modifications are

- Severe itching at night: Add Ji Xue Teng, He Huan Pi, and Ye Jiao Teng.
- Heart restlessness and dry mouth: Add Zi Cao, Mu Dan Pi, and Shi Gao.
- Heart restlessness and insomnia: Add Dai Zhe Shi and Ci Shi.
- Thickening of the local skin: Add Sheng Mu Li.
- Thinning of the local skin: Add Chan Tui.
- Local dry pale skin: Add Hu Tao Ren, Huo Ma Ren, and Bai Zi Ren.

Wind-Evil Chronic Stagnation, and Failing to Nourish Qi and Blood Syndrome

Treatment Principle: Nourish blood, invigorate blood, expel wind, and stop itching.

Use Si Wu Tang [Four-Substance Decoction] plus Jing Jie, Fang Feng, Dan Shen, Chan Tui, Wu Shao She, and Chi Shao Yao.

Related conditions and modifications are

- Rough scale-like skin: Add Huang Jing and Ji Xue Teng.
- Local dark brown skin, rough to the touch: Add Tao Ren, Hong Hua, and San Leng.

Furthermore, local skin thickening like ox skin can also be treated with Bo Fu Gao [Thinning the Skin Ointment].

Chuan Shan Jia 10 g, fried San Leng 15 g, Dang Gui 15 g, E Zhu 15 g, Bai Xian Pi 12 g, Wu Bei Zi 20 g, Bing Pian 6 g, Zhang Nao 5 g, and Lu Gan Shi 30 g.

Grind the above herbs into fine powder and mix with sesame oil to form ointment. Apply topically two to three times a day.

PSORIASIS

This condition is often caused by blood-heat, wind-dryness, and heat stasis accumulating and obstructing the skin. In Western medicine, it is a common recurrent chronic inflammation of the skin with unknown reasons. It is often seen in children and young adults. The clinical characteristics during the progression period are irregular red papules covered by multilayers of white silver scaling, and blood spots under the scales when their thin membrane is ruptured with scratching. Patient feels itchy. The shape of the rash can be round, coin shaped, or maplike. During the silent period, the affected area is rough with scaling. If the scaling is removed, then one can see dark red skin with bleeding spots underneath. During the regression period, the inflammation is reduced at the local area. There is also decreased scaling, rash becomes smoother and pigmentation of surrounding area is decreased. During the progression period, TCM treatment focuses on clearing heat, relieving toxicity, cooling blood, and expelling wind. During the silent period, the treatments focuses on invigorating blood, transforming stasis, unblocking collaterals, and scattering accumulations. During regression period, the treatments focus on nourishing blood, moistening dryness, expelling wind, and stopping itch.

Progression Period

Wind–Heat Accumulation at Skin Internally Attacking Ying Level Syndrome

Treatment Principle: Clear heat, cool blood, disperse wind, and stop itching.

Use Yin Qiao San [Honeysuckle and Forsythia Powder] plus Ban Lan Gen, Huang Qin, Zi Cao, Bai Xian Pi, Shan Dou Gen, and Chi Shao Yao.

Internal Accumulation of Blood–Heat, Transforming into Dryness and Wind Syndrome

Treatment Principle: Clear heat, cool blood, relieve toxicity, and expel wind.

Use Xi Jiao Di Huang Tang [Rhinoceros Horn and Rehmannia Decoction] plus Zi Cao, Da Qing Ye, Bai Mao Gen, Ren Dong Teng, and Ku Shen.

Above two syndromes can be modified according to presenting symptoms as the following:

- More upper body psoriasis: Add Da Qing Ye.
- More lower body psoriasis: Add Ban Lan Gen.
- More at head: Add Qian Huo and Sheng Ma.
- More at back: Add Chan Tui and Fu Ping.
- More at chest: Add Zi Cao and Bai Zhi.
- Swollen painful throat: Add Shan Zhi Zi and Ma Bo.
- Thick and dry scaling: Add 30 g of Sheng Mu Li.
- Dark urine: Add Hua Shi and Mu Tong.
- Severe itching: Add Bai Xian Pi and Chan Tui.

Furthermore, one can also decoct 10 g of Jin Yin Hua and 6 g of Zi Cao and drink frequently.

Silent Period

Geographic distribution of patchy, thick, hard skin in coin-like shapes.

Blood Stasis Obstructing Collaterals Syndrome

Treatment Principle: Invigorate blood, transform stasis, and unblock collaterals and channels.

Use Tao Hong Si Wu Tang [Four-Substance Decoction with Peach Pit and Safflower] remove Sheng Di Huang, Bai Shao Yao, and add Chi Shao Yao, Dan Shen, San Leng, E Zhu, Ru Xiang, Mu Yao, Zhi Da Huang, and Tu Bie Chong.

Related conditions and modifications are

- If new lesions are seen: Add Da Qing Ye, Jin Yin Hua, and Lian Qiao.
- Thick scaling: Add Chuan Shan Jia and Lu Feng Fang.
- Severe itching: Add Bai Xian Pi and Fang Feng.
- Darkened and hard skin lesions: Add Sheng Mu Li and Xia Ku Cao.
- Severe dry, red, and itchy skin lesions: Add Shui Niu Jiao powder, Sheng Di Huang, and Zi Cao.
- Pale skin lesions with scaling and slight itching: Add Dang Gui, Dan Shen, and Shu Di Huang.

Furthermore, one can also add food therapy.

Hua Sheng Zhou [Peanut Congee]: Cook peanuts without removing the red skin 45 g, Chi Xiao Dou 30 g, Gao Liang Mi (sorghum) 30 g, Sheng Shan Zha 15 g, and brown sugar 15 g to form congee. Eat every day.

Regression Period

Blood-Deficiency and Wind-Dryness Syndrome

Treatment Principle: Nourish blood, moisten dryness, expel wind, and stop itching.

Use Dang Gui, Bai Shao Yao, Shu Di Huang, Mai Men Dong, Ji Xue Teng, E Jiao, Huo Ma Ren, Xuan Shen, and Huang Qi.

Blood-Dryness Damaging Yin Syndrome

Treatment Principle: Moisten dryness and nourish blood.

Use Si Wu Tang [Four-Substance Decoction], remove Chuan Xiong, add Tian Men Dong, Huang Jing, Sha Shen, Sang Shen, Huo Ma Ren, and Yu Zhu.

Above two syndromes can be modified accordingly.

- Dry skin lesions: Add He Shou Wu.
- Severe itching: Add Zao Jiao Ci and Bai Ji Li.
- Dry stool: Add Yu Li Ren and Tao Ren.
- Dark purple skin lesions: Add Dan Shen and Tao Ren.
- Thickening of scaling like a clam shell: Add Sheng Mu Li and Bai Xian Pi.

Furthermore, one can also take Ou Pian Zhou [Lotus Root Congee].

Fresh sliced lotus root 30 g, Chi Xiao Dou 30 g, Yi Yi Ren 60 g, Bai Mao Gen 15 g, and Geng Mi (rice) 100 g. Decoct Bai Mao Gen first. Use the decoction of Bai Mao Gen to cook with other herbs to make congee.

ROSACEA

This condition is often caused by blood–heat, Lung stagnation, and Stomach fire. In Western medicine, it is a chronic skin condition due to neuro-dysfunction of vasomotor nerve. The characteristics are diffused redness at the center of the face, or accompanied by papules, pustules, and dilation of the capillaries. TCM treatments are cooling blood, clearing heat, invigorating blood, and scattering accumulations.

Blood–Heat Steaming Lung Syndrome

Treatment Principle: Cool blood and clear heat.

Use Jia Jian Yin Qiao San [Modified Honeysuckle and Forsythia Powder] and add Mei Gui Hua (Rose), Sheng Di Huang, Shi Gao, Zi Cao, and Hong Hua.

Accumulated Heat in Lung and Stomach Syndrome

Treatment Principle: Clear and drain accumulated heat from Lung and Stomach.

Use Huang Lian Jie Du Tang [Coptis Decoction to Eliminate Toxin] plus Qing Wei San [Clear Stomach Powder] and Bai Hua She She Cao, Pi Pa Ye, Sang Bai Pi, and Yin Chen Hao.

Qi Stagnation and Blood Stasis Syndrome

Treatment Principle: Move Qi and transform stasis.

Use Tong Qiao Huo Xue Tang [Unblock the Orifices and Invigorate the Blood Decoction] plus Shi Chang Pu, Chen Pi, Sheng Di Huang, and Bai Zhi.

The above three formulas can be modified according to the following:

- Severe pustule: Add Pu Gong Ying and Zi Hua Di Ding.
- Red skin: Add Zao Xiu and Han Shui Shi.
- Red nose and face from overdrinking: Add Ge Hua, Ku Shen, and Zhi Shi.
- Worsening before menstruation: Add Yi Mu Cao and Mu Dan Pi.

Furthermore, one can also add topical application. Grind equal amount of Liu Huang and Da Huang. Mix with fresh Lu Hui juice, mix into paste, and apply topically before going to sleep. It is indicated for acne with red patches, papules, and pustules. It usually takes 2 weeks to see the effect.

URTICARIA

This condition is often caused by external wind–cold, and wind–heat combating internal damp–heat and toxin-fire and then manifesting on skin. In Western medicine, this can be caused often by allergens leading to temporary inflammatory infection at the skin and mucous membrane vessels, and edema and itching at the skin due to large amount of exudates. The clinical manifestations are irregularly shaped or sized bright red or pale wind patches, appearing fast, disappearing fast, with localized itching or burning sensation. Some patients also experience fever, nausea, vomiting, abdominal pain, and diarrhea. It is common in young adults. TCM treatment at acute stage focuses on expelling wind, eliminating damp, and relieving toxins. At the chronic stage, the treatment focuses on nourishing Yin, invigorating blood, expelling wind, and relieving toxicity.

Acute Urticaria

Wind–Heat Syndrome

Treatment Principle: Disperse wind, clear heat, and cool blood.

Use Ma Huang Lian Qiao Chi Xiao Dou Tang [Ephedra, Forsythia, and Aduki Bean Decoction] plus Sang Bai Pi, Jiang Can, Xing Ren, and Chan Tui.

Related conditions and modifications are

- Scratching when the skin itches leading to raised marks right after scratching. This indicates excess heart fire: Remove Ma Huang, and add Sheng Di Huang, Chi Shao Yao, Xuan Shen, Dang Gui, Zhi Mu, Zi Cao, and Shi Gao.
- Abdominal pain: Add Bi Ba, Yan Hu Suo, and fried Chuan Lian Zi.
- Diarrhea: Add Huang Lian, Huang Qin, and Ku Shen.
- Caused by parasites: Add Wu Mei, fried Shi Jun Zi, and Bing Lang.

Wind–Cold Syndrome

Treatment Principle: Disperse wind, scatter cold, and harmonize Ying and Wei.

Use Gui Zhi Tang [Cinnamon Twig Decoction] plus Ma Huang, Jing Jie, Fang Feng, Chan Tui, and Xing Ren.

Related conditions and modifications are

- Easily catch cold: Add Yu Ping Feng San [Jade Windscreen Powder].
- Constipation: Add Da Huang and Zhi Shi.
- Diarrhea: Add fried Bai Zhu, Fu Ling, and Yi Yi Ren.

Chronic Urticaria

Wind–Toxin Stasis and Dampness Syndrome

Treatment Principle: Expel wind, relieve toxicity, and eliminate dampness and blood stasis.

Use Sheng Di Huang, Chi Shao Yao, Mu Dan Pi, Dang Gui, Fang Feng, Jiang Can, She Chuang Zi, Ku Shen, Di Fu Zi, Da Huang, and Gan Cao.

Related conditions and modifications are

- Red burning rash: Add Lian Qiao, Jin Yin Hua, Ju Hua, and Chan Tui.
- White or light red rash: Remove Sheng Di Huang, Mu Dan Pi, and Da Huang, and add Jing Jie, Ma Huang, Gui Zhi, Xing Ren, Sheng Jiang, and Da Zao.
- Severe itching: Add Jing Jie, Chan Tui, Wu Shao She, and Di Long.
- Poor appetite, loose stool, severe lower limbs itching, and thick white tongue coating: Remove Da Huang, and add Cang Zhu, Huang Bai, and Yi Yi Ren.
- Damp–heat: Remove Sheng Di Huang and Mu Dan Pi, and add Xing Ren, Yi Yi Ren, Dong Gua Ren, Bai Jiang Cao, and Ma Chi Xian.
- Blood deficiency: Remove Da Huang, and add Shu Di Huang, Zhi He Shou Wu, E Jiao, and Huo Ma Ren.
- Attacks of wind rash before menstruation: Add Bai Shao Yao, Ji Xue Teng, and Zhi He Shou Wu.
- Heat disturbing Heart meridian, fluid exhaustion, and intestinal dryness: Add Sha Shen, Mai Men Dong, Bai Zi Ren, Suan Zao Ren, Lian Zi Xin, and Shan Zhi Zi.
- Nausea and vomiting: Add Ban Xia, Huo Xiang, Zhu Ru, and Sheng Jiang.
- Damp–heat in intestines: Add Huang Lian and Ge Gen.

VITILIGO

This is often caused by internal injuries of seven emotions, disorders of Qi and blood, obstruction of channels; or deficiency of Qi, blood, Liver, and Kidney. In Western medicine, it is a skin condition with loss of melanocytes (pigmentation) resulting in whitish or milky spots with unknown cause. There is possible genetic correlation and it is often seen in young adults and middle-aged people. The clinical manifestations are irregular white spots or patches on skin with clear outlines. TCM treatments focus on dispersing Liver, regulating Qi, harmonizing blood, and tonifying Liver and Kidney, combined with invigorating blood and transforming stasis.

Treatment According to Syndrome Differentiations

Liver Qi Stagnation Syndrome

Treatment Principle: Disperse Liver, relieve stagnation, invigorate blood, and expel wind.

Use Xiao Yao San [Relaxed Wanderer Powder] plus Yu Jin, Yi Mu Cao, Bai Ji Li, and Chuan Xiong.

Related conditions and modifications are

- Dry mouth, bitter taste, and hypochondriac distention: Add Mu Dan Pi, Xiang Fu, and Xia Ku Cao.
- Easily angered: Add Shan Zhi Zi and Long Dan Cao.
- Irritability: Add Huang Qin.
- Dry stool: Add Da Huang and Tao Ren.
- Blood stasis: Add Dan Shen, Hong Hua, and Tao Ren.

Liver and Kidney Yin Deficiency Syndrome

Treatment Principle: Nourish Liver, tonify Kidney, and transform stasis.

Use Liu Wei Di Huang Wan [Six-Ingredient Pill with Rehmannia] plus Zhi He Shou Wu, Gou Qi Zi, Tu Si Zi, Huang Qi, Huang Jing, and Chuan Xiong.

Related conditions and modifications are

- Depression: Add Chai Hu and Xiang Fu.
- Dizziness, shortness of breath, and pale face: Add E Jiao, Dang Gui, Dang Shen, and Huang Qi.
- Aversion to cold, cold limbs, loose stool, and clear copious urine: Add Fu Zi, Rou Gui, and Xian Mao.
- Itch and dark purple tongue: Add Dan Shen, Di Long, and Tu Bie Chong.

Qi Stagnation and Blood Stasis Syndrome

Treatment Principle: Invigorate blood, transform stasis, activate channels, and unblock collaterals.

Use Tong Jing Zhu Yu Tang [Unblock the Channels and Drive Out Stasis Decoction] plus Dang Gui, Zi Cao, Chuan Xiong, and Lu Lu Tong.

When part or all of the skin lesions turn to light or dark brown color, then one can decrease the dosage or change to other patent medicines, such as Xiao Yao Wan [Relax Wanderer Pill], Shu Gan Wan [Spread the Liver Pill], Liu Wei Di Huang Wan [Six-Ingredient Pill with Rehmannia], Shou Wu Pian [Polygoni Pill], and Fu Fang Dan Shen Pian [Revised Salvia Tablet].

External Application

Use Bu Gu Zhi 200 g, Tu Si Zi 300 g, Hong Hua 100 g, Bai Ji Li 50 g, and Zi Cao 100 g.

Soak the herbs in 75% alcohol for 7 days and remove the dregs. Apply topically one to three times a day and take a moderate amount of sunlight.

Food Therapy

1. Gou Qi Mu Er Tang [Lycium Fruit and Wood Ear Soup]: Hei Mu Er (black fungus) 15 g, Gou Qi Zi 10 g, and brown sugar 10 g. Cook the above herbs with water and take once a day.
2. Sang Shen Hu Tao Zhou [Mulberry and Walnut Congee]: Sang Shen Zi 20 g and Hu Tao Rou 30 g. Cut into small pieces and cook with 60 g of Geng Mi (rice) to make congee. Eat three times a week.

CHAPTER 15

Endocrine Disorders

ADIPOSIS (OBESITY)

This is caused by greasy lipid, turbid phlegm, and water dampness obstructing and stagnating in the body. The condition discussed here is simple obesity without metabolic, endocrine, and hypothalamus pathologies. In general, when the body weight is 20% (some use 10%) higher than the average weight of people with same age and sex, then it is referred to as obesity. TCM treatment focuses on strengthening Spleen, tonifying Qi, draining damp, promoting urination, expelling phlegm, and transforming turbidity. It assists in nourishing Kidney and regulating Liver.

Treatment According to Syndrome Differentiation

Qi Stagnation and Damp Obstruction Syndrome

Treatment Principle: Move Qi and transform dampness.

Use Si Ni San [Frigid Extremities Powder] plus San Ren Tang [Three-Nut Decoction] plus Hua Shi, Dong Gua Ren, Wang Bu Liu Xing, Zhi Ke, Xiang Fu, Yi Yi Ren, and Chuan Xiong.

Related conditions and modifications are

- Copious cough and phlegm: Add Jie Geng and Niu Bang Zi.
- Epigastric and hypochondriac distention: Add Yu Jin and Qing Pi.
- Constipation, ulcerations in mouth and on tongue: Add Sheng Da Huang and replace Zhi Ke with Zhi Shi.

Phlegm–Damp Accumulation and Obstruction Syndrome

Treatment Principle: Strengthen Spleen, dry dampness, and transform phlegm.

Use Er Chen Tang [Two-Cured Decoction] plus Cang Zhu, fried Bai Zhu, Dong Gua Ren, and Gan Jiang.

Related conditions and modifications are

- Unproductive cough with scanty yellow phlegm: Add Gua Lou Pi and Yu Xing Cao.
- Focal distention at epigastrium, loose stool, and poor appetite: Add Xiao Hui Xiang and increase Gan Jiang by 15 g.

Yang Deficiency and Damp Accumulation Syndrome

Treatment Principle: Warm Yang and drain damp.

Use Wu Ling San [Five-Ingredient Powder with Poria] plus Gan Jiang and Cang Zhu. Related conditions and modifications are

- Cold limbs: Add Zhi Fu Zi.
- Fatigue sensation in the chest, shortness of breath: Add Hong Ren Shen.
- Numbness of the four limbs, sharp chest and abdominal pain, and scanty menstrual flow with clots: Add Tao Ren, Hong Hua, and Yi Mu Cao.

Tea Therapy

1. Shan Zha, He Ye, and Lu Cha (green tea). Drink frequently to calm Liver, drain heat, promote digestion, and reduce fat.
2. Shan Zha, Ju Hua, Chen Pi, Mai Ya, and Lu Cha (green tea). Promote digestion, relieve stagnation, clear heat, and drain damp.
3. Da Huang, Shan Zha, and Lu Cha (green tea). Unblock stool, clear heat, and reduce fat.
4. Gou Qi Zi, He Ye, Shan Zha, Lu Dou (mung bean), and Yi Yi Ren. Strengthen Spleen, drain damp, and reduce fat.

Exercise and Food Therapy

1. Exercise everyday.
2. Avoid sitting or lying for too long, or sleeping right after a meal.
3. Control diet. Eat regular amounts at regular times.
4. Low salt, no sugar, and avoid coffee and alcohol.

CHRONIC ADRENOCORTICAL HYPOFUNCTION (ADDISON'S DISEASE)

This condition belongs in consumptive diseases in TCM. Kidney and Ming Men are involved. The pathogeneses are Kidney deficiency and declined Ming Men fire. In Western medicine, it is described as being caused by tuberculosis of adrenal gland, autoimmune atrophy of the adrenal gland, hypernephroma or infection; or chronic use of corticosteroids causing insufficient secretion of the cortical hormones. The clinical manifestations at the beginning stage are a subtle onset of fatigue, poor appetite, emaciation, and pigment precipitation. As the condition progresses, the symptoms also become more severe. At the middle or late stages, there can be dizziness, hypotension, nausea, vomiting, abdominal distention, abdominal pain, diarrhea, craving for salty and sweet food, depression, insomnia, anger that comes quickly, aversion to cold, shortness of breath, decreased libido, loss of hair from armpit and genital area, and irregular menstruation. If it is caused by infection, surgery, trauma, delivery, overexhaustion, heavy sweating, diarrhea, and loss of water, a critical condition may exist at the cortex of the adrenal gland. Patients presenting with this condition should be sent to the emergency room. TCM treatments focus on warming the Kidney and strengthening Ming Men fire.

Treatment According to Syndrome Differentiation

Kidney Deficiency and Essence Exhaustion Syndrome

Treatment Principle: Supplement essence and warm Kidney.

Use Zuo Gui Wan [Restore the Left (Kidney) Pill] plus Hong Ren Shen, Rou Cong Rong, Du Zhong, Dang Gui, Xian Mao, Rou Gui, Zhi Gan Cao, and Bu Gu Zhi.

Heart and Kidney Deficiency Syndrome

Treatment Principle: Warm Kidney, support Yang, tonify Heart, and calm Shen.

Use You Gui Wan [Restore the Right (Kidney) Pill] plus Dang Gui Bu Xue Tang [Tangkuei Decoction to Tonify the Blood] and Dan Shen, Suan Zao Ren, Bai Zi Ren, Wu Wei Zi, and Hong Ren Shen.

Spleen and Kidney Yang Deficiency

Treatment Principle: Warm and tonify Spleen and Kidney.

Use Gui Fu Li Zhong Wan [Cinnamon and Prepared Aconite Decoction to Regulate the Middle Jiao] plus Huang Qi, Dan Shen, Fu Ling, Tu Si Zi, Yin Yang Huo, and Lu Jiao Jiao.

Liver and Kidney Yin Deficiency Syndrome

Treatment Principle: Nourish Liver and Kidney and assist Yang.

Use Mai Wei Di Huang Wan [Ophiogonis, Schisandra, and Rehmannia Pill] plus Gou Qi Zi, Huang Jing, Zhi Gan Cao, Zhi Gui Ban, and Tu Si Zi.

Modifications of the Above Syndromes

- Edema: Remove Gan Cao, increase Huang Qi by 50 g, and add Zhu Ling, Ze Xie, Che Qian Zi, Fu Ling, and Bai Zhu.
- Diarrhea: Remove Dang Gui and add Yi Yi Ren, Qian Shi, and Wu Mei.
- Insomnia: Add Zhi He Shou Wu, Ye Jiao Teng, and Suan Zao Ren.
- Low blood pressure and palpitation: Add Huang Jing, Zhi Shi, and Sheng Mai San [Generate the Pulse Powder].
- Decreased libido: Add Ba Ji Tian, She Chuang Zi, Gou Qi Zi, Yin Yang Huo, and Bu Gu Zhi.
- Dark purple lips and eyelids: Add Dang Gui, Dan Shen, Ji Xue Teng, and Tu Bie Chong.
- Dry mouth/tongue and five-center heat: Remove Gui Zhi, Fu Zi, and add Zhi Mu, Huang Bai, and Di Gu Pi.
- Severe aversion to cold and cold limbs: Add Lu Rong powder 1 g in the morning with water.

Once the condition has been relieved, change to patent medicines according to syndrome differentiations, such as He Che Da Zao Wan [Great Construction Pill with Placenta], Gui Ling Ji [Collection of Turtle Spirit], Ren Shen Yang Ying Wan [Ginseng Pill to Nourish the Nutritive Qi], Zuo Gui Wan [Restore the Left (Kidney) Pill], and You Gui Wan [Restore the Right (Kidney) Pill]. This should be taken for 3 months as one course.

Tonify and Support with Herbal Food Therapy

1. Fu Zi Sheng Jiang Yang Rou Tang [Prepared Aconite, Fresh Ginger, and Mutton Soup]: Cook Zhi Fu Zi 15 g and Sheng Jiang 15 g, and fresh mutton 500 to 1000 g together and drink the soup. It is especially good during winter. It is indicated for Spleen and Kidney Yang deficiency syndrome.
2. Dang Gui Huang Qi Sheng Jiang Yang Rou Tang [Tangkuei, Astragalus, Fresh Ginger, and Mutton Soup]: Cook Dang Gui 30 g, Huang Qi 30 g, Sheng Jiang 15 g, and fresh mutton 500 to 1000 g together and drink the soup. It is good for Qi and blood deficient patients.
3. Hai Shen Dan Cai Yang Shen Tang [Sea Cucumber, Mussels, and Goat Kidney Soup]: Cut sea cucumber 100 g, Dan Cai 100 g, and goat kidney 400 g into small pieces and cook. Eat the kidney and drink the soup. It is indicated for Liver and Kidney Yin deficiency syndrome.

DIABETES MELLITUS

This condition belongs to Xiao Ke in TCM. The pathogeneses are Yin deficiency and dry heat which, in the long run, damage Qi and Yin, and damaged Yin affects Yang and at the end presents as Qi and Yin deficiency. Fluid congestion, damp–heat, blood stasis, and heat toxins can also occur along the disease process. In Western medicine, the pathogenesis of this condition is described as absolutely or relatively insufficient insulin secretion and insulin resistance. These lead to metabolic dysfunction of sugar, protein, fat, and electrolytes. The main clinical manifestation is high blood sugar. The symptoms include excessive hunger, excessive thirst, copious urination, emaciation, fatigue, sugar in urine, elevated blood sugar on empty stomach or 2 hours after meal. TCM treatments focuses on nourishing Yin, clearing heat, tonifying Qi, and invigorating blood. TCM treatment is not very effective in lowering the blood sugar, thus one should treat the syndrome according to differentiation in combination with the insulin therapy.

Treatment According to the Three Xiao

Shang Xiao (Upper)

Treatment Principle: Clear heat, moisten Lung, generate fluid, and stop thirst.

Use Mai Men Dong Tang [Ophiopogonis Decoction] plus Tian Hua Fen, Zhi Mu, Ge Gen, Wu Wei Zi, Dan Zhu Ye, Fu Shen, and Sheng Di Huang, and replace Ren Shen with Tai Zi Shen.

Zhong Xiao (Middle)

Treatment Principle: Clear heat, drain fire, nourish Yin, and replenish fluid.

Use Bai Hu Jia Ren Shen Tang [White Tiger plus Ginseng Decoction] plus Mai Men Dong, Huang Lian, and Xuan Shen.

Related condition and modification are

- Constipation: Add Tiao Wei Cheng Qi Tang [Regulate the Stomach and Order the Qi Decoction].

Xia Xiao (Lower)

Treatment Principle: Nourish Yin and consolidate Kidney.

Use Liu Wei Di Huang Wan [Six-Ingredient Pill with Rehmannia] plus Tu Si Zi and Fu Pen Zi, and remove Ze Xie and Fu Ling.

Related condition and modification are

- Slight Kidney Yang deficiency: Use Jin Gui Shen Qi Wan [Kidney Qi Pill from the Golden Cabinet] instead.

Treatment According to Yin, Yang, Qi, and Blood

Qi and Yin Deficiency Syndrome

Treatment Principle: Tonify Qi, generate fluid, nourish Yin, and tonify Kidney.

Use Sheng Mai San [Generate the Pulse Powder] plus Liu Wei Di Huang Wan [Six-Ingredient Pill with Rehmannia] and Tian Hua Fen and Huang Qi.

Related conditions and modifications are

- Insomnia: Add Er Zhi Wan [Two-Ultimate Pill].
- Night sweats and tidal fever: Add Zhi Mu and Huang Bai.
- Lumbar and knee soreness and weakness: Add Du Zhong and Sang Shen.

Dry Heat Syndrome

Treatment Principle: Nourish Yin, moisten Lung, clear Stomach, and drain fire.

Use Yi Wei Tang [Benefit the Stomach Decoction] plus Bai Hu Tang [White Tiger Decoction] and Tian Hua Fen, Ge Gen, and Tai Zi Shen.

Related conditions and modifications are

- Severe thirst: Add Shi Hu and Wu Mei.
- Ulcerations on tongue and mouth and toothache: Add Xuan Shen, Chuan Niu Xi, and Huang Lian.
- Constipation: Add Ma Zi Ren. In severe cases: Steep 5 g of Fan Xie Ye and drink as tea.

Yin and Yang Deficiency Syndrome

Treatment Principle: Warm Yang and nourish Yin.

Use Jin Gui Shen Qi Wan [Kidney Qi Pill from the Golden Cabinet] remove Ze Xie, Mu Dan Pi, and add calcined Long Gu, calcined Mu Li, and Du Zhong.

Related conditions and modifications are

- Nocturnal emission or early ejaculation: Add Tu Si Zi, Jin Ying Zi, and Nu Zhen Zi.
- Copious nocturia: Add Yi Zhi Ren, Fu Pen Zi, and Sang Piao Xiao.
- Edema: Add Yi Mu Cao, Che Qian Zi, and Yu Mi Xu.
- Red face, red eyes, headache, and yellow urine: Remove Zhi Fu Zi and Rou Gui, and add Zhi Mu and Huang Bai.

Internal Blood Stasis Obstruction Syndrome

Treatment Principle: Invigorate blood, transform stasis, tonify Qi, and nourish Yin.

Use Tao Hong Si Wu Tang [Four-Substance Decoction with Peach Pit and Safflower] plus Sheng Di Huang, Dan Shen, Huang Qi, and Yi Mu Cao, and remove Chuan Xiong.

Related conditions and modifications are

- Severe dizziness and headache: Add Gou Teng and Cao Jue Ming.
- Stifling sensational chest and pain: Add San Qi and Zhi Ke.
- Painful limbs and body: Add Ji Xue Teng and Wei Ling Xian.

Treatments for Complications

Peripheral Neuropathy

Numbness, burning, tingling sensation, and blood stasis at the vessels.

Formula includes Wei Ling Xian, Ji Xue Teng, Luo Shi Teng, Hai Feng Teng, Gou Teng, Dan Shen, and Tao Ren.

Related conditions and modifications are

- Sensory dysfunction, painful limbs, and Bi syndrome: Use Huang Qi Gui Zhi Wu Wu Tang [Astragalus and Cinnamon Twig Five-Substance Decoction].
- Motor dysfunction, muscle atrophy, and Lung heat damaging fluid: Use Bai Hu Jia Ren Shen Tang [White Tiger and Ginseng Decoction].
- Chronic stasis and damaged Yin affecting Yang: Add Zhi Fu Zi, Yin Yang Huo, and Bu Gu Zhi.

Itching

- Whole body itching: Add Di Fu Zi, Ku Shen, and Fang Feng.
- Genital itching: Add Huang Bai and Zhi Mu. One can also decoct She Chuang Zi, Di Fu Zi, Jing Jie, and Ku Fan to use as an external wash, two to three times a day.

Vasculitis

- This belongs to blocked vessels: Add Su Mu, Hong Hua, Ji Xue Teng, and Liu Ji Nu, or use Si Miao Yong An Tang [Four-Valiant Decoction for Well-Being].

Urinary Tract Infection

- Add Che Qian Cao, Pu Gong Ying, and Bei Xie.

Skin Infection

- This belongs to heat toxin accumulating in blood. Add Huang Qin, Huang Lian, Huang Bai, Lian Qiao, Pu Gong Ying, Zi Hua Di Ding, Zi Bei Tian Kui, Jin Yin Hua, Chi Shao Yao, Xuan Shen, and Sheng Di Huang, or use Wu Wei Xiao Du Yin [Five-Ingredient Decoction to Eliminate Toxin].

Retinopathy, Cataract

- Liver and Kidney deficiency: Add Yu Zhu, Gou Qi Zi, and Chong Wei Zi.
- Heat in Liver channel: Add Cao Jue Ming, Gu Jing Cao, and Ju Hua.
- Bleeding at fundus which belongs to stasis obstruction of blood vessels: Add Da Ji, Xiao Ji, Hong Hua, and San Qi powder.
- Cataract: Add Shi Hu Ye Guang Wan [Dendrobium Pill for Night Vision].

Ketosis

- Tonify Qi, nourish Yin, clear heat, harmonize blood, descend rebellious Qi, and transform turbidity.
- Use Huang Lian Jie Du Tang [Coptis Decoction to Relieve Toxicity] plus Chi Shao Yao, Chuan Xiong, Sheng Di Huang, Huang Qi, Shan Yao, Xuan Shen, Fu Ling, Ze Xie, Ban Xia, Zhu Ru, and Sheng Mu Li.

GOUT

Gout is called Tong Feng (painful wind) or painful Bi in TCM. It is often seen in middle-aged or senior obese males. It is closely related to dysfunctions of separating clear from turbid and metabolism of spleen and kidney. This causes internal generation of phlegm turbid which causes obstruction in channels and collaterals. In Western medicine, this relates to dysfunction of purine metabolism and increased uric acid. The clinical manifestations are recurrent joint pain, especially common at the interphalangeal joint of the big toe, more in lower joints than joints of the upper limbs, and more in small joints than big joints. One can also see gout precipitation at joints or subcutaneous, chronic arthritis, and deformity of the joints.

During attack period, main treatment is to clear heat and drain damp or warm the channels and scatter cold for some patients. The treatment between attacks focuses on draining damp and expelling turbid phlegm. One can invigorate blood and transform stasis through the whole disease process. From modern research, Che Qian Zi, Di Long, and Shan Ci Gu have certain effect in expelling uric acid. Qin Jiao and Wei Ling Xian can eliminate the pain due to uric acid. One can add these herbs into the formula according to syndrome differentiation.

During Attack

Treatment Principle: Clear heat, drain damp, stop pain, and invigorate blood.

Use San Miao San [Three-Marvel Pill] plus Yan Hu Suo, Dang Gui Wei, Pu Gong Ying, Di Long, Shan Ci Gu, and Hua Shi.

Related condition and modification are

- Severe pain: Mash equal amount of fresh Ye Ju Hua and fresh Fu Rong Ye (cotton rose leaf) and add 20 g of Sheng Da Huang powder. Mix well and use topically. Change once a day.

Between Attacks

Treatment Principle: Drain damp, expel turbid phlegm, invigorate blood, and unblock collaterals.

Use San Ren Tang [Three-Nut Decoction] plus Tu Fu Ling, Ze Lan, Che Qian Zi, Ji Xue Teng, Chuan Niu Xi, and Lu Lu Tong.

Related conditions and modifications are

- Excess turbidity, damp, and thick, greasy tongue coating: Add Cang Zhu, Fu Ling, Dan Nan Xing, and Jiang Can.
- Severe damp heat: Add Huang Bai, Zhi Mu, and Ren Dong Teng.
- Obvious blood stasis, severe local swelling, and purple dark skin: Add Dan Shen, Tu Bie Cong, Dang Gui, Tao Ren, and Hong Hua.
- Severe Pain: Add Quan Xie, Wu Gong, and Yan Hu Suo.
- Pain in upper limb: Remove Chuan Niu Xi, and add Sang Zhi, Qin Jiao, and Wei Ling Xian.

Chronic Period

Treatment Principle: Drain turbid damp, invigorate blood, and expel stasis.

Use Bei Xie Fen Qing Yin [Dioscorea Hypoglauca Decoction to Separate the Clear] plus Tu Fu Ling, Qin Jiao, Fang Ji, Wu Gong, Tao Ren, Di Long, and Tu Bie Chong.

Related conditions and modifications are

- Local heat sensation, red tongue, and yellow tongue coating: Add Hu Zhang, Ren Dong Teng, and Huang Bai.
- Local cold pain and pale tongue with white coating: Add Zhi Chuan Wu and Yin Yang Huo.
- Joint stiffness and deformity: Add Zhi Chuan Shan Jia, Gui Zhi, and Zao Jiao Ci.

HYPERCORTISOLISM (CUSHING'S SYNDROME)

This condition is thought to be caused by emotions damaging the interior, internal accumulation of damp–heat, Kidney essence deficiency with a common cause of constitution deficiency, and internal heat. According to descriptions in Western medicine, corticosuprarenoma, adrenal cortical hyperplasia, and pituitary microadenoma can cause dysfunction of the pituitary gland. Increased secretion stimulates the cortical hormone and hyperplasia of the cortex leading to elevated cortical hormone in the blood. The clinical manifestations are concentric weight accumulation especially at the face (full moon face), posterior part of the neck and back (buffalo back), chest and abdomen; relatively slim four limbs, thinning or transparency of the skin, back and lumbar pain, acne, increased hair, decreased libido, irritability, agitation, easily angered, insomnia, depression, and elevated blood pressure. TCM treatment focuses on dispersing Liver, strengthening Spleen, tonifying Qi, transforming dampness, and tonifying Kidney. It is contraindicated using Gan Cao in this disease.

Excess Heat in Liver and Stomach Syndrome

Treatment Principle: Clear Liver and Stomach and drain heat.

Use Er Zhi Wan [Two-Ultimate Pill] plus Ge Gen Huang Qin Huang Lian Tang [Kudzu, Coptis, and Scutellaria Decoction] plus Chai Hu, Ban Xia, and Hou Po.

Damp–Heat in Middle Jiao Syndrome

Treatment Principle: Clear and transform damp heat from the Middle Jiao.

Use Xie Huang San [Drain the Yellow Powder] plus Long Dan Cao, Huang Qin, Cang Zhu, Da Fu Pi, Yi Yi Ren, Sheng Di Huang, and Sang Shen.

Damaged Yin Transforms into Dryness Syndrome

Treatment Principle: Drain and clear heat and fire and nourish Yin.

Use Tiao Wei Cheng Qi Tang [Regulate the Stomach and Order the Qi Decoction] plus Yu Zhu, Long Dan Cao, Sheng He Shou Wu, and Huang Jing.

Liver and Kidney Yin Deficiency Syndrome

Treatment Principle: Nourish Liver and Kidney.

Use Liu Wei Di Huang Wan [Six-Ingredient Pill with Rehmannia] plus Gui Ban, Bie Jia, and Huang Jing.

Turbid Damp Obstructing Middle Jiao Syndrome

Treatment Principle: Strengthen Spleen and transform dampness.

Use Huo Xiang Zheng Qi San [Agastache Powder to Rectify the Qi] plus Bing Lang, Pei Lan, and Fang Ji.

Patients who overuse steroids for a long time can also present with similar manifestations as this syndrome. They can be treated according to this condition.

HYPERLIPIDEMIA

It is caused by phlegm turbidity, blood stasis, and water dampness accumulating in the body obstructing the channels and collaterals. Spleen, liver, and stomach are involved. In Western medicine, this refers to elevated concentration of cholesterol and triglyceride level in blood with fasting. Clinical manifestations are arteriosclerosis and thrombosis which is often seen in middle age or in seniors. TCM treatment focuses on expelling turbid phlegm, invigorating blood, augmenting Qi, and tonifying Kidney. Herbs that can decrease blood lipid are Shan Zha, Cao Jue Ming, Ze Xie, Yin Chen Hao, Da Huang, Huang Qi, He Shou Wu, Gou Qi Zi, Sang Ji Sheng, Qian Shi, Yu Zhu, Huang Jing, Ling Zhi, Jin Ying Zi, Dang Gui, Du Zhong, Hu Zhang, Pu Huang, San Qi, Chong Wei Zi, Jiang Huang, Bei Xie, Shui Zhi, He Ye, Chuan Xiong, and Jue Ming Zi. These herbs can be added to the formula according to syndrome differentiation.

Treatments According to Syndrome Differentiation

Spleen and Kidney Deficiency Syndrome

Treatment Principle: Tonify Qi, nourish Yin, and nourish Kidney.

Formula consists of Zhi He Shou Wu, Huang Jing, Sang Ji Sheng, and Shan Zha.

Internal Excess of Phlegm Damp

Treatment Principle: Expel phlegm, drain damp, and transform turbidity.

Formula consists of Yin Chen Hao, Ze Xie, Hu Zhang, He Ye, Gua Lou Pi, and Chen Pi.

Blood Stasis and Phlegm Turbidity Syndrome

Treatment Principle: Invigorate blood, transform stasis, and expel phlegm.

Formula consists of Chuan Xiong, Hong Hua, Jiang Huang, Xie Bai, Chi Shao Yao, Gua Lou Pi, Dan Shen, and Ge Gen.

Liver Qi Stagnation Syndrome

Treatment Principle: Disperse Liver and regulate Qi.

Formula consists of Chai Hu, Cao Jue Ming, Ji Nei Jin, Yu Jin, Chen Pi, Shan Zha, Chuan Xiong, and He Ye.

Food Therapy

1. Low fat, low cholesterol, and low sugar diet.
2. Reduce carbohydrate intake and avoid alcohol.
3. Increase intake of onion, garlic, bean sprouts, carrot, sea cucumber, mushroom, turnip, and mung beans.
4. Steep Shan Zha, Yin Chen Hao, Ze Xie, Da Huang, Hu Zhang, and He Shou Wu and drink as tea.
5. Long-term consumption of Da Shan Zha Wan [Big Hawthorn Pill], Shan Zha Jing Jiang Zhi Pian [Hawthorn Extract Tablet to Lower Lipid], and Shou Wu Pian [Polygoni Tablet].

HYPERTHYROIDISM

This condition is similar to goiter (Ying syndrome) and palpitation. It is often caused by emotional stress leading to Qi stagnation, phlegm damp, blood stasis, fire transformation, and damaged Yin. In severe cases, there can be wind and damaged kidney from chronic condition. In Western medicine, this syndrome is caused by excessive secretions from the thyroid gland. The clinical manifestations are enlarged thyroid; or with nodular formation; or diffused swelling. Increased metabolic rate manifests as easily hungry, excessive eating, aversion to heat, excessive sweating, fatigue, lack of strength, and reduced weight. There can also be unstable emotions, irritation, agitation, inability to concentrate, insomnia, tachycardia; and increased pressure difference between systole and diastole and trembling while extending hands. Early stage of TCM treatment focuses on clearing heat, draining fire, dispersing Liver, relieving stagnation, expelling stasis, and transforming phlegm to attack the excess. At middle and late stages, the treatments focus on tonifying Qi and nourishing blood and Yin to support the deficient constitution or a combination of treating the excess and deficiency at the same time. TCM treatments are better for the mild or medium type of patients.

For patients with decreased white blood cells due to long-term drug use, one should add Huang Qi, Bai Shao Yao, Mai Men Dong, Wu Wei Zi, and Nu Zhen Zi according to syndrome differentiation to tonify Qi and nourish blood. After taking the formula for 2 to 4 weeks, white blood cell count may start rising.

Treatment According to Syndrome Differentiation

Liver Qi Stagnation and Phlegm Accumulation Syndrome

Treatment Principle: Spread Liver, regulate Qi, transform phlegm, and scatter accumulation.

Use Xiao Chai Hu Tang [Minor Bupleurum Decoction] or Xiao Yao San [Relaxed Wanderer Powder], plus Er Chen Tang [Two-Cured Decoction] and Sheng Mu Li.

Related conditions and modifications are

- Severe phlegm and obstructed sensation at throat: Add Zhi Nan Xing and She Gan.
- Bitter dry mouth, insomnia, red tongue tip and sides: Add Xia Ku Cao and Suan Zao Ren.
- Palpitation: Add Bai Zi Ren and Long Gu.
- Pain and swollen thyroid: Add Pu Gong Ying and Zi Hua Di Ding.
- Loose stool and abdominal fullness and distention: Add Hou Po, Cang Zhu, and Sha Ren.

Excess Stomach Fire Syndrome

Treatment Principle: Sweet cold to clear Stomach, generate fluid, and stop thirst.

Use Bai Hu Jia Ren Shen Tang [White Tiger Decoction plus Ginseng Decoction], Yu Nu Jian [Jade Women Decoction] plus Tian Hua Fen, and replace Ren Shen with Tai Zi Shen.

Related conditions and modifications are

- Constipation: Add Huo Ma Ren and Yu Li Ren.
- Soreness and distention of the eyes: Add Xia Ku Cao and Shi Chang Pu.
- Discomfort of the swollen thyroid: Add Sheng Mu Li and Zhe Bei Mu.

Stomach Excess and Spleen Deficiency Syndrome

Treatment Principle: Clear Stomach and support Spleen.

Use Bai Hu Jia Ren Shen Tang [White Tiger Decoction plus Ginseng Decoction] and Si Jun Zi Tang [Four-Gentleman Decoction] plus Shan Yao, Mai Men Dong, Tian Hua Fen, and Bai Bian Dou, and replace Ren Shen with Tai Zi Shen.

Related conditions and modifications are

- Bad breath and abdominal distention: Add Shen Qu and Lai Fu Zi.
- Greasy tongue coating: Add Huang Lian and Huo Xiang.
- Thirst with desire to drink: Add Shi Hu and Ge Gen.

Spleen Deficiency and Liver Excess Syndrome

Treatment Principle: Clear Liver and strengthen Spleen.

Use Dan Zhi Xiao Yao San [Relax Wanderer Powder with Moutan and Gardenia] plus Si Jun Zi Tang [Four-Gentleman Decoction] plus Bie Jia, Sheng Mu Li, and fried Bai Bian Dou; and change Ren Shen with Tai Zi Shen.

Related conditions and modifications are

- Swollen thyroid gland with nodules: Add Hong Hua, Tao Ren, and Dan Shen.
- Poor appetite and abdominal distention: Add Sha Ren, Bai Zhu, and fried Mai Ya.
- Diarrhea and abdominal pain: Remove Shan Zhi Zi and add Mu Xiang, Sha Ren, and Cang Zhu.

Both Qi and Yin Deficiency Syndrome

Treatment Principle: Tonify Qi and nourish Yin.

Use Sheng Mai San [Generate the Pulse Powder] plus Suan Zao Ren, Fu Ling, Shan Yao, Fu Xiao Mai, and Sheng Mu Li.

Related conditions and modifications are

- Severe diarrhea: Add Cang Zhu, Bai Zhu, and fried Bian Dou.
- Diffused swelling of the thyroid gland: Add Zhe Bei Mu, Ban Xia, and Yu Jin.

Liver and Kidney Yin Deficiency Syndrome

Treatment Principle: Nourish Liver and Kidney.

Use Liu Wei Di Huang Wan [Six-Ingredient Pill with Rehmannia] plus Er Zhi Wan [Two-Ultimate Pill], remove Ze Xie and add Mai Men Dong.

Related conditions and modifications are

- Five-center heat, restlessness, yellow tongue coating, and rapid pulse: Add Zhi Mu, Yin Chai Hu, and Di Gu Pi.
- Low grade fever: Add Bai Wei.

Yin Deficiency with Hyperactive Fire Syndrome

Treatment Principle: Nourish Yin, descend fire, scour phlegm, and scatter accumulation.

Use Zhi Bai Di Huang Wan [Anemarrhena, Phellodendron, and Rehmannia Pill] plus Xia Ku Cao, Han Lian Cao, Mu Li, Huang Yao Zi, Zhe Bei Mu, and Xuan Shen.

Related conditions and modifications are

- Excess Heart fire: Add Deng Xin Cao and Huang Lian.
- Excess Stomach fire: Add Sheng Shi Gao.

Hyperactive Yang with Stirring Wind Syndrome

Treatment Principle: Nourish Yin, calm Liver, subdue Yang, extinguish wind, and scour phlegm.

Use Zhen Gan Xi Feng Tang [Sedate the Liver and Extinguish Wind Decoction] plus Sheng Di Huang, Xia Ku Cao, Huang Yao Zi, Zhe Bei Mu, Mu Dan Pi, Yu Jin, Zhi Mu, Suan Zao Ren, and Fu Shen.

Related conditions and modifications are

- Dry mouth with desire to drink: Add Sha Shen and Yu Zhu.
- Tinnitus and deafness: Add Gou Qi Zi and Nu Zhen Zi.
- Trembling hands and feet: Add Bie Jia.

Key Points to Note during Treatment

1. Protruding eyes can exist concomitantly with hyperthyroidism; it can also occur after the hyperthyroidism is being controlled. Both are difficult for Chinese and Western medical treatments. Western medicine uses large amounts of steroids or immunosuppressive drugs that do not have stable results and also have side effects. TCM treatment focuses on clearing Liver, brightening the eyes, transforming phlegm, and scattering accumulations. The formula is composed of Xia Ku Cao, Ju Hua, Qian Li Guang, Jue Ming Zi, Zhe Bei Mu, Ze Lan, and Dan Shen in the amount of 12 g each.
   - Related conditions and modifications are
     Liver and Kidney Yin deficiency, dry eyes, and blurry vision: Add 12 g each of Gou Qi Zi, Mi Meng Hua, and Sang Shen and 30 g of Sheng Mu Li.
   - Damp phlegm which is seen as swollen eyelids: Add 12 g each of Ban Xia and Fu Ling.
   - Heart and spirit not nourished, and insomnia: Add 15 g of Suan Zao Ren and 30 g of Ye Jiao Teng.
   - Qi deficiency: Add 30 g of Tai Zi Shen and 12 g of Shan Yao.

   This formula not only is effective in treating protruding eyes in hyperthyroidism, it can also prevent the condition from worsening and even developing.
2. Hyperthyroidism with both excess and deficiency, and mixed cold and heat. Syndrome needs to be carefully differentiated in order to prescribe treatments.
   - If patient has excess Liver fire, then one should use bitter–cold herbs to descend fire. However, if patient often presents with loose stool, emaciation, and fatigue, which are signs of Liver overacting on Spleen, then bitter–cold herbs will further damage the Spleen.
   - Patients with severe diarrhea should have their Spleen supported. However, overuse of Ren Shen, Bai Zhu, and Huang Qi can assist the fire.
   - Irritability, agitation, five-center heat, restlessness, red tongue scanty with coating, and thin rapid pulse are signs of Yin deficiency. Overuse of Yin and softening herbs will affect the transformation function of Spleen.

   Therefore, one has to focus on Liver, which is the cause of this condition, and treat the condition by spreading and calming the Liver and assist by supporting earth, clearing heat, nourishing Kidney, and softening hardness. One should avoid overusing bitter–cold, warm–dry, and ascending herbs.
3. Small numbers of patients overuse bitter–cold herbs or take Western medications for a long time, resulting in hypofunction of the thyroid. The manifestations are edema, aversion to cold, and other Yang deficiency symptoms. At this time, one should change to Ba Ji Tian, Huang Qi, Yin Yang Huo, Bu Gu Zhi, and other warm Kidney herbs.
4. For goiter, TCM always uses Hai Zao and Kun Bu which contains iodine. From the clinical results, patients with hyperthyroid can feel better with increased intake of high iodine herbs for a long time. However, the condition is still recurrent, and T3 and T4 levels are still high in the blood. This is because iodine-containing herbs cannot directly inhibit the formation of thyroid hormones but do inhibit the release of these hormones. Once the patient stops taking the herbs, then the thyroid hormones are released in the blood, leading to recurrent or worsened symptoms of hyperthyroid. Thus, one should be cautious with herbs containing iodine.

HYPOGLYCEMIA

This belongs to deficient syndrome, Jue syndrome, and exhausted sweating. The pathogeneses are Spleen and Stomach deficiency, Qi and blood exhaustion, and Yang Qi not consolidating. Western medicine refers to a syndrome caused by various reasons leading to low glucose concentration in blood which is lower than 2.8 mmol/L or 50 mg/dL. The most common cause is functional hypoglycemia with unknown reasons, which is also called autoimmune hypoglycemia. This condition often happens in adults and young females without a family history of diabetes and often occurs 2 to 4 hours after a meal, with hunger, weakness, sweating, tachycardia, nervousness, agitation, and trembling of the four limbs. TCM treatment focuses on strengthening Spleen, tonifying Qi, nourishing Heart, and calming spirit. It is assisted by dispersing Liver, clearing heat, and transforming turbidity.

Liver Qi Stagnation and Spleen Deficiency Syndrome

Treatment Principle: Disperse Liver, relieve stagnation, strengthen Spleen, and nourish blood.
Use Xiao Yao San [Relax Wanderer Powder] plus Suan Zao Ren, Huang Qi, and Dang Shen.

Heart and Spleen Deficiency Syndrome

Treatment Principle: Strengthen Spleen, tonify Qi, nourish Heart, and calm spirit.
Use Gui Pi Tang [Restore the Spleen Decoction] plus calcined Mu Li, Bai Shao Yao, and Chen Pi.

Damp–Heat Obstructing the Orifices Syndrome

Treatment Principle: lear heat, transform turbidity, and open orifices.
Use Qing Dan Li Shi Tang [Clear the Gallbladder and Drain Damp Decoction], remove Da Huang, Chai Hu, and add Zhu Li, Lian Qiao, Zhu Ye, Shi Chang Pu, and Mu Dan Pi.

Collapsing Yin and Yang Syndrome

Treatment Principle: Rescue Yin and Yang.
Use Sheng Mai San [Generate the Pulse Powder] plus Zhi Fu Zi.
Related condition and modification are

- Copious sweating: Add calcined Long Gu and calcined Mu Li.

HYPOTHYROIDISM

This condition belongs to consumptive disease in TCM. The pathogeneses are Spleen, Kidney, and Yang deficiency, and Qi and blood deficiency. In Western medicine, this refers to decreased formation and secretion of thyroid hormones caused by various reasons. Decreased thyroid hormones lead to decreased metabolism. The manifestations in mild cases include fatigue, aversion to cold, scanty sweat, poor appetite, and abdominal distention. In severe cases, there can be apathy, dullness, swelling of face and eyelids, pale or sallow complexion and skin tone, dry flaky skin, full body swelling without pitting edema, weight gain, loss of hair, flaccidity and weakness of the muscles, aversion to talk, flabby big tongue, abdominal distention, no appetite, decreased libido, impotence in male, and excessive bleeding or amenorrhea in female. The treatments in TCM are to warm the Kidney, support Yang, tonify Spleen, and augment Qi. Even though the formulas do not contain thyroid hormone, from the clinical results, they can increase levels of T3 and T4 in blood and decrease TSH. Thus, the functions of the thyroid gland and anterior pituitary can be regulated. If one can take small dose of thyroid hormone with the herbal formula, then this can be more effective than using the thyroid hormone alone. TCM treatment is especially effective in treating patients with slow onset or mild condition.

The following formula is a base to tonify and warm the Spleen and Kidney. Modifications are done according to syndrome differentiation. The formula consists of Zhi Fu Zi, Yin Yang Huo, Xian Mao, Rou Cong Rong, Lu Jiao Jiao, Zhi Huang Qi, Fu Ling, Ban Xia, Bu Gu Zhi, and Bai Shao Yao.

- Severe Spleen Yang deficiency: Remove Rou Cong Rong, Lu Jiao Jiao, Xian Mao, and Bai Shao Yao, and add Bai Zhu, Sha Ren, Mu Xiang, Pao Jiang, and Shen Qu.
- Heart and Kidney Yang deficiency: Add Gui Zhi, Dang Gui, Shi Chang Pu, and Long Yan Rou.
- Yin and Yang deficiency: Remove Zhi Fu Zi, Rou Cong Rong, Lu Jiao Jiao, and Ban Xia, and add Dang Gui, Chuan Xiong, Shu Di Huang, Hei Zhi Ma, Zhi He Shou Wu, and Long Yan Rou.
- Skin dryness, excess flaking and without cold: Add Mu Dan Pi and Sang Bai Pi.
- Blood stasis: Add Dang Gui, Chuan Xiong, and Yi Mu Cao. In severe cases, add Tao Ren and Hong Hua.
- Damp turbid obstructing interior: Remove Bai Shao Yao and Rou Cong Rong, and add Dang Shen, Cang Zhu, Yi Yi Ren, and Ze Xie.
- Severe swelling: Add Che Qian Zi and Zhu Ling.
- Phlegm congealed and Qi stagnation, and swelling at anterior neck, soft in quality: Add Zhe Bei Mu, Bai Jie Zi, San Leng, and E Zhu.
- Damp–heat obstructing interior: Take Sheng Shi Gao, Zhi Mu, Fu Ling, Ze Xie, Dang Gui, Pei Lan, Tong Cao, and Chen Pi first. When damp–heat is cleared, then continue to use the base formula.
- Liver fire excess: Take Bai Shao Yao, Chai Hu, Dang Gui, Shan Zhi Zi, Mu Dan Pi, Huang Qin, Fu Ling, Bai Zhu, Dang Shen, and Ye Jiao Teng. After the Liver fire is cleared, then continue to use the base formula.
- Constipation: Add Ma Zi Ren, Huai Niu Xi, Rou Cong Rong, Dang Gui, and Tao Ren.

CHAPTER 16

Ear, Eye, Nose, and Throat Disorders

HEARING LOSS AND DEAFNESS

This is often caused by Kidney deficiency, stagnant Liver and Gallbladder fire, Spleen deficiency, and Heart fire. In Western medicine deafness is described as an abnormality of the conduction, sensory, or analyzing function of the ear, which leads to different degrees of hearing loss, classified as conduction hearing loss, sensorineural hearing loss, functional hearing loss, and mixed type of hearing loss. TCM treatment focuses on tonifying Kidney, invigorating blood, and transforming stasis. Commonly used herbs are Ci Shi, Ge Gen, Bu Gu Zhi, and Huang Jing to improve ear congestion, tinnitus, dizziness, and vertigo. It is more effective in treating conductive and sudden sensory hearing loss.

Conductive Hearing Loss

This is a decrease in hearing caused by mechanical problems of external ear, ear canal, or middle ear, which block sounds that can reach the inner ear. If it is caused by otitis media, one can use the formulas below for treatment:

Treatment Principle:

1. Disperse Lung, open orifices, transform stasis, and open collaterals.
   Formula consists of Ma Huang, Xing Ren, Jing Jie, Lu Lu Tong, Ma Dou Ling, Chuan Xiong, Chai Hu, Ge Gen, Gan Cao, and Shi Chang Pu.
2. Move Qi and invigorate blood.
   Formula consists of Chuan Xiong, Chuan Niu Xi, Dang Gui Wei, Sang Leng, E Zhu, Chong Wei Zi, Hong Hua, Qing Pi, Yu Jin, Zhi Fu Zi, and Gan Cao.

Related conditions and modifications are

- Severe tinnitus: Add Ci Shi, Bai Shao Yao, and Wu Wei Zi.
- Severe blood stasis: Add Shui Zhi, Wu Shao She, and Tu Bie Chong.

Sensory Hearing Loss (Most Common Type of Hearing Loss)

Sudden Hearing Loss

This often occurs in one ear with sudden development (24 to 48 hours) to severe hearing loss. It is accompanied with tinnitus, vertigo, nausea, and more hearing damages of higher pitch than lower ones. Research shows this is related to obstruction in the blood supply of the inner ear.

Treatment Principle: Nourish blood, invigorate blood, and open orifices.

Use Tong Qiao Huo Xue Tang [Unblock the Orifices and Invigorate the Blood Decoction] and Dan Shen, Ge Gen, Ci Shi, and Zhi Da Huang.

Related conditions and modifications are

- Congested obstruction of the ear after a common cold: Remove Da Huang and add Ma Huang, Xing Ren, and Chan Tui.
- History of hypertension, irritability, agitation, and easily angered: Add Ju Hua and Gou Teng.
- After hearing has improved: Change the formula to Dan Shen, Hong Hua, Huang Qi, Dang Shen, Fu Ling, Huang Jing, Ge Gen, Shi Chang Pu, Gan Cao, or Bu Zhong Yi Qi Wan [Tonify the Middle and Augment the Qi Pill] for 1 to 2 months.

Hearing Loss Due to Trauma

Very loud noise and chronic noise irritation.

Treatment Principle: Move Qi, invigorate blood, strengthen Spleen, and tonify Kidney.

Formula consists of Tao Ren, Hong Hua, Chuan Xiong, Su Mu, Yan Hu Suo, Wu Ling Zhi, Bu Gu Zhi, Tai Zi Shen, and Huang Jing.

Related conditions and modifications are

- Unhealed new injury of the ear: Add Qian Cao and Pu Huang.
- Decreased appetite after taking this formula: Remove Wu Ling Zhi, and add Sheng Shan Zha and Chen Pi.
- Caused by loud noise: Add Lu Jiao Jiao, Gui Ban, and Gou Qi Zi.
- If no results after taking the above formula: Add Shui Zhi and Chuan Shan Jia.

Hearing Loss Due to Infectious Diseases

This occurs after epidemic encephalitis, scarlet fever, diphtheria, typhoid, influenza, parotitis, measles, chicken pox, herpes zoster, or syphilis. It is often seen in children.

Treatment Principle:

1. Nourish Kidney and benefit the ear.
   Formula consists of Shu Di Huang, Gou Qi Zi, Tu Si Zi, Rou Gui, Ci Shi, fried Shan Zha, and Ge Gen.
2. Invigorate blood and open orifices.
   Formula consists of Chi Shao Yao, Bai Shao Yao, Chuan Xiong, Lu Lu Tong, Ma Dou Ling, Ge Gen, and Gan Cao.

Related conditions and modifications are

- Strong constitution: Add Chuan Shan Jia and Hong Hua.
- Weak constitution: Add Dang Shen and Huang Qi.

Drug-Induced Hearing Loss

It is often caused by long-term use of streptomycin, kanamycin, and gentamicin. The hearing loss occurs in both ears; more in high-frequency range, tinnitus, dizziness, and vertigo.

Treatment Principle I: Relieve toxicity and open orifices (drug toxins affect ear).

Formula consists of Lu Dou (mung bean), Gan Cao, Chuan Niu Xi, Chai Hu, Xia Ku Cao, Shi Chang Pu, Ge Gen, and Jin Yin Hua.

Treatment Principle II: Drain Heart fire and nourish Kidney Yin (Heart and Kidney disharmony syndrome).

Formula consists of Huang Lian, Fu Shen, Bai Mao Gen, Mu Dan Pi, Sheng Di Huang, Xuan Shen, Mai Men Dong, Ci Shi, and Gan Cao. For young children, use Deng Xin Cao instead of Huang Lian.

Treatment Principle III: Strengthen Spleen and ascending Yang (clear Yang Failing to Ascend Syndrome).

Formula consists of Dang Shen, Huang Qi, Bai Zhu, Fu Ling, Huang Jing, Ge Gen, Sheng Ma, and Chai Hu. One can also take Er Long Zuo Ci Wan [Pill for Deafness that is Kind to the Left (Kidney)], Liu Wei Di Huang Wan [Six-Ingredient Pill with Rehmannia], Yi Qi Cong Ming Wan [Augment the Qi and Increase Acuity Decoction] (Huang Qi, Ren Shen, Sheng Ma, Ge Gen, Man Jing Zi, Bai Shao Yao, Huang Bai, and Zhi Gan Cao).

Age-Associated Hearing Loss

This is due to degeneration of hearing apparatus, often occurs at the inner ear, thus it is a type of sensorineural hearing loss.

Treatment Principle: Tonify Kidney and supplement essence.

Zuo Gui Wan [Restore the Left (Kidney) Pill] plus Bu Gu Zhi, Du Zhong, Ci Shi, Sang Shen Zi, and Shi Chang Pu. One can also take Er Long Zuo Ci Wan [Augment the Qi and Increase Acuity Pill].

Furthermore, one can eat 15 g of walnuts every night and 5 g of black sesame seeds every morning and night.

FUNCTIONAL HEARING LOSS

This is also called psychic deafness, often caused by emotional disturbance. It can occur in one or both ears with sudden hearing loss. It is often accompanied with being quiet, fullness and distention of the chest and hypochondriac region, sighing, or with trembling or numbness of the four extremities.

Treatment Principle: Disperse Liver and relieve stagnation.

Use Xiao Yao San [Relax Wanderer Powder] plus Su Geng, Qing Pi, Yu Jin, Chuan Xiong, Dai Zhe Shi, Bai Ji Li, and Shi Chang Pu.

MENIERE'S DISEASE

This condition is discussed under vertigo and dizziness in TCM. It is often caused by wind-phlegm upward disturbing the clear orifices and Liver, Spleen, and Kidney pathologies. In Western medicine it is described as an effect of lymphedema in the inner ear. The clinical manifestations are attacks of dizziness and vertigo, decreased hearing, tinnitus, or with nausea, vomiting, sweating, sallow complexion, and twitching of the eyeballs. In severe cases, patient has to close eyes and lie still, otherwise symptoms become worse. TCM treatment focuses on calming Liver, extinguishing wind, transforming phlegm, and descending rebellion. They are assisted by promoting urination, strengthening Spleen, and tonifying Kidney.

Attack Period

Dizziness and Vertigo

Treatment Principle: Expel phlegm, calm Liver, and promote urination.
Formula consists of Xia Ku Cao, Dai Zhe Shi, Jiang Ban Xia, Ze Xie, and Che Qian Zi.

Dizziness, Vertigo, Nausea, and Vomiting Thin Saliva

Treatment Principle: Expel wind phlegm, warm Middle Jiao, and strengthen Spleen.
Formula consists of Gui Zhi, Bai Zhu, Ze Xie, Fu Ling, Zhu Ling, Jiang Ban Xia, Chen Pi, Zhi Shi, Zhu Ru, Long Gu, and Mu Li.

Between Attacks

Liver Qi Stagnation and Hyperactive Yang Syndrome

Treatment Principle: Disperse Liver and calm Liver.
Use Xiao Chai Hu Tang [Minor Bupleurum Decoction] plus Fu Ling, Shi Jue Ming, Long Gu, Mu Li, and Gou Teng.

Spleen Deficiency with Water Accumulation Syndrome

Treatment Principle: Strengthen Spleen and drain water.
Use Ling Gui Zhu Gan Tang [Poria, Cinnamon Twig, and Atractylodes Macrocephala Decoction] plus Tian Ma, Tai Zi Shen, and Huang Qi.

Yang Deficiency with Water Accumulation Syndrome

Treatment Principle: Warm Yang and drain water.
Use Zhen Wu Tang [True Warrior Decoction] plus Tian Ma and Gou Teng.
Related condition and modification are

- Aversion to cold, cold limbs, frequent nocturia: Add Rou Gui, Ba Ji Tian, and Yin Yang Huo.

Stomach Deficiency with Phlegm Obstruction Syndrome

Treatment Principle: Descend rebellious Qi, harmonize Stomach, regulate Qi, and transform phlegm.
Use Xuan Fu Dai Zhe Tang [Inula and Haematite Decoction] plus Bai Zhu, Tian Ma, and Fu Ling.

SUPPURATIVE OTITIS MEDIA

Acute suppurative otitis media is caused by wind–heat stirring Liver and Gallbladder fire and heat, upward blazing to the ear, and rotting the flesh and blood. Chronic type is caused by Spleen deficiency, excess dampness, or Yin deficiency with deficient fire. In Western medicine this is called a suppurative inflammatory syndrome of the mucous membrane in the middle ear. The acute type often occurs in young children with fever and earache. The symptoms worsen in a day or two, with pulsating pain and decreased hearing. After the eardrum ruptures and discharges the pus, fever and pain are then alleviated. Chronic type manifests as chronic or intermittent discharge of pus, perforated and unhealed eardrum, and decreased hearing. TCM treatments for acute type focus on draining fire and relieving toxicity, and for the chronic type, the focus is on tonifying Qi and nourishing Yin.

Acute Type

Beginning Stage

Treatment Principle: Disperse wind, clear heat, relieve toxicity, and reduce swelling.

Use Man Jing Zi, Bo He, Ju Hua, Huang Qin, Xia Ku Cao, Ge Gen, Jing Jie, and Gan Cao.

Related conditions and modifications are

- With summer dampness: Add Huo Xiang and Pei Lan.
- Ear congestion and discomfort: Add Lu Lu Tong and Ma Dou Ling.

Middle Stage (Suppurative Stage)

Treatment Principle: Clear Liver, drain fire, reduce swelling, and expel pus.

Use Shan Zhi Zi, Huang Qin, Chai Hu, Jin Yin Hua, Chuan Shan Jia, Dang Gui Wei, Chi Shao Yao, and Gan Cao.

Related conditions and modifications are

- Severe pain before rupture of the eardrum: Add Zhi Ru Xiang and Zhi Mo Yao at 3 g each.
- High fever: Add Da Cheng Qi Tang [Major Order of Qi Decoction].
- Residual exterior syndrome: Add Ju Hua and Bo He.
- Ruptured eardrum, pus discharge, alleviated fever, and ear ache: Add Zao Jiao Ci.

Late Stage (Recovery Stage)

Treatment Principle: Support Zheng Qi and expel evil pathogens.

Use Wu Wei Xiao Du Yin [Five-Ingredient Decoction to Eliminate Toxin] plus Dang Gui and Huang Qi.

Related conditions and modifications are

- Deep red tongue and scanty coating: Add Sheng Di Huang and Lu Gen.
- Copious yellow pus discharge: Add Huang Bai and Cang Zhu.
- Copious clear watery pus discharge: Add Fu Ling, Ze Xie, and Gui Zhi.
- Difficulty healing after puncture of the eardrum: Add Fu Ling, Dang Shen, Hong Hua, and Ji Xue Teng.

Chronic Type

Treatment Principle: Tonify Qi, nourish Yin, and eliminate dampness.

Use Er Miao San [Two-Marvel Powder] plus Ze Xie, Che Qian Cao, Deng Xin Cao, Xia Ku Cao, Mu Dan Pi, Hua Shi, and Gan Cao.

Related conditions and modifications are

- Scanty white pus discharge: Remove Huang Bai, and add Fu Ling and Yi Yi Ren.
- Qi Deficiency: Add Dang Shen and Huang Qi.
- Spleen deficiency with loose stool, poor appetite: Remove Mu Dan Pi and Deng Xin Cao, and add fried Shan Zha, Ge Gen, and Shan Yao.
- Yin deficiency: Add Zhi Mu and Xuan Shen.

Furthermore, combine with an external herbal preparation: Huang Bai Di Er Ye [Ear Drops with Phellodendron] Huang Bai 30 g, Bing Pian 0.3 g, Peng Suan (boric acid) 4 g, and distilled water 100 mL. After cleaning and drying the ear canal, put three to five drops of the preparation, two to three times a day into the ear. Discontinue when the pus in the ear has decreased.

TINNITUS

This is also called ringing in the ear as per patient report. It can be seen in many conditions, often caused by Liver and Kidney insufficiency, Qi, and blood deficiency. It can also be caused by wind, fire, phlegm, and stasis. When the ringing is low pitch and it comes and goes, the condition is a deficient type. If the ringing is high pitch, constant, and loud, then it is an excess type.

Wind–Heat Attacking Lung Syndrome

Treatment Principle: Disperse wind and clear heat.

Use Yin Qiao San [Honeysuckle and Forsythia Powder].

Related conditions and modifications are

- Severe fever, sore throat: Add Da Qing Ye and Ban Lan Gen.
- Interior heat and constipation: Change to Fang Feng Tong Sheng San [Ledebouriella Powder that Sagely Unblocks].

Excess Liver and Gallbladder Fire Syndrome

Treatment Principle: Clear Liver and drain Gallbladder.

Use Long Dan Xie Gan Tang [Gentiana Longdancao Decoction to Drain the Liver] and Dang Gui Long Hui Wan [Tangkuei, Gentiana Longdancao, and Aloe Pill].

Knotted Phlegm Fire Stagnation Syndrome

Treatment Principle: Transform phlegm and clear fire.

Use Meng Shi Gun Tan Wan [Vaporize Phlegm Pill with Chlorite Schist] which is also called Gun Tan Wan [Vaporize Phlegm Pill].

Blood Stasis at Ear Meridians Syndrome

Treatment Principle: Open the orifices and invigorate blood.

Use Tong Qiao Huo Xue Tang [Unblock the Orifices and Invigorate the Blood Decoction] plus Zhe Bei Mu, Hai Zao, and Kun Bu.

Liver and Kidney Deficiency Syndrome

Treatment Principle: Tonify Liver and Kidney.

Use Er Long Zuo Ci Wan [Pill for Deafness that is Kind to the Left (Kidney)].

Related conditions and modifications are

- Severe Liver and Kidney deficiency: Add Gou Qi Zi and Er Zhi Wan [Two-Ultimate Pill].
- Upward blazing of deficient fire: Add Huang Bai and Huang Lian.
- Ear congestion: Add Shi Chang Pu and Mu Tong.

Qi and Blood Deficiency Syndrome

Treatment Principle: Tonify Qi and benefit blood.

Use Ba Zhen Tang [Eight-Treasure Decoction] plus Suan Zao Ren, Yi Zhi Ren, Mai Men Dong, and Long Yan Rou.

Related conditions and modifications are

- Liver blood deficiency: Add Nu Zhen Zi and Mu Gua.
- Blood stasis: Add Tao Ren, Dan Shen, and Hong Hua.

ACUTE CONJUNCTIVITIS

This condition is often caused by wind–heat invading Lung, rising upward, and attacking the eyes. It can also be caused by constitutional excess Lung and Stomach heat combined with external pathogenic factors. Clinical manifestations are acute onset, obvious conjunctiva infection, or with mucous or purulent discharge. In severe cases, there is a sensation of foreign body in eyes, light sensitivity, tearing, burning sensation; in general there is no sharp, severe pain unless cornea is affected. TCM treatments focus on dispersing Lung, expelling wind, clearing heat, and draining Lung. One can use the following basic formula and modify it according to the manifestations.

Use Sang Dan Xie Bai San [Mulberry Leaf and Moutan Powder to Drain the White] plus Jing Jie, Fang Feng, Huang Qin, Shan Zhi Zi, Lian Qiao, Mu Tong, Jie Geng, Zhi Ke, Chi Shao Yao, and Sang Ye.

Related conditions and modifications are

- Wind more severe than heat: Remove Lian Qiao and Jin Yin Hua, and add Qiang Huo and Bo He.
- Heat more than wind: Add Shi Gao, Da Huang, and Zhi Mu.
- Excess congestion of Lung heat: Remove Jing Jie and Fang Feng, and add Huang Lian, Sheng Di Huang, and Da Huang.
- Transformation to chronic conjunctivitis: Add Huang Qi, Bai Zhu, Dang Shen, Sheng Di Huang, Mai Men Dong, Xuan Shen, and Sha Shen.
- Dark nebula: Add Mu Zei, Chan Tui, Gu Jing Cao, Cao Jue Ming, and Xia Ku Cao.

CHRONIC GLAUCOMA

This condition is caused by Liver Qi stagnation, Liver fire, phlegm stasis causing hesitant flow of Qi and blood, and stasis and obstruction inside the eyes. The clinical manifestations are recurrent attacks, discomfort at the back of the eyes, paroxysmal rainbow vision, blurry vision, or asymptomatic. However, there is decreased vision or blindness of the affected side under eye exam. TCM treatment focuses on dispersing Liver, relieving stagnation, moving Qi, invigorating blood, transforming phlegm, and draining dampness.

Liver Qi Stagnation Syndrome

Treatment Principle: Disperse Liver, clear heat, invigorate blood, and transform stasis.

Use Dan Zhi Xiao Yao San [Relax Wanderer Powder with Moutan and Gardenia] plus Yu Jin, Xia Ku Cao, and Dan Shen.

Related condition and modification are

- Severe distention and eye pain: Add Xia Ku Cao, Mu Tong, and Ze Xie.

Phlegm Fire Upward Attacking Syndrome

Treatment Principle: Clear heat, expel phlegm, harmonize Stomach, and descend rebellious Qi.

Use Huang Lian Wen Dan Tang [Coptis Decoction to Warm the Gallbladder] plus Pei Lan, Chuan Bei Mu, and Huang Qin.

Related condition and modification are

- Severe phlegm heat: Add Jiang Can, Meng Shi, and Bo He.

Liver and Stomach Deficient Cold Syndrome

Treatment Principle: Warm Liver and Stomach, descend rebellious Qi, and stop vomiting.

Use Wu Zhu Yu Tang [Evodia Decoction] plus Bai Zhi, Chen Pi, Chuan Xiong, Fu Ling, Ban Xia, and Gan Cao.

Liver and Kidney Yin Deficiency Syndrome

Treatment Principle: Nourish Yin and descend fire.

Use Zhi Bai Di Huang Wan [Anemarrhena, Phellodendron, and Rehmannia Pill] plus Ci Shi, Shi Jue Ming, Nu Zhen Zi, Shi Hu, and Ju Hua.

(Caution: Chronic increase of intraocular pressure should be treated with surgery, then with herbs to prevent recurrence.)

HERPES SIMPLEX KERATITIS

This condition is caused by external wind–heat toxins invasion combining with internal Liver fire and then upward-attacking the pupil. In Western medicine it is attributed to infections of herpes simplex I, which occurs more in youth and middle-aged people and is usually unilateral. The damage of the cornea can appear to be dendritic, geographic, and discoid shape. The disease progression is long, often recurrent, and difficult to treat. If the treatment is not given early enough, vision can be affected or even blindness may occur. TCM treatments focus on clearing heat, relieving toxicity, expelling wind, draining damp, and at the late stage, on supporting Zheng Qi. Herbs that can inhibit herpes simplex I virus are Zi Cao, Chai Hu, Bo He, Jin Yin Hua, and Pu Gong Ying. Herbs that can increase immune system are Huang Qi, Dang Shen, Shu Di Huang, Bai Zhu, and Nu Zhen Zi. These can be added to the formula according to syndrome differentiation. During symptom period, the manifestations are painful eye with same-sided headache, aversion to light, and tearing.

Wind–Heat Attacking Eye Syndrome

Treatment Principle: Disperse wind and scatter heat.

Use Si Ni San [Frigid Extremities Powder] plus Jing Jie, Fang Feng, Huang Qin, Long Dan Cao, Man Jing Zi, and Da Qing Ye, and substitute Bai Shao Yao with Chi Shao Yao.

Related conditions and modifications are

- Frontal headache and nasal congestion: Add Bai Zhi.
- Bitter taste, yellow urine, and wiry rapid pulse: Add Huang Lian and Shan Zhi Zi.
- Severe infection at conjunctiva: Add Tao Ren, Hong Hua, Mu Dan Pi, and Chong Wei Zi.

Damp–Heat Accumulation and Steaming Syndrome

Treatment Principle: Transform dampness, clear heat, expel wind, and relieve toxicity.

Use Da Qing Ye, Hua Shi, Mu Tong, Huang Qin, Lian Qiao, Jing Jie, Fang Feng, Huo Xiang, Tu Fu Ling, and Gan Cao.

Related conditions and modifications are

- Severe swelling at cornea: Remove Huang Qin, add Cang Zhu, Yi Yi Ren, and Bai Dou Kou.
- Poor appetite and apathy: Remove Hua Shi, and add Tai Zi Shen, Bai Zhu, and Shen Qu.

Zheng Qi Deficiency and Residual Evil Pathogens

Treatment Principle: Tonify and attack at the same time.

Use Huang Qi, Bai Zhu, Zi Cao, Da Qing Ye, Chi Shao Yao, Fang Feng, Dang Gui, Chai Hu, Mu Tong, Qiang Huo, and Gan Cao.

Related conditions and modifications are

- Eye pain, aversion to light, and tearing: Minus Huang Qi and Dang Gui, and add Jing Jie, Man Jing Zi, and Jin Yin Hua.
- Red tongue, yellow coating, and wiry rapid pulse: Add Huang Qin and Long Dan Cao.

Zheng Qi Deficiency with Blood Stasis Syndrome

Treatment Principle: Tonify Qi, invigorate blood, and expel stasis.

Use Yu Ping Feng San [Jade Windscreen Powder] plus Dan Shen, Sheng Di Huang, Nu Zhen Zi, Chong Wei Zi, Dang Gui, Chai Hu, Mu Zei, Mi Meng Hua, and Chan Tui.

Related conditions and modifications are

- Chronic unhealed cornea ulcer: Add Dang Shen, Bai Lian, and increase Huang Qi to 60 g.
- Middle Jiao Qi sinking: Combine with Bu Zhong Yi Qi Tang [Tonify the Middle and Augment the Qi Decoction].

HORDEOLUM (STY)

Sty is called "needle eye" in TCM. It is often caused by wind evil and heat toxins invading the body, or heat accumulation at Spleen, Stomach, or Heart, upward-attacking the eyelids with Qi and blood stagnation and congested heat forming pus. The characteristics are acute onset, fast progression, local swelling, redness and burning pain due to infection, and pus with ulceration. TCM treatment should be given as early as possible to control, eliminate, and scatter the development of nodular swelling. The treatment principles are clearing heat, dispersing wind, relieving toxicity, and reducing swelling.

Wind-Heat Toxins Attack (Before Pus Formation)

Treatment Principle: Clear heat, disperse wind, relieve toxicity, and reduce swelling.

Use Jin Yin Hua, Lian Qiao, Ye Ju Hua, Pu Gong Ying, Zi Hua Di Ding, Zi Bei Tian Kui, Niu Bang Zi, and Jie Geng.

Related conditions and modifications are

- Local swelling, redness, and itching: Add Jing Jie, Fang Feng, and Bo He.
- Severe local redness, swelling, and pain: Add Dang Gui Wei, Chi Shao Yao, and Dan Zhu Ye.

Exuberant Fire Heat Accumulation (Pus Formation Period)

Treatment Principle: Clear heat, drain fire, and reduce swelling.

Use Jin Yin Hua, Fang Feng, Bai Zhi, Dang Gui Wei, Chi Shao Yao, Xia Ku Cao, Mu Tong, Dan Zhu Ye, Huang Qin, and Huang Lian.

Related conditions and modifications are

- Appearing pus head, not yet obvious: Add Shan Zhi Zi and Sheng Di Huang.
- Constipation: Add Da Huang.
- Small pus head that is hard to ripen: Add Zao Jiao Ci and Jie Geng.

OPTIC ATROPHY

This condition is called "blue-green blindness" in TCM. It is often caused by Liver and Kidney essence deficiency, Qi, and blood deficiency. In Western medicine it is a result of any severe retinopathy and optic neuropathy. The clinical manifestations are dysopsia, decreased visual field, and at the end, blindness. TCM treatments focus on tonifying Liver and Kidney, tonifying Qi, and nourishing blood. Treatments should be long term, lasting at least 3 months.

Liver and Kidney Deficiency Syndrome

Treatment Principle: Tonify Liver and Kidney, benefit essence, and brighten the eyes.

Use Ming Mu Di Huang Wan [Improve Vision Pill with Rehmannia] plus Chai Hu and Quan Xie.

Related conditions and modifications are

- Liver Qi stagnation: Add Bai Shao Yao, Zhi Ke, and Bai Ji Li.
- Yin deficiency and internal heat: Add Sheng Di Huang, Mu Dan Pi, and Zhi Mu.
- Kidney Yang deficiency: Add Rou Gui, Tu Si Zi, and Zi He Che.

Qi and Blood Deficiency Syndrome

Treatment Principle: Tonify Qi and nourish blood.

Use Ren Shen Yang Ying Tang [Ginseng Decoction to Nourish the Nutritive Qi].

Heat Condition Damaging Yin Syndrome

Treatmen Principle: Nourish Yin and benefit Stomach.

Use Yi Guan Jian [Linking Decoction] plus Shan Yao, Wu Wei Zi, and Bai Ji Li.

Qi Stagnation and Blood Stasis Syndrome

Treatment Principle: Invigorate blood and transform stasis.

Use Tao Hong Si Wu Tang [Four-Substance Decoction with Peach Pit and Safflower] plus Mu Dan Pi, Chong Wei Zi, Che Qian Zi, Yu Jin, and Quan Xie.

All of the above formulas can add two to three of the following herbs: She Xiang, Chai Hu, Dan Shen, Yu Jin, Dang Gui, Chuan Xiong, and Quan Xie. These herbs open and unblock the eyes, disperse the stagnation, and unblock the stasis to nourish the eye and to improve vision.

OPTIC NEURITIS

This condition is discussed under sudden blindness and blurry vision in TCM. The pathogeneses are Liver Qi stagnation transforming into fire and upward-attacking of the eyes, or Qi and blood deficiency failing to nourish the eyes. The clinical manifestations are acute onset, sudden and fast decrease in vision or even blindness, or with pain when moving the eyes, or with big central black spot at visual field or concentric decrease of the visual field. TCM treatment focuses on dispersing Liver, relieving stagnation, clearing heat, draining fire, and opening and unblocking the eyes.

Stagnant Heat at Liver Channel Syndrome

Treatment Principle: Disperse Liver, relieve stagnation, and clear heat.

Use Dan Zhi Xiao Yao San [Relax Wanderer Powder with Moutan and Gardenia] plus Xia Ku Cao, Bai Ji Li, and Ju Hua.

Related conditions and modifications are

- Distending pain at orbit, dizziness, tinnitus, red face, and red eyes: Add Shi Jue Ming, Cao Jue Ming, and Gou Teng.
- Stifling sensation of the head, eye distention, and circuitous vein at fundus: Add Dan Shen, Tao Ren, Hong Hua, and Bai Mao Gen.

Heat Condition Damaging Yin Syndrome

Treatment Principle: Clear heat and nourish Yin.

Use Sha Shen Mai Men Dong Tang [Glehnia and Ophiopogonis Decoction] plus Jin Yin Hua, Lian Qiao, Sheng Di Huang, Gou Qi Zi, Zhi Mu, and Mu Dan Pi.

Related conditions and modifications are

- Severe heat: Add Shan Zhi Zi and Huang Qin.
- Mild heat: Remove Zhi Mu and Lian Qiao, and add Shan Yao and Tu Si Zi.

Qi and Blood Deficiency Syndrome

Treatment Principle: Tonify Qi and blood.

Use Ren Shen Yang Ying Tang [Ginseng Decoction to Nourish the Nutritive Qi].

Related conditions and modifications are

- Liver Qi stagnation: Add Bai Shao Yao, Yu Jin, and Chai Hu.
- Kidney deficiency: Add Gou Qi Zi, Tu Si Zi, and Chong Wei Xin.

Liver and Kidney Deficiency Syndrome

Treatment Principle: Tonify Liver and Kidney.

Use Zhu Jing Wan Jia Jian [Modified Preserve Vistas Pill].

Related conditions and modifications are

- Severe Kidney Yang deficiency: Add Rou Gui and Zi He Che.
- Severe Liver Yin deficiency: Add Shan Zhu Yu and Bai Shao Yao.

Pterygium

Pterygium is a fleshy growth in the eyes, especially at inner or outer canthus, from conjunctiva to cornea. It is caused by wind–heat, excess heat, or deficient fire attacking the canthus; obstructing channels and collaterals, and with Qi stagnation and blood stasis. It is often seen in people who work outside, especially fishermen. If it is small and smooth, then there is no obvious discomfort. If it is big, thick, and infected on the surface, then one can have an uncomfortable sensation or it can cause astigmatism. When the fleshy growth enters the pupil, the vision can be affected. Most of the pterygium grow slowly. TCM treatments focus on eliminating discomfort and preventing or delaying progression.

Wind–Heat at Lung and Heart Syndrome

Treatment Principle: Expel wind and clear heat.

Use: Formula consists of Fang Feng, Jing Jie, Qiang Huo, Bai Ji Li, Gu Jing Cao, Cao Jue Ming, Ju Hua, Mi Meng Hua, Man Jing Zi, Shan Zhi Zi, Huang Qin, Mu Zei, Chuan Xiong, Chan Tui, and Gan Cao.

Related conditions and modifications are

- Red eyes, copious eye discharge, and flooding big pulse: Remove Mi Meng Hua and Qiang Huo, and add Da Huang.

Excess Heat at Spleen and Stomach Syndrome

Treatment Principle: Drain heat and unblock Fu organs.

Use Huang Qin, Che Qian Zi, Chong Wei Zi, Fang Feng, Da Huang, Mang Xiao, Jie Geng, and Huang Lian.

Related conditions and modifications are

- Copious sticky eye discharge: Add Xuan Shen and Xia Ku Cao.
- For fast growth of pterygium: Add Shan Zha and Bing Lang.

Yin Deficiency with Deficient Fire Syndrome

Treatment Principle: Nourish Yin and descend fire.

Use Zhi Bai Di Huang Wan [Anemarrhena, Phellodendron, and Rehamnnia Pill].

Related condition and modification are

- Irritability, insomnia: Add Mai Men Dong, Wu Wei Zi, and Suan Zao Ren.

SENILE CATARACT

This is often caused by Qi and blood deficiency, essence exhaustion, Liver and Kidney deficiency, and inability to nourish eyes. It can also be caused by damp–heat and Qi stagnation disturbing eyes. It is more common in people older than 50 years of age and often occurs in both eyes simultaneously or one after another. The clinical manifestations are progressive decrease in vision, fixed black spots, or double vision. With the increased cloudiness of the lens, cataract is mature, visibility can decrease to not being able to recognize people or things with only sensation to light left. TCM treatment at the beginning stage can control the clouding of the lens and increase visibility which has better effect.

Liver Kidney Yin Deficiency Syndrome

Treatment Principle: Tonify Liver and Kidney, brightens the eyes.

Use Zhi Shou Wu, Huang Jing, Shu Di Huang, Tu Si Zi, Ci Shi, Gou Qi Zi, Yu Zhu, Zhi Ke, and Shen Qu.

Related conditions and modifications are

- Tired vision: Add Bai Zhu and Huang Qi.
- Distending sensation at eyes: Add Xia Ku Cao and Shi Jue Ming.

Spleen Deficiency Syndrome

Treatment Principle: Strengthen Spleen and tonify Qi.

Use Bu Zhong Yi Qi Tang [Tonify the Middle and Augment the Qi Decoction] plus Qing Xiang Zi and Shi Jue Ming.

Related condition and modification are

- Loose stool: Minus Dang Gui, and add Yi Yi Ren, Fu Ling, and Bian Dou.

Stagnant Heat at Liver Channel Syndrome

Treatment Principle: Disperse Liver, clear heat, and brighten the eyes.

Use Dan Zhi Xiao Yao San [Relax Wanderer Powder with Moutan and Gardenia].

Related conditions and modifications are

- Long-term stagnation turning into fire: Add Long Dan Cao, Huang Qin, and Xia Ku Cao.
- Liver heat damaging Yin: Add Xuan Shen, Shi Hu, Mai Men Dong, and Yu Zhu.

Yin Deficiency with Damp–Heat Syndrome

Treatment Principle: Nourish Yin and clear heat.

Use Shu Di Huang, Sheng Di Huang, Mai Men Dong, Tian Men Dong, Huang Qin, Yin Chen Hao, Zhi Ke, Pi Pa Ye, and Shi Hu.

Related conditions and modifications are

- Excess dampness: Remove Shu Di Huang, and add Yi Yi Ren, and Fu Ling.
- Severe distention at eyes: Add Bing Lang, Che Qian Zi, and Ze Xie.

SENILE MACULAR DEGENERATION

This condition belongs to blurry vision and vitreous opacity in TCM. It is caused by deficiency of the Zheng Qi leading to Spleen deficiency, Liver Qi stagnation, blood stasis, phlegm damp, and Kidney deficiency. In Western medicine it is described as often occurring in people 45 years or older with regressive pathological condition at the macula lutea of the retina. At an early stage, vision may not be affected or there is mild damage. At middle stage, there is an obvious decrease in vision and distorted vision. At late stage, patient can have complete blindness. TCM treatment at early stage focuses on tonifying Qi, strengthening Spleen, draining dampness, and transforming turbidity. At middle stage it should focus on expelling phlegm, transforming dampness, dispersing Liver, and invigorating blood. At late stage, the treatment should focus on tonifying Liver and Kidney, benefiting the essence and blood. Invigorating blood and transforming stasis should happen at every stage of the condition to increase effectiveness.

Early Stage

Mild decrease in vision, vitreous membrane wart at macula lutea, or increased pigmentation at macula lutea area.

Treatment Principle: Tonify Qi, strengthen Spleen, drain damp, and transform turbidity.

Use Dan Bai Si Ni San [Moutan and Phellodendron Powder for Frigid Extremities] plus Bai Zhu, Ju Hua, Che Qian Zi, Chong Wei Zi, Dang Gui, Dang Shen, Fu Ling, Huang Qi, Dan Shen, Sheng Ma, Ling Zhi, and Zhi Shou Wu.

Middle Stage

Severe decrease in vision, swelling at macula lutea area with gray white exudates.

Strengthen Spleen, tonify Qi, expel phlegm, transform dampness, disperse Liver, and invigorate blood.

Use Si Jun Zi Tang [Four-Gentleman Decoction] plus Er Chen Tang [Two-Cured Decoction] and Che Qian Zi, Dan Shen, Zhe Bei Mu, Yi Yi Ren, Ze Xie, and Ji Xue Teng.

Related conditions and modifications are

- Copious exudates: Add San Leng, E Zhu, Kun Bu, and Hai Zao.
- Severe swelling: Add zhu Ling, Yi Mu Cao, and Gui Zhi.
- Vitreous opacity and bleeding: Add Xian He Cao, Han Lian Cao, Bai Ji, Sheng Pu Huang, Nu Zhen Zi, Ling Zhi, and San Qi powder.

Late Stage

Severe damage or loss of vision and formation of discoid scar at macula lutea area.

Treatment Principle: Benefit Qi, nourish blood, tonify Liver and Kidney, soften hardness, and scatter accumulation.

Use Si Jun Zi Tang [Four-Gentleman Decoction] plus Er Zhi Wan [Two-Ultimate Pill] and Huang Qi, Zhi He Shou Wu, Bai Shao Yao, Kun Bu, Huang Jing, Zhe Bei Mu, Sha Yuan Ji Li, Sheng Ma, Chu Shi Zi, and Gou Qi Zi.

UVITIS

This condition belongs to vitreous opacity, shrinking of the pupil, dry pupil, and blurry vision. The pupil belongs to the Kidney. Iris and the ciliary body belong to the Liver. The choroid belongs to the Heart. The exudates and swelling during inflammation process belong to phlegm damp which is related to the Spleen. At early stage, it is caused by wind–heat in the Liver channel, wind–damp with heat damaging eyes, or excess fire of Liver and Gallbladder attacking the eyes. These are excess and heat syndromes. During the middle stage, it is related to heat evil damaging Yin, upward blazing of deficient fire, damp obstruction due to Spleen deficiency, and mixed deficiency and excess patterns. At late stage, there are Liver and Kidney Yin deficiency, damaged collaterals and blood stasis, and knotted phlegm and blood stasis. These are root deficiencies with excess manifestations. In Western medicine it is described as an inflammatory condition of uvea (including iris, ciliary body, and choroid) due to various reasons. It often occurs in youth and middle-aged people, and often both eyes are involved. It is recurrent and is a common reason for the cause of blindness. The clinical manifestations are severe eye pain, aversion to light, tearing, quick worsening of vision, cloudy vitreous humor, and decreased pupil size. TCM treatments during acute stage focus on clearing heat, relieving toxicity, cooling blood, invigorating blood, expelling wind, and eliminating dampness. During middle and late stage, the treatments focus on nourishing Liver and Kidney, and are supported by transforming dampness and stasis.

Wind–Heat in Liver Channel Syndrome

Treatment Principle: Disperse wind and clear heat.

Use Si Ni San [Frigid Extremities Powder] plus Huang Qin, Shan Zhi Zi, Qing Xiang Zi, Man Jing Zi, Jing Jie, Fang Feng, Ci Ji Li, and Long Dan Cao.

Related conditions and modifications are

- Dry mouth: Add Lu Gen.
- Severe headache and eye pain: Add Bai Zhi and Qiang Huo.
- Hypopyon–pus accumulation at anterior chamber: Add Pu Gong Ying and Shi Gao.

Excess Fire of Liver and Gallbladder Syndrome

Treatmen Principle: Clear Liver and drain Gallbladder.

Use Long Dan Xie Gan Tang [Gentiana Longdancao Decoction to Drain the Liver] plus Xia Ku Cao and Zi Hua Di Ding.

Related condition and modification are

- Hyphema–blood accumulation at anterior chamber: Add Sheng Pu Huang, Bai Mao Gen, and San Qi powder.

Wind–Damp with Heat Syndrome

Treatment Principle: Expel wind, eliminate dampness, and clear heat.

Use Fang Feng, Qiang Huo, Du Huo, Zhi Mu, Huang Lian, Fang Ji, Huang Qin, Huang Bai, Qin Jiao, Shan Zhi Zi, Ze Xie, Yi Yi Ren, and Sheng Di Huang.

Related condition and modification are

- Swollen, red, and painful joints: Add Ren Dong Teng and Sang Zhi.

Yin Deficiency with Deficient Fire Syndrome

Treatment Principle: Nourish Yin, descend fire, and transform stasis.

Use Zhi Bai Di Huang Wan [Anemarrhena, Phellodendron, and Rehmannia Pill] plus Ju Hua, Xuan Shen, Chong Wei Zi, Nu Zhen Zi, and Han Lian Cao.

Related conditions and modifications are

- Eye pain, severe red eyes, sudden decrease in vision, headache, bitter taste, and yellow tongue coating: Add Long Dan Cao, Xia Ku Cao, Huang Qin, and Qing Xiang Zi.
- Copious exudates at ocular fundus, cloudy vitreous humor, increased deposits behind the cornea, and thick, greasy, tongue coating: Add Yi Yi Ren, Yin Chen Hao, and Pei Lan.

Liver and Kidney Deficiency with Blood Stasis Syndrome

Treatment Principle: Tonify Kidney and Liver, transform stasis, and scatter accumulation.

Use Zhu Jing Wan [Preserve Vistas Pill] plus Er Chen Tang [Two-Cured Decoction] and Si Wu Tang [Four-Substance Decoction].

Related conditions and modifications are

- Severe eye pain and increased intraocular pressure: Add Chuan Niu Xi, Shi Jue Ming, and Quan Xie.
- Yellow spots and edema: Add Yi Yi Ren and Ze Xie.
- Seeing streaks due to hardening of the vitreous body: Add San Leng, E Zhu, and Kun Bu.

NASAL POLYP

Nasal polyp is often caused by wind, damp, and heat invasion, damp–heat accumulation, and congelation at the nasal cavity. In Western medicine, many factors are attributed to mild edema or thickening of the mucous membrane at the nasal cavity or sinuses. It is related to pathological reaction to chronic inflammatory irritations. It often occurs in adults 20 to 30 years of age. Main symptoms are nasal congestion, disturbance of sense of smell, increased nasal discharge, breathing with mouth open, dizziness, and headache. TCM treatments focus on clearing and transforming damp–heat, and dispersing and unblocking nasal orifices.

Lung Qi Congestion and Stagnation Syndrome

Treatment Principle: Disperse Lung and scatter accumulation.

Use Cang Er Zi San [Xanthium Powder] plus Huang Qin, Ting Li Zi, Sang Bai Pi, Yu Xing Cao, Ma Dou Ling, Che Qian Zi, and Jiang Can.

Related conditions and modifications are

- Severe cold: Add Jing Jie, Xi Xin, and Bi Cheng Qie.
- Severe heat: Add Jin Yin Hua, Lian Qiao, and Bo He.

Spleen Dampness Syndrome

Treatment Principle: Strengthen Spleen, transform turbidity, and open orifices.

Use Liu Jun Zi Tang [Six-Gentleman Decoction] plus Huo Xiang, Pei Lan, Xin Yi Hua, Cang Er Zi, Shi Chang Pu, and Huang Lian.

Related conditions and modifications are

- Qi stagnation at Middle Jiao: Add Mu Xiang, Sha Ren, and Fo Shou.
- Poor appetite and loose stool: Add Ji Nei Jin, fried Shen Qu, and fried Mai Ya.
- Headache: Add Chuan Xiong and Bai Zhi.
- Recurrent polyps after surgery: Add Cang Zhu, Wu Wei Zi, Wu Mei, Zao Jiao Ci, and Shi Liu Pi.

RHINITIS (ACUTE)

This condition is referred to as nasal congestion due to wind injuries. It is caused by wind–cold heat invasion, Lung failing to disperse and descend, and nasal blockage. In Western medicine it is described as an acute inflammation of the nasal mucous membrane. The clinical manifestations are nasal congestion, discharge, or with fever, aversion to wind, or aversion to cold. TCM treatments focus on releasing wind, dispersing Lung, and open orifices.

Exterior Wind–Cold Syndrome

Treatment Principle: Disperse wind, scatter cold, and use acrid warm to open the orifices.

Use Jing Fang Bai Du San [Schizonepeta and Ledebouriella Powder to Overcome Pathogenic Influences].

Related conditions and modifications are

- Distending pain at forehead: Add Bai Zhi and Gao Ben.
- Severe nasal congestion: Add Cong Bai and Sheng Jiang.

Exterior Wind–Heat Syndrome

Treatment Principle: Disperse wind, clear heat, and acrid cool to open the orifices.

Use Sang Ju Yin [Mulberry Leaf and Chrysanthemum Decoction].

Related conditions and modifications are

- Nasal congestion with yellow discharge: Add Jin Yin Hua, Huang Qin, and Guan Zhong.
- Sore throat: Add Niu Bang Zi and Xuan Shen.
- Productive cough: Add Qian Hu and Gua Lou Pi.

RHINITIS (ATROPHIC)

This is called Bi Gao in TCM. It is caused by dry Qi damaging fluid, Lung and Kidney Yin deficiency, and Spleen failing to transport and transform. In Western medicine it is seen as a pathological atrophy of the mucous membrane in the nasal cavity and turbinate. It is more common in young females. The clinical characteristics are dry nasal cavity, scanty nasal discharge or dry scab, and easily having nose bleed. TCM treatments are nourishing Yin, moistening dryness, and benefiting Qi.

Yin Deficiency and Lung Dryness Syndrome

Treatment Principle: Nourish Yin and moisten dryness.

Use Qing Zao Jiu Fei Tang [Eliminate Dryness and Rescue the Lungs Decoction]

Related conditions and modifications are

- With nose bleed: Add Bai Mao Gen and Xian He Cao.
- Nasal dryness and pain: Add Tian Hua Fen, Lu Gen, and Zhi Mu.

Spleen and Stomach Deficiency Syndrome

Treatment Principle: "Support the earth to generate metal."

Use Shen Ling Bai Zhu San [Ginseng, Poria, and Atractylodes Macrocephala Powder] plus Sha Shen Mai Men Dong Tang [Glehnia and Ophiopogonis Decoction].

Kidney Yin Deficiency Syndrome

Treatment Principle: Nourish Yin and descend fire.

Use Zhi Bai Di Huang Wan [Anemarrhena, Phelloendron, and Rehmannia Pill] plus Bai Mao Gen, Gui Ban, Huai Niu Xi, and Sheng Ma.

RHINITIS (CHRONIC)

This is under nasal congestion in TCM. It is caused by clear Yang failing to reach nasal orifices, Qi stagnation, and blood stasis. It is also closely related with Lung and Spleen. In Western medicine, it is seen as a chronic inflammation of the nasal mucous membrane and submucous layer which includes chronic simple rhinitis and chronic hypertrophic rhinitis. The clinical characteristics are intermittent nasal congestion, mucous and nasal discharge, throat discomfort with post nasal drip, copious sputum, heavy nasal voice, and decreased sense of smell. TCM treatments focus on dispersing Lung, opening orifices, invigorating blood, and transforming phlegm.

Chronic Simple Rhinitis

Treatment Principle: Tonify and benefit Spleen and Lung, disperse Lung, and open orifices.

Use Si Jun Zi Tang [Four-Gentleman Decoction], Yu Ping Feng San [Jade Windscreen Powder], and Jie Geng Gan Cao Tang [Platycodon and Licorice Decoction] with Xi Xin and Xin Yi Hua.

Related conditions and modifications are

- Severe Spleen deficiency: Add Chen Pi and Shan Yao.
- Copious nasal discharge and distending headache: Remove Dang Shen and add Bai Zhi, Bo He, and Cang Er Zi.
- Nasal congestion with yellow sticky discharge: Change to Xie Bai San [Drain the White Powder] plus Jin Yin Hua, Lu Gen, Xin Yi Hua, Cang Er Zi, Jie Geng, and Huang Qin.

Chronic Hypertrophic Rhinitis

Treatment Principle: Open orifices and transform stasis.

Use Tong Qiao Huo Xue Tang [Unblock the Orifices and Invigorate the Blood Decoction] plus Tao Hong Si Wu Tang [Four-Substance Decoction with Peach Pit and Safflower] and San Leng, E Zhu, Lu Lu Tong, Shi Chang Pu, Xin Yi Hua, and Cang Er Zi.

Related conditions and modifications are

- Distending painful forehead: Add Bai Zhi, Gao Ben, and Bo He.
- Yellow, sticky nasal discharge: Add Gua Lou Pi, Zhu Ru, Huang Qin, and Long Dan Cao.
- Qi Deficiency: Add Huang Qi, Dang Shen, and Bai Zhu.

SINUSITIS

Acute sinusitis is often caused by disturbances from wind–heat invading Lung channel, stagnant heat in Gallbladder channel, and damp–heat in the Spleen and Stomach. In chronic sinusitis, it is often caused by deficiency and cold in the Lung, Spleen Qi deficiency, and evil toxins accumulating in Lung and nasal orifices. In Western medicine, it is seen as an inflammatory disease caused by pathogenic infections of the mucous membrane in the sinus cavities. The manifestations of acute sinusitis are fever and chills, nasal congestion, thick nasal discharge, headache, whole body discomfort, decreased appetite, and constipation. For chronic stage, whole body manifestations are not obvious. However, there are sometimes dizziness, distending headache, thick nasal discharge or nasal congestion, decreased concentration, fatigue, and apathy. TCM treatments focus on clearing heat, relieving toxicity, discharging pus, and unblocking orifices. For acute sinusitis, integrated treatments of pharmaceutical medication and Chinese herbal therapy are effective. For chronic sinusitis, Chinese herbal therapy is more effective.

Acute Sinusitis

Wind–Heat Invading Lung Channel

Yellow and white sticky nasal discharge, nasal congestion, decreased sense of smell, and forehead or maxillary pain.

Treatment Principle: Disperse wind, clear heat, disperse Lung Qi, and unblock orifices.

Use Sang Ju Yin [Mulberry Leaf and Chrysanthemum Decoction] plus Cang Er Zi San [Xanthium Powder], Sang Bai Pi, and Huang Qin.

Related conditions and modifications are

- Copious yellow nasal discharge: Add Shan Zhi Zi and Yu Xing Cao.
- Severe fever: Add Jin Yin Hua and Pu Gong Ying.

Stagnant Heat at Gallbladder Channel

Yellow, thick, sticky nasal discharge, copious with foul smell, severe pain at forehead, bitter taste in mouth, dry throat, agitation, and irritability.

Treatment Principle: Clear heat from Gallbladder channel, relieve toxicity, and unblock orifices.

Use Long Dan Xie Gan Tang [Gentiana Longdancao Decoction to Drain the Liver] plus Mu Dan Pi, Bai Zhi, Shi Chang Pu, and Chi Shao Yao.

Related conditions and modifications are

- Severe heat: Add Qing Hao, Sang Bai Pi, Zhi Mu, Shi Gao, and Ban Lan Gen.
- Copious sticky nasal discharge: Add Zhe Bei Mu, Gua Lou Pi, Jie Geng, Dan Nan Xing, and Bo He.

Damp–Heat in Spleen and Stomach

Copious, yellow, and turbid nasal discharge, severe and persistent nasal congestion, and decreased sense of smell.

Treatment Principle: Drain heat, eliminate damp, disperse Lung Qi, and unblock orifices.

Use Xie Huang San [Drain the Yellow Powder] plus Jia Wei Cang Er San [Augmented Xanthium Powder] and Huang Qin, Hua Shi, Mu Tong, Fu Ling, Shi Chang Pu, and Ya Zhi Cao.

Chronic Sinusitis

Lung Qi Deficiency and Cold

Treatment Principle: Warm and tonify Lung Qi, and scatter and disperse wind–cold.

Use Yu Ping Feng San [Jade Windscreen Powder], Si Jun Zi Tang [Four-Gentleman Decoction], and Gui Zhi Tang [Cinnamon Twig Decoction] plus He Zi, Xin Yi Hua, Lu Lu Tong, and Bai Zhi.

Related condition and modification are

- Nasal congestion and headache: Cang Er Zi San [Xanthium Powder].

Spleen Qi Deficiency

Treatment Principle: Strengthen Spleen, tonify Qi, transform damp, and benefit the orifices.

Use Shen Ling Bai Zhu San [Ginseng, Poria, and Atractylodes Macrocephala Powder] plus Huang Qi, Huo Xiang, Shi Chang Pu, and Xin Yi Hua.

Related condition and modification are

- Poor appetite: Add Shen Qu and Mai Ya.
- Abdominal distention: Add Mu Xiang and Sha Ren.

(Note: After the symptoms of acute sinusitis are alleviated, patient should continue to take the herbal formula for about 1 week. This can prevent acute sinusitis from progressing into chronic sinusitis, which is difficult to heal. Patient with chronic sinusitis should take the herbs long term, generally 1 to 3 months or even longer.)

APHTHA

Aphtha is also called canker sores. This is often caused by fire heat upward-blazing. In Western medicine it refers to ulcerations of the mucous membrane of the oral cavity. The clinical manifestations are round or oval shape of painful ulcerations with a diameter of 2 to 3 mm at the mucous membrane of mouth, cheeks, and lips. It can appear as a single or multiple spots. It heals in about 7 to 10 days. TCM treatments focus on clearing heat and nourishing Yin.

Upward-Blazing Heart Fire Syndrome

Treatment Principle: Clear Heart, descend fire, cool blood, and stop pain.

Use Jia Wei Dao Chi San [Augmented Guide Out the Red Powder] plus Huang Qin, Huang Lian, and Shan Zhi Zi.

Related conditions and modifications are

- Yellow dark urine: Add Che Qian Zi and Bai Mao Gen.
- Obvious red swelling around the ulcerations: Add Chi Shao Yao and Mu Dan Pi.
- Constipation: Add Sheng Da Huang.

Spleen and Stomach Heat Accumulation Syndrome

Treatment Principle: Clear heat, drain fire, reduce swelling, and stop pain.

Use Liang Ge San [Cool the Diaphragm Powder] and Yu Nu Jian [Jade Woman Decoction].

Related conditions and modifications are

- Obvious redness and swelling of the ulcers: Add Jin Yin Hua, Pu Gong Ying, and Chi Shao Yao.
- Restlessness, agitation, and irritability: Add Xia Ku Cao and Mu Dan Pi.

Yin Deficiency with Deficient Fire Syndrome

Treatment Principle: Nourish Yin and descend fire.

Use Zhi Bai Di Huang Wan [Anemarrhena, Phellodendron, and Rehmannia Pill] plus Xuan Shen, Mu Dan Pi, and Chuan Niu Xi.

Related conditions and modifications are

- Restlessness and insomnia: Add Huang Lian and E Jiao.
- Tidal fever and night sweats, and lower back and knee soreness and ache: Add Sang Ji Sheng and Wu Wei Zi.
- Mouth ulcers, bitter taste, and irritability before menstruation: Add Huang Qin and Xia Ku Cao.
- Dry stool: Add He Shou Wu and Nu Zhen Zi.

Spleen Deficiency with Damp Retention Syndrome

Treatment Principle: Strengthen Spleen and transform dampness.

Use Shen Ling Bai Zhu San [Ginseng, Poria, and Atractylodes Macrocephala Powder] plus Hua Shi.

Related conditions and modifications are

- Sticky mucous and sweet taste in mouth: Add Huo Xiang and Pei Lan.
- No taste in mouth: Add Sha Ren and Hou Po.
- Epigastric distention and stifling sensation: Add Da Fu Pi and Zhi Ke.
- Qi deficiency: Add Huang Qi and Shan Yao.

Additionally, one can add external applications.

Use Yang Yin Sheng Ji San [Nourish Yin and Generate the Flesh Powder]: Shi Gao 10 g, Xiong Huang 3 g, Qing Dai 1 to 5 g, Long Dan Cao 1 g, Huang Bai 1 g, Bo He 1 g, Pu Huang 1 g, Gan Cao 2 g, and Bing Pian 2 g. This can nourish Yin, generate flesh, reduce swelling, and stop pain.

One can also use decoction of Jin Yin Hua, Zhu Ye, Bai Zhi, and Bo He to rinse the mouth. It has the functions of clearing heat, relieving toxicity, reducing swelling, and stopping pain.

HERPETIC STOMATITIS

This condition is caused by accumulated heat in Heart and Spleen, external wind–damp heat invasion, upward-steaming of damp–heat to the mouth leading to ulceration of the muscle membrane. In Western medicine, herpetic stomatitis is seen as an inflammation of the mucous membrane of the oral cavity caused by herpes simplex virus. The manifestations are scattered or clustered small blisters at the mucous membrane inside the oral cavity. There is redness around the blisters which quickly rupture, forming ulcers. Small ulcers can form into a big ulcer. Pain is noticeable, and the course of development is about 7 to 14 days. It is often accompanied by fever and swollen lymph nodes.

External Wind–Heat Syndrome

Treatment Principle: Disperse wind and clear heat.

Use Yin Qiao San [Honeysuckle and Forsythia Powder] plus Ban Lan Gen and Pu Gong Ying.

Related condition and modification are

- Severe pain: Add Chi Shao Yao and Xuan Shen.

Accumulated Heat in Heart and Spleen Syndrome

Treatment Principle: Clear and drain accumulated heat from Heart and Spleen.

Use Dao Chi San [Guide Out the Red Powder] plus Bai Hu Tang [White Tiger Decoction], plus Lu Gen, Ban Lan Gen, and Huang Lian.

Related conditions and modifications are

- Constipation: Add Da Huang and Mang Xiao.
- Yellow, greasy tongue coating: Add Huo Xiang, Pei Lan, Yi Yi Ren, and Bian Dou.

Furthermore, one can also decoct Jin Yin Hua, Huang Qin, and Dan Zhu Ye to rinse the mouth.

LARYNGITIS

This condition is discussed under loss of voice in TCM. The acute type is often caused by wind–cold, wind–heat, and external dryness pathogen, Lung failing to purify and descend, disorder of airway, or knotted phlegm heat in the throat obstructing the airway. The chronic type is often caused by Qi and Yin deficiency. Chronic hypertrophic laryngitis is caused by Qi accumulation, phlegm obstruction, and blood stasis. In Western medicine, laryngitis is seen as an inflammatory pathological change of the mucous membrane at larynx. It can affect the vocal cord. The clinical manifestations for acute type are dry throat, itching, slight pain, foreign body sensation at throat, follow by hoarseness, or even loss of voice, and intermittent cough. These can be accompanied by headache, aversion to cold, fever, and whole body discomfort. For chronic type, at the beginning stage there is throat discomfort or scanty, sticky phlegm. After using the voice, the voice becomes lower and thicker, followed by pain or intermittent hoarseness. Then symptoms become continuous hoarseness, dry throat, itching, slight cough, and frequent clearing the throat without coughing out any sputum.

Acute Laryngitis

Wind–Cold Constrict Lung Syndrome

Treatment Principle: Disperse wind, scatter cold, disperse Lung, and unblock stagnation.

Use Xiang Su Cong Chi Tang [Cyperus, Perilla Leaf, Scallion, and Prepared Soybean Decoction] plus Qian Hu, Xing Ren, Jing Jie, Ban Xia, and Fu Ling.

Related conditions and modifications are

- Constitution Yang deficiency: Add Ma Huang Fu Zi Xi Xin Tang [Ephedra, Prepared Aconite, and Asarum Decoction].
- Exterior condition already resolved: Change to Gui Zhi, Ban Xia, and Gan Cao.
- Chronic phlegm in Lung, recurrent wind–cold transforming into heat when entering interior: Change to Da Qing Long Tang [Major Blue-Green Dragon Decoction].

Wind–Heat Obstructing the Orifices Syndrome

Treatment Principle: Disperse wind, clear heat, transform phlegm, and benefit the throat.

Use Lian Qiao, Huang Qin, Shan Zhi Zi, Zhi Mu, She Gan, Zao Xiu, Niu Bang Zi, Sang Ye, Zhe Bei Mu, Jie Geng, Zhi Pi Pa Ye, Pang Da Hai, Chan Tui, and Gan Cao.

Related conditions and modifications are

- Constitution Yin deficiency with recurrent wind–heat: Add Mai Men Dong, Sha Shen, Yu Zhu, and Xuan Shen.
- Evil heat entering interior and excess internal heat: Change to Liang Ge San [Cool the Diaphragm Powder].

Dry Heat Damaging Lung Syndrome

Treatment Principle: Clear heat, moisten dryness, nourish Yin, and generate fluid.

Use Qing Zao Jiu Fei Tang [Eliminate Dryness and Rescue the Lungs Decoction] plus Sang Xing Tang [Mulberry Leaf and Apricot Kernel Decoction].

Related conditions and modifications are

- Severe dry painful throat: Add She Gan, Pang Da Hai, and Mu Hu Die.
- Sticky sputum that is hard to expectorate: Add Gua Lou Pi and Ma Dou Ling.
- Severe thirst: Add Tian Hua Fen and Zhi Mu.
- Constipation: Add Gua Lou Ren and Yu Li Ren.

Chronic Laryngitis

Chronic Simple Laryngitis

Treatment Principle: Scatter accumulation, scour out phlegm, invigorate blood, benefit Qi, and nourish Yin.

Use Zeng Ye Tang [Increase the Fluids Decoction] plus Sheng Mai San [Generate the Pulse Powder] and Tao Ren, Hong Hua, Jie Geng, Chai Hu, Chi Shao Yao, Chuan Bei Mu, Tai Zi Shen, and Hai Zao.

Chronic Hypertrophic Laryngitis

Treatment Principle: Invigorate blood, transform stasis, scour out phlegm, and scatter accumulation.

Use Dan Shen, Huang Qi, Xuan Shen, Hai Zao, Zao Jiao Ci, Chi Shao Yao, Chuan Shan Jia, Zhe Bei Mu, Tao Ren, and Hong Hua.

Above two syndromes can be modified with the following:

- Red swollen painful vocal cord: Add Mu Dan Pi, Jin Yin Hua, Pu Gong Ying, She Gan, Shan Dou Gen, and Sheng Gan Cao.
- Grey white polyps with semiclear blisters: Add Fu Ling, Ze Xie, Yi Yi Ren, Bai Zhu, and Ban Xia.
- Hard nodules: Add Shan Dou Gen, Xia Ku Cao, and Zhong Jie Feng.

PHARYNGITIS

This condition is called throat Bi or sore throat in TCM. It is often caused by wind–heat attacking Lung, accumulating at throat, and obstructing the channels and collaterals. In Western medicine, this refers to inflammatory changes that occur at mucous membrane of the pharynx. At the beginning stage, the symptoms include dryness, burning, or slight painful sensation at throat, followed by obvious sore throat worse with swallowing; itchy throat, cough, or with fever, chills, headache, and body ache. At acute stage, TCM treatment focuses on dispersing wind, clearing heat, relieving toxicity, eliminating swelling, scattering accumulation, and benefiting the throat. At chronic stage, the focus is on nourishing Yin, clearing heat, invigorating blood, and transforming phlegm.

Acute Pharyngitis

Mild Case

Treatment Principle: Disperse wind, clear heat, scatter accumulation, and benefit the throat.

Use Yin Qiao Ma Bo San [Honeysuckle, Forsythia, and Puffball Powder] plus She Gan, Ban Lan Gen, and Huang Qin.

Severe Case

Treatment Principle: Clear heat, relieve toxicity, eliminate swelling, and scatter accumulation.
Use Pu Ji Xiao Du Yin [Universal Benefit Decoction to Eliminate Toxin].
Modification of above two formulas:

- Exterior cold: Add Fang Feng and Qiang Huo.
- Yellow hot urine: Add Mu Tong, Hua Shi, and Che Qian Zi.
- Constipation: Add Mang Xiao or combine with Huang Lian Shang Qing Wan [Coptis Pill to Clear the Upper Jiao].
- Severe swelling and distention of the throat: Add Qian Li Guan, Pu Gong Ying, Ye Ju Hua, and Zi Hua Di Ding.

Chronic Pharyngitis

Chronic Simple Pharyngitis

Treatment Principle: Nourish Yin, clear heat, invigorate blood, and scatter stasis.

Use Yang Yin Qing Fei Tang [Nourish the Yin and Clear the Lungs Decoction] plus Tao Ren, Hong Hua, Chuan Shan Jia, Chi Shao Yao, and Zao Jiao Ci.

Related conditions and modifications are

- Severe dry throat: Add Tian Men Dong, Lu Gen, and Ge Gen.
- Dry cough and itchy throat: Add Pi Pa Ye, Sang Ye, and Bo He.
- Red dry purple throat: Add Dan Shen and Si Gua Lou.

Chronic Hypertrophic Pharyngitis

Treatment Principle: Invigorate blood, transform stasis, scour out phlegm, and scatter accumulation.

Use Tao Hong Si Wu Tang [Four-Substance Decoction with Peach Pit and Safflower] plus Yang He Tang [Yang-Heartening Decoction] minus Shu Di Huang, and add Huang Qi, Dan Shen, Chuan Shan Jia, Zao Jiao Ci, Zhe Bei Mu, Xuan Shen, and Hai Zao.

Related conditions and modifications are

- Sticky sputum hard to expectorate: Add Gua Lou Ren and Hai Fu Shi.
- Sharp pain in throat: Add Yan Hu Suo and Bai Zhi.
- Nausea: Add Ban Xia and Zhu Ru.
- Severe hypertrophy of the pharynx: Add San Leng and E Zhu.

Chronic Atrophic Pharyngitis

Treatment Principle: Nourish Lung and Kidney, and generate fluid.

Use Sha Shen Mai Men Dong Tang [Glehnia and Ophiopogonis Decoction] and Liu Wei Di Huang Wan [Six-Ingredient Pill with Rehmannia], and Xuan Shen, Zhe Bei Mu, Bo He, and Dan Shen.

Related conditions and modifications are

- Severe dry throat and pain: Add Tian Men Dong, Shi Hu, or steep tea with Xuan Shen, Mai Men Dong, Gan Cao, and Jie Geng.
- Copious dry scabs: Add Zhu Ru and Hai Fu Shi.

TONSILLITIS (ACUTE)

This condition is often caused by wind–heat, heat toxins, and knotted external and internal heat. The clinical manifestations are acute onset, chills and fever, headache, whole body ache, sore throat, cough, ear congestion, earache, swollen tonsils, and injection, with purulent secretion at the surface. TCM treatment focuses on dispersing wind, clearing heat, relieving toxicity, and clearing and draining Lung and stomach heat toxins.

Warm Evil Attacking Lung, Stagnant Heat at Lung Channel Syndrome

Treatment Principle: Acrid cool to release exterior, benefit throat, and reduce swelling.

Use Yin Qiao San [Honeysuckle and Forsythia Powder] plus Pu Gong Ying and Zi Hua Di Ding.

Related conditions and modifications are

- Severe chills: Add Jing Jie.
- Severe swelling of the tonsils: Add Ye Ju Hua and Shan Dou Gen.
- Cough: Add Sang Bai Pi and Zhe Bei Mu.

Evil Heat Entering Interior, Excess Heat in Lung and Stomach Syndrome

Treatment Principle: Clear heat toxins, drain Lung and stomach, benefit the throat, and reduce swelling and pain.

Use Qing Fei Jie Du Tang [Clear the Lungs and Relieve Toxicity Decoction] plus Da Huang, Xuan Shen, Jin Yin Hua, and Huang Lian.

Related conditions and modifications are

- High fever: Add Shi Gao and Tian Hua Fen.
- Excessive purulent secretion: Add Ma Bo, Zhe Bei Mu, and Dan Nan Xing.
- Severe swelling and pain: Add Liu Shen Wan [Six-Miracle Pill].
- Swollen, painful cervical and submaxillary lymph nodes: Add Shan Ci Gu and Gua Lou Pi.

TONSILLITIS (CHRONIC)

This condition often progresses from acute tonsillitis or Qi and blood insufficiency, Lung and Kidney Yin deficiency, and deficient fire upward-burning. The clinical manifestation are recurrent sore throat, dry throat, discomfort foreign body sensation at throat, frequent common colds, slightly swollen or swollen tonsils, or tonsils with white and yellow pus spots. TCM treatments focus on strengthening Spleen and tonifying Qi. One can use the following formula with modifications.

Use Shen Ling Bai Zhu San [Ginseng, Poria, and *Atractylodes Macrocephala* Powder] plus Lu Dou Yi, Jin Yin Hua, and Jie Geng.

Related conditions and modifications are

- Easily catch colds: Add Huang Qi and Fang Feng.
- Dry throat, dry cough, and scanty sputum: Add Sha Shen, Bai He, and Xuan Shen.
- Dry, painful throat, worse at night: Add Huang Bai, Zhi Mu, and Sheng Di Huang.
- Severe swollen tonsils, pale and tender looking: Add Jiang Can and Chuan Bei Mu.
- Dark purple, big tonsils with adhesion and scar: Add Tao Ren, Hong Hua, Ru Xiang, and Mo Yao.
- Big tonsils that are hard: Add Kun Bu and Dan Shen.
- Bad breath: Add Huo Xiang and Shi Chang Pu.

CHAPTER 17

Gastrointestinal Disorders

CHOLECYSTITIS

This condition is discussed under hypochondriac pain, epigastric pain, and jaundice in TCM. It is caused by Liver and Gallbladder Qi stagnation, internal excess of damp–heat, Qi stagnation, and blood stasis. The clinical manifestations of chronic cholecystitis are not typical. There is often a history of gripping pain at the Gallbladder area with aversion to oily food, abdominal distention, and belching. The manifestations for acute cholecystitis are intermittent severe pain at right upper quadrant, radiation to right shoulder area, nausea and vomiting, or with fever, chills, and jaundice. The treatments in TCM include clearing heat, draining damp, dispersing Liver, promoting bile discharge, moving Qi, invigorating blood, strengthening Spleen, and harmonizing Stomach.

Liver and Gallbladder Qi Stagnation Syndrome

Treatment Principle: Regulate Qi, stop pain, and relieve stagnation.

Use Shu Yu Qing Gan Tang [Spread the Depression and Clear the Liver Decoction] plus Jin Qian Cao, Pu Gong Ying, remove Mu Dan Pi and Shan Zhi Zi.

Related conditions and modifications are

- Pain upon palpation: Add Chuan Xiong and Dan Shen.
- Nausea: Add Zhu Ru.

Liver and Gallbladder Damp–Heat Syndrome

Treatment Principle: Disperse Liver, promote bile discharge, clear heat, and drain damp.

Use Jia Jian Xiao Yao San [Modified Relaxed Wanderer Powder] plus Da Huang, Jin Qian Cao, Hu Zhang, Yu Jin, and Bai Jiang Cao.

Related conditions and modifications are

- Nausea and vomiting: Add Huang Lian, Ban Xia, and Sheng Jiang.
- Spleen deficiency, poor appetite, and loose stool: Add Si Jun Zi Tang [Four-Gentleman Decoction] or Shen Ling Bai Zhu San [Ginseng, Poria, and Atractylodes Macrocephala Powder].
- Yin deficiency: Add Yi Guan Jian [Linking Decoction].

Blood Stasis in Gallbladder Syndrome

Treatment Principle: Promote bile discharge, invigorate blood, and regulate Qi.

Use Shu Gan Jie Yu Tang [Spread Liver and Relieve Stagnation Decoction] remove Yan Hu Suo, and add Bai Shao Yao, Chuan Shan Jia, Dang Gui, and Zhi Ke.

Related conditions and modifications are

- Constipation: Add Da Huang.
- Jaundice: Add Yin Chen Hao and Jin Qian Cao.
- With stones in Gallbladder symptoms: Add Hai Jin Sha, Ji Nei Jin, and Jin Qian Cao.

CHOLELITHIASIS

This condition is discussed under hypochondriac pain, jaundice, epigastric pain, and pain in epigastrium. The pathogenesis is bile stagnation which generates damp–heat and formation of sand and stones. This condition has double the occurrences in females than in males. Ultrasound is the first choice for examining this condition. There are mild digestive complaints or none for cholelithiasis. When the stone is stuck at the neck of the Gallbladder, then there can be wretching pain at the Gallbladder region which radiates to the right shoulder. If the stone is located at the bile duct of the Liver, then there can be distending pain at the Liver area, fever, pain with palpation, and striking at the Liver area. If the stone is located at the bile duct outside the Liver, such as obstruction of the bile duct leading to cholangitis, then there are abdominal pains, alternating chills and fever, and jaundice. TCM treatments during the acute stage are clearing heat, draining damp, dispersing Liver, promoting bile discharge, unblocking Fu organs, and attacking downward. After attacks, the treatments are dispersing Liver, promoting bile discharge, moving Qi, invigorating blood, and regulating Spleen and Stomach. Commonly used herbs are Huang Qin, Shan Zhi Zi, and Jin Qian Cao to clear heat and promote bile discharge; Da Huang to purge downward; Chuan Lian Zi, Yan Hu Suo, Mu Xiang, Zhi Ke, and Yu Jin to regulate Qi and disperse Liver; Bai Shao Yao, Gan Cao, and Yan Hu Suo to stop pain; and Yin Chen Hao and Hu Zhang to reduce jaundice and drain damp.

Damp–Heat Accumulation Syndrome

Treatment Principle: Clear heat, transform dampness, disperse Liver, promote bile discharge, unblock the interior, and purge downward.

Use Qing Dan Li Shi Tang Jia Wei [Augmented Clear the Gallbladder and Drain Damp Decoction] plus Jin Qian Cao, Che Qian Cao, Yin Chen Hao, Zhi Ke, Shan Zhi Zi, Yu Jin, Yan Hu Suo, Hu Zhang, Huang Qin, Jin Yin Hua, Chai Hu, Da Huang, Ji Nei Jin, and Mu Xiang.

Related condition and modification are

- Constipation: Add Xiao Cheng Qi Tang [Minor Order the Qi Decoction].

Accumulated Heat Toxins Syndrome

Treatment Principle: Clear heat, disperse Liver, drain fire, cool blood, relieve toxicity, and open orifices.

Use Qing Dan Xie Huo Tang [Clear the Gallbladder and Drain Fire Decoction] plus Dan Dao Pai Shi Tang [Expel Stone from Biliary Tract Decoction] and Jin Qian Cao, Shui Niu Jiao, Sheng Di Huang, Chi Shao Yao, Jin Yin Hua, Lian Qiao, Yin Chen Hao, Da Huang, Huang Lian, Mu Dan Pi, and Chai Hu.

Liver Qi Stagnation Syndrome

Treatment Principle: Disperse Liver, regulate Qi, promote bile discharge, and expel stones.

Use Qing Dan Xing Qi Tang Jia Wei [Augmented Clear the Gallbladder and Move Qi Decoction] plus Jin Qian Cao, Ji Nei Jin, Chai Hu, Yu Jin, Xiang Fu, Mu Xiang, Zhi Ke, Chuan Lian Zi, Hou Po, and Yan Hu Suo.

Qi Stagnation and Blood Stasis Syndrome

Treatment Principle: Disperse Liver, promote bile discharge, invigorate blood, transform stasis, soften hardness, and scatter accumulations.

Use Ge Xia Zhu Yu Tang [Drive Out Blood Stasis Below the Diaphragm Decoction] plus San Leng, E Zhu, Yu Jin, and Yan Hu Suo.

Related conditions and modifications are

- Enlarged Liver and Spleen: Add Dan Shen, Bie Jia, and Chuan Shan Jia.
- Abnormal function of the Liver: Add Huang Qi, Dang Shen, Bai Zhu, and Fu Ling.

CONSTIPATION

Constipation refers to bowel movement that is less than three times per week, or 2 or more days without a bowel movement with dry stool that is difficult to pass. The main cause is decreased motility of the large intestine. The treatment focuses on unblocking the large intestine and recovering the transportation function. Furthermore, opening Lung Qi with Xing Ren, Gua Lou Ren, Sheng Ma, Jie Geng, Qian Hu, Lai Fu Zi, and Sang Bai Pi can assist the descending function. In addition, dispersing Liver herbs such as Bai Shao Yao, Chai Hu, Zhi Ke, and Xiang Fu; and nourishing Liver and tonifying Kidney herbs such as Dang Gui, Zhi He Shou Wu, Ji Xue Teng, Bai Zi Ren, Hei Zhi Ma, and Rou Cong Rong can also be added to formulas to help treat the condition.

Heat Accumulation Type of Constipation

Treatment Principle: Clear heat, moisten intestines, and unblock stools.

Use Tiao Wei Cheng Qi Tang [Regulate the Stomach and Order the Qi Decoction] plus Xing Ren, Sheng Di Huang, and Huo Ma Ren.

Related conditions and modifications are

- Heat damaging fluid: Add Xuan Shen and Mai Men Dong.
- Severe constipation: Also steep 5 g of Fan Xie Ye and drink frequently.

Qi Stagnation Type of Constipation

Treatment Principle: Smooth Qi, open the stagnation, and unblock stools.

Use Liu Mo Tang [Six Milled Herbs Decoction] plus Lai Fu Zi and Da Fu Pi.

Related conditions and modifications are

- Stagnation transform into fire: Add Shan Zhi Zi and Huang Qin.
- Yin damage: Add Xuan Shen, Sheng Di Huang, and Mai Men Dong.

Knotted Phlegm Type of Constipation

Treatment Principle: Disperse Lung, transform phlegm, moisten intestine, and unblock stools.

Use Ma Zi Ren Wan [Hemp Seed Pill] plus Zi Wan, Xuan Shen and Feng Mi. It is also good to drink 3 to 5 pieces of Pang Da Hai steeped in hot water as tea.

Stasis and Stagnation Type of Constipation

Treatment Principle: Invigorate blood, expel stasis, moisten intestines, and unblock stool.

Use Tao He Cheng Qi Tang [Peach Pit Decoction to Order the Qi], add Dang Gui, Chi Shao Yao, Mu Dan Pi, Jie Geng, and remove Mang Xiao.

Qi Deficient Type of Constipation

Treatment Principle: Strengthen Spleen, benefit Qi, and moisten stools.

Use Bu Zhong Yi Qi Tang [Tonify the Middle Jiao and Augment the Qi Decoction] with increased dosage of Bai Zhu (60 to 80 g), and add Huo Ma Ren and Zhi Shi.

Blood Deficient Type of Constipation

Treatment Principle: Nourish blood, moisten intestine, and unblock stool.

Use Dang Gui Bu Xue Tang [Tangkuei Decoction to Tonify the Blood] plus Shu Di Huang, Zhi He Shou Wu, Huo Ma Ren, Tao Ren, and Zhi Ke.

Related condition and modification are

- Qi deficiency: Add Huang Qi at 50 g; add Dang Shen and Bai Zhu.

Yin Deficient Type of Constipation

Treatment Principle: Nourish Yin, clear heat, moisten intestine, and unblock stool.

Use Zeng Ye Cheng Qi Tang [Increase the Fluids and Order the Qi Decoction] plus Zhi He Shou Wu, Zhi Mu, and Mu Dan Pi.

Yang Deficient Type of Constipation

Treatment Principle: Warm Yang and moisten intestines.

Use Jin Gui Shen Qi Wan [Kidney Qi Pill from the Golden Cabinet] plus Rou Cong Rong, Dang Gui, and Bai Zhu.

DRUG-INDUCED LIVER DISEASE

This is discussed under jaundice, hypochondriac pain, abdominal mass, and tympanites. The pathogeneses are invasion of drug toxins, obstruction and stagnation of damp–heat and blood and stasis damaged Zheng Qi. As the condition becomes chronic, there can be Spleen deficiency with concomitant damp obstruction and blood stasis, or even tympanites, and abdominal mass formation. In Western medicine, this is described as liver toxicity caused by drug and medication or liver damages due to allergic reactions to the medication. The characteristics are pathological changes or necrosis of liver cells, bile accumulation and stasis, chronic hepatitis, and liver cirrhosis. The manifestations of mild case are similar to acute viral hepatitis. In severe cases, there are symptoms of liver failure, coma, jaundice, ascites, and bleeding. TCM treatment includes dispersing Liver, clearing heat, relieving toxicity, draining damp, transforming turbidity, invigorating blood, and strengthening Spleen.

Liver Qi Stagnation Type

Treatment Principle: Disperse Liver and relieve stagnation and toxicity.

Use Xiao Chai Hu Tang [Minor Bupleurum Decoction] plus Yin Chen Hao, Lian Qiao, Dan Shen, Qing Pi, Yu Jin, and Wu Wei Zi.

Related conditions and modifications are

- Abdominal distention and turbid greasy tongue coating: Add Ban Xia, Mu Xiang, Cang Zhu, and Ze Xie.
- Constipation: Add Sheng Da Huang.
- Fever and rash: Add Bo He, Fang Feng, and Chan Tui.
- Persistent elevated SGPT and SGOT: Add Sheng Ma 20 g.

Damp–Heat with Wind Type

Treatment Principle: Disperse wind, clear heat, drain damp, cool blood, and relieve toxicity.

Use Xiao Feng San [Eliminate Wind Powder] plus Lian Qiao, Yin Chai Hu, Mu Dan Pi, Yin Chen Hao, Wu Wei Zi, and Bai Xian Pi.

Damp–Heat Type

Treatment Principle: Clear heat, relieve toxicity, drain damp, transform turbidity, and invigorate blood.

Use Yin Chen Hao Tang [Artemisia Decoction] plus Bai Hua She She Cao, Ban Zhi Lian, Che Qian Cao, Huang Bai, Lian Qiao, Chi Shao Yao, and Fu Ling.

Related conditions and modifications are

- Whole body heaviness and poor appetite: Add Chen Pi, Cang Zhu, and Yi Yi Ren.
- Severe jaundice and heat: Add Ku Shen, Pu Gong Ying, and Jin Qian Cao.
- Concomitant phlegm and stasis: Add Shi Xiao San [Sudden Smile Powder], Chen Pi, Yi Yi Ren, Ban Xia, and Ge Gen.

Stasis in Gallbladder Type

Treatment Principle: Disperse Liver, promote Gallbladder, transform stasis, expel phlegm, cool blood, and invigorate blood.

Use Qing Dai, Ming Fan, Dan Shen, Chi Shao Yao, Hu Zhang, and Ge Gen.

Related conditions and modifications are

- Constipation: Add Da Huang and Mang Xiao
- Enlarged Liver and Spleen: Add Tao Ren, Hong Hua, E Zhu, and Qian Cao.
- Water sounds in epigastrium: Add Gui Zhi, Fu Ling, Sheng Jiang, and Ban Xia.
- Yin type of jaundice, and pale purple flabby tongue with stasis spots: Add Rou Gui.

ACUTE ENTERITIS

Acute enteritis is discussed under sudden diarrhea in TCM. It is caused by wind, cold, summer heat, or damp evil invasion, or improper diet, ingesting cold, raw, and uncleaned food. These damage the Spleen and Stomach which cannot separate the clear from turbid which is sent down to the intestines and causes sudden diarrhea. The characteristics are sudden acute onset, abdominal pain, and diarrhea. If Stomach is involved, then there is nausea and vomiting. If colon is involved, then there is mucous in the stool. If rectum is involved, then there can be tenesmus or fever. In severe cases, there can be dehydration, acidosis, lethargy, or shock. TCM treatment focuses on controlling the diarrhea. When clearing heat, be careful with very bitter and cold herbs because they can damage Stomach Qi. Astringent herbs should not be used too early to prevent retention of evil pathogens. In order to protect Qi and Yin when dispersing and promoting urination, use caution with aromatic, dry, and Qi invigorating herbs. To prevent production of dampness and sensation of fullness, avoid use of sweet and greasy herbs to tonify deficiency; instead, opt for sweet, bland, and neutral herbs. The main formula consists of Cang Zhu, Hou Po, Fu Ling, Shen Qu, Chen Pi, Zhu Ling, Ze Xie, and Yi Yi Ren. Also take 15 g of Da Suan (garlic) separately.

Related conditions and modifications are

- Wind–cold exterior syndrome: Add Qiang Huo, Chuan Xiong, Fang Feng, and Bai Zhi. Or plus Huo Xiang Zheng Qi San [Agastache Powder to Rectify the Qi].
- Summer heat: Add Bian Dou, Xiang Ru, Lian Qiao, and Hua Shi.
- Upward disturbing of damp turbid: Add Huo Xiang, Ban Xia, Zi Su Geng, and Zhi Ke.
- Cold–damp harass Spleen: Add Gan Jiang. In severe cases, add Zhi Fu Zi.
- Internal excess of damp–heat: Add Ge Gen, Huang Qin, Huang Lian, Lu Gen, and Ma Chi Xian.
- Tenesmus: Add Bai Tou Weng, Bai Shao Yao, Qin Pi, and Huang Bai.
- High fever: Add Shi Gao.
- Epigastric and abdominal distention, belching, and yellow, greasy tongue coating: Add Jiu Da Huang and Bing Lang.
- Dry tongue and scanty fluid: Add Wu Mei, Shi Hu, and Lu Gen.
- Spleen and Stomach Qi deficiency: Remove Cang Zhu, and add Dang Shen and Huang Qi.
- Qi sinking and rectal prolapse: Add Sheng Ma and Chai Hu.
- Frequent bowel movement, watery stool, and absence of heavy sensation at anus: Add He Zi, Sheng Jiang, and Rou Dou Kou.

During remission period, Shen Ling Bai Zhu San [Ginseng, Poria, and Atractylodes Macrocephala Powder] is used as a base with appropriate modification.

FATTY LIVER

Fatty Liver is discussed under hypochondriac pain, abdominal mass, and accumulated lumps in TCM. It is often caused by Liver failing to disperse and circulate, Qi stagnation, and blood stasis. It can also be caused by Spleen deficiency generating phlegm–damp, and congestion in earth with stagnation in wood. In Western medicine, fatty liver refers to accumulation of fats (triglyceride, cholesterol, and phospolipid) which exceeds 5% of the weight of liver, or more than 50% of the Liver cells with lipid deposit. This condition often occurs at the beginning stage of liver cirrhosis. The clinical manifestations include asymptomatic during mild cases, enlarged liver, smooth liver border and surface, denser with absence of pain upon palpation in moderate and severe cases. If inflammation occurs due to rupture of lipocysts, then there can be pain at liver area or upon palpation, mild case of jaundice, fever, poor appetite, nausea, vomiting, abdominal distention, and increase in white blood cells. The clinical diagnosis should be based on an ultrasound. TCM treatment focuses on scouring phlegm, draining turbid, and transforming stasis. One can choose Fu Yuan Huo Xue Tang [Revive Health by Invigorating the Blood Decoction], Qi Ju Di Huang Wan [Lycium Fruit, Chrysanthemum, and Rehmannia Pill], or Gua Lou Xie Bai Tang Jia Jian [Modified Trichosanthes Fruit and Chinese Chive Decoction]. Then one will take one of the above formula with Da Shan Zha Wan [Major Hawthorn Pill].

Core formula: Chai Hu, Quan Gua Lou, Shan Zha, Chi Shao Yao, Dan Shen, Yu Jin, He Shou Wu, Nu Zhen Zi, Xie Bai, Cang Zhu, Hong Hua, Pu Huang, Gou Qi Zi, Ju Hua, Ze Xie, and Zhe Chong.

Related conditions and modifications are

- Liver and Gallbladder damp–heat, and yellow, thick tongue coating: Add Hou Po, Yin Chen Hao, Jin Qian Cao, Yu Jin, and Hu Zhang.
- Excess Liver fire: Jue Ming Zi, Cao Jue Ming, Chong Wei Zi, and Ju Hua.
- Flabby tongue with white, slippery and thick tongue coating: Add Lu Dou, Yi Yi Ren, Zhu Ling, Che Qian Zi, and Ze Xie.
- Constipation: Add Sheng He Shou Wu, Jue Ming Zi, and Sheng Da Huang.
- Liver and Kidney Yin deficiency: Add Nu Zhen Zi, Sheng He Shou Wu, Ling Zhi or combine with Liu Wei Di Huang Wan [Six-Ingredient Pill with Rehmannia].

Additionally, patients with fatty Liver often have high lipid in blood, especially triglycerides, thus one should also combine herbs to lower the lipid in blood, such as a tea made from Ze Xie, Shan Zha, He Ye, Pu Huang, and Ju Hua.

FUNCTIONAL INDIGESTION

This condition is discussed under focal distention and fullness, stomach pain, hiccups, and rumbling sounds in stomach. It is often caused by emotional stress such as nervousness, stress, and worries. There can be Liver Qi overacting on Stomach, stagnant heat at Liver and Stomach, or ingesting greasy fatty foods that are difficult to digest and can cause food accumulation and obstruction of Qi mechanism in the stomach and epigastrium. In Western medicine, functional indigestion is not described as having significant organic changes but only functional changes of the stomach activity. This causes dysfunction of the upper digestive tract, which manifests as fullness and distention of the epigastrium, pain, burning sensation, reverse flow, and other indigestion symptoms. TCM treatments focus on dispersing Liver, relieving stagnation, descending rebellious Qi, harmonizing Stomach, clearing and draining stagnant heat, clearing and transforming damp–heat, and eliminating food stagnation. According to research results, the following herbs can increase Stomach function: Zhi Ke, Zhi Shi, Bing Lang, Chai Hu, Lai Fu Zi, and Ji Nei Jin. The following herbs can inhibit gastric kinetic functions: Wu Yao, Chen Xiang, Bai Shao Yao, Gan Cao, Xiao Hui Xiang, Gui Zhi, Rou Gui, Yan Hu Suo, and Yang Jin Hua. These herbs can be added to the base formula according to syndrome differentiation.

Reverse Flow Type

Treatment Principle: Disperse Liver, relieve stagnation, descend rebellious Qi, and harmonize Stomach.

Use Ban Xia Hou Po Tang [Pinellia and Magnolia Bark Decoction], Xuan Fu Dai Zhe Shi Tang [Inula and Haematite Decoction], or Qu Mai Zhi Zhu Tang [Medicated Leaven, Barley Sprout, Immature Bitter Orange, and Atractylodes Macrocephala Pill].

Related conditions and modifications are

- Severe belching: Add Huo Xiang, Ding Xiang, and Shi Di.
- Severe acid regurgitation: Add Wu Zei Gu, Wa Leng Zi, and Huang Lian.
- Bitter regurgitation: Add Huang Qin, Chai Hu, Shan Zhi Zi, and Zhu Ru.
- Incomplete bowel movement: Add Bing Lang and Da Fu Pi.
- Food stagnation: Add Shan Zha and Lai Fu Zi.
- Damp obstruction and white greasy tongue coating: Add Huo Xiang, Pei Lan, Cang Zhu, and Xia Ku Cao.

Decrease Kinetic Function Type

Treatment Principle: Descend Qi, reduce focal distention, reduce and guide food stagnation.

Use Zhi Shi Xiao Pi Wan [Immature Bitter Orange Pill to Reduce Focal Distention] and Bao He Wan [Preserve Harmony Pill].

Related conditions and modifications are

- Transformation into heat and dry yellow tongue coating: Add Lian Qiao, Ge Gen, and Huang Qin.
- Abdominal fullness and constipation: Add Bing Lang, Da Fu Pi, and Jiu Da Huang.

Ulceration Type

Treatment Principle: Disperse Liver, harmonize Stomach, and clear stagnant heat.

Use Si Ni San [Frigid Extremities Powder] and Xiang Su San [Cyperus and Perilla Leaf Powder] plus Mu Dan Pi, Shan Zhi Zi, Qing Pi, and Zhe Bei Mu.

Related conditions and modifications are

- Severe epigastric pain: Add Mu Xiang and Yan Hu Suo.
- Phlegm–heat: Add Huang Lian, Fa Xia, and Quan Gua Lou.
- Damp–heat and yellow, greasy tongue coating: Add Ge Gen, Huang Qin, Huang Bai, Cang Zhu, Huo Xiang, and Pei Lan.
- Damaged Yin, red tongue, and scanty tongue coating: Add Xuan Shen, Yu Zhu, Bei Sha Shen, and Mai Men Dong.

GASTRITIS (ACUTE)

Acute gastritis is discussed under cholera, nausea, and vomiting in TCM. Ingesting cold, uncleaned food causes Spleen failing to transport and Stomach losing harmonization and descending function. These lead to nausea, vomiting, and pain in the stomach and epigastrium. In Western medicine, it is attributed various physio-chemical reasons, bacteria, and/or toxins that irritate mucous membrane of the stomach leading to diffused or local inflammatory responses and even bleeding or acute ulcerations. Harmonizing Stomach, descending rebellious Qi, stopping vomiting, and strengthening Spleen are the treatment principles.

Nausea and vomiting are important symptoms of this condition. Sheng Jiang is an important herb to stop nausea. Thus, it is added to various syndromes in this condition.

Cold–Damp Stagnation in the Middle Jiao Syndrome

Treatment Principle: With aromatic herbs, transform damp, regulate Qi, and harmonize Stomach.

Use Huo Xiang Zheng Qi San [Agastache Powder to Rectify the Qi].

Damp–Heat Congestion and Stagnation Syndrome

Treatment Principle: Transform damp, clear heat, descend rebellious Qi, and harmonize Stomach.

Use Ban Xia Xie Xin Tang [Pinellia and Magnolia Bark Decoction]; replace Gan Jiang with Sheng Jiang and add Zhi Shi, Bai Dou Kou, Zhu Ru, and Chen Pi.

Food Stagnation Syndrome

Treatment Principle: Clear accumulation, guide stagnation, harmonize Stomach, and stop nausea.

Use Bao He Wan [Preserve Harmony Pill] plus Huang Lian, Sha Ren, Ji Nei Jin, Mai Ya, and Zhi Shi.

Food Poisoning Syndrome

Treatment Principle: Relieve toxicity and harmonize Stomach.

Decoct Gan Cao and Lu Dou; add Sheng Jiang juice and drink frequently.

GASTRITIS (CHRONIC)

This condition is discussed under epigastric pain, focal distention, and fullness. It is caused by various factors that damage Spleen and Stomach function, disturb normal flow of Qi, generate dampness, stagnate Qi, and obstruct the collaterals due to phlegm and stasis. The main manifestations are fullness, distention and discomfort at the epigastrium, or pain, poor appetite, belching, nausea, and irregular bowel movements. During the inflammatory stage, TCM treatment focuses on regulating Qi, invigorating blood, eliminating dampness, clearing heat, and stopping pain. During remission period, the focus is on regulating Spleen and Stomach.

Since the most common infection is caused by Helicobacter pylori, one should add herbs that can inhibit this bacterial growth to the formula, which is selected according to syndrome differentiation. These herbs include Da Huang, Huang Lian, Pu Gong Ying, Shi Liu Pi, Bing Lang, Gui Zhi, Wu Mei, Yan Hu Suo, Chen Pi, Hou Po, Bai Hua She She Cao, Wu Zhu Yu, Lian Qiao, Zhi Mu, Chai Hu, San Qi, Shi Hu, and Dang Shen.

Discussion According to Syndrome Differentiations

Liver and Stomach Qi Stagnation Syndrome

Treatment Principle: Disperse Liver, harmonize Stomach, move Qi, and reduce distention.

Use Chai Hu Shu Gan San [Bupleurum Powder to Spread the Liver].

Related conditions and modifications are

- Acid regurgitation: Add Wa Leng Zi, and Hai Piao Xiao or take with Zuo Jin Wan [Left Metal Pill].
- Food stagnation, poor appetite and greasy thick tongue coating: Add Shen Qu, Mai Ya, Lai Fu Zi, and Chen Pi.
- Abdominal distention: Add Da Fu Pi, Su Geng, Zhi Shi, and Hou Po.

Spleen Deficiency Stomach Heat Syndrome

Treatment Principle: Strengthen Spleen, clear heat, and harmonize Stomach.

Use Gan Cao Xie Xin Tang [Licorice Decoction to Drain the Epigastrium] plus Sha Ren.

Related conditions and modifications are

- Severe epigastric distention: Add Mu Xiang, Xiang Fu, and Yan Hu Suo.
- Loose stool: Add Bai Zhu and Shan Yao.
- Constipation: Add Hou Po and Zhi Shi.

Spleen and Stomach Deficiency Syndrome

Treatment Principle: Strengthen Spleen, warm Middle Jiao, and harmonize Stomach.

Use Xiang Sha Liu Jun Zi Tang [Six-Gentleman Decoction with Aucklandia and Amomum] plus Huang Qi and Xiang Fu.

Related conditions and modifications are

- Distention and fullness after meal: Add Mai Ya and Ji Nei Jin.
- Sallow complexion, pale tongue and lip: Add Dang Gui and He Shou Wu.
- Vomiting clear saliva and cold limbs: Add Wu Zhu Yu, Fa Xia, Gui Zhi, and Gan Jiang.

Stomach Yin Deficiency Syndrome

Treatment Principle: Nourish Yin, benefit Stomach, and stop pain.

Use Mai Men Dong Tang [Ophiopogonis Decoction] and Shao Yao Gan Cao Tang [Peony and Licorice Decoction] plus Sha Shen, Yu Zhu, Shi Hu, Tian Hua Fen, and Ji Nei Jin.

Blood Stasis Obstructing Collaterals Syndrome

Treatment Principle: Move Qi, invigorate blood, and transform stasis.

Use Dan Shen Yin [Salvia Decoction] plus Chuan Lian Zi, Fo Shou, E Zhu, Yan Hu Suo, and Gan Cao.

Discussion According to Diseases

Erosive Gastritis

This is often due to Liver heat attacking Stomach. Treatment focuses on clearing Liver, harmonizing Stomach, and reducing distention.

Use Zuo Jin Wan [Left Metal Pill] plus Da Huang, Mu Zei, Pu Gong Ying, Dai Zhe Shi, and Gan Cao.

Bile Reflux Type Gastritis

The reverse bile flow is often caused by dysfunction of the pyloric sphincter after gastrectomy. The treatment should disperse Liver, promote bile, harmonize Stomach, and reduce distention.

Use Long Dan Xie Gan Tang [Gentiana Longdancao Decoction to Drain the Liver] plus Yu Jin, Mu Xiang, Da Huang, Jin Qian Cao, Pu Gong Ying, and Dai Zhe Shi.

Anastomotic Inflammation or Residual Gastritis

These two are common complications of total or partial gastrectomy. They should be treated by strengthening Spleen, clearing Stomach, moving Qi, and transforming stasis.

Use Jia Wei Xiang Lian Wan [Augmented Aucklandia and Coptis Pill] plus Fa Xia, Bai Zhu, Chi Shao Yao, Gan Jiang, Dang Shen, Pu Gong Ying, Sha Ren, Ji Nei Jin, and Zhi Shi.

Atrophic Gastritis with Intestinal Metaplasia

The symptoms of this condition are similar to gastritis, which can be determined through gastroscope. It is often a deficient Spleen, Stomach heat with blood stasis syndrome. Thus, treatment should strengthen Spleen, tonify Qi, clear heat, and transform stasis.

Use Si Jun Zi Tang [Four-Gentleman Decoction] plus Huang Qi, Huang Jing, Ban Zhi Lian, Dang Gui, E Zhu, Qi Ye Yi Zhi Hua, and Bai Hua She She Cao.

Prognosis from Tongue Coating

From clinical observations, if tongue coating changes from thick, greasy, and white to thin and moist, this indicates that the condition has improved. If the tongue coating changes from white to yellow, from thin to thick and greasy, then the condition is worsening and pathogens are active. If the tongue coating remains thick and the tongue proper is purple dark and lusterless, this indicates that the condition is persistent, difficult to treat, or worsening. One should pay close attention. In summary, if the disease duration has been long, then the treatment should last more than 3 months, and preferably 6 months, to see the results.

HICCUPING

Hiccuping is often caused by some irritations that lead to spasms of the diaphragm. In TCM, they are caused by various reasons that disturb the functions of Lung and Stomach Qi. The diaphragm is located between the lung and stomach. When Qi rebels, it disturbs the diaphragm and leads to hiccuping. The treatment principles are to harmonize Stomach, descend rebellious Qi and calm hiccuping. Herbs that can calm and descend rebellious Qi are Ding Xiang, Shi Di, Ju Pi, Zhu Ru, Xuan Fu Hua, Dai Zhe Shi, Chen Xiang, and Dao Dou. They can be selected according to the syndrome differentiation.

Treatments According to Syndrome Differentiation

Stomach Cold Syndrome

Treatment Principle: Warm the Middle Jiao, scatter cold, and stop hiccuping.

Use Ding Xiang Shi Di Tang [Clove and Persimmon Calyx Decoction] plus Hou Po, Zhi Shi, Gao Liang Jiang, Wu Zhu Yu, Rou Gui, and Zhi Gan Cao.

Stomach Heat Syndrome

Treatment Principle: Clear Stomach, descend rebellious Qi, and stop hiccuping.

Use Xuan Bai Cheng Qi Tang [Order the Qi Decoction to Disperse White] plus Xuan Shen, Mai Men Dong, Lai Fu Zi, Fa Xia, Shi Di, and Hou Po.

Phlegm Stagnation Syndrome

Treatment Principle: Regulate Qi, transform phlegm, descend rebellious Qi, and stop hiccupping.

Use Er Chen Tang [Two-Cured Decoction] plus Mu Xiang, Chen Xiang, Zhi Shi, Wu Yao, Bing Lang, Yu Jin, and Dai Zhe Shi.

Spleen Yang Deficiency Syndrome

Treatment Principle: Strengthen Spleen, warm the Middle Jiao, descend rebellious Qi, and stop hiccupping.

Use Fu Zi Li Zhong Wan [Prepared Aconite Pill to Regulate the Middle] plus Ding Xiang and Wu Zhu Yu.

Stomach Yin Deficiency Syndrome

Treatment Principle: Nourish Yin, generate fluid, and stop hiccupping.

Use Yi Wei Tang [Benefit the Stomach Decoction] plus Zhu Ru, Pi Pa Ye, Shi Di, Chen Pi, and Gan Cao.

Treatments According to Types

Acute Attack of Hiccupping

Treatment Principle: Drain and purge heat accumulation, descend rebellious Qi, and stop hiccupping.

Use Da Chai Hu Tang [Major Bupleurum Decoction] plus Dai Zhe Shi, Gan Cao, and Jiang Xiang.

Related condition and modification are

- If only constipation, abdominal distention, and hiccupping but the tongue coating is not yellow or thick: Change to Zi Su Zi, Lai Fu Zi, Tao Ren, and Yu Li Ren.

Neurotic Type of Hiccupping

Treatment Principle: Disperse Liver, harmonize Stomach, transform phlegm, and stop hiccupping.

Use Xuan Fu Dai Zhe Tang [Inula and Haematite Decoction] plus Fu Ling, Pi Pa Ye, Yu Jin, and Shi Di.

Stubborn Type of Hiccupping

Treatment Principle: Warm Spleen, tonify Kidney, ascend clear, and descend rebellious Qi.

Use Bu Zhong Yi Qi Tang [Tonify the Middle Jiao and Augment the Qi Decoction] plus Bu Gu Zhi and Shi Di.

INTESTINAL ADHESION

This condition is discussed under abdominal pain in TCM. It is caused by Qi stagnation, blood stasis, and dysfunction of Qi flow in Fu organs with rebellion after abdominal surgeries or traumatic injuries. The clinical manifestations are abdominal pain and distention and borborygmus, which is alleviated by belching or passing gas, occasional nausea, vomiting, protruding abdomen, or visible intestinal peristalsis and mild pain with palpation. The treatment focuses on moving Qi, guiding the stagnation, alleviating spasms, and stopping pain. One should also refrain from eating and drinking to reduce the pressure in the stomach and intestine. Patient is then supported by IV. This can prevent adhesive ileus.

Qi Stagnation and Blood Stasis Syndrome

Treatment Principle: Invigorate blood, expel stasis, move Qi, and stop pain.

Use Shao Fu Zhu Yu Tang [Drive Out Stasis in the Lower Abdomen Decoction] plus Tao Ren and Hong Hua.

Related conditions and modifications are

- Severe Qi stagnation: Add Chai Hu, Xiang Fu, Zhi Ke, and Bai Shao Yao.
- Severe blood stasis: Add Ze Lan and San Qi.
- Inability to pass accumulated dry stool: Add Mang Xiao to the decoction.

Damp–Heat Congestion and Stagnation Syndrome

Treatment Principle: Drain heat and unblock Fu organs.

Use Xiao Cheng Qi Tang [Minor Order the Qi Decoction] plus Huang Qin, Shan Zhi Zi, and Yan Hu Suo.

Related condition and modification are

- Abdominal and flank pain: Add Chai Hu and Yu Jin.

Mixed Deficiency and Excess Syndrome

Treatment Principle: Warm the Middle Jiao, tonify deficiency, moderate the acute situation, and stop pain.

Use Xiao Jian Zhong Tang [Minor Construct the Middle Decoction] plus Xi Xin, Dang Gui, and Huang Qi.

Related condition and modification are

- Constipation: Add Da Huang.

Additionally, if patient also has severe abdominal distention, nausea, vomiting, and constipation, enema should also be used as part of the treatment.

IRRITABLE BOWEL SYNDROME

Irritable bowel syndrome is discussed under abdominal pain, constipation, and diarrhea in TCM. The pathology occurs at Liver and Spleen, and in chronic cases, Kidney can be involved. The pathogeneses are Liver overacting on Spleen, or irregular food intake, and disharmony of cold and warm which damage the Spleen. If the condition lasts for a long time, it can cause ulcerative colitis, hemorrhoids, and enteritis. In Western medicine, the intestines (mainly colon or small intestines) overreact or have abnormal reaction to stimulations. The manifestations are symptoms of abnormal motility or secretion. It is often seen in young to middle-aged people, more commonly in men. There can be family problems, emotional factors, and a diet containing too many refined foods. The clinical manifestations are abdominal pain especially lower left abdomen, diarrhea, or constipation or alternating constipation and diarrhea, clear jelly-like mucous in the stool, small amount of white blood cell in the stool, and negative bacterial culture. It is difficult to treat the root of this condition. However, Chinese herbs can alleviate the symptoms and attacks.

Treatments According to Syndrome Differentiations

Liver Stagnation and Spleen Deficiency Syndrome

Treatment Principle: Disperse Liver and strengthen Spleen.

Use Tong Xie Yao Fang [Important Formula for Painful Diarrhea] plus Mu Xiang, burned Shan Zha, fried Mai Ya, Fu Ling, and Shen Qu.

Related conditions and modifications are

- Constipation: Remove Fu Ling, add Dang Gui, Tao Ren, Xing Ren, and Huo Ma Ren.
- Diarrhea: Add Shan Yao, Lian Zi, Dang Shen, Yi Yi Ren, and Zhi He Zi.
- Severe abdominal distention: Add Hou Po, Zhi Ke, and Bing Lang.
- Tenesmus: Add Huang Lian and Qin Pi.
- Apathy and fatigue: Add Tai Zi Shen and Huang Qi.
- Hypochondriac fullness and easily angered: Add Chai Hu and Xiang Fu.

Cold–Damp Harass Spleen Syndrome

Treatment Principle: Dry dampness and strengthen Spleen.

Use Wei Ling Tang [Calm the Stomach and Poria Decoction] plus Chi Xiao Dou and Huo Xiang.

Related conditions and modifications are

- Aversion to cold and cold limbs: Add Gan Jiang and Gao Liang Jiang.
- Copious jelly-like mucous, abdominal pain, tenesmus, and greasy and dirty tongue coating: Add San Wu Bei Ji Wan [Three-Substance Pill for Emergencies]. Discontinue once the symptoms are cleared.
- Scanty urination and edema at face or lower limbs: Add Gui Zhi.

Spleen and Stomach Deficiency Syndrome

Treatment Principle: Benefit Qi and strengthen Spleen.

Use Shen Ling Bai Zhu San [Ginseng, Poria, and Atractylodes Macrocephala Powder] plus Zhi He Zi, burned Shan Zha, and fried Mai Ya.

Related conditions and modifications are

- Spleen Qi sinking: Add Sheng Ma, Chai Hu, and Huang Qi.
- Morning diarrhea, dizziness, and tinnitus: Add Bu Gu Zhi, Tu Si Zi, Wu Zhu Yu, Wu Wei Zi, and Zhi Rou Dou Kou.

Intestinal Dryness and Qi Stagnation Syndrome

Treatment Principle: Moisten dryness and move Qi.

Use Ma Zi Ren Wan [Hemp Seed Pill] plus Tao Ren, Huang Qin, Rou Cong Rong, Chai Hu, Dang Gui, Bai Zi Ren, and processed Da Huang.

Related conditions and modifications are

- Fluid exhaustion and dry tongue: Add Yu Zhu, Mai Men Dong, and Tian Hua Fen and remove Tao Ren, Xing Ren, and processed Da Huang.
- Absence of obvious heat and thin pulse: Remove Huang Qin.

Qi Stagnation and Blood Stasis Syndrome

Treatment Principle: Move Qi and transform stasis.

Use Shi Xiao San [Sudden Smile Powder] plus Si Mo Tang [Four Milled Herb Decoction], remove Ren Shen, and add Mu Xiang, Zhi Shi, processed Da Huang, and Tao Ren.

Related conditions and modifications are

- Bitter taste, irritability, and heat sensation: Add Huang Qin and Shan Zhi Zi.
- Qi and blood deficiency: Add Dang Shen and Dang Gui, and remove Da Huang and Bing Lang.

Common Patent Herbal Supports

Liver Qi Stagnation and Spleen Deficiency

Use Xiao Yao Wan [Relaxed Wanderer Pill] and Shu Gan Wan [Spread the Liver Pill].

Cold–Damp Harass Spleen

Use Huo Xiang Zheng Qi Wan [Agastache Pill to Rectify the Qi], Xiang Sha Liu Jun Zi Wan [Six-Gentleman Pill with Aucklandia and Amomum], and Xiang Sha Yang Wei Wan [Nourish the Stomach Pill with Aucklandia and Momum].

Spleen and Stomach Deficiency

Use Ren Shen Jian Pi Wan [Strengthen the Spleen Pill with Ginseng], Shen Ling Bai Zhu Wan [Ginseng, Poria, and Atractylodes Macrocephala Pill], and Bu Zhong Yi Qi Wan [Tonify the Middle Jiao and Augment the Qi Pill].

Intestine Dryness and Qi Stagnation

Use Run Chang Wan [Moisten the Intestine Pill], Tao Ren Wan [Peach Pit Pill], and Jin Gui Shen Qi Wan [Kidney Qi Pill from the Golden Cabinet].

Qi Stagnation and Blood Stasis

Use Fu Fang Dan Shen Pian [Compound Salvia Tablet] plus Jia Wei Xiao Yao Wan [Augmented Relaxed Wanderer Pill].

LIVER CIRRHOSIS

Liver cirrhosis is discussed under accumulated lumps or tympanite in TCM. At the beginning, Liver is involved, then Spleen, and later Heart and Kidney are involved. Thus, the symptoms begin with swelling in Liver and Spleen and symptoms of digestive tract, then advance to difficult urination and ascites. The pathological changes are Qi stagnation, blood stasis, and fluid congestion. It is often caused by viral hepatitis, alcoholic poisoning, schistosomiasis, chemical or medication toxins accumulation, and stasis of bile, blocked circulation, and malnutrition, in addition to unknown reasons. During the compensatory period, there may not be obvious symptoms, or there can be fatigue, poor appetite, aversion to greasy food, abdominal distention, nausea, and loose stool. During the decompensation period, other than above symptoms, there can also be bleeding, low-grade fever, jaundice, skin itching, anemia, ascites, liver complexion, spider veins, liver palm, swelling and hardening of the Spleen, lower body edema, development of breasts in males, varicosity in lower portion of the esophagus fundus of Stomach. During the compensatory period (accumulated lumps), the focus of the treatment is to regulate Qi, invigorate blood, expel stasis, and assist by softening the hardness. During the decompensation period (tympanites), one should differentiate the deficiency or excess. If it is excess, then the focus is on transforming stasis and regulating Qi. If it is deficient, then one should use formulas to support the Zheng Qi.

Accumulated Lumps (Compensatory Period)

Liver Qi Stagnation and Spleen Deficiency Syndrome

Treatment Principle: Disperse Liver, strengthen Spleen, invigorate blood, and soften the hardness.

Use Liu Jun Zi Tang [Six-Gentleman Decoction] plus Xiang Fu, Mu Xiang, Tao Ren, Ji Nei Jin, and Bie Jia.

Related conditions and modifications are

- Thick greasy tongue coating: Add Cang Zhu and Huo Xiang.
- Apathy and shortness of breath: Add Huang Qi.

Qi Stagnation and Blood Stasis Syndrome

Treatment Principle: Disperse Liver, regulate Qi, invigorate blood, and transform stasis.

Use Ge Xia Zhu Yu Tang [Drive Out Blood Stasis Below the Diaphragm Decoction] plus Yan Hu Suo.

Related conditions and modifications are

- Severe jaundice: Add Yin Chen Hao, Lian Qiao, and increase Chi Shao Yao to 60 to 120 g.
- Hyperactivity of Spleen: Add E Jiao, Shu Di Huang, and Da Zao.
- Bleeding tendency: Add Da Ji, Xiao Ji, Xian He Cao, and San Qi powder.
- Low grade fever: Add Chai Hu, Bai Wei, and Di Gu Pi.
- Fever due to inflammation: Add Hong Teng and Bai Jiang Cao.

Liver and Kidney Yin Deficiency Syndrome

Treatment Principle: Nourish Yin, soften Liver, soften the hardness, and scatter accumulation.

Use Yi Guan Jian [Linking Decoction] plus Yu Jin, He Shou Wu, Bai Shao Yao, Dan Shen, Sheng Mu Li, and processed Bie Jia and Ju Ye.

Related conditions and modifications are

- Epistaxis: Add Hei Zhi Zi (charred Zhi Zi).
- Bleeding gums: 50 g of Di Gu Pi, soak in water and rinse mouth with it.

Tympanites Type (Decompensation Period)

Qi Stagnation and Damp Obstruction Syndrome

Treatment Principle: Disperse Liver, regulate Qi, expel dampness, and eliminate fullness.

Use Chai Hu Shu Gan San [Bupleurum Powder to Spread the Liver] plus Ping Wei San [Calm the Stomach Powder].

Related conditions and modifications are

- Vomiting clear water: Add Gan Jiang and Ban Xia.
- Severe abdominal distention: Add Sha Ren and Mu Xiang.

Damp–Heat Accumulation Type

Treatment Principle: Clear heat, transform dampness, and reduce distention.

Use Gan Lu Xiao Du Dan [Sweat Dew Special Pill to Eliminate Toxin] plus Si Ling San [Four-Ingredient Poria Powder] and Jin Qian Cao.

Related conditions and modifications are

- Constipation: Add Shang Lu, Da Huang, and Qian Niu Zi.
- Severe jaundice: Add Qing Hao and increase Chi Shao Yao to 60 to 120 g.
- Abdominal distention: Add Lai Fu Zi and Chen Xiang.

Liver and Spleen Blood Stasis Type

Treatment Principle: Invigorate blood, transform stasis, move Qi, and promote urination.

Use Jia Jian Huo Luo Xiao Ling Dan [Modified Fantastically Effective Pill to Invigorate the Collaterals] plus Hong Hua, Qing Pi, Yu Jin, Sheng Mu Li, Sheng Bai Zhu, Zhi Chuan Shan Jia, and Dang Gui.

Liver and Kidney Yin Deficiency Type

Treatment Principle: Nourish Yin and promote urination.

Use Bie Jia Yang Yin Jian [Nourish Yin Decoction with Soft-Shell Turtle Shell], remove Sheng Di Huang, Bai Shao Yao, and Gou Qi Zi, and add Ze Lan, Lu Lu Tong, Hai Zao, and Chu Shi Zi.

Related conditions and modifications are

- Abdominal distention: Add Mei Gui Hua and Fo Shou.
- Scanty urination: Add Bai Mao Gen and Sha Shen.
- Bleeding tendency: Add Er Zhi Wan [Two-Ultimate Pill] plus Xian He Cao.
- Loose stool: Add Bai Zhu and Shan Yao.

Spleen and Kidney Yang Deficiency Type

Treatment Principle: Warm and tonify Spleen and Kidney, transform Qi, and move water.

Use Fu Zi Li Zhong Wan [Prepared Aconite Pill to Regulate the Middle].

Related condition and modification are

- In severe cases: Add Zi He Che and Lu Rong powder.

Furthermore, one can assist the condition with food therapy.

1. Take one soft-shell turtle and coat it with soil. Roast until golden then remove the soil. Grind the shell into powder. Take 5 g each time with brown sugar, three times a day. This can treat enlarged Spleen.
2. Li Yu Chi Dou Tang [Carp and Adzuki Bean Soup]: Cook carp (500 g), adzuki bean (30 g), a small amount of vinegar, sugar, and fresh ginger for 1 hour. Then add small amount of alcohol to get rid of the fish smell. Drink this soup once a day. This is good for the deficient type of Liver cirrhosis ascites which can promote urination and reduce edema.
3. Dong Gua Zhou [Wax Gourd Congee]: Cook wax gourd with rind (wintermelon) 60 g, with rice 60 g. Eat on an empty stomach, once to twice a day to treat mild edema.

MALABSORPTION

This syndrome is discussed under chronic diarrhea in TCM. It is caused by damp harassing the Middle Jiao. This damages Spleen, Stomach, and Kidney. In Western medicine, malabsorption is seen as caused by various reasons that affect the digestive function or absorption function. Thus, one or multiple nutrients cannot be absorbed through the mucosal layer into the body and is eliminated through bowel movements. This results in symptoms of nutrient deficiencies. Clinical manifestations include chronic diarrhea, emaciation, fatigue, abdominal distention, stomatitis, or anemia. This condition has two different types: primary and secondary. Secondary malabsorption occurs after pathological conditions of small intestine, liver, Gallbladder, or pancreas, or diabetes, hyperthyroidism, uremia, Lupus, or taking purgatives, weight-loss pills, or antibiotics for a long time. The treatment of TCM focuses on stopping diarrhea and strengthening Spleen.

Diarrhea Period

Cold–Damp Syndrome

Treatment Principle: Dry dampness, warm Middle Jiao, leach out dampness, and promote urination.

Use Shen Ling Bai Zhu San [Ginseng, Poria, and Atractylodes Macrocephala Powder] plus Gan Jiang, Shen Qu, Zhu Ling, and Zhi Rou Dou Kou.

Related conditions and modifications are

- Qi deficiency: Add Hong Shen.
- Edema: Add Huang Qi and Gui Zhi.
- Kidney Yang deficiency and intestinal hernia: Add Zhi Fu Zi, Ying Su Ke, and He Zi.
- White, thick, greasy tongue coating: Add Cao Guo, Fa Xia, Bing Lang, and Hou Po.
- Severe Qi stagnation: Add Hou Po, Huo Xiang Geng, Yu Jin, and Mu Xiang.
- Dark stagnant tongue, dull pain at the sides of the lower abdomen: Add Shi Xiao San [Sudden Smile Powder] and San Qi powder.
- Wind–cold exterior syndrome: Add Huo Xiang Zheng Qi San [Agastache Powder to Rectify the Qi].
- Cold pain at lower abdomen: Add Rou Gui and Ding Xiang.
- White sticky tongue coating or yellow at the center: Add Huang Lian, Hou Po, and Bai Shao Yao.

Damp–Heat Syndrome

Treatment Principle: Clear heat and drain damp.

Use Ge Gen Huang Qin Huang Lian Tang [Kudzu, Coptis, and Scutellaria Decoction] plus Mu Xiang, Chi Shao Yao, Shan Zha, Che Qian Zi, Fu Ling, and Gan Cao.

Related conditions and modifications are

- Severe Qi stagnation: Add Da Fu Pi and Hou Po.
- Food stagnation: Add burned Shan Zha, fried Mai Ya, Shen Qu, Fa Ban Xia, and Lai Fu Zi.
- Damp worse than heat: Remove Huang Lian and Chi Shao Yao, and add Huo Xiang, Fa Xia, and Hou Po.
- Severe dampness: Remove Gan Cao, add Cang Zhu and Sheng Jiang.
- Heat worse than dampness: Remove Fu Ling and add Shi Gao and Lu Gen.
- Stomach Yin deficiency: Remove Che Qian Zi, Fu Ling, and Mu Xiang, and add Wu Mei, Shan Yao, Sha Shen, and Bian Dou.
- Abdominal distention: Add Fo Shou and Chen Pi.

Qi Stagnation Syndrome

Treatment Principle: Disperse Liver and strengthen Spleen.

Use Chai Hu Shu Gan San [Bupleurum Powder to Spread the Liver] plus Bai Zhu, Yu Jin, and Fang Feng.

Related conditions and modifications are

- Severe Spleen Qi deficiency: Add Tai Zi Shen, Dang Shen, Lian Zi, and Huang Qi.
- Spleen Yang deficiency: Add Gan Jiang and Ren Shen.
- Chronic Liver disorders and Liver blood stasis: Add Tao Ren, Hong Hua, Dan Shen, Bie Jia, and Zhe Chong.
- Damp–heat in the Liver and Gallbladder: Add Huang Qin and Qing Hao.
- Cholelithiasis: Add Jin Qian Cao and Yu Jin.
- Diarrhea right after abdominal pain or with emotional causes: Add Tong Xie Yao Fang [Important Formula for Painful Diarrhea].

Spleen Deficiency Syndrome

Treatment Principle: Strengthen Spleen and harmonize Stomach.

Use Xiang Sha Liu Jun Zi Tang [Six-Gentleman Decoction with Aucklandia and Amomum] plus Huang Qi, Lian Zi, Shan Yao, Shen Qu, and Yi Yi Ren.

Related conditions and modifications are

- Sinking Middle Jiao Qi: Add Jie Geng, Chai Hu, Sheng Ma, and Hong Shen, and increase Huang Qi.
- Frequent diarrhea: Add He Zi, Zhi Rou Dou Kou, and Chi Shi Zhi.

Kidney Deficiency Syndrome

Treatment Principle: Warm Kidney, strengthen Spleen, and astringe and stop diarrhea.

Use Si Shen Wan [Four-Miracle Pill] plus Hong Shen, Bai Zhu, Fu Ling, Gan Jiang, Mu Xiang, Zhi He Zi, and Gan Cao.

Related conditions and modifications are

- Severe aversion to cold and cold limbs: Add Zhi Fu Zi.
- Chronic diarrhea, weak constitution, old age, apathy, and shortness of breath: Add Huang Qi 30 g, Sheng Ma 9 g, and Ying Su Ke 12 g.

Remission Period

The condition is relatively stable with unformed bowel movements two to three times a day. In general, congee food therapy is used in this case.

1. Fu Ling Qian Shi Zhou [Poria and Gordon Euryale Seed Congee]: To Geng Mi 60 g add an appropriate amount of water. Add 10 pieces of Hong Zao (Chinese date) after the rice is boiled. Add 15 g of Fu Ling powder and 20 g of Qian Shi powder to the congee and mix well. Good for Spleen and Stomach deficiency.
2. Shuang Ren Zhou [Two-Seeds Congee]: Wrap Che Qian Zi 30 g in cloth; decoct and remove from the decoction. Add 30 g of Yi Yi Ren and Geng Mi 60 g to make congee. Good for Spleen deficiency with damp obstruction.
3. Cao Zhu Zhou [Licorice and Atractylodes Congee]: Zhi Gan Cao 50 g, Cang Zhu 500 g, Hui Xiang 30 g, and dry Hong Zao (Chinese date) 500 g (remove pit). Roast the herbs, grind into powder. Make Geng Mi 500 g and salt 20 g into congee, then add the powder. Good for cold–damp obstructing Middle Jiao.

NONSPECIFIC ULCERATIVE COLITIS (CHRONIC)

This condition is discussed under diarrhea, intestinal wind, and dysentery. During an acute period, the main pattern is damp–heat in large intestine. During remission period, one can see Spleen and Kidney deficiency with Liver Qi stagnation and blood stasis. Fulminant type is often due to heat toxins. In Western medicine, this is described as a superficial, nonspecific inflammation syndrome of the mucous membrane of the colon and rectum. It mainly affects the rectum and sigmoid colon. In severe cases, the whole colon is involved. During the active inflammatory period, there is erosion and ulceration of the mucous membrane. During remission period, the mucous membrane appears to be pale, rough, thickened, or with polyps formation. This condition often occurs in patients 20 to 40 years old. The manifestations are mucous and blood in the stool, abdominal pain, diarrhea, or tenesmus, with repeated occurrences and chronic condition. At the beginning stage, the onset is gradual and seldom sudden or acute. Other than taking the herbs orally in TCM, enema is also an important method to apply the herbs locally. Colonoscopy can assist the application of herb by observing the changes of mucous membrane at the rectum and colon. From long-term clinical observations, swelling of the mucous membrane with obvious oozing (damp syndrome) can benefit from Cang Zhu, Ku Fan, Bai Zhi, He Zi, and Wu Bei Zi. If there is congestion at the mucous membranes and copious pus discharge (evil heat syndrome), then Huang Lian, Huang Qin, Da Huang, Pu Gong Ying, Bai Jiang Cao, Ma Chi Xian, Jin Yin Hua, Huang Bai, Qin Pi, Bai Tou Weng, and Zi Cao are used together. For erosion of the mucous membrane with some blood spots or ulceration (heat toxins damaging collaterals), one should use Qing Dai, Xue Jie, Zhen Zhu, Bai Ji, Pu Huang, Chi Shi Zhi, Bing Pian, and Yun Nan Bai Yao. For thickened and red mucous membrane, rough, grainy, stasis spots, and easily bleeding, use Huai Hua, E Zhu, Ru Xiang, Mo yao, Dan Shen, Zi Cao, Qian Cao, San Qi, Xue Jie, and Yun Nan Bai Yao. If there are a lot of polyps (congealed phlegm and stasis), one should use fresh turnip juice. At late stage, when the mucous membrane becomes pale and there is atrophy of the spots (deficient cold), use Dang Shen, Chuan Xiong, Huang Qi, Hong Hua, Bai Zhi, Dang Gui, and Gan Jiang. One can also modify the formula according to the symptoms.

Treatment According to Syndrome Differentiation

Internal Accumulation of Damp–Heat Syndrome

Treatment Principle: Clear heat, drain damp, regulate Qi, and stop pain.

Use Bai Tou Weng Tang [Pulsatilla Decoction] plus Ge Gen Huang Qin Huang Lian Tang [Kudzu, Coptis, and Scutellaria Decoction] plus Mu Xiang, Bai Shao Yao, Bing Lang, and Gan Cao.

Related conditions and modifications are

- Excess accumulation of heat toxins: Add Shui Niu Jiao, Shi Gao, Jin Yin Hua, Bai Jiang Cao, Sheng Di Huang, Di Yu, Xian He Cao, and E Jiao.
- Qi and Yin damage: Add Tai Zi Shen, Wu Wei Zi, and Mai Men Dong.
- Thick, yellow tongue coating, slippery rapid pulse, and food stagnation: Add Sheng Da Huang, and fried Shan Zha; substitute Mu Xiang with Zhi Shi.
- Damp worse than heat: Replace Mu Xiang and Bing Lang with Hou Po, remove Huang Qin, Bai Tou Weng, and add Huo Xiang, Ban Xia, Bian Dou, Bai Zhu, Fu Ling, and Che Qian Zi.
- Swollen joint pain: Add Han Fang Ji, Qin Jiao, Jiang Huang, and Hai Tong Pi.
- Red nodules on skin: Add Chi Shao Yao, Hong Hua, Tao Ren, Zi Cao, and Zao Jiao Ci.
- Photophobia (iritis): Add Xia Ku Cao, Ju Hua, and Long Dan Cao.

Qi Stagnation and Blood Stasis Syndrome

Treatment Principle: Move Qi, invigorate blood, strengthen Spleen, and harmonize Stomach.

Use Shi Xiao San [Sudden Smile Powder] plus Tao Hong Si Wu Tang [Four-Substance Decoction with Peach Pit and Safflower], remove Shu Di Huang, and add Zhi Ke and Wu Yao.

Related conditions and modifications are

- For stasis and heat: Add Hong Teng, Bai Jiang Cao, Lian Qiao, and Jin Yin Hua.
- Jelly-like mucous in stool, white tongue coating, and deep pulse: Add Xiao Hui Xiang, Gan Jiang, and Rou Gui.

Spleen and Kidney Deficiency

Treatment Principle: Strengthen Spleen, tonify Kidney, astringe the intestines, and stop diarrhea.

Use Li Zhong Wan [Regulate the Middle Jiao Pill] plus Si Shen Wan [Four-Miracle Pill] and Dang Gui, Fu Ling, Shan Yao, Huang Qi, Gou Qi Zi, Tu Si Zi, and Zhi He Zi.

Related conditions and modifications are

- For excess damp and greasy tongue coating: Remove Dang Gui and Gou Qi Zi, and add Huo Xiang, Sha Ren, Yi Yi Ren, Cang Zhu, and Bai Dou Kou.
- Liver Qi stagnation: Add Chai Hu, Bai Shao Yao, and Zhi Ke.
- Damp–heat: Remove Bu Gu Zhi, Tu Si Zi, and Gou Qi Zi, and add Huo Xiang Geng, Huang Lian, burned Shan Zha, and Zhi Da Huang.
- Sinking Middle Jiao Qi: Add Chai Hu and Sheng Ma.
- Spleen and Kidney Yang deficiency: Add Zhi Fu Zi and Wu Wei Zi.
- Blood in stool: Add Sheng Di Huang, Chi Shi Zhi, and E Jiao.

Yin and Blood Deficiency Syndrome

Treatment Principle: Soften Liver and nourish Kidney.

Use Yi Guan Jian [Linking Decoction] plus Bai Shao Yao, Bie Jia, Shan Yao, Lian Zi, Fo Shou, and Sheng Di Tan.

Related conditions and modifications are

- Anemia: Add E Jiao and Huang Qi.
- Blood in stool: Add Wu Mei, Ce Bai Ye, and Di Yu Tan.
- Dry stool: Add Huo Ma Ren, Xing Ren, Bai Zi Ren, Yu Li Ren, Rou Cong Rong, and Sheng Di Huang.
- Yin deficiency with excess fire: Add Zhi Mu, Huang Bai, and Mu Dan Pi.
- Apathy, fatigue, and shortness of breath: Add Tai Zi Shen and Wu Wei Zi.

Post-Disease Congee Food Therapy

1. Huang Qi Bian Dou Zhou [Astragalus and Dolichoris Bean Congee]: Decoct Huang Qi 30 g with Bian Dou 30 g. Cook 150 g of rice with the decoction into congee. It is indicated for Spleen deficiency and dysfunctional immune system.
2. Shuang Ren Bai He Zhou [Two-Seeds and Lily Bulb Congee]: Decoct wrapped Che Qian Zi 30 g. Cook Yi Yi Ren 30 g, Bai He 30 g, and Geng Mi 150 g with the decoction to make congee. Take the congee frequently. It is indicated for Spleen and Stomach deficiency and water dampness accumulation.

PANCREATITIS (ACUTE)

This condition is discussed under abdominal pain, accumulation of pathogens in the chest, jaundice, and collapse syndrome. It is an acute chemical inflammation caused by the active form of pancreatic enzyme digesting the pancreatic tissues. The clinical manifestations are sudden attack of persistent severe pain at upper abdomen, nausea, vomiting, and fever. In TCM, this condition is thought to be caused by improper diet, such as overeating, overindulgence in greasy, fatty foods and alcohol leading to damp–heat obstructing Liver and Gallbladder, interference with Qi mechanism, and blood stasis. At the beginning stage, mild syndrome and edema type can be treated with Chinese herbal medicine. The severe type and hemorrhagic necrotic type should be sent to an emergency room. Da Huang is the key herb. It is antiinflammatory, antibacterial, bile promoting, and inhibitory to the activity of pancreatic enzymes which unblock the Fu organ and clear internal heat.

Qi Stagnation and Heat Accumulation Syndrome

Treatment Principle: Disperse Liver, regulate Qi, clear heat, and unblock stool.

Use Qing Yi Tang [Clear the Pancreas Decoction].

Related condition and modification are

- Food stagnation: Add Shan Zha, Lai Fu Zi, and Ji Nei Jin.

Liver and Gallbladder Damp–Heat Syndrome

Treatment Principle: Clear Liver and promote bile secretion.

Use Da Chai Hu Tang [Major Bupleurum Decoction] plus Chuan Lian Zi, Long Dan Cao, Yin Chen Hao, and Pu Gong Ying.

Related conditions and modifications are

- Dry stool: Add Mang Xiao and Hou Po.
- Irritability, agitation, stifling sensation at chest, and dry heaves: Add Zhu Ru and Shan Zhi Zi.

Excess Heat in Stomach and Intestines Syndrome

Treatment Principle: Purge accumulated heat downward, move Qi, and unblock stool.

Use Da Cheng Qi Tang [Major Order the Qi Decoction] plus Hong Teng, Chi Shao Yao, Tao Ren, Chuan Lian Zi, and Bai Jiang Cao.

Blood Stasis and Qi Stagnation Syndrome

Treatment Principle: Regulate Qi, invigorate blood, and transform stasis.
Use Ge Xia Zhu Yu Tang [Drive Out Stasis Below the Diaphragm Decoction] plus Zao Jiao Ci, E Zhu, and San Leng.

PANCREATITIS (CHRONIC)

This condition is discussed under abdominal pain and diarrhea in TCM. The pathogenesis is damp–heat phlegm stasis obstructing the Middle Jiao. In Western medicine, it is seen as a chronic inflammation of the pancreas which is often caused by acute pancreatitis. Cholelithiasis and infection of the biliary tract causing reverse flow of the bile or narrowed pancreatic duct can cause repeated attacks of pancreatitis. These are the most common reasons that cause chronic pancreatitis. At the later stage, it can lead to diabetes or pancreatic cancer. TCM treatment focuses on strengthening Spleen, transporting dampness, invigorating blood, and clearing heat, which can effectively control the symptoms such as diarrhea, fatty diarrhea, emaciation, malnutrition, and abdominal distending pain.

Treatments According to Syndrome Differentiation

Spleen Deficiency with Dampness Syndrome

Treatment Principle: Strengthen Spleen and transform dampness.
Use Shen Ling Bai Zhu San [Ginseng, Poria, and Atractylodes Macrocephala Powder] plus Huo Xiang and Cang Zhu.
Related conditions and modifications are

- Aversion to fat or oil: Add Shan Zha, Gu Ya, and Mai Ya.
- Abdominal distention after meal: Add Ji Nei Jin, Shan Zha, and Shen Qu.

Damp–Heat Obstruction Syndrome

Treatment Principle: Clear heat and drain damp.
Use Yin Chen Hao Tang [Artemisia Yinchenhao Decoction] plus Huang Bai, Fu Ling, Che Qian Zi, Jin Qian Cao, Yu Jin, and Chuan Xiong.
Related conditions and modifications are

- Severe abdominal pain: Add Mu Xiang and Chai Hu.
- Fever: Add Huang Qin, Huang Lian, and Zhi Mu.
- Constipation: Add Da Huang.

Kidney Deficiency with Blood Stasis Syndrome

Treatment Principle: Tonify Kidney and invigorate blood.
Use Zuo Gui Yin [Restore the Left (Kidney) Decoction] plus Dan Shen and Dang Gui.
Related conditions and modifications are

- Deficient heat: Add Huang Bai and Gui Ban.
- Kidney Yang deficiency: Add Yin Yang Huo and Xian Mao.
- Severe blood stasis spots on skin and rough skin: Add Chi Shao Yao, Mu Dan Pi, and Xian He Cao.

Modification of a Common Formula for Chronic Pancreatitis

Common formula: Chai Hu, Yu Jin, Jin Yin Hua, Tian Hua Fen, Chi Shao Yao, Chen Pi, Bai Zhi, Dang Gui, Zao Jiao Ci, Chuan Shan Jia, Ru Xiang, Tao Ren, and Mo Yao.

Related conditions and modifications are

- Severe abdominal pain: Add Mu Xiang and Xiang Fu.
- Diarrhea: Remove Tao Ren and Dang Gui, add Zhu Ling, Fu Ling, and Ji Nei Jin.
- Fatigue and apathy: Add Huang Qi and Dang Shen.
- Epigastric fullness and distention: Add Zhi Ke and Ban Xia.
- Thick, greasy tongue coating: Add Cang Zhu and Bai Dou Kou.
- Frequent sighing and hypochondriac discomfort: Add Xiang Fu and Yu Jin.
- Constipation: Add Da Huang. Once the stool has passed, use Lai Fu Zi and Bing Lang to maintain the smooth flow of bowel movements.

PEPTIC ULCER

This condition is discussed under Stomach pain, abdominal pain, regurgitation, and bleeding syndrome. The formation is related to the digestive function of gastric acid and gastric protease. The pathological location is at Stomach, Spleen, and Liver. The pathogenesis is weak Spleen and Stomach, with Qi stagnation, stagnant heat, turbid phlegm, and blood stasis. The manifestations include epigastric pain, belching, acid regurgitation, nausea, vomiting, abdominal distention, or constipation, or diarrhea, or tarry stool. The treatment during active ulcer period should clear heat, promote discharge, stop bleeding, and generate flesh. Treatment during remission period focuses on strengthening Spleen and tonifying Qi. After selecting the treatment according to syndrome differentiation, herbs are added that control acidity such as Wu Zei Gu, Wa Leng Zi, Zhe Bei Mu, calcined Mu Li, calcined Long Gu, and Zuo Jin Wan [Left Metal Pill], and herbs that astringen and generate flesh such as Bai Ji, Chi Shi Zhi, Zhen Zhu powder, and Huang Qi.

Active Stage

Stomach Heat Syndrome

Treatment Principle: Clear Stomach and drain heat.

Use Qing Wei San [Clear the Stomach Powder] plus Huang Qin, Bai Shao Yao, Pu Gong Ying, and Wu Zhu Yu.

Related conditions and modifications are

- Dry mouth, red tongue, and scanty fluid: Add Yu Zhu, Mai Men Dong, and Sha Shen.
- Dry stool and abdominal distention: Add Hou Po and Zhi Shi.

Qi Stagnation Syndrome

Treatment Principle: Disperse Liver, regulate Qi, harmonize Stomach, and stop pain.

Use Chai Hu Shu Gan San [Bupleurum Powder to Spread the Liver] plus Chuan Lian Zi.

Related condition and modification are

- Frequent belching: Add Xuan Fu Hua and Dai Zhe Shi.

Blood Stasis Syndrome

Treatment Principle: Move Qi, invigorate blood, expel stasis, and stop pain.

Use Shi Xiao San [Sudden Smile Powder] and Dan Shen Yin [Salvia Decoction].

Related conditions and modifications are

- Severe pain: Add San Qi powder.
- Qi deficiency, and pale and flabby tongue: Add Huang Qi, Bai Zhu, and Dang Shen.

Remission Period

Deficient Cold Syndrome

Treatment Principle: Warm Middle Jiao, strengthen Spleen, scatter cold, and stop pain.

Use Li Zhong Wan [Regulate the Middle Pill] plus Zhi Fu Zi, Gui Zhi, Sha Ren, and Xiang Fu.

Related conditions and modifications are

- Acid regurgitation: Add Wa Leng Zi and calcined Mu Li.
- Sounds in stomach: Add Fu Ling and Ban Xia.
- Severe pain due to cold invasion or ingesting cold foods: Add Gao Liang Jiang, Xiang Fu, and Zi Su Ye.

Yin Deficiency Syndrome

Treatment Principle: Nourish Yin, benefit Stomach, regulate Qi, and stop pain.

Use Yi Wei Tang [Benefit Stomach Decoction] plus Shan Yao, He Huan Hua, Chuan Lian Zi, Bai Shao Yao, and Gan Cao.

Related conditions and modifications are

- Insufficient gastric acid and poor appetite: Add Wu Mei and Shan Zha.
- Constipation: Add Gua Lou Ren, Huo Ma Ren, and Yu Li Ren.

CHAPTER 18

Genitourinary Disorders

GLOMERULONEPHRITIS (ACUTE)

This condition is discussed under Feng Shui (wind edema), Pi Shui (skin edema), and Yang Shui (Yang edema) in TCM. Manifestation of edema occurs when external evil pathogens attack the lungs, resulting in dispersing and descending failure. It can also be caused by damp–heat accumulation in the urinary bladder, burning blood collaterals leading to hematuria, or blood heat disturbing Liver causing upward movement of Qi and blood and manifesting as elevated blood pressure. The clinical manifestations are acute onset, hematuria, proteinuria, edema, and hypertension. The treatment focuses on promoting sweat, unblocking stool and promoting urination to reduce edema, lowering blood pressure, and preventing further complications.

Acute Stage

Feng Shui (Wind Water) Syndrome

Treatment Principle: Disperse wind, scatter heat, disperse Lung, and promote urination.

Use Yue Bi Tang [Maidservant from Yue Decoction] plus Ma Xing Shi Gan Tang [Ephedra, Apricot Kernel, Gypsum, and Licorice Decoction] and Si Ling San [Four-Ingredient Powder with Poria].

Related conditions and modifications are

- Severe exterior cold: Remove Shi Gao, and add Fang Feng and Qiang Huo.
- Severe wind–heat: Add Bo He, Lian Qiao, Jin Yin Hua, Shan Dou Gen, and Jie Geng.
- Phlegm–damp obstructing lung: Add Ban Xia, Chen Pi, and Xing Ren.

Damp Toxin Syndrome

Treatment Principle: Drain damp, expel toxin, strengthen Spleen, and separate water.

Use Wu Ling San [Five-Ingredient Powder with Poria] plus Hou Po, Sheng Jiang Pi, Chi Xiao Dou, and Dang Gui.

Related conditions and modifications are

- Distention and fullness at chest and abdomen and unsmooth urination and bowel movement: Add Ting Li Zi and Da Zao.
- Internal excess damp–heat: Remove Dang Gui and Sheng Jiang Pi, and add Huang Qin, Huang Lian, Yin Chen Hao, and Yi Mu Cao.
- Hematuria: Remove Gui Zhi.
- Ulcerations, and pus and swelling on skin: Remove Gui Zhi, and add Pu Gong Ying, Lian Qiao, Hua Shi, Jin Yin Hua, Tu Fu Ling, and Zi Hua Di Ding.
- Sallow complexion and apathy: Add Dang Shen and Huang Qi.

Excess Heat Syndrome

Treatment Principle: Clear heat, drain fire, cool blood, and stop bleeding.

Use Xiao Ji Yin Zi [Cephalanoplos Decoction] plus Huang Bai, Da Ji, Bai Mao Gen, Che Qian Cao, Nu Zhen Zi, and Han Lian Cao.

Related conditions and modifications are

- Elevated level of RBC in urine: San Qi powder.
- Yin deficiency with excess fire: Add Bie Jia, Shan Zhu Yu, and Xuan Shen.
- Hypertension and yellow greasy tongue coating: Add Gou Teng, Sang Ji Sheng, and Shi Jue Ming.

Recovery Period

At this stage symptoms disappear and only red blood cells are found in urine analysis. This can last from a few months to a few years.

Treatment Principle: Sweet and bland herbs bolster Spleen and with aromatic herbs to transform dampness and nourish Kidney.

Use Mai Wei Di Huang Wan [Ophiopogonis, Schisandra, and Rehmannia Pill] plus Tai Zi Shen, Qian Shi, Sang Ji Sheng, Huo Xiang, Yi Mu Cao, and Bai Mao Gen.

Related conditions and modifications are

- Proteinuria: Add Huang Qi, Dang Shen, fried Mai Ya, burned Shan Zha, and remove Mu Dan Pi and Sheng Di Huang.
- High level of RBC in urine: Remove Huo Xiang and Fu Ling, and add Er Zhi Wan [Two-Ultimate Pill], fried Shan Zhi Zi, Ou Jie, Da Ji, and Xiao Ji.
- Persistent edema for a long period: Add Qian Cao, Yi Mu Cao, Chi Shao Yao, Dan Shen, Ji Xue Teng, Dang Gui, and Chuan Xiong.

GLOMERULONEPHRITIS (CHRONIC)

This condition is discussed under Yin Shui (Yin edema) and deficient damage. The pathogeneses are Kidney Qi, Yang, or Yin deficiency with water dampness, heat toxins, and stasis. Even though acute glomerulonephritis can develop into chronic type, most of the cases are chronic type right from the beginning. The manifestations are chronic, slow development, frequent proteinuria, hematuria, and cylindruria. At late stage, there often is edema, anemia, hypertension, and incomplete kidney function. TCM treatment focuses on supporting the Zheng Qi, expelling evil pathogens, eliminating proteinuria, and at the same time invigorating blood and transforming stasis. Dan Shen, Yi Mu Cao, Dang Gui, Chuan Xiong, Tao Ren, Hong Hua, Ze Lan, Da Huang, and Shui Zhi all have anticoagulant, inhibitory functions of platelet aggregation and kidney improving functions.

The herbs that have better function in eliminating proteinuria are Huang Qi, Dang Shen, Shi Wei, Jiang Can, Chan Tui, Wu Mei, Shan Zha, Ge Gen, and Yu Mi Xu. Eliminating edema herbs are Ze Xie, Zhu Ling, Mu Tong, Che Qian Zi, processed Fu Zi, Xian Mao, Yin Yang Huo, Hu Lu Ba, Ba Ji Tian, Fu Ling, and Bai Zhu. The herbs that can reduce hematuria are Bai Mao Gen, Xian He Cao, Han Lian Cao, San Qi, Sheng Di Huang, Di Yu, Huai Hua Mi, Da Ji, Xiao Ji, Ce Bai Ye, and Liu Ji Nu. One can select these herbs according to syndrome differentiations. Furthermore, Lei Gong Teng preparation has functions that are similar to steroid but without the side effects of steroid. One can use this herb in combination with other herbs.

Treatment According to Syndrome Differentiation

Spleen and Kidney Yang Deficiency Syndrome

Treatment Principle: Warm and tonify Spleen and Kidney.

Prepare Jin Gui Shen Qi Wan [Kidney Qi Pill from the Golden Cabinet] in decoction and take with 8 pills of Bu Zhong Yi Qi Wan [Tonify the Middle and Augment the Qi Pill].

Related condition and modification are

- Severe condition: Add Che Qian Zi, Ba Ji Tian, and Huang Qi.

Liver and Kidney Yin Deficiency Syndrome

Treatment Principle: Nourish Liver and Kidney.

Use Zhi Bai Di Huang Wan [Anemarrhena, Phellodendron, and Rehmannia Pill] plus Sang Ji Sheng, Zhi He Shou Wu, Du Zhong, and Jue Ming Zi.

Accumulation of Damp–Turbid Syndrome

Excess damp: Aromatic herbs transform turbid, harmonize Stomach, and descend rebellious Qi.

Use Huo Po Xia Ling Tang [Agastache, Magnolia Bark, Pinellia, and Poria Decoction].

Excess heat: Bitter cold drain heat, bland leach out dampness, and promote urine.

Use Ge Gen Huang Qin Huang Lian Tang [Kudzu, Coptis, and Scutellaria Decoction] plus Huang Lian Wen Dan Tang [Coptis Decoction to Warm the Gallbladder Decoction].

Modifications

Related conditions and modifications are

- Exterior cold: Xing Su San [Apricot Kernel and Perilla Leaf Powder] plus Yu Xing Cao, Xi Xin, Fang Feng, and Chan Tui.
- Exterior heat: Add Sang Ju Yin [Mulberry Leaf and Chrysanthemum Decoction] plus Jie Geng, Chan Tui, Jiang Can, Niu Bang Zi, Yu Xing Cao, and Jin Yin Hua.
- Severe edema and puffiness: Add Dong Gua Pi (wax gourd), Zhu Ling, Da Fu Pi, Che Qian Zi, Fu Ling, Huang Qi, Dan Shen, and Yi Mu Cao.
- Hypertension: Add Gou Teng, Du Zhong, Sang Ji Sheng, Di Long, Di Gu Pi, and Ci Ji Li.
- Proteinuria: Add Qian Shi, Yi Zhi Ren, and Jin Ying Zi.
- Hematuria: Add Bai Mao Gen, Han Lian Cao, and Hu Po.
- Cylindruria: Increase dosage of Shu Di Huang, Shan Zhu Yu, and add Gou Qi Zi, Zhi He Shou Wu, and Hong Hua.

- Anemia: Add Dang Gui, Huang Qi, Dang Shen, E Jiao, and Gui Ban.
- Yang deficiency: Add Lu Rong and Bu Gu Zhi.
- High cholesterol: Add Shan Zha, Jue Ming Zi, increase dosage of Ze Xie.
- Obvious elevation of nonprotein nitrogen and creatinine: Add Dan Shen and Da Huang.
- Damaged kidney functions: Add Dong Chong Xia Cao.

Medicinal Food Therapy Support

1. Ji Xue Teng E Jiao Zhou [Jixueteng and Ass-Hide Gelatin Congee]: Cook 30 g of Ji Xue Teng, 15 g of E Jiao, and 50 g of Nuo Mi (glutinous rice) to make congee. This is good for blood deficient syndrome.
2. Gou Qi Sang Shen Zhou [Lycium Fruit and Mulberry Fruit Congee]: Cook 15 g of Gou Qi Zi, 15 g of Sang Shen, and 50 g of Nuo Mi to make congee. It is good for Kidney and Liver Yin deficiency.
3. Yi Ren Fu Ling Zhou [Coix Seeds and Poria Congee]: Cook 30 g of Yi Yi Ren, 15 g of Fu Ling, and 50 g of Nuo Mi (glutinous rice). It is good for damp turbid obstructing internally.

IMPOTENCE

Impotence is called Yang Wei or Yin Wei in TCM. The pathogeneses are kidney deficiency with declined fire or Lower Jiao damp–heat. Impotence is inability to initiate or maintain a firm erection in at least 50% of the attempts at intercourse or the cessation of attempts at intercourse. It sometimes accompanied by decreased libido. The treatment in TCM focuses on strengthening Yang. If it is caused by damp–heat, then the treatment focuses on clearing and draining Lower Jiao damp–heat.

Kidney Yang Deficiency Syndrome

Treatment Principle: Tonify Kidney and strengthen Yang.
Use Zan Yu Dan [Special Pill to Aid Fertility].
Related condition and modification are

- Severe Yang deficiency: Add Lu Rong and Ren Shen.

Liver Qi Stagnation Syndrome

Treatment Principle: Spread Liver and relieve stagnation.
Use Dan Zhi Xiao Yao San [Relax Wanderer Powder with Moutan and Gardenia].
Related condition and modification are

- Insomnia or excessive dreaming: Add Suan Zao Ren and Zhi He Shou Wu.

Lower Jiao Damp–Heat Syndrome

Treatment Principle: Clear and drain damp–heat.
Use Long Dan Xie Gan Tang [Gentiana Longdancao Decoction to Drain the Liver] plus Er Miao San [Two-Marvel Powder].

Cold Stagnation in Liver Channel Syndrome

Treatment Principle: Warm Liver and move Qi.

Use Nuan Gan Jian [Warm the Liver Decoction].

Related condition and modification are

- Severe cold pain at sides of the lower abdomen: Add Wu Zhu Yu, Xi Xin and Bi Ba.

Blood Stasis Due to Cold at Liver Channel Syndrome

Treatment Principle: Warm the channel, invigorate blood, and scatter cold.

Use Dang Gui Si Ni Jia Wu Zhu Yu Sheng Jiang Tang [Tangkuei Decoction for Frigid Extremities plus Evodia and Fresh Ginger].

This formula can be used to treat cases without any results from all other formulas and without heat manifestations.

Related conditions and modifications are

- Nocturnal emission with dreams: Add Suan Zao Ren, Huang Lian, calcined Mu Li, and Ye Jiao Teng.
- Nocturnal emission without dreams: Add Qian Shi, Lian Xu, Yi Zhi Ren, Jin Ying Zi, Wu Wei Zi, and Wu Zei Gu.

PROSTATITIS (ACUTE)

This condition is discussed under turbid Lin in TCM. It is often caused by damp–heat accumulation in the Lower Jiao. The manifestations are acute onset, high fever, chills, whole body discomfort, fatigue, nausea, vomiting; frequent, urgent, hesitant, and painful urination; turbid urine with mucous or with blood and essence; difficulty urinating or even urinary retention; or pain at pubic bone. From the physical examination, one finds enlarged prostate gland with pain upon palpation. TCM treatment focuses on clearing and draining Lower Jiao damp–heat.

Lower Jiao Damp–Heat Syndrome

Treatment Principle: Clear and drain damp–heat.

Use Long Dan Xie Gan Tang [Gentiana Longdancao Decoction to Drain the Liver] plus Bei Xie and Bai Jiang Cao.

Related conditions and modifications are

- Severe lower abdominal pain: Add Chi Shao Yao and Tao Ren.
- Urinary difficulty: Add Dan Zhu Ye, Huang Bai, and Wang Bu Liu Xing.
- Severe testicle pain: Add Ju He and Li Zhi He.

Yin Deficiency and Damp–Heat Syndrome

Treatment Principle: Nourish Yin, clear heat, and drain damp.

Use Zhi Bai Di Huang Wan [Anemarrhena, Phellodendron, and Rehmannia Pill] plus Che Qian Zi and Bei Xie.

PROSTATITIS (CHRONIC)

Chronic prostatitis can be asymptomatic or with increased urinary frequency, burning urination, dribbling urination, or turbid white mucous discharge with urination; and distending pain or discomfort at lower abdomen, genital area, or testicles. Some people also experience decreased libido, early ejaculation, and nocturnal emission. The pathogenesis is kidney deficiency failing to astringe with Lower Jiao damp–heat and there can also be Qi stagnation and blood stasis when the condition lasts a long time.

Use Bei Xie Fen Qing Yin Jia Jian [Modified Dioscorea Hypoglauca Decoction to Separate the Clear from Medical Revelations].

Related conditions and modifications are

- Slight Kidney Yang deficiency: Remove Huang Bai and add Zhi Fu Zi, Rou Gui, Suo Yang, Jin Ying Zi, Wu Wei Zi, and Yin Yang Huo.
- Slight Kidney Yin deficiency: Add Zhi Mu, Sheng Di Huang, Shan Zhu Yu, and Er Zhi Wan [Two-Ultimate Pill].
- Spleen deficiency: Add Huang Qi and Dang Shen.
- Qi stagnation and blood stasis: Add Tao Ren, Chi Shao Yao, Wang Bu Liu Xing, and Li Zhi He.
- Severe damp–heat: Add Guan Zhong, Ku Shen, Cang Zhu, Ma Chi Xian, and Tu Niu Xi.

PYELONEPHRITIS

Acute pyelonephritis is discussed under heat Lin and blood Lin syndrome. The pathogeneses are damp–heat accumulating in the Lower Jiao and disturbance in Qi transformation at urinary bladder. Chronic type of pyelonephritis is discussed under Qi stagnation and exhausted Lin syndrome. It is often caused by acute unhealed pyelonephritis, accumulation of damp–heat, or overuse of bitter cold herbs damaging Middle Jiao Qi. The manifestations of acute type are chills and high fever, nausea, vomiting, poor appetite, frequent, urgent and painful urination, and discomfort or pain at lumbar region or pain upon palpation or percussion at the kidney area. Manifestations of chronic type are not typical, and show as white blood cells or with bacteria in urine. TCM treatment focuses on clearing heat, relieving toxicity, unblocking Lin syndrome, and draining damp. TCM treatment is even more effective in treating chronic pyelonephritis. It often tonifies spleen, benefits kidney, drains damp, and unblocks Lin. Commonly used anti-infectious herbs are Jin Qian Cao, Huang Qin, Huang Bai, and Huang Lian.

Acute Pyelonephritis

Lower Jiao Damp–Heat Syndrome

Treatment Principle: Unblock Lin syndrome, drain dampness, clear heat, and drain fire.

Use Ba Zheng San [Eight-Herb Powder for Rectification] plus Shi Wei, Jin Qian Cao, and Tu Fu Ling.

Related conditions and modifications are

- Abdominal distention and constipation: Add Zhi Shi, and increase dosage of Da Huang.
- Distending sinking sensation at lower abdomen: Add Wu Yao and Tao Ren.
- Dark urine: Add Bai Mao Gen, Qian Cao, and Di Yu Tan.
- Exterior syndrome: Add Jin Yin Hua, Bo He, and Lian Qiao.

Liver and Gallbladder Damp–Heat Syndrome

Treatment Principle: Clear Liver, promote Gallbladder, unblock Lin, and drain dampness.

Use Long Dan Xie Gan Tang [Gentiana Longdancao Decoction to Drain the Liver] plus Jin Qian Cao.

Damp–Heat Damaging Collaterals Syndrome

Treatment Principle: Clear heat, drain damp, cool blood, and stop bleeding.

Use Dao Chi San [Guide Out the Red Powder] plus Mai Men Dong, Han Lian Cao, Shi Wei, Huang Bai, Pu Gong Ying, Jin Qian Cao, Bai Mao Gen, Tu Fu Ling, and Ou Jie.

Furthermore, one should remind the patient to drink more water and urinate more and also keep the genital area clean.

Chronic Pyelonephritis

Yin Deficiency with Damp–Heat Syndrome

Treatment Principle: Nourish Yin, tonify Kidney, clear heat, and drain damp.

Use Zhi Bai Di Huang Wan [Anemarrhena, Phellodendron, and Rehmannia Pill] plus Tu Fu Ling, Huang Qin, Jin Qian Cao, and fried Chuan Lian Zi.

Related conditions and modifications are

- Severe heat: Add Bai Jiang Cao and Zi Hua Di Ding.
- Fatigue and shortness of breath: Add Huang Qi and Sha Shen.
- Lower abdominal distention: Add Tao Ren.

Spleen and Kidney Yang Deficiency Syndrome

Treatment Principle: Warm Kidney and Spleen.

Use Jin Gui Shen Qi Wan [Kidney Qi Pill from the Golden Cabinet].

Related conditions and modifications are

- Distending and sinking sensation at the lower abdomen and dribbling urination: Add Bu Zhong Yi Qi Tang [Tonify the Middle and Augment the Qi Decoction].
- Severe Kidney Yang deficiency and exhaustion: Add You Gui Wan [Restore the Right (Kidney) Pill].

Treatments According to Symptoms

1. Severe edema: Use Wu Pi Yin [Five-Peel Decoction] plus Bai Zhu, Tu Fu Ling, Cang Zhu, Gui Zhi, Che Qian Zi, and Zhu Ling. Return to the core formula after the edema has cleared.
2. Hypertension: Add Tian Ma Gou Teng Yin [Gastrodia and Uncaria Decoction] plus Sheng Di Huang, Bai Shao Yao, and Ze Xie.
3. Pyuria, or urine color like pus: Add Bei Xie, Yi Yi Ren, Ze Xie, and Fu Ling.
4. Stones in urinary tract: Add Chuan Niu Xi, Jin Qian Cao, Hai Jin Sha, Gan Cao, and Bai Shao Yao.

RENAL FAILURE (CHRONIC)

Chronic renal failure is discussed under urinary retention, obstruction and rejection (guan ge), and deficiency due to exhaustion. The pathogeneses are lungs failing to disperse and descend, spleen failing to transform and transport, and kidneys failing to open and close which leads to damp–turbid urinary toxins accumulating internally, failing to descend, and stagnating at blood vessels. In Western medicine, various reasons can cause organic damage of the kidneys. When the condition worsens, the kidneys fail to maintain its basic functions. Then, there is accumulation of metabolic waste products, imbalance of the electrolytes and acidity, dysfunction of the endocrine system, and damages to the other systems. Chronic glomerulonephritis is most commonly seen at the beginning stage, followed by chronic pyelonephritis. The manifestations are fatigue, poor appetite, nausea, vomiting, headache, dizziness, palpitation, heart fatigue, aloofness, elevated blood pressure, skin itching, anemia, increased nocturia, scanty urination, edema, labored breathing or even difficulty, heart exhaustion, coma, and subcutaneous or digestive tract bleeding. TCM treatment focuses on supporting Zheng Qi, expelling the evil and treating both root and branches at the same time to improve kidney function and prolong life. In severe cases, one should be referred to his or her medical doctor. Da Huang, Huang Qi, Dong Chong Xia Cao, Chuan Xiong, and San Qi can stimulate kidney functions and inhibit kidney damages.

Treatment According to Syndrome Differentiation

Spleen and Kidney Qi and Yang Deficiency Syndrome

Treatment Principle: Tonify Spleen and Kidney, assist Yang, and promote urination.
Use Zhen Wu Tang [True Warrior Decoction] plus Jiu Da Huang.
Related conditions and modifications are

- Severe nausea and vomiting: Remove Fu Zi, and add Chen Pi and Zi Su Ye.
- Blood stasis: Add Hong Hua, Dang Gui, and San Qi.

Spleen and Kidney Qi and Yin Deficiency Syndrome

Treatment Principle: Tonify Spleen and Kidney, tonify Qi, and nourish Yin.
Use Liu Wei Di Huang Wan [Six-Ingredient Pill with Rehmannia] plus Huang Qi and Ren Shen.
Related conditions and modifications are

- Severe Yin deficiency: Add Er Zhi Wan [Two-Ultimate Pill].
- Severe Qi deficiency: Add Bai Zhu and Tu Si Zi.
- Edema: Add Zhu Ling and Che Qian Zi or change the formula to Ji Sheng Shen Qi Wan [Kidney Qi Pill from Formulas that Aid the Living].
- Blood stasis: Add Hong Hua, Dan Shen, Yi Mu Cao, Tao Ren, and Chuan Xiong.

Liver and Kidney Yin Deficiency Syndrome

Treatment Principle: Nourish Liver and Kidney.
Use Qi Ju Di Huang Wan [Lycium Fruit, Chrysanthemum, and Rehmannia Pill].

Both Yin and Yang Deficiency Syndrome

Treatment Principle: Tonify Yang and nourish Yin.
Use Di Huang Yin Zi [Rehmannia Decoction] plus Shi Hu.
Related condition and modification are

- Edema: Add Che Qian Zi, Zhu Ling, and Fu Ling Pi.

Treatments of Complications

Hypertension

Use Ban Xia Bai Zhu Tian Ma Tang [Pinellia, Atractylodes Macrocephala, and Gastodia Decoction] plus Zhi Ke, Mai Men Dong, Ze Xie, Dan Shen, Ju Hua, Che Qian Zi, and Yi Mu Cao.

Heart Failure

Use Ling Gui Zhu Gan Tang [Poria, Cinnamon Twig, Atractylodes Macrocephala, and Licorice Decoction] plus Zhen Wu Tang [True Warrior Decoction].

Anemia

Use Gui Pi Tang [Restore the Spleen Decoction] plus Shi Quan Da Bu Tang [All-Inclusive Great Tonifying Decoction].

Skin Itching

Add Di Fu Zi, Bai Xian Pi, and Hong Zao.

SPERMACRASIA

This condition refers to sperm count that is below 0.6×10^6/mL or normal sperm that is lower than 60% or seminal fluid that is less than 1.5 mL. These are all possible causes of decreased fertility. In TCM, this condition is caused by insufficient Kidney essence, Qi and blood exhaustion and deficiency, and lack of postnatal proper care. These patients do not have obvious signs. It is often determined through examination of the seminal fluids. Small number of patients can experience fatigue, apathy, dizziness, tinnitus, soreness and weakness of the lumbar region, forgetfulness, dampness at scrotum, low libido, and decreased volume and thin quality of semen per patient report. TCM treatment focuses on tonifying Kidney, supporting essence and assists by tonifying Qi and nourishing blood. In this case, it is contraindicated to use very bitter, cold, warm, and dry formulas.

In general, one can use the following formulas as a base and modify according to the syndrome differentiation: Wu Zi Yan Zong Wan [Five-Seeds Pill to Continue the Lineage] plus Yin Yang Huo, Xu Duan, Zhi He Shou Wu, and Shu Di Huang.

Related conditions and modifications are

- Severe Kidney Yin deficiency: Add Gui Ban, Zhi Mu, and Di Gu Pi.
- Severe Kidney Yang deficiency: Add Lu Jiao Jiao, Ba Ji Tian, and Suo Yang.
- Severe Qi and blood deficiency: Add Dang Shen, Huang Qi, E Jiao, and Ji Xue Teng.
- Liver Qi stagnation: Add Chai Hu, Yu Jin, and Xiang Fu.
- Spleen deficiency: Add Shan Yao, Bai Zhu, and Lian Zi.
- Cold congealed in liver channel, retraction of the scrotum due to cold, clear semen, and cold in the sides of the lower abdomen: Add Zhi Fu Zi, Gui Zhi, Xiao Hui Xiang, and Li Zhi He.
- Decreased motility of sperms: Add Huang Jing, Zi He Che, and Sha Yuan Ji Li.
- High viscosity of semen and slow liquefaction: Add Bei Xie, Fu Ling, Ze Xie, Gui Ban, Dan Shen, and Huang Bai.
- Impotence: Add Bai Ji Li, Jiu Zi, Hong Shen, and Rou Gui.

- Seminal emission, early ejaculation: Add Qian Shi, Lian Xu, and Mu Li.
- Excess dead sperms (more than 40% of the sperms are immobile): Add Sheng Di Huang, Zhi Mu, Chi Shao Yao, Mu Dan Pi, Jin Yin Hua, Pu Gong Ying, and Yin Chen Hao.

URINARY STONES

This condition is also called Shi Lin (Stone Lin) and Sha Lin (Sandy Lin). Damp–heat accumulation in the Lower Jiao turning into fire, damaging Yin, and burning urine into sand and stones. About 90% of stone composition is calcium oxalate and less than 10% is calcium phosphate. According to the location of the stone, it can be classified as kidney stone, ureter stone, urinary bladder stone, or urethra stone. TCM treatment focuses on clearing heat, draining damp, expelling stone, and unblocking the Lin syndrome. Herbs, such as Jin Qian Cao, Hai Jin Sha, Ji Nei Jin, Chuan Shan Jia, Mang Xiao, and Hu Po can dissolve stones. Dong Kui Zi, Hua Shi, and Bai Mao Gen have the function of promoting urination and expelling stone.

Damp–Heat Accumulation in the Lower Jiao Syndrome

Treatment Principle: Clear heat, drain damp, unblock Lin, and expel stones.

Use Ba Zheng San [Eight-Herb Powder for Rectification] plus Jin Qian Cao, Hai Jin Sha, Shi Wei, Bai Mao Gen, Ji Nei Jin, Wang Bu Liu Xing, and Dong Kui Zi.

Related conditions and modifications are

- Severe spasms and pain at the sides of the lower abdomen: Add Chuan Lian Zi, Yan Hu Suo, and Wu Yao.
- Lower abdominal distention and pain: Add Chen Xiang and Qing Pi.
- Painful burning and hesitant urination: Add Huang Bai and Pu Gong Ying.
- Itching of external genitalia: Add Bai Xian Pi.

Qi Stagnation and Blood Stasis Syndrome

Treatment Principle: Invigorate blood, transform stasis, soften hardness, and expel stones.

Use Pai Shi Tang [Expel Stone Decoction] plus Dan Shen, Hong Hua, Chuan Shan Jia, San Leng, E Zhu, Wu Yao, Wang Bu Liu Xing, Jin Qian Cao, and Ji Nei Jin.

Related conditions and modifications are

- Severe pain: Add Zhi Ru Xiang and Zhi Mo Yao.
- Hematuria: Add San Qi powder and Hu Po powder.
- Painful, hesitant, and dribbling urination, and yellow tongue coating: Add Che Qian Cao, Bian Xu, and Huang Bai.

Spleen and Kidney Yang Deficiency Syndrome

Treatment Principle: Warm Spleen and Kidney, expel stone, and unblock Lin.

Use You Gui Wan [Restore the Right (Kidney) Pill] plus Chuan Shan Jia, Jin Qian Cao, and Ji Nei Jin.

Related conditions and modifications are

- Severe Yang deficiency: Add Zhi Fu Zi and Rou Gui.
- Forceless urination: Add Dang Shen and increase Huang Qi dosage to 30 g.
- Radiating pain to sides of the abdomen and genital area and is better with warmth: Add Wu Yao, Xiao Hui Xiang, and Chai Hu.

Liver and Kidney Yin Deficiency Syndrome

Treatment Principle: Nourish Yin, increase fluid, unblock Lin, and expel stone.

Use Zeng Ye Tang [Increase the Fluids Decoction] plus Yu Zhu, Jin Qian Cao, Hai Jin Sha, Mu Dan Pi, Cu Bie Jia, Tu Si Zi, Zhi Shou Wu, and Nu Zhen Zi.

Related conditions and modifications are

- Soreness and weakness of the lumbar region: Add Gou Qi Zi, Shan Zhu Yu, and Huai Niu Xi.
- Hematuria: Add Bai Mao Gen and Han Lian Cao.
- Night sweats: Add Fu Xiao Mai and Duan Mu Li.

If the stone is greater than 1 cm in diameter, then patients should be referred to their M.D.s.

CHAPTER 19

Gynecological Disorders

AMENORRHEA

There are two types of pathogenic factors for amenorrhea: excess and deficiency. Deficiencies often involve Spleen, Kidney, Qi, and blood. Excess type is obstruction of channels and collaterals. In Western medicine, if period has never started, then it is primary amenorrhea. If period ceases to occur for more than 6 months due to any reason, then it is called secondary amenorrhea. TCM treatments focus on tonifying and unblocking the two conditions, respectively.

Uterine Amenorrhea

Blood Stasis at Uterus, Blockage at Chong and Ren Channels

Amenorrhea due to adhesion of cervix and uterine cavity. This often occurs after surgery, artificial abortion, and delivery.

Treatment Principle: Move Qi, invigorate blood, transform stasis, and unblock collaterals.

Use Xue Fu Zhu Yu Tang [Drive Out Stasis in the Mansion of Blood Decoction] plus Dan Shen, Xiang Fu, Wu Ling Zhi, E Zhu, Tu Bie Chong, and Chuan Shan Jia.

Related conditions and modifications are

- More cold and lower abdominal cold pain: Add Gui Zhi and Wu Zhu Yu.
- More heat with damp: Add Huang Qin, Che Qian Zi, and Yin Chen Hao.

Damaged Chong and Ren Channels, Insufficient Essence and Blood, and Empty Deficiency at the Collaterals of Uterus

Frequent artificial abortion with deep damage to the endometrium due to deep curettage.

Treatment Principle: Tonify Kidney, supplement essence, benefit Qi, nourish blood, regulate, and tonify Chong and Ren channels.

Use Zuo Gui Wan [Restore the Left (Kidney) Pill] plus Si Wu Tang [Four-Substance Decoction] and Xiang Fu, Huang Qi, and Bai Dou Kou.

Related conditions and modifications are

- More cold: Add Gui Zhi.
- More heat: Remove Huang Qi, and add Mai Men Dong and Xuan Shen.
- Liver stagnation: Remove Huang Qi, and add Yu Jin and Chuan Lian Zi.

Take the above formula for 20 days which is about 10 bags of herb. If there is increased abdominal distention and increased vaginal discharge, then one should change the formula to Tao Hong Si Wu Tang [Four-Substance Decoction with Peach Pit and Safflower] with Gou Qi Zi, Ji Xue Teng, Chong Wei Zi, Chai Hu, Zhi Shi, and Chuan Niu Xi. Continuously take five bags of the formula. If menstruation has not yet restarted in 1 week, then repeat the above treatment for one more course.

Yin Deficiency with Internal Heat, Heat Burned Yin Fluid, and Blood Dryness Leading to Amenorrhea

Tuberculosis of the endometrium.

Treatment Principle: Nourish Yin, clear heat, nourish blood, invigorate blood, and regulate Chong and Ren channels.

Use Qing Hao Bie Jia Tang [Artemisia Annua and Soft-Shelled Turtle Shell Decoction] plus Di Gu Pi, Yin Chai Hu, Mai Men Dong, Bai Shao Yao, Dang Gui, Ji Xue Teng, Shan Yao, and Dan Shen.

If symptoms of Yin deficiency and internal heat are better after taking the formula, then change to Si Wu Tang [Four-Substance Decoction] plus Ji Xue Teng, Chong Wei Zi, Nu Zhen Zi, Dan Shen, Tao Ren, Ze Lan, Xiang Fu, and Mai Men Dong. One can also combine with antituberculosis pharmaceutical drugs.

Kidney Qi Deficiency, Empty Chong Channel, Blocked Ren Channel, and Tian Kui Not Yet Arrived

Maldevelopment of the uterus (infantile uterus).

Treatment Principle: Tonify Kidney, nourish blood, regulate Chong, and Ren channels.

Use Tu Si Zi Wan [Cuscuta Seed Pill], remove Fu Zi, Mu Li, Sang Piao Xiao, and add Fu Pen Zi, Yin Yang Huo, Zi Shi Ying, Shu Di Huang, Gou Qi Zi, Dang Gui, and Xiang Fu.

Related conditions and modifications are

- Kidney Yin deficiency: Remove Lu Rong, Zi Shi Ying, and Yin Yang Huo, and add Nu Zhen Zi, He Shou Wu, and Mai Men Dong.
- Liver stagnation: Add Chai Hu and Shan Zhi Zi.

After taking the above formula for few months, if the vaginal discharge increases with distending sensation at breast and lower abdomen, one should change to Tao Hong Si Wu Tang [Four-Substance Decoction with Peach Pit and Safflower] with Ji Xue Teng, Chong Wei Zi, Gou Qi Zi, Nu Zhen Zi, Chi Shao Yao, Xiang Fu, Ze Lan, and Sheng Di Huang.

Ovarian Amenorrhea

This amenorrhea is caused by decreased ovarian hormones leading to absence of cyclical changes of the endometrium.

Kidney Qi Deficiency, Tian Kui Not Arrived Yet, and Blocked Chong and Ren Channels

Amenorrhea caused by maldevelopment of the ovaries.

Please refer to the treatment for maldevelopment of uterus.

Kidney and Blood Deficiency, Early Decline of Chong and Ren Channels, Empty Sea of Blood, and Dried Blood Leading to Amenorrhea

Early decline of ovaries leading to amenorrhea. The manifestations occur before 40 years of age or even before age 30. There are four steps to the treatment:

- Step 1: Nourish Yin, clear heat, nourish blood, and soften Liver.
  Use Zhi Bai Di Huang Wan [Anemarrhena, Phellodendron, and Rehmannia Pill] plus Er Zhi Wan [Two-Ultimate Pill] and Bai Shao Yao, Gui Ban, and He Shou Wu. Take five bags per week.
- Setp 2: After few weeks of above treatment, when symptoms of Yin deficiency and internal heat is reduced, then change to tonify Kidney, supplement essence, nourish blood, and regulate Chong channel.
  Use Zuo Gui Wan [Restore the Left (Kidney) Pill] plus Yin Yang Huo, Sang Shen, Nu Zhen Zi, Fu Pen Zi, Dang Gui, Mai Men Dong, and Bai Dou Kou. Take five bags per week.
- Step 3: During one or more weeks of taking the above formula, the estrogen level and vaginal discharge should be monitored. If the estrogen level has increased, then one should tonify Kidney, nourish blood, and invigorate blood.
  Use Qi Bao Mei Ran Dan [Seven-Treasure Special Pill for Beautiful Whiskers] plus Yin Yang Huo, Huang Jing, Chong Wei Zi, Chuan Xiong, Tao Ren, Hong Hua, and Xiang Fu. Take five bags per week.
- Step 4: Take the above formula until basal body temperature has increased. Continue by taking three to four more bags. Then change to invigorating blood and unblocking menstruation formula.
  Use Tao Hong Si Wu Tang [Four-Substance Decoction with Peach Pit and Safflower] plus Ji Xue Teng, Ze Lan, Chuan Niu Xi, Zhi Ke, Xiang Fu, Dan Shen, and Qu Mai. In general, one should see effect in 3 to 5 weeks. If this is not effective, one should combine it with pharmaceutical drugs.

Pituitary Amenorrhea

Pathological changes of the anterior lobe of the pituitary gland causes decrease in the level of gonadotrophin hormone, which in turns leads to decreased ovarian function and amenorrhea.

Greatly Deficient Qi and Blood, Failed to Nourish Zang and Fu Organs, Yang Qi Deficiency, and Failed to Warm and Nourish Eight Extra Meridians

Damage or trauma to the pituitary, leading to dysfunction (Sheehan's syndrome) in hormonal regulation leading to amenorrhea. It is often seen in postpartum hemorragic bleeding with shock.

Treatment Principle: Greatly tonify Qi and blood, arouse Yang Qi, warm and nourish Eight Extra Meridians, and stimulate Zang and Fu organ functions.

Use Si Wu Tang [Four-Substance Decoction] plus Ren Shen, Huang Qi, Lu Rong, Ba Ji Tian, Yin Yang Huo, Zi He Che powder, Bu Gu Zhi, Shan Yao, and Chen Pi.

After the patient takes the above formula and there is reduced vaginal discharge, then remove Lu Rong and Zi He Che powder, and add Tao Ren and Hong Hua. In general, one needs to take 20 bags to see increased levels of estrogen.

Kidney Deficiency and Damaged Qi and Blood

Amenorrhea after surgery and chemotherapy for pituitary tumor.

Treatment Principle: Tonify Kidney, invigorate blood, and regulate Chong channel.

Use Tu Si Zi Wan [Cuscuta Seed Pill], remove Fu Zi, Sang Piao Xiao, calcined Mu Li, and Wu Wei Zi, and add Bu Gu Zhi, Yin Yang Huo, Chong Wei Zi, Dang Gui, Chuan Xiong, Tao Ren, Hong Hua, Nu Zhen Zi, Dan Shen, Xiang Fu, Ji Xue Teng, and Fu Pen Zi.

Related conditions and modifications are

- Yin deficiency with heat: Add Huang Bai, Zhi Mu, Mai Men Dong, and Shi Hu.
- Spleen deficiency: Add Bai Dou Kou, Bai Zhu, and Hou Po.
- Obesity: Add Fu Ling, Ban Xia, Ze Xie, and Cang Zhu.
- High blood pressure: Add Xia Ku Cao, Gou Teng, Huang Qin, Sang Ji Sheng, and Du Zhong.
- Liver Qi stagnation: Add Chai Hu and Yu Jin.

Hypothalamic Amenorrhea

Dysfunction of hypothalamus affecting the pituitary, ovaries, and uterus leading to amenorrhea.

Liver Qi Stagnation, Dysfunction of Chong and Ren Channels

This is caused by disturbances of seven emotions.

Treatment Principle: Disperse Liver, regulate Qi, invigorate blood, and regulate Chong channel.

Use Xiao Yao San [Relax Wanderer Powder] plus Xiang Fu, Yu Jin, Ji Xue Teng, Chong Wei Zi, Tu Si Zi, Tai Zi Shen, and Bai Dou Kou.

- After taking four to six bags of herbs and there is improvement of symptoms, then one should disperse Liver and unblock meridians. Change the formula to Chai Hu Shu Gan San [Bupleurum Powder to Spread the Liver] and add E Zhu, Ji Xue Teng, Dan Shen, Tao Ren, Hong Hua, and Chuan Niu Xi.
- After taking four to six bags, one should expect the onset of menstruation. If menstruation is absent, add Tu Si Zi, Bu Gu Zhi, and Yin Yang Huo.

Liver Fire Due to Long-Term Liver Qi Stagnation (Stagnant Qi Leading to Fire, Damaging Yin)

Treatment Principle: Nourish Yin, clear heat, disperse Liver, invigorate blood, and regulate Chong channel.

Use Dan Zhi Xiao Yao San [Relax Wanderer Powder with Moutan and Gardenia] plus Zhi Mu, Mai Men Dong, Xuan Shen, Nu Zhen Zi, Chong Wei Zi, Shi Jue Ming, Zhi He Shou Wu, and Ji Xue Teng.

- After taking six to eight bags and improvement of symptoms, then one should disperse Liver and unblock meridians. The formula is the same as listed above (Xiao Yao San Jia Wei).

Liver Overacting on Spleen (Psychogenic Anorexia)

Treatment Principle: Strengthen Spleen, disperse Liver, nourish blood, and tonify Chong channel.

Use Si Jun Zi Tang [Four-Gentleman Decoction] plus Xiao Yao San [Relax Wanderer Powder] and Bai Dou Kou, Xiang Fu, Tai Zi Shen, Ji Xue Teng, Gou Qi Zi, Tu Si Zi, Shen Qu, and Shan Zha.

- After taking four to six bags of the above formula with stronger Spleen and Stomach and increased appetite, then one can use Ba Zhen Tang [Eight-Treasure Decoction] plus Xiang Fu and Ze Lan.

Malnutritional Amenorrhea

Qi and Blood Deficiency, Empty Blood Sea, and Scanty Source

Treatment Principle: Malnutrition, also seen in consumptive disease, gastrointestinal disease, severe tuberculosis, malaria, and severe anemia.

Strengthen Spleen, benefit Qi, nourish blood, and regulate Chong channel.

Use Ba Zhen Tang [Eight-Treasure Decoction] plus Zhi He Shou Wu, Huang Jing, Bu Gu Zhi, Bai Dou Kou, and Chen Pi.

Take the above formula until Qi and blood are recovered and Spleen and Stomach are stronger, then change to Tao Hong Si Wu Tang [Four-Substance Decoction with Peach Pit and Safflower] plus Dang Shen, Xiang Fu, and Ze Lan to benefit Qi, nourish blood, invigorate blood, and unblock channels.

Drug-Induced Amenorrhea

Drug Interference and Chong and Ren Disorder

This amenorrhea is caused by drug inhibition. It is often seen in long-term use of oral contraceptives, leading to inhibition of gonadotrophin hormone and amenorrhea.

Treatment Principle: Tonify Kidney, invigorate blood, and regulate Chong channel.

Use Zuo Gui Wan [Restore the Left (Kidney) Pill] plus Yin Yang Huo, Zhi He Shou Wu, Huang Jing, Chong Wei Zi, Dang Gui, Xiang Fu, Tao Ren, and Hong Hua.

Internal Obstruction of Phlegm–Damp and Blocked Chong and Ren Channels

Polycystic Ovarian Syndrome

It is often seen in females of 20 to 30 years old. Symptoms other than amenorrhea include infertility, hirsutism, weight gain, and persistent unovulation.

Yang Deficiency and Phlegm Damp

Treatment Principle: Warm Kidney Yang, strengthen Spleen, eliminate phlegm, and regulate Chong channel.

Use Dao Tan Tang [Guide Out Phlegm Decoction] plus Cang Zhu, Xiang Fu, Shan Zha, Yin Yang Huo, Ba Ji Tian, Lu Jiao Jiao, and Tu Si Zi.

Phlegm–Damp Transform into Heat

Treatment Principle: Drain heat and damp.

Use Long Dan Xie Gan Tang [Gentiana Longdancao Decoction to Drain the Liver] plus Cang Zhu, Shan Zha, and Zhe Bei Mu.

After taking the formula with alleviated symptoms and raised estrogen, then one should change to Tao Hong Si Wu Tang [Four-Substance Decoction with Peach Pit and Safflower] plus Xu Duan, Xiang Fu, Dan Shen, Ze Lan, Chuan Niu Xi, and Bu Gu Zhi to invigorate blood, unblock meridians, and stimulate ovulation.

Related conditions and modifications are

- Severe damp–phlegm: Add Gua Lou Pi, Ban Xia, and Dan Nan Xing.
- If menstruation still does not occur, one can repeat the above formulas for one or two more cycles.

CHRONIC CERVICITIS

This condition is discussed under leukorrhea (Dai Xia) under TCM. It is mainly caused by damp accumulation in the Lower Jiao obstructing Ren and Dai meridians. Spleen deficiency with damp–heat downward-pouring can also cause this condition. In Western medicine, more than 50% of married women are reported to have chronic cervicitis. The pathogens often invading after delivery, miscarriage, surgery, or traumatic clinical manifestations increase leukorrhea, which is milky and sticky or light yellow pus, with foul smell, lower back pain, infertility, and sinking sensation in the pelvic cavity or with pain during intercourse. TCM treatments focus on strengthening Spleen, ascending Yang, eliminating damp, and clearing heat.

Internal Treatment

Treatment Principle: Strengthen Spleen, benefit Qi, clear heat, and eliminate damp.

Use Wan Dai Tang [End Discharge Decoction] plus Yi Yi Ren, Guan Zhong, Huang Bai, and Jin Yin Hua.

Related conditions and modifications are

- Copious leukorrhea, yellow, and sticky: Remove Dang Shen and Shan Yao, and add Bei Xie, Ze Xie, and Yin Chen Hao.
- Lower abdominal and lumbar distending pain, yellow and white leukorrhea, and sticky and foul smell: Remove Dang Shen, and add Zhi Ke, Chuan Niu Xi, Chuan Lian Zi, Tu Fu Ling, and Pu Gong Ying.
- Blood-tinged leukorrhea: Add Di Yu and Qian Cao.
- Copious leukorrhea with white and watery quality and lower abdominal coldness: Remove Guan Zhong and Huang Bai, and add Jin Ying Zi, Ai Ye, and Lu Jiao Jiao.

External Application

Apply powder topically.

Treatment Principle: Clear heat, relieve toxicity, invigorate blood, and eliminate the rotten and generate flesh.

Use Hai Ge Ke powder 30 g, Zhang Nao 15 g, Ru Xiang 5 g, Mo Yao 5 g, and Bing Pian 5 g. Grind all ingredients into powder. Follow routine of cleaning the vaginal area, then spray the powder at the cervix where there is inflammation. Apply once every three days, six to seven times as one course. Avoid using the powder during menstruation. Avoid sexual activity while using the herbs.

CHRONIC PELVIC INFLAMMATION

There is no corresponding pathological name in TCM. It is discussed under leukorrhea, irregular menstruation, and infertility. Damp–heat and heat–toxins invade the body while there is deficiency after menstruation and delivery. The pathogen attacks the uterus directly, obstructs Qi and blood, and damages Chong and Ren channels. In Western medicine, this refers to inflammations of the female internal genital organs including uterus, fallopian tubes, ovaries and the surrounding tissues, and pelvic peritoneum. The clinical characteristics are lower abdominal pain, increased leukorrhea, or abnormal bleeding from the vagina, or with low-grade fever and fatigue. TCM treatments focus on clearing heat, draining damp, invigorating blood, and transforming stasis.

Damp–Heat and Blood Stasis Accumulating in Lower Jiao

Treatment Principle: Clear heat, drain damp, expel stasis, and scatter accumulation.

Use Qing Re Zhi Dai Tang [Clear Heat and Stop Excessive Uterine Bleeding Decoction] plus Yi Mu Cao and Pu Huang.

Related conditions and modifications are

- Constipation: Add Tao Ren and Da Huang.
- Copious yellow, sticky, and foul leukorrhea: Add Dong Gua Ren and Chi Xiao Dou.

Qi Stagnation and Blood Stasis

Treatment Principle: Move Qi, invigorate blood, transform stasis, and stop pain.

Use Ge Xia Zhu Yu Tang [Drive Out Stasis Below the Diaphragm Decoction] plus Pu Huang and Dan Shen.

Related conditions and modifications are

- Restlessness, agitation, and easily angered: Add Chai Hu and Yu Jin.
- Breast tenderness before menstruation: Add Si Gua Luo and Ju He.
- Copious leukorrhea: Add Yi Yi Ren and Che Qian Zi.
- Abdominal mass: Add San Leng, E Zhu, and Zao Jiao Ci.
- Infertility: Add Wang Bu Liu Xing, Ji Xue Teng, and Lu Lu Tong.

Qi Deficiency and Blood Stasis

Treatment Principle: Benefit Qi, strengthen Spleen, nourish blood, and invigorate blood.

Use Si Jun Zi Tang [Four-Gentleman Decoction] plus Tao Hong Si Wu Tang [Four-Substance Decoction with Peach Pit and Safflower] plus Huang Qi and Yan Hu Suo.

Related conditions and modifications are

- Copious leukorrhea: Add Qian Shi and Cang Zhu.
- Lumps and mass: Add Ji Xue Teng and Ji Nei Jin.

Kidney Deficiency and Blood Stasis and Qi Stagnation

Treatment Principle: Tonify Kidney and expel stasis.

Use Zuo Gui Wan [Restore the Left (Kidney) Pill] plus Dan Shen, Ji Xue Teng, and Du Zhong.

Related conditions and modifications are

- Severe distention and soreness at lower back: Add Sang Ji Sheng, Xu Duan, and Gou Ji.
- Qi deficiency: Add Huang Qi and Dang Shen.

Qi Deficiency with Blood Stasis and Heat–Damp

Chronic long-term pelvic inflammation with lower abdominal pain with palpation or painful lumps and mass.

Treatment Principle: Tonify Qi, invigorate blood, transform stasis, clear heat, and eliminate dampness.

Use Yin Jia Wan [Silver Armor Pill] plus Si Jun Zi Tang [Four-Gentleman Decoction].

DYSMENORRHEA

Dysmenorrhea refers to pain during or right before the onset of menstruation. Six pathogenic factors, seven emotions, and internal organ injuries can cause disorders of the Chong and Ren channels, obstruct Qi and blood flow, or Chong, Ren, and uterus failing to be nourished can lead to pain. In Western medicine, this refers to lower abdominal pain, distention, and lumbar soreness or discomfort before, after, or during menstruation. In severe cases, there is excruciating pain in the lower abdomen and lumbar region, pale complexion, cold sweats, cold limbs, nausea and vomiting, whole body weakness, and even fainting. These symptoms occur with or sometimes following the menstrual cycle. TCM treatments focus on stopping pain when pain occurs by focusing on unblocking the channels and collaterals. During other times, the treatments are regulating Liver, tonifying Kidney, and regulating and supporting Chong and Ren channels.

Menstrual Pain Period

Liver Qi Stagnation Syndrome

Treatment Principle: Disperse Liver, scatter cold, and stop pain.

Use Si Ni San [Frigid Extremities Powder] plus Shi Xiao San [Sudden Smile Powder], Jin Ling Zi San [Melia Toosendan Powder] and Dang Gui, Chuan Xiong, Xiang Fu, Ai Ye, and Ban Xia.

Related conditions and modifications are

- Congealed cold and blood stasis: Remove Chuan Lian Zi, and add Gui Zhi, Wu Zhu Yu, and Xue Jie.
- Qi deficiency and blood stasis: Remove Chuan Lian Zi and Shi Xiao San [Sudden Smile Powder], and add Huang Qi, Tai Zi Shen, and Ji Xue Teng.
- Kidney Yang deficiency: Remove Chuan Lian Zi and Shi Xiao San [Sudden Smile Powder], and add Rou Gui, Xian Mao, Yin Yang Huo, and Bu Gu Zhi.
- Mix with damp–heat: Remove Dang Gui, Chuan Xiong, and Ai Ye, and add Mu Dan Pi, Huang Bai, and fried Guan Zhong.
- Excessive menstrual flow: Remove Dang Gui and Chuan Xiong, and add fried Di Yu and Hai Piao Xiao.

Congealed Cold and Blood Stasis Syndrome

Treatment Principle: Warm channels, scatter cold, invigorate blood, and transform stasis.

Use Wen Jing Tang [Warm the Menses Decoction], add San Leng, E Zhu, Ai Ye, and Xue Jie, and replace Sheng Jiang with Pao Jiang.

Related conditions and modifications are

- Qi stagnation and damp obstruction: Remove Mai Men Dong and Mu Dan Pi, and add Bing Lang, Bai Dou Kou, and Yi Yi Ren.
- Kidney Yang deficiency: Remove Mu Dan Pi and Mai Men Dong, and add Yin Yang Huo, Ba Ji Tian, and Bu Gu Zhi.
- Blood stasis with heat: Remove Dang Gui, Chuan Xiong, and Pao Jiang, add fried Di Yu, and increase Mai Men Dong by 20 g.

During Normal Period

This is a treatment to be used when there is no pain to prevent and reduce pain at next menstruation. In general, one should take three to six bags of the herbs during the 10-day period before menstruation.

Liver Qi Stagnation and Blood Stasis, Unregulated Chong, Ren Channels, and Qi and Blood

Treatment Principle: Disperse Liver and regulate Chong and Ren.

Use Chai Hu Shu Gan San [Bupleurum Powder to Spread Liver], add Dang Gui, Ji Xue Teng, Chong Wei Zi, Yu Jin, Tu Si Zi, Nu Zhen Zi, Fu Pen Zi, Tao Ren, and Yan Hu Suo.

Liver Qi Stagnation Transforming into Fire, Unregulated Chong, Ren Channels, and Qi and Blood

Treatment Principle: Clear Liver, drain fire, and regulate Chong and Ren Channels.

Dan Bai Si Ni San [Frigid Extremities Powder with Moutan and Phellodendron] plus Er Zhi Wan [Two-Ultimate Pill] and Shan Zhi Zi, Xiang Fu, Yu Jin, Ju He, Xuan Shen, Mai Men Dong, and He Shou Wu.

Damp–Heat Stagnation and Stasis

Treatment Principle: Clear heat, transform damp, and remove stasis.

Use Si Miao Wan [Four-Marvel Pill] plus Guan Zhong, Yin Chen Hao, Zhi Ke, Chong Wei Zi, Sheng Pu Huang, Tao Ren, Chuan Lian Zi, Dan Shen, and Xu Duan.

Blood Stasis Syndrome (Membranous Dysmenorrhea Blood Clots can be Observed with Endometrial Discharge)

Treatment Principle: Transform stasis and stop pain.

Use Wen Jing Tang [Warm the Menses Decoction] plus Bu Gu Zhi, Yin Yang Huo, Xue Jie, San Leng, and E Zhu.

ECLAMPSIA

This is discussed under swelling, dizziness, and epilepsy all due to fetus. There are Liver, Spleen, and Kidney deficiency with wind, fire, and phlegm. In Western medicine, this refers to hypertension, edema, and proteinuria that occurs after 20 weeks of pregnancy. In severe cases, there can be spasms, convulsions, and loss of consciousness. The cause of the condition is not clearly understood. TCM treatments are effective for patients at early stage and mild stages. The method is to nourish Yin, subdue Yang, strengthen Spleen, drain water, warm Kidney, and move water. One should be aware that eclampsia can occur 24 hours to 5 days after delivery.

Treatments According to Syndrome Differentiation for Mild Early Signs of Preclampsia

Spleen and Kidney Yang Deficiency Syndrome

Treatment Principle: Warm Kidney, strengthen Spleen, transform Qi, and move water.

Use Zhen Wu Tang [True Warrior Decoction] plus Si Ling San [Four-Ingredient Powder with Poria].

Related conditions and modifications are

- Mild Kidney Yang deficiency: Use Gui Zhi instead of Fu Zi.
- Patient with history of abortion or miscarriage: Remove Fu Zi, and add Ba Ji Tian, Xu Duan, and Tu Si Zi.
- Dizziness and vertigo: Add Gou Teng and Tian Ma.
- Poor appetite and loose stool: Add Shan Yao and Qian Shi.

Kidney and Liver Yin Deficiency with Hyperactive Liver Yang Syndrome

Treatment Principle: Nourish Yin, descend fire, and subdue Yang.

Use Qi Ju Di Huang Wan [Lycium Fruit, Chrysanthemum, and Rehmannia Pill] plus Xia Ku Cao, Mu Dan Pi, Gou Teng, Shi Jue Ming, and Gui Ban Jiao.

Related conditions and modifications are

- Severe headache and dizziness: Add Shan Yang Jiao powder and Bai Ji Li.
- Phlegm–heat: Add Zhu Ru and Dan Nan Xing.
- Severe edema: Add Bai Zhu, Zhu Ling, and Da Fu Pi.
- Lumbar soreness: Add Du Zhong and Sang Ji Sheng.

Spleen Deficiency with Excess Liver Syndrome

Treatment Principle: Strengthen Spleen, drain water, and calm Liver.

Use Ban Xia Bai Zhu Tian Ma Tang [Pinellia, Atractylodes Macrocephala, and Gastrodia Decoction] plus Gou Teng, Shi Jue Ming, Dan Shen, and Da Fu Pi.

Related conditions and modifications are

- Severe swelling and scanty urine: Add Zhu Ling, Chi Xiao Dou, and Ze Xie.
- Headache and distention: Add Bai Ji Li and Zhen Zhu Mu.
- Apathy: Add Dang Shen and Huang Qi.
- Copious phlegm with rapid breathing: Add Ban Xia and Zhu Ru.

Modern TCM Treatments for Mild Case of Preclampsia

Modern research believes that preclamsia belongs to blood stasis syndrome in TCM. Herbs that invigorate blood and transforms stasis can improve ischemia of the uterus, poor microcirculation of the placenta, and reduce spasms of whole body arterioles. Then, symptoms are alleviated and the progression of this condition is controlled.

Furthermore, if there is lower abdominal pain, limb pain, thirst but with no desire to drink. dark tongue with stasis spots, and accompanying mild preclamsia in the clinic, one should add blood invigorating and transforming stasis herbs after syndrome differentiation. This can increase the effectiveness. These herbs include Dang Gui, Dan Shen, Bai Shao Yao, Chi Shao Yao, Tao Ren, Chuan Xiong, Ru Xiang, Jiang Can, and Wu Gong.

Also, one can take Shen Ling Bai Zhu Wan [Ginseng, Poria, and Atractylodes Macrocephala Pill] together for edema. For headache and dizziness, one can add Qi Ju Di Huang Wan [Lycium Fruit, Chrysanthemum, and Rehmannia Pill] and Tian Ma Tou Teng Wan [Gastrodia Pill for Headache]. For blood stasis one can add Fu Fang Dan Shen Pian [Compound Salvia Tablet].

ENDOMETRIOSIS

This is discussed in TCM under painful menstruation and abdominal mass. It is caused by blood stasis and obstruction. In Western medicine, this refers to endometrial tissues which are found only in the uterine lining growing outside the uterus. It is seen more often in women between 30 and 40 years of age. It is often accompanied with irregular menstruation, infertility, pain during intercourse, and cyclic rectal irritation symptoms such as distention and sinking sensation at anus; or irritated urinary bladder symptoms such as frequent, urgent urination, or blood in the urine. TCM treatments focus on invigorating blood and transforming stasis. The goal of the treatment is to alleviate pain and eliminate abdominal mass. In general, one should take blood invigorating and transforming stasis formula seven days before menstruation because the onset of menstrual flow will remove the chronic stasis thus relieving pain. Even if the menstrual flow is excessive, one should still invigorate blood and stop bleeding.

External Endometriosis (Pelvic Cavity Endometriosis)

Progressive lower abdominal pain during menstruation; or with severe sinking sensation at the anus, urge for bowel movement; or with excess menstrual flow, copious blood clots or unceasing spotting.

Internal Treatment

Treatment Principle: Move Qi and invigorate blood.

Use Jia Jian Huo Luo Xiao Ling Dan [Modified Fantastically Effective Pill to Invigorate the Collaterals] plus Chai Hu, Zhi Ke, Yan Hu Suo, Xue Jie, Dang Gui, Chuan Xiong, and Xiang Fu.

Related conditions and modifications are

- Qi deficiency: Add Tai Zi Shen and Huang Qi.
- Kidney deficiency: Add Tu Si Zi and Yin Yang Huo.
- Cold–damp: Add Wu Zhu Yu, Gui Zhi, and Yi Yi Ren.
- Damp–heat: Remove Dang Gui and Chuan Xiong, and add Yin Chen Hao, Guan Zhong, Yi Yi Ren, and Huang Bai.

- Excess menstrual flow: Remove Dang Gui and Chuan Xiong, and add fried Pu Huang, Wu Ling Zhi, and fried Di Yu.
- Lower abdominal pain with palpation: Add Zao Jiao Ci, Xia Ku Cao, Tu Bie Chong, and Tao Ren.
- Ovarian cysts, adhesion, and painful lumps and mass: Add Li Zhi He, Kun Bu, Hai Zao, and fried Chuan Shan Jia.

External Applications

Hot poultice formula: Gui Zhi, Xi Xin, Ru Xiang, Mo Yao, Ju He, Bai Zhi, and Dang Gui. Grind all ingredients into powder and add Can Sha and place into a cloth bag. Steam until warm and place on lower abdomen, one to two times a day.

Internal Endometriosis

Infertility for many years, increased menstrual flow, prolonged menstruation, and progressively painful menstruation. One can palpate an enlarged and hardened uterus with pain upon palpation, especially worse during menstruation.

Treatment Principle: Move Qi, invigorate blood, and transform stasis.

Use Ge Xia Zhu Yu Tang [Drive Out Blood Stasis Below the Diaphragm Decoction] plus Yan Hu Suo, Ru Xiang, Mo Yao, Xue Jie, and Dan Shen.

Related conditions and modifications are

- After menstruation is complete: Add Gui Zhi Fu Ling Wan [Cinnamon Twig and Poria Pill].
- Severe heat: Remove Gui Zhi, and add Xia Ku Cao, Xuan Shen, and Jiao Shan Zhi Zi.

HYPERPLASIA OF BREAST (MAMMARY GLAND)

This condition is often caused by emotional disturbance and disorder of the Chong and Ren channels. Nipples belong to Stomach meridian and breasts belong to Liver channel. Thus the pathogeneses include Liver Qi stagnation and Spleen deficiency leading to Qi stagnation, phlegm accumulation, and blood stasis. In Western medicine it is that at the beginning stage, there are mild hyperplasia of surrounding lobules of mammary gland and swelling of the connective tissues at mammary gland. This painful breast condition is a mild case and can be self-limiting. For chronic types, there is recurrent edema and hyperplasia leading to irreversible structural changes that are mammary pathological conditions. In some people, it can also be self-eliminating after delivery and menopause. Some develop into cystic hyperplasia or even abnormal activity of hyperplasia duct of mammary glands leading to hyperplasia of breast. This condition often occurs in women who are young or middle-aged. The clinical characteristics are cyclic breast tenderness, which is often worse right before menstruation and better after menstruation. Pain can be sharp or dull with local painful spots upon palpation, radiating to armpit, shoulder, or hip area. Some patients find round, hard, small and scattered lumps in the breast mostly with an unclear border at upper lateral quadrant, and other patients may be asymptomatic. The size of the lumps can change with menstrual cycle, but the lumps do not disappear. TCM treatments focus on spreading Liver Qi, scattering phlegm accumulation, and unblocking stagnation and stasis and regulating Chong and Ren channels. One can use the formula below as the base and modify according to the presenting symptoms.

Use Chai Hu Shu Gan San Jia Jian [Modified Bupleurum Powder to Spread the Liver] plus Dang Gui, Chi Shao Yao, Gua Lou Pi, Zhe Bei Mu, Ju He, Yu Jin, Ban Xia, Xia Ku Cao, and Pu Gong Ying.

Related conditions and modifications are

- Chong and Ren channel disorders and unsmooth menstrual flow with scanty menstruation: Add Yi Mu Cao, Ji Xue Teng, and Chong Wei Zi.
- Excess menstruation: Add Er Zhi Wan [Two-Ultimate Pill] plus Guan Zhong.
- Delayed menstruation: Add Lu Jiao Shuang, Yin Yang Huo, and Tu Si Zi.
- Early menstruation: Add Mu Dan Pi, Shan Zhi Zi, and Di Gu Pi.
- Severe Qi stagnation and blood stasis and severe pain and distention: Add Chuan Lian Zi, Lu Lu Tong, and Yan Hu Suo.
- Hard lumps and severe phlegm and stasis: Add Mu Li, Zao Jiao Ci, and Chuan Shan Jia.
- Purulent and/or bloody discharge from nipple: Add Han Lian Cao and Sheng Di Huang.

After few weeks of treatment, if pain has decreased and lumps are softer and smaller, then one should continue with the same formula until swelling and lumps are eliminated and scattered. If there is no change, then one should search for other treatments.

INFERTILITY

Infertility is often caused by Kidney deficiency, damaged Chong and Ren channels, blood deficiency, Liver Qi stagnation, phlegm–damp, blood stasis, and damp–heat accumulation and stagnation. In Western medicine, this refers to childbearing women with normal sex life but unable to conceive for two years or two years from last pregnancy. This can be caused by abnormal ovary, fallopian tubes, and uterus or cervix. There can also be psychological and emotional factors. TCM treatments are effective for women under the age of 40 by focusing on tonifying the Kidney, benefiting essence and blood, regulating Chong and Ren channels, regulating menstruation, and unblocking channels and collaterals, and also treating Liver Qi stagnation, phlegm stasis, and damp–heat at the same time.

Infertility Due to Disorder of Ovarian Functions

Chong and Ren Blood Deficiency, and Failing to Nourish the Uterus

Treatment Principle: Nourish blood, nourish Kidney, and regulate menstruation.

Use Zuo Gui Wan [Restore the Left (Kidney) Pill] plus Dang Gui, Bai Shao Yao, Zi He Che, Du Zong, Xu Duan, Ji Xue Teng, Bai Dou Kou, and Xiang Fu. Continuously take 8 to 10 bags of this formula.

Phlegm–Damp Obstructing Chong and Ren Channels, and Obstructing the Uterine Channels

Treatment Principle: Dry damp, transform phlegm, regulate Qi, and regulate menstruation.

Use Dao Tan Tang [Guide the Phlegm Decoction] plus Cang Zhu, Xiang Fu, Shan Zha, Dang Gui, and Chuan Xiong.

Infertility Due to Obstruction in Fallopian Tubes (Postsurgery or After Inflammation)

Treatment Principle: Liver Qi stagnation, obstruction in uterine channels, and internal accumulation of damp–heat. This can be discussed in two conditions:

After Menstruation, before Rising of Basal Body Temperature

Treatment Principle: Spread Liver, unblock collaterals, eliminate damp, and clear heat.

Herbal formula consists of Chai Hu, Tu Bie Chong, fried Chuan Lian Zi, Zhi Chuan Shan Jia, Zao Jiao Ci, Ji Xue Teng, Lu Lu Tong, Sheng Pu Huang, Dan Shen, Chi Shao Yao, Bai Shao Yao, Guan Zhong, Yin Chen Hao, and Yi Yi Ren. Take 8 to 10 bags continuously.

Related conditions and modifications are

- Severe damp–heat: Add Huang Bai, Jin Yin Hua, and Pu Gong Ying.
- Severe cold–damp: Remove Chuan Lian Zi and Guan Zhong, and add Ai Ye, Lu Jiao Jiao, and Xiao Hui Xiang.
- Water accumulation at fallopian tube: Add Zhu Ling, Ze Xie, and Hu Po powder.
- Tuberculosis at fallopian tube: Add Zhe Bei Mu, Xuan Shen, Mu Li, and Xia Ku Cao.
- If there is ovulation after increased basal body temperature: Remove Chuan Shan Jia, Tu Bie Chong, Zao Jiao Ci, Sheng Pu Huang, and Dan Shen, and add herbs that tonify Kidney and supplement essence, such as Tu Si Zi, Zhi He Shou Wu, Gou Qi Zi, Nu Zhen Zi, and Sang Ji Sheng.

Before or during Menstruation

Treatment Principle: Spread Liver, nourish blood, invigorate blood, and regulate Chong and Ren channels.

Use Si Ni San [Frigid Extremities Powder] plus Si Wu Tang [Four-Substance Decoction] plus Yi Mu Cao and Qian Cao to expel stasis, generate new blood, and unblock the channels.

Infertility Due to Uterine Pathologies

This includes adhesion in the uterus (metrosynizesis), uterine leiomyoma, nodules formation at endometrium, and uterine hypoplasia. Please refer to treatments for amenorrhea and uterine amenorrhea.

Infertility Due to Pathologies of Uterine Cervix

Chronic Cervicitis

Sticky mucous at the cervix with high quantity of white blood cells, which obstruct the way of sperms to passing through. This belongs to damp–heat congesting and obstructing the uterus.

Clear heat and eliminate damp.

Formula consists of Huang Bai, Shan Zhi Zi, Mu Dan Pi, Fu Ling, Ze Xie, Yi Yi Ren, Yin Chen Hao, Guan Zhong, Chuan Niu Xi, and Zhi Ke. Take six to eight bags continuously.

NOTE: **Leiomyoma of the uterine cervix and polyps at cervix obstructing the opening of the uterus should be treated by Western medicine.**

Infertility Due to Immunity

In recent years, antisperm antibodies (ASAB) play an important role in infertility. According to clinical observations, this type of infertility is often related to Yin deficiency with hyperactive Yang and Kidney deficiency and blood stasis.

Yin Deficiency with Hyperactive Yang

Treatment Principle: Nourish Yin and guide the fire back to its source.

Use Liu Wei Di Huang Wan [Six-Ingredient Pill with Rehmannia] plus Dang Gui, Chi Shao Yao, Bai Shao Yao, and Gou Teng. Take one bag of this formula a day after finishing menstruation.

Kidney Deficiency and Blood Stasis

Treatment Principle: Tonify Kidney, supplement essence, invigorate blood, and transform stasis.

Use Gui Lu Er Xian Jiao [Tortoise Shell and Deer Antler Syrup] plus Xu Duan, Ba Ji Tian, Huang Qi, Xian Mao, Dan Shen, Chi Shao Yao, and Chong Wei Zi. One should take one bag of the formula per day after ovulation and before menstruation.

MENOPAUSAL SYNDROME

There is no corresponding TCM disease term to menopausal syndrome. However, from the clinical manifestations, the pathogeneses often include disorder of Kidney Yin and Yang leading to Liver and Kidney, Spleen and Kidney, and Heart and Kidney pathologies, which lead to various and complex manifestations of this condition. Menopause age is generally between 40 to 60 years. About one third of the females go through menopause without any symptoms. The other two thirds of women present with symptoms due to decreased sexual hormone level. On average, the symptoms can last from 1 to 4 years and can sometimes be 10 years long. The main manifestations are delayed irregular menstrual cycle, decreased menstrual flow, hot flushes, increased sweating, depression, easily angered, and uncontrolled emotions. TCM treatments focus on tonifying Kidney and regulating Heart, Liver, and Spleen to balance Yin and Yang. However, cold, cool, and greasy herbs are not appropriate to nourish Yin. Acrid dry herbs are not appropriate to support Yang. One should select neutral to moderate herbs in this case.

Liver and Kidney Yin Deficiency Syndrome

Treatment Principle: Nourish tonify Liver and Kidney, nourish Yin, and clear heat.

Use Zuo Gui Yin [Restore the Left (Kidney) Decoction] plus Er Zhi Wan [Two-Ultimate Pill] and Zhi He Shou Wu, Mai Men Dong, Zhi Mu, Huang Bai, and Rou Cong Rong.

Related conditions and modifications are

- Five-center heat and excessive sweating: Add Di Gu Pi, Bai Wei, and Qing Hao.
- Headache and vertigo: Add Tian Ma, Shi Jue Ming, and Gou Teng.
- Depression and hypochondriac distending pain: Add Yu Jin, Xiang Fu, and Chuan Lian Zi.
- Lumbar soreness and leg weakness: Add Du Zhong and Huai Niu Xi.

Heart and Kidney Disharmony Syndrome

Treatment Principle: Nourish Yin, descend fire, and harmonize Heart and Kidney.

Use Si Wu Tang [Four-Substance Decoction] plus Jiao Tai Tang [Grand Communication Pill], remove Chuan Xiong, and add Mai Men Dong, Sha Shen, Fu Shen, Yuan Zhi, Ye Jiao Teng, and Wu Wei Zi.

Related conditions and modifications are

- Insomnia and night sweats: Add Bai Zi Ren, Fu Xiao Mai, and Mu Li.
- Hyperactive Heart fire, yellow dark urine, and ulcerations on tongue and mouth: Add Mu Tong, Che Qian Zi, and Dan Zhu Ye.
- Sadness, crying, and abnormal emotional state: Change to Liu Wei Di Huang Wan [Six-Ingredient Pill with Rehamannia] plus Gan Mai Da Zao Tang [Licorice, Wheat, and Jujube Decoction].

Essence and Blood Deficiency Syndrome

Treatment Principle: Benefit essence and nourish blood.

Use Xiao Ying Jian [Minor Nourishing Decoction] plus Zi He Che, E Jiao, Bai Ji Li, He Shou Wu, and Xu Duan.

Related conditions and modifications are

- Severe skin itch: Add Mu Dan Pi, Bai Xian Pi, fried Jing Jie, and Chan Tui.
- Numbness and tingly muscle and skin: Add Ji Xue Teng and Si Gua Luo.
- Heel pain: Add Du Zhong, Sang Ji Sheng, and Chuan Niu Xi.

Kidney Yang Deficiency Syndrome

Treatment Principle: Warm Kidney, benefit Qi, and leach out damp.

Use You Gui Wan [Restore the Right (Kidney) Pill] plus fried Bai Zhu, Fu Ling, Bu Gu Zhi, and Dang Shen.

Related conditions and modifications are

- Apathy and poor appetite: Add Huang Qi, Mu Xiang, and Sha Ren.
- Swelling of the face and limbs, loose stool, and difficulty urinating: Add Ze Xie, Dong Gua Pi, and Yi Yi Ren.
- Frequent nocturia and urinary incontinence: Remove Fu Ling, and add Yi Zhi Ren, Sang Piao Xiao, and Wu Yao.
- Copious, clear, thin leukorrhea: Add Qian Shi, Lian Xu, and Hai Piao Xiao.

Kidney Yin and Yang Deficiency Syndrome

Treatment Principle: Tonify both Yin and Yang.

Use Er Xian Tang [Two-Immortal Decoction] plus Shan Zhu Yu, Ba Ji Tian, Fu Shen, Shan Zhi Zi, fried Bai Zhu, and Dang Shen.

The effect is better when combined with Tian Wang Bu Xin Dan [Emperor of Heaven's Special Pill to Tonify the Heart], once in the morning and at night.

All deficient syndromes above can have Liu Wei Di Huang Wan [Six-Ingredient Pill with Rehmannia] or Zhi Bai Di Huang Wan [Anemarrhena, Phellodendron, and Rehmannia Pill] added to those formulas.

MENSTRUAL IRREGULARITY AND UTERINE BLEEDING

This is discussed under early menstruation, late menstruation, early or late menstruation, excessive bleeding, and Beng and Lou in TCM. The root cause of the condition is at Kidney and Liver. The pathological changes occur at Qi and blood and the location of disease is at Chong and Ren channel. It often occurs with Liver stagnation, Spleen deficiency, blood stasis, and blood heat. In Western medicine, irregular menstruation and dysfunctional uterine bleeding are attributed to abnormal neuroendocrine function in charge of regulating the reproductive system. There are no organic changes of the whole body or reproductive organs. Conditions can be divided into anovulatory and ovulatory types. The former is often seen in puberty and menopause. The manifestations are irregular menstrual cycle or amenorrhea for a few weeks or a few months, followed by heavy bleeding which can last for a few weeks. The bleeding can also be intermittent, sometimes heavy, and sometimes scanty without pain. The latter is often seen in women during childbearing age, which can be further divided into inadequate luteal function and atrophy of the corpus luteum. With inadequate luteal function, the menstrual cycle is often shortened, with infertility or easily miscarriage. The latter has regular menstrual cycle and prolonged menstrual flow for about 10 days with heavy bleeding. In TCM, the treatments are regulating Chong and Ren channels, restoring the menstrual cycle (regulated with hormones), tonifying Kidney and supplementing essence (stimulating ovulation), and stopping heavy bleeding. Commonly used herbs to tonify Kidney and supplement essence are Yin Yang Huo, Fu Pen Zi, Tu Si Zi, Bu Gu Zhi, Shu Di, Gou Qi Zi, Shan Zhu Yu, Nu Zhen Zi, Lu Jiao Jiao, Xian Mao, Sang Ji Sheng, Xu Duan, and Ba Ji Tian. Herbs that can regulate menstruation and stop bleeding are Yi Mu Cao, Ai Ye, San Qi, Dang Gui, Chuan Xiong, Huang Qi, Shan Yao, and Xian He Cao.

Treatment during Menstrual Period

Menorrhagia, Irregular Menstrual Cycle, and Prolonged Menstruation

Kidney Yin Deficiency

Treatment Principle: Nourish Kidney and stabilize Chong channel

Use Liang Di Tang [Two-Rehmannia Decoction] plus Er Zhi Wan [Two-Ultimate Pill] and fried Di Yu, Hai Piao Xiao, Yi Mu Cao, and fried Guan Zhong.

Kidney Yang Deficiency

Treatment Principle: Warm Kidney and stabilize Chong channel

Use Gu Yin Jian [Stabilize the Yin Decoction] plus Lu Jiao Jiao, Bu Gu Zhi, Chi Shi Zhi, fried Ai Ye, and Xian Ling Pi.

Excessive, Unceasing, Watery Blood Flow

The bleeding time is long or without clear menstrual cycle (bleeding every day).

Spleen Deficiency

Treatment Principle: Tonify Spleen and stabilize Chong channel.

Use Ju Yuan Jian [Lift the Source Decoction] plus Shan Yao, Bai Dou Kou, Chen Pi, calcined Long Gu, calcined Mu Li, Hai Piao Xiao, and Jiang Tan.

Bright Red, Sticky Blood Flow with Burning Sensation

Early menstruation or prolonged menstrual bleeding.

Blood–Heat

Treatment Principle: Cool the blood and stabilize Chong channel.

Use Gu Jing Wan [Stabilize the Menses Pill] plus Sheng Di Huang, Mu Dan Pi, burned Shan Zhi Zi, fried Di Yu, fried Guan Zhong, Mai Men Dong, and Zong Lu Tan.

Heavy Flow or Irregular Flow (Sometimes Heavy and Sometimes Scanty)

Purple dark-colored blood, blood clots, abdominal distention and sharp pain which is better after passing the clots, irregular menstruation, and spotting.

Blood Stasis

Treatment Principle: Transform stasis and regulate Chong channel.

Use Si Wu Tang [Four-Substance Decoction] plus Shi Xiao San [Sudden Smile Powder] and Yi Mu Cao, Zhi Ke, and San Qi powder.

One can modify the above four types with the following:

- Distention and fullness at chest, hypochondriac, and breasts: Add Chai Hu, Yu Jin, and Zhi Ke.
- Cold pain at the sides of the lower abdomen and lumbar region: Add Gui Zhi, fried Ai Ye, and Ba Ji Tian.
- Damp–heat, red tongue, and yellow, thick, and greasy tongue coating: Add Yin Chen Hao, Guan Zhong, and Yi Yi Ren.

Treatment for Unceasing Spotting after Menstruation and Unclear Menstrual Cycle

Anovulatory Type

Unceasing bleeding or spotting which can last for a few decades or days. This belongs to blood stasis due to chronic leakage.

Treatment Principle: Transform stasis and stop bleeding.

Use Tao Hong Si Wu Tang [Four-Substance Decoction with Peach Pit and Safflower], remove Shu Di, and add Dan Shen, E Zhu, Qian Cao, Zhi Ke, Pu Huang, Yi Mu Cao, Guan Zhong, and Di Yu.

Ovulatory Type

Uterine bleeding during ovulation which may last until the onset of menstruation.

Yin Deficiency with Blood–Heat

Treatment Principle: Nourish Yin, cool blood, and stabilize Chong channel.

Use Liang Di Tang [Two-Rehmannia Decoction] plus Er Zhi Wan [Two-Ultimate Pill] and Di Gu Pi, Sang Ji Sheng, and Xu Duan.

Liver Qi Stagnation and Blood–Heat

Treatment Principle: Disperse Liver, clear heat, and regulate Chong channel.

Use Dan Zhi Xiao Yao San [Relax Wanderer Powder with Moutan and Gardenia] plus Xiang Fu, fried Chuan Lian Zi, and fried Di Yu.

Internal Accumulation of Damp–Heat

Treatment Principle: Clear heat, drain damp, and regulate the Middle Jiao.

Use Yi Huang Tang [Benefit the Yellow Powder] with Yin Chen Hao, Fu Ling, Ze Xie, Chi Shao Yao, Yi Yi Ren, Guan Zhong, and Qian Cao.

Maintenance during Normal Days

These are treatments for spotting which occurs after stopping the heavy bleeding with slight irregular menstrual cycle. In order to completely stop bleeding and restore regular menstrual cycle, one should tonify Kidney, nourish blood, and tonify Qi 5 to 11 days after menstrual period has stopped.

Use Zuo Gui Wan [Restore the Left (Kidney) Pill] plus Huang Qi, Chen Pi, and Fu Pen Zi.

Related conditions and modifications are

- More Yin deficiency: Add Mai Men Dong, Nu Zhen Zi, and Di Gu Pi.
- More Yang deficiency: Add Xian Mao and Yin Yang Huo.
- Spleen deficiency: Add Dang Shen and Bai Zhu.
- Blood-heat: Remove Shu Di Huang, Lu Jiao Jiao, Huang Qi, Chen Pi, and add Sheng Di Huang, Mu Dan Pi, and Di Gu Pi.
- Liver Qi stagnation: Add Chai Hu and Xiang Fu.
- With uterine myoma or pelvic inflammation: Change to Gui Zhi Fu Ling Wan [Cinnamon Twig and Poria Pill] plus Xia Ku Cao, Zhe Bei Mu, Pu Gong Ying, Guan Zhong, and Jin Yin Hua to invigorate the blood, soften hardness, clear heat, and eliminate damp.
- After 16 to 26 days of cessation of menstruation, one needs to tonify Kidney, invigorate blood, and regulate Liver: Hei Xiao Yao San [Black Relax Wanderer Powder] plus Ba Ji Tian, Bu Gu Zhi, Chuan Xiong, Tu Si Zi, Tao Ren, Hong Hua, and Xiang Fu.

MORNING SICKNESS

This is often caused by Stomach Qi unable to descend and rebellious Chong Qi. In Western medicine, this refers to a small number of pregnant women who react severely during the early stages of pregnancy causing persistent nausea, vomiting, and inability to eat or drink, leading to fluid loss and metabolic disorders. TCM treatments focus on harmonizing Stomach, descending rebellious Qi, strengthening Spleen, calming Chong channel, and calming the fetus.

Spleen and Stomach Deficiency Syndrome

Treatment Principle: Strengthen Spleen, harmonize Stomach, descend rebellious Qi, and stop nausea.

Use Xiang Sha Liu Jun Zi Tang [Cyperus and Amomum Powder to Calm the Stomach] plus Sheng Jiang.

Disharmony of Liver and Stomach Syndrome

Treatment Principle: Clear Liver, harmonize Stomach, and descend rebellious Qi.

Use Qin Lian Ju Ru Tang [Scutellaria, Coptis, Tangerine Peel, and Bamboo Shaving Decoction] plus Mai Men Dong and Sheng Jiang.

Qi and Yin Deficiency Syndrome

Treatment Principle: Benefit Qi, nourish Yin, harmonize Stomach, and stop nausea.

Use Sheng Mai San [Generate the Pulse Powder] plus Zeng Ye Tang [Increase the Fluids Decoction] and Lu Gen, Su Geng, Huang Qin, Shi Hu, and Yu Zhu.

All of the above three syndromes can be modified according to the presenting symptoms as follows:

- Spleen and Stomach deficiency with cold: Use Gan Jiang instead of Sheng Jiang.
- Severe phlegm damp: Add Huo Xiang, Gua Lou Pi, and Zhi Ke.
- Liver Qi stagnation: Add Xiang Fu and Yu Jin.
- Restless fetus, threatened miscarriage, bleeding or spotting during pregnancy: Remove Ban Xia, add Du Zhong, Sang Ji Sheng, Tu Si Zi, Bai Shao Yao, Xian He Cao, and E Jiao.
- If the vomit is dark brown in color: Add Ou Jie Tan.

Furthermore, from lab testing of the herbs, it can be seen that processed Ban Xia is not toxic. Thus, Ban Xia can be used to treat nausea and vomiting in pregnant women without a history of miscarriage. If the woman has a history of miscarriage or weak constitution, then one should use Ban Xia carefully or replace with Sheng Jiang.

PREMENSTRUAL SYNDROME (PMS)

Premenstrual syndrome is often caused by Liver Qi stagnation, insufficient Liver blood, and Spleen deficiency with damp obstruction or Kidney deficiency. In Western medicine, premenstrual syndrome refers to physical, emotional, and behavioral changes and symptoms that occur between ovulation and menstruation. Generally, it occurs 7 to 10 days before menstruation. There are 150 possible symptoms. Main symptoms are abdominal distention, breast tenderness and distention, headache, weight gain, body swelling and distention sensation; irritability, agitation, depression, nervousness, worry, insomnia, somnolence, changes in appetite and libido, emotional disturbance, and easily prone to crying. TCM treatments focus on dispersing Liver, regulating Qi, expelling stasis, transforming phlegm, nourishing blood, strengthening Spleen, and regulating Chong and Ren channels.

Liver Qi Stagnation Syndrome

Disperse Liver, relieve stagnation, regulate Qi, and stop pain.

Breast Distention and Tenderness

Use Chai Hu Shu Gan San [Bupleurum Powder to Spread the Liver] plus Chuan Lian Zi, Yu Jin, Yan Hu Suo, Chi Shao Yao, Lu Lu Tong, and Xia Ku Cao.

Abnormal Emotions

Use Xiao Yao San [Relax Wanderer Powder] plus He Huan Pi, Yu Jin, Yuan Zhi, Shan Zhi Zi, and Ye Jiao Teng.

Absent-Mindedness, Melancholy, and Crying

Use Gan Mai Da Zao Tang [Licorice, Wheat, and Jujube Decoction] plus Bai He and Sheng Di Huang.

Spleen Fail to Transport and Transform

Strengthen Spleen, tonify Qi, and leach out damp.

Diarrhea before and during Menstruation

Use Shen Ling Bai Zhu San [Ginseng, Poria, and Atractylodes Macrocephala Powder] plus Pao Jiang and Chen Pi.

Related conditions and modifications are

- Severe diarrhea: Add Wu Zhu Yu and Wu Wei Zi.
- Diarrhea triggered by pain and is relieved after diarrhea: Add Tong Xie Yao Fang [Important Formula for Painful Diarrhea].

Swelling and Edema before and during Menstruation

Use Si Jun Zi Tang [Four-Gentleman Decoction] with Wu Pi Yin [Five-Peel Decoction] and Dong Gua Pi.

Related conditions and modifications are

- Aversion to cold, soreness, and weak lower back and legs: Add Ba Ji Tian and Gui Zhi.
- Severe breast distention and pain: Add Lu Lu Tong and Yu Jin.
- Severe lower limb swelling: Add Fang Ji.

UTERINE FIBROID

This condition belongs to abdominal mass. It is often caused by Qi stagnation, blood stasis, and phlegm accumulation in the uterus. In Western medicine, this is the most common benign tumor condition in female reproductive system. It often occurs in women between 30 to 50 years and the frequency of occurrence is 20%. It was commonly thought that uterine fibroid was related to long-term excessive use of estrogen. The clinical characteristics for most patients are increased menstrual flow, prolonged menstruation, shortened cycle, and irregular vaginal bleeding leading to anemia, infertility, and compressed sensation on the urinary bladder or rectum. Upon palpation, doctors find a lump or mass in the lower abdominal region. This condition is chronic with slow changes, thus patients need to take the herbs for a long period of time. TCM treatments focus on invigorating blood, regulating Qi, breaking up stasis, eliminating abdominal mass, softening hardness, and scattering accumulation. If the uterus size is bigger than the size of a 12-week pregnant woman and there is excessive menstrual flow, one should be treated with surgery.

Treatments According to Stages

Treatments during Menstruation

Treatment Principle: Benefit Qi, nourish blood, stop bleeding, and transform stasis.

Formula consists of Dang Shen, Huang Qi, Zhi He Shou Wu, Mu Li, Yi Mu Cao, Li Zhi He, Bai Zhu, Xue Yu Tan, fried Pu Huang, Sheng Ma, Qian Cao Gen, and Wu Zei Gu.

One can also make this into pills, which are easier to take long term.

Related conditions and modifications are

- Severe abdominal pain: Add Yan Hu Suo and Wu Ling Zhi.
- Lumbar soreness: Add Tu Si Zi and Xu Duan.
- Cold abdominal pain: Add Gui Zhi, Xiao Hui Xiang, and Wu Yao.
- Yin deficiency: Remove Huang Qi and add Er Zhi Wan [Two-Ultimate Pill] and Mai Men Dong.

Treatments for Nonmenstrual Period

Treatment Principle: Regulate Qi, invigorate blood, break up stasis, eliminate mass, soften hardness, and scatter accumulation.

Formula consists of Li Zhi He, Mu Li, Bie Jia, Ju He, Zhi He Shou Wu, Zhe Bei Mu, Hai Zao, Wu Yao, San Leng, E Zhu, Wu Ling Zhi, and Qing Pi.

One can also make this into pills which is easier to take long term.

Related conditions and modifications are

- Qi and blood deficiency: Add Dang Shen, Ji Xue Teng, and Bai Zhu.
- Kidney deficiency: Add Du Zhong, Xu Duan, and Sang Ji Sheng.
- Severe cold: Add Xiao Hui Xiang, Wu Zhu Yu, Jin Ying Zi, and Qian Shi.
- Severe damp–heat: Add Huang Bai, Hong Teng, and Yi Yi Ren.

Treatments According to Syndrome Differentiation

Qi Stagnation and Blood Stasis Syndrome

Treatment Principle: Regulate Qi, invigorate blood, eliminate mass, and scatter accumulation.

Use Ge Xia Zhu Yu Tang [Drive Out Blood Stasis Below the Diaphragm Decoction] plus Lu Lu Tong, Yan Hu Suo, and Li Zhi He.

Congealed Cold and Blood Stasis Syndrome

Treatment Principle: Warm Yang, scatter cold, transform stasis, and eliminate mass.

Use Gui Zhi Fu Ling Wan [Cinnamon Twig and Poria Pill] plus E Zhu, Qing Pi, Wu Yao, Huang Qi, Mu Li, Bie Jia, and Xiao Hui Xiang.

Qi Deficiency and Blood Stasis Syndrome

Treatment Principle: Tonify Qi, strengthen Spleen, transform stasis, and scatter accumulation.

Use Ba Zhen Tang [Eight-Treasure Decoction] plus Dan Shen, Ji Nei Jin, Sheng Ma, Huang Qi, and E Zhu.

Yin Deficiency and Blood Stasis Syndrome

Treatment Principle: Nourish Yin, clear heat, transform stasis, and eliminate mass.

Use Liu Wei Di Huang Wan [Six-Ingredient Pill with Rehmannia] plus Xuan Shen, Mai Men Dong, Sha Shen, Nu Zhen Zi, Gui Ban, Bai Shao Yao, Di Gu Pi, and Mu Li.

Food Therapy

Steamed Egg with San Qi

Use 3 g of San Qi powder, one small cup of Ou Zhi (lotus juice), and one egg. Mix all ingredients and steam until cooked. This can tonify Qi, nourish blood, transform stasis, and stop bleeding.

Tao Ren Zhou [Peach Pit Congee]

Use decoct of 10 g Tao Ren and remove dreg. Cook 30 g of rice with the decoction to make congee. This can eliminate mass and scatter accumulation.

VAGINITIS

This condition is discussed under vaginal itching and leukorrhea in TCM. It is mainly caused by damp–heat pouring downward. The pathological locations are Liver, Spleen, and Dai channel. In Western medicine, vaginitis is an inflammatory condition caused by invasion of various pathogens. The common types are trichomonas vaginitis, candida vaginitis, and senile vaginitis. TCM treatments focus on clearing heat, draining damp, killing pathogens, and stopping itch.

Trichomonal Vaginitis

This is caused by trichomonas vaginalis. It often reproduces before and after menstruation leading to inflammatory symptoms. The manifestations are thin, bubbly, and white vaginal discharge, itching or burning sensation at the vulva, or pain during intercourse.

Clear heat, drain damp, kill pathogen, and stop itch.

Internal Treatment

Use Yi Huang Tang [Benefit the Yellow Decoction] plus Shan Zhi Zi, Mu Dan Pi, Yi Yi Ren, Fu Ling, Bai Xian Pi, Guan Zhong, Pu Gong Ying, and Ze Xie.

External Applications

1. Sitz bath: Wu Mei, Huang Bai, Pu Gong Ying, She Chuang Zi, Bai Xian Pi, and Qin Pi, 15 g each. Remove dregs from decoction. Fumigate the area first followed by sitz bath, one to two times a day.
2. Vaginal wash: Sheng Wu Mei 10 g. Make decoction to be used as a vaginal wash. Wash the area once a day for 10 days as one course.

Candidal Vaginitis

This is often seen after long-term use of antibiotics, oral contraceptives, or in diabetic patients. The manifestations are severe unbearable itching of the vulva, burning pain, copious leukorrhea, and restlessness while sitting or lying down. The discharge is milky white like tofu curd.

Clear heat, drain damp, kill pathogens, and stop itching.

Internal Treatment

Same as treatment for trichomonal vaginitis.

External Applications

1. Vaginal wash: She Chuang Zi 15 g. Use the decoction to wash vaginal area. Twice a day for 10 days as one course.
2. Sitz bath: She Chuang Zi, Ku Shen, Mu Jin Pi, Xiao Ji, Chuan Jiao, and Ai Ye, 10 g each. Remove dregs from decoction. Fumigate area first then follow by sitz bath, once a day for 7 days as one course.

Postmenopausal Vaginitis

This is more often seen in menopausal women. The manifestations are itching at vulva or burning pain, increased vaginal discharge which is light yellow in color, or bloody purulent leukorrhea.

Nourish Kidney Yin, clear heat, and eliminate damp.

Internal Treatment

Use Zhi Bai Di Huang Wan [Anemarrhena, Phellodendron, and Rehmannia Pill] plus Jin Yin Hua, Guan Zhong, Che Qian Zi, and Bai Xian Pi.

Related conditions and modifications are

- Constipation: Add Huo Ma Ren and Xing Ren.
- Leukorrhea with blood: Add Han Lian Cao, Qian Cao Gen, Du Zhong, and Xu Duan.
- Dry mouth, throat, and vagina: Add Mai Men Dong, Xuan Shen, Shi Hu, and Xi Yang Shen.

External Applications

1. External application of Zi Cao You [Arnebia Ointment]. Two to three times a day to treat burning pain at vulva.
2. Sitz bath: She Chuang Zi, Bai Xian Pi, Di Fu Zi, Huang Bai, and Wu Wei Zi, 15 g each. Remove dregs from decoction. Fumigate area first then follow with sitz bath, once a day for 10 days as one course.

CHAPTER 20

Immune System Disorders

AIDS (ACQUIRED IMMUNODEFICIENCY SYNDROME)

This condition is similar to consumptive and epidemic disease. It is an infectious disease caused by the HIV virus. At stage one, there can be asymptomatic or chronic lymphadenopathy. At stage two, there is irregular fever, night sweats, diarrhea, emaciation, and swelling of the lymph nodes all over the body. At stage three, there are whole body symptoms and severe infection. TCM treatment principles benefit Qi, nourish Yin, clear heat, and relieve toxicity, mainly acting to improve symptoms, stabilize the condition, and prolong and improve quality of life. Tian Hua Fen and its extract is the main herb for anti-HIV virus and can be added into any formulas.

Treatments According to Stages

Beginning Stage (Hidden Evil Pathogens)

Treatment Principle: Separate, eliminate, and expel evil pathogens.

Use Sheng Ma Ge Gen Tang [Cimicifuga and Kudzu Decoction] plus Sheng Jiang San [Ascending and Descending Powder] and Cang Zhu and Fang Feng.

Late Stage (Zheng Qi Deficiency)

Treatment Principle: Support the source and consolidate the root.

Use Gui Lu Er Xian Jiao [Tortoise Shell and Deer Antler Syrup] plus Huang Qi, Tu Si Zi, Ku Shen, Nu Zhen Zi, Zhi Gan Cao, Dong Chong Xia Cao, and Tian Hua Fen.

Treatment According to Syndrome Differentiation

Lung and Stomach Yin Deficiency

Treatment Principle: Benefit Qi, nourish Yin, clear heat, and transform phlegm.

Use Shen Ling Bai Zhu San [Ginseng, Poria, and Atractylodes Macrocephala Powder] plus Bai He Gu Jin Tang [Lily Bulb Decoction to Preserve the Metal].

Spleen and Stomach Deficiency Syndrome

Treatment Principle: Strengthen Spleen, benefit Qi, harmonize Stomach, and stop diarrhea.

Use Bu Zhong Yi Qi Tang [Tonify the Center and Augment the Qi Decoction], Wen Dan Tang [Warm the Gallbladder Decoction], and Xiao Chai Hu Tang [Minor Bupleurum Decoction].

Spleen and Kidney Deficiency Syndrome

Treatment Principle: Benefit Qi, strengthen Spleen, and warm Kidney.

Use Si Jun Zi Tang [Four-Gentleman Decoction], Si Shen Wan [Four-Miracle Pill] and Huang Qi, Tu Si Zi, and Huang Lian.

Excess Heat with Phlegm Misting Syndrome

Treatment Principle: Clear heat, transform phlegm, extinguish wind, and open orifices.

Use Ling Jiao Gou Teng Yin [Antelope Horn and Uncaria Decoction] plus An Gong Niu Huang Wan [Calm the Palace Pill with Cattle Gallstone]. Take one pill, twice a day.

Treatment According to Symptoms

Fever

- Middle Jiao Qi deficiency: Use Bu Zhong Yi Qi Tang [Tonify the Middle and Augment the Qi Decoction].
- Yin damage: Qing Hao Bie Jia Tang [Artemisia Annua and Soft-Shelled Turtle Shell Decoction], Qing Gu San [Cool the Bone Powder], and Qin Jiao Bie Jia Tang [Gentiana Qinjiao and Soft-Shelled Turtle Shell Decoction].
- Exterior syndrome: Use Sheng Ma Ge Gen Tang [Cimicifuga and Kudzu Decoction] or Yin Qiao San [Honeysuckle and Forsythia Powder] and Sheng Jiang San [Ascending and Descending Powder].
- Evil pathogens at Shao Yang: Use Xiao Chai Hu Tang [Minor Bupleurum Decoction] plus Sheng Jiang San [Ascending and Descending Powder].
- Damp–heat obstructing San Jiao: Use Gan Lu Xiao Du Dan [Sweet Dew Special Pill to Eliminate Toxins] plus Sheng Jiang San [Ascending and Descending Powder].

Diarrhea

- Closed interior due to hidden evil pathogens: Use Sheng Ma Ge Gen Tang [Cimicifuga and Kudzu Decoction] plus Wu Ling San [Five-Ingredient Powder with Poria].
- Damp–heat obstructing Middle Jiao: Use Ban Xia Xie Xin Tang [Pinellia Decoction to Drain the Epigastrium] plus Gan Lu Xiao Du Dan [Sweet Dew Special Pill to Eliminate Toxins].
- Spleen and Stomach deficiency: Use Li Zhong Wan [Regulate the Middle Pill] plus Si Shen Wan [Four-Miracle Pill].
- Chronic diarrhea with mixed hot and cold: Use Wu Mei Wan [Mume Pill].

LUPUS ERYTHEMATOSUS

There is no equivalent disease name in ancient Chinese. In modern times, this condition is caused by internal emotional damages, Liver Qi stagnation, Qi stagnation, and congealed blood manifesting as red spots in TCM. At the late stage, there is Qi and Yin deficiency, Liver and Kidney exhaustion, mixed deficiency, and excess and mixed cold and heat. Western medicine believes dysfunction of the autoimmune system affect connective tissues and multiple organs in the body. It is more common in females. The ratio between female and male patients is about 8:1. It is commonly seen between the age of 20 to 40. The clinical manifestations are slow onset with various symptoms. Discoid erythema occurs on the face and the ears and is worse with sunlight and fatigue. Systematic erythema is most commonly seen as butterfly erythema on the face. Fever is most often seen as chronic low-grade fever. There is joint pain which often affects fingers and toes, kidney damage such as nephritis, pericarditis, myocarditis, pleurisy, and anemia. TCM treatment of acute stage focuses on clearing Ying and relieving toxicity. At chronic stage, treatment focuses on benefiting Qi and nourishing Liver and Kidney. TCM treatment is more effective in treating discoid erythema. In recent years, Lei Gong Teng and its prepared formulas have been used to treat Lupus erythematosus. It can control the condition from further worsening and also protect the joints and Zang organs.

Treatments According to Syndrome Differentiation

Qi Stagnation and Blood Stasis Syndrome

Treatment Principle: Disperse Liver, regulate Qi, invigorate blood, and transform stasis.

Use Xiao Yao San [Relax Wanderer Powder] plus Tao Hong Si Wu Tang [Four-Substance Decoction with Peach Pit and Safflower] plus Zhi Ke, Chen Pi, and Yu Jin.

Excess Heat Toxin Syndrome

Treatment Principle: Cool blood, relieve toxicity, clear heat, and generate fluid.

Use Qing Wen Bai Du Yin [Clear Epidemics and Overcome Toxin Decoction] plus Bai Mao Gen, Lu Gen, and Jin Yin Hua.

Qi and Yin Deficiency Syndrome

Treatment Principle: Benefit Qi, nourish Yin, and tonify Liver and Kidney.

Use Mai Wei Di Huang Wan [Ophiopogonis, Schisandra, and Rehmannia Pill] plus Sha Shen, Dan Shen, Han Lian Cao, Gou Qi Zi, Zhi Mu, Huang Bai, Di Gu Pi, and Sang Shen.

Spleen and Heart Deficiency Syndrome

Treatment Principle: Nourish Heart and strengthen Spleen.

Use Gui Pi Tang [Restore the Spleen Decoction] plus Shu Di Huang and Shan Zhu Yu.

Evil Heat Damaging Liver Syndrome

Treatment Principle: Clear heat, nourish Liver, regulate Qi, and invigorate blood.

Use Yi Guan Jian [Linking Decoction] plus Dan Shen, Chi Shao Yao, He Shou Wu, Yu Zhu, Shi Hu, Gou Qi Zi, Bie Jia, Bai Zhu, Fu Ling, and Yan Hu Suo.

Internal Disturbance of Liver Wind Syndrome

Treatment Principle: Cool Liver and extinguish wind.

Use Ling Jiao Gou Teng Tang [Antelope Horn and Uncaria Decoction] plus Mu Dan Pi, Shi Chang Pu, Lian Qiao, Huang Lian, and Yuan Zhi.

Spleen and Kidney Yang Deficiency Syndrome

Treatment Principle: Warm and tonify Spleen and Kidney and promote urination.

Use Shi Pi Yin [Bolster the Spleen Decoction] plus Zhi Huang Qi, Dang Shen, and Xian Ling Pi.

Maintenance

1. Gradually reduce the herbal dosage as the condition improves. For example, one can reduce from one dose a day to one dose every other day and then every 3 days.
2. During remission period, one can also change to patent medicine, such as Liu Wei Di Huang Wan [Six-Flavor Rehmannia Pill], Zhi Bai Di Huang Wan [Anemarrhena, Phellodendron, and Rehmannia Pill], Mai Wei Di Huang Wan [Ophiopogonis, Schisandra, and Rehmannia Pill], Jin Gui Shen Qi Wan [Kidney Qi Pill from Golden Cabinet], Gui Pi Wan [Restore the Spleen Pill], Xiao Yao Wan [Relax Wanderer Pill], and Da Bu Yin Wan [Great Tonify the Yin Pill].
3. Avoid direct sun and ultraviolet exposure.
4. Women of child-bearing age should be on birth control during the active stage of this condition.

RHEUMATIC ARTHRITIS

This condition belongs to heat Bi, fixed Bi, painful Bi, and wandering Bi. It is often caused by wind–cold damp and heat evil invading the body, accumulating at the joints, and obstructing Qi and blood flow. In Western medicine, this is a whole body pathological reaction to external streptococcus. There might also be correlation of viral infection to this condition. TCM treatments are expelling wind, clearing heat, transforming dampness, scattering cold, and unblocking collaterals. From modern research, herbs that have antistreptococcus functions are Ju Hua, Jin Yin Hua, Lian Qiao, Da Qing Ye, Ban Lan Gen, Xia Ku Cao, Bai Jiang Cao, Chuan Xin Lian, Huang Qin, Huang Lian, Huang Bai, Da Huang, Shan Zhi Zi, Chi Shao Yao, Bai Shao Yao, Zhi Mu, Tian Men Dong, Hou Po, and Dang Gui. Herbs that have steroid functions are Qin Pi, Qin Jiao, Fang Ji, Hai Feng Teng, Xu Chang Qing, She Xiang, Jiang Can, Gan Cao, Huang Qi, Ren Shen Ye, San Qi, Ci Wu Jia, He Shou Wu, Wu Wei Zi, Yu Zhu, Fu Zi, Wu Tou, and Ba Ji Tian. Herbs that can reduce heat and stop pain are Sheng Di Huang, Wu Jia Pi, Xu Duan, Du Zhong, Niu Xi, Fang Feng, Qiang Huo, Du Huo, Xi Xin, and Shi Nan Teng. One can add these herbs to the main formula according to the presentations.

Damp–Heat Type

Treatment Principle: Clear heat, expel dampness, disperse wind, and unblock collaterals.

Use Xuan Bi Tang [Disband Painful Obstruction Decoction] plus Fu Ling, Dang Gui, Huang Bai, Cang Zhu, Hai Tong Pi, Xi Xian Cao, and Qin Jiao.

Related conditions and modifications are

- Severe heat: Remove Dang Gui, and add Huang Qin and Ren Dong Teng.
- Severe dampness: Add Bei Xie and Wu Jia Pi.
- Severe painful limbs: Add Wei Ling Xian and Sang Zhi. If severe at upper limbs, add Gui Zhi and Jiang Huang. If severe at lower limbs, add Sang Ji Sheng and Chuan Niu Xi.
- Red erythema at skin: Add Mu Dan Pi and Chi Shao Yao.
- Subcutaneous small nodules: Add Mu Tong, Chi Shao Yao, and Tao Ren.

Wind–Cold–Damp–Heat Syndrome

Treatment Principle: Expel wind, scatter cold, clear heat, eliminate dampness, invigorate collaterals, and stop pain.

Use Juan Bi Tang [Remove Painful Obstruction Decoction] plus Du Huo, Gui Zhi, Chuan Xiong, Qin Jiao, Hai Feng Teng, Sang Zhi, and Ji Xue Teng.

Related conditions and modifications are

- Swollen and heavy joints with fixed pain: Add Fang Ji, Yi Yi Ren, and Mu Gua.
- Wandering pain: Add Jing Jie. In severe cases, add Du Huo and Qin Jiao up to 20 g.
- Severe, red, and swollen painful joints: Add Ren Dong Teng, Huang Bai, and Zhi Mu.
- Severe and sharp pain improving with heat: Increase Gui Zhi to 20 g and add Xi Xin and Ma Huang.
- Chronic and recurrently painful joints: Add Dang Shen, Huang Qi, Du Zhong, Sang Ji Sheng, Yin Yang Huo, and Ba Ji Tian.

Furthermore, when the condition improves, one can also take Yu Ping Feng San [Jade Windscreen Powder] or Si Jun Zi Tang [Four-Gentleman Decoction] for 1 to 2 months to strengthen the constitution, and prevent cold and recurrence of the condition.

RHEUMATOID ARTHRITIS

This condition is discussed under Bi syndrome and Li Jie Feng. It is caused by repeated attacks of wind–cold damp evil to patients with weak Zheng Qi leading to obstruction of the tendons and channels. The pathogeneses are Qi stagnation, congealed blood and joints failing to be nourished. In Western medicine, the characteristic of this whole body autoimmune disease is chronic inflammation of the joints and tissue. The cause and pathology are not fully understood at this time. The onset of this condition is often latent, often with fatigue, weight loss, poor appetite, low-grade fever, and numbness or sharp pain of hands and feet. Following these symptoms are joint stiffness, especially worse in the morning, multiple joint swelling and pain, and symmetrical, wandering pain. From the lab examination, patients show positive rheumatoid factor. This condition occurs more in females than males and the ratio is about 3:1. It often occurs in people older than 30 years of age and peaks around menopausal period. TCM treatments are expelling wind, eliminating dampness, warming the channels, scattering cold, invigorating blood, unblocking collaterals, nourishing Yin, clearing heat, tonifying Qi, nourishing blood, and tonifying and nourishing Liver and Kidney. During remission, one should protect the muscles and joints from irreversible pathological changes from happening. Thus, take the herbs internally, and combine with external wash or paste and acupuncture. By adding Lei Gong Teng (15 to 30 g) to the internal formula, one can increase the effectiveness.

Treatments According to Different Stages

Active Stage

Damp–Heat Accumulation Syndrome

Treatment Principle: Clear heat, relieve toxicity, drain damp, and unblock collaterals.

Use Si Miao San [Four-Marvel Powder] plus Jiang Huang, Wei Ling Xian, Yu Zhu, Ren Dong Teng, Yin Chen Hao, Hai Tong Pi, and Sheng Gan Cao.

Related conditions and modifications are

- Severe swollen burning pain of the joints, worse at night and better during the day: Add San Qi powder, Chuan Shan Jia, and Shui Niu Jiao.
- Red erythema at skin: Add Mu Dan Pi, Sheng Di Huang, Chi Shao Yao, and Di Fu Zi.
- Excess heat damaging Yin: Add Sheng Di Huang, Xuan Shen, and Mai Men Dong.

Mixed Cold and Heat Syndrome

Treatment Principle: Warm the channels, scatter cold, clear heat, and unblock collaterals.

Use Gui Zhi Shao Yao Zhi Mu Tang [Cinnamon Twig, Peony, and Anemarrhena Decoction] plus Xi Xin, Wei Ling Xian, and Dang Gui.

Related conditions and modifications are

- Headache: Add Jing Jie and Bai Zhi.
- Cold pain at joints with difficulty in extension or flexion: Zhi Fu Zi and Quan Xie.
- Red, swollen, and hot joints: Add Sheng Shi Gao and Huang Bai.
- Severe swollen joints: Cang Zhu and Fang Ji.

Yin Deficiency, Blood Stasis and Heat Syndrome

Nourish Yin, clear heat, invigorate blood, and unblock collaterals.

Use Shen Tong Zhu Yu Tang [Drive Out Blood Stasis from a Painful Body] plus Sha Shen, Mai Men Dong, and Di Gu Pi.

Remission Period

Liver and Kidney Exhaustion with Stagnation Syndrome

Treatment Principle: Nourish Liver and Kidney, expel stasis, and unblock collaterals.

Use Du Huo Ji Sheng Tang [Angelica Pubescens and Sangjisheng Decoction].

Related conditions and modifications are

- Dizziness and vertigo: Add Gou Qi Zi and Ju Hua.
- Sore and painful lower back: Add Gou Ji and Xu Duan.
- Deformity of the joints and difficult to move: Add Wu Gong, Bai Hua She, and Tu Bie Chong.

Qi and Blood Insufficiency and Phlegm Stasis Syndrome

Treatment Principle: Tonify Qi, nourish blood, drive out phlegm, and expel stasis.

Use Dang Gui Bu Xue Tang [Tangkuei Decoction to Tonify the Blood] plus Shu Di Huang, Bai Shao Yao, Ji Xue Teng, Dang Shen, Chen Pi, Bai Jie Zi, Chuan Xiong, and Di Long.

External Herbal Wash

Fumigation and Wash Formula

Use Sang Zhi, Dang Gui, Gui Zhi, Ji Xue Teng, Si Gua Luo, Qiang Huo, Du Huo, Ru Xiang, Mo Yao, Chuan Xiong, and Xi Xin. Decoct. Fumigate first then wash, twice a day.

Herbal Rubbing Alcohol Formula

Use Chuan Wu, Dang Gui, Chi Shao Yao, Dan Shen, Gan Jiang, Xi Xin, and Hong Hua, soak in 50% alcohol for 2 weeks. Then the herbal alcohol can be used to rub the affected area, two to three times a day.

SJOGREN'S SYNDROME

This is discussed under consumptive disease. It is caused by damaged true Yin, internal deficiency of essence and blood, dryness, and exhaustion of the blood sea. These all lead to dryness symptoms. In Western medicine, most people believe that it is an autoimmune disease, which most often attacks the salivary and lacrimal glands. This condition is seen more in women. The clinical manifestations are dry eyes, dry mouth, and also dry nose, trachea, bronchi, vaginal dryness, and dry skin. There is often kerato–conjunctivitis sicca and symmetrical chronic swelling of the parotid glands. TCM treatments focus on generating fluid and moistening dryness. It is effective in treating primary drying syndrome.

Both Qi and Yin Deficiency

Treatment Principle: Nourish Yin, benefit Qi, and generate fluid.

Use Zuo Gui Yin [Restore the Left (Kidney) Pill] plus Tai Zi Shen, Sha Shen, Huang Jing, Huang Qi, Bai Zhu, Mai Men Dong, Shi Hu, and Xuan Shen.

Yin Deficiency with Internal Heat Syndrome

Treatment Principle: Nourish Yin, clear heat, generate fluid, and moisten dryness.

Use Yi Guan Jian [Linking Decoction] plus Tai Zi Shen, Mai Ya, Zhi Mu, Shi Hu, Tian Hua Fen, Tian Men Dong, Huang Lian, and Gan Cao.

Damp–Heat Obstruction and Accumulation Syndrome

Treatment Principle: Clear heat, drain damp, strengthen Spleen, and harmonize Stomach.

Use Er Miao San [Two-Marvel Powder] plus Ping Wei San [Calm the Stomach Powder], Er Chen Tang [Two-Cured Decoction] plus Jin Yin Hua, Lian Qiao, Bei Xie, Huo Xiang, Xia Ku Cao, Tu Fu Ling, and Yin Chen Hao.

Related condition and modification are

- Constipation: Add Da Huang.

Wind–Heat Transforming into Toxins Syndrome

Treatment Principle: Clear heat, relieve toxicity, disperse wind, and generate fluid.

Use Yi Wei Tang [Benefit the Stomach Decoction] plus Yin Qiao San [Honeysuckle and Forsythia Powder] and Da Qing Ye, Ban Lan Gen, Jiang Can, Shi Hu, and Ge Gen.

Related condition and modification are

- High fever: Add Shi Gao.

Yang Deficiency and Internal Cold Syndrome

Treatment Principle: Warm Yang, expel cold, and transform water.

Use Jin Gui Shen Qi Wan [Kidney Qi Pill from the Golden Cabinet] plus Li Zhong Wan [Regulate the Middle Pill].

One can modify the above formulas with the following according to syndrome differentiation.

- Dry eyes with ulcerations: Add Gu Jing Cao, Pu Gong Ying, Ju Hua, and Chan Tui.
- Scanty menstrual flow and severe sharp pain in the lower abdomen: Add Yan Hu Suo, Xiang Fu, Wu Ling Zhi, and Pu Huang.
- Yin deficiency and internal heat: Add Di Gu Pi, Bai Wei, Zhi Mu, and Gui Ban.
- Rashes on skin, dry nose with cracks and bleeding, and excessive menstrual flow: Add Ou Jie, Xian He Cao, Bai Mao Gen, Zi Cao, Qian Cao, and Xiao Ji.
- Enlarged Liver and Spleen: Add Bie Jia, Dan Shen, Da Huang, Yu Jin, and Tu Bie Chong.
- Dark complexion, and scaly skin with purple spots: Add Dan Shen, Chi Shao Yao, Tao Ren, Hong Hua, Chuan Xiong, and Su Mu.

CHAPTER 21

Infectious Disorders

ANCYLOSTOMIASIS

Hookworm is called Fu Chong in TCM. Ancylostomiasis, hookworm infestation in the West, is called Huang Pang Bing or Gan Huang in TCM. Hookworm toxins invading the skin and muscle, obstructing Ying and Wei, and body fluids leading to generation of damp–heat cause this condition. The symptoms include severe itching, burning sensation of the skin, and eruptions of red dots, papule, or small vesicles. After 3 to 5 days, hookworms travel to the lung through the bloodstream, thus there can be cough, itchy throat, hoarseness, coughing of blood-tinged sputum, and even asthma. When the hookworms are swallowed into the stomach and intestines, they attach to the lining of the intestinal wall and suck the blood to grow into the adult worm. At this stage, black stools, dizziness, blurry vision, fatigue, palpitation, shortness of breath, scanty food intake, diarrhea, or constipation can be observed. The Chinese herbs that expel parasites are Bing Lang, Lei Wan, Fei Zi, Ku Lian Gen Pi, Guan Zhong, and Ma Chi Xian.

Expel Worm

Choose one of the following formulas:

- Lei Wan powder 25 g. Mix with small quantity of white sugar and water. Take before bedtime continuously for 5 days.
- Fresh Ma Chi Xian 90 g. Add small quantity of sugar and vinegar to the decoction. Take before bedtime continuously for 5 days.
- Gou Chong Wan [Hookworm Pill]: Take once in the morning and once at night for 5 days.

Pharmaceutical drugs to expel parasites are stronger than Chinese herbs and they also have greater side effects and toxicity. Thus, for a person with a strong constitution, pharmaceutical drugs are indicated; however, for a weak constitution person and pregnant women, one should use pharmaceutical drugs with caution or use Chinese herbs.

Support the Zheng Qi

Mild Case of Anemia

Use Xiang Sha Liu Jun Zi Tang [Six-Gentleman Decoction with Aucklandia and Amomum].

Moderate Case of Anemia

Use Shi Quan Da Bu Tang [All-Inclusive Great Tonifying Decoction] plus Gou Qi Zi, Shan Yao, and Da Zao.

Severe Case of Anemia

Use Sheng Mai San [Generate the Pulse Powder] plus E Jiao and Shi Quan Da Bu Tang [All-Inclusive Great Tonifying Decoction].

BACILLARY DYSENTERY

Bacillary dysentery happens more often in the summer and fall seasons. It is due to over consumption of raw, cold, or unclean food which damages the stomach and intestines. Damp–heat toxin enters the body and accumulates in the intestinal tract, obstructing Qi and blood, eroding intestinal membrane, and damaging channels and collaterals, thus transforming into red and white pus with blood to create damp–heat type dysentery. The clinical manifestations include whole body poison symptoms, lumbar pain, diarrhea, tenesmus, mucous, pus or blood in stool. If damp–heat toxin evils transform into fire attacking Heart and Ying, then blood stasis occurs due to the toxins, and Qi becomes stagnant because the blood leads to epidemic toxic dysentery. In TCM, the treatment focuses on clearing heat, eliminating dampness, regulating Qi, and invigorating blood. From modern research, it has been found that Huang Lian, Huang Bai, Huang Qin, Da Huang, Ku Shen, Mu Xiang, Di Yu, Qin Pi, Da Suan, Ma Chi Xian, Bai Tou Weng, Bai Jiang Cao, Chuan Xin Lian, San Ke Zhen, and Shi Da Gong Lao Ye have inhibitory function to *Bacillus dysenteriae*.

Acute Dysentery

Mild Type

Treatment Principle: Clear heat, eliminate dampness, and regulate Qi.

Use Modified Xiang Lian Wan [Aucklandia and Coptis Pill] plus Ping Wei San [Calm the Stomach Powder].

Common Type

Treatment Principle: Clear heat, eliminate dampness, regulate Qi, and harmonize blood.

Use Shao Yao Tang Jia Jian [Modified Peony Decoction].

Related conditions and modifications are

- Severe tenesmus: Add Da Huang to the first bag and remove it from the second bag.
- Dysentery with more pus than blood: Minus Da Huang, and add Di Yu and Jin Yin Hua.
- Epigastric and stomach discomfort, nausea with urge to vomit: Add Ban Xia, Chen Pi, and Zhu Ru.
- Epigastric and stomach distention/fullness with sour regurgitation: Add Hou Po, Shen Qu, and Lai Fu Zi.

Severe Type

Treatment Principle: Clear heat, cool blood, drain damp, and relieve toxins.

Use Bai Tou Weng Tang Jia Wei [Augmented Pulsatilla Decoction]: Bai Tou Weng, Jin Yin Hua, Qin Pi, Huang Bai, Huang Lian, Chi Shao Yao, Mu Dan Pi, Ku Shen, and Ma Chi Xian.

Toxic Type

This is a severe acute syndrome that should be treated by combining Chinese and Western medicine. In TCM, Bai Tou Weng Tang Jia Wei [Augmented Pulsatilla Decoction] is used with Da Huang and Xuan Ming Fen to quickly drain the toxins downward. If there is high fever with wind, loss of consciousness, and syncope, it is treated according to the acute syndromes. An Gong Niu Huang Wan [Calm the Palace Pill with Cattle Gallstone], Zhi Bao Dan [Greatest Treasure Special Pill], and Zi Xue Dan [Purple Snow Special Pill] can be used according to the syndrome.

Chronic Dysentery

This syndrome is often more deficiency and slight excess, severe cold, and mild heat. One should eat foods that are easy to digest and avoid overeating or excessive hunger.

Chronic Hidden Type (Often Qi Deficiencywith Damp–Heat)

Treatment Principle: Strengthen Spleen and Stomach, eliminate dampness, regulate Qi, and move stagnation.

Use Shen Ling Bai Zhu San [Ginseng, Poria, and Atractylodes Macrocephala Powder] plus Pao Jiang and fried Huang Lian.

Related conditions and modifications are

- Focal distention at epigastrium and abdomen, incomplete sensation with bowel movements and white slippery tongue coating: Add Huo Xiang and Hou Po.
- Abdominal coldness and cold limbs with preference for heat: Add Fu Zi and Rou Gui.
- Burning heat in abdomen and burning sensation with bowel movement and urination, red tongue, yellow tongue coating, and thin, rapid pulse: Add Ma Chi Xian and Wu Mei.

Chronic Persisting Type (Often Spleen Kidney Yang Deficiency with Damp–Heat)

Treatment Principle: Use both heat and cold to treat excess and deficiency.

Use Wu Mei Wan [Mume Pill].

Related conditions and modifications are

- Diarrhea with more blood than mucous and severe abdominal pain: Remove Gui Zhi and Chuan Jiao, add Di Yu, Bai Shao Yao, and Jin Yin Hua.
- Epigastric focal distention, abdominal distention, poor appetite, and white, slippery tongue coating: Add Cang Zhu, Hou Po, Mu Xiang, and Chen Pi.

BACTERIAL FOOD POISONING

This syndrome is caused by ingesting bacteria contaminated food or bacterial toxins which leads to acute infection and poisoning conditions. It is often seen in summer and fall seasons. In TCM, this syndrome is thought to be caused by ingesting unclean food or external cold–damp, or summer damp–turbid pathogenic factors. The clinical manifestations include fever, nausea, vomiting, abdominal pain, diarrhea, and other acute gastroenteritis symptoms. Diarrhea is often watery and frequency of diarrhea can vary from a few bouts to as many as 20 bouts or more. Some patients may have stools with pus and blood. Severe vomiting and diarrhea can lead to dehydration and shock. Treatment focuses on regulating the ascending and descending functions of spleen and stomach, transforming dampness and expelling turbidity. Treatments for different patterns are as follows:

Damp–Heat Type

Treatment Principle: Clear heat, transform damp, separating clear from turbid, and expelling turbid.

Use Ge Gen Huang Qin Huang Lian Tang Jia Wei [Augemented Kudzu, Coptis, and Scutellaria Decoction]: Ge Gen, Huang Qin, Huang Lian, Ban Xia, Hou Po, Bai Dou Kou, Yi Yi Ren, Fu Ling, and Gan Cao.

Related conditions and modifications are

- Severe fever: Add Jin Yin Hua and Lian Qiao.
- Severe vomiting and rebellious Qi: Add Zhu Ru and Sheng Jiang juice.
- Thirst and scanty urination: Add Lu Gen and Wu Mei.
- Food stagnation: Add Shen Qu and Jiao Shan Zha.
- Spasms of the extremities, especially of the gastrocnemius muscle: Add Bai Shao Yao, Mu Gua, and Can Sha.
- Blood in watery stool: Add Di Yu and Huai Hua Mi.

Cold–Damp Syndrome

Treatment Principle: Aromatic to transform damp, warm Middle Jiao, and scatter the cold.

Use Huo Xiang Zheng Qi San [Agastache Powder to Rectify the Qi] plus Gan Jiang and Bai Dou Kou.

Related conditions and modifications are

- Severe vomiting and rebellious Qi: Add Wu Zhu Yu and Sheng Jiang juice.
- Severe internal cold with abdominal pain and cold extremities: Add Rou Gui and Ding Xiang.
- Heat, restlessness, thirst, and yellow tongue: Add Huang Lian.
- Spasms of muscles especially at gastrocnemius muscle: Add Mu Gua, Bai Shao Yao, and Wu Zhu Yu.

CHOLERA

Cholera (Huo Luan) is a disease that occurs during summer and fall seasons. Pathogenic infectious agents and toxins invade the body with food, damaging Spleen and Stomach and causing a disruption in the ascending and descending function, as well as mixing clear and turbid. The main manifestations are severe and frequent vomiting, diarrhea, and presence or absence of abdominal pain. Furthermore, even if it is not caused by epidemic pathogenic influences, the presentation of acute alternating vomiting and diarrhea is also diagnosed as Huo Luan.

Cold Type

Mild Case

Treatment Principle: Aromatic to transform dampness, warm Middle Jiao, and scatter cold.

- Beginning stage
  Use Huo Xiang Zheng Qi San [Agastache Powder to Rectify the Qi].
- Cold damages Yang in Middle Jiao
  Use Li Zhong Wan [Regulate the Middle Pill].

Severe Case

Treatment Principle: Rescue Yang, revive desertion, tonify deficiency, and nourish Yin.

Use Jiang Fu Li Zhong Wan [Regulate the Middle Pill with Ginger and Prepared Aconite] or Si Ni Jia Ren Shen Tang [Frigid Extremities Decoction plus Ginseng] or Si Jun Zi Tang [Four-Gentleman Decoction] plus Shao Yao and Gao Liang Jiang.

Heat Type

Treatment Principle: Clear heat, transform dampness, descend and drain turbid.

Use Ran Zhao Tang [Lighting Decoction].

Related conditions and modifications are

- Severe heat: Gui Ling Gan Lu Yin [Cinnamon and Poria Sweet Dew Decoction], Bai Hu Tang [White Tiger Decoction], Zhu Ye Shi Gao Tang [Luphatherus and Gypsum Decoction].
- Soreness of extremities and hypertonicity of sinews: Can Shi Tang [Silk Worm Feces Decoction].
- Acute coldness of extremities: Bai Hu Tang [White Tiger Decoction] with Zi Xue Dan [Purple Snow Special Pill], Bi Xue Dan [Green Snow Special Pill], or Jiang Xue Dan [Dark Red Snow Special Pill].

Dryness Type

Treatment Principle: Descend turbid, disinhibit Qi, and unblock the congestion.

Use Yu Shu Dan [Jade Pivot Special Pill], Fei Long Duo Ming Dan [Flying Dragon Life-Clutching Special Pill], and Xing Jun San [Troop-Marching Powder]. Can also use salt water to induce vomiting or Gua Di San [Melon Pedicle Powder] and San Sheng San [Three-Sage Powder] to promote vomiting.

COMMON COLD

The common cold is a disease caused by wind evil pathogens invading the body. Its main symptoms include headache, nasal congestion, nasal discharge, sneezing, chills, fever, and floating pulse. If the condition is more severe and spreads quickly within a period of time with similar presentations, it is referred to as influenza.

Wind–Cold Common Cold

Treatment Principle: Acrid, warm formulas to release exterior, diffuse Lung Qi, and scatter cold.

Mild Case

Use Cong Chi Tang [Scallion and Prepared Soybean Decoction].

Severe Case

Use Jing Fang Bai Du San [Schizonepeta and Ledebouriella Powder to Overcome Pathogenic Influences].

With Dampness

Use Qiang Huo Sheng Shi Tang [Notopterygium Decoction to Overcome Dampness].

Wind–Heat Common Cold

Treatment Principle: Acrid, cool formulas to release exterior, expel wind, and clear heat.

Mild Case with Cough

Use Sang Ju Yin [Mulberry Leaf and Chrysanthemum Decoction].

Severe Case with Sore Throat

Use Yin Qiao San [Honeysuckle and Forsythia Powder].

Summer Wind–Heat with Dampness

Use Xin Jia Xiang Ru Yin [Newly-Augmented Elsholtzia Decoction].

Summer Dampness More than Heat Case

Use Huo Xiang Zheng Qi San [Agastache Powder to Rectify the Qi].

Heat and Cold Mixed Presentations

Exterior Cold and Internal Heat (Cold Wrapping Fire)

Use Ma Xing Shi Gan Tang [Ephedra, Apricot Kernel, Gypsum, and Licorice Decoction] with Qiang Huo and Yu Xing Cao.

Related conditions and modifications are

- Severe exterior cold, chills, and joint pain: Add Zi Su Ye and Gui Zhi.
- Severe internal heat and sore, painful, swollen throat: Add Ban Lan Gen and Huang Qin.

Both Exterior and Interior Excess with Constipation and Fever

Use Fang Feng Tong Sheng San [Ledebouriella Powder that Sagely Unblocks].

Common Cold with Qi Deficiency

Treatment Principle: Tonify Qi, release exterior, and harmonize Ying and Wei.

Severe Case of Qi Deficiency

Use Shen Su Yin [Ginseng and Perilla Leaf Decoction].

Severe Case of Ying and Wei Disharmony

Use Huang Qi Gui Zhi Wu Wu Tang [Astragalus and Cinnamon Twig Five-Substance Decoction].

Qi Deficiency with Frequent Colds

Take Yu Ping Feng San [Jade Windscreen Powder] for the long term.

Common Cold with Yang Deficiency

Treatment Principle: Warm Yang and release exterior.
Use Gui Zhi Jia Fu Zi Tang [Cinnamon Twig Decoction plus Aconite].
Related condition and modification are

- Watery diarrhea and dull abdominal pain: Add Pao Jiang and Rou Gui.

Common Cold with Blood Deficiency

Treatment Principle: Nourish blood and release exterior.
Use Cong Bai Qi Wei Yin [Scallion Decoction with Seven Ingredients].
Related conditions and modifications are

- More coldness: Add Huang Qi, Fang Feng, and Jing Jie.
- More heat: Add Jin Yin Hua and Lian Qiao.
- Blood deficiency with blockage in channels and knotted or intermittent pulse: Add Gui Zhi, Hong Hua, and Dan Shen.

Common Cold with Yin Deficiency

Treatment Principle: Nourish Yin and release exterior.
Use Jia Jian Wei Rei Tang [Modified Polygonatum Odoratum Decoction].
Related conditions and modifications are

- Severe Yin deficiency: Add Zhu Ye, Xian Mao Gen, and Tian Hua Fen.
- Cough and dry throat: Add Niu Bang Zi, She Gan, and Gua Lou Pi.
- Cough, chest pain, and sputum with blood: Add Pu Huang and Ou Jie.

CYSTICERCOSIS

This is caused by cysticercus living parasitically in various parts of the body, such as muscles, subcutaneous layer, brain, eye, Heart, Lung, tongue, and mouth. It is most commonly seen in the brain, muscles, and subcutaneous layer. Cysticercosis commonly manifests as epileptic attacks, mental aberration, headache, nausea, vomiting, blurry vision, staggering gait, or temporary weakness and paralysis of the body. For muscle and subcutaneous cysticercosis, one can palpate one to a few dozen nodules formed by cysticercus; patient will experience muscle soreness or jumping pain. Brain cysticercosis often occurs with subcutaneous nodules at the same time. Eye cysticercosis can occur in one or both eyes and can be in single or multiples. The manifestations are decreased vision or blindness. There is no clear record of this syndrome in old Chinese classical texts. The modern Chinese medicine treatment focuses on killing parasites, expelling wind, invigorating blood, and eliminating phlegm.

Kill Parasites

This is indicated for brain, muscle, and subcutaneous cysticercosis. It cannot be used for eye cysticercosis.

Use Lei Wan 200 g, Pao Chuan Shan Jia 50 g, Shui Zhi 50 g, Gan Qi 50 g, Bai Jie Zi 100 g, Mu Li 100 g, Jin Yin Hua 100 g, Dan Shen 100 g, Fu Ling 200 g, Yi Yi Ren 200 g, and Jiang Can 200 g.

Grind all herbs into powder, form into the size of mung beans pills and take 5 g, twice a day, 30 minutes before meals.

To treat brain cysticercosis, take the herbs for 3 to 6 months. Some people may experience increased headache or frequent epileptic attacks when beginning the herbs, but the symptoms gradually improve over time. After taking the herbs for 2 to 3 months, subcutaneous nodules will become bigger and softer and then slowly disappear after another 2 months.

Treatment for Complications of Epilepsy

During Epileptic Attack

Acupuncture Treatment: Du 26, Du 20, and K 1.

Related conditions and modifications are

- Attacks during day: Add UB 62.
- Attacks during night: Add K6.

Between Epileptic Attack

Treatment Principle: Subdue Yang, transform phlegm, and expel water.

Use Decoction of Da Ji 1g, Shang Lu 2 g, Ban Xia 15 g, Bai Zhi 15 g, Chuan Xiong 15 g, Jing Jie 15 g, Tu Fu Ling 20 g, Fu Ling 20 g, Long Gu 30 g, Mu Li 30 g, and Xi Xin 5 g.

Related conditions and modifications are

- Headache and severe nausea and vomiting: Add Xuan Fu Hua, Quan Xie, and Dai Zhe Shi.
- Insomnia and poor memory: Add Shi Chang Pu and Zhi Suan Zao Ren.
- Decreased vision: Add Shi Jue Ming and Jue Ming Zi.
- Mental aberration, irritability, red tongue, and dry yellow tongue coating: Add An Gong Niu Huang Wan [Calm the Palace Pill with Cattle Gallstone], one pill twice a day.

DIPHTHERIA

Diphtheria is called Suo Hou Feng, Chan Hou Feng, or Bai Hou in TCM. It is caused by dry heat epidemic toxin (*Corynebacterium diphtheriae*) that enters the body through the nose, accumulates in the throat, damages mucous membrane by forming a false membrane and then releases toxins into the body through the blood. The manifestations include fever, fatigue, nausea, vomiting, headache or myocarditis and in severe cases delirium, palpitation, purple lips, sallow complexion, and minute thin or knotted intermittent pulse. The main treatment principle in Chinese medicine is to clear heat and relieve toxicity. The secondary treatment principles are to nourish Yin and expel that which is rotten. Acrid warm herbs are scattering, and dry hot herbs can damage Yin so they should be used with caution.

Pharyngeal Diphtheria (Yan Bai Hou—About 80% of the Cases)

Mild Case

Treatment Principle: Release wind, clear heat, and relieve toxicity.

Use Yin Qiao San [Honeysuckle and Forsythia Powder] plus Sang Ju Yin [Mulberry Leaf and Chrysanthemum Decoction] plus Qing Dai, Mu Tong, Ge Gen, Pi Pa Ye, Chuan Bei Mu, and Tu Niu Xi.

Related conditions and modifications are

- Severe heat: Add Huang Qin and Shi Gao
- Dry throat and thirst: Add Mai Men Dong, Xuan Shen, and Sha Shen.
- Constipation and abdominal distention: Add Da Huang and Lai Fu Zi.
- Yellow dark urine: Add Bai Mao Gen, Long Dan Cao, and Che Qian Zi.
- Red, swollen throat and tonsils: Add Liu Shen Wan [Six-Miracle Pill], 10 pills, three times per day.

Common Type

Treatment Principle: Clear heat, relieve toxin, expel rotten, and scatter the accumulation.

Use Huang Lian Jie Du Tang [Coptis Decoction to Relieve Toxicity] plus Shi Gao, Niu Bang Zi, Jie Geng, Bo He, Xuan Shen, Jin Yin Hua, Lian Qiao, and Chi Shao Yao.

Related conditions and modifications are

- Dry mouth and throat: Add Sheng Di Huang and Tian Hua Fen.
- Nausea and vomiting: Add Zhu Ru, Ban Xia, Sheng Jiang, and Dai Zhe Shi.
- Constipation and abdominal pain: Add Jiu Da Huang and Mang Xiao.
- False membrane appears to be dark red, black, or mixed with blood: Add San Qi Fen [Pseudo-ginseng Powder].

Severe Type

Treatment Principle: Drain heat, relieve toxin, nourish Yin, and cool blood.

Use Qing Wen Bai Du Yin [Clear Epidemics and Overcome Toxin Decoction] plus Tu Niu Xi.

Related conditions and modifications are

- Constipation: Add Da Huang and Mang Xiao.
- Burned mouth and dry throat: Add Mai Men Dong and Yu Zhu.

Extreme Severe Type

Treatment Principle: Clear heat, relieve toxin, disperse Lung, and reduce swelling.

Use Qing Wen Bai Du Yin [Clear Epidemics and Overcome Toxin Decoction] plus Ma Bo, Chi Shao Yao, Long Dan Cao, Jiang Can, Zhe Bei Mu, and Mu Tong.

Related condition and modification are

- Severe swelling of lymph nodes below the jaw and neck: Add Jie Geng, Xia Ku Cao, Dan Shen, and Zao Jiao Ci.

Laryngeal Diphtheria (Often Spread from Diphtheria at Pharynx)

The main manifestations are obstruction of the airway and cough that sounds like barking.

Treatment Principle: Clear heat, eliminate phlegm, open the obstruction, and scatter accumulation.

Use Xian Fang Huo Ming Yin [Sublime Formula for Sustaining Life] plus Jiang Can, Qing Dai, and Ting Li Zi.

Related conditions and modifications are

- Severe phlegm sounds in throat: Add Mang Xiao, Sang Bai Pi, and Bai Jie Zi.
- Breathing difficulty, purple lips, and dark tongue: Add Xing Ren, Tao Ren, Dan Shen, and Zhi Ke.
- If individual has obstructed airway, the patient should be sent to the emergency room.

Nasal Diphtheria (Rarely Seen, Often Spread from Diphtheria at Pharynx)

Treatment Principle: Clear heat, relieve toxins, scatter accumulations, and open orifices.

Use Yin Qiao San [Honeysuckle and Forsythia Powder] plus Huang Qin, Ye Ju Hua, Xin Yi Hua, Shan Zhi Zi, Shi Chang Pu, and Sheng Di Huang.

Related conditions and modifications are

- Bloody nasal discharge: Add Bai Mao Gen and Ce Bai Tan.
- Nasal discharge with pus: Add Xia Ku Cao, Long Dan Cao, and Zhe Bei Mu.
- Dry scabs in nasal cavity: Add Xuan Shen and Mai Men Dong.

FASCIOLOPSIASIS

This condition belongs to worm syndromes in TCM. It is caused by ingesting unclean food containing worm evil infection (*Fasciolopsis buski*) which grows parasitically in the small intestine, causing irregular ascending and descending Qi mechanism of Spleen and Stomach function and affecting Qi and blood generation and transformation. In mild cases, symptoms can range from none at all to having alternating diarrhea and constipation, nausea, vomiting, becomes hungry easily, has strange cravings, and undigested food in stool with extremely foul smell. In severe cases, there can be emaciation, edema, ascites, and malnutrition. The treatment focuses on expelling the parasites and strengthening the spleen.

Expel Parasites

Choose one of the formulas:

1. Bing Lang 30 g, Nan Gua Zi 20 g, and Qian Niu Zi (black and white) 3 g each. Take the decoction after waking on empty stomach continuously for 3 days.
2. Bing Lang 45 g, Fei Zi 40 g, Da Huang 5 g, and Mu Xiang 10 g. Take the decoction twice a day continuously for 3 days.
3. Bing Lang 10 g, Da Huang 6 g, and Qian Niu Zi (black and white) 1.5 g each. Grind into powder and divide into two doses. Take once in the morning and night with sugar water on an empty stomach continuously for 3 to 5 days.

After taking the formula for 3 to 5 days, if parasite eggs still remain in stool culture, then rest for 2 to 3 days before taking the formula again for 3 to 5 days until the stool culture becomes negative.

Strengthen Spleen and Support Zheng Qi

Treatment Principle: Use before or after expelling the parasites.
Use Xiang Sha Liu Jun Zi Tang [Six-Gentleman Decoction with Aucklandia and Amomum].

INFECTIOUS MONONUCLEOSIS

This syndrome is due to warm toxins belonging to warm febrile diseases in TCM. Warm, heat, epidemic toxin (Epstein–Barr Virus) invades the lung and Wei level. Symptoms include fever, cough, sore throat, skin rash, swollen liver, spleen, or lymph node, increase lymphocytes in blood, antibodies for Epstein–Barr virus found from blood test, or nausea, vomiting, jaundice, epigastric and abdominal fullness, and focal distention. In severe cases, presentation of high fever, delirium and convulsion. The focus of TCM treatment is on heat and toxins. At the beginning stage, treatment focuses on clearing heat and relieving toxins. At the late stage, treatment principles are tonifying Qi and nourishing Yin. Commonly used herbs are Jin Yin Hua, Lian Qiao, Chi Shao Yao, Ban Lan Gen, Qing Dai, Tu Niu Xi, and Zhe Bei Mu.

Evil Pathogens Attacking Lung and Wei Syndrome

Treatment Principle: Clear heat, relieve toxins and acrid cool to release exterior.
Use Yin Qiao San [Honeysuckle and Forsythia Decoction] plus Xia Ku Cao, Huang Qin, Xuan Shen, Zhe Bei Mu, and She Gan.
Related conditions and modifications are

- High fever: Add Shi Gao, Chai Hu, Qing Dai, and Qing Hao.
- Red skin rash: Add Mu Dan Pi, Sheng Di Huang, Shui Niu Jiao, and Chan Tui.
- Red eyes and bitter taste in mouth: Add Shan Zhi Zi, Ju Hua, and Jue Ming Zi.
- Cough: Add Sang Ye and Gua Lou Pi.
- Severe swollen lymph node: Add Dan Shen, Tao Ren, and Zao Jiao Ci.

Excess Evil Toxins Syndrome

Treatment Principle: Drain fire, relieve toxins, clear Ying, and cool blood.
Use Qing Wen Bai Du Yin [Clear Epidemics and Overcome Toxin Decoction].
Related conditions and modifications are

- Chills, high fever, bitter taste, and dark urine: Add Hua Shi, Zhu Ru, Qing Hao, and Qing Dai.
- Thirst with desire to drink and dry yellow coat: Add Mai Men Dong, Tian Hua Fen, and Lu Gen.
- Chest pain and cough with yellow sticky phlegm: Add Gua Lou, Zhe Bei Mu, and Xing Ren.
- Constipation: Add Sheng Da Huang and Mang Xiao.
- Delirium and delirious speech: Add Zi Xue Dan [Purple Snow Special Pill], one pill three times a day.

Heat Toxins in San Jiao with Damp-Turbid Syndrome

Treatment Principle: Clear heat, relieve toxins, transform dampness, and drive out turbidity.
Use Gan Lu Xiao Du Dan [Sweet Dew Special Pill to Eliminate Toxin].

Related conditions and modifications are

- Enlarged Spleen and Liver: Add Yu Jin, Hu Zhang, E Zhu, and Dan Shen.
- Dry painful throat: Add Xuan Shen and Mai Men Dong.
- Jaundice: Add Yin Chen Hao, Hu Zhang, and Che Qian Zi.

Qi and Yin Deficiency Syndrome

Treatment Principle: Benefit Qi, nourish Yin, clear and relieve remaining evil pathogens.
Use Zhu Ye Shi Gao Tang [Lophatherus and Gypsum Decoction].
Related conditions and modifications are

- Severe thirst: Add Yu Zhu and Tian Hua Fen.
- Constipation: Add Sheng Di Huang and Huo Ma Ren.
- Epigastric and abdominal distention and bland taste in mouth: Add Da Fu Pi, Huo Xiang, Shan Zha, and Mai Ya.

Recovery Stage (Damaged Lung and Stomach Yin)

Treatment Principle: Nourish Lung and Stomach.
Use Sha Shen Mai Dong Tang [Glehnia and Ophiopogonis Decoction] plus Chuan Bei Mu.
Related conditions and modifications are

- Fatigue and poor appetite: Add Tai Zi Shen and Shan Yao.
- Irritability and insomnia: Add Zhu Ye, Suan Zao Ren, and Ye Jiao Teng.

INTESTINAL AMEBIASIS

This syndrome is under the scope of TCM dysentery. It is caused by external damp–heat and insect toxins in the Intestines and Stomach leading to internal accumulation of damp–heat, Qi stagnation, blood stasis, damaged intestinal collaterals, and formation of pus and blood. Chronic condition can damage Middle Jiao Yang leading to internal cold–damp obstruction. From modern medicine, ingesting *Entamoeba histolytica* where the parasites infect the large intestine wall leading to dysenteric symptoms causes this condition. In acute stage, classical symptoms begin with abdominal pain and diarrhea which varies from a few bouts to a couple dozen bouts with mucous and blood in the stool which present as jelly color, foul fetid smell, different degree of tenesmus or with fever, and abdominal tenderness worse at sides of the lower abdomen. If the treatment during acute stage is not complete, the condition will progress into the chronic stage with mild diarrhea, three to five times a day with some mucous. The treatment should focus on relieving toxicity and killing parasite. After the clinical symptoms are cleared, one should continue taking the herbs for 2 to 4 weeks. Modern research shows that Bai Tou Weng, Ya Dan Zi, and Da Suan have inhibitory and killing functions to combat *Entamoeba histolytica*.

Acute Stage

Damp–Heat Accumulating in Intestines Syndrome (Common Type)

Treatment Principle: Clear heat, drain damp, and relieve toxins.

Use Bai Tou Weng Tang [Pulsatilla Decoction] plus Ku Shen, Mu Xiang, Di Yu, Chi Shao Yao, and Jin Yin Hua.

Related conditions and modifications are

- Fever, chills, headache, and body ache: Add Jing Jie and Ge Gen.
- Severe dampness and white, greasy tongue coating: Add Cang Zhu and Huo Xiang.
- Abdominal pain refusing pressure, incomplete sensation with bowel movement, and foul fetid smell: Add Da Huang, Zhi Shi, and Lai Fu Zi.
- Nausea and vomiting: Add Zhu Ru and Ban Xia.

Excess Parasite Toxin Syndrome (Fulminant)

Treatment Principle: Clear heat, relieve toxin, cool blood, and stop dysentery.

Use Bai Tou Weng Tang [Pulsatilla Decoction] plus Xi Jiao Di Huang Tang [Rhinoceros Horn and Rehmannia Decoction] and Di Yu.

Related conditions and modifications are

- Delirium and delirious speech: Add one pill of Shen Xi Dan [Magical Rhinoceros Special Pill], twice a day.
- Qi deficiency, closed interior, and collapsed exterior: Add Xi Yang Shen 30 g. Take the decoction frequently.

Chronic Stage

Spleen and Stomach Deficient Cold Syndrome

Treatment Principle: Warm Middle Jiao, strengthen Spleen, astringe, and rescue collapse.

Use Tao Hua Tang [Peach Blossom Decoction] plus Shan Yao, Long Gu, Mu Li, Di Yu, Qin Pi, and Ya Dan Zi.

Zheng Qi Deficiency with Lingering Evil Pathogens Syndrome

Treatment Principle: Strengthen Spleen, benefit Qi, and expel evil.

Zi Sheng Wan [Assist the Life Pill] plus Bai Bian Dou and Bai Tou Weng.

LEPROSY

This syndrome occurs in people with Zheng Qi deficiency. The pathogen (Mycobacterium leprae) invades the body through skin and mucous membrane. It leads to stagnation in the channels and collaterals and dysfunction of the Zang Fu organs, especially Liver and Spleen. The main clinical manifestations include damages of skin, mucous membrane and nerves, such as irregular-shaped macules, papules, tubercles, plaques, infiltrations, ulcerations, peripheral nerve damage especially peripheral cutaneous nerves causing decreased sensation with pain and pustular ulcerations of nerves and decreased mobility. At late stage, there can be loss of sensation of the whole hand up to the wrist. In TCM, primary treatment focus is to attack toxins and kill bacteria. Commonly used herbs are Da Feng Zi, Cang Er Zi, Ku Shen, Wu Shao She, Wu Gong, Zao Jiao Ci, Pi Shuang, and Ban Mao. Secondary treatment focus depends on syndrome differentiation to boost the immune function of the body. For best results pharmaceutical drugs should be taken in conjunction with the herbs.

Blood-Heat, Blood Stasis Syndrome

Treatment Principle: Clear heat, cool blood, and transform stasis.

Use Qing Ying Tang [Clear the Nutritive Level Decoction] minus Huang Lian, Mai Men Dong, plus Jing Jie, Huang Qin, Zi Cao, Bai Ji Li, Di Long, and Chan Tui.

Damp–Heat Obstruction Syndrome

Treatment Principle: Clear heat, drain damp, transform stasis, and unblock collaterals.

Use Wu Wei Xiao Du Yin [Five-Ingredient Decoction to Eliminate Toxin] plus Shi Gao, Tu Fu Ling, Da Qing Ye, and Yi Yi Ren.

Yin Deficiency and Blood Heat Syndrome

Treatment Principle: Nourish Yin, clear heat, and transform stasis.

Use Xi Jiao Di Huang Tang [Rhinoceros Horn and Rehmannia Decoction] plus Sheng Ma Ge Gen Tang [Cimicifuga and Kudzu Decoction] plus Bie Jia, Zi Cao, Huang Qin, Lian Qiao, and Zhi Mu.

Related conditions and modifications are

- Excess heat toxins: Add Da Qing Ye, Jin Yin Hua, Pu Gong Ying, and Zi Hua Di Ding.
- Swelling and ulcerations of the extremities: Add Yi Yi Ren, Ze Xie, and Tu Fu Ling.
- Qi and Yin deficiency: Change to Sheng Mai San [Generate the Pulse Powder] plus Jia Jian Wei Rui Tang [Modified Polygonatum Odoratum Decoction].
- Kidney Yin exhaustion: Add Er Zhi Wan [Two-Ultimate Pill] and Shu Di Huang, Gou Qi Zi, and He Shou Wu.

Liver Qi Stagnation and Blood Stasis Syndrome

Treatment Principle: Smooth Liver and invigorate blood.

Use Shu Gan Jie Yu Tang [Smooth Liver and Resolve Stagnation Decoction] plus Chi Shao Yao, Bai Shao Yao, Yin Chen Hao, Zhi Ke, Sheng Ma, Ge Gen, San Leng, E Zhu, Fu Ling, and Cang Zhu.

Qi and Blood Deficiency Syndrome

Treatment Principle: Tonify Qi, nourish blood, and transform stasis.

Use Shi Quan Da Bu Tang [All-Inclusive Great Tonifying Decoction] plus Dan Shen, Huai Niu Xi, Xiang Fu, Ze Lan, and Bai Ji Li.

Qi Stagnation and Blood Stasis Syndrome

Treatment Principle: Move Qi, invigorate blood, unblock collaterals, and stop pain.

Use Huo Luo Xiao Ling Dan Jia Jian [Modified Fantastically Effective Pill to Invigorate the Collaterals].

Related conditions and modifications are

- Slight cold: Add Zhi Chuan Wu, Gui Zhi, Gan Jiang, and Xi Xin.
- Slight heat: Add Zi Hua Di Ding, Pu Gong Ying, Huang Qin, Mu Dan Pi, Di Long, and Ren Dong Teng.
- Disease in upper extremities: Add Qiang Huo and Sang Zhi.
- Disease in lower extremities: Add Chuan Niu Xi, Mu Gua, and Qin Jiao.
- Disease in whole body: Add Wei Ling Xian and Fang Ji.

LEPTOSPIROSIS

This syndrome belongs to summer–warm and damp–warm categories in the study of warm febrile diseases in TCM. It often occurs in fall season and is caused by an invasion of summer–damp epidemic toxins (Leptospira interrogans) through the skin surface. In the early stage, it inhibits Wei Qi and can cause septicemia. In the middle stage, the damages occur at Lung, Spleen, Stomach, and Gallbladder organs. The symptoms of this stage include headache, nausea, vomiting, jaundice, and bleeding. In the late stage, Yin and Qi are damaged. The clinical manifestations are fever, headache, red conjunctiva, swollen lymph nodes, pain upon palpation, pain at gastrocnemius muscle, bleeding, Liver and Kidney damages, and sequela of allergic reactions. Severe critical condition patients should be sent to the emergency room. Clinically, Jin Yin Hua, Lian Qiao, Lu Gen, Bai Mao Gen, Ban Lan Gen, Huang Qin, Shan Zhi Zi, Huo Xiang, Tong Cao, Ye Ju Hua, Yin Chen Hao, Qing Hao, and Tu Fu Ling are effective in treating this syndrome.

Infectious Toxin Type

Damp Obstructing the Upper Jiao Syndrome

Treatment Principle: Disperse Lung and transport Spleen.

Use San Ren Tang [Three-Nut Decoction] plus Bo He, Huang Lian, Jin Yin Hua, and Huang Qin.

Related conditions and modifications are

- Body ache and gastrocnemius pain: Add Qin Jiao, Sang Zhi, and Fang Ji.
- Epigastric and abdominal distention/fullness, nausea, and vomiting: Add Su Geng, Zhu Ru, Huo Xiang, and Ji Shi Teng.
- Severe diarrhea: Minus Xing Ren and Hua Shi, and add Fu Ling, Ze Xie and Zhu Ling.

Evil Pathogens Accumulating at Qi Level Syndrome

Treatment Principle: Clear heat, relieve toxicity, and transform dampness.

Use Yin Qiao Bai Hu Tang [White Tiger Decoction with Honeysuckle and Forsythia] plus Huang Qin, Bai Mao Gen, Da Qing Ye, Huo Xiang, and Tong Cao.

Related conditions and modifications are

- Absence of sweat or scanty sweat: Add Jing Jie and Chai Hu.
- Red tongue and thin dry yellow coat: Add Tian Hua Fen and Mai Men Dong.
- Greasy tongue coat, no taste in mouth, and scanty yellow urine: Add Deng Xin Cao, Che Qian Zi, Fu Ling, and Zhu Ru.
- Epistaxis: Add Ce Bai Tan, Bai Ji, Di Yu Tan, Mu Dan Pi, Ou Jie, and fried Qian Cao Gen.
- Heat excess generating wind, and headache and stiff neck: Add Ling Yang Jiao, Sheng Ju Hua, and Di Long.
- After treatment where fever has subsided and fluid was damaged: Change to Yi Wei Tang [Benefit the Stomach Decoction].

Jaundice, Bleeding Type

Treatment Principle: Clear heat, relieve toxins, cool blood, and reduce jaundice.

Use Yin Chen Hao Tang [Artemisia Yinchenhao Decoction] plus Shen Xi Dan [Magical Rhinoceros Special Pill].

Related conditions and modifications are

- Abdominal distention: Add Huo Geng, Bing Lang, and Lai Fu Zi.
- Bleeding: Take with San Qi Fen [Pseudoginseng Powder] and Bai Mao Gen.
- Delirium and delirious speech: Add An Gong Niu Huang Wan [Calm the Palace Pill with Cattle Gallstone], one pill twice a day. Feed the pill through nose.

Pulmonary Hemorrhage Type

Treatment Principle: Clear heat, relieve toxicity, cool blood, and stop bleeding.

Xi Jiao Di Huang Tang [Rhinoceros Horn and Rehmannia Decoction] plus Shan Zhi Zi, Huang Qin, Bai Mao Gen, Qian Cao Gen, San Qi powder, fried Di Yu, Sheng Shi Gao, and Huang Lian.

Meningoencephalitis Type

Treatment Principle: Clear Ying, open orifices, cool liver, and extinguish wind.

Use Ling Yang Gou Teng Tang [Antelope Horn and Uncaria Decoction] plus Qing Re Xi Feng Tang [Clear Heat Extinguish Wind Decoction] and Shui Niu Jiao.

Related condition and modification are

- Delirium and delirious speech: Add Zi Xue Dan [Purple Snow Special Pill], one pill three times a day via nose.

Renal Failure Type

Treatment Principle: Clear heat, drain damp, disperse, and unblock San Jiao.

Use Dao Chi San [Guide Out the Red Powder] plus Sheng Shi Gao, Han Shui Shi, Hua Shi, Che Qian Zi, Fu Ling, Bai Mao Gen, and Deng Xin Cao.

MALARIA

Malaria is caused by invasion of malarial infectious agents or wind, cold, summer heat, and dampness. These pathogenic factors reside in Shao Yang battling with Zheng Qi between Ying and Wei level. The symptoms are shivering chills followed by fever, hot body, headache, restlessness and thirst, then followed by sweating which breaks the fever. Alternating chills and fever can occur every day, every other day or every 3 days, which is characteristic of this disease. Once diagnosed with malaria, one should be treated with the "Interrupt Malaria" method. Chang Shan Yin [Dichroa Decoction] or Jie Nue Qi Bao Dan [Malaria-Interrupting Seven-Treasure Decoction] can be taken 4 hours, 2 hours, and 1 hour before the attack. It is also treated according to syndrome differentiation.

Zheng Nue (Primary Malaria)

Treatment Principle: Harmonize Shao Yang, release exterior, and expel evil.

Use Xiao Chai Hu Tang and Da Yuan Yin Jia Jian [Modified Minor Bupleurum Decoction and Reach the Membrane Source Decoction].

Related conditions and modifications are

- Exterior excess with scanty sweat, chills, and greasy coating: Add Gui Zhi and Qiang Huo.
- Dry mouth with desire to drink: Add Ge Gen and Shi Hu.
- Excess phlegm, stifling sensation at chest, and greasy tongue coating: Qing Pi Yin [Clear the Spleen Decoction].

Wen Nue (Warm Malaria)

Treatment Principle: Clear heat and expel evil.

Use Bai Hu Jia Gui Zhi Tang [White Tiger Plus Cinnamon Twig Decoction].

Related conditions and modifications are

- Constipation and yellow, greasy tongue coating: Add Da Huang.
- Heat accumulation in Shao Yang and jaundice: Add Yin Chen Hao, Shan Zhi Zi, Huang Bai, and Mu Dan Pi.
- Delirium: Add Shi Chang Pu, Dan Nan Xing, and An Gong Niu Huang Wan [Calm the Palace Pill with Cattle Gallstone].
- Chest stifling sensation and nausea: Add Huang Lian, Hua Shi, and Fu Ling.
- Excess sputum: Add Gua Lou, Zhi Shi, Zhu Ru, and Chen Pi.
- Fever without chills, thirst with desire to drink, and frequent urge to vomit: Bai Hu Jia Ren Shen Tang [White Tiger plus Ginseng Decoction] plus Mai Men Dong, Sheng Di Huang, and Sha Shen.
- Chronic condition with damages of Yin fluid, emaciation, and mild low-grade fever: Qing Hao Bie Jia Tang [Artemisia Annua and Soft-Shelled Turtle Shell Decoction].

Han Nue (Cold Malaria)

Treatment Principle: Harmonize Shao Yang, warm, and transform to reach evil.

Use Chai Hu Gui Jiang Tang [Bupleurum, Cinnamon, and Ginger decoction].

Related conditions and modifications are

- Severe cold due to Spleen Yang deficiency: Fu Zi Li Zhong Tang [Prepared Aconite Decoction to Regulate the Middle].
- Food accumulation, phlegm, and damp: Chang Shan Cao Guo Yin [Dichroa and Amomi Tsao-ko Decoction].

Shi Nue (Damp Malaria)

Treatment Principle: Drain heat, clear summer heat, and transform damp.

Use Jia Wei Xiang RuYin [Augmented Elsholtzia Decoction] and Yi Yuan San [Benefit Source Powder].

Related condition and modification are

- Severe summer damp: Chai Ping San [Bupleurum and Stomach-Calming Decoction] plus Huo Xiang and Pei Lan.

Zhang Nue (Miasmic Malaria)

Treatment Principle: Descend turbid, clear toxins, and transform turbid.

Use severe heat toxin: Qing Zhang Tang [Clear Miasma Decoction].

Related conditions and modifications are

- Excess heat damaging Jin fluid with dark red tongue: Add Sheng Di Huang, Xuan Shen, and Shi Hu.
- Dry stool and burned tongue coating: Add fresh Da Huang and Mang Xiao.
- Severe vomiting: Yu Shu Dan [Jade Pivot Special Pill].
- High fever, Shen disturbance, and delirious speech: Zi Xue Dan [Purple Snow Special Pill].
- Severe cold–damp: Augmented Bu Huan Jin Zheng Qi San [Rectify the Qi Powder Worth More than Gold Decoction].
- Phlegm misting the mind: Su He Xiang Wan [Liquid Styrax Pill].

Lao Nue (Taxation Malaria)

Treatment Principle: Tonify Zheng Qi.

Use Bu Zhong Yi Qi Tang [Tonify the Middle Jiao and Augment the Qi Decoction].

Related conditions and modifications are

- Qi and blood deficiency: He Ren Yin [Polygonum Multiflorum Root and Ginseng Decoction] plus Wu Fu Yin [Five-Fortune Decoction].
- Yin and blood deficiency: Xiao Ying Jian [Minor Construction Decoction].
- Yin deficiency with tidal fever: Da Bu Yin Wan [Great Tonify the Yin Pill].

Li Nue (Dysentery Malaria)

Treatment Principle: Harmonize Shao Yang, reach evil, and clear and transform damp–heat.

Use Chai Qin Jian [Bupleurum and Scutellaria Decoction].

Related conditions and modifications are

- Clear watery diarrhea and greasy white tongue coating: Jia Wei Bu Huan Jin Zheng Qi San [Newly-Augmented Rectify the Qi Powder Worth More than Gold] plus Chai Hu and Chang Shan.

Nue Mu (Mother of Malaria)

Treatment Principle: Tonify and regulate Qi and blood, crack stasis, and unblock channels.

Bie Jia Jian Wan [Turtle Shell Decoction Pill]: This pill can also be taken with Qi tonic and Yin-nourishing decoction to achieve simultaneous draining and tonifying effect.

SCARLET FEVER

Scarlet fever is called Lan Hou Sha or Yi Hou Sha in Chinese medicine. It is caused by warm heat epidemic toxin that enters the body from mouth and nose, directly attacks Lung and Stomach, and disturbs the muscles and blood vessels. The main manifestations are fever, angina, widespread, fresh, red, small-bumps skin rash, and peeling of the skin after the fever subsides. A small number of patients may have Heart and Kidney conditions 2 to 3 weeks after the fever. The tongue is dry and dark red like a plum, without coating. There can also be symptoms that are chronic and difficult to treat such as red swollen joints, Heart palpitation, and edema. Chinese medicine focuses on clearing heat and relieving toxins as the basic treatment principle. At the same time, according to the manifestations at different stages, assisting principles are acrid-cool to vent the exterior, clearing Qi, cooling Ying, nourishing Yin, and clearing heat.

Common Type

Evil Pathogen Attacking the Lung and Wei Syndrome

Treatment Principle: Vent exterior, drain heat, clear heat, and benefit the throat.

Use Yin Qiao San [Honeysuckle and Forsythia Powder] plus Shan Zhi Zi, Shui Niu Jiao, Chi Shao Yao, Xuan Shen, Zhe Bei Mu, Jiang Can, and Mu Dan Pi.

Toxins Accumulate and Congested at Qi Level Syndrome

Treatment Principle: Clear Qi and relieve toxins.

Use Qing Xin Liang Ge San [Clear the Heart and Cool the Diaphragm Powder].

Related condition and modification are

- Severe epidemic toxins, severe painful throat, and swollen neck: Add Pu Gong Ying, Ban Lan Gen, and Qing Dai.

Scorching Qi and Ying Levels Syndrome

Treatment Principle: Clear Qi and cool Ying.

Use Qing Ying Tang [Clear Ying Level Decoction] plus Bo He, Shi Gao, and Bai Mao Gen.

Residual Evil Pathogens Damaged Yin Syndrome

Treatment Principle: Nourish Yin, generate fluid, and clear remnant heat.

Use: Xi Yang Shen, Bai Shao Yao, Mai Men Dong, Sheng Di Huang Tian Men Dong, Tian Hua Fen, Xuan Shen, Zhi Mu, and Gan Cao.

Mild Type

Treatment Principle: Acrid cool to release exterior, disperse Lung, and vent rash.

Use: Yin Qiao San [Honeysuckle and Forsythia Powder] plus Da Qing Ye, Sheng Di Huang, and Mu Dan Pi.

Poisoning Type

Treatment Principle: Drain fire, relieve toxins, clear heat, and open orifices.

Use: Qing Wen Bai Du Yin [Clear Epidemics and Overcome Toxin Decoction] plus An Gong Niu Huang Wan [Calm the Palace Pill with Cattle Gallstone], one pill twice a day. One needs to open internal close and rescue external collapse. Use Shen Fu Tang [Ginseng and Prepared Aconite Decoction]. Take the decoction frequently or feed through nose.

Toxic Pus Type

Treatment Principle: Clear heat, relieve toxins, and expel pus.

Huang Lian Jie Du Tang [Coptis Decoction to Relieve Toxicity] plus Wu Wei Xiao Du Yin [Five-Ingredient Decoction to Eliminate Toxin].

Furthermore, one should also treat complications of the main syndrome:

Myocarditis

Treatment Principle: Benefit Qi, invigorate blood, transform stasis, and recover pulse.

Use Sheng Mai San [Generate the Pulse Powder] plus Huang Qi, Huang Lian, Hong Hua, Suan Zao Ren, Dan Shen, and Zhi Gan Cao.

Arthritis

Treatment Principle: Clear heat, expel wind, eliminate dampness, and disperse Bi pain.

Use Xuan Bi Tang [Disband Painful Obstruction Decoction] plus Huang Bai, Qin Jiao, and Sang Zhi.

Glomerulonephritis

Treatment Principle: Clear heat, drain damp, cool blood, and stop bleeding.

Use Er Miao San [Two-Marvel Powder] plus Xiao Ji Yin Zi [Cephalanoplos Decoction].

SCHISTOSOMIASIS

This syndrome belongs to Gu Du or Gu Zhang in TCM. It is caused by worm infection (blood fluke). The pathogenic factor is damp–heat and the condition is lingering and hard to recover from. At the acute stage, the manifestations are fever, urticaria, dermatitis, tender areas in Liver region, enlargement of Liver, and increased acid granulocytes in blood. At chronic stage, there are dysenteric diarrhea, repeated attacks, and enlargement of Liver and Spleen. At the late stage, the main characteristic is Liver cirrhosis with extremely enlarged Spleen and ascites. There are no satisfactory Chinese herbs that can kill the worms, but herbs can reduce the toxic reaction, control fever, improve the digestive system, and prevent or improve fibrosis. The treatment principles are to transform damp and clear heat during acute stage, smooth Liver and strengthen Spleen during chronic stage, and support Zheng Qi, reduce ascites, and remove accumulations and clumps during the late stage.

Acute Stage

Treatment Principle: Transform damp and clear heat.

Use Qing Pi Yin [Clear the Spleen Decoction] plus Liu Yi San [Six-to-One Powder].

Related conditions and modifications are

- Severe dampness: Add Bai Dou Kou, Yi Yi Ren, and Xing Ren.
- Stifling sensation at chest and cough with sticky phlegm: Add Huang Lian, Huang Qin, and Ma Chi Xian.
- Jaundice: Add Yin Chen Hao and Lian Qiao.
- Fever and chills: Add Jin Yin Hua and Lian Qiao.

Chronic Stage

Treatment Principle: Smooth liver, strengthen spleen, move Qi, invigorate blood, and scatter accumulation.

Use Chai Hu Shu Gan San [Bupleurum Powder to Spread the Liver] plus Dang Gui, Tao Ren, Chong Wei Zi, and Bie Jia.

Related conditions and modifications are

- Damp–heat loose stool: Add Huang Qin, Huang Lian, and Mu Xiang.
- Tenesmus, pus, and blood in stool: Add Di Yu Tan, Huang Bai, Qin Pi, and Bai Tou Weng.
- Severe abdominal pain: Add Chuan Lian Zi and Xiao Hui Xiang.
- Abdominal distention: Add Hou Po and Da Fu Pi.
- Poor appetite: Add fried Bai Zhu, Ji Nei Jin, and Shan Zha.
- Ascites: Add Jiao Mu and Qian Niu Zi.
- Hypochondriac distention and pain: Add Chai Hu and Yan Hu Suo.

Late Stage

Please refer to the sections on liver cirrhosis.

TAENIASIS

Tapeworm is called Cun Bai Chong (white inchworm). Clinically, the characteristics of taeniasis include a history of ingesting raw pork or beef, dull abdominal pain, indigestion, itching of the anus, proglottids (section of the tapeworm containing reproductive organs) found in stool, diarrhea, emaciation, and fatigue. Chinese herbs are effective to expel tapeworms.

Expel Parasites

Choose one of the following formulas:

1. Nan Gua Zi Fen Bing Lang Jian [Pumpkin Seed Powder and Betel Nut Decoction]: Take the decoction in the morning on an empty stomach. Wait for 1 hour, then take Xuan Ming powder 20 g with water. One should have bowel movement a while later and the tapeworm should come out with the stool. Nan Gua Zi acts on the middle and tail end of the tapeworm and Bing Lang mainly acts on the head section of the tapeworm.
2. Ku Lian Pi 60 g, Bing Lang 180 g. Decoct with 1000 mL of water until 250 mL left. Take the whole decoction at once on an empty stomach in the morning.

Treatment After Expelling the Parasites

Use Xiang Sha Liu Jun Zi Tang [Six-Gentleman Decoction with Aucklandia and Amomum] plus Shen Qu and Da Zao.

TETANUS

This is also called Po Shang Feng in TCM. It is caused by trauma and wind evil toxin invasion (toxin from *Clostridium tetani*) obstructing the muscles, channels, and Ying and Wei, causing spasms of the tendons. The evil toxins can transform into heat and enter the interior and generate wind. Evil pathogens in both interior and exterior level can cause muscle spasms and opisthotonos. The treatment principle in TCM focuses on expelling wind, relieving toxicity, and calming spasms. Yu Zhen San [True Jade Powder] is used topically.

Mild Case

Treatment Principle: Expel wind, unblock channels, relieve toxicity, and stop spasms.

Use Wu Hu Zhui Feng Tang [Five-Tiger Decoction to Pursue Wind].

Related conditions and modifications are

- Severe chills: Add Fang Feng, Jing Jie, and Bai Zhi.
- Fever with sweating and slight yellow coat: Add Jin Yin Hua, Lian Qiao, and Ju Hua.

Severe Case

Treatment Principle: Extinguish wind, relieve spasm, clear heat, and relieve toxicity.

Use Mu Gua, Fang Feng, Jiang Can, Tian Ma, Wu Zhu Yu, Dan Nan Xing, Bai Ji Li, Quan Xie, Chan Tui, Wu Gong, and Zhu Dan Zhi (pig bile).

Related conditions and modifications are

- Severe heat: Add Jin Yin Hua, Lian Qiao, Pu Gong Ying, Shi Gao, Huang Lian, and Huang Qin.
- Frequent spasms: Add Di Long and Gou Teng.
- Irritability and thirst: Xi Yang Shen and Mai Men Dong.
- Postpartum Qi and blood deficiency: Add Dang Shen, Dang Gui, Huang Qi, and Sheng Di Huang.
- Bluish-purple complexion, copious sweating, floating, big and rapid pulse: Add Shen Fu Tang [Ginseng and Prepared Aconite Decoction] and Sheng Mai San [Generate the Pulse Powder].

Zheng Qi Deficiency with Lingering Evil Pathogens

Treatment Principle: Tonify Qi, nourish blood, and unblock channels.

Use Dang Gui, Fang Feng, Bai Zhi, Dang Shen, Gao Ben, Sheng Di Huang, Bai Shao Yao, Mu Gua, Chuan Xiong, Huang Qi, Lu Lu Tong, and Si Gua Luo.

TYPHOID FEVER AND PARATYPHOID FEVER

These conditions belong to damp warm in TCM. It is caused by external summer dampness or internal injuries due to food and diet that causes internal accumulation of damp–heat obstructing the normal functions of Spleen and Stomach. Generally, it follows the progression of Wei, Qi, Ying, and Xue. The main manifestation of the syndromes is heat-damaging intestinal collaterals leading to bleeding. In severe cases, there can be Qi collapse due to excessive bleeding which drains the heat. Clinically, the manifestations include slow onset, increased body temperature in step-like fashion, then constant high fever, sallow complexion, dull expression, poor appetite, abdominal distention, borborygmus, constipation, and other poisoning symptoms. From physical examination, there are moderate pulse, enlarged Spleen, and possible slight enlargement of the Liver in children. At the beginning stage, dampness is at Wei level, thus the treatment focuses on using aromatic herbs to transform and disperse dampness. At middle stage, the dampness is at Qi level, thus one should separate and reduce damp–heat. At the late stage, one needs to cool blood and stop bleeding. For concomitant intestinal bleeding, treatment in TCM is effective.

Beginning Stage

Treatment Principle: Damp obstruction at Wei Qi syndrome.

Use Modified San Ren Tang [Modified Three-Nut Decoction].

Peak Stage

Treatment Principle: Damp–heat accumulating at Qi level syndrome.

Use Lian Po Yin [Coptis and Magnolia Bark Decoction].

Related conditions and modifications are

- Severe dampness: Add Pei Lan, Huo Xiang, and Cang Zhu.
- Severe heat: Add Shi Gao and Zhi Mu.
- Severe heat causing thirst: Add Lu Gen, Tian Hua Fen, and Shi Hu.
- Abdominal distention, constipation, or incomplete sensation with loose stool: Add Da Huang and Zhi Shi.
- Damp–heat stagnation manifesting as sudamina crystalline: Add Yi Yi Ren and Hua Shi.
- Damp–heat manifesting as jaundice: Add Yin Chen Hao, Shan Zhi Zi, and Da Huang.
- Internal obstruction of turbid phlegm, confused consciousness: Add Shi Chang Pu, Yu Jin, and Zhu Li.

Relieving Stage with Intestinal Bleeding

Evil pathogens at blood level syndrome.

Excess Blood Heat Syndrome

Treatment Principle: Cool blood, relieve toxins, and stop bleeding.

Xi Jiao Di Huang Tang [Rhinoceros Horn and Rehmannia Decoction] plus Di Yu, Zi Cao, Huai Hua Mi, Qian Cao Tan, Xian He Cao, and Ce Bai Tan.

Deficient Collapse Syndrome

Treatment Principle: Benefit Qi and relieving collapse.

Du Shen Tang [Unaccompanied Ginseng Decoction] plus Sheng Mai San [Generate the Pulse Powder].

VIRAL HEPATITIS

Viral hepatitis is caused by invasion of damp–heat epidemic toxic pathogens. The accumulated damp–heat pathogens manifest as acute hepatitis. Chronic hepatitis manifests when damp–heat epidemic toxic pathogens hide deeply in the blood level causing Zang Fu pathological changes, Qi stagnation, blood stasis, and mass formulations, thus making this disease difficult to treat. The main clinical manifestations are fatigue, decreased appetite, nausea, discomfort in upper abdomen, enlargement of Liver, and damaged Liver function with jaundice and fever. In the acute stage, the treatment should focus on clearing heat, draining damp, relieving toxicity, smoothing Liver, cooling blood, and invigorating blood. One should also pay attention to free movement of stool and urine in order to expel toxins out the body. In chronic stage, expel evil and support Zheng Qi simultaneously. Thus, the treatment consists of combining clearing heat, relieving toxicity, draining damp, smoothing Liver, tonifying Qi, strengthening Spleen, and tonifying Liver and Kidney. At the same time, invigorating blood, unblocking channels, transforming phlegm, and expelling stasis are also combined. From the modern research, it is found that Ban Lan Gen, Shan Dou Gen, Guan Zhong, Zao Xiu, Zi Cao, Hu Zhang, Da Huang, Yi Yi Ren, Tu Fu Ling, and Ban Zhi Lian have antiviral functions. Mu Dan Pi, Chi Shao Yao, Sheng Di Huang, Hong Hua, Tao Ren, and Yi Mu Cao have inhibitory function to the immunity. Xian Ling Pi, Dong Chong Xia Cao, Huang Qi, Ren Shen, Ba Ji Tian, Bai Zhu, Dang Gui, and Gou Qi Zi have stimulatory function to immune system. Wu Wei Zi, Shan Dou Gen, and Chui Pen Cao can decrease transaminase (aminotransferase). Tao Ren, Hong Hua, Dan Shen, San Qi, Chai Hu, Shan Ci Gu, Bie Jia, and Zhe Chong have antifibrosis function.

Acute Hepatitis with Jaundice

Yang Jaundice

Heat Worse than Dampness Type: Clear Heat and Drain Damp

Use Yin Chen Hao Tang [Artemisia Yinchenhao Decoction] plus Zhi Zi Bai Pi Tang [Gardenia and Phellodendron Decoction].

- Severe heat: Add Ku Shen, Pu Gong Ying, Tu Fu Ling, and Jin Qian Cao.

Dampness Worse than Heat Type: Aromatic to Transform Turbid, Drain Damp, and Clear Heat

Use Yin Chen Wu Ling San [Artemisia Yinchenhao and Five-Ingredient Powder with Poria Decoction] plus San Ren Tang [Three-Nut Decoction].

- Severe dampness: Add Cang Zhu, Dou Juan, Yi Yi Ren, and Sha Ren.

Yin Jaundice

Treatment Principle: Warm and transform cold damp.

Use Yin Chen Zhu Fu Tang [Artemisia Yinchenhao, Atractylodes, and Prepared Aconite Decoction]. Related conditions and modifications are

- Severe cold–damp with Qi stagnation and blood stasis: Add Gui Zhi, Xiang Fu, Dang Gui, Chi Shao Yao, and Bai Shao Yao.
- Mild jaundice after treatment for acute stage of hepatitis: Add Chai Shao Liu Jun Zi Tang Jia Jian [Modified Six-Gentleman Decoction with Bupleurum and Peony].

Acute Hepatitis C with Jaundice

Treatment Principle: Clear heat, drain damp, and relieve toxicity.

Use Yin Chen Hao Decoction [Artemisia Yinchenhao Decoction] with Bai Hua She She Cao, Pu Gong Ying, Yi Yi Ren, Fu Ling, Chi Shao Yao, Zi Cao, Xiang Fu, Che Qian Cao, Jin Yin Hua, and Hu Zhang.

Acute Hepatitis without Jaundice

Liver Qi Stagnation with Spleen Damp Syndrome (Often Seen in Hepatitis A & B)

Treatment Principle: Clear heat, drain damp, smooth Liver, regulate Qi, harmonize Stomach, and transport Spleen.

Use Chai Hu Shu Gan San [Bupleurum Powder to Spread the Liver] and Huo Pu Xia Ling Tang [Agastache, Magnolia Bark, Pinellia, and Poria Decoction].

Liver Qi Stagnation with Phlegm and Toxin Stasis Syndrome (Often Seen in Hepatitis C)

Treatment Principle: Relieve toxins, invigorate blood, smooth Liver, regulate Qi, and transform phlegm.

Use Xiao Chai Hu Tang [Minor Bupleurum Decoction] minus Ren Shen plus Yu Jin, Gua Lou, Bai Hua She She Cao, Jin Yin Hua, Qing Pi, Chen Pi, Fu Ling, Zhe Bei Mu, and Xiang Fu.

Related conditions and modifications are

- Severe epigastric and abdominal distention, fullness, and pain: Remove Huang Qin, add Huang Lian, Zhi Shi, and Zi Cao.
- Lumbar and knee weakness and soreness, cold limbs, apathy, and dull expression: Add Yin Yang Huo, Ba Ji Tian, and Huang Qi.
- Persistent high transaminase, purple dark tongue or with purple spots: Add Pu Huang, Wu Ling Zhi, or San Qi.

Chronic Hepatitis

Damp–Heat Obstructing Middle Jiao Syndrome

Treatment Principle: Clear heat, drain damp, cool blood, and relieve toxins.

Use Xiao Chai Hu Tang [Minor Bupleurum Decoction] minus Ren Shen, Da Zao, add Yin Chen Hao, Bai Hua She She Cao, Hu Zhang, Pu Gong Ying, Yi Yi Ren, Chi Shao Yao, and Yu Jin.

Related conditions and modifications are

- Obvious jaundice: Add Jin Qian Cao and Da Huang.
- Severe jaundice and damp-heat stagnation: Remove Ban Xia, add Jin Qian Cao, and increase dosage of Chi Shao Yao to 50 g.
- Hepatitis B with blood heat: Add Bai Mao Gen and Mu Dan Pi.
- Hepatitis C with excess heat: Add Jin Yin Hua, Ku Shen, and Zi Cao.
- Enlarged spleen, abnormal albumin-globulin ratio: Remove Ban Xia, and add Zhi Bie Jia, Mu Li, Qian Cao, and Dan Shen.
- High level of SGOT and SGPT which is due to:
  - Internal accumulation of heat toxins: Use Tu Fu Ling, Yin Chen Hao, Qian Cao, Chai Hu, Bai Shao Yao, Zhi Shi, Gan Cao, Pu Gong Ying, Bai Hua She She Cao, Pu Huang, and Wu Ling Zhi.
  - Internal stagnation of damp toxins: Use Huang Lian, Pu Gong Ying, Hu Zhang, Chi Shao Yao, Ku Shen, Gua Lou, Ban Xia, Yu Jin, Mu Xiang, Zhi Shi, Chen Pi, and Che Qian Cao.

Liver Qi Stagnation and Spleen Deficiency Syndrome

Treatment Principle: Smooth Liver, relieve stagnation, strengthen Spleen, and harmonize Middle Jiao.

Use Xiao Yao San [Relax Wanderer Powder] with Shan Yao, Yu Jin, Ji Nei Jin, and Xiang Fu.

Related conditions and modifications are

- Liver Qi stagnation turning into fire: Remove Ji Nei Jin and Shan Yao, add Xia Ku Cao, Pu Gong Ying, Bai Hua She She Cao, Mu Dan Pi, and Dan Shen.
- Liver fire disturbing the Heart: Remove Chai Hu, Bai Zhu, and Shan Yao. Add Huang Lian, Pu Gong Ying, Suan Zao Ren, Bai Hua She She Cao, Dan Shen, Yuan Zhi, Long Gu, Mu Li, and Ye Jiao Teng.
- Spleen deficiency with internal obstruction of phlegm damp: Use Er Chen Tang [Two-Cured Decoction] with Shi Xiao San [Sudden Smile Powder].
- Chronic hepatitis C which belongs to Liver Qi stagnation and blood stasis: Use Chai Hu, Jin Yin Hua, Bai Hua She She Cao, Bai Xian Pi, Zi Cao, Dan Shen, Yu Jin, and Sheng Huang Qi.

Liver and Kidney Yin Deficiency Syndrome

Treatment Principle: Nourish Yin, tonify Kidney, nourish blood, soften Liver, and assist by cooling blood and relieving toxins.

Use Yi Guan Jian Jia Wei [Augmented Linking Decoction]: Sheng Di Huang, Dang Gui, Sha Shen, Mai Men Dong, Gou Qi Zi, Bai Shao Yao, Shan Zhu Yu, Mu Dan Pi, Huang Jing, Chuan Lian Zi, Shan Yao, Sheng Mai Ya, and Sheng Gu Ya.

Related conditions and modifications are

- Stomach Yin deficiency: Remove Huang Jing, add Shi Hu, Yu Zhu, and Wu Mei.
- Liver Qi stagnation, blood heat, and blood stasis: Remove Sha Shen, Mai Men Dong, Shan Zhu Yu, and Shan Yao, and add Chai Hu, Dan Shen, Yu Jin, Qian Cao, and Bai Mao Gen.
  - If positive for Hepatitis C: Add Jin Yin Hua and Zhi Bie Jia.
  - If positive for Hepatitis B: Add Pu Gong Ying, Ye Ju Hua, and Zi Cao.
- Repeated increase in SGPT level or persistently high: Remove Sha Shen and Shan Zhu Yu, add Ren Shen, Bai Mao Gen, and San Qi powder.
  - If it is chronic hepatitis C: Add Sheng Huang Qi, Jin Yin Hua, Bai Xian Pi, and Liu Ji Nu.
- Qi deficiency: Remove Sha Shen, Huang Jing, and Shan Zhu Yu, add Dang Shen, Fu Ling, Huang Qi, Zi Cao, Dan Shen, and Bai Hua She She Cao.

Spleen and Kidney Yang Deficiency Syndrome

Treatment Principle: Strengthen Spleen, tonify Qi, warm Kidney, and support Yang.

Use Fu Zi Li Zhong Wan [Prepared Aconite Pill to Regulate the Middle] plus Wu Ling San [Five-Ingredient Powder with Poria].

Related conditions and modifications are

- Yang deficiency with severe cold; Loose stool and abdominal cold pain: Add Gan Jiang, Bu Gu Zhi, Ba Ji Tian, fried Shan Zha, fried Gu Ya, and fried Mai Ya.
- Yang deficiency with edema: Taken with Ji Sheng Shen Qi Wan [Kidney Qi Pill from Formulas that Aid the Living]
- Cold–damp: Remove Fu Zi, and add Yi Yi Ren, Jin Qian Cao, Xian Mao, Yin Yang Huo, and Dang Gui.
- Blood stasis: Add Tao Ren, Dang Gui, San Qi, and Dan Shen.
- Kidney essence deficiency and exhaustion: Add Gou Qi Zi, Tu Si Zi, Sang Ji Sheng, Ba Ji Tian, Nu Zhen Zi, Sheng Di Huang, and Niu Xi.
- Blood stasis and hypochondriac pain: Add Tao Ren, Dang Gui, and Ze Lan.

Blood Stasis and Obstruction of the Channels and Collaterals Syndrome

Treatment Principle: Invigorate blood, transform stasis, eliminating phlegm, and open the channels.

Use Modified Xue Fu Zhu Yu Tang [Drive Out Stasis in the Mansion of Blood Decoction] plus Tao Hong Si Wu Tang [Four-Substance Decoction with Safflower and Peach Pit]. Or with Da Huang Zhe Chong Wan [Rhubarb and Eupolyphaga Pill].

Related conditions and modifications are

- Damp–heat: Remove Sheng Di Huang and Pu Huang, add Chi Shao Yao, Ku Shen, Jin Qian Cao, and Ge Gen.
- Qi deficiency: Remove Hong Hua, Sheng Di Huang, and Mu Dan Pi, and add Huang Qi, Dang Shen, Bai Zhu, Fu Ling, Shan Yao, Ji Nei Jin, and Sha Ren.

- Yang deficiency: Remove Hong Hua, Sheng Di Huang, and Mu Dan Pi, and add Gui Zhi, Zhu Ling, Fu Zi, Ba Ji Tian, and Zhi Gan Cao.
- Yin deficiency: Remove Tao Ren, Hong Hua, and Pu Huang, and add Sha Shen, Bai Shao Yao, Gou Qi Zi, Nu Zhen Zi, Gui Ban, and Qian Cao.

WHOOPING COUGH

Whooping cough is called Bai Ri Ke or Dun Ke in Chinese medicine. It often occurs at the end of winter and beginning of spring. It is caused by seasonal evil pathogen (*Bordetella pertussis*) attacking Lung, phlegm obstruction, and Qi rebellion. Clinical manifestations include paroxysmal spasmodic cough, ending in a prolonged crowing or whooping respiration worse at night. The treatment principles are to descend Qi, drain Lung, transform phlegm, and stop cough. Commonly used herbs are Bai Jie Zi, fried Lai Fu Zi, Zi Su Zi, Ting Li Zi, Da Zao, and Da Qing Ye.

Catarrh Period (Mucositis, about 3 to 6 weeks)

Wind Attacking Lung Syndrome

Treatment Principle: Disperse wind, clear heat, transform phlegm, and descend Qi.

Use Zhi Sou San [Stop Coughing Powder] plus Sang Ye, Che Qian Zi, Chuan Bei Mu, Bo He, Lian Qiao, and Lu Gen.

Related conditions and modifications are

- External cold with internal fluid congestion: Add Xiao Qing Long Tang [Minor Blue-Green Dragon Decoction].
- Equally severe wind and Lung heat: Use Ma Xing Shi Gan Tang [Ephedra, Apricot Kernel, Gypsum, and Licorice Decoction] plus Yin Qiao San [Honeysuckle and Forsythia Powder].

Upper Jiao Damp–Heat Syndrome

Treatment Principle: Disband Bi pain, transform dampness, expel phlegm, and stop cough.

Use Xuan Bi Tang [Disband the Painful Obstruction Decoction] plus Gua Lou Ren, Zi Wan, Kuan Dong Hua, Dan Nan Xing, Dong Gua Ren, Chuan Bei Mu, and Qing Dai.

Combating Phlegm and Qi Syndrome

Treatment Principle: Clear Lung, transform phlegm, disperse wind, clear heat, and scatter accumulation.

Use decoct of Niu Bang Zi, Xing Ren, Bai Jie Zi, fried Zhi Ke, Zhi Sang Bai Pi, Mai Men Dong, Bai Bu, Zhi Kuan Dong Hua, and Mu Hu Die together. Dissolve 3 g of rock sugar to the decoction.

Spasmodic Cough Period (about 3 to 6 weeks)

Lung–Heat Scorched Metal Syndrome

Treatment Principle: Clear heat, drain Lung, transform phlegm, and stop cough.

Use Sha Shen, Chuan Bei Mu, Bai Jie Zi, Ting Li Zi, Bai Qian, Bai Bu, Xing Ren, Bai Guo, Zi Wan, and Qing Dai.

Related conditions and modifications are

- Dry mouth and throat: Add Mai Men Dong.
- Nausea and vomiting: Add Zhi Pi Pa Ye and Zhu Ru.
- Hemoptysis and epistaxis: Add Bai Mao Gen and Ce Bai Ye.
- Swollen eyelids and face: Add Che Qian Zi and Fu Ling.
- Chest and hypochondriac pain: Add Yu Jin and Jie Geng.
- Low-grade fever: Add Di Gu Pi and Hu Huang Lian.

Liver Fire Causing Wind Syndrome

Treatment Principle: Clear Liver, drain fire, and stop cough.

Use Long Dan Cao, Bai Shao Yao, Chai Hu, Gou Teng, Gou Qi Zi, Dai Zhe Shi, Dan Nan Xing, Qing Dai, Gan Cao, Jiang Can, Wu Gong, Bai Mao Gen, and Jiang Ban Xia.

Related condition and modification are

- Chest and hypochondriac pain: Add Yu Jin and Qing Pi.

Gallbladder Heat Attacking Lung Syndrome

Treatment Principle: Clear Gallbladder, calm cough, relieve spasm, and stop cough.

Use Decoction of Huang Qin, Di Long, Fu Ling, Ban Xia, Bai Bu, Hai Ge Ke, Zhu Ru, Jiang Can, Gan Cao, Zhi Ke, Chen Pi, and Qing Dai. Take five bags.

Spleen Deficiency with Excess Dampness Syndrome

Treatment Principle: Strengthen Spleen, clear heat, and drain damp.

Use Liu Yi San [Six-to-One Powder] plus Si Jun Zi Tang [Four-Gentleman Decoction].

Stomach Heat Attacking Lung Syndrome

Treatment Principle: Clear Stomach, drain Lung, cool blood, and relieve toxicity.

Use Bai Hu Jia Ren Shen Tang [White Tiger plus Ginseng Decoction] plus Wei Jing Tang [Reed Decoction].

Stagnation and Stasis of the Blood Vessels Syndrome

Treatment Principle: Invigorate blood, stop bleeding, disperse Lung, and transform dampness.

Use Xian He Cao, Bai Mao Gen, Dong Gua Ren, Zhi Ke, Chen Pi, Er Cha, Dang Gui, Tao Ren, Jie Geng, Xing Ren, Ban Xia, and Ma Huang.

Recovery Stage (about 2 to 3 weeks)

Spleen and Lung Deficiency Syndrome

Treatment Principle: Strengthen Spleen, benefit Lung, and stop cough.

Use Si Jun Zi Tang [Four-Gentleman Decoction] plus Wu Wei Zi, Sheng Jiang, and Da Zao, and replace Ren Shen with Xi Yang Shen.

Lung Yin Deficiency Syndrome

Treatment Principle: Nourish Lung and stop cough.

Use Mai Men Dong Tang [Ophiopogonis Decoction] plus E Jiao, Zhi Ma Dou Ling, Niu Bang Zi, Wu Wei Zi, and Xing Ren.

CHAPTER 22

Miscellaneous Disorders

CHRONIC FATIGUE SYNDROME

This condition often caused by improper diet, overexertion, and overindulgence in sexual activity which damages Heart, Liver, Spleen, and Kidney. In Western medicine this is a chronic disease with unknown etiology, however medical doctors use it commonly as an assessment after ruling out all other possible conditions. The clinical manifestations include persistent and recurrent fatigue, becoming easily tired, which does not improve after sleep or rest, and decreased daily activity to less than 50% of the normal activities. TCM treatments focus on ascending and generating Yang.

Spleen Yang not Transporting Syndrome

Treatment Principle: Warm and transport Spleen Yang.

Use Xiang Sha Liu Jun Zi Tang [Six-Gentleman Decoction with Aucklandia and Amomum] plus Yin Yang Huo, Gu Ya, Xian Mao, Mu Xiang, Gui Zhi, and Sheng Ma.

Related conditions and modifications are

- Fever due to Qi deficiency: Change to Bu Zhong Yi Qi Tang [Tonify the Middle and Augment the Qi Decoction].
- Sore throat: Add Hu Zhang and Jie Geng.
- Unrelieved low-grade fever: Add Qing Hao and Bie Jia.

Insufficient Heart Yang Syndrome

Treatment Principle: Warm Heart Yang and tonify Heart Qi.

Use Sheng Mai San [Generate the Pulse Powder] plus Rou Gui, Xian Mao, Yin Yang Huo, Shu Di Huang, Wei Ling Xian, Xian He Cao, and Zhi Gan Cao.

Related condition and modification are

- Palpitations, knotted or intermittent pulse: Change To Zhi Gan Cao Tang [Honey-Fried Licorice Decoction] plus Xian He Cao.

Liver Qi Stagnation Syndrome

Treatment Principle: Disperse Liver, regulate Qi, tonify and benefit the essence.

Use Dang Gui Shao Yao San [Tangkuei and Peony Powder] plus Ren Shen, Shu Di Huang, Xian Mao, Xian Ling Pi, Xian He Cao, Chai Hu, and Sheng Ma.

Heart and Spleen Deficiency Syndrome

Treatment Principle: Tonify and benefit Heart and Spleen.

Use Gui Pi Tang [Restore the Spleen Decoction] plus Mai Men Dong and Wu Wei Zi.

Kidney Yang Deficiency and Decline Syndrome

Treatment Principle: Warm and tonify Kidney Yang.

Use You Gui Wan [Restore the Right (Kidney) Pill] plus Rou Cong Rong and Ba Ji Tian.

HEADACHE

The basic etiologies of headache are wind, phlegm, stasis, and deficiency. It is often caused by dysfunctions of Liver, Spleen, and Kidney. Chronic intermittent headache that is severe in quality is called Head–Wind. If it only occurs at one side, then it is called one-sided Head–Wind. The headache that is located at both Tai Yang acupuncture points and is caused by wind mixed with phlegm–stasis obstructing the head orifices. In Western medicine, headaches are categorized into vascular, neurological, migraine, changes of intracranial pressure, tension, and traumatic type. Among these different types, neurological, tension, and migraine are the most common. TCM treatments during the acute stage focus on the manifestations by treating Wind–Yang, turbid phlegm, and blood stasis. During chronic stage or between attacks, the treatment principle is to tonify Spleen, Liver, and Kidney.

Caution: Herbs that guide to particular meridians are added to the formulas according to syndrome differentiation to strengthen effectiveness or:

- Pain along Tai-Yang meridian, headache with neck stiffness, lumbar, and back pain: Add Qiang Huo and Gao Ben.
- Pain along Shao-Yang meridian, headache at temple, ear, and corner of the head: Add Chuan Xiong and Huang Qin.
- Pain along Yang-Ming meridian, headache at forehead and eyebrow: Add Bai Zhi, Sheng Ma, and Ge Gen.
- Vertex headache, cold extremities in severe cases, and wiry pulse which is Jue-Yin headache: Add Wu Zhu Yu and Chai Hu.
- Headache with toothache, cold feet, rebellious Qi, cold at chest and back, and deep thin pulse which is Shao Yin headache: Add Xi Xin and Fu Zi.
- Headache with dizziness and heaviness, copious phlegm, and deep, moderate pulse which is Tai Yin headache: Use Cang Zhu and Ban Xia.

Liver Yang Headache Syndrome

Treatment Principle: Calm Liver and subdue Yang.

Use Tian Ma Gou Teng Yin [Gastrodia and Uncaria Decoction] plus Sheng Mu Li and Sheng Long Gu.

Related conditions and modifications are

- Liver and Kidney Yin deficiency and headache that is better in the morning and worse in the evening, and worse with exertion: Add Bai Shao Yao, Nu Zhen Zi, and Shi Hu.
- Hyperactive Liver fire: Add Long Dan Cao, Xia Ku Cao, and Chai Hu.

Turbid Phlegm Headache Syndrome

Treatment Principle: Transform phlegm and descend rebellion.

Use Ban Xia Bai Zhu Tian Ma Tang [Pinellia, Atractylodes Macrocephala, and Gastrodia Decoction] plus Man Jing Zi and Shi Chang Pu.

Related conditions and modifications are

- Phlegm stagnation transforming into heat: Remove Bai Zhu, and add Huang Lian, Zhi Shi, and Zhu Ru.

Blood Stasis Headache Syndrome

Treatment Principle: Invigorate blood and transform stasis.

Use Tong Qiao Huo Xue tang [Unblock the Orifices and Invigorate the Blood Decoction] plus Mao Dong Qing and Yu Jin.

Related conditions and modifications are

- Severe pain: Add Quan Xie, Wu Gong, and Di Long.
- With cold: Add Xi Xin, Bai Zhi, and Gui Zhi.
- Qi and blood deficiency: Add Dang Gui, Huang Qi, and Dang Shen.

Qi and Blood Deficiency Headache Syndrome

Treatment Principle: Tonify Qi and nourish blood.

Use Ba Zhen Tang [Eight-Treasure Decoction] plus Huang Qi, Gou Qi Zi, Man Jing Zi, and Chai Hu.

Related conditions and modifications are

- Palpitation with or without fright: Add Long Yan Rou, Suan Zao Ren, and Bai Zi Ren.
- Severe Qi deficiency: Remove Shu Di Huang, and add Sheng Ma and Ge Gen.

Kidney Deficiency Headache Syndrome

Treatment Principle: Tonify Kidney and supplement marrow.

Use Da Bu Yuan Jian [Great Tonify the Basal Decoction].

Related conditions and modifications are

- Kidney Yang deficiency: Add Lu Jiao and Fu Zi.
- Kidney Yin deficiency: Add Gui Ban, Bie Jia, and Nu Zhen Zi.

Atypical Symptoms for the Headache

Symptoms are atypical, which makes syndrome differentiation difficult. The following formula can be used as a base with modification.

Formula consists of Bai Shao Yao, Chuan Xiong, Zhi Nan Xing, Ju Hua, Xiang Fu, Bai Zhi, and Gan Cao.

Related conditions and modifications are

- Headache that is worse with wind–cold: Add Gui Zhi and Fang Feng.
- Headache occurs with heat, worse in the summer: Remove Zhi Nan Xing and Xiang Fu, and add Shi Gao and Huang Qin.
- Headache with heavy sensation, and heavy limbs: Remove Bai Shao Yao, and add Qiang Huo, Cang Zhu, and Man Jing Zi.

- Headache caused by emotional disturbance: Remove Zhi Nan Xing and Chuan Xiong, and add Shi Jue Ming, Tian Ma, and Jiang Can.
- Headache that is sharp, pulsating, and headache occurs or worsens before menstruation: Remove Bai Zhi and Zhi Nan Xing, and add Tao Ren, Hong Hua, Chi Shao Yao, and Dang Gui.
- Yang deficient constitution, aversion to cold, and cold hands and limbs: Add Fu Zi, Gui Zhi, and Dang Gui.
- Chronic headache, stubborn pain due to wind, phlegm, stasis, and stagnation obstructing the orifices: Add Quan Xie, Jiang Can, Di Long, and Yan Hu Suo.

RESTLESS LEG

There is no equivalent condition in TCM. However, after syndrome analysis, the condition is caused by Qi and blood disharmony, blood deficiency failing to nourish muscle, tendons, and channels, and unsmooth flow of Qi which fails to control and regulate normal movements and resting state. In Western medicine, the cause is not clear, but some doctors believe it is due to increased excitation of sympathetic nerves, spasms of vessels, and ischemia and hypoxia of the tissues. Patient often reports soreness, distention, itching, and "stranded" sensation or discomfort that is difficult to describe at the lower part of the leg. Some people can also have upper leg or upper limbs involvement. It often occurs bilaterally, better during the day and worse at night. All lab tests appear normal. TCM treatments focus on regulating Qi and blood and unblocking channels and collaterals. The following is the base formula to use and modified according to presenting symptoms.

Formula: Shao Yao Gan Cao Tang [Peony and Licorice Decoction] plus Mu Gua and Chuan Niu Xi. The dosage for Bai Shao Yao can be up to 30 g.

Related conditions and modifications are

- Worse with fatigue, apathy, tiredness; dark red tongue and thin, choppy pulse: Use Bu Yang Huan Wu Tang [Tonify the Yang to Restore Five-Tenths Decoction] plus Mu Gua and Chuan Niu Xi.
- Five-center-heat, irritability, dry mouth, dry throat, red tongue, scanty tongue coating, and thin, rapid pulse: Add Sheng Di Huang, Zhi Mu, Bie Jia, and Hu Huang Lian.
- Blood deficiency: Add Dang Gui, Huang Qi, Ji Xue Teng, Dan Shen, and Lu Lu Tong.

SENILE DEMENTIA

This condition is often caused by Kidney deficiency and marrow emptiness; Spleen deficiency leading to internal phlegm–damp generation; and Liver deficiency failing to regulate blood and causing stagnation and stasis in TCM. In Western medicine, senile dementia refers to progressive decline in several dimensions of intellectual function at old age, which mainly includes Alzheimer's disease and cerebrovascular dementia. The manifestations include decline in memory, judgment, calculation, recognition, speech and thinking, as well as a change in personality. This often occurs in people older than 60. TCM treatments focus on tonifying Spleen and Kidney, nourishing Liver blood, transforming stasis, expelling phlegm, and calming the Shen.

Insufficient Bone Marrow Syndrome

Treatment Principle: Tonify Kidney, supplement marrow, and benefit intelligence.

Use Zuo Gui Wan [Restore the Left (Kidney) Pill] plus Zhi He Shou Wu, Shi Chang Pu, Yuan Zhi, Yu Jin, Dan Shen, Xu Duan, Chuan Xiong, Zi He Che, and Tian Men Dong.

Spleen and Kidney Deficiency Syndrome

Treatment Principle: Strengthen Spleen and tonify Kidney.

Use Gui Pi Tang [Restore the Spleen Decoction] plus Jin Gui Shen Qi Wan [Kidney Qi Pill from Golden Cabinet].

Phlegm Stasis Obstructing Orifices Syndrome

Treatment Principle: Strengthen Spleen, scour out phlegm, invigorate blood, and transform stasis.

Use Dao Tan Tang [Guide Out the Phlegm Decoction] plus Tong Qiao Huo Xue Tang [Unblock the Orifices and Invigorate the Blood Decoction].

Qi and Blood Deficiency Syndrome

Treatment Principle: Tonify Qi and blood and nourish Heart and Shen.

Use Ba Zhen Tang [Eight-Treasure Decoction] plus Chai Hu, fried Suan Zao Ren, and Zhi He Shou Wu.

Above syndromes can occur independently or in combination with other syndromes. Liver Qi stagnation, Yin deficiency, and Yang deficiency or hyperactive Yang can also occur. Thus, above formulas can be modified accordingly.

SMOKING ADDICTION

In TCM, when a person is addicted to smoking, the smoke toxins accumulate in the body, damage Zang and Fu organs, disturb blood flow, and lead to Heart and Lung Qi deficiency, Lung and Kidney Yin deficiency, weak Spleen and Stomach, and damp–heat in the Middle Jiao. Chinese herbal treatment in combination with acupuncture can relieve toxicity, expel toxins, and restore normal functions of Qi and blood, thus helping smokers to eliminate the addiction to cigarettes. The treatment principles are to clear and dissolve nicotine, reduce craving, change the taste of cigarettes in mouth (more bitter, spicier, or bland without taste), and increase will power.

Lung and Heart Qi Deficiency

Smoking for long term and weak constitution.

Treatment Principle: Tonify Qi and calm Shen.

Use Jiu Xian San [Nine-Immortals Powder] plus Xing Ren, Sang Ye, Jin Yin Hua, Dan Zhu Ye, and Mai Men Dong.

Related conditions and modifications are

- Shortness of breath, easily catching colds, and spontaneous sweating: Add Yu Ping Feng San [Jade Windscreen Powder].
- Palpitation and restlessness: Combine with Suan Zao Ren Tang [Sour Jujube Decoction].
- Insomnia: Add Ye Jiao Teng and Zhi He Shou Wu.

Lung and Stomach Damp–Heat

Heavy smoker, damp–heat constitution, and increased craving after meals or with coffee.

Treatment Principle: Clear and drain damp–heat.

Use Qing Wei San [Clear the Stomach Powder] plus Qing Fei Yin [Clear the Lung Decoction].

Related conditions and modifications are

- Severe heat: Add Lian Qiao, Shan Zhi Zi, Shi Gao, and Huang Qin.
- Severe damp: Add Ze Xie and Yi Yi Ren.
- Constipation: Add Da Huang and Fan Xie Ye.

Spleen and Stomach Weakness

Weak digestive system and cravings increased with hunger.

Treatment Principle: Tonify and benefit Spleen and Stomach.

Use Shen Ling Bai Zhu San [Ginseng, Poria, and Atractylodes Macrocephala Powder] and Yi Wei Tang [Benefit the Stomach Decoction].

Related conditions and modifications are

- Shortness of breath, too lazy to talk, and sinking sensation: Change to Bu Zhong Yi Qi Tang [Tonify the Middle and Augment the Qi Decoction].
- Spleen deficiency with damp, nausea and vomiting phlegm and saliva, and loose stool: Huang Lian Wen Dan Tang [Coptis Decoction to Warm the Gallbladder] plus Zi Su Ye.

Liver and Kidney Yin Deficiency with Blazing Deficient Heat

Long-term light smoker and weak and warm constitution. Strong cravings in the evening and night.

Treatment Principle: Nourish Liver and Kidney and guide heat downwards.

Use Da Bu Yin Wan [Great Tonify the Yin Pill] plus Yi Guan Jian [Linking Decoction].

Related conditions and modifications are

- Severe Yin deficiency with dry mouth and dry tongue: Add Er Zhi Wan [Two-Ultimate Pill].
- Irritability, agitation, night sweats, insomnia, and difficulty focusing: Add Wu Wei Zi, Di Gu Pi, Ye Jiao Teng, Suan Zao Ren, and Bai Zi Ren.
- Frequently stops smoking and picks up smoking again: Add Yuan Zhi, He Huan Pi, Rou Cong Rong, Tu Si Zi, and Zhi Da Huang.

Furthermore, patient should not drink coffee, tea, and other foods or drinks that contain nicotine or caffeine.

CHAPTER 23

Musculoskeletal Disorders

CERVICAL SPONDYLOSIS

This condition is discussed under Bi syndrome, neck pain, and shoulder pain in TCM. It is caused by traumatic injury and wind–cold–damp invasion with Liver and Kidney deficiency failing to nourish tendons and bones. It often occurs after 40 years of age. In Western medicine, cervical spondylosis describes symptoms caused by cervical vertebrae hyperplasia, cervical intervertebral disc degeneration, and irritation or compression of the nearby tissues and nerves. Clinical characteristics are C5 and C6 involvement, soreness, and pain at neck radiating down to arm; vertigo, dizziness, numbness, pain, or mild degree of weakness or paralysis of the affected upper extremity. TCM treatments focus on tonifying Liver and Kidney, unblocking channels and collaterals, invigorating blood, and stopping pain. One can use the following formula as base formula and modify accordingly: Shao Yao Gan Cao Tang [Peony and Licorice Decoction] plus Mu Gua, Ji Xue Teng, Wei Ling Xian, and Ge Gen.

Related conditions and modifications are

- Severe cold–damp, severe pain: Add Gui Zhi, Qiang Huo, and Xi Xin.
- Liver and Kidney deficiency: Add Xu Duan, Gou Qi Zi, Shu Di Huang, and Gu Sui Bu.
- Qi and blood deficiency: Add Dang Gui, Huang Qi, Gui Zhi, and Gou Qi Zi.
- Severe blood stasis: Add Dan Shen, Chuan Xiong, and San Qi powder.
- Spleen Deficiency: Add Sha Ren, Fu Ling, Bai Zhu, and Shan Zha.

Further, one can also combine patent medicines such as Jian Bu Hu Qian Wan [Hidden Tiger Pill to Strengthen Stride] and Gu Zhi Zeng Sheng Pian [Bone Hyperplasia Tablet].

FEMORAL HEAD NECROSIS

This condition is discussed under Bi syndrome in TCM. Qi and blood deficiency, Qi and blood stasis, Kidney deficiency, and decreased bone density can cause this condition. In Western medicine, it is seen as being caused by traumatic injury and long-term use of steroids or infections (TB or osteomyelitis) which lead to insufficient blood supply to the femoral head, interruption of nourishment supply to bone tissues, and loss of activity of bone tissue. Main manifestations include persistent or intermittent pain at hip joint, difficulty squatting, increased pain sitting in cross-legged position, or external rotation of the hip joint. TCM treatment at the beginning stage focuses on invigorating blood, moving Qi, supplementing bone and marrow, and nourishing Qi and blood. One should also combine with acupuncture, Tui Na, physical therapy, and muscle strengthening exercises.

Qi Stagnation, Blood Stasis Syndrome

Treatment Principle: Move Qi, invigorate blood, and transform stasis.

Use Shao Fu Zhu Yu Tang [Drive Out Blood Stasis in the Lower Abdomen Decoction] plus Tao Ren, Xu Duan, Wu Yao, Ru Xiang, Su Mu, and Dan Shen.

Qi Deficiency and Blood Stasis Syndrome

Treatment Principle: Benefit Qi, invigorate blood, and strengthen tendons and bones.

Use Tao Hong Si Wu Tang [Four-Substance Decoction with Peach Pit] plus Dang Shen, Huang Qi, Xu Duan, and Zi Ran Tong.

Qi and Blood Deficiency Syndrome

Treatment Principle: Tonify Qi, strengthen Spleen, invigorate blood, and nourish blood.

Use Ba Zhen Tang [Eight-Treasure Decoction] plus Huang Qi, Xu Duan, and Huai Niu Xi.

Liver and Kidney Deficiency Syndrome

Treatment Principle: Nourish Liver and Kidney, invigorate blood, and nourish blood.

Use Yi Guan Jian [Linking Decoction] plus Zhi Mu, Gui Ban, Bai Shao Yao, and Huai Niu Xi. Related condition and modification are

- Yang deficiency: Add Rou Gui, Zhi Fu Zi, and calcined Zi Ran Tong.

Damp–Heat Syndrome

Treatment Principle: Clear heat, drain damp, invigorate blood, and unblock collaterals.

Use Si Miao Wan [Four-Marvel Pill] plus Fu Ling, Dang Gui, Chuan Xiong, Lu Lu Tong, and Tu Bie Chong.

For severe case of femoral head necrosis, one should also take Qi Li San [Seven Thousandths of a Tael Powder] or Die Da Wan [Trauma Pill] or Jie Gu Wan [Set the Bone Pill].

FIBROMYALGIA SYNDROME

Fibromyalgia is discussed under muscle Bi syndrome, flesh Bi syndrome, overexertion, and depression syndrome. It is often caused by external wind–cold–damp and seven emotional disturbances leading to Qi stagnation, blood stasis, and channel obstruction. In chronic cases, Liver, Spleen, and Kidney are involved as the root cause and become more deficient. In Western medicine, the AMA recognized fibromyalgia syndrome in 1987. It occurs more in females and in people between the ages of 20 and 50. The clinical characteristics are chronic extensive muscle pain, soreness, sharp pain or stiffness with insomnia, depression, and fatigue. TCM treatments focus on unblocking meridians, invigorating collaterals, dispersing and nourishing Liver, and tonifying Qi and blood. Secondary treatment principles focus on transforming phlegm, invigorating blood, clearing heat, and draining damp.

Liver Qi Stagnation

Treatment Principle: Disperse Liver, regulate Qi, and stop pain.

Use Xiao Yao San [Relax Wanderer Powder] or Chai Hu Shu Gan San [Bupleurum Powder to Spread the Liver].

Related condition and modification are

- Severe pain: Add Qin Jiao and Mu Gua.
- Liver stagnation transforming into heat: Add Huang Qin and Shan Zhi Zi.
- Blood stasis with sharp pain: Add Ji Xue Teng and Dan Shen.

Liver Yin Deficiency with Liver Qi Stagnation

Treatment Principle: Nourish Yin, soften the Liver, disperse Liver, and stop pain.

Use Yi Guan Jian [Linking Decoction] or Zi Shui Qing Gan Yin [Nourish Water and Clear the Liver Decoction].

Related conditions and modifications are

- Yin deficiency with deficient heat: Add Zhi Mu and Chuan Niu Xi.
- Severe Liver Qi stagnation: Add Chai Hu, Qing Pi, and Bai Shao Yao.

Liver Blood Deficiency with Liver Qi Stagnation

Treatment Principle: Nourish blood, soften Liver, and stop pain.

Use Dan Gui Shao Yao San [Angelica and Peony Powder] or Hei Xiao Yao San [Black Relax Wanderer Powder].

Related conditions and modifications are

- Severe blood deficiency: Add Ji Xue Teng, Dan Shen, and Da Zao.
- Excess damp: Add Mu Xiang, Chen Pi, Xiang Fu, and Yi Yi Ren.

Congealed Cold in Liver Channel

Treatment Principle: Warm Yang, expel cold, relax tendons, and invigorate collaterals.

Use Bu Gan Tang [Tonify the Liver Decoction].

Related conditions and modifications are

- Tight tendons and muscle spasms: Add Bai Shao Yao and Shen Jin Cao.
- Blurry vision: Add Qing Xiang Zi and Cao Jue Ming.

Obstruction due to Phlegm Stasis

Treatment Principle: Invigorate blood, move stasis, transform phlegm, and unblock collaterals.

Use Shen Tong Zhu Yu Tang [Drive Out Blood Stasis from a Painful Body Decoction] and Er Chen Tang [Two-Cured Decoction].

Related conditions and modifications are

- Hypochondriac pain: Add fried Chuan Shan Jia.
- Cold presentations: Add Gui Zhi.
- Heat presentations: Add Ren Dong Teng.
- Qi deficiency: Add Huang Qi, Dang Shen, and Bai Zhu.
- Blood deficiency: Add Bai Shao Yao and Shu Di Huang.

Obstruction due to Damp–Heat

Treatment Principle: Clear and drain damp–heat, disperse painful obstruction, and unblock collaterals.

Use Long Dan Xie Gan Tang [Gentiana Longdancao Decoction to Drain the Liver] plus Fang Ji, Yi Yi Ren, Mu Gua, and Sang Zhi.

Related conditions and modifications are

- Severe pain: Add Di Long and Yan Hu Suo.
- Severe heat: Add Shi Gao and Ren Dong Teng.
- Damages of Yin: Remove Che Qian Zi and Mu Tong, and add Sheng Di Huang and Bai Shao Yao.

Heart and Spleen Deficiency

Treatment Principle: Nourish blood, tonify Qi, strengthen Spleen, and calm Shen.

Use Gui Pi Tang [Restore the Spleen Decoction] plus Suan Zao Ren Tang [Sour Jujube Decoction].

Related conditions and modifications are

- Severe Spleen deficiency: Add Sha Ren, Chen Pi, Yi Yi Ren, and Bai Bian Dou.
- Severe Heart blood deficiency: Add Bai Zi Ren, Dan Shen, and He Shou Wu.

LUMBAR SPRAIN: ACUTE

This is often caused by external injuries, motor vehicle accidents, overexertion, and sports injuries, which lead to Qi stagnation, blood stasis, and obstructed channels and collaterals. In Western medicine, sudden violent force is said to cause strain, sprain, and tear of the lower back muscles, fascia, tendons, and small joints between the vertebrae, leading to lower back pain, resulting in limited movement and difficulty walking. TCM treatments focus on reducing swelling, stopping pain, moving Qi, and invigorating blood. At the same time, acupuncture treatment and Tui Na are combined.

Mild Case: Qi Stagnation Obstructing Collaterals Syndrome

Treatment Principle: Move Qi, unblock collaterals, invigorate blood, and stop pain.

Use Chai Hu Shu Gan San [Bupleurum Powder to Spread the Liver] with Mu Xiang, Chuan Lian Zi, Yan Hu Suo, Lu Lu Tong, Ru Xiang, and Mo Yao.

Severe Case: Obstruction due to Blood Stasis Syndrome

Treatment Principle: Invigorate blood, transform stasis, move Qi, and stop pain.

Use Die Da Wan [Trauma Pill] plus Chuan Shan Jia, Tao Ren, and Jiu Da Huang.

Related conditions and modifications are

- Qi stagnation: Add Yan Hu Suo and Qing Pi.
- Severe stasis: Add Huo Xue Jiu [Invigorate Blood Wine] and Xiao Huo Luo Dan [Minor Invigorate the Collaterals Special Pill].
- Constipation: Add Tao He Cheng Qi Tang [Peach Pit Decoction to Order the Qi].

LUMBAR STRAIN: CHRONIC

This condition is discussed under Bi syndrome and lower back pain in TCM. It is often caused by Kidney deficiency, with invasion of wind–cold–damp evil pathogens, or long-term overexertion or delayed treatment for acute lower back sprain leading to Qi stagnation, blood stasis, and cold congeal. The manifestations are slow onset, soreness or distending pain at lumbar area, worse after exertion, better with rest, or pain related to weather changes. TCM treatments focus on moving Qi, invigorating blood, tonifying Liver and Kidney, expelling wind, scattering cold, and eliminating damp. One should also combine with acupuncture and Tui Na.

Kidney Deficiency Syndrome

Kidney Yang Deficiency

Treatment Principle: Tonify Kidney, strengthen Yang, invigorate blood, and stop pain.

Use You Gui Wan [Restore the Right (Kidney) Pill] or Jian Bu Hu Qian Wan [Hidden Tiger Pill to Strengthen Stride].

Kidney Yin Deficiency

Treatment Principle: Nourish Kidney Yin and invigorate blood.

Use Liu Wei Di Huang Wan [Six-Ingredient Pill with Rehmannia] or Zuo Gui Wan [Restore the Left (Kidney) Pill].

The above two cases can be modified according to presenting symptoms.

- Wind–cold: Add Qiang Huo, Du Huo, and Ge Gen.
- Wind–damp: Add Mu Gua, Cang Zhu and Fang Ji.
- Sore and weak lower back: Add Xu Duan, Gou Ji, and Yin Yang Huo.

Kidney Deficiency with Cold–Damp Syndrome

Treatment Principle: Tonify Kidney, move Qi, invigorate blood, scatter cold, and eliminate damp.

Use Du Huo Ji Sheng Tang [Angelica Pubescens and Sangjisheng Decoction].

Related conditions and modifications are

- Severe cold and damp: Add Wei Ling Xian, Ba Ji Tian, Yin Yang Huo, and Du Zhong.
- Kidney deficiency: Add Gou Qi Zi, Xu Duan, Shu Di Huang, and Rou Cong Rong.
- Severe blood stasis: Add Dan Shen, San Qi, and Zhe Chong.

Blood Stasis Syndrome

Treatment Principle: Invigorate blood, transform stasis, regulate Qi, and stop pain.

Use Shen Tong Zhu Yu Tang [Drive Out Stasis from Painful Body Decoction] and Tao Hong Si Wu Tang [Four-Substance Decoction with Peach Pit and Safflower].

One can also add Jin Gui Shen Qi Wan [Kidney Qi Pill from Golden Cabinet], Zuo Gui Wan [Restore the Left (Kidney) Pill], You Gui Wan [Restore the Right (Kidney) Pill], Hu Qian Wan [Hidden Tiger Pill], Liu Wei Di Huang Wan [Six-Ingredient Pill with Rehmannia], and Yao Tong Pian [Lumbago Pill] according to the syndrome.

METAPLASTIC SPONDYLITIS

This belongs to lumbar pain and Bi syndrome. It is due to Liver and Kidney deficiency, wind, cold, damp evil invasions, and blood stasis internally obstructing channels. In Western medicine, hyperplacia rachitis, hypertrophic rachitis, senile rachitis, and degeneration of lumbar rachitis are all types of osteoarthritis with bone hyperplacia and degeneration of the cartilage. The pathological changes include degeneration of the vertebral body, intervertebral discs, and surrounding soft tissues, narrowing of the vertebral space, compressed vertebrae, and formation of bone spurs along the upper or lower rims of vertebral body. Clinical manifestations include progressive lumbar pain without history of traumatic injuries; soreness, dull pain, and limited range of motion, and occasional pain radiating to hip and thigh. TCM treatment focuses on tonifying Liver and Kidney, transforming stasis, stopping pain, dispersing wind, and scattering cold.

Beginning Stage (Severe Pain)

Treatment Principle: Invigorate blood, transform stasis, regulate Qi, stop pain, expel wind, and scatter cold.

Use Shen Tong Zhu Yu Tang [Drive Out Stasis from a Painful Body Decoction] plus Du Huo, Gui Zhi, Chi Shao Yao, Ru Xiang, Zhi Chuan Wu, Ze Xie, and Yu Jin.

Middle Stage (Endless Pain)

Treatment Principle: Regulate Qi, invigorate blood, tonify Liver and Kidney, expel wind, and unblock collaterals.

Use Du Huo Ji Sheng Tang [Angelica Pubescens and Sangjisheng Decoction] plus Xu Duan, Gu Sui Bu, and Gou Qi Zi.

Related conditions and modifications are

- Liver and Kidney Yin deficiency: Remove Rou Gui and Xi Xin and add Er Zhi Wan [Two-Ultimate Pill].
- Severe cold and damp: Add Wei Ling Xian.
- Cold–damp transforming into heat: Remove Rou Gui and Xi Xin, and add Jin Yin Hua, Lian Qiao, and Huang Bai.
- Phlegm obstruction: Add Dan Shen and Shi Xiao San [Sudden Smile Powder].

Late Stage (Pain Greatly Reduced, Only Dull Pain and Deficient Manifestations)

Treatment Principle: Tonify Liver and Kidney, benefit Qi and blood, and unblock collaterals.

Use Shi Quan Da Bu Tang [All-Inclusive Great Tonifying Decoction] plus Sang Ji Sheng, Xu Duan, Huai Niu Xi, Gou Qi Zi, Qin Jiao, Wei Ling Xian, and Du Huo.

Furthermore, one can add Zuo Gui Wan [Restore the Left (Kidney) Pill], You Gui Wan [Restore the Right (Kidney) Pill], Gu Zhi Zeng Sheng Pian [Bone Hyperplacia Tablet], Xiao Huo Luo Dan [Minor Invigorate the Collaterals Special Pill], and Da Huo Luo Dan [Major Invigorate the Collaterals Special Pill].

SOFT TISSUE NECK INJURY

This condition is discussed under tendon injury in TCM. It is often caused by overexertion, traumatic injury, wind, cold, damp evil invasion leading to Qi stagnation, blood stasis, and blocked channels and collaterals at the neck. In Western medicine, this is described as damage of muscle, fascia, and nerves of the neck due to trauma, motor vehicle accidents, or sports activities. After the soft tissue neck injury, there is local congestion, effusion, and edema. The manifestations are pain, swelling, distention, limited range of motion, stiffness, or fixed at certain posture. TCM treatment focuses on moving Qi, invigorating blood, reducing swelling, and stopping pain. One should also combine with acupuncture and Tui Na to treat the condition internally and externally.

Beginning Stage

Treatment Principle: Invigorate blood, scatter stasis, reduce swelling, and stop pain.

Use Jia Jian Huo Luo Xiao Ling Dan [Modified Fantastically Effective Pill to Invigorate the Collaterals] plus Qiang Huo, Wei Ling Xian, Chuan Niu Xi, Xiang Fu, Mu Tong, Yi Yi Ren, Di Long, Ji Xue Teng, Mu Dan Pi, Tu Bie Chong, and Ge Gen.

Related conditions and modifications are

- Headache: Add Chuan Xiong and Fang Feng.
- Severe blood stasis: Add San Qi powder.

Middle and Late Stage

Treatment Principle: Expel stasis, generate the new, invigorate blood, and stop pain.

Use Shu Jin Huo Xue Tang [Relax the Tendon and Invigorate the Blood Decoction] plus Chi Shao Yao, Su Mu, Wu Yao, Mu Tong, and Chuan Xiong.

Related conditions and modifications are

- Liver and Kidney deficiency: Add Gou Qi Zi, Gou Ji, and Shu Di Huang.
- Stiff neck and limited range of motion: Add Ge Gen, Lu Lu Tong, and San Qi powder.

Furthermore, one can also apply plasters locally to stop pain and invigorate blood which can increase the effectiveness, such as Gou Pi Gao [Dog Skin Plaster], She Xiang Zhuang Gu Gao [Musk Plaster to Strengthen the Bone], Shang Shi Zhi Tong Gao [Stop Pain Plaster due to Injury, Damp], Wai Ca Zheng Gu Shui [External Use Correct the Bone Liniment], and Zheng Hong Hua You [Origin Safflower Oil].

OSTEOPOROSIS

This condition is discussed under Bi syndrome and deficient exhaustion in TCM. It is often caused by Spleen and Kidney deficiency and exhausted essence and blood failing to nourish bone. In Western medicine, it is seen that as a person grows older, bone density and mass decreases, and per unit bone tissue content decreases. The bone becomes weak, thin, and easily fractured. It is more often seen in senior people, after menopause, and more often in females than in males. TCM treatments focus on tonifying Spleen and Kidney, regulating Qi, stopping pain, relaxing tendons and invigorating collaterals.

Spleen Qi Deficiency Syndrome

Treatment Principle: Strengthen Spleen and benefit Qi.

Use Shen Ling Bai Zhu San [Ginseng, Poria, and Atractylodes Macrocephala Powder] with Ji Nei Jin, Shan Zha, Du Zhong, and Xu Duan.

Kidney Deficiency Syndrome

Treatment Principle: Tonify Kidney and strengthen bone.

Use Tu Si Zi Wan [Cuscuta Seed Pill] plus Dang Gui, Gou Qi, Huai Niu Xi, and Xu Duan.

Spleen and Kidney Deficiency Syndrome

Treatment Principle: Tonify Spleen and Kidney.

Use Si Jun Zi Tang [Four-Gentleman Decoction] plus Zuo Gui Wan [Restore the Left (Kidney) Pill] and Gu Sui Bu, Dan Shen, Xu Duan, and Dang Gui.

Related conditions and modifications are

- Severe Kidney Yang deficiency: Add Rou Gui, Yin Yang Huo, and Ba Ji Tian.
- Severe Kidney Yin deficiency: Add Er Zhi Wan [Two-Ultimate Pill] and Shi Hu.
- Wind–damp: Add Du Huo, Qin Jiao, and Wei Ling Xian.

Furthermore, one can often use food therapy to increase the effectiveness.

Pai Gu Tang [Pork Rib Soup]: Pork ribs 1000 g, Sheng Jiang 12 g, and Chen Pi 10 g. Cook together to make soup. Take often.

Related conditions and modifications are

- Spleen Qi deficiency: Add Dang Shen, Yi Yi Ren, and Huang Qi, 20 g each.
- Kidney deficiency: Add Shan Yao, Dang Gui, and Gou Qi Zi, 20 g each.

CHAPTER 24

Neurological Disorders

BRAIN ATROPHY

This disorder shows as forgetfulness, amnesia, and atrophy syndrome in TCM. The pathogeneses are damages to the five Zang organs, and empty marrow and essence. Some scholars suggest Kidney deficiency with blood stasis. It often occurs in people over 50 years of age. Sometimes it is asymptomatic. Some have decreased memory, thinking, calculating, or motor functions. TCM treatment tonifies Kidney, supplements essence, expels stasis, unblocks collaterals, scours out phlegm, and opens orifices. Use Zuo Gui Wan [Restore the Left (Kidney) Pill] as the base formula and modify accordingly: Remove Chuan Niu Xi, and add Ren Sheng, Huang Qi, Mai Men Dong, Wu Wei Zi, and Fu Ling.

Related conditions and modifications are

- Decreased intelligence: Remove Mai Men Dong and Fu Ling, and add Yi Zhi Ren, Zhi He Shou Wu, and Sang Shen Zi.
- Insomnia with excessive dreaming: Remove Shu Di Huang, Mai Men Dong, and Lu Jiao Jiao, and add Suan Zao Ren, Ye Jiao Teng, and He Huan Pi.
- Headache, dizziness, and heavy sensation of the head: Remove Mai Men Dong and Fu Ling, and add Chuan Xiong, Tian Ma, Sheng Ma, and Yan Hu Suo.
- Flaccid extremities, unstable gait, and speech difficulty: Remove Gou Qi Zi, Mai Men Dong, and Fu Ling, and add Dan Shen, Shi Chang Pu, Jiang Can, Di Long, Xue Tong, and Jiang Xiang.

Furthermore, one can combine patent medicine, such as Liu Wei Di Huang Wan [Six-Ingredient Pill with Rehmannia], Qi Ju Di Huang Wan [Lycium Fruit, Chrysanthemum, and Rehmannia Pill], Ba Xian Chang Shou Wan [Eight-Immortals Pill for Longevity], Jin Gui Shen Qi Wan [Kidney Qi Pill from Golden Cabinet], Gui Pi Wan [Restore the Spleen Pill], Shou Wu Pian [Polygoni Tablet], and Gui Ling Ji [Aged Turtle Formula].

CARPAL TUNNEL SYNDROME

This condition belongs to tendon Bi in TCM. It is caused by wind–cold–damp evil invading the body with weak constitution, obstruction of channels and collaterals, and unsmooth flow of Qi and blood. In Western medicine, this refers to numbness and tingling of the fingers due to compression of the median nerve in the wrist area and especially common from repeated motion of the wrist. The characteristics are numbness, tingling, and sharp pain in the wrist radiating to the elbow and down to middle and index finger and thumb. The pain is more noticeable in early morning and at night. TCM treatments focus on expelling wind, eliminating damp, invigorating blood, unblocking collaterals, and stopping pain.

Cold–Damp Syndrome

Treatment Principle: Warm the channels, scatter cold, and dry damp.
Use Wu Tou Tang [Aconite Decoction] plus Can Zhu, Qiang Huo, Jiang Huang, Dang Gui, and Bai Zhu.

Wind–Cold Syndrome

Treatment Principle: Disperse wind and scatter cold.
Use Du Huo Ji Sheng Tang [Angelica Pubescens and Sangjisheng Decoction].

Damp–Heat Syndrome

Treatment Principle: Clear heat and eliminate damp.
Use Xuan Bi Tang [Disband Painful Obstruction Decoction] plus Shen Tong Zhu Yu Tang.

Blood Stasis Syndrome

Treatment Principle: Invigorate blood, transform stasis, relax tendons, and activate collaterals.
Use Huo Luo Xiao Ling Dan [Fantastically Effective Pill to Invigorate the Collaterals] plus Shen Tong Zhu Yu Tang [Drive Out Blood Stasis from the Painful Body Decoction].

Liver and Kidney Deficiency Syndrome

Treatment Principle: Nourish Liver and Kidney and strengthen tendons and bones.
Use Hu Qian Wan [Hidden Tiger Pill].

CEREBRAL THROMBOSIS (RECOVERY PERIOD)

Cerebral thrombosis belongs to wind stroke in TCM. Six months after an acute attack is the key recovery period for cerebral thrombosis. During the recovery period, the TCM presentations are Qi deficiency, blood stasis, and obstructed brain orifices. Thus, treatment focuses on tonifying Qi, invigorating blood, transforming stasis, and opening the orifices. Commonly used herbs are Dan Shen, Chi Shao Yao, Chuan Xiong, and Hong Hua. These herbs can improve microcirculation in brain, increase blood flow to brain, and inhibit platelet aggregation.

Qi Deficiency, Blood Stasis Syndrome

Treatment Principle: Tonify Qi, invigorate blood, and expel stasis.
Use Tao Hong Si Wu Tang [Four-Substance Decoction with Peach Pit and Safflower], remove Shu Di, and add Huang Qi, Dang Shen, Chi Shao Yao, and Chuan Shan Jia.

Blood Stasis Obstructing Collaterals Syndrome

Treatment Principle: Invigorate blood, expel stasis, and unblock collaterals.

Use Tong Jing Zhu Yu Tang [Unblock the Meridians and Drive Out Stasis Decoction] plus E Zhu and Qin Jiao.

Qi and Blood Deficiency with Stasis at Brain Orifices

Treatment Principle: Tonify Qi and blood and open orifices.

Use Ba Zhen Tang [Eight-Treasure Decoction] plus Tong Qiao Huo Xue Tang [Unblock the Orifices and Invigorate the Blood Decoction] and Chuan Shan Jia, Lu Lu Tong, and Mang Chong.

Above formulas can be modified according the following presenting symptoms:

- Deviation of the mouth and eye: Add Qian Zheng San [Lead to Symmetry Powder].
- Speech difficulties: Add Shi Chang Pu, Bai Fu Zi, Tian Ma, Yuan Zhi, Quan Xie, and Dan Nan Xing.
- Paralysis of upper limbs: Add Gui Zhi and Sang Zhi.
- Paralysis of lower limbs: Add Huai Niu Xi, Xu Duan, Chuan Niu Xi, and Sang Ji Sheng.
- Urinary incontinence: Add Yi Zhi Ren and Wu Yao.
- Hypertension, dizziness, and vertigo due to Liver Yang hyperactivity: Add Dai Zhe Shi, Cao Jue Ming, Ju Hua, and Gou Teng.
- Somnolence: Add Zhu Li and Shi Chang Pu.

Furthermore, one can also combine with patent medicine, such as Ren Shen Zai Zao Wan [Rebuild Pill with Ginseng], Xiao Huo Luo Dan [Minor Invigorate the Collateral Special Pill], and Fu Fang Dan Shen Pian [Newly Revised Salvia Tablet].

One can also combine with food therapy.

Excess Heat in Liver Channel

Use Tao Ren Jue Ming Cha [Peach Pit and Cassia Seeds Tea]: Decoct Tao Ren 10 g and Cao Jue Ming 12 g, and add honey and drink as tea.

Yin Deficiency with Yang Hyperactivity

Use Zhen Zhu Mu Zhou [Mother of Pearl Congee]: Decoct Zhen Zhu Mu 50 g and Sheng Mu Li 60 g in 500 mL of water for 2 hours. Remove dregs and add 100 g of Geng Mi to make congee.

Liver and Kidney Yin Deficiency

Use Gou Mai Yin [Lycium Fruit and Ophiopogonis Congee]: Decoct 30 g of Gou Qi Zi and 30 g of Mai Men Dong. Drink as tea.

Qi and Blood Deficiency with Blood Stasis

Use Huang Qi Zhu Rou Tang [Astragalus Pork Soup]: Huang Qi 30 g, Da Zao 10 pieces, Dang Gui 10 g, lean pork 100 g, and salt 3 g. Make into soup.

EPILEPSY (BETWEEN ATTACKS)

The treatment for epilepsy between attacks is prevention. During an acute attack the patient should be sent to the emergency room immediately. During the remission period, the presentations are wind, fire, phlegm, and stasis in the body. At early stage, TCM treatments focus on extinguishing wind, clearing fire, and scouring out phlegm. At middle and late stage, the treatments focus on tonifying Qi, extinguishing wind, transforming phlegm, and invigorating blood.

Wind–Fire, Turbid Phlegm Syndrome (Early Stage Excess Syndrome)

Treatment Principle: Extinguish wind, clear fire, and scour out phlegm.

Use Jia Wei Wen Dan Tang [Augmented Warm the Gallbladder Decoction] plus Huang Qin, Jiang Can, Long Dan Cao, Di Long, Shan Yang Jiao, and Dan Nan Xing.

Liver Excess and Spleen Deficiency Syndrome (Often Seen in Pediatric Epilepsy)

Treatment Principle: Calm Liver, strengthen Spleen, and scour out phlegm.

Use Dao Tan Tang [Guide the Phlegm Decoction] plus Gui Zhi, Bai Shao Yao, Jiang Can, Bai Zhu, Gou Teng, Huang Qi, and Bai Ji Li.

Qi Deficiency, Blood Stasis, and Phlegm Obstruction Syndrome (Often Seen in Middle or Late Stage)

Treatment Principle: Tonfiy Qi, transform stasis, and eliminate phlegm.

Use Xiong Gui Er Chen Tang [Two-Cured Decoction with Ligusticum Chuanxiong and Angelica] plus Huang Qi, Tai Zi Shen, Dan Nan Xing, Jiang Can, and Shui Zhi.

Furthermore, one can also add food therapy.

- Slight excess wind–phlegm: Tian Ma Zhu Li Zhou [Gastrodia and Bamboo Sap Congee].
- Slight excess damp–heat: Yi Ren Lu Dou Zhou [Coix Seed and Mung Bean Congee].
- Qi deficiency and Kidney deficiency: Shan Yao Gou Qi Huang Qi Zhou [Dioscorea, Lycium Fruit, and Astragalus Congee].
- Qi deficiency and blood stasis: Dang Gui Shan Yao Huang Qi Zhou [Angelica, Dioscorea, and Astragalus Congee].

INSOMNIA

This is often caused by Liver Qi stagnation, fire–heat, phlegm–heat, Yin deficiency, and blood deficiency, which fail to nourish Heart and ground the mind. Insomnia includes lack of duration, depth of sleep, and not feeling rested in the morning. The manifestations are difficulty falling asleep, frequently waking up, excessive dreaming, and waking early. TCM treatments focus on removing the cause, nourishing Heart, and calming the spirit.

Liver Qi Stagnation Transforming into Fire

Treatment Principle: Clear Liver and drain fire.

Use Dan Zhi Xiao Yao San [Relax Wanderer Powder with Mutan and Gardenia] plus Huang Qin, Sheng Long Gu, and Suan Zao Ren.

Internal Hyperactive Phlegm Fire

Treatment Principle: Clear heat and transform phlegm.

Use Huang Lian Wen Dan Tang [Coptis Decoction to Warm the Gallbladder] plus Sheng Long Gu, Ye Jiao Teng, and Zhu Li.

Yin Deficiency with Deficient Fire

Treatment Principle: Nourish Yin, clear fire, and calm Shen.

Use Zhi Bai Di Huang Wan [Anemarrhena, Phellodendron, and Rehmannia Pill] plus Huang Lian E Jiao Tang [Coptis and Ass-Hide Gelatin Decoction] plus Ji Zi Huang. Or use Jiao Tai Wan [Grand Communication Pill].

Heart and Spleen Deficiency

Treatment Principle: Benefit Qi and nourish blood.

Use Gui Pi Tang [Restore the Spleen Decoction].

Heart and Gallbladder Qi Deficiency

Treatment Principle: Benefit Qi, anchor fright, and calm spirit.

Use Wen Dan Tang [Warm the Gallbladder Decoction] plus Dang Shen, Yuan Zhi, Wu Wei Zi, and fried Suan Zao Ren.

Disorder of Stomach Qi

Treatment Principle: Harmonize Stomach and regulate Qi.

Use Yue Ju Wan [Escape Restraint Pill].

INTERCOSTAL NEURALGIA

This condition is discussed under chest pain and hypochondriac pain in TCM. It is often caused by Liver and Gallbladder Qi and blood stagnation, obstruction of the channels and collaterals, and Liver and Kidney Yin deficiency. In Western medicine, this refers to pain in the area where one or more intercostal nerves innervate. The characteristics are severe and sharp pain attacks. TCM treatments focus on dispersing Liver and Gallbladder, moving Qi, and invigorating blood.

Liver Qi Stagnation Syndrome

Treatment Principle: Disperse Liver and relieve stagnation.
Use Si Ni San [Frigid Extremities Powder] plus Ban Xia.
Related conditions and modifications are

- Chest pain: Add Gua Lou Xie Bai Bai Jiu Tang [Trichosanthes Fruit, Chinese Chive, and Pinellia Decoction].
- Vertigo, bitter taste in mouth, and wiry pulse: Change to Xiao Yao San [Relax Wanderer Powder].

Blood Stasis Leading to Stagnation Syndrome

Treatment Principle: Invigorate blood, move Qi, and stop pain.
Use Shi Xiao San [Sudden Smile Powder] plus Jin Ling Zi San [Melia Toosendan Powder].
Related condition and modification are

- Sharp and distending pain: Change to Xue Fu Zhu Yu Tang [Drive Out Stasis in the Mansion of Blood Decoction].

Qi Stagnation and Blood Stasis Syndrome

Treatment Principle: Move Qi, invigorate blood, and stop pain.
Use Xuan Fu Hua Tang [Inula Decoction] plus Yan Hu Suo, Wei Ling Xian, Wu Ling Zhi, Gui Zhi, Gua Lou, Bai Jie Zi, and Chuan Xiong.
Related conditions and modifications are

- More pain at hypochondrium: Add Chai Hu.
- More pain at chest: Add Yu Jin.

Liver and Gallbladder Damp–Heat Syndrome

Treatment Principle: Disperse Liver and Gallbladder, clear heat, and eliminate damp.
Use Long Dan Xie Gan Tang [Gentiana Longdancao Decoction to Drain the Liver].

Liver and Kidney Yin Deficiency Syndrome

Treatment Principle: Nourish Liver and Kidney.
Use Yi Guan Jian [Linking Decoction] plus He Huan Hua, Suan Zao Ren, Nu Zhen Zi, and Mei Gui Hua.

MULTIPLE SCLEROSIS

Multiple Sclerosis is discussed under Wei (atrophy) syndrome in TCM. It is often caused by wind, cold, and damp; and heat invading tendons and muscles, damaging body fluid and exhausting Qi. This leads to Qi and Yin deficiency, malnourishment of tendons and vessels, Qi stagnation and blood stasis over time, affecting Kidney and causing atrophy of the tendons and muscles and damaged bone. In Western medicine, this is described as a chronic progressive autoimmune disease with unknown etiology, but many researchers believe that MS is related to chronic malnutrition. MS is caused by disseminated patches of demyelination in brain and spinal cord and is seen more often in females than males and more in people 20 to 40 years of age. At the beginning stage, the symptoms are not clear and gradually there are developments of numbness and weakness of the extremities, face and body, clumsy movements, visual disturbance such as double vision, and eye pain or blurriness. At late stage, there can be paralysis, muscular atrophy, inability to walk, and inability to manage daily activities. TCM treatments focuses on strengthening Spleen, dispersing Liver, clearing heat, draining damp, and scouring out phlegm at early stage. At late stage, the treatment focuses on tonifying Liver and Kidney, invigorating blood, and transforming stasis.

Wind–Cold–Damp Syndrome

Treatment Principle: Expel wind, scatter cold, eliminate damp, and unblock collaterals.

Use Juan Bi Tang [Remove Painful Obstruction Decoction] or San Bi Tang [Three Painful Obstructions Decoction].

- Numbness of upper limb: Increase Qiang Huo and Gui Zhi to 15 to 25 g.
- Numbness of lower limb: Add Fang Ji and increase Chuan Niu Xi to 15 g.
- Severe cold–damp: Add Zhi Chuan Wu.
- Transformation of heat: Add Huang Bai, Dan Nan Xing, and Tu Fu Ling.

Liver Deficiency and Cold in Du Channel Syndrome

Treatment Principle: Tonify Kidney, strengthen Du channel, warm the meridian, scatter cold, and invigorate blood stasis.

Use You Gui Wan [Restore the Right (Kidney) Pill] plus Xu Duan, Gou Ji, Yin Yang Huo, Yi Yi Ren, Jiang Can, Gan Jiang, Du Huo, Fang Feng, and Huai Niu Xi.

Related conditions and modifications are

- Numbness and weakness of upper limb: Add Sang Ye and Gui Zhi.
- Numbness and weakness of lumbar and spine: Add Sang Ji Sheng and Qiang Huo.
- Chronic sickness: Ze Lan and San Qi powder.

Kidney Deficiency with Empty Marrow Syndrome

Treatment Principle: Tonify Kidney, supplement marrow, invigorate blood, and unblock collaterals.

Use San Bi Tang [Three Painful Obstructions Decoction] plus Wu Gong, Gu Sui Bu, Lu Feng Fang, Bu Gu Zhi, Hong Hua, Ji Xue Teng, Zhi He Shou Wu, and Shu Di Huang.

Obstruction due to Blood Stasis Syndrome

Treatment Principle: Invigorate blood, transform stasis, and unblock collaterals.

Use Tao Hong Si Wu Tang [Four-Substance Decoction Peach Pit and Safflower] plus Shi Xiao San [Sudden Smile Powder] and add Qiang Huo, Mo Yao, Huai Niu Xi, Quan Xie, Wu Gong, Ru Xiang, and Lu Lu Tong.

Related conditions and modifications are

- Severe above the waist: Remove Huai Niu Xi and add Jiang Huang.
- Severe below the waist: Add Chuan Niu Xi and Wu Shao She.

Liver and Kidney Yin Deficiency with Deficient Heat

Treatment Principle: Nourish and tonify Liver and Kidney and clear deficient heat.

Use Zhi Bai Di Huang Wan [Anemarrhena, Phellodendron, and Rehmannia Pill] plus Shen Jin Cao, Nu Zhen Zi, Shi Hu, and Tian Men Dong.

Related conditions and modifications are

- Numbness and heaviness: Add Cang Zhu and Tu Fu Ling.
- Muscular atrophy: Increase Huang Qi by 30 to 50 g and add Lu Feng Fang and Dan Shen.

Spleen and Kidney Yang Deficiency Syndrome

Treatment Principle: Warm and tonify Spleen and Kidney, benefit Qi, invigorate blood, and unblock collaterals.

Use You Gui Wan [Store the Right (Kidney) Pill] plus Si Jun Zi Tang [Four-Gentleman Decoction] and Yin Yang Huo, Gan Jiang, Dan Shen, Tu Bie Chong, Bai Jie Zi, and Gu Sui Bu.

Related conditions and modifications are

- Blood stasis: Add Ru Xiang, Mo Yao, Wu Gong, Xue Jie, and Yan Hu Suo.
- Diarrhea, loose stool, and poor appetite: Add Sha Ren, Mu Xiang, Yi Yi Ren, Shan Zha, and Shen Qu.

MYASTHENIA GRAVIS

This condition is similar to atrophy syndrome and drooping eyelids. It is caused by Spleen and Stomach deficiency failing to nourish muscles leading to weakness and atrophy. In Western medicine, this is a chronic condition due to autoimmune disorder causing transmission obstruction of motor neurons. The clinical manifestations are easy fatigue of affected muscles, most often affecting the eye muscles first, then throat, and tongue muscles; followed by neck, shoulder, upper limbs, trunk, and lower limb muscle groups. Symptoms are usually better in the morning, worse at night, and better with rest. In mild cases, there are drooping eyelids, double vision or squint, and limited range of motion of the eyeballs. In severe cases, there are weakness of the four limbs, whole body fatigue, difficulty swallowing, choking with drinking water, soft neck, tilted head, weak mastication, shortness of breath, mumbled speech or even inability to manage daily activity, and breathing difficulty and in critical conditions. TCM treatment uses big dosages of sweet and warm herbs to tonify Spleen and strengthen Qi, combined with very acrid and very warm herbs to tonify Spleen and strengthen Yang. One can use the following formulas and modify them accordingly.

Treatment Principle: Tonify Spleen, benefit the Qi, and ascend Yang.

Use Bu Zhong Yi Qi Tang [Tonify the Middle and Benefit the Qi Decoction] plus Fu Ling and Huang Qin.

Related conditions and modifications are

- Liver blood deficiency: Add Zhi He Shou Wu and Gou Qi Zi.
- Soft neck and weak lower back: Add Gou Qi Zi, Tu Si Zi, and Xu Duan.
- Frequent nocturia: Ad Du Zhong and Sang Piao Xiao.
- Kidney Yang deficiency, aversion to cold, and cold limbs: Add Ba Ji Tian and Zhi Fu Zi.
- Thick white tongue coating: Add Yi Yi Ren and Huo Xiang.
- Difficulty swallowing: Remove Chen Pi, and add Zhi Shi and Jie Geng.

- Sticky and thick phlegm: Add Zhe Bei Mu, Yu Xing Cao, and Ting Li Zi.
- Severe cases of myasthenia gravis: In acute cases use Zhi Fu Zi, Gan Jaing, Hong Shen, Zhi Gan Cao, and Huang Qi to make decoction. Take the decoction orally or through nose; or add ground Ma Qian Zi 0.5 g and take once a day. Patient then is closely monitored. If clenched jaws and spasms of the body occur, then one should stop taking Ma Qian Zi.

Note: Ma Qian Zi is extremely toxic and one should be careful while taking it.

Preparation of Ma Qian Zi: Take appropriate amount of herb, soak in water for 7 days, and change water every day. Remove the herbs in 7 days, remove the skin, and cut into thin slices. Dry under the sun. Fry with oil until golden and grind into powder to be used.

PARKINSON'S DISEASE

This is often caused by Liver and Kidney Yin deficiency, failing to nourish tendons, Qi and blood deficiency, and spasms of channels and muscles, generated fire–phlegm, and fire and stasis from deficiency which worsen the deficiency. In Western medicine, this is an extra pyramidal pathology, which occurs in senior people. The clinical characteristics include slow onset, progressively worsening with manifestations of tremors, stiffness, slow movement, mask-like face, head tilted forward, stooped posture, difficulty maintaining balance, shuffling, short-stepped gait, and absence of arm swing with stride. TCM treatment nourishes Liver and Kidney, tonifies Qi, generates blood, extinguishes wind, and unblocks collaterals. Commonly used herbs are Shu Di Huang, Gou Qi Zi, Shan Zhu Yu, Sang Ji Sheng, He Shou Wu, Gui Ban, Bie Jia, Dang Gui, Xu Duan, Ji Xue Teng, Huang Qi, Dang Shen, Huang Jing, Zhen Zhu Mu, Long Gu, Mu Li, Gou Teng, Tian Ma, Jian Can, and Bai Ji Li. Among these herbs, Gou Teng and Zhen Zhu Mu are first choices. The dosage can vary from 30 to 50 g. This formula should combine with blood-nourishing and blood-invigorating herbs, such as Dang Gui, Chi Shao Yao, Bai Shao Yao, Dan Shen, Chuan Xiong, Hong Hua, Chuan Niu Xi, and Tao Ren, and also insects such as Quan Xie, Di Long, Wu Gong, and Jiang Can to search, chase, and disperse wind; and unblock collaterals, extinguish wind, and stop spasms.

Liver and Kidney Yin Deficiency Syndrome

Treatment Principle: Nourish Liver and Kidney and support Yin to extinguish wind.

Use Da Ding Feng Zhu [Major Arrest Wind Pearl] plus Gui Ban, Wu Gong, Jiang Can, and Gou Teng.

Related conditions and modifications are

- Yin deficiency with deficient fire: Add Zhi Mu, Huang Bai, and Xuan Shen.
- Blood stasis obstructing collaterals: Add Chi Shao Yao, Ji Xue Teng, Dang Gui, Dan Shen, and Chuan Niu Xi.
- Damaged Yin affecting Yang: Add Ba Ji Tian, Rou Gui, and Rou Cong Rong.

Qi and Blood Deficiency Syndrome

Treatment Principle: Tonify Qi, nourish blood, and extinguish wind.

Use Ba Zhen Tang [Eight-Treasure Decoction] plus Huang Qi, Tian Ma, Qin Jiao, Quan Xie, Fang Feng, Jing Jie, Xi Xin, and Wei Ling Xian.

Related conditions and modifications are

- Severe trembling and spasms of tendons: Add Jiang Can and Wu Gong.
- Insomnia: Add fried Suan Zao Ren.

Spleen Deficiency and Phlegm Stasis Syndrome

Treatment Principle: Strengthen Spleen, scour out phlegm, and unblock collaterals.

Use Dao Tan Tang [Guide the Phlegm Decoction] plus Tian Ma Gou Teng Yin [Gastrodia and Uncaria Decoction].

Related condition and modification are

- Abdominal distention and poor appetite: Add Mu Xiang, Shan Zha, and Hou Po.

Qi Stagnation and Blood Stasis Syndrome

Treatment Principle: Move Qi, invigorate blood, and unblock collaterals.

Use Shen Tong Zhu Yu Tang [Drive Out Blood Stasis from a Painful Body Decoction].

Related conditions and modifications are

- Numbness of the limbs and body: Add Ji Xue Teng and He Shou Wu.
- Qi deficiency: Add Dang Shen and Huang Qi.
- Unceasing trembling: Add Quan Xie, Wu Gong, and Jiang Can.

POLYNEURITIS

This condition is discussed under numbness, tingling, and atrophy syndrome in TCM. It is often caused by internal invasion of damp–heat, Bi obstructing channels and collaterals, and in chronic condition there is Liver and Kidney Yin deficiency and insufficient Jing and Qi, which fails to nourish muscles, tendons, and bones. In Western medicine, inflammations occur simultaneously at multiple peripheral nerves leading to symmetrical sensory disturbance and flaccid paralysis. Clinical characteristics at the beginning stage are numbness, tingling or pain at the end of four limbs, hyperalgesia or abnormal sensation such as ant crawling sensation, sharp pain, and symmetrical glove- or sock-like distributions. In severe cases, the pathological changes progress from distal to proximal part of the limbs. At the same time, there is decreased muscle tone and decreased or disappearing tendon reflexes. At the late stage, there is muscle atrophy with dropping of hands and feet. TCM treatments are clearing heat, draining damp, invigorating blood, and unblocking collaterals. One can use the following formula as the core formula and modify according to symptom presentations.

Core formula: Huang Bai, Cang Zhu, Chuan Niu Xi, Bei Xie, Qin Jiao, and Yi Yi Ren.

Related conditions and modifications are

- Fever and thirst: Add Dan Zhu Ye and Shi Gao.
- Severe pain at the end of extremities: Add Tao Ren, Hong Hua, and Mu Gua.
- Itching: Add Wu Shao She and Ci Ji Li.
- Stifling sensation and difficulty breathing: Add Gua Lou, Hou Po, Xie Bai, and Gui Zhi.
- Urinary retention: Add Che Qian Zi and Bai Mao Gen.
- Urinary incontinence: Add Wu Yao, Yi Zhi Ren, and Shan Yao.
- Atrophy and weakness: Add Dang Shen, Bai Zhu, and Fu Ling.
- Worse at the upper limb: Add Sang Zhi.
- Worse at the lower limb: Add Mu Gua.
- Damp–heat damaging Yin: Add Sheng Di Huang and Zhi Mu.
- Severe heat damaging Yin: Remove Cang Zhu and add Sheng Di Huang, Gui Ban, Mai Men Dong, and Shi Hu.
- Cold extremities: Remove Huang Bai and add Gui Zhi.
- Aversion to cold and deep weak pulse: Remove Huang Bai and Cang Zhu, and add Lu Jiao Jiao, Bu Gu Zhi, and Yin Yang Huo.

SCIATICA

This is discussed under Bi syndrome in TCM. It is often caused by wind–cold–damp evil pathogens invading the body that has a deficiency, obstructing the channels and collaterals, or due to traumatic injuries damaging the channels, collaterals, and causing Qi stagnation and blood stasis. In Western medicine, it is described as pain occurring along the sciatic nerve. Primary sciatica is pain at distal end of the sciatic nerve due to inflammation of the nerve with pathological change at the nerve trunk. Secondary sciatica is due to pathological changes at intervertebral discs or intervertebral tumor causing pain at the proximal end of the nerve. TCM treatment unblocks meridian, invigorates blood, and stops pain.

Wind–Cold–Damp Syndrome

Treatment Principle: Warm meridian and scatter cold.

Use Xiao Huo Luo Dan [Minor Invigorate the Collaterals Special Pill] and Juan Bi Tang [Remove Painful Obstruction Decoction].

Related conditions and modifications are

- Sore and heavy lower limb: Add Yi Yi Ren, Cang Zhu, and Fang Ji.
- Lumbar pain: Add Xu Duan and Du Zhong.

Damp–Heat Syndrome

Treatment Principle: Clear heat and drain damp.

Use Xuan Bi Tang [Disband Painful Obstruction Decoction] and San Miao San [Three-Marvel Powder].

Related conditions and modifications are

- Extreme damp and severe numbness: Add Du Huo, Mu Gua, and Wei Ling Xian.
- Excess heat: Add Shi Gao, Zhi Mu, and Di Long.

Qi Stagnation and Blood Stasis Syndrome

Treatment Principle: Invigorate blood and transform stasis.

Use Shen Tong Zhu Yu Tang [Drive Out Blood Stasis from a Painful Body Decoction].

Related conditions and modifications are

- Blood deficiency: Add Bai Shao Yao and Dan Shen.
- Severe pain: Add Quan Xie, Jiang Can, and Wu Shao She.

Liver and Kidney Yin Deficiency Syndrome

Treatment Principle: Tonify Liver and Kidney, and strengthen tendons and bones.

Use Du Huo Ji Sheng Tang [Angelica Pubescens and Sangjisheng Decoction] plus Gou Qi Zi and Xu Duan.

Related conditions and modifications are

- Yin deficiency: Add Zhi He Shou Wu.
- Yang deficiency: Add Rou Gui, Zhi Fu Zi, and Yin Yang Huo.

Qi and Blood Deficiency Syndrome

Treatment Principle: Tonify Qi and blood, and unblock channels and collaterals.

Use Ba Zhen Tang [Eight-Treasure Decoction] and Juan Bi Tang [Remove Painful Obstruction Decoction].

Related conditions and modifications are

- Slight cold: Add Gui Zhi and Zhi Fu Zi.
- Slight heat: Add Qin Jiao and Sang Zhi.
- Blood stasis: Add Tao Ren and Hong Hua.
- Liver and Kidney deficiency: Add Du Zhong and Xu Duan.

SEQUELAE OF POLIOMYELITIS

This condition is often caused by wind–heat and summer damp invading Lung and Spleen, obstruction of channels and collaterals, disturbed Qi and blood that causes eventual damage to Liver and Kidneys by interrupting nourishment of tendons and bones. In Western medicine, poliomyelitis refers to flaccid paralysis of limbs after the infection of poliovirus. The manifestations are atrophy of the muscles, deformed joints, and limping. TCM treatments focus on nourishing and tonifying Liver and Kidney and warming and unblocking the channels and collaterals. One should also pay attention to the saying that "selecting Yang Ming only to treat atrophy" applies to this disorder. Thus, focus on the postnatal foundation: Spleen and Stomach. Herbs such as Dang Gui, Chuan Xiong, Dan Shen, Ren Shen, Huang Qi, She Xiang, Lu Lu Tong, Xu Duan, Gou Qi Zi, and Du Zhong are often used.

Qi Deficiency and Blood Stasis Syndrome

Treatment Principle: Tonify Qi, invigorate blood, and unblock collaterals.

Use Bu Yang Huan Wu Tang [Tonify the Yang to Restore Five-Tenths Decoction] plus Gui Zhi, Jiang Can, Quan Xie, and Sang Ji Sheng.

Liver and Kidney Deficiency Syndrome

Treatment Principle: Tonify Liver and Kidney, and warm and unblock channels and collaterals.

Use Hu Qian Wan [Hidden Tiger Pill] plus Huang Qi, Du Zhong, Mu Gua, Jiang Can, and Quan Xie.

Furthermore, one can also combine with Huo Xue Jiu [Invigorating Blood Wine]. Drink 5 mL twice a day.

TRANSIENT ISCHEMIC ATTACK (TIA)

This condition belongs to slight wind or small wind stroke. It often occurs in elderly people with excess above and deficient below, which is Liver Yang hyperactive, Kidney Yin deficiency with Qi stagnation, blood stasis, and turbid phlegm. In Western medicine, TIA is seen as a temporary disturbance of the brain function due to insufficient blood supply at local area of the brain. The causes of this condition include artherosclerosis leading to cerebral vasospasm and blood clots causing insufficient blood supply to the brain tissue. This condition often occurs in people above

40, more so in males than females with a history of artherosclcerosis and hypertension. It presents with sudden onset generally shorter than 2 minutes and recovery within 24 hours; it is often recurrent. TCM treatments focus draining above, tonifying below, invigorating blood, and transforming stasis.

Empty Channels and Collaterals, Attacked by Wind

Treatment Principle: Nourish blood and expel wind.
Use Da Qin Jiao Tang [Major Gentiana Qinjiao Decoction] plus Jiang Can, and Chi Shao Yao.

Yin Deficiency and Yang Hyperactivity

Treatment Principle: Nourish Yin and subdue Yang.
Use Zhen Gan Xi Feng Tang [Sedate the Liver and Extinguish Wind Decoction] plus Dan Shen, Zhi He Shou Wu, and Gou Qi Zi.

Knotted Phlegm and Stasis Syndrome

Treatment Principle: Expel phlegm and transform stasis.
Use Tong Qiao Huo Xue Tang [Unblock Orifices and Invigorate Blood Decoction] plus Er Chen Tang [Two-Cured Decoction], Shi Chang Pu, Huang Qi, and Shen Qu.

Qi Deficiency and Blood Stasis Syndrome

Treatment Principle: Tonify Qi, invigorate blood, and transform stasis.
Use Bu Yang Huan Wu Tang [Tonify the Yang to Restore Five-Tenths Decoction] plus Di Long and Huai Niu Xi.

Blood Deficiency Generating Wind Syndrome

Treatment Principle: Nourish blood and extinguish wind.
Use Si Wu Tang [Four-Substance Decoction] plus Gou Teng, Chuan Niu Xi, and Gou Qi Zi.

TRIGEMINAL NEURALGIA

This condition is often caused by wind–cold, wind–heat upward attacking the three Yang meridians at the face; or Liver and Gallbladder wind–fire blazing; ascending Yang Ming dry-heat, or wind mixed with turbid phlegm attacking upward, causing obstruction of the channels and collaterals and blocked Qi and blood circulation, leading to facial pain. In Western medicine it is described as paroxysmal, transient, severe pain along the distribution of the trigeminal nerve on the face. It occurs more in people between 40 to 60 years of age, and more in females than males. Clinical characteristics are sudden onset, excruciating pain of short duration, and suddenly stopping but recurrent. It often occurs on one side especially the second branch or second and third branch of the trigeminal nerve. TCM treatments include expelling wind, clearing fire, scattering cold, scouring out phlegm, and transforming stasis. Acupuncture is effective and efficient in stopping pain.

Wind–Cold Attacking the Collaterals Syndrome

Treatment Principle: Disperse wind, scatter cold, unblock collaterals, and stop pain.

Use Chuan Xiong Cha Tiao San [Ligusticum Chuanxiong Powder to be Taken with Green Tea] plus Bi Ba.

Related conditions and modifications are

- Severe cold: Add Gao Ben and Sheng Jiang.
- Severe pain: Add Chuan Wu.
- Muscle twitching on face: Add Di Long and Jiang Can.

Wind–Heat Damaging Collaterals Syndrome

Treatment Principle: Disperse wind, clear heat, unblock collaterals, and stop pain.

Use Xiong Zhi Shi Gao Tang [Ligusticum Chuanxiong, Angelica Dahurica, and Gypsum Decoction] plus Jin Yin Hua, Shan Zhi Zi, Man Jing Zi, Bo He, Quan Xie, Wu Gong, and Gan Cao.

Related conditions and modifications are

- Excess heat damaging fluids: Add Shi Hu, Lu Gen, and Tian Hua Fen.
- Dark and yellow urine: Add Mu Tong and Dan Zhu Ye.
- Alternating chills and fever: Add Chai Hu and Huang Qin.

Wind–Phlegm Obstructing Collaterals Syndrome

Treatment Principle: Disperse wind, transform phlegm, unblock collaterals, and stop pain.

Use Qian Zheng San [Lead to Symmetry Powder] plus Yu Zhen San [True Jade Powder] plus Jing Jie, Ban Xia, and Gua Lou Pi.

Related conditions and modifications are

- Numbness and tingling of the facial muscle: Add Qin Jiao and Ji Xue Teng.
- Evil pathogens entering blood level: Add Hong Hua and Chi Shao Yao.

Heat Excess in Yang Ming Syndrome

Treatment Principle: Clear and drain Stomach fire, unblock collaterals, and stop pain.

Use Bai Hu Tang [White Tiger Decoction] plus Xie Xin Dao Chi Tang [Drain the Epigstrium and Guide Out the Red Decoction] and Mu Dan Pi, Zhi Mu, Lian Qiao, Di Long, Sheng Da Huang, Sheng Ma, and Chuan Xiong.

Related condition and modification are

- Very thirsty and likes to drink, burning pain at stomach: Add Tian Hua Fen and Ge Gen.

Liver and Gallbladder Fire Blazing Syndrome

Treatment Principle: Clear and drain Liver and Gallbladder, unblock collaterals, and stop pain.

Use Long Dan Xie Gan Tang [Gentiana Longdancao Decoction to Drain the Liver] plus Ju Hua, Tian Ma, Di Long, Gou Teng, and Xia Ku Cao.

Related condition and modification are

- Hypochondriac pain: Add Zhi Ke and Yu Jin.

Blood Stasis Obstructing Collaterals Syndrome

Treatment Principle: Invigorate blood, transform stasis, unblock collaterals, and stop pain.

Use Tong Qiao Huo Xue Tang [Unblock the Orifices and Invigorate the Blood Decoction] plus Dan Shen, Yan Hu Suo, and Jiang Can.

Related conditions and modifications are

- Chronic stasis generating heat: Add Chi Shao Yao and Mu Dan Pi.
- Chronic unhealed condition: Add Shui Zhi and Chuan Shan Jia.

Qi and Blood Deficiency Syndrome

Treatment Principle: Tonify Qi, nourish blood, unblock collaterals, and stop pain.

Use Ba Zhen Tang [Eight-Treasure Decoction] plus Dan Shen, Quan Xie, and Ji Xue Teng.

Related conditions and modifications are

- Severe Qi deficiency: Add Huang Qi.
- Severe blood deficiency: Add Zhi He Shou Wu and Gou Qi Zi.

CHAPTER 25

Pediatric Disorders

Herbal dosage for children:
Above 3 years old, use half of the adult dosage.
For 1 to 3 years old, use one third of the adult dosage.
For 6 months to 1 year old, use one fourth of the adult dosage.
For infants younger than 6 months old, use one fifth of the adult dosage.

CHICKENPOX

This condition is caused by external epidemic toxins invading Lung and Spleen. In Western medicine, it is an acute contagious disease caused by Varicella-Zoster virus. The main characteristics are fever, gradual appearing of macules, papules, blisters, and scabs. It often occurs in the winter and spring. TCM treatments for mild cases are clearing heat and venting evil pathogens. For severe cases, treatments are clearing Ying level and cooling blood. One can use Yin Qiao San Jia Jian [Modified Honeysuckle and Forsythia Powder] as a base formula that consists of Jin Yin Hua, Lian Qiao, Zi Cao, Ban Lan Gen, Yi Yi Ren, Chan Tui, Bo He, and Gan Cao.

Related conditions and modifications are

- Sore throat and cough: Add Niu Ban Zi, Jiang Can, Jie Geng, and Xing Ren.
- Severe skin itching: Add Bai Ji Li, Fang Feng, and Jing Jie.
- High fever: Add Shi Gao and Zhi Mu.
- Dense pox and dark purple or bleeding: Add Huang Lian, Sheng Di Huang, Chi Shao Yao, Mu Dan Pi, and Dan Shen.
- Herpes on tongue and mouth: Add Huang Lian, Mu Tong, and Dan Zhu Ye.
- Excess heat toxins: Add Yu Xing Cao, Ye Jue Hua, Pu Gong Ying, and Zi Hua Di Ding.

ENURESIS

It is caused by insufficient Kidney Qi, deficient cold at lower Yuan (source Qi), Spleen and Lung Qi deficiency, and weakness due to long-term illness. In Western medicine, this refers to children above 3 years of age having repeated or accidental urination during sleep. It has been thought to be caused by dysfunction of the cerebral cortex and subcortex and insufficient tension of the urinary bladder sphincter. TCM treatments focus on supporting Zheng Qi and astringing the bladder. The following formula is the base and should be modified according to symptoms. Jia Wei Suo Quan Wan [Modified Shut the Sluice Pill] plus Wu Yao, Yi Zhi Ren, Shan Yao, Shan Zhu Yu, and Sang Piao Xiao.

Related conditions and modifications are

- Insufficient Middle Jiao Qi: Add Bu Zhong Yi Qi Wan [Tonify the Middle Augment the Qi Pill].
- Ying and Wei disharmony and Yin and Yang disorder: Add Gui Zhi Jia Long Gu Mu Li Tang [Cinnamon Twig Decoction plus Dragon Bone and Oyster Shell].
- Insufficient Kidney Qi: Add Jin Gui Shen Qi Wan [Kidney Qi Pill from the Golden Cabinet].
- Lung Qi failing to disperse: Add Xing Ren, Gui Zhi, Zhi Gan Cao, Jie Geng, and Zhi Ma Huang.
- Damp–heat pouring downward: Use Ba Zheng San [Eight-Herb Powder for Rectification] first, followed by the above formula.
- Liver Qi stagnation: Add Chai Hu Shu Gan San [Bupleurum Powder to Spread the Liver].

Furthermore, one can also incorporate navel application. Grind equal amount of Gan Cao, Bai Shao Yao, and Bai Zhu. Put 0.5 to 1 g of the powder into the navel after cleaning, and cover the area with tape. Change every 3 to 5 days.

Food therapy can also be added. Place 50 g of Shan Yao in a pig urinary bladder and seal well. Cook with water, ginger, and salt. Drink the soup and eat the whole thing once a week. Continue for 5 weeks.

EPIDEMIC MUMPS

This is caused by wind–damp–evil pathogens invasion. In Western medicine, it is an acute respiratory infectious disease caused by epidemic paramyxovirus. The clinical manifestations are nonsuppurative inflammation of the parotid gland, fever, whole body discomfort, and possible orchitis in boys. It often occurs in the winter and spring in kids between 5 to 8 years of age. The incubation period is 14 to 21 days. TCM treatments focus on dispersing wind, clearing heat, relieving toxicity, and scattering accumulations.

Warm Toxins at Exterior

Treatment Principle: Disperse wind, clear heat, scatter accumulation, and reduce swelling.

Use Yin Qiao San [Honeysuckle and Forsythia Powder], remove Jie Geng, and add Ban Lan Gen, Pu Gong Ying, Xia Ku Cao, and Chai Hu.

Heat Toxins Accumulation

Treatment Principle: Clear heat, relieve toxicity, soften hardness, and scatter accumulation.

Use Pu Ji Xiao Du Yin [Universal Benefit Decoction to Eliminate Toxin] plus Shi Gao and Zao Xiu.

Related conditions and modifications are

- Obvious hardness and accumulation at parotid gland: Add Xia Ku Cao, Hai Zao, Kun Bu, and Zhe Bei Mu. One can also use mashed fresh Pu Gong Ying as topical poultice at affected area.
- Constipation: Add Da Huang.
- Orchitis: Change to Long Dan Xie Gan Tang [Gentiana Longdancao Decoction to Drain the Liver] plus Chuan Lian Zi, Tao Ren, Yan Hu Suo, and Ban Lan Gen. Huang Bai and Ce Bai Ye can be powdered and made into paste for external use.

FEVER

There are two types of fever in TCM. One is exterior fever, and the other is interior fever. Exterior fever is caused by invasion of six pathogenic factors or epidemic evil pathogen. Interior fever is caused by excess Yang and Yin deficiency or chronic stagnation and obstruction. In Western medicine, fever is a common clinical symptom in children. The causes can be divided into infectious and noninfectious. TCM treatments are effective in treating low-grade fever or fever that is not effectively treated by antibiotics. Syndrome differentiation is the key in TCM treatment. Even though fever is the complaint, treatment principles are different depending on the syndrome.

Exterior Fever

Treatment Principle: Disperse Lung and release exterior.

Use modified Xing Su San [Apricot Kernel and Perilla Leaf Powder] plus Chai Ge Jie Ji Tang [Bupleurum and Kudzu Decoction to Release the Muscle Layer]: Chai Hu, Huang Qin, Shi Gao, Jiang Can, Bai Zhi, Xing Ren, Zi Su Ye, Chen Pi, Mai Ya, Sheng Jiang, and Da Zao.

Related conditions and modifications are

- Sore and red swollen throat: Add Niu Bang Zi and Bo He.
- Mix with dampness: Remove Shi Gao and add Huo Xiang and Bai Dou Kou.

Damp–Heat Fever

Treatment Principle: Clear and transform damp–heat.

Use Da Yuan Yin Jia Jian [Modified Reach the Membrane Source Decoction]: Chai Hu, Huang Qin, Shan Zhi Zi, Bing Lang, Hou Po, Cao Guo, Zhi Mu, Fu Ling, Zhi Ke, and Huo Geng.

Related conditions and modifications are

- Exterior syndrome: Add Xiang Ru and Bo He.
- Liver and Gallbladder syndrome: Add Yin Chen Hao and Qing Pi.
- Lung Wei syndrome: Add Jie Geng and Xing Ren.
- Lower Jiao syndrome: Add Hua Shi and Bian Xu.

Qi Level Heat Syndrome

Treatment Principle: Clear heat and relieve toxicity.

Use modified Bai Hu Tang [White Tiger Decoction] plus Huang Lian Jie Du Tang [Coptis Decoction to Relieve Toxicity]: Shi Gao, Zhi Mu, Huang Lian, Huang Qin, Shan Zhi Zi, and Chai Hu.

Related conditions and modifications are

- Cough and wheezing: Add Xing Ren, Su Zi, and Zhi Ma Huang.
- Constipation: Combine with Tiao Wei Cheng Qi Tang [Regulate the Stomach and Order the Qi Decoction].

Fever Due to Heat at Ying and Blood Levels

Treatment Principle: Clear Ying level, cool blood, and rescue Yin.

Use Shen Xi Dan Jia Jian [Modified Magical Rhinoceros Special Pill]: Shui Niu Jiao, Sheng Di Huang, Mu Dan Pi, Ban Lan Gen, Zi Cao, Huang Qin, Jin Yin Hua, Lian Qiao, Shi Chang Pu, and Chi Shao Yao.

Food Accumulation Fever

Treatment Principle: Clear heat and reduce accumulation.

Use Xiang Lian Wan Jia Wei [Augmented Aucklandia and Coptis Pill]: Shan Zha, Shen Qu, Bing Lang, Chen Pi, Mu Xiang, Da Huang, Huang Lian, and Huang Qin.

Related conditions and modifications are

- Nausea and vomiting: Add Sheng Jiang and Ban Xia.
- Abdominal pain: Add Yan Hu Suo and Bai Shao Yao.

Yin Deficiency Fever

Treatment Principle: Nourish Yin and clear heat.

Use Qing Hao Bie Jia Tang Jia Jian [Artemisia Annua and Soft-Shelled Turtle Shell Decoction]: Qing Hao, Bie Jia, Di Gu Pi, Sheng Di Huang, Tian Hua Fen, Tai Zi Shen, Mai Men Dong, Qin Jiao, Yin Chai Hu, and Wu Mei.

Related condition and modification are

- Copious sweating: Add Fu Xiao Mai, Long Gu, and Mu Li.

Qi Deficiency Fever

Treatment Principle: Tonify middle and benefit Qi.

Use Bu Zhong Yi Qi Tang Jia Jian [Modified Tonify the Middle and Augment the Qi Decoction]: Huang Qi, Dang Shen, Bai Zhu, Zhi Gan Cao, Dang Gui, Chai Hu, Mai Ya, Bian Dou, and Chen Pi.

Fever after Fright (Fever Due to Floating Heart, and Liver Yang)

Treatment Principle: Clear heat, anchor fright, and calm Shen.

Formula consists of Long Dan Cao, Huang Lian, Lian Zi Xin, Fang Feng, Gou Teng, Qing Dai, Zhi Jiang Can, Bai Zhu, Fu Shen, Long Gu, Gan Cao, and fried Suan Zao Ren.

Furthermore, one can incorporate topical application to the navel region. Grind Qing Hao 50 g, Shi Gao 50 g, Hua Shi 30 g, green tea leaves 20 g, and Bing Pian 2 g into powder, and add glycerin and egg white to make a paste. Apply to navel, cover with a cloth, and change every 8 hours. It is good for any type of fever.

GERMAN MEASLES

German Measles is also called wind rash in TCM. It is caused by epidemic external wind–heat. In Western medicine, it is described as an acute contagious disease caused by a virus, which is often seen in winter and spring in children aged 1 to 5 years. The clinical characteristics are fever, maculo–papule rashes all over the body, and swollen lymph nodes behind the ear, occipital, and back of the neck. The maculo–papule rash is light red in color, 2 mm in diameter, and spreads from face and neck area to the trunk and four limbs in one day. There are no rashes at palms and feet. TCM treatments focus on dispersing wind, clearing heat, clearing Qi, cooling Ying, and relieving toxicity.

One can use Yin Qiao San Jia Jian [Modified Honeysuckle and Forsythia Powder] as the base and modify according to presenting symptoms.

The base formula consists of Jin Yin Hua, Da Qing Ye, Lian Qiao, Zi Cao, Dan Zhu Ye, Bo He, Chan Tui, and Gan Cao.

Related conditions and modifications are

- Cough: Add Xing Ren, Qian Hu, and Jie Geng.
- Severe itching: Add Bai Ji Li and Jing Jie.
- High fever, irritability, and thirst: Add Chai Hu, Shi Gao, Lu Gen, and Zhi Mu.
- Red or purple dense rashes: Add Chi Shao Yao, Mu Dan Pi, and Sheng Di Huang.
- Severe swollen lymph nodes: Add Xia Ku Cao and Yu Xing Cao.
- High fever and dizziness: Add Zi Hua Di Ding, Gou Teng, Shan Yang Jiao, Huang Lian, and Huang Qin.

INFANTILE DIARRHEA

This is caused by external wind, cold, summer heat, damp or improper diet, and weak and deficient Spleen and Stomach. This condition shows as increased bowel movement frequency, and thin watery stool is the key manifestation that often occurs in the summer and fall seasons. This is the most common digestive complaint for children under 2 years. This can be divided into infectious diarrhea and noninfectious diarrhea. TCM treatment should be given after clearly identifying the type of diarrhea in terms of cold and heat, and deficiency or excess in regulating the Spleen and Stomach, as well as ascending clear and descend turbidity. From the lab, Huang Lian, Bai Tou Weng, Qin Pi, Wu Mei, Huang Qin, and Huang Bai have antibacterial function. Bai Shao Yao and Gan Cao can moderate the peristalsis of the intestines. Si Jun Zi Tang [Four-Gentleman Decoction], Shen Ling Bai Zhu San [Ginseng, Poria, and Actratylodes Macrocephala Powder], and other strengthening Spleen formulas can stimulate the activity of digestive enzyme and increase the absorption function of the small intestine. These can be considered in clinical practice.

Treatments According to Type of Diarrhea

Acute Diarrhea

Sudden onset of watery diarrhea with episodes ranging from a few to over a dozen.

Treatment Principle: Regulate Qi, drain damp, and stop diarrhea.

Formula consists of Cang Zhu, Fu Ling, Hou Po, Che Qian Zi, Zhi Shi Tan, Bai Shao Yao, and Gan Cao.

Related conditions and modifications are

- Exterior wind–cold: Add Zi Su Ye and Huo Xiang.
- Damp–heat: Add Huang Qin, Huang Lian, and Bai Tou Weng.
- Damages by food: Add Ji Nei Jin, Mai Ya, and Lai Fu Zi.
- Severe vomiting and nausea: Add Ban Xia, Chen Pi, and Sheng Jiang.
- Fever: Add Chai Hu and Shi Gao.
- Tenesmus: Add Ge Gen and Bing Lang.

Chronic Diarrhea

Long course of disease, recurrent with weak constitution, poor appetite, and emaciation.

Treatment Principle: Strengthen Spleen, drain damp, and stop diarrhea.

Use Si Jun Zi Tang Jia Wei [Augmented Four-Gentleman Decoction] plus Dang Shen, fried Bai Zhu, Yi Yi Ren, Shan Yao, Bai Dou Kou, Fu Ling, Zhi Ke, Lian Zi, and Gan Cao.

Related conditions and modifications are

- Disharmony between Liver and Spleen: Add Bai Shao Yao and Wu Mei.
- Spleen and Kidney Yang Deficiency: Add Zhi Fu Zi and Gan Jiang.
- Unhealed chronic diarrhea and weak constitution: Add Ying Su Ke and He Zi.
- Chronic diarrhea with rectal prolapse: Add Huang Qi, Ge Gen, and Sheng Ma.
- Nausea and vomiting: Add Ban Xia and Sheng Jiang.
- Irritability and agitation: Add Bai Shao Yao and Jiang Can.
- Chronic diarrhea and enlarged vein on abdomen or focal distention and lumps at flank: Add Dan Shen, E Zhu, and Dang Gui.
- Diarrhea in fall: Use Ge Gen Huang Qin Huang Lian Tang [Kudzu, Coptis, and Scutellaria Decoction] plus Chai Hu, Hua Shi, Bing Lang, Da Fu Pi, and Zhu Ru.

Formulas for Diarrhea Determined by Western Diagnosis

Fungal (Candida) Diarrhea Formula

Use Zhi Fu Zi, Dang Shen, Gan Jiang, fried Bai Zhu, Bai Tou Weng, Huang Lian, Er Cha, Ying Su Ke, and Yu Xing Cao.

Diarrhea Due to Pathogenic Colibacillus

Use Bai Zhu, Bing Lang, Hou Po, Dang Gui, Huang Qin, Huang Lian, Guan Zhong, Ku Shen, and Gan Cao.

Hypokalemia Formula

Use Huang Qi, Sha Shen, Bai Zhu, Bai Bian Dou, Yi Yi Ren, Mai Men Dong, and Mu Xiang.

Food Therapy

Lian Zi Shan Yao Zhou [Lotus Seed and Chinese Yam Congee]

Grind Lian Zi 15 g and Shan Yao 15 g into powder. Cook with 50 g of Geng Mi to make congee. Eat frequently. It is good for chronic diarrhea.

Da Zao Yi Ren Tang [Jujube and Coix Seed Soup]

Cook 10 pieces of Da Zao and 30 g of Yi Yi Ren until the seed splits slightly, then add 10 g of brown sugar, and 3 slices of ginger for 5 more minutes. At the beginning, drink the soup two to three times a day. When the digestive function is stronger, then one can consume Da Zao and Yi Yi Ren as well. It is suitable for recovery stage of diarrhea.

MEASLES

This is caused by epidemic measles virus invading the Lung and Spleen through mouth and nose. It is an acute respiratory infectious disease, which can occur all year round. It has higher occurrence in children between 6 months to 5 years of age. The clinical manifestations are fever, symptoms of upper respiratory infection, skin rash which first occur behind the ear and spreads all over the body after 2 or 3 days and then disappears according to the order of appearance. Brown pigmented spots are left on skin with desquamation. The whole disease course is 10 to 14 days. TCM treatments focus on dispersing rash, venting toxins, clearing and releasing heat, nourishing Yin, and preventing measles toxin from attacking deeper.

Treatments According to Periods

Before Rash Period

Exterior condition and tiny white or gray spots occurring in mucous membrane close to the wisdom tooth.

Treatment Principle: Acrid cool to vent the exterior and clear and disperse Lung and Wei Qi.

Use Yin Qiao San [Honeysuckle and Forsythia Powder] plus Ge Gen, Xing Ren, Fang Feng, Chan Tui, and Ban Lan Gen.

Related conditions and modifications are

- Severe heat: Add Jiang Can and Gou Teng.
- Sore throat: Add She Gan and Ma Bo.
- Severe chills: Add Jing Fang Bai Du San [Schizonepeta and Ledebouriella Powder to Overcome Pathogenic Influences].
- Weak constitution: Add Tai Zi Shen.
- Rash unable to surface: Add Xi He Liu, Zu Ling, and Yuan Sui.

Rash Period

Treatment Principle: Clear heat, relieve toxicity, and vent rash.

Use Zhu Ye Shi Gao Tang [Lophatherus, Willow, and Burdock Seeds Decoction] plus Jin Yin Hua, Lian Qiao and Da Qing Ye, Shi Gao, Sang Ye, Zi Cao, and Sheng Ma.

Related conditions and modifications are

- Red and purple spots and stasis: Add Sheng Di Huang, Chi Shao Yao, and Mu Dan Pi.
- Severe coughing: Add Sang Bai Pi and Xing Ren.
- Sticky and yellow phlegm: Add Zhe Bei Mu and Dan Nan Xing.
- High fever: Combine with Bai Hu Tang [White Tiger Decoction] and Shan Zhi Zi.

Recovery Period

Treatment Principle: Clear and relieve residual toxins, nourish Yin, and generate fluids.

Use Sha Shen Mai Dong Tang [Glehnia and Ophiopogonis Decoction] and Ren Shen Ye, Pi Pa Ye, and Xing Ren.

Related conditions and modifications are

- Persisting low-grade fever: Add Di Gu Pi, Yin Chai Hu, and Bai Wei.
- Dry cough: Add Chuan Bei Mu, Bai He, and Bai Bu.
- Hoarseness of the voice: Add Mu Hu Die.
- Poor appetite: Add Shan Zha, Mai Ya, and Ji Nei Jin.
- Apathy and shortness of breath: Add Tai Zi Shen and Gan Cao.

MENTAL RETARDATION

This is discussed in TCM under the "Five Slows" syndrome, i.e., slow in standing, walking, hair growth, teeth development, and talking. It is often caused by congenital insufficiency, Liver and Kidney deficiency; postnatal malnourishment, Qi and blood deficiency, Heart and Shen not nourished; phlegm stasis obstructing the Heart orifices and Shen losing its support. In Western medicine, this is also called hypophrenia and refers to children with significantly lower intelligence than the average children of the same age. Thus their abilities in recognition, memory, comprehension, imagination, thinking, judgment, and adaptation to social environment are significantly behind children of the same age. Combination of psychological treatment and herbal formula sometimes has better results.

Treatments According to Syndrome Differentiation

Heart and Shen Failed to be Nourished

Treatment Principle: Tonify Heart and nourish blood.

Use Yang Xin Tang [Nourish the Heart Decoction], remove Sheng Di Huang and Shu Di Huang, and add Chuan Xiong, Yuan Zhi, Shi Chang Pu, Dan Shen, and Bai Zhu.

Liver and Kidney Deficiency Syndrome

Treatment Principle: Tonify and benefit Liver and Kidney.

Use You Gui Wan [Restore the Right (Kidney) Pill], remove Zhi Fu Zi, and add Zi He Che and Sha Ren.

Phlegm Stasis Obstructing Orifices Syndrome

Treatment Principle: Expel phlegm, move stasis, and open orifices.

Use Tong Qiao Huo Xue Tang plus Di Long, Ban Xia, Dan Nan Xing, and Shi Chang Pu.

Kidney Essence Deficiency Syndrome (Congenital Insufficiency)

Treatment Principle: Tonify Kidney supplement essence.

Use He Che Ba Wei Wan [Eight-Flavor Pill with Placenta], remove Fu Zi and Rou Gui, and add Shi Chang Pu and Yuan Zhi.

Food Therapy

1. Mash Hu Tao Rou 30 g, Hei Zhi Ma 10 g, and mix with 5 g of sugar. Take once a day for a long period of time.
2. Decoct Lian Zi Xin 5 g, pig bone marrow 15 g, and rock sugar 10 g. Drink often.

NEPHROTIC SYNDROME

Nephrotic syndrome belongs to Yin Water in TCM. It is caused by weak functions of Lung, Spleen, and Kidney, disorders of Qi transformation, congestion in San Jiao, blocked waterways, and overflowing of water evil. At the beginning stage, there is water–damp obstruction. As the disease becomes chronic, there is heat mixed with damp. In Western medicine, there are many reasons advanced for the cause of increased permeability of glomeruli which leads to large amount of plasma protein lost through the urine. The clinical characteristics are high level of proteinuria, low protidemia, high cholesterolemia, and edema. This occurs more in boys than girls. TCM treatments focus on dispersing Lung, strengthening Spleen, tonifying Kidney, transforming damp, leaching out damp, and promoting urination.

Treatments According to Syndrome Differentiation

Spleen Qi Deficiency and Damp Accumulation

Treatment Principle: Strengthen Spleen, tonify Qi, transform damp, and drain water.

Use Shen Ling Bai Zhu San [Ginseng, Poria, and Atractylodes Macrocephala Powder] plus Cang Zhu, Da Fu Pi, Zhu Ling, and Ze Xie.

Related condition and modification are

- Damp stagnation transforming into heat: Add Huang Qin, Huang Lian, and Shan Zhi Zi.

Spleen and Kidney Yang Deficiency

Treatment Principle: Warm Yang and promote urination.

Use Zhen Wu Tang [True Warrior Decoction] plus Rou Gui, Huang Qi, Yin Yang Huo, Dang Gui, Bu Gu Zhi, and Yu Mi Xu.

Yin Deficiency with Damp–Heat Syndrome

Treatment Principle: Nourish Yin, clear heat, and transform damp.

Use Zhu Jing Wan [Preserve Vistas Pill] and Ji Sheng Shen Qi Wan [Kidney Qi Pill from Formulas that Aid the Living], remove Fu Zi, Rou Gui, replace Shu Di Huang with Sheng Di Huang, and add Gou Qi Zi, Tu Si Zi, Xian He Cao, and Huang Bai.

Treatment According to Western Index

1. Simple Type of Nephrotic Syndrome
   After using high doses of steroids to control the pathological condition, if proteinuria is lower than + and the use of steroids is decreasing, one can combine with the following formula.
   Formula consists of Jiang Can, Huang Qi, Bai Zhu, Xi Yang Shen, Yin Yang Huo, Shu Di Huang, Gou Qi Zi, Nu Zhen Zi, Ling Zhi, Yi Mu Cao, San Qi powder, Sha Ren, Shan Yao, Bu Gu Zhi, Hei Nuo Mi, Hei Zhi Ma, Hei Da Dou, Hu Tao Ren, Zi He Che, and Huang Bai.
   Grind all of the ingredients into powder and encapsulate into pills of 1.5 g each. Take one in morning and one at night with ginger decoction.
2. In recent years, clinical results show that it is more effective treating this condition with invigorating blood and transforming stasis methods. Commonly used herbs include Dan Shen, Chuan Xiong, Tao Ren, Hong Hua, Mu Dan Pi, San Leng, E Zhu, San Qi, and Yi Mu Cao. With syndrome differentiation as the foundation, if any of the following symptoms occur, these herbs can be added in the formula accordingly.
   - Nephritis type of nephritic syndrome. Increased RBC in urine or hypertension.
   - Dark complexion and tongue proper.
   - Dry, itchy, or scaly skin.
   - Stubborn condition. Chronic use of corticosteroid.
3. Huang Qi, Dang Shen, and Yu Mi Xu can reduce proteinuria. These can be added to the formula according to syndrome differentiation.

4. If level of RBC is high in urine, Bai Mao Gen, San Qi, Xiao Ji, and Yi Mu Cao can be added to the formula according to syndrome differentiation. Chi Shao Yao and Mu Dan Pi should be used with caution.
5. Jiang Can and Chan Tui can reduce the side effects of corticosteroids. They can be added to assist steroids treatment.
6. In recent years, doctors are using prepared Lei Gong Teng extract to treat this condition. It is effective in treating especially stubborn kind of recurrent nephrotic syndrome. It is also especially effective in lowering proteinuria.

UPPER RESPIRATORY INFECTION

This is caused by external invasion of wind, cold, heat, damp, dryness, and summer heat to the Lung and Wei. In Western medicine, this is often caused by viruses and sometimes by bacteria. The clinical characteristics are acute onset, running nose, nasal congestion, sneezing, coughing, sore throat, fever, and full body ache. Sometimes, there can also be nausea, vomiting, diarrhea, and lack of appetite. If it is not properly treated or treated in time, then it can progress into trachitis, bronchitis, pneumonia, nephritis, or convulsion. TCM treatments focus on releasing exterior, dispersing Lung, transforming phlegm, draining damp, eliminating food stagnation, clearing heat, and relieving toxicity and modified according to each syndrome.

Treatments according to types:

Common Cold

Wind–Cold Syndrome

Treatment Principle: Acrid warm herbs to release exterior.

Formula consists of Jing Jie, Fang Feng, Zi Su Ye, Qian Huo, Lian Qiao, Xing Ren, Jie Geng, Qian Hu, Zhi Ke, Bo He, and Gan Cao.

For mixed cold and heat pattern, change to Chai Ge Jie Ji Tang [Bupleurum and Kudzu Decoction to Release the Muscle Layer].

Wind–Heat Syndrome

Treatment Principle: Acrid cool herbs to release exterior.

Formula consists of Jin Yin Hua, Lian Qiao, Niu Bang Zi, Bo He, Dan Zhu Ye, Dan Dou Chi, Huang Qin, Jing Jie, Ban Lan Gen, Sheng Shi Gao, and Jiang Can.

Related condition and modification are

- Cough: Change to Sang Ju Yin [Mulberry Leaf and Chrysanthemum Decoction].

Summer Damp Syndrome

Treatment Principle: Release summer heat and expel damp.

Formula consists of Xiang Ru, Hou Po, Bian Dou, Huo Xiang, Su Ye, Jiang Ban Xia, Fu Ling, Chen Pi, and Gan Cao.

Influenza

Wind–Warm Toxin Syndrome

Treatment Principle: Disperse wind, clear heat, and relieve toxicity.

Formula consists of Jin Yin Hua, Lian Qiao, Niu Bang Zi, Huang Qin, Da Qing Ye, Jing Jie, Yu Xing Cao, Qian Hu, and Gan Cao.

Tonsillitis

Wind–Heat Transferring Internally and Transforming into Toxins

Treatment Principle: Clear Lung heat, relieve toxicity, and reduce swelling.

Formula consists of Ban Lan Gen, Jie Geng, Xuan Shen, She Gan, Zi Hua Di Ding, Huang Qin, Ma Bo, Sheng Shi Gao, Jin Yin Hua, and Lian Qiao.

CHAPTER 26

Psychological Disorders

DEPRESSION

Excessive worrying can often cause depression, which affects the Heart and Spleen. The pathogeneses are Qi stagnation and phlegm accumulation obstructing the Heart orifices. In Western medicine, this is described as a mood disorder with the key characteristic of having a depressed or moody feeling. Main manifestations are low emotions, slow mental activity, a sense of inferiority, self-criticism, decreased physical activities, dull expression, dislike for talking, and sluggishness. This condition often has a slow onset and progression. TCM treatment focuses on moving Qi, relieving stagnation, expelling phlegm, and opening orifices.

Knotted Phlegm and Qi Stagnation Syndrome

Treatment Principle: Disperse Liver, relieve stagnation, transform stasis, and open orifices.

Use Chai Hu Shu Gan San [Bupleurum Powder to Spread the Liver] plus Ban Xia, Yu Jin, Zhu Ru, Dan Nan Xing, Shi Chang Pu, and Yuan Zhi.

Qi Deficiency and Phlegm Accumulation Syndrome

Treatment Principle: Tonify Qi, strengthen Spleen, scour phlegm, and open orifices.

Use Si Jun Zi Tang [Four-Gentleman Decoction] and Di Tan Tang [Scour Out Phlegm Decoction] plus Yuan Zhi, Yu Jin, and Tian Zhu Huang.

Related condition and modification are

- Dull mental state: Add Su He Xiang Wan [Liquid Styrax Pill].

Heart and Spleen Deficiency Syndrome

Treatment Principle: Strengthen Spleen and nourish Heart.

Use Gui Pi Tang [Restore the Spleen Decoction] plus Dan Nan Xing, Zhu Ru, Yu Jin, and Bai Zi Ren.

HYSTERIA

This condition is caused by internal injury from seven emotions and a reverse flow of Qi. There is no particular pattern of disease progress. In Western medicine, this is described as a mood disorder due to psychoemotional stimulation or negative trigger. It often occurs in youth, more prevalent in females than male. The onset is sudden. The manifestations are temporary amentia or sensory and motor disturbances. The symptoms can occur under negative triggers and can also disappear with stimuli. The pathological duration is short but recurrent and with good prognosis. TCM focuses on regulating Qi mechanism. One can use the following base formula and modify according to presenting symptoms. Usually, Gan Mai Da Zao Tang [Licorice, Wheat, and Red Date Decoction] is used as a basic formula with appropriate modification according to syndrome differentiation.

Disturbance of Psychoconsciousness

Symptoms: Agitation, restlessness, quick anger, sudden fainting, and stiffness or spasms of the limbs and trunk.

Add Suan Zao Ren, Yu Jin, Dan Nan Xing, Long Gu, and Mu Li.

Related condition and modification are

- Severe twitching: Add Quan Xie.

Disturbance of Psychomotors

Symptoms: Depressed, sad, worried, and cries easily; aphasia, or paralysis.

Add Yu Jin, Shi Chang Pu, Long Gu, Mu Li, Di Long, and Chuan Niu Xi.

Related conditions and modifications are

- Loss of hearing: Add Ci Shi and Wu Wei Zi.
- Aphasia: Add Bai Zi Ren and Dan Shen.
- Blindness: Add Ju Hua and Gou Qi Zi.

Functional Disturbance of Internal Organs and Vegetative Nerve

Symptoms: Nausea, vomiting, hiccup, palpitation, wheezing, and difficulty swallowing.

Add Ban Xia, Sheng Jiang, Dai Zhe Shi, Zhu Ru, Shi Di, and Suan Zao Ren.

Related conditions and modifications are

- Severe wheezing: Add Zi Su Zi and Chen Pi.
- Aversion to food: Add Shan Zha, Ji Nei Jin, and Gu Ya.

Plum Pit Qi

Symptoms: Discomfort in the throat with a sense of obstruction, unable to get rid of the sensation by swallowing, and sometimes experiencing Qi sensation rushing upward to the larynx.

Add Hou Po, Ban Xia, Zi Su Geng, Dai Zhe Shi, Xiang Fu, and Gua Lou Pi.

MANIA (MANIC ATTACK)

This condition belongs to Kuang syndrome in TCM. It is often caused by fright and excessive worrying that generates fire and phlegm, leading to phlegm–fire obstructing the Heart orifices and disturbing Shen (spirit). The organs involved are Heart, Liver, and Stomach. In Western medicine mania is a mood disorder with key characteristics of extreme elation or easy irritability. The main manifestations are elation, mental hyperactivity, and increased physical activity. TCM treatments focuses on scouring phlegm, opening orifices, draining fire, relieving toxicity, tranquilizing Heart, and calming Shen (spirit).

Liver Stagnation Transforming into Fire Syndrome

Treatment Principle: Clear heat, smooth Liver, and calm Shen.

Use Long Dan Xie Gan Tang [Gentiana Longdancao Decoction to Drain the Liver] plus Suan Zao Ren, Dan Nan Xing, Long Gu, and Yu Jin.

Accumulated Heat in Stomach and Intestines Syndrome

Treatment Principle: Unblock and drain Stomach and Intestines.

Use Da Cheng Qi Tang [Major Order of Qi Decoction] plus Sheng Shi Gao, Sheng Tie Luo, and Yuan Zhi. Or one can start with Da Cheng Qi Tang [Major Order of Qi Decoction] to quickly drain but preserve Yin, followed by Liang Ge San [Cool the Diaphragm Powder] to clear residual heat.

Phlegm–Fire Disturbing the Heart Syndrome

Treatment Principle: Drain fire, scour out phlegm, and open orifices.

Use Xie Xin Dao Chi San [Drain the Epigastrium and Guide Out the Red Decoction] remove Ren Shen, and add Sheng Tie Luo and Sheng Da Huang.

Yin Deficiency with Deficient Fire Syndrome

Treatment Principle: Nourish Yin, descend fire, tranquilize the Heart, and calm Shen.

Use Xie Xin Dao Chi San [Drain the Epigastrium and Guide Out the Red Decoction] and Suan Zao Ren Tang [Sour Jujube Decoction] plus Xuan Shen, Mu Dan Pi, Shi Chang Pu, and Dan Nan Xing.

SCHIZOPHRENIA

This condition is discussed under Dian and Kuang in TCM. It is caused by internal damages of Heart, Spleen, Liver, and Gallbladder due to seven emotions, causing Qi stagnation, phlegm accumulation, fire stagnation, and blood stasis. These factors disturb the Heart and Shen and mist the Heart orifice leading to abnormal mental state. Dian syndrome (psychosis) is often caused by knotted phlegm and Qi stagnation with characteristics of not liking to talk, dementia, rambling, incoherent speech, silence, or excessive, inappropriate laughter or smiling. Kuang syndrome (manic type) is often caused by excess congestion of phlegm–fire and manifests as making a lot of noise, agitation, hitting or yelling, and moving a lot and often irritable. In Western medicine, the cause is unknown. It often begins in youth, and the clinical manifestations include abnormal thinking, emotions, perception, and behavior. It is a disorder of unconsciousness and intelligence. Schizophrenia is possibly related to genetic and social psychological factors. TCM treatments are clearing Liver, draining fire, scouring phlegm, opening orifices, regulating Qi, relieving stagnation, invigorating blood, and transforming stasis. Commonly used herbs are Shui Niu Jiao, Sang Ji Sheng, Di Long, Yang Jin Hua, Yuan Hua, and Da Ji.

Treatment According to Syndrome Differentiation

Liver Stagnation and Excess Fire Syndrome

Treatment Principle: Drain Liver, clear fire, scour out phlegm, and calm Shen.

Use Tiao Wei Cheng Qi Tang [Regulate the Stomach and Order of Qi Decoction] plus Gun Tan Wan [Vaporize Phlegm Pill] and Sheng Shi Gao, Sheng Tie Luo, Long Gu, Huang Bai, Huang Lian, and Yu Jin.

Blood Deficiency and Phlegm–Fire Syndrome

Treatment Principle: Nourish blood, calm Shen, clear heat, and transform phlegm.

Use Tiao Wei Cheng Qi Tang [Regulate the Stomach and Order of Qi Decoction] plus Zhi Mi Fu Ling Wan [Poria Pill to Point Our Confusion] and Shi Chang Pu, Huang Qin, Sheng Shi Gao, Lu Hui, Bo He, Ji Xue Teng, and Niu Huang.

Yang Ming Excess Heat Syndrome

Treatment Principle: Clear and drain Yang Ming, purge and scour out the stasis and accumulation.

Use Da Cheng Qi Tang [Major Order of Qi Decoction] plus Dang Gui Long Hui Wan [Tangkuei, Gentiana Longdancao, and Aloe Pill] and Sheng Shi Gao.

Qi Stagnation and Blood Stasis Syndrome

Treatment Principle: Regulate Qi, relieve stagnation, invigorate blood, and transform stasis.

Use Chai Hu Shu Gan San [Bupleurum Powder to Spread the Liver] plus Xiao Cheng Qi Tang [Minor Order of Qi Decoction] and Tao Ren, Hong Hua, Dan Shen, Long Gu, and Mu Li.

Related condition and modification are

- Severe blood stasis: Add Da Huang Zhe Chong Wan [Rhubarb and Eupolyphaga Pill].

Knotted Phlegm and Qi Stagnation Syndrome

Treatment Principle: Regulate Qi, relieve stagnation, transform phlegm, and open orifices.

Use Jia Wei Dao Tan Tang [Augmented Guide Out Phlegm Decoction] plus Mu Xiang, Xiang Fu, Yuan Zhi, Yu Jin, and Shi Chang Pu.

Furthermore, for agitated and excited patients, one can add small dose of Yang Jin Hua because of its sedative effect.

Posttreatment

This condition has high recurrent rate; one should also focus on treatments during recovery stage to strengthen the effect and prevent recurrence.

Dian Syndrome (Psychosis)

Use Yang Xin Tang [Nourish the Heart Decoction] plus Yuan Zhi, Huang Qi, Suan Zao Ren, Chuan Xiong, and Rou Gui. One can also add Su He Xiang Wan [Liquid Styrax Pill].

Kuang Syndrome (Mania)

Use Chai Hu Jia Long Gu Mu Li Tang [Bupleurum Plus Dragon Bone and Oyster Shell Decoction], remove Ren Shen, and add Sheng Tie Luo.

STRESS

This condition is caused by internal injuries due to seven emotions (anger, worry, grief, joy, thought, fright, and fear), which causes Qi stagnation, phlegm accumulation, and transformation of fire and blood stasis. These pathologic factors obstruct the Heart orifice disturbing Heart and Shen. In Western medicine, this is described as an abnormal mental reaction to strong or chronic emotional irritation. The symptoms are closely related to emotional irritations. During attacks there are two types of conditions. One is agitation and a manic condition that manifests as extreme nervousness, fright, worries, nonsensical speech, delusion, and hallucination. Another type is an inhibited and depressed state that manifests as low-spiritedness, inhibited emotions, and low emotional state. TCM treatments focus on dispersing Liver, relieving stagnation, transforming phlegm, and opening orifices.

Liver Qi Stagnation and Phlegm Obstruction Syndrome

Treatment Principle: Disperse Liver, regulate Qi, eliminate phlegm, and open orifices.

Use Dao Tan Tang [Guide Out Phlegm Decoction] plus Zhu Ru, Chai Hu, Yu Jin, Shi Chang Pu, Yuan Zhi, and Tian Zhu Huang.

Related conditions and modifications are

- Focal distention and stifling sensation at epigastrium and abdomen, poor appetite: Add Shen Qu and Hou Po.
- Insomnia: Add Ye Jiao Teng, He Huan Pi, and Long Gu.

Spleen Deficiency and Damp Excess Syndrome

Treatment Principle: Strengthen Spleen and transform damp.

Use Shen Ling Bai Zhu San [Ginseng, Poria, and Atractylodes Macrocephala Powder] plus Yu Jin, Dan Nan Xing, Shi Chang Pu, and Yuan Zhi.

Related conditions and modifications are

- Severe damp: Add Zhu Li.
- Chest stifling sensation: Add Fo Shou and Zi Su Geng.
- Palpitation and shortness of breath: Add Gui Zhi and Xi Xin.

Yin Deficiency and Deficient Fire Syndrome

Treatment Principle: Nourish Yin and descend fire.

Use Zhi Bai Di Huang Wan [Anemarrhena, Phellodendron, and Rehmannia Pill] plus Mu Tong, Dan Zhu Ye, Suan Zao Ren, Shi Chang Pu, Zhen Zhu Mu, Long Gu, and Mu Li.

Furthermore, one should combine Tui Na, Qi Gong, and Tai Ji Quan to treat the condition.

CHAPTER 27

Respiratory Disorders

ACUTE TRACHEOBRONCHITIS

This disease in TCM can be discussed in terms of cough. It is caused by external pathogens constricting the surface, lung failing to disperse and descend, and phlegm turbid obstructing the lungs. In Western medicine, it described as acute inflammatory syndrome of the trachea and bronchi that is often caused by virus or bacteria infecting the entire respiratory tract. It can also be caused by allergic or physicochemical stimulation. The main treatment principles are releasing exterior, expelling phlegm, and stopping cough. For symptoms of dry cough, dry throat, sore throat, and itchy throat, Bo He, Ban Lan Gen, and She Gan are used. For severe cough with phlegm, add Wu Wei Zi, Xing Ren, Bai Bu, and Ying Su Ke. If the sputum is sticky and hard to expectorate, then add Gua Lou Pi, Hai Fu Shi, and Zhu Li.

Wind–Cold Syndrome

Treatment Principle: Release exterior, scatter cold, disperse Lung, and transform phlegm.

Use Zhi Sou San [Stop Coughing Powder] plus Xing Ren and Zi Su Ye.

Related conditions and modifications are

- Phlegm changing from thin and white to yellow and sticky: Remove Bai Bu and add Yu Xing Cao and Huang Qin.
- Dry cough with scanty phlegm and dry throat and lips: Remove Chen Pi and Zi Su Ye and add Sang Bai Pi and Kuan Dong Hua.

Wind–Heat Syndrome

Treatment Principle: Disperse wind, release exterior, clear Lung, and transform phlegm.

Use Sang Ju Yin [Mulberry Leaf and Chrysanthemum Decoction], remove Lu Gen, and add Jin Yin Hua, Niu Bang Zi, and Huang Qin.

Related conditions and modifications are

- Severe heat: Add Chai Hu, Shi Gao, and Sang Bai Pi.
- Blood-tinged phlegm: Add Ce Bai Ye, Gua Lou Pi, and Bai Mao Gen.

Wind–Dryness Syndrome

Treatment Principle: Disperse wind, moisten dryness, and stop cough.

Use Qing Zao Jiu Fei Tang [Eliminate Dryness and Rescue the Lungs Decoction], remove Ren Shen, and add Zhe Bei Mu and Bo He.

Related conditions and modifications are

- Cough due to itchy throat: Add Jiang Can, She Gan, Chan Tui, and Xuan Shen.

Phlegm–Heat Syndrome

Treatment Principle: Clear heat, scorch out phlegm, and stop cough.

Use Xiao Xian Xiong Tang [Minor Sinking into the Chest Decoction] plus Jie Geng, Xing Ren, Lian Qiao, Zhi Ke, Zhe Bei Mu, Yu Xing Cao, and Huang Qin.

BRONCHIAL ASTHMA

This condition is commonly called "asthma." In TCM, it is under Xiao (wheezing) and Chuan (breathlessness) syndrome. The basic pathogenesis is hidden phlegm obstructing the lung. During acute attack, there is Qi stagnation, phlegm obstruction, and phlegm following the Qi rebellion. The clinical characteristics are difficulty breathing especially exhalation during attack, vibrating sound heard during respiration (stridor), stifling sensation of the chest, cough with phlegm, and diffused wheezing sounds from both lungs. In TCM, the treatment during acute attack is to scorch out phlegm, descend Qi, and calm wheezing. During remission stage, the treatment focuses on regulating Lung, Spleen, and Kidney.

Acute Attack Period

Treatment Principle: Clear heat, transform phlegm, diffuse Lung, and calm wheezing.

Use Ding Chuan Tang [Arrest Wheezing Decoction] plus Chen Pi, Ting Li Zi, and Wu Wei Zi.

Related conditions and modifications are

- Slight cold, clear watery phlegm, and pale tongue with white moist coating: Add Gui Zhi, Gan Jiang, and Xi Xin.
- Slight heat, yellow sticky phlegm, red tongue with yellow tongue coating: Add Huang Qin and Yu Xing Cao.
- Copious phlegm difficult to expectorate: Add Gua Lou Pi, Bai Jie Zi, Zhe Bei Mu, and Zi Wan.

Remission Period

Lung Qi Deficiency Syndrome

Treatment Principle: Tonify Lung and consolidate Wei Qi.

Use Yu Ping Feng San [Jade Windscreen Powder] plus Gui Zhi, Bai Shao Yao, and Dang Shen.

Spleen Qi Deficiency Syndrome

Treatment Principle: Tonify Qi and strengthen Spleen.

Use Bu Zhong Yi Qi Tang [Tonify the Middle and Augment the Qi Decoction]

Kidney Qi Deficiency Syndrome

Treatment Principle: Warm Kidney and tonify Qi.

Use Ren Shen Hu Tao Tang [Ginseng and Walnut Decoction] plus Jin Gui Shen Qi Wan [Kidney Qi Pill from the Golden Cabinet] and Wu Wei Zi.

Kidney Yin Deficiency Syndrome

Treatment Principle: Tonify Kidney and nourish Yin.

Use Zhi Bai Di Huang Wan [Anemarrhena, Phellodendron, and Rehmannia Pill] plus Xian Mao, Xian Ling Pi, Gou Qi Zi, Tai Zi Shen, and Wu Wei Zi.

Related conditions and modifications are

- Chronic condition with stasis: Add San Qi, Dan Shen, and Tao Ren.

BRONCHIECTASIS

This condition belongs to cough, Tan Yin (phlegm and fluid congestion syndrome), cough with blood, and lung abscess in TCM. The main pathogeneses are evil heat attacking lung, internal accumulation of phlegm–heat, phlegm obstruction leading to Qi rebellion, and damage to the channels and collaterals. In Western medical pathology, it refers to chronic inflammation of the bronchi and surrounding tissues damaging the walls of the bronchi leading to dilation and deformation. Main manifestations are productive cough with copious thick phlegm and symptoms that do not change with the seasons. The secretions due to bronchitis remain at the dilated area of the bronchi that flows out when there are postural changes. This causes repeated coughing of sputum and often is present with blood-tinged sputum or hemoptysis. It can also be accompanied with night sweat, fever, emaciation, poor appetite, and anemia. During periods of acute attack, the treatment focuses on the manifestations and the damages of the airway, whereas during the chronic stage, the treatment focuses on the root deficiencies. Thus, treatment principles for acute stage are to clear heat and transform phlegm, and regulate and tonify Lung, Spleen, and Kidney deficiencies during chronic stage.

Acute Attack Period

Wind–Heat Attacking Lung Syndrome

Treatment Principle: Disperse wind, clear heat, and transform phlegm.

Use Sang Ju Yin [Mulberry Leaf and Chrysanthemum Decoction] plus Zhe Bei Mu and Zi Wan.

Related conditions and modifications are

- High fever: Add Huang Qin and Shi Gao.
- Hemoptysis: Add Xiao Ji and Xian He Cao.

Internal Accumulation of Phlegm–Heat Syndrome

Treatment Principle: Clear heat, relieve toxins, transform phlegm, stop bleeding, and move stasis.

Use Shi Hui San [Ten-Partially-Charred Substances Powder] plus Yu Xing Cao, Huang Qin, Xian He Cao, Huang Lian, and San Qi powder.

Liver Fire Attacking the Lung Syndrome

Treatment Principle: Clear Liver fire, benefit the Lung, and stop cough.

Use Ke Xue Fang [Cough Blood Formula] plus Chai Hu, Huang Qin, Sang Bai Pi, Sheng Da Huang, Ce Bai Ye, and Jie Geng.

Heat Damage Lung Collaterals Syndrome

Treatment Principle: Clear heat, drain fire, cool blood, and stop bleeding.

Use Sheng Pu Huang Tang [Fresh Cattail Pollen Decoction] plus Qing Dai, San Qi powder, Shan Zhi Zi, Da Huang, Xian He Cao, and Bai Ji.

Persisting Stage (Remaining Cough, Sometimes with Scanty Hemoptysis)

Yin Deficiency Lung Heat Syndrome

Tonify Qi, nourish Yin, clear heat, and stop cough.

Use Sha Shen Mai Dong Tang [Glehnia and Ophiopogonis Decoction] plus Shan Yao, Bai He, Chuan Bei Mu, Tao Ren, and Chi Shao Yao.

Spleen and Kidney Yang Deficiency Syndrome

Strengthen Spleen, tonify Kidney, transform phlegm, and move stasis.

Use Zhen Wu Tang [True Warrior Decoction] plus Huang Qi, Dan Shen, Ban Xia, Dang Gui, and Chen Pi.

- Poor appetite: Add Shan Zha, Mai Ya, and Gu Ya.

CHRONIC BRONCHITIS

This condition belongs to Tan Yin (phlegm congested fluid syndrome) and cough in TCM. Cough is triggered when phlegm is in contact with bronchi. Wheezing occurs when phlegm obstructs the bronchi. When Qi does not return to the root, then there is shortness of breath. Coughing and wheezing for a long period of time spreads the Lung pathology to Spleen pathology. Spleen deficiency will generate damp causing it to move upward and attack the Lungs. Chronic cough will also affect Kidney pathology because Kidney grasps the Qi in respiration. The treatment during attack focuses on treating cough and phlegm by descending Lung Qi, draining damp, and transforming phlegm. Treatment between attacks focuses on treating the root by tonifying Spleen and Kidney.

Acute Attack Period

Wheezing Type

Scorch out phlegm and calm wheezing. Use the following:

- Cold syndrome: Xiao Qing Long Tang [Minor Bluegreen Dragon Decoction]
- Heat syndrome: Ma Xing Shi Gan Tang [Ephedra, Apricot Kernel, Gypsum, and Licorice Decoction]

Related conditions and modifications are

- Severe cough, wheezing: Add Di Long, Quan Xie, and Jiang Can.
- Hypertension: Remove Ma Huang and add Zi Su Zi, Di Long, and Xie Bai.

Simple Type

Treatment Principle: Disperse Lung, transform phlegm, and stop cough. Use the following:

- Cold syndrome: Ling Gan Wu Wei Jiang Xin Tang [Poria, Licorice, Schisandra, Ginger, and Asarum Decoction].
- Heat syndrome: Xiao Xian Xiong Tang [Minor Sinking into the Chest Decoction] plus Wei Jing Tang [Reed Decoction].
- Heat with exterior syndrome: Xiao Chai Hu Tang [Minor Bupleurum Decoction] plus Xiao Xian Xiong Tang [Minor Sinking into the Chest Decoction].
- Heat syndrome transform into dryness: Qing Zao Jiu Fei Tang [Eliminate Dryness and Rescue the Lungs Decoction].

Related conditions and modifications are

- Aversion to wind and headache: Add Gui Zhi and Bai Shao Yao.
- Severe chest pain: Add Yu Jin and Lian Qiao.
- Severe fever, sticky phlegm, and white blood cells not elevated: Add Ban Lan Gen, Jin Yin Hua, Lian Qiao, Zhi Mu, Gu Ya, and Sheng Jiang.
- Chronic unceasing cough: Change to Wu Wei Zi, Ying Su Ke, Ku Fan, and Xing Ren. Grind into powder, mix with honey to form pills, and take 10 to 15 g, twice a day.

Heat Damage Lung Collaterals Syndrome (Bronchitis with Bronchiectasis)

Treatment Principle: Clear heat, moisten dryness, transform phlegm, and stop bleeding.

Use Huang Qin, Sheng Di Huang, Ou Jie, Mai Men Dong, Bai Ji, Gua Lou Pi, Tao Ren, Xue Yu Tan, E Jiao, Xian He Cao, Huang Lian, Wu Wei Zi, and Bai Bu. In case of a severe syndrome, add Yun Nan Bai Yao [White Powder from Yun Nan].

Chronic Persisting Type

Mild symptoms with unclear heat or cold signs. Main manifestation is phlegm–damp. One should treat both root and branch manifestations.

Spleen Deficiency with Phlegm–Damp Syndrome

Treatment Principle: Strengthen Spleen and transform dampness.

Use Liu Jun Zi Tang [Six-Gentleman Decoction] plus San Zi Yang Qin Tang [Three-Seed Decoction to Nourish One's Parents].

Related conditions and modifications are

- Spontaneous sweating and catching cold easily: Add Huang Qi and Fang Feng.
- Poor appetite and indigestion: Shen Qu and Gu Ya.

Lung Deficiency with Phlegm–Heat Syndrome

Treatment Principle: Tonify Qi, clear heat, and expel phlegm.

Use Sheng Mai San [Generate the Pulse Powder], Ting Li Da Zao Xie Fei Tang [Descurainia and Jujube Decoction to Drain the Lungs], and Xiao Xian Xiong Tang [Minor Sinking into the Chest Decoction].

Related conditions and modifications are

- Wheezing and chest tightness: Add Ma Huang, Zi Su Zi, and Di Long.

Remission Period

Treatment Principle: Support the Zheng Qi, stabilize the root, and tonify both Qi and Yin.

Use Ren Shen Ge Jie San [Ginseng and Gecko Powder] plus Dong Chong Xia Cao, Gou Qi Zi, Huang Qi, Shu Di Huang, Bai Guo, and Bai Zhu.

CHRONIC OBSTRUCTIVE PULMONARY EMPHYSEMA

This condition is discussed under Lung distention, Tan Yin (phlegm fluid congestion syndrome), and breathlessness syndromes in TCM. In the pathological progression, chronic cough exhausts Qi, phlegm causes obstruction and Qi rebellion, which then damages Lung, Spleen, and Kidney. The manifestations are productive cough, breathlessness, swelling, and stasis. The root of the condition is at Lung, Spleen, and Kidney. In Western pathology, the condition is described as chronic cough causing overinflation of the Lung, overdistended, ruptured, and fused bronchioles and alveoli. It will often happen after chronic bronchitis and bronchial asthma. Main manifestations are cough for many years, cough with sputum, attacks during winter which improve during summer season, and gradual increase in severity of exhalation difficulties through the years. Physical examination will present with symptoms of barrel chest, decreased respiration movement, decreased vocal fremitus, high clear percussion sound, lowered Lung border, and decreased Heart boundary. In TCM, if there is an exterior syndrome, then the treatment principle is to release the exterior, expel phlegm, and stop cough. It there is no exterior syndrome, then transform phlegm, warm Yang, and tonify Qi at the same time.

During Attack Infections Period

Phlegm–Heat Type

Treatment Principle: Clear heat, expel phlegm, disperse Lung, and release exterior.

Use Yin Qiao Ma Xing Shi Gan Tang [Honeysuckle, Forsythia, Ephedra, Apricot Kernel, Gypsum, and Licorice Decoction] plus Huang Qin, Yu Xing Cao, Jie Geng, Ban Xia, Gua Lou Pi, Dong Gua Ren, and Sang Bai Pi.

Related conditions and modifications are

- After fever has reduced: Remove Shi Gao and Huang Qin, and add Dang Shen.

Excess Wheezing Syndrome

Treatment Principle: Clear heat, calm wheezing, release exterior, and expel phlegm.

Use Ting Li Ma Xing Shi Gan Tang [Descurainia, Ephedra, Apricot Kernel, Gypsum, and Licorice Decoction] plus Di Long, Huang Qin, Lai Fu Zi, Gua Lou Pi, Sang Bai Pi, Jie Geng, and Lu Gen.

Stabilized Period

Treatment Principle: Tonify Lung, Spleen, and Kidney, transform phlegm, and expel stasis.

Use Ren Shen Ge Jie San [Ginseng and Gecko Powder] plus Dong Chong Xia Cao, Gou Qi Zi, Huang Qi, Bai Zhu, Dang Gui, Tu Si Zi, Lu Jiao Jiao, and Quan Gua Lou.

Treatment According to Symptoms

Wheezing

Di Long has the function to relieve spasms of the bronchi. One can take 2 g of Di Long powder twice a day. Or grind equal amount of Di Long and Gan Cao, and take 3 to 5 g of the powder three times a day.

Copious Phlegm

Related conditions and modifications are

- Clear, watery, and white phlegm: Add Ban Xia, Bai Jie Zi, Gan Jiang, Zi Su Zi, Jie Geng, Yuan Zhi, Zao Jiao Ci, Zi Wan, and Kuan Dong Hua.
- Yellow, sticky, and thick phlegm: Add Zhe Bei Mu, Gua Lou, Ting Li Zi, Hai Fu Shi, and Zhu Ru.
- Sticky phlegm slightly blood-tinged: Add Mai Men Dong, Pi Pa Ye, Sha Shen, and Li Pi.
- White, sticky, and copious phlegm: Add Ban Xia, Fu Ling, Chen Pi, and Jie Geng.
- Copious phlegm difficult to expectorate and stifling sensation of the chest: Add Dan Nan Xing, Ju Hong, Fu Ling, Zhi Shi, Shi Chang Pu, Zhu Ru, Ma Huang, and Ban Xia.

Cough

Add Chuan Bei Mu, Zhe Bei Mu, Xing Ren, Wu Wei Zi, and Xi Xin to the base formulas. Related condition and modification are

- Severe cough: Add Ying Su Ke.

Edema

In this syndrome, it is often mild edema of the face or lower extremities.

Use Jia Wei Ting Li Da Zao Xie Fei Tang [Augmented Descurainia and Jujube Decoction to Drain the Lung] plus Huang Qi, Ze Xie, Fu Ling, Ren Shen, Dan Shen, and Wu Wei Zi.

CHRONIC PULMONARY HEART DISEASE

This condition belongs to Lung distention and Tan Yin (phlegm congested fluid syndrome) in TCM. Lingering cough, wheezing, Tan Yin, and repeated attacks often cause this condition. Lung Qi becomes more deficient, and Lung turbid becomes more obstructed which manifests as cough with or without phlegm, shortness of breath, stifling sensation of the chest, palpitation, distention and fullness of the epigastric and abdominal region, and cyanosis. The primary treatment principle for acute stage is to expel evil pathogens by treating the exterior and phlegm stasis, secondary to giving support by tonifying the deficiency (Lung, Kidney, Heart, and Spleen). During remission period, the primary focus is to tonify the deficiency and, secondarily, to support by transforming phlegm fluid congestion. Invigorating blood, transforming stasis, and unblocking the Qi flow are also included as part of the treatment.

Exterior Cold Phlegm Syndrome

Treatment Principle: Scatter cold, transform congested fluid, stop cough, and calm wheezing.

Use Xiao Qing Long Tang [Minor Blue-Green Dragon Decoction] plus Fang Feng, Bai Qian, Kuan Dong Hua, Zi Su Zi, and Ting Li Zi.

Exterior Heat Phlegm Syndrome

Treatment Principle: Scatter wind, clear heat, transform phlegm, and calm wheezing.

Use Ma Xing Shi Gan Tang [Ephedra, Apricot Kernel, Gypsum, and Licorice Decoction] plus Ting Li Da Zao Xie Fei Tang [Descurainia and Jujube Decoction to Drain the Lungs] and Huang Qin, Sang Bai Pi, Gua Lou Pi, Ban Lan Gen, Lu Gen, and Shi Gao.

Lung Deficiency with Phlegm Turbid Syndrome

Treatment Principle: Warm Lung, transform phlegm, descend Qi, and calm wheezing.

Use Su Zi Jiang Qi Tang [Perilla Fruit Decoction for Directing Qi Downward] plus San Zi Yang Qin Tang [Three-Seed Decoction to Nourish One's Parents] and Fu Ling, Cang Zhu, and Bai Zhu.

Related condition and modifications are

- Purple lips and palpitation: Add Dan Shen, Hong Hua, and Chi Shao Yao.
- Spontaneous sweating: Add Huang Qi, Fang Feng, and Dang Shen.

Yang Deficiency with Edema Syndrome

Treatment Principle: Warm Kidney, strengthen Spleen, transform congested fluid, and drain water.

Use Zhen Wu Tang [True Warrior Decoction] plus Wu Ling San [Five-Ingredient Powder with Poria].

Related condition and modifications are

- Cold limbs and minute pulse: Add Ren Shen and Zhi Huang Qi.

Phlegm Obstructing the Shen and Orifices Syndrome

Treatment Principle: Clear heat, eliminate phlegm, open orifices, and awaken Shen.

Use Di Tan Tang [Scour Phlegm Decoction], remove Ren Shen, and add Jiang Can, Yu Jin, and Zhi Yuan Zhi.

For closed cold type: Combine with Su He Xiang Wan [Liquid Styrax Pill] one pill twice a day.

For closed heat type: Combine with Zhi Bao Dan [Greatest Treasure Special Pill] 0.7 g twice a day.

Heart–Qi Deficiency Syndrome

Treatment Principle: Tonify Qi and recover pulse.

Use Zhi Gan Cao Tang [Honey-Fried Licorice Decoction] plus Ge Jie, Hu Tao Ren, and Dong Chong Xia Cao.

Yin Deficiency with Phlegm–Heat Syndrome

Treatment Principle: Nourish Yin, clear heat, and expel phlegm.

Use Yang Yin Qing Fei Tang [Nourish the Yin and Clear the Lungs Decoction] plus Huang Qin and Sang Bai Pi.

LUNG ABSCESS

This condition is referred to as Fei Yong in TCM. Exterior wind–cold or wind–heat pathogenic factors invade the Lung generating phlegm that transforms into heat over time. Heat congestion, blood stasis, and rotten flesh form abscesses in the Lung. From Western medical point of view, Lung abscess and swelling is a suppurative pathological change of the Lung tissues due to various germs. At the early stage, there is suppurative inflammation of the Lung tissue followed by pus and swelling formation due to necrosis. Then there is pus discharge, which manifests as purulent sputum, sometimes with blood. The clinical manifestations are acute onset, aversion to cold, high fever, chest pain, cough, and vomiting of purulent sputum with foul smell. The TCM treatment focuses on expelling phlegm, discharging pus, and releasing exterior. At the later stages, one should tonify Qi and nourish Yin. From the laboratory study, equal dosages of Jie Geng and Gan Cao together can increase the discharge of purulent fluid.

Exterior Syndrome Period (Beginning Stage)

Wind–Heat Attacking Lung Syndrome

Treatment Principle: Release wind, clear heat, disperse Lung, and transform phlegm.

Use Yin Qiao San [Honeysuckle and Forsythia Powder] plus Zhe Bei Mu, Quan Gua Lou, and Pu Gong Ying.

Related conditions and modifications are

- Vomiting brown phlegm: Remove Bo He and Niu Bang Zi, and add Huang Qin, Yu Jin, Tao Ren, Dong Gua Ren, and Yi Yi Ren.
- Severe heat: Add Yu Xing Cao and Shi Gao.

Wind–Cold Constricting Lung Syndrome

Treatment Principle: Expel wind, scatter cold, disperse Lung, and transform phlegm.

Use She Gan Ma Huang Tang [Belamcanda and Ephedra Decoction] plus Yu Xing Cao and Gua Lou Pi.

Pus Formation (Abscess Formation) Period

Knotted Phlegm and Stasis Syndrome

Treatment Principle: Transform stasis and disperse accumulation.

Use Wei Jing Tang [Reed Decoction].

Related conditions and modifications are

- Vomiting foul and fetid phlegm: Add Jie Geng and Sheng Gan Cao.
- Chest and hypochondriac pain: Add Yu Jin, Chi Shao Yao, and Mu Dan Pi.

Phlegm–Heat Accumulation and Stagnation Syndrome

Treatment Principle: Drain Lung, expel phlegm, clear heat, and relieve toxicity.

Use Ting Li Da Zao Xie Fei Tang [Descurainia and Jujube Decoction to Drain the Lungs] plus Jin Yin Hua, Lian Qiao, Huang Qin, Yu Xing Cao, and Xian Lu Gen.

Related conditions and modifications are

- More heat than phlegm–damp: Change to Yue Bi Jia Ban Xia Tang [Maidservant from Yue Decoction plus Pinellia].
- Yellow coating all over the tongue, foul breath with heat, and dark urine: Add Pu Gong Ying and Zi Hua Di Ding.
- Copious phlegm: Add Zhe Bei Mu and Zhu Li.

Purulent Ulceration Period

Purulent Ulceration and Phlegm Congestion Syndrome

Treatment Principle: Transform stasis, expel pus, clear heat, and relieve toxicity.

Use Jie Geng Tang [Platycodi Decoction] (Jie Geng and Gan Cao) plus Wei Jing Tang [Reed Decoction] and Ban Lan Gen, Da Huang, and Shi Gao.

Blazing Heat Toxin Syndrome

Treatment Principle: Clear heat, relieve toxicity, and expel pus.

Use Huang Lian Jie Du Tang [Coptis Decoction to Relieve Toxicity] plus Wei Jing Tang [Reed Decoction] and Hong Teng, Bai Jiang Cao, Da Qing Ye, and Zhe Bei Mu.

Related conditions and modifications are

- Dry mouth: Add Sha Shen, Tian Hua Fen, and Mai Men Dong.
- Hemoptysis: Remove Jie Geng, and add Mu Dan Pi and Bai Mao Gen.

Lung Excess and Spleen Deficiency Syndrome

Treatment Principle: Transform stasis, clear heat, tonify Qi, and expel pus.

Use Wei Jing Tang [Reed Decoction] plus Bai Mao Gen, Yu Xing Cao, Jin Yin Hua, Lian Qiao, Mu Dan Pi, Jie Geng, Gan Cao, Huang Qin, Xing Ren, Gu Ya, and Bai Zhu.

Related conditions and modifications are

- Copious phlegm: Add Sang Bai Pi, Ting Li Zi, and fried Lai Fu Zi.
- Hemoptysis: Add San Qi powder.

Recovery Period

Lung Yin Deficiency Syndrome

Treatment Principle: Tonify Qi and nourish Yin.

Use Sha Shen Mai Dong Tang [Glehnia and Ophiopogonis Decoction] plus Tai Zi Shen, Yi Yi Ren, Huang Qi, Dong Gua Ren, and Jie Geng.

Related conditions and modifications are

- Chest pain: Add Yu Jin and Si Gua Luo.
- Cough: Add Bai Bu and Xing Ren.
- Hot flashes and night sweats: Add Qing Hao and Di Gu Pi.
- Greasy tongue coating and poor appetite: Add Chen Pi, fried Gu Ya, and Zhi Ke.
- Constipation: Add Zhi Shi and Gua Lou Ren.

Kidney Yin Deficiency Syndrome

Treatment Principle: Tonify Qi, nourish Yin, and transform phlegm.

Use Mai Wei Di Huang Wan [Glehnia, Schisandra, and Rehmannia Pill] plus Sha Shen and Zhe Bei Mu.

PLEURISY

This condition is called Xuan Yin (suspended congested fluid syndrome) in TCM. The external pathogens cause internal fluid congestion to accumulate in the Chest and hypochondriac region. Western medicine describes this is an inflammation of the pleura in the Lung and Chest wall, which can be divided into dry and exudative pleurisy. Clinical manifestations are Chest pain, cough, rapid breathing, fever, chills, and various degrees of restriction in respiration or even difficulty breathing and cyanosis. Chinese medicine is effective in reducing Chest pain, preventing adhesion of the pleura, and controlling the infection.

Beginning Stage (Early Stage of Dry or Exudative Pleurisy)

Treatment Principle: Harmonize exterior and interior, and disperse and benefit the Qi mechanism.

Use Xiao Chai Hu Tang [Minor Bupleurum Decoction] and Xiao Xian Xiong Tang [Minor Sinking into the Chest Decoction] plus Yu Jin, Qing Hao, and Yu Xing Cao.

Related conditions and modifications are

- Internal excess of phlegm–heat: Remove Chai Hu and add Ma Xing Shi Gan Tang [Ephedra, Apricot Kernel, Gypsum, and Licorice Decoction].
- Constipation: Add Da Huang and drink separately. Stop once bowel movements start.
- Congestion and stagnation of damp–heat: Add Yin Chen Hao, Yi Yi Ren, Long Dan Cao, and Bai Dou Kou.

Fluid Congestion Period

Treatment Principle: Drain damp and attack congested fluid.

Use Shi Zao Tang [Ten-Jujubes Decoction] or Kong Yan Dan [Control Mucous Special Pill].

Related conditions and modifications are

- Mild case of fluid congestion: Use Bai Jie Zi, Ting Li Zi, Zi Su Zi, Gua Lou Pi, Xing Ren, Sang Bai Pi, Zhi Ke, Ze Xie, Fu Ling, Che Qian Zi, Dong Gua Pi, and Jiao Mu.
- Fullness and stifling sensation of the chest and hypochondriac region and copious phlegm: Add Ban Xia and Xie Bai.
- Chronic accumulation of water in the chest, focal distention and fullness of the chest and epigastric region, scanty food intake, and loose stool: Add Gui Zhi, Bai Zhu, Tong Cao, and Lu Lu Tong.

Recovery Period

Treatment Principle: Regulate Qi, activate collaterals, and expel congested fluid.

Use Xiang Fu Xuan Fu Hua Tang [Cyperus and Inula Decoction] plus Xing Ren,Yi Yi Ren, and Zhi Ke.

Related conditions and modifications are

- Yin deficiency and dry cough: Use Sha Shen Mai Dong Tang [Glehnia and Ophiopogonis Decoction] and Xie Bai San [Drain the White Powder].
- Severe hot flashes and night sweats: Add Yin Chai Hu, Qing Hao, and Bie Jia.
- Severe chest and hypochondriac pain: Add Yu Jin, Jiang Xiang, Zhi Ke, and Ban Xia.
- Small quantity of residual fluid in chest cavity: Add Bai Jie Zi, Ze Xie, Fu Ling, and Lu Lu Tong.
- Shortness of breath, apathy, and pale complexion: Add Huang Qi, Dang Shen, and Wu Wei Zi.

PNEUMONIA

This syndrome is discussed under warm febrile diseases in TCM. It is often diagnosed as cough, Chest pain, and fever. The wind–warm or warm pathogens (bacteria, mycoplasms, virus, and fungi) invade the Lung through the mouth and nose and cause parenchymatous inflammatory degeneration of the Lung.

SARS (Severe Acute Respiratory Syndrome)

It is a warm febrile disease caused by epidemic toxins (virus). The pathogeneses are heat toxin stasis, congestion obstructing Lung collaterals, and excess evil heat pathogens. In severe cases, there is high fever, cough, and wheezing, which damages both Qi and Yin and leads to collapse (Tuo) syndrome.

Early Stage

Heat toxins attack the Lungs and damp–heat obstructing the Lungs.

Heat Toxins Attack the Lungs

Treatment Principle: Clear heat, disperse Lung, release exterior, and unblock collaterals.

Use Yin Qiao San [Honeysuckle and Forsythia Powder] plus Ma Xing Shi Gan Tang [Ephedra, Apricot, Kernel, Gypsum, and Licorice Decoction] and Guan Zhong.

Damp–Heat Obstructing the Lungs

Treatment Principle: Disperse heat, transform damp, and vent evils to the exterior.

Use Huo Po Xia Ling Tang [Agastache, Magnolia Bark, Pinellia, and Poria Decoction] plus Sheng Jiang San [Ascending and Descending Powder], San Ren Tang [Three-Nut Decoction], and Da Qing Ye.

Exterior Cold and Interior Heat with Damp

Treatment Principle: Release exterior, clear interior, disperse Lung, and transform damp.

Use Ma Xing Shi Gan Tang [Ephedra, Apricot Kernel, Gypsum, and Licorice Decoction] plus Sheng Jiang San [Ascending and Descending Powder], Huo Po Xia Ling Tang [Agastache, Magnolia Bark, Pinellia, and Poria Decoction], and Cang Zhu.

Middle Stage

Damp–heat epidemic toxins and blazing excess in exterior and interior Shao Yang syndrome.

Epidemic Toxins Invade Lung with Excess in Exterior and Interior

Treatment Principle: Clear heat, relieve toxicity, drain Lung, and descend rebellious Qi.
Use Qing Fei Jie Du Tang [Clear the Lungs and Relieve Toxicity Decoction].

Damp–Heat and Toxins Accumulation

Treatment Principle: Transform dampness, separate the clear from turbid, clear heat, and relieve toxicity.
Use Gan Lu Xiao Du Yin [Sweet Dew Decoction to Eliminate Toxin].

Damp–Heat Obstruction in Shao Yang

Treatment Principle: Clear heat, drain Shao Yang, separate, and transform damp–heat.
Use Hao Qin Qing Dan Tang [Artemisia Annua and Scutellaria Decoction to Clear the Gallbladder].

Blazing Heat Toxins

Treatment Principle: Clear heat, cool blood, drain fire, and relieve toxicity.
Use Qing Wen Bai Du Yin [Clear Epidemics and Overcome Toxin Decoction] plus Jiang Can.

Extreme Period

Accumulated excess heat toxins, excess evil pathogens, deficient Zheng Qi, both Qi and Yin deficiency, and close internally and collapse externally.

Phlegm–Heat Toxins Stasis and Congested and Obstructed Lung Collaterals

Treatment Principle: Relieve toxicity, transform phlegm, cool blood, and unblock collaterals.
Use Jia Jian Pu Ji Xiao Du Yin [Modified Universal Benefit Decoction to Eliminate Toxin] plus Sang Dan Xie Bai San [Mulberry Leaf and Moutan Decoction to Drain the White] and Qing Dai, Ban Lan Gen, and Dan Shen.

Damp–Heat Congesting Lung and Both Qi and Yin Damage

Treatment Principle: Clear heat, drain damp, tonify Qi, and nourish Yin.
Use Qing Fei Yin [Clear the Lungs Decoction] plus Mai Men Dong, Zhu Ru, Tai Zi Shen, and Dang Shen.

Excess Evil Pathogens and Deficient Zheng Qi, Closes Internally and Collapses Externally

Treatment Principle: Tonify Qi and rescue the collapse.
Use Shen Fu Tang [Ginseng and Prepared Aconite Decoction].

Recovery Period

Related conditions and modifications are

Qi and Yin Deficiency and Residual Evil Pathogens

Treatment Principle: Tonify Qi, nourish Yin, transform dampness, and unblock collaterals.

Qing Shu Yi Qi Tang [Clear Summer Heat and Augment the Qi Decoction] plus Ge Gen and Pei Lan.

Lung and Spleen Deficiency

Treatment Principle: Tonify Qi and strengthen Spleen.

Shen Ling Bai Zhu San [Ginseng, Poria, and Atractylodes Macrocephala Powder] plus Ge Gen Huang Qin Huang Lian Tang [Kudzu, Scutellaria, and Coptis Decoction].

Furthermore, people who come in contact with SARS patients may take this formula to increase their Wei Qi (defensive Qi):

Sheng Huang Qi 15 g	Cang Zhu 9 g
Jin Yin Hua 15 g	Yi Yi Ren 15 g
Chai Hu 9 g	Huo Xiang 9 g
Huang Qin 9 g	Fang Feng 9 g
Ban Lan Gen 15 g	Sheng Gan Cao 6 g
Guan Zhong 15 g	

Take the decoction twice a day for 10 to 14 days.

Pneumococcal Pneumonia

Clinically, this condition happens suddenly with chills and high fever, chest pain, cough with rust color phlegm, shallow rapid breathing, ala nasi flaring, and rapid or irregular heartbeats, or even shock. Chinese medicine focuses on clearing heat, relieving toxicity, dispersing Lung, transforming phlegm, and nourishing Yin. Even though the progression of the disease follows the Four Stages—Wei, Qi, Ying, and Blood—the extensive use of antibiotics will block the evil pathogens from penetrating into Ying and Blood level. Thus, most patients only manifest with symptoms in Wei and Qi stage and recover after these. From modern research, Jin Yin Hua, Huang Qin, Yu Xing Cao, Jing Qiao Mai, Pu Gong Ying, Chai Hu, and Ban Zhi Lian have antibacterial and antiinflammatory functions that can be used according to syndrome differentiation.

Heat at Lung and Wei Syndrome

Treatment Principle: Acrid cool herbs to release exterior, disperse Lung, and stop cough.

Use Yin Qiao San [Honeysuckle and Forsythia Powder] plus Sang Ju Yin [Mulberry Leaf and Chrysanthemum Decoction].

Related conditions and modifications are

- Severe cough: Add Zhe Bei Mu.
- Swelling and painful throat: Add Shan Dou Gen and Pang Da Hai.
- Severe heat: Add Sang Bai Pi, Huang Qin, and Yu Xing Cao.

Phlegm–Heat Congesting the Lungs Syndrome

Treatment Principle: Clear heat, transform phlegm, disperse Lung, and stop cough.
Use Qing Jin Hua Tan Tang [Clear the Metal and Transform the Phlegm Decoction]
Related conditions and modifications are

- Sticky, thick, and copious yellow sputum: Add Yu Xing Cao, Wei Jing, and Pu Gong Ying.
- Chest pain: Add Zhi Ke and Yu Jin.

Heat Attacks Pericardium Syndrome

Treatment Principle: Clear Lung, drain heat, vent Ying level, and cool blood.
Use Qing Ying Tang [Clear the Nutritive Level Decoction].
Related conditions and modifications are

- Severe heat toxins: Also take An Gong Niu Huang Wan [Calm the Palace with Gallstone Pill].
- Delirium and closed orifices: Add Zhi Bao Dan [Greatest Treasure Special Pill].
- Convulsion and opisthotonos: Add Zi Xue Dan [Purple Snow Special Pill].

Exhausted Yin and Collapsing Yang Syndrome

Treatment Principle: Tonify Qi and rescue collapse.
Use Sheng Mai San [Generate the Pulse Decoction].

Qi and Yin Both Damaged with Residual Heat Syndrome

Treatment Principle: Nourish Yin, clear heat, moisten Lung, and transform phlegm.
Use Zhu Ye Shi Tao Tang [Lophatherus and Gypsum Decoction], replace Ren Shen with Xi Yang Shen, and add Tian Hua Fen, Jie Geng, and Chuan Bei Mu.
Related conditions and modifications are

- No clear heat manifestations: Use Yang Yin Qing Fei Tang [Nourish Yin and Clear the Lungs Decoction].

Staphylococcal Pneumonia

There are two types of staphylococcal pneumonia that are inhaled and hematogenous types. In general, there is respiratory tract infection. Shortly after, there are chills, high fever, cough, rapid breathing, and yellow sticky phlegm changing to bloody sputum with pus. In severe cases, there are also muscle and joint soreness, weakness of the whole body, and other whole body poisoning symptoms. The TCM syndrome differentiation follows the Wei, Qi, Ying, and Blood levels.

Lung Wei Syndrome

Treatment Principle: Acrid cool herbs to release the exterior, disperse Lung, and transform phlegm.
Use Yin Qiao San [Honeysuckle and Forsythia Powder] plus Pu Gong Ying and Yu Xing Cao.
Related condition and modification are

- If Qi level symptoms are also seen at this time: Add Shi Gao, Zhi Mu, Huang Qin, and Shan Zhi Zi.

Phlegm–Heat Congesting the Lung Syndrome

Treatment Principle: Clear heat, disperse Lung, transform phlegm, and stop cough.

Use Wei Jing Tang [Reed Decoction] plus Pu Gong Ying, Yu Xing Cao, Ye Ju Hua, Zhe Bei Mu, Jin Yin Hua, Lian Qiao, Jie Geng, Gua Lou Pi, and Qian Hu.

Related conditions and modifications are

- Sore throat: Add Niu Bang Zi and She Gan.
- Severe chest pain: Add Mu Dan Pi, Chi Shao Yao, and Dan Shen.

Heat Trapped in Ying and Blood Syndrome

Treatment Principle: Clear Ying, cool blood, expel evil from Qi level, and transform phlegm.

Use Qing Ying Tang [Clear the Nutritive Level Decoction] plus Xi Jiao Di Huang Tang [Rhinoceros Horn and Rehmannia Decoction].

Related condition and modifications are

- Delirium and delirious speech: Add An Gong Niu Huang Wan [Calm the Palace with Cattle Gallstone].

Qi and Yin Deficient Syndrome

Treatment Principle: Tonify Qi and nourish Yin.

Use Sheng Mai San [Generate the Pulse Powder] plus Sheng Di Huang, Shu Di Huang, Xuan Shen, Huang Qi, Xi Yang Shen, Huang Qin, Zhi Mu, Dan Shen, and Gan Cao.

Pulmonary Mycoses (Fungal Pneumonia)

This condition often occurs in people with defect in immune function, critical or chronic conditions such as diabetes, cancer, AIDS, and chronic consumption of antibiotics or steroids. The treatment focus is on strengthening the immune system and inhibiting the growth of fungi.

Phlegm–Heat Obstructing Lung Syndrome

Treatment Principle: Clear heat, transform phlegm, disperse Lung, and stop cough.

Use Qing Jin Hua Tan Tang [Clear the Metal Decoction to Transform Phlegm] plus Zhi Sou San [Stop Cough Powder].

Phlegm–Damp Obstructing Lung Syndrome

Treatment Principle: Dry damp and transform phlegm.

Er Chen Tang [Two-Cured Decoction] plus Jie Geng, Qian Hu, Bai Bu, Kuan Dong Hua, Chuan Bei Mu, and Yu Xing Cao.

Both Qi and Yin Deficiency Syndrome

Treatment Principle: Tonify Qi and nourish Yin.

Related conditions and modifications are

- More Qi deficiency: Shen Ling Bai Zhu San [Ginseng, Poria, and Atractylodes Macrocephala Powder].
- More Yin deficiency: Bai He Gu Jin Tang [Lily Bulb Decoction to Preserve the Metal] plus Xing Ren and Bai Bu.

Mycoplasmal Pneumonia

This syndrome is caused by mycoplasms. It often occurs in summer and fall seasons. Main clinical manifestations are cough, spitting white and sticky or yellow and sticky sputum, breathlessness, chest pain, and sore throat. In few cases, there can also be erythema circinatum, hemolysis, and thrombocytopenia. The TCM syndrome differentiation follows Wei, Qi, Ying, and Blood levels. One can refer to the discussion and formulas for Pneumococcal Pneumonia. If there is erythema circinatum, hemolysis, or thrombocytopenia, then the following herbs can be used:

Chi Xiao Dou, Huang Qin, Huang Lian, Huang Bai, Yi Yi Ren, Fang Feng, Tu Fu Ling, Zao Xiu, Lian Qiao, Bai Xian Pi, and Gan Cao.

PULMONARY TUBERCULOSIS

This condition is referred to as consumptive disease of the Lung (Fei Lao) in TCM. It is caused by "consumptive bacteria" (bacillus tubercle) invading the Lungs. The clinical manifestations are chronic low-grade fever, emaciation, fatigue, night sweats, anemia, poor appetite, chest pain, palpitation, cough with sputum, and hemoptysis or irregular menstruation. Yin deficiency with Lung heat is the pathogenesis of this condition. There are two parts to the treatment in TCM. One is to kill the consumptive bacteria and the other is to tonify the deficiency. Da Suan, Huang Lian, Shi Liu Pi, Di Yu, Jin Yin Hua, Huang Qin, and Bai Guo are used to kill the bacteria.

Yin Deficiency and Lung Dryness Syndrome

Treatment Principle: Nourish Yin and moisten dryness.

Use Yue Hua Wan [Moonlight Pill] plus Dong Chong Xia Cao.

Yin Deficiency and Spleen Weakness Syndrome

Treatment Principle: Nourish Yin and strengthen Spleen.

Use Shi Quan Da Bu Tang [All-Inclusive Great Tonifying Decoction] plus Tian Men Dong, Mai Men Dong, Sheng Di Huang, Yin Chai Hu, Di Gu Pi, Wu Wei Zi, Lian Xu, Chen Pi, Huang Bai, and Zhi Mu.

Lung and Kidney Yin Deficiency Syndrome

Treatment Principle: Moisten Lung and nourish Yin.

Use Zhi Bai Di Huang Wan [Anemarrhena, Phellodendron, and Rehmannia Pill] plus Bai Zi Yang Xin Wan [Biota Seed Pill to Nourish the Heart] and Ren Shen, Long Gu, and Gui Ban.

Related conditions and modifications are

- Persisting tidal fever with fluctuating temperature: One should stop taking the core formula and change to Qing Gu San [Clear the Bone Powder].
- Hemoptysis: Add Bai Ji, Xian He Cao, and Xiao Ji.
- Frequent dry cough: Add Xing Ren, Wu Wei Zi, Ma Dou Ling, Bai Bu, and Chuan Bei Mu.
- Night sweats: Add Ma Huang Gen, Fu Xiao Mai, Wu Wei Zi, and Bai Shao Yao. In severe cases, stop the core formula and change to Mu Li San Jia Wei [Augmented Oyster Shell Powder] (Mu Li, Huang Qi, Fu Xiao Mai, Ma Huang Gen, Wu Wei Zi, Mai Men Dong, and Gui Ban).
- Insomnia: Add Suan Zao Ren, Bai Zi Ren, Fu Shen, Ye Jiao Teng, and Wu Wei Zi.
- Poor appetite: Add Shan Zha, Shen Qu, Gu Ya, Chen Pi, and Zhi Ke.
- Diarrhea: Add Shen Ling Bai Zhu San [Ginseng, Poria, and Atractylodes Macrocephala].
- Hoarseness: Add Yu Zhu, Shi Hu, Xing Ren, Chuan Bei Mu, Zhu Ye, Xi Yang Shen, Sheng Di Huang, and Mai Men Dong.

SILICOSIS

This condition is called stone consumptive disease (Shi Lao) in TCM. It is caused by inhaling silica dust for many years, which caused fibrosis and silica nodule formation in the Lungs. The manifestations are cough, expectoration of sputum, stifling sensation at chest, chest pain, rapid breathing, and cyanosis due to obstruction of Lung Qi function and emphysema. The pathogenesis of this condition is internal obstruction of phlegm, turbid, Qi stagnation, blood stasis, and deficiency caused by excess evil pathogens. The excess evil pathogens are the quartz dust, phlegm, and stasis. The deficiencies are Lung, Spleen, Kidney, and Heart. Treatment principles are to clear dryness, moisten Lung, transform phlegm, scatter accumulation, invigorate blood, unblock collaterals, and regulate and tonify Spleen and Kidney. Herbs that can dissolve silica dust and expel silica are Dan Shen, Tao Ren, Hong Hua, Chi Shao Yao, Liu Ji Nu, San Qi, Gua Lou, Ban Xia, Zhi Nan Xing, Zhe Bei Mu, Hai Fu Shi, Mang Xiao, Ji Nei Jin, and Xia Ku Cao. Herbs that contain aluminum are Chang Shan, Wu Gong, Ban Bian Lian, and Xiang Fu. They have the function of controlling the progression of silicosis. In general, stages I and II of silicosis are often excess syndromes and stages III and IV are often deficiency or mixed syndromes.

Dryness Damaging Lung Syndrome

Treatment Principle: Nourish Yin, moisten dryness, disperse Lung, and transform phlegm.

Use Qing Zao Jiu Fei Tang [Eliminate Dryness and Rescue the Lungs Decoction] remove Ren Shen, and add Jie Geng, Niu Bang Zi, Sha Shen, Gua Lou Ren, Shi Gao, and Shen Qu.

Related conditions and modifications are

- Severe wheezing and cough: Add Zi Su Zi, Lai Fu Zi, and Sang Bai Pi.
- Phlegm with blood: Add Huang Qin, Bai Mao Gen, and Xian He Cao.

Phlegm and Qi Stagnation and Nodule Formation Syndrome

Treatment Principle: Disperse Lung and expel phlegm.

Use Qing Qi Hua Tan Tang [Clear Qi Decoction to Transform Phlegm] and Wei Jing Tang [Reed Decoction].

Related conditions and modifications are

- Sharp pain at chest: Add Chuan Xiong, Yu Jin, Ju Luo, and Bai Jie Zi.
- Stifling sensation of the chest, dizziness, and yellow, thick and greasy tongue coating: Remove Chuan Bei Mu, and add Cang Zhu, Hou Po, and Huo Xiang.

Blood Stasis Obstructing the Collaterals Syndrome

Treatment Principle: Transform stasis, unblock collaterals, disperse accumulations, and dissolve silica.

Use Bu Yang Huan Wu Tang [Tonify the Yang to Restore Five-Tenths Decoction] plus Xue Fu Zhu Yu Tang [Drive Out Stasis in the Mansion of Blood Decoction].

Lung Yin Deficiency Syndrome

Treatment Principle: Nourish Yin, moisten Lungs, dissolve silica, and disperse accumulation.

Use Bai He Gu Jin Tang [Lily Bulb Decoction to Preserve the Metal] remove Shu Di Huang, and add Mu Li, He Ye, Gu Ya, Qing Hao, Bie Jia, and Yin Chai Hu.

Related conditions and modifications are

- Copious hemoptysis: Add Gua Lou Ren, Huang Qin, Shan Zhi Zi, and Hai Fu Shi.
- Palpitation, shortness of breath, and fatigue: Add Ren Shen and Wu Wei Zi.

Lung and Spleen Qi Deficiency Syndrome

Treatment Principle: Tonify Spleen, benefit Lung, expel evil, and dissolve silica.

Use Liu Jun Zi Tang [Six-Gentleman Decoction] plus Shan Yao, Dan Shen, Huang Qi, Shan Zha, Ji Nei Jin, Jie Geng, Dang Gui, Hong Hua, and Gua Lou Pi.

Related conditions and modifications are

- Swelling and distention of lower limbs: Add Gui Zhi, Zhu Ling, and Ze Xie.
- Abdominal distention: Add Hou Po and Zhi Shi.

Lung and Kidney Qi and Yin Deficiency Syndrome

Treatment Principle: Nourish Kidney, benefit Lung, and dissolve silica.

Use Sheng Mai San [Generate the Pulse Powder] and Liu Wei Di Huang Wan [Six-Ingredient Pill with Rehmannia].

Heart Yang Deficiency Syndrome

Treatment Principle: Stimulate Heart Yang, unblock Yang, transform stasis, and disperse accumulation.

Use Zhi Gan Cao Tang [Honey-Fried Licorice Decoction] plus Wu Wei Zi, Gua Lou Pi, Dan Shen, and Hong Hua.

Related conditions and modifications are

- Aversion to cold with cold limbs: Add Zhi Fu Zi, Gan Jiang, and double the dosage of Gui Zhi.
- Swollen body: Add Zi Su Zi, Sheng Jiang, Hou Po, Da Fu Pi, and Ze Xie.
- Severe cough and wheezing: Remove Mai Men Dong, Wu Wei Zi, and Gua Lou Pi, and add Fu Ling, Bai Zhu, Ting Li Zi, and Xuan Fu Hua.

PART III

Single Herbs Classification

TOXIC HERBS

(Please refer to individual herb listing for exact dosage amounts)

Extremely Toxic: Dosages between 0.03 and 1.5 g in Pill/Powder

- Ba Dou (*Croton Seed*)
- Ban Mao (*Cantharides*)
- Cao Wu (*Kusnezoff Monkshood Root*)
- Chuan Wu (*Sichuan Monkshood Root*)
- Li Lu (*Mountain Onion*)
- Nong Ji Li (*Crotalaria Plant*)
- Pi Shi (*Arsenolite*)
- Shui Yin (*Mercury*)
- Tian Xian Zi (*Henbane Seed*)
- Xue Shang Yi Zhi Hao (*Brachypodum Monkshood Root*)
- Yang Zhi Zhu (*Yellow Azalea*)

Toxic: Dosages between 3 and 12 g in Decoction and 0.3 and 3 g in Pill/Powder

- Ba Jiao Feng (*Chinese Alangium Root*)
- Bai Fu Zi (*Giant Typhonium Tuber*)
- Bai Hua She (*Agkistrodon Snake*)
- Chan Su (*Toad Venom*)
- Chang Chun Hua (*Madagascar Periwinkle*)
- Chang Shan (*Dichroa Root*)
- Chen Shi Hui (*Old Lime*)
- Da Feng Zi (*Chaulmoogra Seed*)
- Da Ji (*Peking Spurge Root*)
- Dan Fan/Shi Dan (*Crystallized Blue Vitriol*)
- Fu Zi (*Aconite*)
- Gan Sui (*Kan-Sui Root*)
- Gua Di (*Melon Pedicle*)
- Ku Lian Gen Pi (*China Tree Root Bark*)
- Lei Gong Teng (*Tripterygium Wilfordii Hook*)
- Liu Huang (*Sulphur*)
- Lu Feng Fang (*Hornets' Nest*)
- Man Tuo Luo/Yang Jin Hua (*Jamestown Weed Seed/Leaf/Flower*)
- Mao Zhua Cao (*Buttercup Root*)
- Mi Tuo Seng (*Litharge or Galena*)
- Mu Bie Zi (*Momordica Seed*)
- Nao Sha (*Sal Ammoniac*)

- Qian Dan (*Lead Oxide* Pb_3O_4)
- Qian Niu Zi (*Morning Glory Seed*)
- Qiang Lang (*Dung Beetle*)
- Qing Fen (*Calomel*)
- Quan Xie (*Scorpion*)
- Shang Lu (*India Pokeberry Root*)
- Shi Liu Gen Pi (*Bark of Pomegranate Tree*)
- Shi Liu Pi (*Pomegranate Husk*)
- Shui Zhi (*Leech*)
- Tian Nan Xing (*Jack-in-the-Pulpit*)
- Tu Bie Chong (*Wingless Cockroach*)
- Tu Jin Pi (*Golden Larch Bark*)
- Wu Gong (*Centipede*)
- Xi Shu (*Happy Tree Fruit Bark*)
- Xiong Huang (*Realgar*)
- Xu Sui Zi (*Moleplant Seed*)
- Ya Dan Zi (*Java Brucea Fruit*)
- Ying Su Ke (*Opium Poppy Husk*)
- Yuan Hua (*Genkwa Flower*)
- Zao Jiao (*Chinese Honey-Locust Fruit*)
- Zhang Nao (*Camphor*)
- Zhu Jie Xiang Fu/Lian Tou Jian (*Anemone*)

Slightly Toxic: Dosages not More than 15 g per Day in Decoction

- Bai Guo (*Ginkgo Nut*)
- Ban Xia (*Pinellia Tuber*)
- Chuan Jiao/Shu Jiao (*Fruit of Szechuan Prickly Ash Pepper*)
- Chuan Lian Zi (*Chinaberry Fruit*)
- Gan Qi (*Dry Lacquer Sap*)
- Guan Zhong (*Dryopteris Root*)
- He Shi (*Starwort Fruit*)
- Hu Er Cao (*Saxifrage*)
- Huang Du (*Air Potato Yam*)
- Jiu Jie Cha (*Sarcandrae Plant*)
- Lei Wan (*Fruiting Body of Stone-Like Omphalia*)
- Liang Mian Zhen (*Prickly Ash Root*)
- Long Kui (*Black Nightshade*)
- Mang Chong (*Gadfly*)
- Wan Nian Qing (*Japanese Lily Bulb*)
- Wu Zhu Yu (*Evodia Fruit*)
- Xian Mao (*Rhizome of Golden-Eye Grass*)
- Xing Ren (*Apricot Seed*)
- Ze Qi (*Spurge*)
- Zi Bei Tian Kui (*Begonia*)

Herbs That are Contraindicated during Pregnancy

- Ba Dou (*Croton Seed*)
- Ban Mao (*Cantharides*)
- Da Ji (*Peking Spurge Root*)
- Mang Chong (*Gadfly*)

- Qian Niu Zi (*Morning Glory Seed*)
- San Leng (*Bur-Reed Rhizome*)
- Shang Lu (*India Pokeberry Root*)
- She Xiang (*Musk of Deer*)
- Shui Zhi (*Leech*)

Herbs Used with Caution during Pregnancy

- Da Huang (*Rhubarb Root Rhizome*)
- Fu Zi (*Aconite*)
- Gan Jiang (*Dried Ginger*)
- Hong Hua (*Safflower Flower*)
- Rou Gui (*Cinnamon Bark*)
- Tao Ren (*Peach Kernel*)
- Zhi Shi (*Immature Fruit of Bitter Orange*)
- And all toxic medicinal substances

Traditional Incompatibilities between Herbs and Food

- Bo He (*Peppermint*) — sea turtle meat
- Chang Shan (*Dichroa Root*) — fresh chive and scallion
- Feng Mi (*Honey*) — fresh scallion
- Fu Ling (*White Poria*) — vinegar
- Han Cai (*Nasturtium*) — sea turtle meat
- He Shou Wu (*Fleeceflower Root*) — fresh chive, garlic, scallion, and turnip
- Sheng Di Huang (*Fresh Rehmannia*) — fresh chive, garlic, scallion, and turnip
- Shu Di Huang (*Cooked Rehmannia*) — fresh chive, garlic, scallion, and turnip

THE EIGHTEEN INCOMPATIBILITIES

- Gan Cao (*Rx. Glycyrrhizae*) incompatible with:
  1. Gan Sui (*Euphorbiae kansui*)
  2. Da Ji (*Rx. Euphorbiae seu Knoxiae*)
  3. Yuan Hua (*Flos Daphnis*)
  4. Hai Zao (*Hb. Sargassii*)
- Wu Tou (*Rx. Aconiti*) incompatible with:
  1. Bai Ji (*Rz. Bletillae*)
  2. Gua Lou (*Fr. Trichosanthis*)
  3. Bei Mu (*Bulbus Fritillariae*)
  4. Ban Xia (*Rz. Pinelliae*)
  5. Bai Lian (*Rx. Ampelopsis*)
- Li Lu (*Rz. et Rx. Veratri*) incompatible with:
  1. Ren Shen (*Rx. Panacis ginsengis*)
  2. Sha Shen (*Rx. Glehniae*)
  3. Dan Shen (*Rx. Salviae miltiorrhizae*)
  4. Xuan Shen (*Rx. Scrophulariae*)
  5. Xi Xin (*Hb. Asari cun radice*)
  6. Bai Shao (*Rx. Paeoniae albae*)
  7. Chi Shao (*Rx. Paeoniae Rubra*)

THE NINETEEN ANTAGONISMS

- Liu Huang (*Sulphur*) — Po Xiao (*Sal Glauberis*)
- Shui Yin (*Hydragyrum*) — Pi Shuang (*Arsenicum*)
- Lang Du (*Rx. Euphorbiae Fischerianae*) — Mi Tuo Seng (*Lithargyrum*)
- Ba Dou (*Sm. Crotonis*) — Qian Niu Zi (*Sm. Pharbitis*)
- Ya Xiao (*Nitrum*) — San Leng (*Rx. Sparganii*)
- Ding Xiang (*Flos Eugeniae*) — Yu Jin (*Rx. Curcumae*)
- Fu Zi/Chuan Wu/Cao Wu (*Rx. Aconiti*) — Xi Jiao (*Cornu Rhinoceri*)
- Rou Gui (*Cx. Cinnamomi Cassiae*) — Chi Shi Zhi (*Halloysitum Rubrum*)
- Ren Shen (*Rx. Panacis Ginsengis*) — Wu Ling Zhi (*Excrementum Trogopteri*)

Table of Abbreviations

Lu	Lung
Sp	Spleen
Liv	Liver
Ht	Heart
Kid	Kidney
Pc	Pericardium
LI	Large Intestine
ST	Stomach
GB	Gallbladder
SI	Small Intestine
UB	Urinary Bladder
SJ	San Jiao
g	Gram
Sl	Slightly
Ex	Extremely
V	Very
C/I	Contraindicated

CHAPTER **28**

Release Exterior Herbs

WARM AND ACRID HERBS

Herb Name	Common Name	Temperature & Tastes	Channels	Functions	Indications	Dosages & Notes
Bai Zhi	Angelica Root	Warm Acrid	Lu ST	*Expels wind–damp *Relieves pain *Reduces swelling *Opens nasal passage	*Yang Ming headache *Toothache *Sores/carbuncles *Damp–cold leukorrhea *Sinus congestion	3–9 g
Cang Er Zi	Cocklebur Fruit	Warm Acrid Sweet Bitter	Lu	*Expels wind–cold *Opens nasal passage *Dispels wind–damp	*Headache at back of neck *Sinus congestion with thick mucous discharge *Bi syndrome arthritis *Itchy skin disorder	3–9 g
Cheng Liu (*Xi He Liu*)	Chinese Tamarisk Tops	Warm Acrid Sweet	Ht Lu ST	*Releases exterior *Encourages rashes to the surface *Reduces toxicity of herbs	*Rash with wind–cold	3–15 g
Cong Bai	Scallion	Warm Acrid	Lu ST	*Expels wind–cold *Promotes sweating *Circulates Yang Qi *Reduces toxin (poultice)	*Acute stage of common cold *Nasal congestion *Sores/abscesses	6–15 g
Fang Feng	Ledebouriella Root	Sl Warm Acrid Sweet	Sp Liv UB	*Expels wind–cold *Dispels wind–damp *Expels internal wind *Expels intestinal wind	*Headaches/migraines *Body aches *Convulsions/tremors *Recurrent and painful diarrhea	3–9 g
Gao Ben	Chinese Lovage Root	Warm Acrid	UB	*Expels wind–cold–damp *Relieves pain	*Vertex headache *Acute low back pain	3–9 g
Gui Zhi	Cinnamon Twig	Warm Acrid Sweet	Lu Ht UB	*Balances Ying/Wei Qi *Warms channel *Disperses wind–cold–damp *Circulates Yang Qi *Strengthens Heart Yang *Warms uterus	*Exterior cold with underlying deficiency *Painful shoulder joints *Palpitations *Connects Heart fire and Kidney water *Painful menses due to cold	3–9 g

(continued)

WARM AND ACRID HERBS (continued)

Herb Name	Common Name	Temperature & Tastes	Channels	Functions	Indications	Dosages & Notes
Jing Jie	Schizonepeta Stem	Sl Warm Acrid Aromatic	Lu Liv	*Releases exterior wind *Releases skin rash *Stops itch *Stops bleeding (charred)	*Wind–heat/wind–cold *Early stage measles or boils *Hemafecia/uterine bleeding	3–9 g
Ma Huang	Ephedra Stem	Warm Acrid Bitter	Lu UB	*Expels wind–cold *Circulates Lung Qi *Promotes urination	*Exterior excess wind–cold *Asthma/cough/ wheezing *Edema with EPI	3–9 g Very hot
Qiang Huo	Notopterygium Root	Warm Acrid Bitter Aromatic	Kid UB	*Expels wind–cold *Dispels wind–cold–damp *Guides Qi to Tai Yang and Du Mai	*Occipital headaches *Painful obstructions in upper back and limbs	6–15 g
Sheng Jiang	Fresh Ginger Rhizome	Sl Warm Acrid	Lu Sp ST	*Releases exterior *Warms Middle Jiao *Disperses cold *Transforms phlegm *Balances Ying/Wei Qi *Reduces toxicity of other herbs	*Exterior wind–cold *Stops nausea/ vomiting/morning sickness *Stops acute/chronic cough *Exterior deficiency cold *Detoxifies Fu Zi/ Ban Xia	3–9 g
Sheng Jiang Zhi	Fresh Ginger Juice	Sl Warm Acrid	Lu Sp ST	*Releases exterior *Warms Middle Jiao *Disperses cold *Transforms phlegm *Balances Ying/Wei Qi *Reduces toxicity of other herbs	*Damp–phlegm Upper and Middle Jiao	5–12 drops per day Stronger than Sheng Jiang in resolving phlegm and stop vomiting
Xi Xin	Chinese Wild Ginger	Warm Acrid	Lu Ht Kid	*Expels wind–cold *Warms Lung *Transforms phlegm *Stops pain *Opens nasal orifice	*Head/body aches *Cough with copious sputum *Arthritis pain due to wind–cold *Nasal congestion	Less than 4.5 g per day Very hot
Xiang Ru	Aromatic Madder	Sl Warm Acrid Aromatic	Lu ST	*Expels wind–cold–damp *Promotes urination *Reduces swelling	*Exterior summer heat with internal turbid dampness *Edema/urinary difficulty	3–9 g
Xin Yi Hua	Magnolia Flower	Warm Acrid	Lu ST	*Disperses wind–cold *Opens nasal orifice	*Sinusitis with headache *Loss sense of smell	3–9 g
Zi Su Geng	Perilla Stem	Sl Warm Acrid Sweet	Lu Sp ST	*Moves Qi *Calms fetus	*Chest and abdominal distention *Restless fetus	9–15 g
Zi Su Ye	Perilla Leaf	Warm Acrid Aromatic	Lu Sp	*Dispels wind–cold *Harmonizes Middle Jiao *Circulates Qi *Reduces seafood poisoning	*Headache/nasal congestion *Nausea/vomiting/ morning sickness/poor appetite *Plum-pit Qi	3–9 g Short cooking time

COOL AND ACRID HERBS

Herb Name	Common Name	Temperature & Tastes	Channels	Functions	Indications	Dosages & Notes
Bo He	Peppermint	Cool Acrid Aromatic	Lu Liv	*Disperses wind–heat *Releases skin rash *Circulates Liver Qi	*Fever/headache/ cough *Sore throat/red eyes *Early stage measles *Hypochondriac pain	2–9 g Add at last 5 min of cooking
Chai Hu	Thorowax Root	Cool Acrid Bitter	GB Liv Pc SJ	*Clears Shao Yang disorders *Releases wind–heat *Smoothes Liver Qi *Ascends Yang Qi	*Alternating chills/ fevers *Chest/flank pain *Menses disorders *Hemorrhoids/ prolapse of anus/ rectum/uterus	3–12 g
Chan Tui	Cicada Molting	Cold Sweet	Lu Liv	*Disperses wind–heat *Brightens eyes *Releases skin rash *Stabilizes spasm	*Loss of voice/sore throat *Eye infections *Early stage measles *Childhood febrile diseases	3–9 g
Dan Dou Chi	Prepared Soybean	Cold Acrid Bitter Sweet	Lu ST	*Releases exterior *Eliminates restlessness	*Wind–cold/wind–he at *Reduces irritability from febrile disorders *Helps with digestion	6–15 g
Fu Ping	Duckweed	Cool Acrid	Lu UB	*Releases exterior *Releases skin rash *Reduces swelling	*Exterior heat with head and body aches *Measles/wind rash *Upper body edema	3–6 g
Ge Gen	Kudzu Root	Cool Acrid Sweet	Sp ST	*Relieves wind–heat *Generates fluid *Releases skin rash *Stops diarrhea	*Fever/headache/ upper back tightness/stiff neck *Stomach heat thirst *Measles *Diarrhea/dysentery	6–12 g
Ge Hua	Kudzu Flower	Sweet Neutral	ST	*Relieves wind–heat *Relieves hangover *Generates fluid *Releases skin rash *Stops diarrhea	*Fever/headache/ upper back tightness/stiff neck *Alcohol intoxication *Stomach heat thirst *Measles *Diarrhea/dysentary	3–12 g
Ju Hua	Chrysanthemum Flower	Sl Cold Acrid Bitter Sweet	Lu Liv	*Disperses wind–heat *Clears Liver heat *Subdues Liver wind *Subdues Liver Yang *Detoxifies Liver	*Exterior wind–heat with fever/headache *Eye infections *Blurry vision/floaters	4–15 g
Man Jing Zi	Vitex Fruit	Cool Acrid Bitter	Liv ST UB	*Expels wind–heat *Drains damp	*Headache *Red, painful, and puffy eyes *Arthritic limbs/joints	6–12 g
Niu Bang Zi	Great Burdock Fruit	Cold Acrid Bitter	Lu ST	*Disperses wind–heat *Reduces swelling *Releases skin rash *Moistens Intestine	*Fever/cough/sore throat *Carbuncles/measles *Mumps *Wind–heat constipation	3–9 g
Sang Ye	Mulberry Leaf	Cold Acrid Bitter Sweet	Lu Liv	*Expels wind–heat *Clears Lung heat *Clears Liver heat *Benefits eyes	*Fever/headache *Sore throat/cough *Red and dry eyes	4–15 g

(continued)

COOL AND ACRID HERBS (continued)

Herb Name	Common Name	Temperature & Tastes	Channels	Functions	Indications	Dosages & Notes
Sheng Jiang Pi	Ginger Peel	Cool Acrid	Lu Sp	*Promotes urination *Reduces swelling *Harmonizes Middle Jiao	*Treats Feng Shui *Reduces superficial edema	3–9 g
Sheng Ma	Black Kohosh	Sl Cold Acrid Sweet	Lu LI Sp ST	*Releases exterior *Releases skin rash *Clears heat *Reduces toxicity *Raises Yang	*Wind–heat headache *Measles *Swollen gums/ ulcerated lips/ canker ulcers/sore throat *Fatigue/shortness of breath	2–9 g

CHAPTER 29

Clear Heat Herbs

CLEAR EXCESS HEAT FROM QI LEVEL

Herb Name	Common Name	Temperature & Tastes	Channels	Functions	Indications	Dosages & Notes
Bi Qi	Water Chestnut	Cool Sweet	Lu LI ST	*Clears heat *Generates fluid *Transform phlegm *Reduces accumulation	*Thirst *Irritability *Dry and hard stools *Cough *Nodules	15–60 g
Dan Zhu Ye	Bamboo Leaves	Cold Sweet	Ht SI ST	*Clears heat *Drains Heart and Stomach fire *Promotes urination *Clears damp–heat	*Irritability/ restlessness *Thirst/insomnia *Canker sores *Heat in small Intestine Mai *Heat Lin Syndrome	6–9 g
He Tao Shu Zhi	Walnut Twig	Cold Bitter Acrid	Lu Liv ST GB	*Clears heat *Relieves toxin *Stops diarrhea *Brightens eye	*Diarrhea/dysentery *Excessive vaginal discharge *Toxin accumulation *Red and swollen eyes due to damp–heat *Treats some forms of cancer	4.5–9 g Up to 30 g when treating cancer
Lei Gong Teng	Tripterygium Wilfordii Hook	Cool Bitter	Ht Kid UB	*Clears heat *Relieves toxin *Kills insects	*Rheumatoid arthritis *Nephritic symptoms *Lupus and other autoimmune diseases	6–12 g Toxic
Li Pi	Asian Pear Peel	Cool Sweet Astringent	Lu Ht	*Clears heat *Moistens Lung *Generates fluid	*Dry cough *Thirst due to summer heat	9–15 g in dried form 30–60 g in fresh form
Lian Zi Xin	Lotus Plumule	Cold Bitter	Ht	*Drains Heart fire *Calms spirit	*Insomnia/irritability *Mental confusion/ delirium *Deficient type diarrhea	2–6 g
Lu Gen	Reed Rhizome	Cold Sweet	Lu ST	*Clears Lung and Stomach heat *Generates fluid *Promotes urination	*Fever/irritability/ thirst *Acute stage cough with thick yellow sputum *Acid reflux/belching *Dark, scanty, and bloody urine	15–30 g

(continued)

CLEAR EXCESS HEAT FROM QI LEVEL (continued)

Herb Name	Common Name	Temperature & Tastes	Channels	Functions	Indications	Dosages & Notes
Shan Zhi Zi	Gardenia Fruit	Cold Bitter	Ht Liv Lu SJ ST	*Clears heat in San Jiao *Drains damp–heat	*Fever/irritability/ insomnia *Delirious speech/restless *Damp–heat Lin syndrome *Bitter taste in mouth	3–12 g
Shi Gao	Gypsum	Very Cold Sweet Acrid	Lu ST	*Clears Lung heat *Drains Stomach fire	*Cough with thick sputum *High fever/extreme thirst *Headache/toothache *Topical use for eczema, burns, and ulcerated sores	9–30 g
Tian Hua Fen	Trichosantes Root	Cold Bitter Sweet	Lu ST	*Drains Lung heat *Transforms phlegm *Generates fluid *Reduces toxicity *Expels pus	*Cough with thick sputum or blood streaked sputum *Wasting/thirsting disorders *Mastitis/carbuncles/ sores *Lobule pneumonia *Beneficial for HIV patients	9–15 g
Ya Zhi Cao	Dayflower	Cold Bitter Sweet	Lu ST SI	*Clears heat *Promotes urination	*Sores/carbuncles *Sty *Lin syndrome	9–15 g
Zhi Mu	Anemarrhena Rhizome	Cold Bitter	Lu ST Kid	*Clears Lung and Stomach heat *Nourishes Lung and Kidney Yin *Generates fluid	*Fever/irritability/ thirst *Dry cough with residual heat *Night sweat/steaming bones disorder	6–12 g
Zi Bai Pi	Chinese Catalpa Bark	Cold Bitter	ST GB	*Clears heat *Detoxifies other herbs *Kills parasite	*Fever *Jaundice *Gastric reflux *Skin itch	9–15 g

COOL BLOOD HERBS

Herb Name	Common Name	Temperature & Tastes	Channels	Functions	Indications	Dosages & Notes
Chi Shao Yao	Red Peony Root	Sl Cold Sour Bitter	Liv Sp	*Moves blood *Clears heat *Cools blood *Drains Liver fire	*Dysmenorrhea *Abdominal pain *Blood level heat *Red, swollen, painful eyes *Diarrhea/dysentery	4–9 g
Han Shui Shi	Calcitum	Cold Acrid Salty	Ht ST Kid	*Clears heat *Drains fire *Cools blood	*Irritability/thirst/fever *Sore throat/mouth ulcers *Skin sores/burns	9–30 g
Mu Dan Pi	Cortex of Peony Tree	Cool Acrid Bitter	Ht Liv Kid	*Clears heat *Invigorates blood *Clears deficient fire *Clears Liver fire *Expels pus	*Blood level heat *Dysmenorrhea/ flank pain *Steaming bones disorder *Headache/eye pain *Intestinal abscess	6–12 g
Qing Dai	Indigo	Cold Salty	Liv Lu ST	*Clears heat *Relieves toxin *Cools blood *Reduces swelling	*Bleeding with fever, convulsion, and delirium	1–3 g in powder or dissolve into decoction

(continued)

COOL BLOOD HERBS (continued)

Herb Name	Common Name	Temperature & Tastes	Channels	Functions	Indications	Dosages & Notes
Sheng Di Huang	Fresh Rehmannia	Cold Bitter Sweet	Ht Kid Liv	*Clears heat *Nourishes Yin *Drains Heart fire	*Nutritive level heat *Low-grade fever/dry mouth, dry throat/ constipation *Tongue sores/ insomnia	9–30 g
Shui Niu Jiao	Water Buffalo Horn	Cold Bitter Salty	Ht Liv ST	*Clears heat *Cools blood *Reduces toxin	*Nutritive/blood level heat *Headache/infantile fever due to epidemic disease	15–30 g Decoct first, then substitute with Xi Jiao
Xi Jiao	Rhinoceros Horn	Cold Bitter Salty	Ht Liv ST	*Clears heat *Cools blood *Reduces toxin	*Nutritive/blood level heat *Convulsion/delirium	Banned due to endangered species
Xuan Shen (*Yuan Shen*)	Ningpo Figwort Root	Cold Bitter Sweet Salty	Lu Kid ST	*Clears heat *Nourishes Yin *Drains fire *Resolves nodule	*Blood level heat *Constipation/ irritability *Red and swollen eyes *Severe sore throat *Boils/skin diseases	9–30 g
Zi Bei Tian Kui	Begonia	Cool Sweet	Liv ST	*Clears heat *Moves blood *Cools blood *Reduces toxin	*Cough due to Lung heat *Topical use for bug/ snake bites and skin diseases	3–9 g Sl Toxic
Zi Cao	Groomwell Root	Cold Sweet	Ht Liv	*Drains fire *Clears damp–heat *Relieves toxin	*Measles/chicken pox *Eczema/genital lesion *Baby milk rash/ diaper rash	3–9 g

DRY DAMPNESS HERBS

Herb Name	Common Name	Temperature & Tastes	Channels	Functions	Indications	Dosages & Notes
Huang Bai	Amur Cork-Tree Bark	Cold Bitter	Kid UB	*Clears heat *Dries damp *Drains Kidney fire *Reduces toxin	*Lower Jiao damp–heat *Steaming bones disorder *Skin lesions/genital sores *Lin syndrome	3–12 g
Huang Lian	Coptis Rhizome	Cold Bitter	Ht LI Liv ST	*Clears heat *Drains damp *Clears Heart and Stomach fire	*Middle Jiao damp–heat *Dysentery/diarrhea *Irritability/insomnia *Acid reflux/heartburn *Canker sores	2–9 g
Huang Qin	Skullcap Root	Cold Bitter	GB LI Lu ST	*Clears heat *Dries damp *Stops bleeding *Calms fetus *Subdues Liver Yang	*Upper Jiao damp–heat *Reckless bleeding disorder *Headache/red eyes/ bitter taste in mouth/irritability	6–15 g
Ku Shen	Sophora Root	Cold Ex Bitter	UB Ht Liv LI SI	*Clears heat *Dries damp *Disperses wind *Stops itch *Kills parasite	*Dysenteric disorder *Genital itch/vaginal discharge *Urinary tract infection *Skin lesion or infestation	3–15 g
Long Dan Cao	Gentiana Root	Cold Bitter	GB Liv ST	*Drains Liver and Gallbladder damp–heat *Clears Liver fire	*Swollen sore throat *Red eyes/ear pain *Foul-smelling vaginal discharge *Headache/flank pain	3–9 g

(continued)

DRY DAMPNESS HERBS (continued)

Herb Name	Common Name	Temperature & Tastes	Channels	Functions	Indications	Dosages & Notes
San Ke Zhen	Berberis Root	Cold Bitter	Liv ST LI	*Clears heat *Relieves toxin *Drains damp–heat	*Swollen gums/sore throat/conjunctivitis *Dysenteric disorders *Eczema/skin sores	9–15 g
Shi Da Gong Lao	Chinese Mahonia	Cold Bitter	Liv ST LI	*Clears Liver heat *Reduces toxin *Drains damp–heat	*Headache/swollen eyes due to Liver heat *Diarrhea/jaundice *Reckless bleeding disorders	9–15 g
Shi Shang Bai	Greater Selaginella	Cold Sweet Acrid Bitter	Liv Lu	*Clears heat *Relieves toxin *Drains damp–heat	*Cough due to Lung heat *Swollen and painful throat *Jaundice/ascites	15–30 g
Tang Song Cao	Thalictrum	Cold Bitter	Ht Lu LI	*Clears heat *Relieves toxin *Drains damp–heat	*Cough due to Lung heat *Skin lesions/sores *Diarrhea/hepatitis	9–15 g

CLEAR DEFICIENT HEAT HERBS

Herb Name	Common Name	Temperature & Tastes	Channels	Functions	Indications	Dosages & Notes
Bai Wei	Swallowwort Root	Cold Bitter Salty	Lu ST Kid	*Clears heat *Promotes urination *Reduces toxin	*Various types of fever *Damp–heat Lin syndrome *Swollen and painful throat	6–15 g
Di Gu Pi	Lycium Bark	Cold Sweet	Lu Liv Kid	*Drains fire *Clears deficiency heat	*Night sweat/ irritability/thirst/ low-grade fever *Steaming bones disorder *Lung heat cough/wheezing *Reckless bleeding disorder *Yin deficiency toothache	6–15 g
Hu Huang Lian	Picrorhiza Rhizome	Cold Bitter	Liv ST LI Ht	*Drains damp–heat *Strongly clears deficiency heat	*Dysentery/hemorrhoid *Afternoon fever/abdominal distention due to childhood malnutrition	3–9 g
Yin Chai Hu	Stellaria Root	Cool Sweet	Liv ST	*Clears deficiency heat *Cools blood	*Steaming bones disorder *Reckless bleeding disorders *Low-grade fever due to malnutrition	3–9 g

RELIEVE TOXIN HERBS (WARM FEBRILE DISEASE)

Herb Name	Common Name	Temperature & Tastes	Channels	Functions	Indications	Dosages & Notes
Ban Lan Gen	Woad Root	Cold Bitter	Ht Lu ST	*Drains heat *Relieves fire toxicity *Cools blood	*Mumps *Damp–heat jaundice *Benefits throat	15–30 g
Chuan Xin Lian	Green Chiretta	Cold Bitter	LI Lu SI ST	*Clears heat *Reduces toxin *Dries damp	*Painful and swollen throat *Diarrhea *Damp–heat Lin syndrome *Eczema	9–15 g
Da Qing Ye	Woad Leaf	Ex Cold Bitter	Ht Lu ST	*Clears heat *Cools blood *Reduces toxin *Reduces inflammation	*Clears Stomach/Heart fire *Febrile/epidemic diseases	9–30 g
Guan Zhong	Dryopteris Root	Cold Bitter	Liv Sp	*Kills parasite *Drains heat *Relieves toxin *Cools blood *Stops bleeding	*Jaundice *Skin lesions/sores *Febrile/epidemic diseases *Reckless bleeding disorders	3–15 g Sl Toxic
Jin Yin Hua	Honeysuckle Flower	Cold Sweet	LI Lu ST	*Clears heat *Relieves fire toxicity *Expels wind–heat	*Swollen and sore throat *Headache/fever *Dysentery/diarrhea *Skin lesions/rash/sores	9–15 g Up to 60 g for severe cases
Jin Zhi	Purified Feces	Cold Bitter	ST	*Clears heat *Relieves toxin	*Fever due to toxin accumulation from warm febrile disease	9–15 g
Lian Qiao	Forsythia Fruit	Cool Bitter Sl Acrid	Ht Liv GB	*Clears heat *Relieves toxin *Reduces swelling *Expels wind–heat	*Clears Heart heat *Dissipates nodules *Headache/fever *Skin lesions/rashes/sores	6–15 g
Sheng Tou Cao	Sweet Basil	Cold Bitter Sweet	Lu LI ST	*Clears heat *Relieves fire toxicity *Expels wind–heat	*Swollen sore throat *Headache/fever *Dysentary/diarrhea *Skinlesions/rash/sores	9—15 g
Zao Xiu (*Chong Lou*)	Paris Rhizome	Cool Bitter	Ht Liv	*Clears heat *Relieves toxin *Reduces mass	*Skin lesions/sores due to heat toxin *Convulsion due to high fever	6–9 g

RELIEVE TOXIN HERBS (ULCER AND ABSCESS SWELLING)

Herb Name	Common Name	Temperature & Tastes	Channels	Functions	Indications	Dosages & Notes
Bai Jiang Cao	Patrinia	Sl Cold Bitter Acrid	LI Liv ST	*Clears heat *Reduces toxin *Dispels blood stasis *Resolves abscess *Expels pus *Stops pain	*Skin lesions/sores *Dysmenorrhea *Postpartum pain	9–15 g
Bai Lian	Ampelopsis	Sl Cold Bitter Acrid	Ht Liv Sp ST	*Clears heat *Reduces mass *Promotes healing of wound	*Skin lesions/sores due to heat toxin	3–9 g
Bai Xian Pi	Dictamnus Root Bark	Cold Bitter	Sp ST	*Clears heat *Reduces toxin *Expels wind–damp *Stops itch	*Eczema/carbuncles *Skin rash/lesions *Sexually transmitted diseases	6–9 g

(continued)

RELIEVE TOXIN HERBS (ULCER AND ABSCESS SWELLING) (continued)

Herb Name	Common Name	Temperature & Tastes	Channels	Functions	Indications	Dosages & Notes
Cao He Che (*Zi Shen*) (*Quan Shen*)	Snakeweed	Cool Bitter	Lu Sp ST Ht	*Clears heat *Reduces toxin *Transforms damp *Reduces abscess *Decreases swelling	*Dysentery with tenesmus *Bloody stools *Acute appendicitis *Fever *Tongue sores *Anticancer treatment *Edema	6–15 g
Guan Ye Liao	Perfoliatum	Neutral Sour	Lu SI	*Clears heat *Reduces toxin *Promotes urination	*Sore throat *Herpes/eczema *Urinary tract infection	9–15 g
Hong Teng	Sargentodoxa Vine	Neutral Bitter	LI Liv	*Clears heat *Reduces toxin *Invigorates blood *Stops pain	*Appendicitis *Dysmenorrhea *Musculoskeletal injury *Bi syndrome	15–30 g
Hu Er Cao	Saxifrage	Cold Bitter Acrid	Lu Liv GB	*Clears heat *Reduces toxin	*Menorrhagia *Otitis media *Urticaria	9–15 g SI Toxic
Jin Qiao Mai	Wild Buckwheat	Neutral Bitter	Lu Liv	*Clears heat *Relieves toxin	*Swollen sore throat *Dysmenorrhea due to blood stasis *Bi syndrome	9–15 g
Lou Lu	Rhaponticum Root	Cold Bitter Salty	LI ST	*Clears heat *Relieves toxin *Reduces swelling *Promotes lactation	*Mastitis *Skin sores/lesions	3–12 g
Mao Dong Qing	Hairy Holly Root	Neutral SI Bitter Sweet	Ht Pc Liv Lu	*Clears heat toxin *Invigorates blood *Dredges meridian *Reduces swelling *Relieves pus	*Congestive heart disease *Chest Bi syndrome *Lung excess heat *Burns and skin disease due to heat toxin	9–15 g 30–60 g in severe cases
Pu Gong Ying	Dandelion	Cold Bitter Sweet	Liv ST	*Clears heat *Relieves fire toxin *Resolves damp	*Mastitis *Jaundice/acute hepatitis *Damp–heat Lin syndrome	9–30 g
Qian Li Guang (*Jiu Li Ming*)	Climbing Groundsel	Cold Bitter	Lu Liv LI	*Clears heat *Reduces toxin *Brightens eye	*Appendicitis *Red, pink, and swollen eyes *Skin sores/lesions	9–15 g
Si Ji Qing	Chinese Holly	Cold Bitter	Ht Lu	*Clears heat *Reduces toxin *Stops bleeding	*Skin ulcers *Skin burns	9–15 g
Tong Bian	Boy Urine	Salty Neutral	Ht Liv ST	*Clears heat *Relieves toxin *Moves blood	*Used in clear heat and relieve toxins formulas	4–8 oz added into decoction or apply topically
Tu Fu Ling	Glabrous Greenbrier Rhizome	Neutral Sweet	Liv ST	*Drains damp *Reduces toxin *Benefits joints	*Syphilis *Detoxifies mercury toxins	15–60 g
Tu Niu Xi	Tuniuxi Root	Neutral Bitter Sour	Liv Kid	*Clears heat toxin *Drains fire *Invigorates blood *Promotes urination	*Diphtheria *Bi syndrome *Amenorrhea *Hematuria	15–30 g
Xiong Dan	Bear Gallbladder	Cold Bitter	GB Ht Liv	*Clears heat *Relieves fire toxicity *Benefits eyes *Extinguishes internal wind	*Skin sores/furuncles *Hemorrhoids *Eye infections *Convulsions/delirium	1–2.5 g

(continued)

RELIEVE TOXIN HERBS (ULCER AND ABSCESS SWELLING) (continued)

Herb Name	Common Name	Temperature & Tastes	Channels	Functions	Indications	Dosages & Notes
Ye Ju Hua	Wild Chrysanthe-mum Flower	Sl Cold Bitter Acrid	Lu Liv	*Drains fire *Reduces toxin	*Furuncles/carbuncles *Stops itch (topical use) *Sore throat/conjunctivitis	6–12 g
Yu Xing Cao	Houttuynia	Cool Acrid	Lu LI	*Clears heat *Reduces toxin *Reduces abscesses *Expels pus *Drains damp–heat	*Damp–heat Lin syndrome *Cough *Lung infections *Carbuncles *Diarrhea	15–60 g Short cooking time
Zi Hua Di Ding	Yedeon's Violet	Cold Bitter Acrid	Ht Liv	*Clears heat *Relieves fire toxicity	*Furuncles/sores/ abscess *Treats acute skin infections by cooling blood	9–15 g

RELIEVE TOXIN HERBS (DIARRHEA AND DYSENTERY)

Herb Name	Common Name	Temperature & Tastes	Channels	Functions	Indications	Dosages & Notes
Bai Tou Weng	Chinese Anemone Root	Cold Bitter	LI Liv ST	*Clears heat *Reduces toxin *Cools blood	*Dysenteric disorders *Reckless bleeding disorders	6–15 g
Fan Bai Cao	Cinquefoil	Neutral Sl Bitter Sweet	Liv LI	*Clears heat *Reduces toxin *Stops bleeding	*Dysenteric disorders *Hemorrhoid bleeding *Uterine bleeding	9–15 g
Feng Wei Cao	Bracken	Cold	LI UB	*Clears heat *Reduces toxin *Cools blood	*Hematuria *Hemafecia *Hemorrhoid bleeding	9–18 g
Ma Chi Xian	Purslane	Cold Sour	LI Liv	*Clears heat toxin *Cools blood *Stops bleeding	*Furuncles/sores/ abscess *Dysenteric disorder *Postpartum bleeding *Uterine bleeding *Red/white leukorrhea	15–60 g
Qin Pi	Bark of Korean Ash Branch	Cold Bitter	GB LI Liv ST	*Drains damp–heat *Clears Liver fire *Dispels phlegm *Stops cough	*Dysenteric disorders *Red and swollen eyes *Chronic bronchitis/asthma	4–15 g
Tie Xian	Copperleaf	Cold Bitter	Ht Lu LI SI	*Drains damp–heat *Stops bleeding	*Dysenteric disorders *Hematuria/hemafecia *Hemoptysis *Trauma bleeding (topically)	9–15 g

RELIEVE TOXIN HERBS (SORE AND SWOLLEN THROAT)

Herb Name	Common Name	Temperature & Tastes	Channels	Functions	Indications	Dosages & Notes
Dian Di Mei	Androsace Plant	Cold Bitter Acrid	Lu Ht	*Drains heat toxin	*Swollen and sore throat *Edema *Trauma	9–15 g
Hai Jin Sha Teng	Japanese Fern	Cold Sweet	UB SI	*Clears heat *Promotes urination *Reduces toxin	*Swollen and sore throat *Mumps *Heat Lin syndrome	15–60 g

(continued)

RELIEVE TOXIN HERBS (SORE AND SWOLLEN THROAT) (continued)

Herb Name	Common Name	Temperature & Tastes	Channels	Functions	Indications	Dosages & Notes
Jin Guo Lan	Tinospora Root	Cold Bitter	Lu	*Clears heat *Reduces toxin	*Sore throat *Diphtheria *Damp–heat diarrhea	2–9 g
Ma Bo	Puffball	Neutral Acrid	Lu	*Clears Lung heat *Stops bleeding	*Sore throat/cough *Aphonia *Trauma/frostbite (topical)	2–5 g Wrap in filter when cooking
Shan Dou Gen	Sophora Root	Cold Bitter	LI Lu	*Clears heat *Relieves toxin	*Painful and swollen throat *Leptospirosis *Early stages of lung/throat/urinary bladder cancer	3–9 g
She Gan	Belamcanda Rhizome	Cold Bitter	Lu	*Clears heat *Relieves toxin *Resolves phlegm	*Painful and sore throat *Cough with phlegm	2–9 g
Suan Jiang	Winter Cherry Fruit	Cold Bitter Sour	Lu Ht	*Clears heat toxin	*Painful and sore throat *Cough with phlegm *Eczema *Urinary tract infection	9–15 g
Wan Nian Qing	Japanese Lily Bulb	Cold Bitter Sl Sweet	Lu	*Clears heat toxin *Promotes urination *Tonifies Heart	*Sore throat *Diphtheria *Edema	9–15 g Sl Toxic
Zhu Sha Gen	Coral Ardisia Root	Cool Bitter Acrid	Lu Liv	*Clears heat *Reduces toxin *Stops pain	*Throat disease/swollen gum due to Stomach fire *Diphtheria *Trauma *Wind–damp Bi syndrome	9–15 g

CLEAR HEAT TO BRIGHTEN THE EYES HERBS

Herb Name	Common Name	Temperature & Tastes	Channels	Functions	Indications	Dosages & Notes
Gu Jing Cao	Pipewort Scapus	Neutral Sweet	Liv ST	*Clears wind–heat *Brightens eyes	*Red eyes/photophobia *Sore throat/painful gums	6–15 g
Jue Ming Zi	Cassia Seeds	Cool Bitter Sweet Salty	Liv Kid LI	*Clears vision *Expels wind–heat *Moistens Intestine	*Eye disease due to Liver heat *Constipation *Lowers high blood pressure *Lowers cholesterol *Headache/dizziness	9–15 g
Mi Meng Hua	Buddleia Flower Bud	Cool Sweet	Liv	*Clears Liver heat *Brightens eyes	*Excessive eye tearing *Photophobia *Eye disease due to excess and deficiency	3–9 g
Mu Zei	Horse Tail	Neutral Sweet Bitter	Lu Liv	*Disperses wind–heat *Clears heat *Stops bleeding	*Eye disease due to Liver heat and wind–heat *Hemafecia	3–9 g
Qing Xiang Zi	Celosia Seeds	Cool Sweet Bitter	Liv	*Clears wind–heat *Drains Liver fire *Improves vision	*Floaters *Blurry vision *Hypertension due to Liver Yang rising	3–15 g Avoid use in patients with glaucoma

(continued)

CLEAR HEAT TO BRIGHTEN THE EYES HERBS (continued)

Herb Name	Common Name	Temperature & Tastes	Channels	Functions	Indications	Dosages & Notes
Xia Ku Cao	Selfheal Spike	Cold Bitter Acrid	GB Liv	*Clears Liver fire *Brightens eyes *Dissipates nodules	*Lowers high blood pressure *Alleviates eye pain *Vertigo/scrofula/goiter	9–15 g
Ye Ming Sha	Bat Feces	Cold Acrid	Liv	*Clears Liver heat *Improves vision	*Childhood malnutrition *Night blindness *Cataract	3–9 g Use caution during pregnancy

CLEAR SUMMER HEAT HERBS

Herb Name	Common Name	Temperature & Tastes	Channels	Functions	Indications	Dosages & Notes
Bian Dou	Hyacinth Bean	Neutral Sweet	Sp ST	*Clears summer heat *Tonifies Spleen	*Damp accumulation diarrhea *Spleen deficient leukorrhea	9–20 g
Dou Juan	Young Soybean Sprout	Neutral Sweet	Sp ST	*Clears summer heat *Drains damp–heat	*Fever *Urinary discomfort *Edema/joint pain *Damp–heat diarrhea	9–15 g
He Ye	Lotus Leaf	Neutral Bitter Sl Sweet	Ht Liv Sp	*Clears summer heat *Raises Spleen Yang *Stops bleeding	*Diarrhea secondary to summerheat *Hemoptysis *Bleeding in Lower Jiao	9–15 g
Lu Dou	Mung Bean	Cool Sweet	Ht ST	*Clears summer heat	*Thirst/irritability/fever *Make into tea to prevent summerheat symptoms	15–30 g Use 120 g plus 60 g Gan Cao as antidote to Fu Zi.
Lu Dou Yi	Mung Bean Skin	Cold Sweet	Ht ST	*Clears summer heat *Drains toxin	*Measles/acne *Benefits vision	9–15 g
Xi Gua	Watermelon Fruit	Cold Sweet	UB Ht ST	*Clears summer heat *Generates fluid *Promotes urination	*Severe thirst *Dry heaves	15–30 g *Not used for cold and Spleen deficiency with damp

ANTI-MALARIAL HERBS

Herb Name	Common Name	Temperature & Tastes	Channels	Functions	Indications	Dosages & Notes
Xi Gua Cui Yi	Watermelon Peel	Cool Sl Bitter Sweet	Sp ST	*Clears heat toxin *Promotes urination	*Severe thirst *Edema *Canker sores/tongue ulcers	9–15 g
Chang Shan	Dichroa Root	Cold Acrid Bitter	Ht Liv Lu	*Kills malaria-causing parasite *Clears heat *Induces vomiting to expel phlegm	*Reduces fever/chills due to malaria	3–9 g Toxic Over 9 g can result in nausea and vomiting

(continued)

ANTI-MALARIAL HERBS (continued)

Herb Name	Common Name	Temperature & Tastes	Channels	Functions	Indications	Dosages & Notes
Ma Bian Cao	European Verbena	Cool Bitter Acrid	Liv Sp UB	*Clears heat *Reduces toxin	*Malarial disorders *Splenomegaly with ascites due to schistosoma *Diarrhea/edema *Amenorrhea due to stasis	9–15 g
Qing Hao	Wormwood	Cold Bitter Acrid	Kid Liv GB	*Clears summer heat *Cools blood *Stops bleeding *Clears deficient heat	*Malarial disorders *Last stage of febrile disease *Jaundice	3–9 g Short cooking time
Tian Ming Jing	Carpesium Root	Cold Bitter Acrid	Lu Liv	*Clears heat *Reduces toxin	*For malarial disorders use 2 hours before attack *Febrile diseases *Swollen sore throat *Skin rash	9–15 g
Xiu Qiu Hua	Hydrangea Flower	Cold Bitter Acrid	Lu Liv	*Clears heat *Reduces toxin	*Malarial disorders *Sore throat *Eczema/scabies	9–15 g
Ya Dan Zi	Java Brucea Fruit	Cold Bitter	LI Liv	*Clears heat toxin *Dries damp	*Malarial disorders *Dysentery *Topical use to treat corns, warts, trichomonas, and vaginitis	9–15 g Toxic High dosage causes stomach irritation
Yan Fu Mu	Chinese Sumac	Cool Sour	Lu Liv	*Clears heat *Reduces toxin *Cools blood	*Malarial disorders *Cough and hemoptysis due to phlegm–heat	9–15 g

CHAPTER 30

Downward Draining Herbs

PURGATIVES

Herb Name	Common Name	Temperature & Tastes	Channels	Functions	Indications	Dosages & Notes
Da Huang	Rhubarb Root Rhizome	Cold Bitter	Ht LI Liv ST	*Clears heat *Drains damp *Purges stagnation *Cools blood *Invigorates blood	*Bleeding due to excess fire *Jaundice *Dysenteric disorder *Intestinal heat excess *Amenorrhea *Abdominal mass	3–12 g Short cooking time for purgative effect C/I during pregnancy and nursing
Fan Xie Ye	Senna Leaf	Cold Sweet Bitter	LI	*Purges stagnation	*Constipation *Ascites	3–9 g Add last 15 minutes of cooking
Lu Hui	Juice of Aloe Leaf	Cold Bitter	LI Liv ST	*Drains fire *Clears heat *Purges stagnation *Kills parasite	*Constipation *Headache/vertigo/ tinnitus *Roundworm/ ringworm *Tinea (topical) *Facial skin enhancement	0.3–1.5 g In pill and powder form only
Mang Xiao	Glauber's Salt	Ex Cold Acrid Bitter Salty	ST LI	*Purges accumulation *Softens hardness *Clears heat *Reduces swelling	*Constipation due to heat in Stomach and Intestine *Skin lesions/sores (topical)	3–9 g Dissolve into strained decoction
Pu Xiao	Less Purfied Mirabilitum	Ex Cold Acrid Bitter Salty	ST LI	*Purges accumulation *Softens hardness *Clears heat *Reduces swelling	*Constipation *Mouth sores/ulcers (topical)	3–9 g Used mostly as a topical
Xuan Ming Fen	Pure Mirabilitum	Ex Cold Acrid Bitter Salty	ST LI	*Mildly purges accumulation *Softens hardness *Clears heat *Reduces swelling	*Constipation *Mouth sores/ulcers (topical)	3–9 g Not as strong as Mang Xiao to move bowels

MOIST LAXATIVES

Herb Name	Common Name	Temperature & Tastes	Channels	Functions	Indications	Dosages & Notes
Feng Mi	Honey	Neutral Sweet	Lu Sp LI	*Moistens Lung *Moistens intestine *Unblocks bowel *Stops cough *Tonifies Middle Jiao	*Constipation due to fluid deficiency *Burns/bee sting (topical) *Postpartum/post surgery	3–30 g Detoxifies aconite poisoning
Huo Ma Ren	Cannabis Seeds	Neutral Sweet	LI Sp ST	*Moistens intestine *Nourishes Yin *Clears heat	*Constipation *Postpartum constipation	9–30 g Must be crushed
Yu Li Ren	Bush Cherry Pit	Neutral Acrid Bitter Sweet	LI SI Sp	*Moistens Intestine *Promotes urination *Reduces edema	*Constipation with edema *Dysuria *Reduces high blood pressure	3–9 g Must be crushed

HARSH EXPELLANTS

Herb Name	Common Name	Temperature & Tastes	Channels	Functions	Indications	Dosages & Notes
Ba Dou	Croton Seed	Hot Acrid	ST LI Lu	*Strongly purges stagnation *Reduces edema *Expels phlegm *Kills parasite	*Ascites due to schistosoma *Intestinal obstruction due to interior cold accumulation *Wheezing/sore throat due to phlegm accumulation	0.1–0.3 g in pill form only EX toxic Overdose can cause severe diarrhea Antagonizes Qian Niu Zi
Da Ji	Peking Spurge Root	Cold Bitter Acrid	Kid LI Lu	*Expels water *Drives out congested fluid *Reduces swelling *Dissipates nodule	*Ascites with wheezing and costal pain *Epilepsy due to phlegm misting the mind *Skin lesions/ sores (topical)	1.5–3 g in decoction 1.0 g in powder Toxic Incompatible with Gan Cao
Gan Sui	Kan-Sui Root	Cold Bitter Sweet	Kid LI Lu	*Expels water *Drives out congested fluid *Clears heat *Reduces swelling	*Constipation *Ascites *Asthma/wheezing *Difficult urination	1.5–3 g in decoction 0.3–1.0 g in powder Toxic Incompatible with Gan Cao
Qian Niu Zi	Morning Glory Seed	Cold Bitter Acrid	Kid LI Lu SI	*Unblocks bowel *Reduces edema *Transforms phlegm *Kills parasite	*Constipation *Congested fluids in lungs *Roundworm/ tapeworm *Food stagnation	4.5–9 g in decoction 1.5–3 g in powder Toxic C/I during pregnancy Antagonizes Ba Dou
Shang Lu	India Pokeberry Root	Cold Bitter Acrid	UB Kid LI Sp	*Expels water *Reduces swelling	*Constipation *Dysuria *Chronic bronchitis due to phlegm accumulation	3–9 g Toxic

(continued)

HARSH EXPELLANTS (continued)

Herb Name	Common Name	Temperature & Tastes	Channels	Functions	Indications	Dosages & Notes
Xu Sui Zi	Moleplant Seed	Warm Acrid	Liv Kid LI	*Expels water *Reduces swelling *Cracks blood stasis	*Constipation *Dysuria with edema *Amenorrhea *Mass in abdomen	0.5–2 g in powder or pill form Toxic C/I during pregnancy
Yuan Hua	Genkwa Flower	Warm Bitter Acrid	Kid LI Lu	*Mildly drains water *Drives out congested fluid *Expels phlegm *Stops cough *Kills parasite	*Phlegm–damp congestion in chest and hypochondriac regions *Chronic bronchitis *Ringworm/tinea infection on scalp	1.5–3 g Toxic C/I during pregnancy Incompatible with Gan Cao

CHAPTER 31

Drain Damp Herbs

PROMOTE URINATION AND REDUCE SWELLING

Herb Name	Common Name	Temperature & Tastes	Channels	Functions	Indications	Dosages & Notes
Ban Bian Lian	Chinese Lobelia	Cool Sweet Acrid	Ht Lu SI	*Promotes urination *Reduces edema *Cools blood *Reduces toxicity	*Ascites from Liver cirrhosis *Snake bite/bee sting/skin swelling (topical) *Combine with other herbs to treat stomach/rectal cancer	15–30 g
Chi Fu Ling	Red Poria	Neutral Sweet Bland	Ht Sp Kid	*Leaches damp–heat (stronger) *Clears heat *Cools blood	*Diarrhea/dysentery *Damp-heat Lin syndrome	9–15 g
Chi Xiao Dou	Aduki Bean	Neutral Sweet Sour	Ht SI Sp	*Clears heat *Promotes urination *Reduces swelling *Drains damp–heat	*Malnutritional edema *Skin swelling (topical) *Jaundice	9–30 g
Dong Gua Pi	Chinese Wax Gourd Peel	Cool Sweet	Lu SI	*Leaches damp *Reduces edema	*Edema *Thirst due to summer heat *Scanty urination	15–30 g
Dong Gua Ren	Chinese Wax Gourd Seed	Cold Sweet	Lu ST LI SI	*Clears heat *Expels phlegm *Promotes pus discharge *Drains damp	*Lung phlegm–heat cough *Pneumonia *Appendicitis *Heat type vaginal discharge	3–12 g
Fu Ling	White Poria	Neutral Sweet Bland	Ht Sp Lu UB	*Promotes urination *Leaches damp *Strongly strengthens Spleen *Calms spirit	*Edema *Diarrhea *Loss of appetite *Abdominal distention	9–15 g 30–45 g for edema
Fu Ling Pi	Poria Skin	Neutral Sweet Bland	Ht Sp Kid	*Leaches damp *Strongly promotes urination	*Edema	9–15 g
Fu Shen	Poria with Wood Root	Neutral Sweet Bland	Ht Sp UB	*Leaches damp *Strengthens Spleen *Calms spirit (strongly)	*Palpitation *Insomnia *Frequent dream	9–15 g

(continued)

PROMOTE URINATION AND REDUCE SWELLING (continued)

Herb Name	Common Name	Temperature & Tastes	Channels	Functions	Indications	Dosages & Notes
Lu Ying	Chinese Sambucus	Sl Cold Bitter Acrid	Liv UB	*Leaches damp *Moves blood *Disperses wind–damp	*Edema *Cough *Asthma *Bi syndrome	9–15 g
Yi Yi Ren	Job's Tears (Coix)	Sl Cold Sweet Bland	Sp Lu Kid LI	*Promotes urination *Leaches damp *Strengthens Spleen *Dissipates pus *Clears heat *Expels wind–damp	*Edema *Bi syndrome *Carbuncles/eczema *Lung abscess *Diarrhea	9–30 g
Yu Mi Xu	Cornsilk	Neutral Sweet	UB GB Liv	*Promotes urination *Leaches damp	*Hot/stone Lin syndrome *Edema from nephritis *Ascites from Liver cirrhosis *Jaundice from hepatitis *Gallbladder stones *Diabetes *Gum and nose bleeding *Reduce hypertension	15–60 g
Ze Xie	Water Plantain	Cold Sweet Bland	Kid UB	*Promotes urination *Leaches damp *Drains Kidney deficiency fire	*Diarrhea *Damp–heat in Lower Jiao *Conduct the Minister fire back to Lower Jiao	6–15 g
Zhu Ling	Umbellate Pore Fungus	Sl Cool Sweet Bland	Sp Kid UB	*Promotes urination *Leaches damp	*Cloudy Lin syndrome *Edema *Leukorrhea	6–15 g Stronger than Fu Ling

PROMOTE URINATION AND UNBLOCK LIN SYNDROME

Herb Name	Common Name	Temperature & Tastes	Channels	Functions	Indications	Dosages & Notes
Bei Xie	Fish-Poison Yam	Sl Cold Bitter	UB Liv ST	*Drains damp *Expels wind *Clears heat	*Cloudy Lin syndrome *Leukorrhea *Bi syndrome	9–15 g Up to 30 g per day
Bian Xu	Knotweed	Sl Cold Bitter	UB	*Clears damp–heat *Promotes urination *Stops itch *Expels parasite	*Damp–heat Lin syndrome *Vaginal/genital itch *Eczema (topical)	9–15 g
Che Qian Zi	Plantago Seeds	Sl Cold Sweet	UB Kid Liv Lu SI	*Promotes urination *Clears heat *Brightens eyes *Expels phlegm	*Damp–heat Lin syndrome *Red, puffy, and painful eyes due to Liver wind–heat *Cough due to Lung phlegm–heat	4.5–9 g Wrap in filter when cooking
Deng Xin Cao	Rush Pith	Sl Cold Sweet Bland	Ht Lu SI	*Promotes urination *Leaches damp *Clears heat	*Hot Lin syndrome *Restless sleep/ insomnia	1.5–4.5 g
Di Fu Zi	Kochia Fruit	Cold Sweet Bitter	UB	*Clears damp–heat *Promotes urination *Stops itch	*Lin syndrome *Damp–heat in Lower Jiao *Skin rash/eczema (topical)	6–15 g

(continued)

PROMOTE URINATION AND UNBLOCK LIN SYNDROME (continued)

Herb Name	Common Name	Temperature & Tastes	Channels	Functions	Indications	Dosages & Notes
Dong Kui Zi	Musk Mallow Seed	Cold Sweet	UB LI SI	*Promotes urination *Moistens Intestine *Promotes lactation	*Lin syndrome with edema *Constipation *Mastitis/breast abscess	6–12 g
Hai Jin Sha	Japanese Fern Spore	Cold Sweet	SI UB	*Clears heat *Expels stones *Promotes urination	*Stone Lin syndrome	6–15 g Wrap in filter when cooking
Han Fang Ji	Stephania Root	Cold Bitter Acrid	UB Kid Sp	*Expels wind–damp *Promotes urination *Reduces edema *Relieves pain	*Lower extremity edema *Swelling from leg Qi *Wind–damp–heat Bi	3–9 g Stronger than Guang Fang Ji
Hua Shi	Talcum	Cold Sweet Bland	ST UB	*Promotes urination *Drains heat from urinary bladder	*Heat Lin syndrome *Thirst/irritability *Eczema/skin rash (topical)	9–18 g Wrap in filter when cooking
Mu Tong	Five Leaf Akebia Stem	Cool Bitter	UB Ht SI	*Promotes urination *Drains Heart heat *Promotes lactation	*Damp–heat Lin syndrome *Irritability/canker sores *Damp–heat Bi syndrome	3–9 g
Qu Mai	Chinese Pink Dianthus	Cold Bitter	UB Ht SI	*Clears damp–heat *Promotes urination *Breaks up blood stasis *Unblocks bowel	*Damp–heat Lin syndrome *Amenorrhea *Constipation	6–12 g Up to 24 g per day
San Bai Cao	Lizard Tail	Cold Bitter Acrid	Lu UB	*Clears heat *Promotes urination *Expels phlegm	*Damp–heat Lin syndrome *Leukorrhea *Edema *Cough	9–15 g
Shi Wei	Felt Fern Leaves	Sl Cold Bitter Sweet	UB Lu	*Clears damp–heat *Stops bleeding *Expels phlegm	*Damp–heat/Bloody Lin *Urinary tract stones *Cough	3–9 g Up to 30 g per day
Tong Cao	Rice Paper Pith	Sl Cold Sweet Bland	Lu ST	*Promotes urination *Clears heat *Promotes lactation	*Cloudy/Bloody Lin *Edema and distention due to damp–heat blockage	3–6 g
Yu Bai Pi	Elm Tree Bark	Neutral Sweet	SI UB LI	*Promotes urination *Reduces edema *Clears heat	*Lin syndrome *Feng Dan *Toxin accumulation	9–15 g

REDUCE JAUNDICE HERBS

Herb Name	Common Name	Temperature & Tastes	Channels	Functions	Indications	Dosages & Notes
Chui Pen Cao	Stringy Stonecrop	Cool Sweet Bland Sour	Ht Liv GB SI	*Clears heat toxin *Promotes urination	*Damp–heat jaundice with difficult urination *Skin swelling/burn (topical) *Snake bites (topical)	9–30 g

(continued)

REDUCE JAUNDICE HERBS (continued)

Herb Name	Common Name	Temperature & Tastes	Channels	Functions	Indications	Dosages & Notes
Di Er Cao	Japanese Saint John's Wort	Neutral Bitter	Liv GB	*Clears heat toxin *Activates blood	*Jaundice/hepatitis *Appendicitis *Eczema *Snake bites *Trauma injuries	9–30 g
Hu Zhang	Giant Knotweed Root & Rhizome	Cold Bitter	Liv GB Lu	*Invigorates blood *Clears heat *Transforms phlegm *Relieves toxin	*Jaundice/leukorrhea/ Lin syndrome/genital itch due to Liver and Gallbladder damp–heat *Thirst/constipation/ skin swellings *Bronchitis/pneumonia *Amenorrhea/ dysmenorrhea/traum a due to blood stasis *Wind–damp Bi syndrome	9–30 g C/I during pregnancy
Jin Qian Cao	Loosestrife Plant	Neutral Sweet Bland Sl Salty	UB Kid GB Liv	*Promotes urination *Clears damp–heat *Reduces toxicity	*Jaundice *Heat/Sandy Lin *Skin swelling (topical) *Snake/bug bites (topical)	9–60 g
Ma Ti Jin	Dichondra Plant	Sl Cold Acrid Bland	Liv Sp	*Clears heat toxin *Drains damp	*Jaundice and Edema *Eczema *Trauma injuries	9–30 g
Tian Hu Sui	Hydrocotyle Plant	Cool Sweet Sl Acrid	Liv Lu	*Clears heat toxin *Drains damp *Stops cough	*Jaundice *Gallstones/kidney stones *Damp-heat Lin syndrome *Influenza *Whooping cough	9–15 g
Yin Chen Hao	Capillaris	Sl Cold Bitter Acrid	Liv GB Sp ST	*Clears damp–heat *Releases exterior	*Main herb for damp–heat type jaundice *Hepatitis *Dysuria *Typhoid	6–15 g

CHAPTER 32

Dispel Wind–Damp Herbs

STOP BI SYNDROME PAIN HERBS

Herb Name	Common Name	Temperature & Tastes	Channels	Functions	Indications	Dosages & Notes
Ba Qia	China Root	Neutral Sweet Sour	Liv Kid	*Dispels wind–damp *Reduces toxicity *Unblocks obstruction	*Bi syndrome *Skin swelling/burn (topical) *Anticancer	9–30 g
Can Sha	Silkworm Feces	Sl Warm Acrid Sweet	Liv Sp ST	*Dispels wind–damp *Harmonizes Stomach *Unblocks obstruction	*Cold–damp Bi syndrome *Itchy rash (topical) *Cholera/abdominal cramp due to turbidity	9–15 g Wrap in filter for cooking
Du Huo	Pubescent Angelica Root	Warm Bitter Acrid	Kid UB	*Dispels wind–cold–damp *Relieves pain *Unblocks obstruction	*Wind–cold–damp Bi below waist *Toothache *Headache	3–9 g
Guang Fang Ji	Aristolochia Root	Cold Bitter Acrid	UB Sp Kid	*Expels wind–damp *Unblocks obstruction *Promotes urination *Reduces edema	*Bi syndrome due to wind–damp–heat	3–15 g
Hai Tong Pi	Erythrina Bark	Neutral Bitter Acrid	Liv Sp Kid	*Dispels wind–damp *Unblocks obstruction *Promotes urination *Stops itch	*Wind–damp Bi syndrome *Scabies/mites/tinea infection (topical)	6–15 g
Qin Jiao	Gentiana Macrophylla Root	Sl Cold Bitter Acrid	GB Liv ST	*Dispels wind–damp *Unblocks obstruction *Clears deficient heat *Moistens Intestine	*Bi syndrome pain *Muscle spasms *Steaming ones symptom *Damp–heat jaundice	4.5–12 g C/I for constipation due to Spleen deficiency
Shi Nan Teng	Chinese Photinia Vine	Neutral Bitter Acrid	Liv Kid	*Dispels wind *Dredges meridian *Benefits Kidney	*Bi syndrome due to Kidney deficiency with cold–damp *Low back pain *Weak lower extremities *Migraines	9–15 g
Song Jie	Knotty Pine Wood	Warm Bitter	Liv Kid	*Dispels wind *Unblocks obstruction *Dries damp	*Bi syndrome *Trauma injuries *Toothache *Abdominal pain due to cold	9–15 g

(continued)

STOP BI SYNDROME PAIN HERBS (continued)

Herb Name	Common Name	Temperature & Tastes	Channels	Functions	Indications	Dosages & Notes
Tou Gu Cao	Garden Balsam	Warm Acrid Sweet	Lu Liv Ht	*Transforms damp *Disperses wind *Moves blood *Relieves pain	*Bi syndrome especially in lower extremities *Spasms *Swellings	9–15 g
Wei Ling Xian	Chinese Clematis Root	Warm Acrid Salty	All 12	*Dispels wind–damp–cold *Unblocks obstruction *Relieves pain	*Softens fish bones stuck in throat *Wind–damp–cold Bi *Edema in lower extremities	6–12 g
Xun Gu Feng	Dutchman's Pipe	Neutral Acrid	Liv	*Dispels wind–damp *Relieves pain *Dredges meridian	*Various types of muscle aches and pains	9–15 g
Zhu Jie Xiang Fu (*Lian Tou Jian*)	Anemone	Warm Acrid	Sp ST Lu	*Dispels wind–damp *Reduces swelling *Relieves pain	*Bi syndrome *Swellings/sores due to traumatic injuries	1.5–3 g in powder Toxic

RELAX TENDONS AND ACTIVATE CHANNELS HERBS

Herb Name	Common Name	Temperature & Tastes	Channels	Functions	Indications	Dosages & Notes
Bai Hua She	Agkistrodon Snake	Warm Sweet Salty	Liv Sp	*Extinguishes wind *Unblocks channel *Stops spasm *Dispels wind from skin	*Poststroke symptoms *Convulsion/delirium/spasm *Skin swelling/sore (topical)	3–10 g in decoction 1–1.5 g in powder Toxic
Chou Wu Tong	Glorybower Leaf	Cool Sweet Acrid Bitter	Liv Sp	*Dispels wind–damp	*Numbness/pain *Wind–damp Bi syndrome *Lowers blood pressure *Eczema (topical)	4.5–15 g Short cooking time
Chuan Shan Long	Japanese Yam	Sl Cold Bitter	Liv Lu	*Dispels wind–cold–damp *Moves blood *Disperses phlegm	*Numbness/pain *Wind–damp Bi syndrome *Cough/wheezing	9–15 g
Hai Feng Teng	Kadsura Stem	Sl Warm Bitter Acrid	Liv	*Dispels wind–damp *Disperses cold *Relieves pain	*Cold–damp Bi syndrome *Asthma/wheezing/cough *Calms fetus due to coldness	6–15 g
Kuan Jin Teng	Chinese Tinospora Stem	Sl Cold Bitter	Liv	*Dispels wind–damp *Unblocks channel *Relaxes sinew *Reduces swelling	*Cramps/stiffness due to wind–damp–heat Bi *Trauma injuries	15–30 g C/I during pregnancy and postpartum
Lao Guan Cao	Heroubill Plant	Neutral Bitter Acrid	Liv LI	*Dispels wind–damp *Activates channel *Relaxes sinew	*Wind–damp Bi syndrome *Damp–heat diarrhea	9–30 g
Lu Lu Tong	Sweetgum Fruit	Neutral Bitter Acrid	Liv ST UB	*Regulates Qi/blood *Unblocks channel *Promotes urination	*Numb/stiff/achy joints *Breast tenderness/distention *Wind type skin rash	3–9 g
Luo Shi Teng	Star Jasmine Stem	Sl Cold Bitter	Liv Ht Kid	*Dispels wind–damp *Cools blood *Reduces swelling	*Wind–damp Bi *Swollen throat with Throat Bi syndrome *Abscess/deep rooted boils	6–15 g

(continued)

RELAX TENDONS AND ACTIVATE CHANNELS HERBS (continued)

Herb Name	Common Name	Temperature & Tastes	Channels	Functions	Indications	Dosages & Notes
Mu Gua	Chinese Quince Fruit	Sl Warm Sour Aromatic	Sp Liv	*Relaxes sinews *Unblocks channel *Harmonizes Stomach *Transforms damp *Reduces food stagnation	*Wind–damp Bi syndrome *Beriberi/calf cramping *Vomiting/intestinal cramping/diarrhea from cholera *Indigestion from overeating and meat intake	4.5–12 g
Sang Zhi	Mulberry Twig	Sl Cold Bitter Sweet	Liv	*Dispels wind *Unblocks channel	*Bi syndrome above waist	10–30 g
Shen Jin Cao	Buck Grass	Warm Bitter Acrid	Liv	*Dispels wind–damp *Activates channel *Relaxes sinews	*Bi syndrome *Trauma injuries (topical)	6–15 g
Si Gua Luo	Suakwa Vegetable Sponge	Neutral Sweet	Lu ST Liv	*Dispels wind *Invigorates channel *Expels phlegm *Clears summer heat *Promotes urination	*Bi syndrome *Chest distention *Breast distention *Insufficient lactation due to Qi and Blood stagnation	6–12 g Up to 30 g Decoct first
Si Gua Zi	Suakwa Seed	Cold Sweet	ST Liv LI	*Expels phlegm *Clears heat *Kills parasite	*Lung phlegm–heat cough *Roundworms	9–15 g Must be crushed
Wu Shao She	Zaocys	Neutral Sweet Salty	Liv Sp	*Extinguishes wind *Unblocks channel *Stops spasm *Dispels wind from skin	*Post-stroke symptoms *Convulsion/delirium/ spasm *Skin swelling/sore (topical)	3–9 g in decoction, 3 g in powder
Xi Xian Cao	Common Saint Paul's Wort	Sl Cold Bitter Acrid	Kid Liv Ht	*Dispels wind–damp *Calms spirit *Clears heat *Pacifies Liver *Transforms damp–heat	*Bi syndrome *Post-stroke symptoms *Jaundice *Skin rash/sores (topical)	6–15 g

STRENGTHEN TENDONS AND BONES HERBS

Herb Name	Common Name	Temperature & Tastes	Channels	Functions	Indications	Dosages & Notes
Hu Gu	Tiger Bone	Warm Acrid Sweet	Liv Kid	*Dispels wind–damp *Strengthens sinews/bones	*Achy joints *Weakness in lower extremities	Banned as endangered species Substitute with bobcat bone, 3–6 g
Qian Nian Jian	Homalomena Rhizome	Warm Acrid Bitter	Liv Kid	*Dispels wind–damp *Strengthens sinews/bones	*Bi syndrome *Weakness of bones/sinews in elderly people *Pain/swelling due to trauma injuries (topical)	6–15 g
She Tui	Snake Skin Slough	Neutral Sweet Salty	Liv	*Dispels wind *Stops spasm *Benefits eye	*Skin lesion *Infantile convulsion *Pterygium/cornea disease	1.5–3 g in decoction, 0.3–0.6 g in powder C/I during pregnancy
Wu Jia Pi	Acanthopanax Root Bark	Warm Acrid Bitter	Liv Kid	*Dispels wind–damp *Strengthens sinews/ bones *Promotes urination	*Chronic wind–cold–damp Bi in elderly people *Slow motor develop-ment in children	4.5–15 g

CHAPTER 33

Transform Phlegm and Stop Cough Herbs

COOL AND TRANSFORM PHLEGM–HEAT HERBS

Herb Name	Common Name	Temperature & Tastes	Channels	Functions	Indications	Dosages & Notes
Chuan Bei Mu	Tendrilled Fritillaria Bulb	Sl Cold Bitter Sweet	Ht Lu	*Clears heat *Transforms phlegm *Dissipates nodules	*Chronic dry cough *Blood-tinged sputum cough *Sores/swellings/scrofula *Lung/breast abscess	3–12 g Not compatible with Wu Tou Counteracts Qin Jiao
Fu Hai Shi	Pumice	Cold Salty	Lu	*Clears heat *Expels phlegm–heat *Softens hardness *Promotes urination	*Phlegm–heat cough *Nodules due to phlegm–heat *Stone Lin syndrome	6–15 g Wrap in filter for cooking
Gua Lou	Mongolian Snakegourd	Cold Sweet	LI Lu ST	*Clears heat *Transforms phlegm–heat *Expands chest *Dissipates nodule	*Heat type cough with thick sticky sputum *Chest tightness/ constriction *Lung/breast abscess	9–30 g Not compatible with Wu Tou
Gua Lou Ren	Mongolian Snakegourd Seed	Cold Sweet	LI Lu ST	*Transforms phlegm–heat *Expands chest *Moistens Intestine *Promotes skin healing	*Heat type cough with thick sticky sputum *Chest tightness/ constriction *Dry constipation with thirst and irritability	9–12 g Not compatible with Wu Tou
Hai Dai	Common Eelgrass	Cold Salty Bitter	Kid Liv Lu ST	*Clears heat *Reduces phlegm nodule *Promotes urination	*Goiter/scrofula *Edema/beriberi *Hernia pain	4–15 g
Hai Ge Ke	Clam Shell	Neutral Bitter Salty	Kid Lu ST	*Clears heat *Disseminates Lung Qi *Transforms phlegm *Softens hardness *Promotes urination	*Heat type cough with thick sticky sputum *Goiter/scrofula *Acid reflux (calcined) *Cloudy Lin syndrome	6–15 g Wrap in filter for cooking
Hai Zao	Seaweed	Cold Salty Bitter	Kid Liv Lu ST	*Clears heat *Reduces phlegm nodules *Promotes urination	*Goiter/scrofula *Edema/beriberi *Hernia pain	4–15 g Not compatible with Gan Cao
Han Cai	Nasturtium	Neutral Acrid Bitter	Lu Liv	*Expels phlegm *Stops cough *Relieves toxin *Drains damp	*Chronic bronchitis *Jaundice due to damp–heat *Swelling/sores/ carbuncles (topical)	Up to 30 g per day
Hou Zao	Macaque Gallstone	Cold Salty Bitter	GB Liv Ht Lu	*Dislodges phlegm *Controls spasm *Arrests wheezing *Clears heat *Reduces toxicity	*Infantile convulsions due to phlegm–heat *Scrofula/swellings (topical)	1–1.5 g in pills or powder

(continued)

COOL AND TRANSFORM PHLEGM–HEAT HERBS (continued)

Herb Name	Common Name	Temperature & Tastes	Channels	Functions	Indications	Dosages & Notes
Huang Yao Zi	Airpotato Yam	Neutral Bitter	Ht Liv Lu	*Dissipates nodule *Cools blood *Reduces toxicity	*Goiter *Gastrointestinal and uterine cancer *Reckless bleeding disorders *Swelling/snakebite/ dog bite (topical)	6–15 g Taken in wine, pill, or powder
Kun Bu	Kelp Thallus	Cold Salty	Kid Liv ST	*Reduces phlegm *Softens hardness *Promotes urination	*Goiter/scrofula *Edema in lower body	4–15 g
Mao Zhua Cao	Buttercup Root	Neutral Acrid Bitter	Lu Liv	*Transforms phlegm *Reduces swelling *Relieves toxin	*Nodules due to toxins *Tuberculosis *Sore throat	5–9 g Toxic
Meng Shi (*Qing Meng Shi*)	Chlorite	Neutral Salty Sweet	Liv Lu	*Guides Qi downward *Eliminates phlegm *Soothes Liver *Calms fright	*Hardened nodules *Infantile convulsion *Constipation due to heat	1.5–3 g in pills or powder. Use caution during pregnancy
Pang Da Hai	Boat Sterculia Seed	Cold Sweet	LI Lu	*Disseminates Lung Qi *Clears Intestine	*Sore throat and dry throat due to Lung heat *Constipation/ headache	3–5 pieces
Pi Pa Ye	Loquat Leaf	Cool Bitter	Lu ST	*Transforms phlegm *Clears heat *Disseminates Lung Qi *Harmonizes Stomach	*Lung and heat cough *Thirst/nausea/ vomiting due to Stomach heat	4–12 g Scrub leaves clean to prevent throat irritation
Qian Hu	Hogfennel Root	Sl Cold Bitter Acrid	Lu	*Disseminates Lung Qi *Expels phlegm *Dispels wind	*Cough with thick sputum *Wheezing *Wind–heat/wind–cold cough	4–9 g
Shi Jian Chuan	Chinese Sage	Neutral Bitter Acrid	Liv GB	*Relieves toxin *Transforms phlegm *Reduces swelling *Relieves pain	*Wheezing with phlegm *Excessive vaginal discharge *Hepatitis *Swellings and sores	9–15 g
Tian Zhu Huang	Tabasheer	Cold Sweet	GB Ht Liv	*Transforms phlegm–heat *Cools the Heart	*Cough with sticky sputum *Infantile spasms/ convulsion/delirium due to phlegm	3–9 g
Wa Leng Zi	Cockle Shell	Neutral Sweet Salty	Lu ST Liv	*Eliminates phlegm *Transforms stasis	*Abdominal mass *Heartburn/acid reflux	6–30 g in decoction 3–9 g in pills or powder
Ze Qi	Spurge	Cool Acrid Bitter	SI LI Lu Sp	*Transforms phlegm *Dissipates nodule *Reduces edema	*Wheezing/scrofula due to Lung phlegm–heat *Edema in upper body *Ascites due to congested fluids	3–15 g Sl Toxic
Zhe Bei Mu	Thunberg Fritillaria Bulb	Cold Bitter	Ht Lu	*Clears heat *Transforms phlegm–heat *Dissipates nodule	*Acute cough with sputum *Neck swellings *Lung/breast abscess	3–9 g Not compatible with Wu Tou
Zhu Li	Dried Bamboo Sap	Ex Cold Sweet	Ht Lu ST Sp	*Clears heat *Transforms phlegm *Stops cough	*Phlegm–heat cough *Fainting/hands and feet numbness/hemiplegia due to phlegm obstruction of orifices	30–60 g 9–15 g to treat cough
Zhu Ru	Bamboo Shavings	Sl Cold Sweet	GB Lu ST	*Transforms phlegm–heat *Clears heat in Lungs *Cools blood *Stops bleeding	*Cough with sticky sputum *Blood-tinged sputum cough *Nausea/vomiting/morning sickness from Stomach heat *Reckless bleeding	4–9 g

WARM AND TRANSFORM PHLEGM–COLD HERBS

Herb Name	Common Name	Temperature & Tastes	Channels	Functions	Indications	Dosages & Notes
Bai Fu Zi	Giant Typhonium Tuber	Warm Acrid Sweet	Liv Sp ST	*Dries dampness *Transforms phlegm *Expels wind *Expels wind–cold–damp *Releases toxicity	*Migraine/headache *Dizziness *Bell's Palsy/hemiplegia *Facial paralysis/sweat blotches (topical)	1.5–6 g Toxic C/I during pregnancy
Bai Jie Zi	White Mustard Seed	Warm Acrid	Lu ST	*Warms Lung *Expels phlegm *Dissipates nodules	*Chronic cough with sputum *Chest congestion *Deep-rooted boils and oozing sores (topical)	3–9 g
Bai Qian	White Swallowwort	Sl Warm Acrid Sweet	Lu	*Disseminates Lung Qi *Expels phlegm	*Cough with white productive sputum *Wheezing *Gurgling sounds in throat due to Lung Qi stagnation	3–9 g Can be used long term because it is not drying
Ban Xia	Pinellia Tuber	Warm Acrid	Lu Sp ST	*Dries dampness *Transforms phlegm *Directs Qi downward *Harmonizes Stomach *Dissipates nodule	*Cough with white productive sputum *Nausea/vomiting due to phlegm in Middle Jiao *Goiter/scrofula	4–12 g Not compatible with Wu Tou
Jie Geng	Balloon flower Root	Neutral Bitter Acrid	Lu	*Disseminates Lung Qi *Expels pus *Benefits throat	*Wind–heat/wind–cold cough *Lung/throat abscess *Sore throat/loss of voice due to heat	3–9 g Guides other herbs toward upper body
Tian Nan Xing	Jack-in-the-Pulpit	Warm Bitter Acrid	Liv Lu Sp	*Strongly dries damp *Expels phlegm *Stops spasms *Reduces swelling *Relieves pain	*Cough/chest congestion due to recalcitrant phlegm *Dizziness/extremities numbness/opisthotonos/ Bell's Palsy/stroke due to obstruction of wind–phlegm	4–9 g Toxic Use caution during pregnancy
Xuan Fu Hua	Inula Flower	Sl Warm Bitter Acrid Salty	LI Lu ST Sp	*Redirects Qi downward *Expels phlegm *Stops vomiting	*Cough with white productive sputum *Wheezing/chest congestion *Vomiting/hiccup	3–12 g Wrap in filter for cooking
Zao Jiao	Chinese Honeylocust Fruit	Warm Acrid	LI Lu	*Expels phlegm *Opens orifices *Reduces clumps *Moves bowels *Kills parasite	*Cough with sticky sputum *Seizure with excess phlegm *Abscess/boils *Impacted stools due to roundworms	1.5–5 g Toxic C/I during pregnancy

RELIEVE COUGH AND WHEEZING HERBS

Herb Name	Common Name	Temperature & Tastes	Channels	Functions	Indications	Dosages & Notes
Bai Bu	Stemona Tuber	Neutral Sweet Bitter	Lu	*Moistens Lung *Stops cough *Expels parasites	*Chronic deficiency cough *Lice/fleas (topical) *Pinworms (enema) *Urticaria/tinea infection/bug bites (topical)	3–9 g
Du Juan Hua Ye	Rhodedendron Leaf	Neutral Sour Sweet	Lu	*Expels phlegm *Stops cough *Calms wheezing	*Cough with sputum *Bronchitis *Skin swelling/deep-rooted sores (topical)	9–15 g

(continued)

RELIEVE COUGH AND WHEEZING HERBS (continued)

Herb Name	Common Name	Temperature & Tastes	Channels	Functions	Indications	Dosages & Notes
Hu Tui Ye	Thorny Elaegnus Leaf	Sl Warm Sl Bitter Sour	Lu	*Stops cough *Calms wheezing	*Chronic deficiency cough *Asthma	2–5 g
Kuan Dong Hua	Coltsfoot Flower	Warm Acrid	Lu	*Directs Qi downward *Stops cough	*All types of cough	2–9 g
Ma Dou Ling	Birthwort Fruit	Cold Bitter Acrid	LI Lu	*Clears Lung heat *Transforms phlegm *Stops cough *Calms wheezing	*Cough due to Lung heat *Hemorrhoid bleeding *Dizziness/headache due to hypertension	3–9 g
Mu Hu Die	India Trumpet-flower Seed	Cool Sweet Bland	Liv Lu	*Moistens Lung *Smoothes Liver Qi	*Cough/sore throat *Hypochondriac pain *Ulcerated sores (topical)	2–6 g
Qian Ri Hong	Globeama-ranth Flower	Neutral Sweet	Liv Lu	*Stops cough *Calms wheezing *Brightens eyes	*Whooping cough *Wheezing *Blurred vision	9–15 g
Sang Bai Pi	Bark of Mulberry Root	Cold Sweet	Lu Sp	*Stops cough *Arrests wheezing *Drains Lung heat *Promotes urination	*Cough/wheezing *Upper body edema *Difficult urination *Lowers mild and high blood pressure	6–15 g
Ting Li Zi	Pepperweed Seed or Flixweed Tansymustard Seed	Very Cold Bitter Acrid	Lu UB	*Redirects Lung Qi down *Reduces phlegm *Calms wheezing *Reduces edema	*Cough with white productive sputum *Gurgling sounds in throat due to Lung Qi stagnation *Pleurisy/pulmonary and cardiovascular disease	3–9 g
Xing Ren (*Ku Xing Ren*) (*Bei Xing Ren*)	Apricot Seed	Sl Warm Bitter	LI Lu	*Stops cough *Calms wheezing *Moistens Intestine	*All types of cough *Dry stools	3–9 g Coarsely crushed for cooking Sl Toxic Overdose will cause nausea and vomiting, dizziness, and headache
Zi Jin Niu (*Ai Di Cha*)	Japanese Ardisia	Sl Cold Bitter	Liv Lu	*Stops cough *Expels phlegm *Promotes urination *Leaches damp *Activates blood *Expels stasis	*Lung heat cough with sticky sputum *Pleurisy/tuberculosis prevention *Jaundice/edema *Bi syndrome/trauma injury *Menstrual cramp	Up to 30 g per day
Zi Su Zi	Purple Perilla Fruit	Warm Acrid	LI Lu	*Stops cough *Arrests wheezing *Directs Qi downward *Moistens Intestine	*Cough/wheezing *Labored exhalation *Chest constriction *Constipation due to dryness	4–9 g Crushed for cooking
Zi Wan	Purple Aster Root	Sl Warm Bitter Sweet	Lu	*Releases cough *Expels phlegm	*Chronic cough with blood-tinged sputum	3–9 g

INDUCE VOMITING TO DISPEL PHLEGM HERBS

Herb Name	Common Name	Temperature & Tastes	Channels	Functions	Indications	Dosages & Notes
Gua Di	Melon Pedicle	Cold Bitter	ST	*Induces vomiting *Clears phlegm–heat *Dispels damp–heat	*Phlegm–heat stuck in chest region *Food stagnation causing abdominal distention *Jaundice/headache due to damp–heat (inhale powder)	2.5–5 g in decoction 0.3–1 g in pills or powder Toxic
Li Lu	Mountain Onion	Cold Bitter Acrid	Liv Lu ST	*Induces vomiting *Dispels wind–phlegm *Kills parasite *Stops itch	*Wind stroke/seizure *Throat Bi syndrome *Scabies/ringworm/lice (topical)	0.3–0.9 g in pills or powder Ex. toxic. Not compatible with Ren Shen, Sha Shen, Dan Shen, Ku Shen, Xi Xin, and Bai Shao

CHAPTER 34

Aromatic Herbs That Transform Damp

Herb Name	Common Name	Temperature & Tastes	Channels	Functions	Indications	Dosages & Notes
Bai Dou Kou	Round Cardamom Fruit	Warm Acrid Aromatic	Lu Sp ST	*Transforms damp *Warms Middle Jiao *Circulates Qi	*Chest constriction *Abdominal distention *Nausea/vomiting due to cold-damp *Infantile vomiting of milk due to Middle Jiao cold	3–6 g Add last 5 min of cooking
Cang Zhu	Chinese Atractylodes	Warm Acrid Bitter Aromatic	Sp ST	*Strongly dries damp *Strengthens Spleen *Forces sweating *Releases exterior	*Diarrhea/nausea/ abdominal distention from damp accumulation *Headache/body ache/oozing Yin sores from wind–damp–cold *Lower Jiao damp–heat *Night blindness *Keratomalacia	4–9 g Very drying Used as prevention for influenza
Cao Dou Kou	Seed of Katsumada's Galangal	Warm Acrid Aromatic	Sp ST	*Dries damp *Warms Middle Jiao	*Vomiting/poor appetite/ diarrhea/abdominal pain due to cold stagnation in Middle Jiao	2–6 g
Cao Guo	Amomum Tsao-Ko	Warm Acrid	Sp ST	*Strongly dries damp *Disperses cold *Reduces food stagnation	*Turbid–damp stagnation in febrile disease *Nausea/abdominal distention from Middle Jiao deficiency cold *Food stagnation from meat *Alternating chills and fever	3–6 g Very drying
Er Cha (*Hai Er Cha*)	Catechu Extract	Bitter Astringent Cool	Lu	*Dries damp *Promotes healing *Stops diarrhea *Stops bleeding *Transforms phlegm	*Skin wounds, ulcers, boils, and sores *Injuries due to trauma *Cough/thirst due to Lung heat *Dysentery	0.1–1 g per day in pill or powder
Hou Po	Magnolia Bark	Warm Acrid Bitter Aromatic	LI Lu Sp ST	*Transforms damp *Circulates Qi *Dissolves stagnation *Directs Qi downward	*Abdominal distention due to Qi or food stagnation *Wheezing/cough with copious mucus from Lung phlegm–cold	3–9 g
Hou Po Hua	Magnolia Flower	Warm Acrid Aromatic	Lu ST	*Transforms damp *Moves Qi	*Epigastric pain *Chest constriction	3–6 g

(continued)

Herb Name	Common Name	Temperature & Tastes	Channels	Functions	Indications	Dosages & Notes
Huo Xiang	Patchouli or Wrinkled Gianthyssop	Sl Warm Acrid Aromatic	Lu Sp ST	*Transforms damp *Harmonizes Middle Jiao *Releases exterior	*Lack of appetite/nausea/vomiting/abdominal discomfort due to damp accumulation *Morning sickness *Wind–cold–damp invasion	4–9 g Short cooking time Double the dosage if fresh
Pei Lan	Eupatorium	Neutral Acrid Aromatic	Sp ST	*Transforms damp *Releases summer heat	*Nausea due to summerheat *Chest constriction *Sweet sticky taste in mouth	4–9 g
Sha Ren	Seed of Grains-of-Paradise	Warm Acrid Aromatic	Sp ST Kid	*Transforms damp *Stops vomiting *Moves Qi *Strengthens Stomach *Awakens Spleen *Calms fetus	*Increase appetite *Promote digestion *Nausea/vomiting *Morning sickness	2–6 g Add last 5 min of cooking
Shi Chang Pu	Sweetflag Rhizome	Sl Warm Acrid Aromatic	Ht ST	*Transforms phlegm *Opens orifice *Calms spirit	*Improves digestion *Increase appetite *Mental clarity *Dizziness/deafness/tinnitus	3–9 g Double the dosage if fresh

CHAPTER 35

Relieve Food Stagnation Herbs

Herb Name	Common Name	Temperature & Tastes	Channels	Functions	Indications	Dosages & Notes
Ge Shan Xiao	Earleaf Cynanchum Root	Neutral Sweet Bitter	Sp ST Liv	*Reduces food stagnation *Strengthens Spleen *Regulates Qi *Stops pain	*Epigastric distention/poor digestion/diarrhea due to Spleen deficiency *Hypochondriac pain	9–15 g in powder
Gu Ya	Rice Sprout	Neutral Sweet	Sp ST	*Reduces food stagnation *Strengthens Stomach	*Food stagnation due to starchy foods	9–15 g Milder than Mai Ya
Ji Nei Jin	Chicken Gizzard Lining	Neutral Sweet	UB SI Sp ST	*Reduces food stagnation *Secures essence *Transforms hardness *Dissolves stones	*Strongly promotes digestion *Enuresis/nocturnal emission *Gallstones/urinary stones	3–9 g in decoction 1–3 g in powder
Ji Shi Teng	Chinese Fevervine	Sl Cool Sweet Bitter	Sp ST Liv Lu	*Reduces food stagnation *Strengthens Spleen *Relieves pain *Transforms phlegm *Clears phlegm–heat *Reduces toxin	*Poor digestion *Infantile malnutrition *Phlegm–heat cough *Hepatitis/dysentery *Sore throat *Skin rash/sores/ burns *Trauma/fracture/ postsurgery pain (topical)	15–60 g
Lai Fu Zi	Chinese Radish Seed	Neutral Sweet Acrid	Lu Sp ST	*Reduces food stagnation *Reduces phlegm	*Epigastric fullness *Cough/wheezing due to phlegm and mucous *Belching with foul smell	6–12 g
Mai Ya	Barley Sprout	Neutral Sweet	Liv Sp ST	*Reduces food stagnation *Strengthens Stomach *Inhibits lactation *Smoothes Liver Qi	*Food stagnation due to starchy foods *Epigastric discomfort *Loss of appetite/ belching	10–15 g 30–120 g per day (toasted) to inhibit lactation

(continued)

Herb Name	Common Name	Temperature & Tastes	Channels	Functions	Indications	Dosages & Notes
Shan Zha	Hawthorn Fruit	Sl Warm Sweet Sour	Liv Sp ST	*Reduces food stagnation due to meat/greasy foods *Breaks blood stasis *Stops diarrhea	*Food stagnation due to meat/ greasy foods *Postpartum abdominal pain *Lowers blood pressure and cholesterol *Chest pain due to blood stasis *Chronic diarrhea (charred)	9–15 g Up to 120 g per day
Shen Qu	Medicated Leaven	Sl Warm Sweet Acrid	Sp ST	*Reduces food stagnation *Harmonizes Stomach *Assists absorption *Promotes digestion	*Abdominal fullness *Borborygmus/ diarrhea *Absorbs mineral/ metal ingredients in formulas	6–15 g

CHAPTER 36

Regulate Qi Herbs

Herb Name	Common Name	Temperature & Tastes	Channels	Functions	Indications	Dosages & Notes
Cha Shu Gen	Tea Tree Root	Neutral Bitter	Ht Lu ST	*Regulates Qi *Moves blood *Relieves toxin *Transforms damp *Dispels wind	*Bi syndrome *Congestive heart disease *Hypertension	30–60 g
Chen Pi	Tangerine Peel	Warm Acrid Bitter Aromatic	Lu Sp ST	*Regulates Qi *Dries damp *Transforms phlegm *Smoothes stagnation	*Cold in Middle Jiao *Cough with copious sputum *Nausea/vomiting/hiccup *Prevents cloying herbs from causing Qi stagnation	3–9 g
Chen Xiang	Aloeswood	Warm Acrid Bitter Aromatic	Kid Sp ST	*Moves Qi *Relieves pain *Directs Qi downward *Aids Kidney in grasping Qi *Warms Middle Jiao	*Epigastric/abdominal pain with distended feeling *Wheezing/asthma due to Kidney deficiency *Nausea/vomiting/hiccup due to deficiency cold in Middle Jiao	1–3 g
Chuan Lian Zi	Chinaberry Fruit	Cold Bitter	Liv SI ST	*Moves Qi *Stops pain *Clears heat *Dries damp *Kills parasite	*Abdominal/hernial pain due to damp–heat stagnation *Hypochondriac pain *Scalp tinea (topical) *Roundworms	Less than 6 g per day Sl Toxic
Da Fu Pi	Betel Husk	Sl Warm Acrid	LI SI Sp ST	*Reduces food stagnation *Moves Qi downward *Expels damp *Promotes urination *Reduces edema	*Epigastric/abdominal pain with distended feeling *Belching/acid reflux *Constipation *Superficial edema and Damp Leg Qi	6–9 g
Dao Dou	Sword Bean	Warm Sweet	ST Kid	*Descends Qi downward *Stops hiccup	*Vomiting due to deficient cold in Middle Jiao	9–15 g
Fo Shou	Finger Citron Fruit	Sl Warm Acrid Bitter	Liv Lu ST Sp	*Spreads Liver Qi *Relieves pain *Harmonizes Middle Jiao *Dries damp *Transforms phlegm	*Hypochondriac pain *Belching/abdominal distention *Lack of appetite/ vomiting *Asthma/cough with sputum	3–9 g

(continued)

Herb Name	Common Name	Temperature & Tastes	Channels	Functions	Indications	Dosages & Notes
Fo Shou Hua	Finger Citron Flower	Sl Warm Acrid Bitter	Liv Sp Lu	*Directs rebellious Qi downward	*Cough/wheezing due to rebellious Lung Qi	3–9 g Milder than Fo Shou
Ju He	Tangerine Seed	Neutral Acrid Bitter	Liv	*Moves Qi *Reduces nodules *Relieves pain	*Hernial swelling *Testicular pain *Mastitis	3–9 g
Ju Hong	Outer Tangerine Peel	Warm Acrid Bitter Aromatic	Sp ST Lu LI	*Regulates Qi *Dries damp *Transforms phlegm *Smoothes stagnation	*Cold in Middle Jiao *Damp–phlegm cough *Nausea/vomiting/hiccup *Prevents cloying herbs from causing Qi stagnation	3–10 g Stronger than Chen Pi
Ju Luo	Tangerine Pith	Neutral Bitter	Liv Lu	*Moves Qi *Dredges meridian *Transforms phlegm	*Cough with sputum *Chest/hypochondriac pain due to phlegm obstruction in channels and collaterals	3–9 g
Ju Ye	Tangerine Leaf	Neutral Acrid Bitter	Liv	*Moves Liver Qi *Reduces swelling	*Hypochondriac pain *Breast swelling and pain	3–9 g
Li Zhi He	Leechee Nut	Warm Acrid Sweet Astringent	Liv ST	*Regulates Qi *Disperses cold *Stops pain *Reduces lump	*Hernial and testicular pain with swelling due to cold in Liver channel *Epigastric discomfort *Menstrual pain	3–15 g Up to 24 g for abdominal pain
Mei Gui Hua	Miniature Chinese Rose	Warm Sl Bitter Sweet	Liv Sp	*Moves Qi *Relieves constraint *Harmonizes blood	*Mild menstrual disorders *Chest constriction *Hypochondriac pain *Belching/epigastric pain due to Liver and Stomach disharmony	1.5–6 g
Mu Xiang	Aucklandia	Warm Acrid Bitter	GB LI Sp ST	*Moves Qi *Relieves pain *Strengthens Spleen *Prevents stagnation	*Hypochondriac discomfort due to Liver Qi stagnation *Nausea/vomiting/poor appetite due to Middle Jiao stagnation *Tenesmus/diarrhea/ abdominal pain/ dysenteric-like disorders due to intestinal Qi blockage *Prevents cloying herbs from causing Qi stagnation *Biliary pain (higher dosage)	3–15 g Add last 5 min of cooking
Qing Pi	Immature Tangerine Peel	Warm Acrid Bitter	GB Liv ST	*Spreads Liver Qi *Reduces food stagnation *Dries dampness *Transforms phlegm	*Various types of pain or hernia-like disorders due to Liver Qi stagnation *Breast abscess from phlegm–damp accumulation	3–9 g
Shi Di	Persimmon Calyx	Neutral Bitter Astringent	Lu ST	*Directs Qi down *Stops hiccup	*Primary herb for hiccup and belching due to cold accumulation in Stomach	6–12 g
Tan Xiang	Sandalwood	Warm Acrid Aromatic	Lu Sp ST Ht	*Moves Qi *Alleviates chest pain *Dispels cold	*Chest pain/cardiac pain due to Qi stagnation and Blood stasis *Epigastric/abdominal pain with poor appetite due to cold accumulation in Stomach	1–6 g

(continued)

Herb Name	Common Name	Temperature & Tastes	Channels	Functions	Indications	Dosages & Notes
Wu Yao	Three-Nerved Spicebush Root	Warm Acrid	UB Kid Lu Sp ST	*Move Qi *Relieves pain *Warms Kidney	*Various pain due to Qi stagnation from cold accumulation *Urinary incontinence and frequency due to deficiency cold	3–9 g
Xiang Fu	Nutgrass	Sl Warm Acrid Bitter	Liv SJ	*Spreads Liver Qi *Regulates menses *Relieves pain	*Hypochondriac pain *Epigastric discomfort *Breast distention/ tenderness *Dymenorrhea	4–12 g
Xiang Yuan	Citron Fruit	Sl Warm Acrid Bitter	Liv Sp Lu	*Spreads Liver Qi *Relieves pain *Harmonizes Middle Jiao *Dries damp *Transforms phlegm	*Hypochondriac pain *Belching/abdominal distention *Lack of appetite/ vomiting *Asthma/cough with sputum	3–9 g
Xie Bai	Bulb of Chinese Chive	Warm Bitter Acrid	LI Lu ST Ht	*Circulates Yang Qi *Disperses cold phlegm *Moves Qi and blood *Relieves pain *Directs Qi downward	*Cough/wheezing/ dyspnea due to cold–turbid–phlegm obstruction *Chest pain due to Heart Blood stasis *Epigastric/abdominal pain due to cold accumulation in Stomach *Dysenteric disorders due to damp stagnation in Large Intestine	4–9 g
Zhi Ke	Ripe Fruit of Bitter Orange	Cool Bitter	Sp ST	*Moves stagnant Qi *Reduces accumulation *Directs Qi downward *Unblocks bowels *Transforms phlegm	*Epigastric distention *Abdominal fullness	3–9 g For weak and deficient type patients Milder than Zhi Shi
Zhi Shi	Immature Fruit of Bitter Orange	Sl Cold Acrid Bitter Sl Sour	LI Sp ST	*Breaks up stagnant Qi *Reduces accumulation *Directs Qi downward *Unblocks bowel *Transforms phlegm	*Epigastric/abdominal pain/constipation with flatulence due to stagnant Qi *Used with Qi tonic herbs for uterine/rectal/ Stomach prolapse *Chest congestion/pain with cough and nausea due to damp or stagnant Qi in Middle Jiao	3–9 g Up to 30 g per day for prolapsed organs Stronger than Zhi Ke

CHAPTER 37

Regulate Blood Herbs

ASTRINGENT TO STOP BLEEDING HERBS

Herb Name	Common Name	Temperature & Tastes	Channels	Functions	Indications	Dosages & Notes
Bai Ji	Hyacinth Bletilla Tuber	Cool Sweet Bitter Astringent	Lu ST Liv	*Stops bleeding *Restrains blood leakage *Reduces swelling *Generates flesh	*Bleeding from Lung and Stomach: vomiting/coughing of blood/nose bleed *Sores/nonhealing ulcers/chapped skin (topical)	2–5 g per dose 5–15 g per day. Not compatible with Wu Tou or Fu Zi
Hua Sheng Yi	Peanut Peel	Neutral Sweet Astringent	Lu Sp Liv	*Restrains blood leakage *Stops bleeding	*Internal/external bleeding	9–15 g
Ji Guan Hua	Cockscomb Flower	Cool Sweet Astringent	Liv LI	*Stops bleeding *Reduces vaginal discharge	*Blood in stools *Hemorrhoid bleeding *Uterine bleeding	6–15 g
Ji Mu	Strap Flower/ Root/Leaf	Neutral Bitter Astringent	Liv ST LI	*Stops bleeding *Stops diarrhea *Clears heat	*Various bleeding cases *Local application for burns	6–9 g: flower 6–15 g: root 15–30 g: leaf
Ou Jie	Node of Lotus Rhizome	Sl Cool Sweet Astringent	Lu ST Liv	*Stops bleeding without creating stasis *Restrains blood leakage	*Bleeding from Lung and Stomach: vomiting/coughing of blood/nose bleed	9–30 g 60 g in squeezed juice
Xian He Cao	Agrimony	Neutral Bitter Acrid	Lu Liv Sp	*Stops bleeding *Clears heat and detoxifies *Alleviates diarrhea *Kills parasite	*Hemoptysis/ hemafecia *Nose bleeding *Uterine bleeding *Bloody Lin syndrome *Chronic dysentery/ diarrhea *Trichomonas vaginitis *Tapeworm	9–30 g 30–60 g for external wash
Xue Yu Tan	Charred Human Hair	Neutral Bitter	Ht Liv Kid	*Stops bleeding without creating stasis *Promotes urination	*Uterine bleeding *Nosebleed *Bloody Lin syndrome	1.5–9 g
Zi Zhu	Beautyberry Leaf	Cool Bitter Astringent	Liv Lu Sp	*Stops bleeding *Detoxifies for skin disease	*Bleeding from Lung and Stomach: vomiting/coughing of blood/nose bleed *Burns/sores/ulcer (topical)	10–15 g 2–3 g in powder

(continued)

ASTRINGENT TO STOP BLEEDING HERBS (continued)

Herb Name	Common Name	Temperature & Tastes	Channels	Functions	Indications	Dosages & Notes
Zong Lu Pi	Stiple Fiber of Fortune Windmill Palm	Neutral Bitter Astringent	Liv LI Lu Sp	*Binds and stops bleeding	*Nose bleed *Uterine bleeding *Hemafecia *Coughing of blood	9–15 g
Zong Lu Tan	Charred Stiple Fiber of Fortune Windmill Palm	Neutral Bitter Astringent	Lu Liv LI	*Binds and stops bleeding	*Nose bleed *Uterine bleeding *Hemafecia	9–15 g

COOL BLOOD TO STOP BLEEDING HERBS

Herb Name	Common Name	Temperature & Tastes	Channels	Functions	Indications	Dosages & Notes
Bai Mao Gen	Woolly Grass Rhizome	Cold Sweet	Lu ST SI UB	*Cools blood *Stops bleeding *Promotes urination *Clears Lung and Stomach heat	*Hemoptysis/ hemafecia *Nose bleeding *Hot Lin syndrome with edema and jaundice *Nausea/thirst/ irritability due to Stomach and Lung heat	10–15 g 30–60 g if fresh
Ce Bai Ye	Oriental Arborvitae Twig	Sl Cold Bitter Astringent	Ht Liv LI	*Cools blood *Stops bleeding *Expels phlegm *Stops cough *Promotes healing of burns *Promotes hair growth	*Hemoptysis/ hemafecia *Nose bleeding *Uterine bleeding *Bloody Lin syndrome *Bleeding gums *Early stage burns (topical ointment) *Cough with blood tinged sputum *Tincture for balding (topical)	6–15 g raw or charred
Da Ji	Japanese Thistle	Cool Sweet	Liv Sp Ht	*Cools blood *Stops bleeding *Reduces swelling	*Hemoptysis/ hemafecia *Nose bleeding *Uterine bleeding *Bloody Lin syndrome *Carbuncles/sores (topical)	10–15 g 30–60 g if fresh
Di Yu	Burnet Bloodwort Root	Sl Cold Bitter Sour	Liv LI ST	*Cools blood *Stops bleeding *Clears heat *Generates flesh	*Hemafecia/bleeding hemorrhoids/ bloody dysentery/ uterine bleeding *Peptic ulcer bleeding *Skin ulcers/burns/ sores (topical)	6–15 g
Huai Hua Mi	Pagoda Tree Flower Bud	Cool Bitter	Liv LI	*Cools blood *Stops bleeding *Clears Liver heat	*Lowers high blood pressure *Hemorrhoids/ bloody dysentery due to damp–heat in Large Intestine *Uterine bleeding *Conjunctivitis/ dizziness due to Liver heat	9–15 g

(continued)

COOL BLOOD TO STOP BLEEDING HERBS (continued)

Herb Name	Common Name	Temperature & Tastes	Channels	Functions	Indications	Dosages & Notes
Huai Jiao	Pagoda Tree Fruit	Cold Bitter	Liv LI	*Stops bleeding *Clears heat (stronger) *Directs Qi downward	*Inflamed hemorrhoids	9–15 g C/I during pregnancy
Ji Cai	Shepherd's Purse	Cool Sweet Bland	Liv ST SI UB	*Cools blood *Stops bleeding *Clears Liver heat	*Lower Jiao bleeding due to heat *Edema *Difficult urination due to damp–heat *Hypertension from Kidney disease	15–30 g in dry form 30–60 g if fresh
Xiao Ji	Small Japanese Thistle	Cool Sweet Bland	Liv Sp Ht	*Cools blood *Stops bleeding *Reduces swelling *Promotes urination	*Blood in urine *Swelling/sores from heat toxin accumulation *Jaundice/edema due to damp–heat in Liver and Kidney *Reduces high blood pressure from Liver heat	10–15 g in dry form 30–60 g if fresh
Zhu Ma Gen	Ramie Root	Cool Sweet	Ht Liv Kid UB	*Stops bleeding *Calms fetus *Drains heat *Promotes urination	*Bleeding from peptic ulcer/ Lung diseases *Terrified fetus with uterine bleeding due to heat *Skin disease and measles from heat toxin *Hot/Bloody Lin with edema	10–30 g

TRANSFORM STASIS TO STOP BLEEDING HERBS

Herb Name	Common Name	Temperature & Tastes	Channels	Functions	Indications	Dosages & Notes
Hua Rui Shi	Ophicalcite	Neutral Sour Astringent	Liv	*Disperses blood stasis *Stops bleeding	*Vomiting of blood *Cough up blood *Postpartum bleeding *Traumatic injury with swelling/pain/bruise	3–9 g in powder
Jiang Xiang	Center of Rosewood	Warm Acrid	Liv Sp ST Ht	*Disperses blood stasis *Stops bleeding *Moves Qi *Relieves pain	*Sprain/contusion/ fracture/laceration (topical) *Stomach/chest pain	3–6 g 1–2 g in powder
Jing Tian San Qi	Aizoon Stonecrop Plant	Neutral Sweet Sl Sour	Ht Liv	*Disperses blood stasis *Stops bleeding *Calms spirit	*Different kinds of bleeding *Thrombocytopenic purpura *Insomnia/palpitation/ irritability *External trauma (topical)	15–30 g in dry form 30–60 g if fresh
Ju Ye San Qi	Canton Tusanqi	Neutral Sl Bitter Sweet	Liv ST	*Disperses blood stasis *Stops bleeding *Reduces swelling *Detoxification	*Mastitis and swelling *Different kinds of bleeding	3–5 g in powder 10–15 g in dry form

(continued)

TRANSFORM STASIS TO STOP BLEEDING HERBS (continued)

Herb Name	Common Name	Temperature & Tastes	Channels	Functions	Indications	Dosages & Notes
Lian Fang	Mature Lotus Receptacle	Warm Bitter Astringent	Liv Sp Kid	*Disperses blood stasis *Stops bleeding *Calms fetus *Dispels summer heat *Drains damp	*Blood in urine *Uterine bleeding *Restless fetus *Diarrhea due to summer heat with dampness	3–9 g
Mao Mei	Japanese Raspberry	Sl Cold Bitter Astringent	Liv	*Disperses blood stasis *Stops bleeding *Reduces swelling *Relieves pain *Clears heat	*Internal/external bleeding *Regulate menses *Dysmenorrhea	30–60 g: root 15–30 g: twig or leaf; 120 g if freshly supplied
Pu Huang	Cattail Pollen	Neutral Sweet Acrid	Liv Ht Sp	*Stops bleeding *Moves blood *Dispels blood stasis *Promotes urination	*Hemoptysis *Hemafecia *Nose bleeding *Uterine bleeding *Bloody Lin syndrome *Postpartum abdominal pain *Dysmenorrhea *External bleeding from traumatic injuries (topical) *Swellings caused by sores	4.5–12 g Use caution during pregnancy Wrap in filter for cooking
Qian Cao Gen	Madder Root	Cold Bitter	Ht Liv	*Cools blood *Stops bleeding *Moves blood *Dispels blood stasis	*Hemoptysis/hemafecia *Nose bleeding *Uterine bleeding *Bloody Lin syndrome *Dysmenorrhea *Chest/hypochondriac pain *Bi syndrome due to blood stagnation *Chronic hepatitis leading to hepatosplenomegaly	10–15 g
San Qi	Pseudoginseng Root	Warm Sweet Bitter	Liv ST LI	*Stops bleeding without causing blood stasis *Transforms blood stasis *Reduces swelling *Relieves pain	*Chest, abdominal, and joint pain *Hemoptysis *Hemafecia *Nose bleeding *Uterine bleeding *Bloody Lin syndrome *Contusions/sprains/falls/fractures	3–9 g 2–5 g in powder

WARM CHANNELS TO STOP BLEEDING HERBS

Herb Name	Common Name	Temperature & Tastes	Channels	Functions	Indications	Dosages & Notes
Ai Ye	Mugwort Leaf	Warm Bitter Acrid	Sp Liv Kid	*Warms uterus *Stops bleeding *Calms fetus *Disperses cold *Relieves pain *Stops cough	*Prolonged menses bleeding *Uterine bleeding *Restless fetus *Threatening miscarriage *Dysmenorrhea *Itchy skin lesion	3–9 g

(continued)

WARM CHANNELS TO STOP BLEEDING HERBS (continued)

Herb Name	Common Name	Temperature & Tastes	Channels	Functions	Indications	Dosages & Notes
Fu Long Gan	Yellow Earth from center of cooking stove	Warm Acrid	Sp ST	*Warms Stomach *Stops bleeding *Reduces nausea *Stops diarrhea	*Nausea/vomiting/ morning sickness/ diarrhea due to Middle Jiao deficiency cold	15–120 g
Pao Jiang	Quick-Fried Ginger Root	Warm Bitter Astringent	Liv Sp	*Stops bleeding *Warms Middle Jiao	*Digestive bleeding due to Middle Jiao deficiency cold *Beng Lou due to Spleen and Kidney Yang deficiency *Abdominal pain and diarrhea from Stomach and Spleen deficiency cold	3–6 g

INVIGORATE THE BLOOD HERBS

Herb Name	Common Name	Temperature & Tastes	Channels	Functions	Indications	Dosages & Notes
Chong Wei Zi	Motherwort Fruit	Sl Cold Acrid Bitter	Ht Liv UB	*Invigorates blood *Builds blood	*Gynecological disorders *Abdominal masses *Systemic edema with blood in urine *Difficult labor and lochioschesis *Traumatic injury (topical)	10–30 g
Chuan Niu Xi	Sichuan Cyathula Root	Neutral Sweet Bitter	Liv Kid	*Invigorates blood *Activates channels *Drains damp	*Lower back pain *Atrophy/spasms in lower limbs *Bloody Lin syndrome *Amenorrhea	4.5–9 g Use caution during pregnancy and menorrhagia Stronger than Huai Niu Xi
Chuan Shan Jia	Pangolin Scales	Cool Salty	Liv ST	*Disperses blood stasis *Expels pus *Dispels wind–damp *Stops bleeding *Promotes lactation	*Amenorrhea *Dysmenorrhea *Abscess/boil *Suppurative lesion (topical) *Bi Syndrome	3–9 g 1–1.5 g in powder
Chuan Xiong	Sichuan Lovage Root	Warm Acrid	Liv GB Pc	*Invigorates blood *Moves Qi *Expels wind *Relieves pain	*Amenorrhea/ dysmenorrhea *Lochioschesis *Chest/hypochon-driac pain *Headache/dizziness *Bi Syndrome	3–6 g Up to 9 g for irregular menses
Dan Shen	Salvia Root	Sl Cold Bitter	Ht Pc Liv	*Invigorates blood *Breaks up blood stasis *Clears heat *Soothes irritability	*Amenorrhea *Dysmenorrhea *Lochioschesis *Chest/hypochon-driac pain *Irritability/palpitation/ insomnia/ restlessness	3–15 g Up to 60 g for vasculitis Not compatible with Li Lu

(continued)

INVIGORATE THE BLOOD HERBS (continued)

Herb Name	Common Name	Temperature & Tastes	Channels	Functions	Indications	Dosages & Notes
E Zhu	Zedoary Rhizome	Warm Bitter Acrid	Liv Sp	*Breaks up blood stasis *Dissolves mass *Moves Qi *Relieves pain	*Dysmenorrhea/ amenorrhea *Abdominal mass *Chest pain/fullness/ constriction due to food accumulation *Childhood Gan Ji symptom *Antitumor effect on skin/cervix/genital regions	3–9 g C/I during pregnancy
Gan Qi	Dry Lacquer Sap	Warm Acrid Bitter	Liv ST	*Breaks up blood stasis *Restores menses *Kills parasite	*Amenorrhea *Abdominal mass *Abdominal pain due to parasite accumulation *Headache/dizziness *Cough/wheezing/ vomiting	0.06–0.1 g in powder Sl toxic Use caution during pregnancy Not compatible with crab
Hong Hua	Safflower Flower	Warm Acrid	Ht Liv	*Invigorates blood *Dispels blood stasis *Relieves pain	*Amenorrhea *Abdominal pain/ postpartum dizziness/abdominal mass/lochioschesis *Wound/sores/ carbuncles *Rash from measles *Chest Bi syndrome	3–9 g Use caution during pregnancy
Huai Niu Xi	Bidentate Achyranthes Root	Neutral Bitter Sour	Liv Kid	*Invigorates blood *Expels blood stasis *Strengthens sinews/bones *Clears damp–heat	*Gynecological disorders *Lower back/knees/ hips/ankles/feet weakness and soreness *Bloody Lin syndrome *Toothache/ headache/dizziness/ hemoptysis/ nosebleed	4.5–9 g Use caution during pregnancy and menorrhagia
Jiang Huang	Turmeric Rhizome	Warm Bitter Acrid	Liv Sp ST	*Invigorates blood *Moves Qi *Unblocks menses *Expels wind	*Dysmenorrhea/ amenorrhea *Abdominal/chest pain *Traumatic injury with swelling/pain/bruise *Bi syndrome due to wind–damp	3–9 g Use caution during pregnancy and menorrhagia
Ling Xiao Hua (*Zi Wei*)	Chinese Trumpetcreeper Flower	Sl Cold Acrid	Liv Pc	*Invigorates blood *Cracks blood stasis *Cools blood *Disperses wind	*Amenorrhea *Lumps due to blood stasis *Internal wind due to extreme heat *Itchy skin due to damp toxins	3–10 g 15–30 g for topical application
Liu Ji Nu	Diverse Wormwood	Warm Bitter	Ht Sp	*Dispels blood stasis *Unblocks menses *Relieves pain	*Bi syndrome *Fall/fracture/burn/ contusion/sprain (topical) *Gynecological disease due to Blood stasis	3–9 g
Mang Chong	Gadfly	Sl Cold Acrid Salty	Liv	*Breaks up blood stasis *Reduces swelling	*Amenorrhea *Mass/lump/swellings from trauma	1–3 g in decoction 0.3–0.5 g in powder Sl toxic C/I during pregnancy

(continued)

INVIGORATE THE BLOOD HERBS (continued)

Herb Name	Common Name	Temperature & Tastes	Channels	Functions	Indications	Dosages & Notes
Mo Yao	Myrrh	Neutral Bitter	Ht Liv Sp	*Invigorates blood *Dispels blood stasis *Reduces swelling *Relieves pain *Promotes healing	*Dysmenorrhea/ amenorrhea *Abdominal/chest pain *Abdominal mass *Chronic nonhealing sores *Sore/swelling/ carbuncles (topical)	3–12 g C/I during pregnancy and menorrhagia
Qi Cao	Grub	Salty Sl Warm	Liv	*Strongly moves blood *Reduces accumulation *Relieves pain *Promotes lactation	*Traumatic injuries *Skin sores/abscess *Hemorrhoid *Feng Dan *Throat Bi syndrome	6–9 g in powder
Ru Xiang	Frankincense or Pistaciacense	Warm Acrid Bitter	Ht Liv Sp	*Invigorates blood *Moves Qi *Relaxes sinews *Relieves pain *Reduces swelling *Generates flesh	*Dysmenorrhea/ amenorrhea *Abdominal/chest pain *Abdominal mass *Chronic nonhealing sores *Sore/swelling/ carbuncles (topical) *Bi syndrome *Swollen gum/ canker sore/ sore throat (oral)	3–9 g C/I during pregnancy Mo Yao and Ru Xiang are used together to promote each other's functions
San Leng	Bur-Reed Rhizome	Neutral Bitter Acrid	Liv Sp	*Strongly breaks blood stasis *Moves Qi *Relieves pain *Dissolves accumulation	*Dysmenorrhea/ amenorrhea *Abdominal pain/ postpartum pain/ abdominal mass *Abdominal pain/ bloating due to food/cold stagnation	3–9 g C/I during pregnancy
Shui Zhi	Leech	Neutral Salty Bitter	Liv UB	*Breaks up blood stasis *Reduces mass	*Dysmenorrhea/ amenorrhea *Abdominal mass *Traumatic injury with swelling/pain/ bruise	3–5 g in decoction 0.3–0.5 g in powder Toxic Don't use on a long-term basis C/I during pregnancy and menstrual cycle
Su Mu	Sappan Wood	Neutral Acrid Sweet Salty	Ht Liv	*Invigorates blood *Reduces swelling *Relieves pain *Stops bleeding	*Gynecological disease due to Blood stasis *Fall/fracture/burn/ contusion/sprain/ swellings *Postpartum bleeding with dizziness/ shortness of breath	3–9 g C/I during pregnancy
Tao Ren	Peach Kernel	Neutral Bitter Sweet	Ht LI Liv Lu	*Breaks up blood stasis *Moistens Intestine *Relieves cough *Alleviates wheezing	*Abdominal/ hypochondriac pain *Lung/intestinal abscess *Constipation *Cough/constipation often used with Xing Ren	6–10 g C/I during pregnancy and hemoptysis

(continued)

INVIGORATE THE BLOOD HERBS (continued)

Herb Name	Common Name	Temperature & Tastes	Channels	Functions	Indications	Dosages & Notes
Tu Bie Chong (*Zhe Chong*)	Wingless Cockroach	Cold Acrid Salty	Liv Ht Sp	*Breaks up blood stasis *Strengthens sinews/bones	*Abdominal mass/pain *Fall/fracture/burn/ contusion/sprain/ swellings *Chronic hepatitis leading to hepatosplenomegaly *Numb/swollen/stiff tongue	6–10 g in decoction 1–1.5 g in powder Toxic C/I during pregnancy
Tu Niu Xi	Prickly Chaff Flower/Root/ Twig	Neutral Bitter Sour	Liv Kid UB	*Moves blood *Disperses stasis *Clears heat *Promotes urination	*Amenorrhea *Bi syndrome *Bloody Lin syndrome *Wind-damp edema *Sore throat/ diphtheria due to heat toxin accumulation *Traumatic injury (topical)	10–15 g in dried form 20–30 g if freshly supplied Use caution during pregnancy and menorrhagia
Wang Bu Liu Xing	Cowherb Seed	Neutral Bitter Acrid	Liv ST	*Invigorates blood *Reduces swelling *Promotes urination *Promotes lactation	*Amenorrhea *Mastitis/lack of lactation *Edema in Lower Jiao *Testicle swelling from damp accumulation and Blood stagnation	3–9 g C/I during pregnancy
Wu Ling Zhi	Flying Squirrel Feces	Warm Bitter Salty	Liv Sp	*Disperses blood stasis *Relieves pain *Stops bleeding	*Dysmenorrhea *Chest/abdominal pain *Uterine bleeding *Lochioschesis *Chest Bi syndrome *Childhood Gan Ji symptom	3–9 g Wrap for cooking Use caution during pregnancy Antagonized by Ren Shen
Xue Jie	Dragon's Blood (Resin)	Neutral Sweet Salty	Ht Liv Pc	*Dispels blood stasis *Relieves pain *Stops bleeding *Promotes healing	*External injuries *Lacerations (topical) *Ulcers/sores/wounds *Raynaud's disease	1–1.5 g in powder C/I during pregnancy
Yan Hu Suo	Corydalis Rhizome	Warm Acrid Bitter	Ht Liv Lu Sp	*Invigorates blood *Moves Qi *Relieves pain	*Dysmenorrhea *Chest/abdominal/ epigastric/hernia pain	3–12 g 1–3 g in powder C/I during pregnancy
Yi Mu Cao	Chinese Motherwort	Sl Cold Acrid Bitter	Ht Liv UB	*Invigorates blood *Regulates menses *Reduces mass *Promotes urination	*Gynecological disorders *Abdominal masses *Systemic edema with blood in urine *Difficult labor and lochioschesis *Traumatic injury (topical)	10–30 g Up to 60 g for edema from glomerulo-nephritis
Yu Jin	Turmeric Tuber	Cool Acrid Bitter	Ht Lu Liv	*Breaks up blood stasis *Moves Qi *Reduces jaundice *Cools blood without causing blood stasis	*Hypochondriac pain *Anxiety/agitation/ seizure *Jaundice from damp–heat *Traumatic injury (topical)	3–9 g Use caution during pregnancy
Yue Ji Hua	Chinese Tea Rose	Warm Sweet	Liv	*Invigorates blood *Regulates menses *Reduces swelling	*Scanty menses/ amenorrhea *Chest/abdominal pain *Neck swelling/scrofula	3–6 g C/I during pregnancy

(continued)

INVIGORATE THE BLOOD HERBS (continued)

Herb Name	Common Name	Temperature & Tastes	Channels	Functions	Indications	Dosages & Notes
Ze Lan	Shiny Bugleweed	Sl Warm Bitter Acrid Aromatic	Liv Sp	*Invigorates blood *Dispels blood stasis *Promotes urination *Reduces edema	*Dysmenorrhea/ amenorrhea *Postpartum abdominal pain/facial edema/dysuria *Traumatic injury (topical)	9–15 g Up to 30 g for topical application Use caution during pregnancy
Zi Jing Mu	Chinese Redbud Wood	Neutral Bitter	Liv Sp	*Moves blood *Dredges meridian *Reduces toxin *Reduces swelling	*Amenorrhea due to Qi stagnation and blood statis *Wind–cold–damp Bi	9–15 g
Zi Jing Pi *(Zi Jing Mu Pi)*	Chinese Redbud Bark	Neutral Bitter	Liv Sp	*Moves blood *Dredges meridian *Reduces toxin *Reduces swelling	*Amenorrhea due to Qi stagnation and blood stasis *Wind–cold–damp Bi	9–15 g
Zi Ran Tong	Pyrite	Neutral Bitter Acrid	Liv Kid	*Dispels blood stasis *Promotes bone/ sinew healing *Relieves pain	*Traumatic injuries and bone fractures	9–15 g 0.3 g in powder

CHAPTER **38**

Warm Interior and Expel Cold Herbs

Herb Name	Common Name	Temperature & Tastes	Channels	Functions	Indications	Dosages & Notes
Ba Jiao Hui Xiang (*Da Hui Xiang*)	Truestar Anisetree Fruit	Warm Acrid	Liv Kid Sp ST	*Warms Kidney *Spreads Liver Qi *Expels cold *Harmonizes Stomach	*Hernia-like pain/ testis disorders in kids due to cold accumulation in Liver meridian *Vomiting/poor appetite due to Stomach cold *Lower abdominal/ hernia pain with testis recession (warm topical)	3–9 g
Bi Ba	Long Pepper Fruit	Hot Acrid	ST LI	*Warms Middle Jiao *Disperses cold from Stomach/ Intestine *Relieves pain	*Vomiting/nausea/ hiccups/abdominal pain/diarrhea due to Stomach cold *Toothache due to cavities (topical)	2–5 g in powder or pill C/I in patients with heat
Bi Cheng Qie	Litse Fruit	Warm Acrid	Sp ST Kid UB	*Warms Middle Jiao *Scatters cold *Moves Qi *Stops pain	*Vomiting/nausea/ hiccups/abdominal pain/diarrhea due to Stomach cold *Hernia-like pain due to cold accumulation *Cold Lin symptom *Infantile turbid urination	2–5 g in powder or pill
Chuan Jiao (*Hua Jiao*)	Fruit of Szechuan Pricklyash Pepper	Hot Acrid Aromatic	Kid Sp ST	*Warms Middle Jiao *Disperses cold *Relieves pain *Kills parasite	*Vomiting/nausea/ hiccups/abdominal pain/diarrhea due to Stomach cold *Hernia-like pain due to cold accumulation *Roundworms *Cold abdomen/ hernia pain (warm topical)	2–5 g SI toxic Use caution during pregnancy
Ding Xiang (*Gong Ding Xiang*)	Clove Flower Bud	Warm Acrid	Kid Sp ST	*Warms Middle Jiao *Directs rebellious Qi downward *Warms Kidney	*Hiccup due to Stomach cold *Nausea/vomiting/ diarrhea *Impotence/ excessive clear leukorrhea due to Kidney Yang deficiency	2–5 g Do not use with Yu Jin

(continued)

Herb Name	Common Name	Temperature & Tastes	Channels	Functions	Indications	Dosages & Notes
Dou Chi Jiang	Aromatic Litsea Root/Stem	Warm Acrid Aromatic	Liv Lu	*Expels wind–damp *Moves Qi and blood *Warms channel *Relieves pain	*Lower back pain/ chills/muscle ache/ headache due to wind–damp–cold *Lower abdominal pain & distension from deficiency cold	1.5–6 g
Fu Zi	Aconite	Very Hot Acrid Sweet	Ht Kid Sp	*Restores devastated Yang *Disperses cold *Warms channel *Relieves pain	*Diarrhea/cold limbs/minute pulse from Kidney and Spleen Yang deficiency *Impotence/ spermatorrhea/ edema due to weak Mingmen fire *Bi syndrome with cold predominance *Spontaneous sweating due to defensive Yang deficiency	1.5–9 g Toxic Decoct for one hour before adding other herbs C/I during pregnancy
Gan Jiang	Dried Ginger	Hot Acrid	Ht Lu Sp ST	*Warms Middle Jiao *Expels interior cold *Rescues devastated Yang *Warms Lung *Warms channel *Stops bleeding	*Diarrhea/vomiting/ poor appetite/cold limbs due to Middle Yang deficiency with cold *For devastated Yang with Fu Zi *Cold-phlegm in Lung *Spontaneous sweating due to defensive Yang deficiency	3–12 g Use caution during pregnancy
Gao Liang Jiang	Lesser Galangal Rhizome	Hot Acrid	Sp ST	*Warms Middle Jiao *Relieves pain	*Vomiting/nausea/ hiccups/abdominal pain/diarrhea due to Stomach cold	3–9 g Used with Dang Shen and Bai Zhu for patients with weak constitution
Guan Gui	Young Cinnamon Bark	Hot Acrid Sweet	Ht Kid Liv Sp	*Warms Kidney *Fortifies Yang *Disperses interior cold *Warms channel *Relieves pain	*Weak legs/lower back pain/ impotence/cold limbs due to Kidney Yang deficiency *Abdominal pain with 5 A.M. diarrhea due to Kidney and Spleen Yang deficiency *Amenorrhea/ dysmenorrheal/ infertility from Mingmen fire fading *Cold below/heat above use with Huang Lian	1–2 g in powder Milder than Rou Gui
Hu Jiao	Black Pepper	Hot Acrid	LI ST	*Warms Middle Jiao *Disperses cold *Promotes appetite	*Vomiting/abdominal pain/diarrhea due to Stomach cold	2–4.5 g Up to 30 g for topical use
Mu Ding Xiang	Clove Fruit	Warm Acrid	Sp ST Kid	*Warms Middle Jiao *Directs rebellious Qi downward *Warms Kidney	*Hiccups due to Stomach cold *Nausea/vomiting/ diarrhea *Impotence/excessive clear leukorrhea due to Kidney Yang deficiency	2–5 g Do not use with Yu Jin Milder than Ding Xiang

(continued)

Herb Name	Common Name	Temperature & Tastes	Channels	Functions	Indications	Dosages & Notes
Rou Gui	Cinnamon Bark	Hot Sweet Acrid	Ht Kid Liv Sp	*Warms Kidney *Fortifies Yang *Disperses interior cold *Warms channel *Relieves pain	*Weak legs/lower back pain/ impotence/cold limbs due to Kidney Yang deficiency *Abdominal pain with 5 A.M. diarrhea due to Kidney and Spleen Yang deficiency *Amenorrhea/ dysmenorrhea/ infertility from Mingmen fire fading *Cold below/heat above use with Huang Lian	2–5 g in powder or pill Short cooking time Use caution during pregnancy
Wu Zhu Yu	Evodia Fruit	Hot Acrid Bitter	Kid Liv Sp ST	*Warms Middle Jiao *Disperses cold *Stops diarrhea	*Nausea/abdominal pain/Jue Yin headache/hernia-like pain due to cold or phlegm obstruction in Stomach and Liver *Hypochondriac/ costal pain/acid reflux due to Liver and Stomach disharmony *Chronic diarrhea due to Kidney and Spleen Yang deficiency *Pinworms	2–5 g Sl toxic Overdose may lead to visual disturbance and hallucinations
Xiao Hui Xiang	Fennel Fruit	Warm Acrid	Liv Kid Sp ST	*Warms Kidney *Spreads Liver Qi *Expels cold *Harmonizes Stomach	*Hernia-like pain/ testis disorders in kids due to cold accumulation in Liver meridian *Vomiting/poor appetite from Stomach cold *Lower abdominal/ hernia pain with testis recession (warm topical)	3–9 g

CHAPTER 39

Tonic Herbs

QI TONIC HERBS

Herb Name	Common Name	Temperature & Tastes	Channels	Functions	Indications	Dosages & Notes
Bai Zhu	Large-headed Atractylodes Rhizome	Warm Bitter Sweet	Sp ST	*Tonifies Spleen (toasted) *Dries damp (raw) *Stops sweating (raw) *Calms fetus *Stops diarrhea (charred)	*Diarrhea/fatigue/poor appetite/vomiting/poor digestion due to Spleen deficiency *Edema with Damp Lin syndrome due to Spleen Qi deficiency *Spontaneous sweating due to Wei Qi deficiency	5–15 g
Bian Dou (*Bai Bian Dou*)	White Hyacinth Dolichos Bean	Sl Warm Sweet	Sp ST	*Tonifies Spleen (toasted) *Transforms summer heat damp (raw) *Stops diarrhea (toasted)	*Poor appetite/loose stool/vomiting/nausea/epigastric distension due to Middle Jiao deficiency with damp *Vomiting/loose stool in summer season in children	10–20 g
Bian Dou Hua	Hyacinth Flower	Sl Cool Sweet Aromatic	Sp ST	*Cools summer heat *Transforms damp	*Irritability/chest distension/nausea/vomiting/diarrhea/feverish sensation due to summer heat	6–12 g Up to 30 g when freshly supplied
Bian Dou Yi	Hyacinth Bean Peel	Sl Warm Sweet	Sp ST	*Strengthens Spleen *Transforms damp	*Loose stool/edema due to Spleen deficiency	6–12 g Milder than Bian Dou
Ci Wu Jia	Siberian Ginseng	Warm Acrid	Liv Kid	*Boosts Qi *Disperses wind–damp *Calms Liver wind	*Fatigue *Lower back pain/arthritis *Dizziness/blurred vision	9–15 g
Da Zao (*Hong Zao*)	Chinese Date	Sl Warm Sweet	Sp ST	*Tonifies Spleen *Nourishes blood *Calms spirit *Protects Stomach	*Poor appetite/fatigue/diarrhea due to the Middle Jiao deficiency *Restless Organ Syndrome (Zang Zao) *Allergic purpura (large dose for long term) *Neutralizes harshness of other herbs	10–30 g Remove peel and core

(continued)

QI TONIC HERBS (continued)

Herb Name	Common Name	Temperature & Tastes	Channels	Functions	Indications	Dosages & Notes
Dang Shen	Codonopsis Root	Neutral Sweet	Lu Sp	*Tonifies Lung *Supports Middle Jiao *Generates fluid	*Fatigue/poor appetite/diarrhea *Chronic cough/clear copious sputum/shortness of breath *Prolapse of Stomach/uterus/anus *Wasting and thirsting syndrome (Xiao Ke) *Leukopenia due to radiation therapy and chemotherapy	10–30 g May be used as a substitute for Ren Shen
Gan Cao	Raw Licorice Root	Neutral Sweet	All 12 Ht Lu Sp ST	*Tonifies Spleen *Moistens Lung *Clears heat *Reduces toxin *Smoothes spasm *Relieves pain	*Irregular/intermittent pulse due to Spleen, Heart, Lung Qi, and blood deficiency *Cough/wheezing *Sore throat/carbuncles *Antidote to toxic substances *Muscle spasms in abdomen and lower extremities *Neutralizes harshness of other herbs	3–12 g Up to 30 g for Addison's disease Not compatible with Da Ji, Yuan Hua, Gan Sui, and Hai Zao
Huang Jing	Siberian Solomon Seal Rhizome	Neutral Sweet	Kid Lu Sp	*Tonifies Kidney *Boosts Spleen *Moistens Lung	*Chronic dry cough due to Lung Qi/Yin deficiency *Lower back pain/weak legs/dizziness/chronic wasting disease due to Kidney Jing deficiency *Poor appetite/lassitude/weak pulse/dry mouth/no taste in mouth/difficult bowel movement due to Spleen Qi & Stomach Yin deficiency	10–20 g in dried form 30–60 g when freshly supplied
Huang Qi	Milk-Vetch Root	Sl Warm Sweet	Lu Sp	*Tonifies Spleen *Benefits Qi (toasted) *Ascends Yang Qi (toasted) *Boosts Wei Qi *Stops sweating *Promotes urination *Generates flesh	*Fatigue/diarrhea/poor appetite due to Spleen Qi deficiency *Prolapse of uterus/stomach/anus/heavy menstrual bleeding due to Yang Qi sinking *Spontaneous sweating/shortness of breath/catching colds frequently due to Wei Qi deficiency *Difficulty urination	10–15 g 30–60 g for severe cases such as paralysis
Mian Hua Gen	Cotton Plant Root	Warm Sweet	Lu Sp	*Tonifies Qi *Ascends Yang *Stops cough *Calms wheezing	*Prolapse of uterus/stomach/anus/heavy menstrual bleeding due to Yang Qi sinking *Fatigue/diarrhea/poor appetite/edema due to Spleen Qi deficiency *Cough/wheezing due to Lung Qi deficiency	15–30 g

(continued)

QI TONIC HERBS (continued)

Herb Name	Common Name	Temperature & Tastes	Channels	Functions	Indications	Dosages & Notes
Ren Shen	Ginseng Root	Sl Warm Sweet Sl Bitter	Lu Sp Kid	*Strongly tonifies Yuan Qi *Tonifies Lung *Boosts Spleen *Benefits Yin *Generates fluid *Calms spirit	*Collapse of Yang/Yin/ exhaustion of Qi and blood and other critical cases *Shortness of breath/ wheezing/labored breathing with exertion due to Lung and Kidney Qi deficiency *Wasting and thirsting syndrome (Xiao Ke) *Insomnia/palpitation/ anxiety/forgetfulness/ restlessness due to Heart Qi and blood deficiency *Immune booster for Wei Qi deficient patients	3–10 g 15–30 g for severe cases 1–2 g in powder Decoct separately Not compatible with Li Lu
Ren Shen Ye	Ginseng Leaf	Neutral Sweet Sl Bitter	Lu Sp Kid	*Nourishes Stomach Yin *Relieves summer heat *Clears deficient heat	*Thirst/dry mouth constipation/poor appetite/sweating due to Yin injury and Qi deficiency *Adrenal gland exhaustion	9–15 g
Shan Yao	Chinese Yam	Neutral Sweet	Kid Lu Sp	*Tonifies Lung *Boosts Spleen/ Stomach *Supports Kidney	*Diarrhea/fatigue/ spontaneous sweating/ poor appetite due to Spleen and Stomach deficiency (toast) *Wasting and thirsting syndrome (Xiao Ke) (raw) *Frequent urination/ leukorrhea/nocturnal emission due to Kidney and Lung Qi deficiency *Chronic cough/ wheezing due to Lung Qi deficiency *Lowers blood sugar for diabetes patient if used long term	10–30 g 60–250 g for severe cases 6–10 g in powder
Tai Zi Shen (*Hai Er Shen*)	Caryophyllaceous Root	Neutral Sweet Sl Bitter	Lu Sp	*Strengthens Spleen *Tonifies Lung *Benefits Qi *Generates fluid	*Lassitude/poor appetite/ excessive sweating *Thirst/dryness after febrile disease/summer heat	10–15 g Up to 30 g for severe cases Stronger Yin nourishing effect
Yi Tang	Barley Malt Sugar	Sl Warm Sweet	Lu Sp ST	*Tonifies Spleen *Moistens Lung *Relieves pain *Resolves cough	*Chronic abdominal pain with excess salivation due to Middle Jiao deficiency cold *Chronic nonproductive cough with shortness of breath/weak voice due to Lung Qi and Yin deficiency	30–60 g Dissolve in strained decoction Overdose leads to abdominal bloating and distension

(continued)

QI TONIC HERBS (continued)

Herb Name	Common Name	Temperature & Tastes	Channels	Functions	Indications	Dosages & Notes
Zhi Gan Cao	Prepared Licorice Root	Warm Sweet	All 12 Ht Lu Sp ST	*Tonifies Spleen *Moistens Lung *Clears heat *Reduces toxin *Smoothes spasm *Relieves pain	*Irregular/intermittent pulse due to Spleen, Heart, Lung Qi, and blood deficiency *Cough/wheezing *Sore throat/carbuncles due to fire poison *Antidote to toxic substances *Muscle spasms in abdomen and lower extremities * Neutralizes harshness of other herbs	2–9 g
Zhi Huang Qi	Prepared Milk-Vetch Root	Warm Sweet	Lu Sp	*Tonifies Spleen (raw) *Benefits Qi *Ascends Yang Qi *Boosts Wei Qi (raw) *Stops sweating (raw) *Promotes urination (raw) *Generates flesh (raw)	*Edema due to Qi deficiency *Chronic ulcers and sores with pus *Postpartum fever/recovery stage due to severe loss of blood and/or Qi *Wasting and thirsting syndrome (Xiao Ke) *Heart failure *Supportive herb for cancer patients undergoing radiation therapy and chemotherapy	10–15 g

BLOOD TONIC HERBS

Herb Name	Common Name	Temperature & Tastes	Channels	Functions	Indications	Dosages & Notes
Bai Shao Yao	White Peony Root	Cool Bitter Sour	Liv Sp	*Nourishes blood (toasted) *Regulates menses *Calms Liver Yang (raw) *Relieves pain *Harmonizes Ying/Wei (toasted) *Softens/soothes Liver	*Menstrual disorders *Leukorrhea *Flank/chest/abdominal pain due to Liver Qi stagnation *Spontaneous/night sweating *Nocturnal emission *Spasm/cramping/headache/dizziness	6–30 g Not compatible with Li Lu Bai Shao is a Yin herb among blood tonic herbs

(continued)

BLOOD TONIC HERBS (continued)

Herb Name	Common Name	Temperature & Tastes	Channels	Functions	Indications	Dosages & Notes
Dang Gui	Chinese Angelica Root	Warm Sweet Acrid Bitter	Ht Liv Sp	*Tonifies blood (main root) *Invigorates blood (root end) *Harmonizes blood (whole root) *Moistens Intestine *Reduces swelling *Relieves pain *Generates flesh	*Menstrual disorders *Pale face/tinnitus/ blurred vision/pale nails/palpitations due to blood deficiency *Abdominal pain/ postpartum/ miscarriage/ postsurgical pain/ trauma with swellings and sores due to blood stasis *Bi syndrome due to blood deficiency/ chronic wind–damp accumulation *Constipation due to blood deficiency	5–15 g Toasted with wine will strongly move blood Use cautiously with patients who have loose stools Dang Gui is Yang herb among blood tonic herbs
E Jiao (*Lu Pi Jiao*)	Donkey-Hide Gelatin	Neutral Sweet	Kid Liv Lu	*Nourishes blood *Moistens Lung (prepared with Ge Jie) *Stops bleeding (prepared with Pu Huang) *Nourishes Yin	*Palpitation/dizziness/ uterine bleeding/ hemoptysis/hemafecia due to Yin/blood deficiency *Vaginal bleeding during pregnancy *Postabortion/ miscarriage bleeding and spotting *Dry cough/insomnia/ thirst/irritability/dry mouth due to Yin deficiency	3–15 g Dissolve and take with wine Use caution in patients with Spleen and Stomach Deficiency
Gou Qi Zi	Chinese Wolfberry Fruit	Neutral Sweet	Liv Kid	*Tonifies Liver/ Kidney *Brightens eye *Benefits Jing *Enriches Yin	*Sore and weak lower back/knees/legs *Impotence/nocturnal emission/dizziness/ tinnitus due to Liver blood and Kidney Jing damage *Blurred vision/ decreased visual acuity due to Liver and Kidney deficiency *Wasting and thirsting syndrome (Xiao Ke)	6–15 g
He Shou Wu	Fleece flower Root	Sl Warm Bitter Sweet	Liv Kid Ht	*Tonifies Liver/ Kidney (toasted) *Nourishes blood (toasted) *Benefits Jing (toasted) *Moistens Intestine (raw) *Detoxifies fire poison (raw)	*Insomnia/Alzheimer's disease/hair loss/ premature gray hair/ early aging due to Liver blood and Kidney Jing deficiency *Malaria/scrofula/goiter/ neck swellings due to fire poison *Skin rash/carbuncles	10–30 g Use fresh quality for best result in treating constipation and fire poison

(continued)

BLOOD TONIC HERBS (continued)

Herb Name	Common Name	Temperature & Tastes	Channels	Functions	Indications	Dosages & Notes
Ji Xue Teng	Leatherleaf Millettia Root/Vine	Warm Sweet Acrid	Ht Liv Sp	*Tonifies blood *Moves blood *Invigorates channel *Promotes movement of Qi	*Irregular menses/ amenorrhea due to blood deficiency *Leukopenia caused by radiation therapy in cancer patients *Lower back pain/knee pain/numbness/ paralysis *Chronic Bi syndrome due to blood deficiency and collateral blockage	9–30 g Weaker than Dang Gui in functions
Long Yan Rou (*Gui Yuan Rou*)	Flesh of Longan Fruit	Warm Sweet	Ht Sp	*Tonifies Heart/ Spleen *Nourishes blood *Calms spirit	*Insomnia/palpitations/ forgetfulness/dizziness/ blurred vision due to Heart and Spleen deficiency *Deficiency after major illness/postpartum and excessive mental work	6–15 g Up to 30 g
Sang Shen	Bud of Mulberry Fruit	Sl Cold Sweet	Ht Liv Kid	*Tonifies blood *Enriches Yin *Lubricates Intestine	*Dizziness/insomnia premature gray hair/ tinnitus/irritability due to Liver/Kidney deficiency *Thirst/dry mouth (Xiao Ke) *Constipation due to blood or Yin deficiency	10–15 g Up to 30 g for severe cases
Shu Di Huang	Cooked Rehmannia Root	Sl Warm Sweet	Ht Kid Liv	*Tonifies blood (chief herb) *Nourishes Yin *Benefits Jing *Stops bleeding (charred)	*Pale face/dizziness/ insomnia/palpitation/ irregular menses/ uterine bleeding/ postpartum bleeding due to blood deficiency *Five-center heat/night sweat/nocturnal emission/steaming bones syndrome/ wasting and thirsting syndrome (Xiao Ke) due to Kidney Yin/Jing deficiency	10–30 g Prepared with Sha Ren and Chen Pi to reduce cloying quality Use caution in patients with Spleen deficiency
Zhe Shi Zi	Paper Mulberry Fruit	Neutral Sweet	Liv Kid Sp	*Tonifies Liver/ Kidney *Brightens eyes *Promotes urination	*Lower back pain/ weakness of knee/ impotence/dizziness/ blurred vision due to Liver/Kidney deficiency *Edema with difficulty urination due to Kidney Qi deficiency	9–15 g

YANG TONIC HERBS

Herb Name	Common Name	Temperature & Tastes	Channels	Functions	Indications	Dosages & Notes
Ba Ji Tian	Indian Mulberry Root	Sl Warm Acrid Sweet	Kid Liv	*Tonifies Kidney *Fortifies Yang (mildly) *Strengthens sinews/ bones *Expels wind–damp–cold	*Impotence/premature ejaculation/nocturnal emission/frequent urination/nocturia/ urinary incontinence/ sore lower back *Paralysis with muscular atrophy *Bi syndrome in lower part of body	10–15 g
Bu Gu Zhi (*Po Gu Zhi*)	Fruit of Scuffy Pea	Warm Acrid Bitter	Kid Sp	*Tonifies Kidney *Fortifies Yang *Benefits Jing *Warms Spleen Yang *Stops diarrhea	*5 A.M. diarrhea with borborygmy and abdominal pain *Impotence/premature ejaculation/nocturnal emission/frequent urination/nocturia/ urinary incontinence/ sore lower back due to Kidney Yang deficiency *Wheezing due to Kidney unable to grasp Lung Qi	3–10 g Used in tincture form to treat tinea, alopecia, beriberi, psoriasis, and vitiligo
Dong Chong Xia Cao (*Chong Cao*)	Cordyceps	Warm Sweet	Lu Kid	*Tonifies Lung/ Kidney *Fortifies Yang *Nourishes Lung Yin	*Impotence/nocturnal emission *Cough/wheezing/night sweat/weakness after major illness *Chronic cough with blood-tinged sputum *Strengthens endurance in athletes	5–15 g 1.5–3 g in powder or taken in broth (cooked with meat)
Du Zhong	Eucommia Bark	Warm Sl Acrid Sweet	Kid Liv	*Tonifies Liver/ Kidney *Strengthens sinews/bones *Promotes circulation *Calms fetus	*Weak/sore/painful lower back and knees *Fatigue/frequent urination/ spermatorrhea Liver and Kidney Yin deficiency *Weakness of sinew/ bone due to poor circulation *Uterine bleeding during pregnancy due to Kidney cold deficiency *Preventively used for miscarriage when fetus is restless or agitated *Dizziness/tinnitus due to hypertension due to Kidney and Liver Yin deficiency	10–15 g; up to 30 g Prepared form has stronger effects
E Guan Shi	Tip of Tubular Stalactites	Warm Sweet	Lu	*Strengthens Yang *Transforms phlegm *Benefits Qi	*Cough/wheezing due to Yang deficiency with phlegm *Insufficient lactation	9–15 g Boil for 30 min then add other herbs

(continued)

YANG TONIC HERBS (continued)

Herb Name	Common Name	Temperature & Tastes	Channels	Functions	Indications	Dosages & Notes
Ge Jie	Gecko	Neutral Salty	Lu Kid	*Benefits Kidney *Tonifies Lung *Arrests wheezing *Fortifies Yang *Benefits Jing	*Wheezing/cough with blood streaked sputum due to Kidney unable to grasp Qi (Use whole Gecko) *Impotence/5 A.M. diarrhea/frequent urination/nocturia due to Kidney Yang deficiency (Use Gecko tail)	9–15 g in decoction 1–2 g in powder or pill Usually used in pairs
Gou Ji (*Jing Mao Gou Ji*)	Chain Fern Rhizome	Warm Bitter Sweet	Kid Liv	*Tonifies Liver/ Kidney *Strengthens sinew/bone *Expels wind–damp	*Weak and sore lower back/spine/lower limbs due to Liver/Kidney deficiency *Urinary incontinence and leukorrhea due to Kidney deficiency *Pain/stiffness/ numbness due to Bi syndrome *Leg edema as an aftermath of illness	10–15 g
Gu Sui Bu (*Hou Jiang*)	Drynaria Rhizome	Warm Bitter	Kid Liv	*Tonifies Kidney *Mends sinews/ bones *Stimulates hair growth	*Weak lower back/knees *Diarrhea/tinnitus/ hearing loss/toothache/ loose teeth/bleeding gums due to Kidney deficiency *Hair loss/baldness (tincture) *Regains strength during healing of fractures/ sprains and ligament damage	10–20 g Topically used as a paste
Hai Gou Shen	Male Seal Sexual Organ	Hot Salty	Kid Liv	*Strengthens Yang *Nourishes Jing	*Impotence/low sexual drive/cold limbs due to Kidney Yang deficiency	3–15 g in powder
Hai Long	Pipe-Fish	Sl Warm Sweet Salty	Kid	*Tonifies Kidney *Strengthens Yang *Reduces swelling	*Impotence/debility in elderly patients *Scrofula/unknown lumps	3–9 g 1.5–2 g in powder or pill C/I during pregnancy Stronger than Hai Ma
Hai Ma	Seahorse	Warm Sweet Salty	Kid Liv	*Tonifies Kidney *Fortifies Yang *Invigorates blood	*Impotence/urinary incontinence/wheezing/ debility due to Kidney Yang and Qi deficiency *Bleeding with pain due to blood stagnation *Swelling due to sores/ boils	5–12 g Steeped in wine C/I during pregnancy
Hai Shen	Sea Cucumber	Warm Salty	Ht Kid	*Tonifies Kidney *Fortifies Yang *Benefits Jing	*Debility/impotence/ nocturnal emission/ urinary frequency/ constipation due to Kidney Jing/Yin/blood deficiency	10–30 g 1–3 g in powder
Hu Lu Ba	Fenugreek Seed	Warm Bitter	Kid Liv	*Warms Kidney *Disperses damp *Relieves pain	*Radiating hernia-like pain due to cold accumulation *Hypochondriac/ abdominal pain	3–10 g

(continued)

YANG TONIC HERBS (continued)

Herb Name	Common Name	Temperature & Tastes	Channels	Functions	Indications	Dosages & Notes
Hu Tao Ren (*Hu Tao Rou*)	Walnut Nut	Warm Sweet	Kid LI Lu	*Tonifies Kidney *Aids Kidney to grasp Qi *Holds Lung Qi *Arrests cough/ wheezing (with peel) *Moistens Intestine (without peel)	*Painful lower back and knees *Nocturnal emission/ frequent urination due to Kidney Yang deficiency *Chronic cough/ wheezing due to Kidney and Lung Qi deficiency *Constipation in elderly patients *Eczema/skin rash (charred) *Urinary stones/early aging with premature gray hair *Slow development in children	10–30 g Use caution in patients with loose stools
Ji Dai (*Kan Qi*)	Human Umbilical Cord	Warm Sweet Salty	Liv Lu Kid	*Tonifies Kidney *Aids Kidney to grasp Qi *Holds in excessive sweat	*Cough/wheezing/night sweat due to Kidney Yin and Jing deficiency *Catch colds frequently due to Wei Qi deficiency	1–2 cords per day; 1.5–3 g in powder or pill 2–3 times per day Boil with Gan Cao, Jin Yin Hua, and wine, then dry
Jiu Xiang Chong (*Qiang Lang Chong*) (*Da Pi Chong*)	Dung Beetle	Warm Salty	Sp Kid Liv	*Tonifies Kidney *Assists Yang *Warms Spleen Yang *Regulates Qi *Stops pain	*Weak/sore lower back/ knees/urinary frequency due to Kidney Yang deficiency *Epigastric bloating/pain due to Spleen Yang deficiency *Flank/lower abdominal pain	3–9 g 1–2 g in powder
Jiu Zi	Seeds of Fragrant-Flowered Garlic	Warm Sweet Acrid	Kid Liv	*Warms Kidney *Fortifies Yang *Secures Jing *Stops vomiting	*Impotence/weak and painful lower back/ knees due to Kidney/ Liver deficiency *Spermatorrhea/urinary frequency/excessive vaginal discharge due to Kidney Qi deficiency *Vomiting due to Stomach cold	5–10 g
Lu Jiao	Deer Horn	Warm Salty	Liv Kid	*Tonify Kidney *Warms Yang	*Impotence/cold limbs/ sore lower back/weak knees/frequent urination due to Kidney Yang deficiency *Delayed mental and physical development in chlildren *Chronic wounds/ ulceration *Uterine bleeding/ leukorrhea	5–10 g Use as a substitution for Lu Rong Milder than Lu Rong

(continued)

YANG TONIC HERBS (continued)

Herb Name	Common Name	Temperature & Tastes	Channels	Functions	Indications	Dosages & Notes
Lu Jiao Jiao	Deer Horn Glue	Sl Warm Sweet Salty	Liv Kid	*Tonifies Liver/ Kidney *Benefits essence/ blood *Stops bleeding *Stabilizes Ren/ Du Mai	*Impotence/cold limbs/ sore lower back/weak knees/frequent urination due to Kidney Yang deficiency *Delayed mental and physical development in children *Chronic wounds/ ulceration *Uterine bleeding/ leukorrhea	5–10 g Use as a substitution for Lu Rong Milder than Lu Rong Used with Gui Ban Jiao to tonify Yin and Yang
Lu Jiao Shuang	Dregs leftover after making Lu Jiao Jiao	Sl Warm Sweet	Liv Kid	*Tonifies Kidney *Fortifies Yang *Benefits essence/ blood	*Nocturnal emission *Leukorrhea/uterine bleeding due to deficiency cold *Trauma bleeding (topical)	10–30 g
Lu Rong	Velvet of Young Deer Horn	Warm Sweet Salty	Kid Liv	*Tonifies Kidney *Strengthens sinew/bone *Nourishes Jing/ blood *Strengthens Du/ Ren/ Dai and Chong Mai	*Impotence/cold limbs/ sore lower back/weak knees/frequent urination due to Kidney Yang deficiency *Delayed mental and physical development in children *Chronic wounds/ ulceration *Uterine bleeding/ leukorrhea	1–3 g in powder 2–3 times per day Also can be soaked in wine
Rou Cong Rong	Fleshy Stem of Broomrape	Warm Sweet Salty	LI Kid	*Tonifies Kidney *Benefits essence *Strengthens Yang *Warms uterus *Moistens Intestine	*Spermatorrhea/ impotence/urinary incontinence/post urination dribbling *Sore lower back/weak knees *Infertility/abnormal uterine bleeding/ leukorrhea due to deficiency cold womb and Jing deficiency *Elderly constipation/ post major chronic illness and leukorrhea due to Kidney Yang deficiency	10–20 g Tonifies Yang without drying Large doses can be used for long duration because effect is mild Use caution in patients with loose stools
Sha Yuan Ji Li	Milk-Vetch Seed	Warm Sweet	Kid Liv	*Fortifies Yang *Secures Jing *Tonifies Liver/ Kidney *Brightens eyes	*Poor/blurred vision due to Liver/Kidney deficiency *Lower back pain/ tinnitus/spermatorrhea/ premature ejaculation/ frequent urination/ urinary incontinence/ leukorrhea due to Kidney Yang deficiency	10–20 g C/I in patients with high sex drive and difficult urination due to Yin deficiency

(continued)

YANG TONIC HERBS (continued)

Herb Name	Common Name	Temperature & Tastes	Channels	Functions	Indications	Dosages & Notes
Suo Yang	Chinese Cynomorium	Warm Sweet	LI Kid Liv	*Tonifies Kidney *Fortifies Yang *Nourishes blood *Moistens Intestine	*Cold sensation and pain in lower back and knees *Spermatorrhea/ impotence/urinary incontinence/post urination dribbling *Infertility/abnormal uterine bleeding/ leukorrhea due to deficiency cold womb and Jing deficiency *Elderly constipation due to Qi and blood deficiency	10–15 g
Tu Si Zi	Chinese Dodder Seeds	Neutral Acrid Sweet	Kid Liv	*Tonifies Kidney/ Liver *Benefits Jing *Brightens eyes *Stops diarrhea *Calms fetus	*Nocturnal emission/ impotence/premature ejaculation/frequent urination *Leukorrhea/tinnitus/ blurred vision/eye floaters *Chief herb for habitual or threatened miscarriage *Loose stools/diarrhea with lack of appetite *Vitiligo (soaked in wine) (topical)	10–15 g Noncloying and neutral Yang tonic
Xian Mao	Rhizome of Golden Eye-Grass	Hot Acrid	Kid Liv	*Tonifies Kidney *Fortifies Yang *Expels cold–damp	*Infertility due to cold womb (female) or cold Jing (male) *Impotence/ spermatorrhea/urinary incontinence/nocturnal emission *Lower back pain/weak knees due to cold–damp Bi syndrome	3–10 g Sl toxic
Xu Duan	Japanese Teasel Root	Sl Warm Sweet Bitter Acrid	Kid Liv	*Tonifies Liver/ Kidney *Strengthens sinew/bone *Stops uterine bleeding (toasted in vinegar) *Calms fetus *Moves blood *Relieves pain *Generates flesh	*Painful/sore lower back *Weak knees/stiff joints *Pain and swelling due to trauma (topical) *Leukorrhea *Restless fetus/uterine bleeding during pregnancy/threatened miscarriage due to Kidney/Chong/Ren deficiency	10–20 g
Yang Qi Shi	Actinolite	Sl Warm Salty	Kid	*Warms Kidney/ Uterus	*Impotence/infertility/ cold in Lower Jiao *Cold sensation in lower back and knees	3–6 g in pill or powder Short-term usage

(continued)

YANG TONIC HERBS (continued)

Herb Name	Common Name	Temperature & Tastes	Channels	Functions	Indications	Dosages & Notes
Yin Yang Huo (*Xian Ling Pi*)	Longspur Epimedium	Warm Acrid Sweet	Kid Liv	*Tonifies Kidney *Fortifies Yang *Enriches Yin *Expels wind–damp–cold *Stops cough/ wheezing	*Impotence/ spermatorrhea/ frequent urination/lower back and knee pain *Wind–cold–damp Bi syndrome *Dizziness/irregular menses *Cough/wheezing with clear copious sputum due to Yang deficiency *Hypertension arisen with menopause	10–15 g Soaked in wine for 10 d take 10–20 g 1–2 times daily to treat Bi syndrome and impotence
Zi He Che (*Tai Yi*) (*Tai Pan*)	Human Placenta	Warm Sweet Salty	Liv Lu Kid	*Tonifies Liver/ Kidney *Augments Qi *Nourishes blood *Benefits Jing	*Night sweats/ emaciation/debility/ impotence/infertility/ habitual miscarriage/ insufficient lactation due to Qi/blood/Yin/ Yang/Jing deficiency *Decreased appetite/ poor digestion *Tinnitus/dizziness/lower back pain *Delayed mental and physical development in children	1.5–5 g in powder 2–3 times daily Double the dosage in severe cases Fresh Zi He Che can be washed and made into soup

YIN TONIC HERBS

Herb Name	Common Name	Temperature & Tastes	Channels	Functions	Indications	Dosages & Notes
Bai He	Lily Bulb	Sl Cold Sl Bitter Sweet	Ht Lu	*Moistens Lung *Clears heat *Clears Heart *Calms spirit *Stops cough	*Dry cough/sore throat due to Lung heat *Insomnia/restlessness/ irritability/low-grade fever/palpitations due to Qi and Yin deficiency or aftermath of a febrile disease	10–30 g C/I with cough due to wind–cold or phlegm
Bai Mu Er	Fruiting Body of Tremella	Neutral Sweet Bland	Lu ST	*Tonifies Lung/ Stomach *Nourishes Yin *Generates fluid	*Dry cough with blood-tinged sputum due to Lung heat *Five-center heat emaciation due to Yin deficiency with Yang rising	3–10 g Soak herb for 1–3 hours, then cook in soup until soft
Bie Jia	Chinese Soft-Shelled Turtle Shell	Sl Cold Salty	Liv Sp Kid	*Nourishes Yin (raw) *Anchors Yang (raw) *Moves blood *Promotes menses *Dissipates nodule (prepared with vinegar)	*Steaming bones syndrome/night sweats/ fever due to Yin deficiency *Excessive menstrual bleeding due to heat in blood *Amenorrhea with pain in hypochondriac region with palpable mass: (Liver cirrhosis or hepatosplenomegal)	10–30 g; decoct first. Use caution during pregnancy

(continued)

YIN TONIC HERBS (continued)

Herb Name	Common Name	Temperature & Tastes	Channels	Functions	Indications	Dosages & Notes
Gui Ban	Land Tortoise Shell	Sl Cold Salty Sweet	Ht Kid Liv	*Nourishes Yin *Anchors Yang *Tonifies Kidney *Strengthens bone *Cools blood *Stops bleeding *Tonifies Heart *Nourishes blood	*Night sweats/dizziness/ tinnitus/steaming bones syndrome due to Yin deficiency with Yang ascending *Hand/feet tremors/facial spasms due to internal wind *Sore lower back/weak knees/failure of fontanel closure and retarded skeletal growth in children *Beng Lou (prepared with vinegar)	10–30 g; decoct first Use caution during pregnancy
Gui Ban Jiao (*Gui Jiao*)	Glue of Land Tortoise Shell	Sl Cold Salty Sweet	Ht Kid Liv	*Nourishes Yin *Anchors Yang *Tonifies Kidney *Strengthens bone *Cools blood *Stops bleeding *Tonifies Heart *Nourishes blood	*Combined with Lu Jiao Jiao to reinforce Yin and Yang, Ren and Du	10–15 g; melt herb and dissolve into strained decoction Stronger than Gui Ban to nourish Yin
Han Lian Cao	Eclipta	Cool Sweet Sour	Kid Liv	*Tonifies Liver/ Kidney *Nourishes Yin *Cools blood *Stops bleeding	*Blurred vision/vertigo *Premature gray hair due to Liver/Kidney Yin deficiency *Reckless bleeding due to blood heat *Trauma bleeding (topical)	10–30 g dried 20–60 g fresh
Hei Zhi Ma	Black Sesame Seeds	Neutral Sweet	Kid Liv	*Tonifies Liver/ Kidney *Nourishes blood *Extinguishes wind *Moistens Intestine *Benefits essence	*Blurred vision/vertigo *Premature gray hair due to Liver/Kidney Yin deficiency *Constipation due to dry Intestines/blood deficiency *Slows down degenerative process (supplement)	10–30 g Use prepared Hei Zhi Ma Use caution in patients with loose stools
Ji Zi Huang	Chicken Egg Yolk	Neutral Sweet	Ht Kid	*Nourishes Yin *Builds blood	*Insomnia *Spasm convulsion *Irritability	1–2 yolks added into decoction
Luo Han Guo	Round Momordica Fruit	Neutral Sweet	Lu Sp	*Moistens Lung *Dissipates nodule	*Heat type cough due to Lung Yin deficiency *Scrofula due to phlegm accumulation	10–15 g
Luo Han Guo Ye	Round Momordica Leaf	Neutral Sweet	Lu Sp	*Moistens Lung *Dissipates nodule *Benefits throat	*Sore throat due to Lung Yin deficiency	10–15 g
Mai Men Dong (*Mai Men Dong*)	Tuber of Dwarf Lilyturf	Sl Cold Sweet Sl Bitter	Ht Lu ST	*Moistens Lung *Generates fluid *Enriches Stomach Yin *Clears Heart	*Irritability/thirst due to Lung and Stomach Yin deficiency due to febrile disease injuring fluids *Cough with thick sputum difficult to expectorate *Hemoptysis *Constipation/dry mouth/ insomnia due to Yin deficiency *Infantile constipation (mild)	6–15 g To tonify Lung and Stomach Yin, use Mai Men Dong without the center To tonify Heart Yin, clear Heart heat, use Mai Men Dong with the center

(continued)

YIN TONIC HERBS (continued)

Herb Name	Common Name	Temperature & Tastes	Channels	Functions	Indications	Dosages & Notes
Ming Dang Shen (*Fen Sha Shen*)	Changium Root	Sl Cold Sweet Sl Bitter	Lu ST	*Moistens Lung *Generates fluid *Harmonizes Stomach *Descends rebellion	*Cough/wheezing due to heat injuring Lung Yin *Nausea/vomiting/ digestive upset due to Stomach Yin deficiency/ Stomach Qi rebellion *Sore/swelling due to toxin accumulation	6–10 g
Nu Zhen Zi	Privet Fruit	Sl Cool Bitter Sweet	Kid Liv	*Tonifies Liver/ Kidney *Clears deficient heat *Improves vision	*Reduced visual acuity/ eye floaters/dizziness/ weak lower back/ premature gray hair/ tinnitus due to Liver/ Kidney Yin deficiency *Early stage cataract and central retinitis	10–15 g
Sang Ji Sheng	Mulberry Mistletoe	Neutral Bitter	Kid Liv	*Tonifies Liver/ Kidney *Strengthens sinew/bone *Expels wind–damp *Nourishes blood *Calms fetus	*Sore lower back/weak knees/joint stiffness/ numbness or atrophy of sinew/bone due to Liver/Kidney deficiency or Bi syndrome *Restless fetus/uterine bleeding during pregnancy due to Chong/Ren disharmony *Dry/scaly skin due to blood deficiency	10–20 g
Sha Shen	Root of Beech Silver-Top	Cool Sl Bitter Sweet	Lu ST	*Moistens Lung *Nourishes Stomach *Generates fluid *Clears heat	*Chronic dry nonproductive cough/ hoarseness due to Lung Yin deficiency *Dry mouth and throat/ constipation due to Stomach Yin *Dry and itchy skin due to cold/dry weather	10–15 g; 15–30 g if fresh supply Not compatible with Li Lu
Shi Hu	Dendrobium Stem	Sl Cold Sweet Bland	Kid ST	*Nourishes Yin *Clears heat *Generates fluids	*Severe thirst/parched mouth/intractable fever due to Yin deficiency due to fluids injury after febrile disease *Wasting and thirsting syndrome (Xiao Ke) *Decreased visual acuity/sore lower back/ weak knees due to Liver/Kidney deficiency *Yin deficiency heat arising due to menopause	10–20 g Up to 30 g if fresh supply Decoct first
Ta Gan	Otter Liver	Neutral Sweet Salty	Liv Kid	*Nourishes Yin *Clears deficient heat *Stops bleeding	*Consumptive diseases with night sweat *Anemia *Night blindness *Cough and wheezing	3–6 g in powder

(continued)

YIN TONIC HERBS (continued)

Herb Name	Common Name	Temperature & Tastes	Channels	Functions	Indications	Dosages & Notes
Tian Men Dong (*Tian Dong*)	Chinese Asparagus Tuber	Very Cold Sweet Bitter	Kid Lu	*Nourishes Kidney Yin *Clears Lung heat *Moistens Lung	*Dry mouth with thick or blood-tinged sputum, difficult to expectorate *Wasting and thirsting syndrome (Xiao Ke) with low-grade afternoon fever due to Lung/Kidney Yin deficiency *Constipation due to Yin fluid deficiency	6–24 g Use caution in patients with loose stools
Xi Yang Shen	American Ginseng Root	Sl Cool Sweet Sl Bitter	Ht Kid Lu ST	*Benefits Qi *Generates fluid *Nourishes Yin *Cools Lung	*Irritability/thirst/fatigue *Weakness/unabating fever/febrile disease/postpartum/post surgery or consumptive disease *Cough with blood/loss of voice due to Lung Yin deficiency *Supplementary boost for patients undergoing chemotherapy/radiation	2–10 g Decoct separately or take as powder
Yu Zhu	Solomon's Seal Rhizome	Sl Cold Sweet	Lu ST	*Nourishes Yin (prepared) *Clears heat (raw) *Extinguishes wind *Generates Stomach fluid (prepared) *Softens sinew	*Dry cough/dry throat/thirst *Steaming bones syndrome *Wasting and thirsting syndrome (Xiao Ke) with intense hunger/constipation due to deficient heat *Pain/spasms in sinews due to fluid deficiency leading to internal wind *Vertigo/dizziness due to internal wind/Meniere's disease *Angina pectoris with abnormal EKG diagnosed as congestive heart disease	10–30 g Use caution in patients with tachycardia or high blood pressure

CHAPTER **40**

Stabilizing and Binding Herbs

STOP SWEATING HERBS

Herb Name	Common Name	Temperature & Tastes	Channels	Functions	Indications	Dosages & Notes
Fu Xiao Mai	Light Wheat Grain	Cool Sweet Salty	Ht	*Stops excessive sweating *Nourishes Heart *Calms spirit	*Spontaneous sweat (raw) *Night sweat/excessive sweat (toasted) *Insomnia/palpitation/irritability/emotional instability due to Heart Shen disturbance (raw) *Childhood bed wetting	15–30 g
Ma Huang Gen	Chinese Ephedra Root	Neutral Sweet	Lu	*Stops sweating	*Sweating due to Qi/Yin deficiency or postpartum *Use powder form for topical use to stop sweating	3–10 g C/I in exterior conditions
Nuo Dao Gen Xu	Glutinous Rice Root	Neutral Sweet	Kid Liv Lu	*Stops sweating	*Spontaneous sweating/night sweats/thirst/fever due to Yin deficiency	15–60 g
Xiao Mai	Wheat Grain	Cool Sweet	Ht	*Nourishes Heart *Calms spirit	*Mood swings and Zang Zao symptoms in women	30–60 g

STOP DIARRHEA HERBS

Herb Name	Common Name	Temperature & Tastes	Channels	Functions	Indications	Dosages & Notes
Chi Shi Zhi	Halloysite	Warm Sweet Sour Astringent	Sp ST LI	*Stops diarrhea *Stops bleeding *Promotes healing of wound	*Chronic diarrhea with mucous/blood in stool due to cold deficiency *Uterine bleeding/rectal prolapse due to cold deficiency *Nonhealing ulcer (topical)	9–30 g Half dose taken in powder Use caution during pregnancy

(continued)

STOP DIARRHEA HERBS (continued)

Herb Name	Common Name	Temperature & Tastes	Channels	Functions	Indications	Dosages & Notes
Chun Pi (*Chun Gen Pi*) (*Chun Bai Pi*)	Bark of Heaven Tree	Cold Bitter Astringent	LI ST Liv	*Clears heat *Dries damp *Stops diarrhea *Kills parasite *Alleviates itch	*Chronic diarrhea/ dysentery with blood *Excessive vaginal discharge due to damp–heat *Hemorrhoid bleeding *Excessive uterine bleeding *Roundworm/tinea infection (topical wash or as a paste)	3–10 g Toast with wheat shell for stronger effects
He Zi (*He Li Le*)	Myrobalan Fruit	Neutral Bitter Sour Astringent	Lu ST LI	*Stops diarrhea *Restrains Lung Qi *Soothes throat Bi	*Chronic diarrhea/ dysentery with blood (toasted) *Chronic cough/ wheezing/loss of voice (raw) *Anal prolapse due to deficiency cold diarrhea	3–10 g
Ming Fan	Purified Alunite	Cold Sour Astringent	Lu LI Liv Sp ST	*Stops bleeding *Clears heat *Stops diarrhea *Reduces toxin *Alleviates itch *Kills insects	*Chronic diarrhea/bloody stool/Beng Lou *Trauma injuries causing bleeding (topical) *Oozing/itchy skin disease due to toxin/ parasite *Seizures due to wind–phlegm misting Heart *Jaundice due to Liver/Gallbladder toxic damp	0.6–3 g
Rou Dou Kou (*Rou Guo*)	Nutmeg Seed	Warm Acrid Aromatic	LI Sp ST	*Stops diarrhea *Warms Middle Jiao *Moves Qi *Relieves pain	*Chronic diarrhea/5 A.M. diarrhea due to Spleen/ Kidney deficiency cold *Epigastric/abdominal pain and distension/ vomiting/poor appetite due to Spleen/Stomach deficiency cold	3–10 g 1.5–3 g in powder Toast herb before decoction to reinforce warming the Middle Jiao and stop diarrhea
Shi Liu Gen Pi	Bark of Pomegranate Tree	Warm Sour Astringent	ST LI	*Kills parasite *Stops diarrhea *Stabilizes Kidney	*Tapeworm/roundworm	3–10 g Toxic
Shi Liu Pi	Pomegranate Husk	Warm Sour Astringent	Kid LI ST	*Stops diarrhea *Stops bleeding (charred) *Stabilizes Kidney *Kills parasite	*Chronic diarrhea/ dysentery/rectal prolapse due to deficiency cold *Tapeworm/roundworm	3–10 g Toxic Use powder or as external wash
Wu Bei Zi	Gallnut of Chinese Sumac	Cold Sour Salty Astringent	Kid LI Lu	*Restrains Lung Qi *Arrests cough *Stops diarrhea *Reduces swelling *Reduces toxin	*Chronic cough due to Lung deficiency *Chronic diarrhea/blood in stool/spermatorrhea/ rectal prolapse *Sores/ringworm/ poisonous swelling/ ulcerated lesion (topical wash) *Scar tissue (topical)	1.5–6 g in pill or powder

(continued)

STOP DIARRHEA HERBS (continued)

Herb Name	Common Name	Temperature & Tastes	Channels	Functions	Indications	Dosages & Notes
Wu Mei	Mume Fruit	Neutral Sour Astringent	LI Liv Lu Sp	*Restrains Lung Qi *Stops diarrhea (charred) *Generates fluid *Expels parasites *Stops bleeding (charred)	*Chronic cough/diarrhea/dysentery-like disorder *Thirst due to Qi/Yin deficiency *Roundworm-induced vomiting/abdominal pain *Excessive uterine bleeding/thirst/parched mouth due to blood deficiency *Corn/wart (topical plaster)	3–10 g Up to 30 g
Ying Su Ke (*Mi Ke*)	Opium Poppy Husk	Neutral Sour Astringent	Kid LI Lu	*Restrains Lung Qi *Stops cough (honey toasted) *Stops diarrhea (raw or toasted with vinegar) *Relieves pain (raw)	*Chronic cough/diarrhea/dysentery-like disorder *Polyuria/spermatorrhea/leukorrhea *Abdominal pain *Joint/sinew/bone pain *Rectal prolapse	3–10 g Toxic Do not use large amounts for long term
Yu Liang Shi (*Yu Yu Liang*)	Limonite	Neutral Sweet Astringent	LI ST Liv Kid	*Stops diarrhea *Stops bleeding	*Chronic intractable diarrhea and elderly diarrhea due to Kidney Yang deficiency and cold in Lower Jiao *Beng Lou/leukorrhea	6–24 g 3–12 g in powder Use caution during pregnancy Do not use large amounts for long term

ASTRINGE ESSENCE/URINE AND STOP LEUKORRHEA HERBS

Herb Name	Common Name	Temperature & Tastes	Channels	Functions	Indications	Dosages & Notes
Bai Guo (*Yin Xing*)	Ginkgo Nut	Neutral Sweet Bitter Astringent	Kid Lu	*Stabilizes Lower Jiao (prepared) *Expels phlegm (raw) *Stops wheezing (prepared) *Leaches damp	*Wheezing/cough with copious sputum *Leukorrhea/turbid urine due to damp–heat *Memory loss/early stage Alzheimer's and other degenerative disease	6–10 g Sl toxic In raw form use smaller dose to prevent poisoning
Bai Guo Ye (*Yin Xing Ye*)	Ginkgo Leaf	Neutral Sweet Sl Bitter	Ht Lu	*Moves Qi *Invigorates blood *Transforms phlegm	*High cholesterol *Hypertension *Angina pectoris	3–6 g
Ci Wei Pi	Hedgehog Skin	Neutral Bitter	ST LI Kid	*Stops bleeding *Astringes essence *Relieves pain *Moves blood	*Bloody stools *Hemorrhoid bleeding *Nocturnal emission and enuresis *Abdominal pain due to Qi and blood stagnation	3–10 g in decoction 1.5–3 g in powder

(continued)

ASTRINGE ESSENCE/URINE AND STOP LEUKORRHEA HERBS (continued)

Herb Name	Common Name	Temperature & Tastes	Channels	Functions	Indications	Dosages & Notes
Fu Pen Zi	Chinese Raspberry	Sl Warm Sweet Sour	Kid Liv	*Stabilizes Kidney *Retains essence *Assists Yang *Brightens eyes	*Frequent nocturia/ excessive wet dreams/ impotence/ spermatorrhea/ premature ejaculation due to Kidney Yang deficiency *Blurred vision/ decreased visual acuity due to Liver/Kidney deficiency	6–15 g
Hai Piao Xiao (*Wu Zei Gu*)	Cuttlefish Bone	Sl Warm Salty Astringent	Kid Liv ST	*Stops bleeding *Retains essence *Controls acidity *Dries damp *Stops diarrhea	*Uterine bleeding/ leukorrhea *Nocturnal emission/ premature ejaculation *Gastric pain/acid reflux/ belching due to Stomach Qi rebellion *Diarrhea/dysentery-like disorder with pain around umbilicus *Chronic/nonhealing skin ulceration (topical)	6–12 g 3–6 g in powder
Jin Ying Zi	Cherokee Rosehip	Neutral Sour Astringent	UB Kid LI	*Stabilizes Kidney *Retains essence *Stops diarrhea	*Spermatorrhea/urinary incontinence/leukorrhea due to Kidney deficiency *Chronic diarrhea/ dysentery-like disorder/ Beng Lou/uterine and anal prolapse	6–18 g C/I in excess heat pattern
Lian Xu	Lotus Stamen	Neutral Sweet Astringent	Ht Kid Sp	*Stabilizes Kidney *Retains essence	*Spermatorrhea/polyuria/ incontinence/leukorrhea due to Kidney deficiency	1.5–6 g
Lian Zi	Lotus Seed	Neutral Sweet Astringent	Ht Kid Sp	*Connects Heart/ Kidney *Tonifies Spleen/ Kidney *Stops diarrhea *Nourishes Heart *Calms spirit	*Irritability/insomnia/ palpitation/dry mouth/ dark yellow urine/wet dream *Chronic diarrhea/ dysentery-like disorders due to Spleen deficiency *Uterine bleeding/ leukorrhea	6–15 g Can be cooked in rice soup C/I in abdominal distension and impacted bowels
Qian Shi	Gordon Euryale Seed	Neutral Sweet Astringent	Kid Sp	*Strengthens Spleen *Stabilizes Kidney *Retains essence *Expels damp	*Chronic diarrhea/poor appetite/slow digestion due to Spleen deficiency *Spermatorrhea/polyuria/ nocturnal emission/ premature ejaculation due to Kidney deficiency *Leukorrhea due to Kidney deficiency or damp–heat	10–15 g Up to 30 g for severe cases Can be cooked in rice soup
Sang Piao Xiao	Praying Mantis Egg-Case laid on Mulberry Leaf	Neutral Sweet Salty Astringent	Kid Liv	*Tonifies Kidney *Retains essence	*Enuresis in children *Nocturnal emission without dreams due to Kidney Yang deficiency *Dribbling urine/polyuria/ leukorrhea/impotence due to Kidney Qi deficiency	3–10 g in powder or pill Raw herb can induce diarrhea

(continued)

ASTRINGE ESSENCE/URINE AND STOP LEUKORRHEA HERBS (continued)

Herb Name	Common Name	Temperature & Tastes	Channels	Functions	Indications	Dosages & Notes
Shan Zhu Yu (*Shan Yu Rou*) (*Zao Pi*)	Asiatic Cornelian Cherry Fruit	Sl Warm Sour	Kid Liv	*Stabilizes Kidney *Tonifies Liver *Stops excessive sweating *Stops bleeding	*Polyuria/urine incontinence/ spermatorrhea/ excessive sweating due to Kidney Qi deficiency *Painful lower back/weak knees/impotence/ reduced hearing acuity/ tinnitus and dizziness due to Liver/Kidney deficiency *Excessive sweating due to devastated Yang and collapsed Qi due to shock	5–12 g 30–60 g in cases of shock C/I in painful and difficult urination
Wu Wei Zi	Schisandra Fruit	Warm Sour	Ht Kid Lu	*Restrains Lung Qi *Tonifies Kidney (prepared with wine) *Retains essence *Inhibits sweating *Stops diarrhea *Generates fluid *Calms spirit	*Chronic cough/ wheezing due to Lung/ Kidney deficiency *Thirst/spontaneous and night sweat/palpitation due to heat damaging Yin fluid and Qi (dried fruit) *Nocturnal emission/ chronic diarrhea due to Spleen and Kidney deficiency *Palpitation/insomnia/ dream-disturbed sleep due to Yin deficiency (dried fruit) *Elevated SGOT/SGPT due to hepatitis	2–9 g 1.5–3 g in powder
Yi Zhi Ren	Galangal Fruit	Warm Acrid	Kid Sp	*Warms Kidney *Retains essence *Astringes saliva/ urine *Warms Spleen *Stops diarrhea	*Spermatorrhea/ premature ejaculation/ polyuria/nocturnal emission due to cold in Lower Jiao *Chronic diarrhea/ abdominal pain/excess salivation due to deficiency cold Spleen *Slow mental development in children	3–10 g C/I in heat type diarrhea, leukorrhea, and spermatorrhea

CHAPTER 41

Calming the Spirit Herbs

ANCHOR AND SETTLE THE SPIRIT HERBS

Herb Name	Common Name	Temperature & Tastes	Channels	Functions	Indications	Dosages & Notes
Ci Shi	Magnetite	Cold Acrid Salty	Kid Liv Ht	*Calms spirit *Pacifies Liver *Restrains floating Yang *Brightens eyes *Aids hearing *Assists Kidney in grasping Qi *Arrests wheezing	*Irritability/palpitation/insomnia/headache/dizziness/seizures due to Yin deficiency with Yang rising *Hearing loss/tinnitus/blurred vision/cataract due to Liver/Kidney Yin deficiency *Wheezing/chronic asthma	10–30 g 1–3 g in powder or pill Decoct first for 20–30 min Use caution with Spleen deficient patients
Dai Zhe Shi	Hematite	Cold Bitter	ST Liv Pc	*Calms Liver *Restrains floating Yang *Descends rebellious Qi *Cools blood *Stops bleeding	*Dizziness/tinnitus/heavy sensation around eyes due to Liver Yang rising *Belching/vomiting/hiccups/cough/wheezing/shortness of breath due to Qi rebellion *Nose bleed/vomiting blood due to heat in blood	10–30 g Decoct first for 20–30 min Use caution during pregnancy
Hu Po	Amber	Neutral Sweet	UB Ht Liv	*Arrests tremor *Calms spirit *Moves blood *Promotes urination *Reduces swelling	*Anxiety/panic attack/dream-disturbed sleep/palpitation/forgetfulness/seizure due to Shen disturbance *Amenorrhea with abdominal pain/mass due to blood stasis *Urinary retention or Bloody Lin syndrome *Skin sores/carbuncles and ulceration (topical) *Painful swollen scrotum	1.5–3 g in powder or pill

(continued)

ANCHOR AND SETTLE THE SPIRIT HERBS (continued)

Herb Name	Common Name	Temperature & Tastes	Channels	Functions	Indications	Dosages & Notes
Long Gu	Fossils of Mammal Vertebrae and Extremity	Neutral Sweet Astringent	Ht Kid Liv	*Calms spirit *Pacifies Liver *Astringes fluid *Restrains floating Yang	*Short tempered/ irritability/dizziness/ vertigo/blurred vision due to Liver Yang floating *Palpitation/insomnia/ anxiety/agitation/ emotional distress due to Heart Shen disturbance *Night and spontaneous sweating/nocturnal emission/spermatorrhea/ leukorrhea/uterine bleeding due to Kidney deficiency *Nonhealing chronic sore/ulcer (calcined form for topical use)	15–30 g Decoct first for 20–30 min Calcined (Duan Long Gu) for astringing fluid and preventing leakage
Mu Li	Oyster Shell	Cool Salty Astringent	Liv Kid	*Calms spirit *Nourishes Yin *Astringes fluid *Softens hardness *Relieves pain	*Restlessness/palpitation/ anxiety/insomnia due to Yin deficiency *Irritability/insomnia/ headache/dizziness/ blurred vision/tinnitus/ short tempered/red face due to Yin deficiency with Yang rising *Excessive sweating/ uterine bleeding/ leukorrhea/nocturnal emission *Swelling/nodules *Acid reflux with abdominal pain/heartburn/sour taste in mouth	15–30 g Decoct first for 20–30 min Calcined (Duan Mu Li) for acid reflux and fluid leakage
Zhen Zhu	Pearl	Cold Sweet Salty	Ht Liv	*Sedates Heart *Settle tremor *Clears Liver *Benefits eyes *Generates flesh	*Infantile convulsion/ seizure/palpitation *Blurred vision due to Pterygium of eyes *Chronic stubborn ulcer *Facial skin enhancement	0.3–1 g in pill or powder
Zi Shi Ying	Fluorite	Warm Sweet	Ht Liv	*Sedates Heart *Settles tremor *Warms Lung/Uterus *Directs Qi downward	*Palpitation/anxiety/ insomnia/disorientation/ infantile convulsion *Cough/wheezing with copious sputum due to Lung deficiency *Uterine bleeding/heavy menstruation/infertility due to deficiency cold in uterus	6–15 g Decoct first for 20–30 min

NOURISHES THE HEART TO CALM SPIRIT

Herb Name	Common Name	Temperature & Tastes	Channels	Functions	Indications	Dosages & Notes
Bai Zi Ren	Arbor-Vitae Seed	Neutral Sweet	Ht Kid LI Sp	*Nourishes Heart *Calms spirit *Moistens intestine *Unblocks bowels	*Insomnia/irritability palpitation/anxiety/ forgetfulness due to Heart blood deficiency *Dry constipation due to Yin and blood deficiency *Night sweats due to Yin deficiency	10–18 g C/I in patients with loose stool or phlegm disorder

(continued)

NOURISHES THE HEART TO CALM SPIRIT (continued)

Herb Name	Common Name	Temperature & Tastes	Channels	Functions	Indications	Dosages & Notes
He Huan Hua	Mimosa Tree Flower	Neutral Sweet	Ht Liv	*Calms spirit *Moves blood *Smoothes Liver Qi *Relieves pain *Dissipates swelling	*Insomnia/poor memory due to constrained emotions with abdominal pain and chest oppression	5–10 g
He Huan Pi	Mimosa Tree Bark	Neutral Sweet	Ht Liv	*Calms spirit *Moves blood *Smoothes Liver Qi *Relieves pain *Dissipates swelling	*Depression/stress/ irritability/insomnia/ forgetfulness due to emotional distress *Pain/swelling due to trauma causing blood stasis *Lung abscess	9–15 g
Ling Zhi	Glossy Ganoderma	Sl Warm Sweet Sl Bitter	Ht Lu Sp Liv Kid	*Nourishes Heart *Calms spirit *Tonifies Qi *Builds blood *Stops cough *Arrests wheezing	*Dream-disturbed sleep palpitation/anxiety/ forgetfulness/mental sluggishness due to Heart Qi and blood deficiency *Wheezing/chronic cough with sputum due to Lung/ Kidney deficiency *Hypertension/ hyperlipidemia/ congestive heart disease/arrhythmia/ leucopenia/weakness after illness	3–15 g Available in tablet, tincture, syrup, or injection form
Qiang Lang	Dung Beetle	Cold Salty	ST LI Liv	*Calms spirit *Moves blood *Reduces toxin *Promotes urination *Moves bowels	*Insanity/schizophrenia (Dian/Kuang) *Skin sores and boils *Lin syndrome *Hemorrhoid *Dysentery *Lumps and clumping	0.9–2.4 g in pill or powder Toxic
Sheng Tie Luo	Iron Filings	Cool Acrid	Ht Liv	*Calms Liver *Sedates spirit	*Insanity/schizophrenia (Dian/Kuang) due to Liver Qi stagnation or Shen disturbance or phlegm–fire disturbance	15–30 g Up to 60 g for severe cases Decoct first for 20–30 min
Suan Zao Ren	Sour Jujube Seed	Neutral Sweet Sour	GB Ht Liv Sp	*Nourishes Heart Yin *Nourishes blood *Calms spirit *Inhibits sweating	*Insomnia/palpitation/ anxiety due to Heart Yin and Liver blood deficiency *Spontaneous/night sweats	10–18 g 1.5–3 g in powder at bedtime Suan Zao Ren is a nourishing sedative
Ye Jiao Teng (*Shou Wu Teng*)	Vine of Solomon's Seal	Neutral Sweet Sl Bitter	Ht Liv	*Nourishes Heart *Calms spirit *Nourishes blood *Alleviates itch *Disperses wind	*Insomnia/irritability/ dream-disturbed sleep due to Yin and blood deficiency *Wind–damp Bi syndrome *Wind–heat skin rash/itch	10–30 g Up to 60 g for severe spirit disturbance
Yuan Zhi	Chinese Senega Root	Sl Warm Acrid Bitter	Ht Lu	*Calms spirit *Opens orifices *Expels phlegm *Reduces abscesses	*Anxiety/disorientation/ restlessness/palpitation/ insomnia due to Heart Shen disturbance due to phlegm misting Heart orifices *Cough with copious sputum *Swelling/abscess (topical)	3–10 g

CHAPTER 42

Aromatic Herbs That Open Orifices

Herb Name	Common Name	Temperature & Tastes	Channels	Functions	Indications	Dosages & Notes
An Xi Xiang	Benzoin	Neutral Acrid Bitter	Ht Liv Sp	*Opens orifices *Expels phlegm *Move Qi *Invigorates blood	*Delirium/stroke/lock-up symptom due to extreme heat and phlegm *Chest/abdominal pain due to Qi stagnation *Nonhealing sore/ulcer (topical)	0.3–1 g in pill or powder C/I during pregnancy
Bing Pian	Borneol	Cool Acrid Bitter	Ht Lu Sp	*Opens orifice *Revives senses *Clears heat *Stops pain *Dissipates nodules *Stops itch	*Unconsciousness/fainting convulsion episodes *Swelling of throat/skin ulcer/ sore/canker sore (topical) *Tearing/photophobia	0.3–1 g in pill or powder C/I during pregnancy
Niu Huang	Cattle Gallstone	Cool Bitter Sweet	Ht Liv	*Clears Heart *Opens orifices *Extinguishes internal wind *Stops tremors *Reduces toxin	*Delirium/coma due to heat obstructing pericardium in febrile disease *Tremor/spasm/coma due to internal wind *Painful swollen throat *Skin ulcer/boil/carbuncle	0.015–0.15 g in pill or powder C/I during pregnancy
She Xiang	Musk of Deer	Warm Acrid Aromatic	Ht Sp Liv	*Opens orifice *Moves blood *Relieves pain *Dredges meridian	*Delirium/coma/convulsion/ lock-up symptom/fainting due to heat blocking pericardium/Heart *Trauma due to blood stasis *Sore/carbuncle/mass due to toxin and stasis (topical) *Descends stillborn/ placenta	0.06–0.15 g in pill or powder C/I during pregnancy
Su He Xiang	Resin of Rose Maloes	Warm Acrid Sweet Aromatic	Ht Sp	*Opens orifice *Expels phlegm	*Lock-up symptom due to cold–turbid phlegm *Epigastric/abdominal pain and distension due to phlegm	0.3–1 g in pill or powder C/I during pregnancy, fever, coma due to heat
Xi Gua Shuang	Frost Powder prepared with Watermelon Peel and Mirabilite	Cool Acrid Bitter	Ht Lu Sp	*Clears heat *Stops pain *Dissipates nodule	*Swelling of throat/skin ulcer/ sore/canker sore (topical)	0.3–1 g in pill or powder

CHAPTER 43

Extinguish Wind and Stop Tremor Herbs

Herb Name	Common Name	Temperature & Tastes	Channels	Functions	Indications	Dosages & Notes
Bai Ji Li (*Ci Ji Li*)	Caltrop Fruit	Neutral Acrid Bitter	Liv Lu	*Calms Liver *Dispels wind–heat *Moves Liver Qi *Brightens eye	*Headache/dizziness *Swollen, red, painful, tearing eyes due to Liver Yang rising *Flank/chest pain/ distension *Insufficient lactation due to Liver Qi stagnation *Wind–heat skin rash/ itch	6–12 g
Di Long	Earthworm	Cold Salty	UB Liv Lu Sp	*Drains heat *Stops spasm *Calms wheezing *Expels wind *Unblocks channel *Promotes urination	*Convulsion/seizure due to high fever *Wheezing with raspy throat *Red, swollen, hot, painful, and stiff joints/ limbs *Hot Lin syndrome with edema *Hypertension due to Liver Yang rising *Acute mumps/burns/ chronic ulcers in lower extremities (topical)	5–15 g dried 10–20 g fresh 1–2 g in powder
Gou Teng	Gambir Vine	Cool Sweet	Liv Pc	*Extinguishes wind *Alleviates spasm *Clears heat *Calm Liver	*Convulsions/seizures/ tremor/eclampsia due to internal Liver wind *Headache/dizziness/ red eyes/irritability *Hypertension due to Liver Yang rising	10–15 g Add last 10 min of decoction
Jiang Can (*Bai Jiang Can*)	Dead Sick Silkworm	Neutral Acrid Salty	Liv Lu ST	*Extinguishes wind *Stops spasm *Releases exterior (raw) *Relieves pain *Transforms phlegm (toasted) *Reduces swelling	*Infantile convulsion/ facial paralysis/ seizures due to wind–phlegm–heat *Headache/red swollen eye/sore throat/ toothache/skin disorder due to wind–heat or Liver fire *Nodule/scrofula due to phlegm–heat	3–10 g 1–1.5 g in powder

(continued)

Herb Name	Common Name	Temperature & Tastes	Channels	Functions	Indications	Dosages & Notes
Ling Yang Jiao	Antelope Horn	Cold Salty	Ht Liv	*Extinguishes wind *Calms Liver *Drains heat *Reduces toxin *Brightens eyes	*Infantile convulsion/red eyes/headache/ dizziness/blurred vision/ photophobia/spasm due to extreme heat leading to internal wind *High fever/delirium/ manic behavior/ unconsciousness due to heat in pericardium due to febrile disease *Wind–damp–heat Bi	1–3 g 0.5–1 g in pill or powder form
Luo Bu Ma	Leaf of Apocynum Venetum	Sl Cold Sweet Bitter Bland Astringent	Liv	*Smoothes Liver *Clears heat *Lowers blood pressure *Strengthens Heart *Promotes urination	*Hypertension due to Liver Yang rising or Liver heat *Edema/difficult urination	3–10 g Add last 10 min of decoction
Quan Xie	Scorpion	Neutral Acrid Salty	Liv	*Extinguishes wind *Stops tremor *Relieves toxin *Reduces swelling *Unblocks channels *Relieves pain	*Infantile convulsion/ spasm/tetany/ opisthotonos/seizure/ poststroke symptom due to Liver wind and phlegm *Severe headache/ migraine/Tou Feng/Bi syndrome *Sore/swelling/scrofula (topical)	3–5 g 0.6–1 g in powder Toxic C/I in patients with internal wind due to blood deficiency
Shi Jue Ming	Abalone Shell	Cold Salty	Kid Liv	*Anchors Yang *Clears Liver heat *Drains fire *Improves vision	*Headache/dizziness due to Liver fire *Photophobia/ pterygium/blurred vision/swollen red eye (fine powder topical use on eyes)	15–30 g Decoct first for 20–30 min
Shou Gong (*Bi Hu*) (*Tian Long*)	House Lizard	Cold Salty	Liv Ht ST	*Extinguishes wind *Relieves toxin *Calms spirit *Reduces swelling	*Li Jie Feng *Stroke and seizure *Unknown swellings and sores	3–12 g in powder
Tian Ma	Gastrodia Rhizome	Neutral Sweet	Liv	*Calms Liver *Anchors Yang *Extinguishes wind *Relieves pain	*Seizure/headache/ dizziness/vertigo/ migraine *Infantile convulsion/ spasm/cramp/ opisthotonos due to Liver wind *Wind-damp Bi syndrome	3–10 g 1–1.5 g in powder
Wu Gong	Centipede	Warm Acrid	Liv	*Extinguishes wind *Stops spasm *Relieves toxin *Reduce swelling *Unblocks channels *Relieves pain	*Infantile convulsion/ seizure/lock jaw/ spasm/opisthotonos due to Liver wind *Headache/migraine *Bi syndrome *Lumps/nodules/sores/ snake bites (topical)	1–3 g 0.6–1 g in powder Toxic C/I during pregnancy

CHAPTER 44

Expel Parasite Herbs

Herb Name	Common Name	Temperature & Tastes	Channels	Functions	Indications	Dosages & Notes
Bing Lang	Betel Nut	Warm Acrid Bitter	LI ST	*Kills parasite *Moves Qi *Promotes urination	*Roundworm/pinworm/fasciolopsis/tapeworm/pig tapeworm *Constipation with food stagnation *Leg Qi with edema *Reduce side effects caused by Shang Shan, e.g., vomit	6–15 g 60–120 g for killing tapeworms and fasciolopsis
Da Suan	Garlic Bulb	Warm Acrid	LI Lu Sp ST	*Kills parasite *Relieves toxin *Reduces swelling	*Pinworm around anus area (topical) *Hookworm/ringworm on scalp *Diarrhea/dysentery *Cough/tuberculosis *Boost immune system *Food poisoning from eating crabs	6–15 g Taken raw, roasted, or as a paste
Fei Zi	Chinese Torreya Seed	Neutral Sweet Astringent	LI Lu ST	*Kills parasite *Moistens Lung *Stops cough	*Tapeworm/hookworm/pinworm/roundworm *Mild constipation/food stagnation *Nonproductive cough	9–15 g Up to 30 g for severe cases More effective when toasted and taken directly
He Cao Ya	Bud of Hairyvein	Cool Bitter Astringent	Liv SI LI	*Kills parasite	*Tapeworm	4.5 g each dose taken in powder
He Shi	Starwort Fruit	Neutral Acrid Bitter	Liv	*Kills parasite *Relieves pain	*Abdominal pain caused by intestinal parasites	6–12 g in pill or powder Sl toxic
Ku Lian Gen Pi	China Tree Root Bark	Cold Bitter	Liv Sp ST	*Kills parasite	*Intestinal/biliary roundworm *Vaginal trichomonas *Tinea/ringworm (topical)	6–15 g dried 15–30 g fresh Toxic Short-term use only
Lei Wan	Fruiting Body of Stone-like Omphalia	Cold Bitter	LI ST	*Kills parasite	*Tapeworm *Hookworm *Roundworm	6–15 g 3–6 g in powder Sl toxic Short-term use only

(continued)

Herb Name	Common Name	Temperature & Tastes	Channels	Functions	Indications	Dosages & Notes
Nan Gua Zi	Pumpkin Seed Husk	Neutral Sweet	LI ST	*Expels parasites *Relieves pain *Reduces edema	*Tapeworm/roundworm (need to take for long period) *Schistosomes *Postpartum swelling of hands and feet *Insufficient lactation	60–120 g
Shi Jun Zi	Rangoon Creeper Fruit with Seed	Warm Sweet	Sp ST	*Kills parasite *Strengthens Spleen *Reduces accumulation	*Roundworm *Infantile Gan Ji (malnutrition/food stagnation/abdominal distension/poor appetite) *Overdose can cause nausea/vomiting/ belching	6–10 g Toast until aromatic For children: eat 1 piece per day
Wu Yi	Stinky Elm Fruit Paste	Warm Acrid Bitter	Sp ST	*Kills parasite	*Roundworm/ tapeworm *Chronic diarrhea with Gan Ji in children	3–10 g

CHAPTER 45

Anesthetic Pain Relieving Herbs

Herb Name	Common Name	Temperature & Tastes	Channels	Functions	Indications	Dosages & Notes
Ba Jiao Feng	Chinese Alangium Root	Warm Acrid	Ht Liv	*Relieves pain *Moves blood *Disperses wind *Stops itch	*Pain/spasm from Bi syndrome *Traumatic injuries *Skin rash/itch (topical wash)	1.5–3 g in powder 6–12 g in bark or leaf form Toxic Use caution during pregnancy
Cao Wu	Kusnezoff Monkshood Root	Hot Acrid Bitter	Ht Liv Sp	*Relieves pain *Expels wind–damp *Warms meridian	*Pain due to wind–damp Bi *Traumatic injuries *Numbness of skin/ muscle from blood stasis	3–9 g; 1–2 g in powder or pill Ex toxic Decoct first for 2 hours C/I during pregnancy Not compatible with Ban Xia, Gua Lou, Bai Ji, Bai Lian, and Bei Mu
Chuan Wu	Sichuan Monkshood Root	Hot Acrid Bitter	Ht Liv Sp	*Relieves pain *Expels wind–damp *Warms meridian	*Pain from cold and/or damp accumulation *Traumatic injuries *Migraine/headache *Yin-type boils/sores	Same as those of Cao Wu
Liang Mian Zhen	Pricklyash Root	Sl Warm Acrid Bitter	Liv ST	*Relieves pain *Moves blood *Circulates Qi *Disperses wind	*Abdominal pain/ epigastric pain/ hernia-like pain *Pain from Bi syndrome *Swelling/pain due to snake-bite	6–15 g Sl toxic Overdose could cause abdominal pain, vomiting, and dizziness
Man Tuo Luo (*Yang Jin Hua*)	Jamestown Weed Seed/Leaf/ Flower	Warm Acrid	Ht Lu Sp	*Relieves pain *Stops spasms *Arrests cough *Suppresses wheezing	*Pain from Bi syndrome *Trauma injuries *Chest/abdominal pain from cold–damp and blood stasis *Seizure/convulsion/ tremor *Cough/wheezing from bronchial spasm	0.3–0.5 g in powder or pill Toxic C/I in patients with EPI and glaucoma

(continued)

Herb Name	Common Name	Temperature & Tastes	Channels	Functions	Indications	Dosages & Notes
Tian Xian Zi	Henbane Seed	Warm Bitter	Ht Liv Lu	*Relieves pain *Stops spasms *Arrests cough *Suppresses wheezing	*Pain from Bi syndrome *Toothache/epigastric pain *Epilepsy *Insanity/schizophrenia (Dian/Kuang) *Cough/wheezing from bronchial spasm	0.6–1.2 g Ex toxic C/I during pregnancy Overdose can lead to dry tongue, dry mouth, and fever
Tie Shu	Sago Cycas	Cool Bitter Sweet	Ht Lu ST	*Clears heat *Reduces toxin *Invigorates blood *Relieves pain	*Pain due to injuries *Skin bruising and swelling *Bleeding due to injuries	9–15 g: flower 15–30 g: leaf
Xia Tian Wu	Corydalis Rhizome	Warm Acrid Bitter	Liv	*Relieves pain *Moves blood	*Pain from Bi syndrome *Trauma/numbness *Hypertension *Paralysis from CAV	5–15 g Used as a substitute for Yan Hu Suo in Japan
Xu Chang Qing	Chinese Cynanchum	SI Warm Acrid	Ht Liv ST	*Relieves pain *Disperses wind *Detoxification	*Lower back pain *Bi syndrome *Epigastric/abdominal pain *Postsurgery pain *Pain due to cancer *Skin rash/swelling *Snake bite *Edema/ascites *Lin syndrome	6–12 g Up to 30 g
Xue Dan	Hemsleya Root	Cold Bitter	Lu ST	*Clears heat *Detoxification *Relieves pain *Reduces swelling	*Pain from heat toxin/ burn *Toothache/sore throat *Toxic reaction to bug bites	0.5–1 g per dose 1.5–3 g per day Overdose can lead to vomiting, sweating, and abdominal bloating C/I in patients with heart disease
Xue Shang Yi Zhi Hao	Brachypodum Monkshood Root	Warm Acrid Bitter	Ht Liv Sp	*Relieves pain *Moves blood *Reduces swelling	*Stronger effect than Chuan Wu and Cao Wu in alleviating pain and reducing sores/ swellings	0.02–0.04 g Ex toxic C/I during pregnancy Overdose can lead to nausea, vomiting, palpitation, pale face and lips, numbness of the limbs, cold skin, and chest tightness
Yang Zhi Zhu	Yellow Azalea	Warm Acrid Bitter	Ht Liv	*Relieves pain *Moves blood *Expels wind–damp	*Pain from Bi syndrome *Trauma injuries *Cough with sputum from cold–damp *Hypertension/ tachycardia	0.3–1.5 g Ex toxic Overdose will lead to loss of consciousnes C/I during pregnancy
Zu Shi Ma	Giraldi Daphne Bark	Warm Acrid Bitter	Ht Liv	*Relieves pain *Moves blood	*Pain from Bi syndrome *Traumatic injuries due to blood stasis *Longer lasting anesthetic effect	3–10 g C/I during pregnancy Topical use may cause blisters

CHAPTER 46

Anti-Tumor Herbs

Herb Name	Common Name	Temperature & Tastes	Channels	Functions	Indications	Dosages & Notes
A Wei	Chinese Asafetida	Warm Bitter Acrid	Liv Sp ST	*Detoxification *Anti-cancer *Softens and reduces lumps and nodules	*Lumps and nodules from blood or phlegm accumulations *Distention from Qi stagnation *Food stagnation in children *Abdominal/ hypochondriac pain *Malaria	0.9–1.5 g in pill or powder
Ba Yue Zha	Akebia Fruit	Neutral Bitter	Liv ST UB	*Detoxification *Anti-cancer *Softens and reduces lumps and nodules	*Cancers of the lung/ esophagus/ stomach	9–30 g or as medicinal wine
Bai Hua She She Cao	Spreading Hedyotis Oldenlandia Plant	Cold Bitter Sweet	Liv ST LI	*Clears heat *Reduces toxin *Reduces abscesses *Promotes urination	*Cancers of the stomach esophagus/ rectum *Abscess/sore/ulcer/ swelling due to heat toxin *Heat Lin syndrome *Snake bites (topical/ internal)	15–60 g Use caution during pregnancy
Bai Ying	Climbing Nightshade Plant	Sl Cold Bitter	Liv GB	*Clears heat–toxins *Reduces abscesses *Expels damp *Disperses wind	*Cancers of uterus/ cervix/liver/urinary bladder *Abscess/swelling/ rash erysipelas *Jaundice/leukorrhea due to damp–heat *Bi syndrome	15–25 g
Ban Mao	Cantharides	Cold Acrid	Liv ST LI Kid	*Detoxifies poison *Reduces abscesses *Breaks blood stasis	*Cancers of liver/ esophagus/stomach/ lung/breast and cardiac region *Swelling/lump/nodule/ scrofula/sore/tinea/ rash	0.03–0.06 g in pill or powder Ex toxic C/I during pregnancy Overdose leads to GI, cardiac, and urinary diseases

(continued)

Herb Name	Common Name	Temperature & Tastes	Channels	Functions	Indications	Dosages & Notes
Ban Zhi Lian	Sun Plant	Cool Sl Bitter Acrid	Liv Lu ST	*Clears heat *Detoxifies poison *Invigorates blood *Reduces swelling	*Cancers of the lung/liver/stomach *Sores/swelling/abscess swollen throat due to heat toxins *Chronic hepatitis/ascites *Traumatic injuries *Difficult urination	15–60 g
Chang Chun Hua	Madagascar Periwinkle	Cool Bitter	Sp Liv UB	*Reduces abscesses *Invigorates blood *Clears heat–toxins *Promotes urination	*Hodgkin's disease *Lymphocytic leukemia *Breast cancer *Choriocarcinona *Hypertension *Hyperglycemia *Skin burns (topical)	9–15 g Toxic Short-term use
Huang Du (*Huang Yao Zi*)	Airpotato Yam	Cold Bitter	Lu Liv	*Detoxifies poison *Cools blood *Reduces swelling	*Cancers of thyroid/stomach/esophagus/liver/rectum *Hyperthyroidism/Graves' disease *Swelling/nodule/sore throat *Bleeding due to blood heat *Snake bites (topical)	9–15 g Sl toxic Short-term use
Ji Xing Zi	Garden Balsam Seed	Warm Sl Bitter Acrid	Liv Sp	*Detoxification *Anti-cancer *Softens and reduces lumps and nodules	*Cancers of the lung/esophagus/stomach	3–4.5 g
Jiu Jie Cha (*Zhong Jie Feng*)	Sarcandrae Plant	Cool Acrid Bitter	Liv Lu ST	*Reduces abscesses *Clears heat–toxins *Disperses wind *Relieves pain	*Cancers of pancreas/rectum/stomach/liver/esophagus *Pain/numbness due to wind–damp Bi syndrome *Pneumonia/appendicitis/gastroenteritis/dysentery/biliary inflammation/abscess	6–15 g Sl toxic C/I during pregnancy
Lang Du	Fischer Euphorbia Root	Neutral Bitter Acrid	Lu Ht	*Detoxification *Anti-cancer *Softens and reduces lumps and nodules	*Anti-tuberculosis *Skin disorders, e.g., eczema *Cancers of the lung/thyroid/liver/stomach	1–2.5 g in pill or powder Not Toxic compatible with Mi Tuo Seng
Ling Jiao	Water Calptrop	Cool Sweet	LI SI ST	*Detoxification *Anti-cancer *Softens and reduces lumps and nodules	*Gastrointestinal cancer (cooked) *Poor appetite, loose stools (cooked)	15–60 g
Long Kui	Black Nightshade	Cold Sl Bitter	Lu UB	*Clears heat toxins *Invigorates blood *Reduces swelling *Promotes urination	*Cancers of uterus/cervix/urinary bladder/throat and liver *Sore throat/abscess/boils and swelling *Itchy skin rash *Damp–heat Lin syndrome with edema	15–30 g Sl toxic Overdose leads to headache, vomiting, diarrhea, and enlarged pupils
Lu Feng Fang	Hornet Nest	Neutral Sweet Acrid	Liv ST Lu	*Detoxifies poison *Expels wind *Relieves pain	*Cancers of the breast/esophagus/nasopharynx *Poisonous lumps/nodules/sores/abscess/tinea *Toothache/chronic sinusitis/rhinitis	3–12 g Toxic Topically as powder or wash

(continued)

Herb Name	Common Name	Temperature & Tastes	Channels	Functions	Indications	Dosages & Notes
Ma Qian Zi (*Fan Mu Bie*)	Nux-Vomica Seed	Cold V Bitter	Liv ST	*Detoxifies poison *Unblocks channels *Reduces swelling *Relieves pain	*Cancers of the lung/ breast/gastrointestine and skin *Psoriasis/sores/boils/ lymphopathy *Traumatic injuries/ spasm *Bi syndrome/paralysis *Also for anesthetic purpose	0.3–0.9 g in pill or powder C/I during pregnancy Overdose leads to convulsion, high blood pressure, dyspnea, and coma
Nao Sha	Sal Ammoniac	Warm Salty Bitter Acrid	ST Lu	*Reduces swelling *Detoxifies poison *Transforms phlegm *Soothes throat	*Cancers of the esophagus/stomach/ rectum/breast/skin *Poisonous sores/boils/ swellings/pterygium (topical) *Sticky difficult to expectorate phlegm with cough/sore throat	0.3–0.6 g in pill or powder Toxic
Nong Ji Li	Crotalaria Plant	Neutral Bitter	ST LI Liv	*Clears heat–toxins *Reduces abscess *Transforms damp	*Cancers of the skin/ uterus/gastrointestine *Dysentery/ gastroenteritis *Abscess/boil/acne/ swellings (topical/ internal)	9–15 g Ex toxic C/I during pregnancy Overdose leads to GI, Liver, and Kidney damage
Shan Ci Gu	Chinese Tulip Bulb	Cold Bitter	Liv ST Lu	*Clears heat–toxins *Reduces nodule *Detoxifies poison	*Cancers of breast/ lung/esophagus/skin/ uterus/cervix/ nasopharynx *Swelling/lumps/ nodules (topical)	0.3–0.6 g in pill or powder Overdose can lead to digestive disorders, leukopenia, and multiple neuritis
She Mei	Mock Strawberry Plant	Cold SI Bitter	Lu ST Liv	*Clears heat–toxins *Reduces nodule *Arrests cough *Stops bleeding	*Cancers of the stomach/uterus/ nasopharynx *Sore throat/diphtheria/ swellings/sores/ eczema/burns (topical) *Cough/whooping cough due to wind–heat *Severe uterine bleeding (Beng Lou) due to blood–heat	9–30 g
Shui Hong Hua Zi	Prince's Feather Fruit	Cold Salty	Liv Sp Lu LI	*Detoxification *Anti-cancer *Softens and reduces lumps and nodules	*Hepatitis *Liver cirrhosis *Cancer of the liver *Abdominal lumps/ masses	6–9 g
Teng Li Gen	Chinese Kiwi Root	Cold Sweet Sour	ST Liv UB	*Clears heat–toxins *Disperses wind *Promotes urination *Transforms damp	*Cancers of the stomach and esophagus *Pain due to Bi syndrome/bone pain/ sinew pain *Lin syndrome/ jaundice	15–30 g
Tian Kui (*Tian Kui Zi*)	Muskroot-like Semiaquilegia Root	Cold SI Bitter	Liv LI Lu	*Clears heat–toxins *Reduces nodules	*Cancers of the breast/ lymph node/urinary bladder/nose *Poisonous lump/ swelling/abscess/ sore throat *Bug and snake bites	3–9 g: root 9–15 g: twig or leaf

(continued)

Herb Name	Common Name	Temperature & Tastes	Channels	Functions	Indications	Dosages & Notes
Tu Bei Mu	Paniculate Boibostemma	Cool Bitter	Lu Ht LI	*Detoxification *Anti-cancer *Softens and reduces lumps and nodules	*Initial stage breast cancer *Stomach cancer *Nasopharyngeal cancer	9–30 g
Xi Shu	Happy Tree Fruit Bark	Cold Bitter	ST LI Liv	*Reduces nodules *Cools blood	*Cancers of the stomach/rectum/ esophagus/lung/ urinary bladder *Psoriasis	3–9 g: fruit 9–15 g: bark Toxic
Xue Jian Chou	Mapleleaf Goosefoot	Neutral Sweet	Liv Sp Lu UB	*Detoxification *Anti-cancer *Softens and reduces lumps and nodules	*Swellings/sores/ lumps (apply fresh herbs topically) *Irregular menses/ uterine bleeding *Gynecological tumors *Bloody Lin symptoms	3–9 g Up to 60 g if use fresh herb

CHAPTER 47

External Application Herbs

Herb Name	Common Name	Temperature & Tastes	Channels	Functions	Indications	Dosages & Notes
Ban Mao	Cantharides	Cold Acrid	LI Liv SI	*Detoxifies poison *Breaks up blood stasis *Disperses clumps	*Nodule/mass due to blood stasis *Ulcer/carbuncle/scrofula *Cancers of stomach/liver	0.03–0.06 g in pill or powder Ex toxic C/I during pregnancy
Chan Su	Toad Venom	Warm Sweet Acrid	Kid ST	*Reduces toxin *Reduces swelling *Relieves pain *Opens orifice *Expels summer heat	*Abdominal pain/vomiting/diarrhea/unconsciousness due to severe turbid evil summer heat (Sha attack) *Swelling/ulcer/carbuncle/sore throat/toothache (internal/topical paste)	0.015–0.03 g in pill or powder Toxic C/I during pregnancy
Chen Shi Hui	Old Lime	Warm Acrid	Liv Sp	*Relieves toxin *Dries damp *Stops bleeding *Relieves pain	*Burns/trauma injury (topical/internal) *Bleeding due to peptic ulcer *Abdominal pain/diarrhea due to Stomach bleeding *Topically used as paste	0.3–1 g in powder Toxic
Da Feng Zi	Chaulmoogra Seed	Hot Acrid	Kid Liv Sp	*Expels wind *Dries damp *Clears toxicity *Kills parasite	*Scabies/tinea *Syphilis/leprosy *Topically used as paste	0.3–0.9 g in pill or powder Toxic Overdose may cause nausea, vomiting, abdominal pain, Liver, and Kidney damage
Dan Fan (*Shi Dan*)	Crystallized Blue Vitriol	Sour Astringent Acrid Cold	Liv	*Detoxify (topical) *Promotes healing *Induces vomiting (oral)	*Topically used for: canker sores, toothache *Throat Bi *Conjunctivitis *Nasal polyps *Orally used to treat epilepsy by inducing vomiting up wind–phlegm	0.3–0.6 g orally Toxic

(continued)

Herb Name	Common Name	Temperature & Tastes	Channels	Functions	Indications	Dosages & Notes
Liu Huang	Sulphur	Hot Sour	Kid LI Pc	*Kills parasite *Stops itch *Clears toxicity *Strengthens Yang	*Ringworm/scabies/ eczema/tinea (topical paste) *Impotence/polyuria/ lower back pain/weak knees due to Kidney Yang deficiency	1–3 g in powder or pill Toxic C/I during pregnancy
Lu Gan Shi	Smithsonite or Calamine	Neural Sweet Astringent	Liv ST	*Clears heat *Dries damp *Speeds wound healing *Stops itch	*Swollen red eyes/ pterygium due to heat or wind–heat *Ulcers with purulent fluid *Open sores *Eczema *Topically used as powder	Do not take internally
Mi Tuo Seng	Litharge or Galena	Neutral Acrid Salty	Liv Sp	*Stops bleeding *Reduces swelling	*Bleeding/swelling/ ulcer/sores/wet skin rash/tinea/leukoderma/ underarm odor *Dysentery-like disorders *Topically used as paste	0.9–3 g in pill or powder Toxic
Mu Bie Zi	Momordica Seed	Warm Bitter Sweet	LI Liv ST	*Reduces toxin *Reduces swelling *Invigorates blood	*Pain due to traumatic injuries *Lower back pain *Hemorrhoid due to blood stasis *Poisonous swelling/ breast abscess/ nonhealing sores and scrofula (topical paste) *Anti-HIV/anti-cancer herb	0.6–1.2 g in pill or powder Toxic C/I during pregnancy
Mu Jin Pi	Shrubalthea Bark	Neutral Sweet	LI SI	*Clears heat *Relieves toxin *Kills parasite *Stops itch	*Itchiness/rash due to scabies/tinea/ ringworm/fungal infection (topical wash/ tincture)	3–10 g
Peng Sha	Borax	Cold Sweet Salty	Lu ST	*Clears heat *Dries damp (toasted) *Reduces toxin (raw)	*Swollen red throat/ canker sores/ candidiasis (topical) *Hot phlegm blockage with hard to expectorate sputum *Toe fungal infection	3–6 g 1–3 g in pill or powder
Pi Shi (*Pi Shuang*)	Arsenolite	Hot Acrid	Lu Liv	*Reduces swelling *Treats malaria *Transforms phlegm *Arrests wheezing	*Malaria *Wheezing/asthma due to cold phlegm obstruction in Lung *Hemorrhoid/fistula (topical) *Nodule/boil/carbuncle *Used to treat various cancers	0.002–0.004 in pill or powder form Ex toxic C/I during pregnancy Short-term use only
Qian Dan (*Huang Dan*)	Lead Oxide (Pb_3O_4)	Cool Acrid	Ht Liv Sp	*Expels toxins *Treats malaria *Generates flesh *Stops itch	*Wet, swollen, poisonous ulcers/sores *Seizures due to wind–phlegm	0.3–0.9 g in pill or powder Toxic Short-term use only
Qing Fen (*Gong Fen*)	Calomel	Cold Acrid	Liv UB Kid	*Kills parasite *Reduces toxin *Promotes urination *Moistens intestine	*Syphilitic chancres/ scabies (topical paste/powder) *Edema/constipation	0.1–0.2 g in powder or pill Toxic C/I during pregnancy

(continued)

Herb Name	Common Name	Temperature & Tastes	Channels	Functions	Indications	Dosages & Notes
She Chuang Zi	Cnidium Seed	Warm Acrid Bitter	Kid	*Dries damp *Kills parasite *Stops itch *Expels wind *Warms Kidney *Enhances Yang	*Infertility/impotence due to Kidney Yang deficiency *Leukorrhea/lower back pain due to damp–cold *Weepy/itchy skin disorders in lower part of body (topical wash/ powder) *Scabies/ringworm	3–10 g Up to 15–30 g
Shui Yin	Mercury	Cold Acrid	Ht Liv Kid	*Clears toxins *Kills parasite	*Syphilitic chancres/ scabies and tinea (topical paste/powder)	Ex toxic C/I during pregnancy Short-term use only
Tu Jin Pi	Goldenlarch Bark	Warm Acrid	LI Lu	*Kills parasite *Relieves itch	*Fungal infection/tinea/ itchy skin (topical wash/tincture)	3–10 g Toxic C/I during pregnancy
Xiong Huang	Realgar	Warm Acrid Bitter	Ht Liv ST	*Kills parasite *Dries damp *Reduces toxin	*Abscess/scabies/ eczema *Snake or bug bites (topical paste) *Roundworm/ schistosoma *Preventive for epidemic disease	0.2–0.4 g in powder or pill Toxic C/I during pregnancy Short-term use only
Zhang Nao	Camphor	Hot Acrid	Ht Sp	*Expels wind–damp *Opens orifice *Invigorates blood *Relieves pain	*Scabies/ringworm/ sores (topical paste) *Delirium/ unconsciousness due to febrile disease *Swelling/traumatic injuries	0.1–0.2 g in pill or powder Toxic C/I during pregnancy

Appendices

APPENDIX 1

Standard Syndrome Differentiation of Traditional Chinese Medicine

Organ System	Pattern	Primary Symptoms	Secondary Symptoms
		Heart System Patterns	
1. Heart patterns	Heart Qi deficiency	Absentmindedness, incoherent speech, feeling sad, on the verge of tears, paranoia, forgetfulness, and palpitation	Spontaneous sweats, pale tongue, and weak or irregular pulse
	Heart Qi close to collapsing	Apathy, weak respiration, spontaneous sweats with big drops, extremely weak, and almost impalpable pulse	Pale complexion, pale and tender tongue
	Heart Yang deficiency	Palpitation, mania triggered by fright, and sore throat	Pale complexion, shortness of breath, spontaneous sweats, pale and tender tongue, slow and forceless pulse, and sore throat alleviated by heat
	Yang deficiency leading to extreme cold	Extremely cold limbs	Mental fatigue, with desire to sleep, incontinence, extremely weak pulse
	Heart Yang obstruction	Angina	May be together with stifling discomfort in front of the Heart, cyanotic lips and purple tongue; may be together with pale tongue with white coating, slow wiry pulse; maybe together with distended discomfort in front of the Heart, shortness of breath, and dyspnea
	Heart Yin deficiency	Palpitation triggered by fright, forgetfulness, insomnia, frequent dreams, and night sweats	Irritability and fever, dry mouth and throat, red tongue with slight coating, and thin and rapid pulse
	Heart blood depletion	Palpitation and forgetfulness	Dull complexion, pale lips and tongue, and thin and weak pulse
	Heart Yin and Yang deficiency	Palpitation and insomnia	Pale tongue with slight coating, deficient and weak pulse, or slow, irregular pulse

(continued)

Organ System	Pattern	Primary Symptoms	Secondary Symptoms
		Heart System Patterns (continued)	
	Heat entering Ying (nutritive) level	Fever worse at night, often with incoherent speech, and faint skin rash	Irritability and restlessness, deep red tongue, and rapid pulse
	Heat entering Xue (blood) level	Skin rash breaking out	Fever worse at night, delirium, deep red tongue, and rapid pulse
	Excessive heat in Heart channel	Inability to stop laughing, mania	Phlegm, red tongue with sticky coating, and rapid pulse
	Heat toxin congesting the collateral channels	Carbuncles and furuncles	Red, swollen, hot, and painful skin, red tongue, and rapid pulse
2. Heart Shen (spirit) patterns	Heat entering pericardium	Delirium	Fever, irritability, deep red tongue with yellow coating, rapid pulse
	Middle Jiao extreme cold leading to Qi blockage	Sudden loss of consciousness and falling	Normal tongue, pulse, and respiration. This is different from wind–stroke, phlegm–stroke, or epilepsy
	Phlegm misting the orifice	Phlegm stroke, phlegm Jue, epilepsy, stupor, and mania	Noisy phlegm in throat, thick greasy tongue coating, and soggy and slippery pulse
		Liver System Patterns	
1. Cold/heat patterns	Deficient cold in Liver channel	Chest pain, hypochondriac pain, abdominal pain, hernia, painful menstruation (dysmenorrhea), irregular menstruation, convulsion, retraction of tongue and genitals, curling up of body, intolerance of fatigue, timidity, dry heave, dizziness, vertigo	Cold hands and feet, cyanotic nails, pale tongue, pale complexion, and pulse is deep, slow, wiry, and thin
	Excess fire in Liver channel	Mania, headache, dizziness, vertigo, red swollen eyes, ringing ears (tinnitus), deafness, hypochondriac pain, jaundice; scanty painful urination with blood, leukorrhagia with foul smell, sweaty scrotum, itchy swollen genitals	Fever, bitter taste, red tongue with yellow coating, and wiry and rapid pulse
2. Liver Qi patterns	Qi flow mechanism dysfunction	Sentimental, depression, and easily enraged, irritability, distention, pain, and discomfort in chest, hypochondria, lower abdomen, lumbar, and sacrum; irregular menstruation, spontaneous sweats, night sweats, difficult and painful scanty urination, red and whitish vaginal discharge, nocturnal emission, and ejaculation during dreams; no appetite, vomiting, abdominal pain, and hiccup	Irregular menstruation, sometimes early, sometimes late; and flow is sometimes heavy, sometimes scanty. Bowel movement is sometimes loose, sometimes hard. Burping, acid reflux, red or pale tongue, and wiry or hesitate pulse
3. Liver blood patterns	Liver blood depletion	Dizziness, blurry vision, oligomenorrhea (abnormally light or infrequent menstruation), and amenorrhea (absence of menstruation)	Pale nails, sallow complexion, pale tongue, and thin and weak pulse

(continued)

Organ System	Pattern	Primary Symptoms	Secondary Symptoms
	Liver System Patterns (continued)		
	Stagnant blood obstruction	Various pain, lumps, dysmenorrhea (painful menstruation), amenorrhea, and external injuries	History of musculoskeletal injuries, stabbing, fixed pain, purple tongue, or spider angioma on face and body, or menstrual clots
	Liver not storing blood	Hemoptysis (spitting blood), nosebleed, coughing blood, bloody stool, bloody urine, metrorrhagia (uterine bleeding), rash, and bruises	Bright color, large quantity, acute onset, fever, bitter taste, red tongue with yellow coating, and rapid pulse
	Liver Yin deficiency	Headache, dizziness, vertigo, blurry vision, tinnitus (ringing ears), night sweats, nocturnal emission, hemoptysis, oligomenorrhea, amenorrhea, dysmenorrhea, and hypochondriac pain	Tidal fever with night sweats, malar flush with bitter taste, red tongue with slight coating, and thin and rapid pulse
4. Liver wind patterns	Seizure due to external wind	Tetanus (opisthotonus, locked jaws), and facial paralysis	With wound, exposure of face to cold wind
	Seizure due to internal wind	Convulsion, spasm, sudden loss of consciousness and fall, and hemiplegia	High fever, dizziness, and vertigo
5. Shaoyang patterns	EPI residing in Shaoyang	Alternating fever and chills, stifling discomfort in chest and hypochondria, bitter taste, dry throat, blurry vision, irritability, with desire to vomit, silence, without appetite, or cough, or palpitation, or abdominal pain	Slightly red-edged tongue with white or yellow coating, wiry and rapid pulse
	EPI hidden in diaphragm	Chills with high fever, malaria-like chills and fever, no fixed timing of occurrence, stuffy chest with nausea and vomiting, and headache with irritability	Deep red edges of tongue, with powder-like coating and wiry and rapid pulse
	Heat stagnation in Gallbladder channel	Hypochondriac pain, jaundice, and acute pain below the Heart	Bitter taste, alternating fever and chills, no appetite, nausea and vomiting, red tongue with yellow coating, and wiry and rapid pulse
	Binding of heat and water in chest	Acute pain below the Heart	Intense and prolonged upper abdominal pain, worse with each episode, muscle tightness and pressuring pain in abdomen, nausea and vomiting, fever, moist tongue with yellow coating, and deep and tight pulse
	Gallbladder Qi deficient cold	Hypochondriac pain	Pale tongue with white coating, slow pulse
	Gallbladder stone formation	Hypochondriac pain	Pain at Gallbladder area after ingesting fatty and greasy food, intense pain as if twisted with a knife, fever, bitter taste, nausea and vomiting, red tongue with yellow coating, and wiry and slippery pulse

(continued)

Organ System	Pattern	Primary Symptoms	Secondary Symptoms
		Liver System Patterns (continued)	
	Roundworm entering bile duct	Right side hypochondriac pain	Sudden and intermittent knife-twisting pain, or head-drilling pain, returning to normal when pain dissipates as if nothing has happened, or together with vomiting roundworms
		Spleen System Patterns	
1. Receiving and transforming mechanism dysfunction	Food stagnation	Stomach pain, vomiting, and diarrhea	Regurgitation of foul smelling food after excessive eating, acid reflux, and diarrhea with foul-smelling stool, abdominal distention, and stifling discomfort, greasy and rotten-smelling tongue coating, and slippery pulse
	Cold–damp restricting the Spleen	Poor appetite, vomiting and nausea, loose stool, abdominal distention, sore and weak limbs with heavy and painful sensation, dizziness, vertigo, and blurry vision	Pale tongue with white coating, slow and moderate pulse
	Damp–heat obstruction	Poor appetite, chest distention, vomiting and nausea, loose stool, or hesitant stool passage, or explosive diarrhea, thirst, without desire to drink, fever, and body emitting foul smell	Pale yellow complexion, red tongue with greasy yellow coating, and soggy and rapid pulse
	Spleen deficiency edema	Edema	Poor appetite, abdominal distention, pale tongue with white coating, and weak pulse
	Spleen–damp producing phlegm	Stifling discomfort and distention, rumbling stomach as if hungry, constant cold feeling on midback, nausea, vomiting, diarrhea; or cough, shortness of breath; or palpitation, cardiac pain, insomnia; or vertigo and dizziness, blurry vision, headache, stroke, aphasia, mania, depression, epilepsy triggered by fright; or sweaty scrotum, leukorrhagia; or numbness of muscles and skin, weak and painful limbs; or tuberculosis of bone and joints, wandering abscess, scrofula, and phlegm node	Copious phlegm, greasy tongue coating, soggy pulse
	Stomach Yin deficiency	Stomach pain and burning sensation	Extremely dry mouth and lips, constipation with dry stool inside intestine, and red tongue with slight coating
	Spleen Qi deficiency	Small appetite with loose stool, epigastric distention	Low voice with shortness of breath, low energy with little desire to talk, fatigue of the four limbs, pale and tender tongue, and weak pulse without force

(continued)

Organ System	Pattern	Primary Symptoms	Secondary Symptoms
		Spleen System Patterns (continued)	
	Middle Jiao deficient cold	Vomiting, diarrhea, and abdominal pain	No thirst, pale tongue with white coating, deep and wiry pulse
	Heat stagnation in Stomach and Intestines	Swollen and painful gums, mouth and lips, bad breath, abdominal pain	Redness, swelling, fever, red tongue with yellow coating, rapid pulse
2. Controlling mechanism dysfunction	Failure of Spleen to control blood	Hemoptysis, nosebleed, coughing blood, bloody stool, bloody urine, metrorrhagia, and rash	Palpitation with shortness of breath, low energy with little desire to talk, dull complexion, or scanty menstrual flow, cold limbs, pale tongue with white coating, and weak pulse
3. Ascending and descending mechanism dysfunction	Middle Jiao Qi collapse	Uterus prolapse, anal prolapse, bloody stool, metrorrhagia, urination difficulty or anuresis (no urination), dribbling painful urination triggered by fatigue, and dizziness	Shortness of breath, lower abdomen feeling empty and dropped, sallow complexion, pale and tender tongue, and weak pulse
	Failure of clear Yang to ascend	Anuresis and diarrhea	Slight red tongue, and soggy and rapid pulse
	Failure of turbid Yin to descend	Vomiting, regurgitation, and hiccup	Stomach distention and discomfort, greasy tongue coating, and soggy pulse
	Disharmony between ascending and descending mechanisms	Vomiting, diarrhea, epigastric distention, or abdominal distention and pain, and cessation of bowel movement	Pale or red tongue with white or yellow coating, or with desire to vomit but without success, desire to have diarrhea but without success
	Spleen and Stomach Qi stagnation	Abdominal pain and distention	Reduced appetite, hesitant stool passage, white tongue coating, hesitant pulse
4. Transferring mechanism dysfunction	Damp–heat pouring downward	Diarrhea and dysentery	Fever, abdominal pain, burning sensation in anus; dysentery with blood and mucous, tenesmus, red tongue with yellow coating, and rapid pulse
	Incontinence due to deficient cold	Chronic diarrhea and chronic dysentery	Chronic nonstop dysentery, poor appetite, mental fatigue, cold limbs, abdominal pain that desires warmth and pressure, pale tongue, and slow pulse
	Intestinal obstruction	Constipation	Constipation with excessive heat; abdominal distention and pain, yellow and dry tongue coating, and excessive and forceful pulse Constipation with fluid deficiency Dry mouth and throat, and red tongue with slight coating Constipation with deficient cold Abdominal pain with slight relief upon warmth, pale tongue, and slow pulse

(continued)

Organ System	Pattern	Primary Symptoms	Secondary Symptoms
		Spleen System Patterns (continued)	
	Parasites gathering in Intestines	Abdominal pain, itchy anus, and desire to ingest inedible objects	Round worm: pain around umbilicus, bug rash on face, little dots with size of a corn kernel on inside of lips, and passing of round worms in stool
		Lung System Patterns	
1. Exterior protective regulation mechanism dysfunction	Exterior wind–cold	Aversion to cold, fever, aches and pain in head and body, without sweats or with spontaneous sweats	Stuffy nose, runny nose, cough, dyspnea, pale tongue with white coating, and floating tight pulse or floating moderate pulse
	Exterior wind–heat	Fever, with sweats, and thirst	Sore throat with little sputum, red tongue tip, thin yellow tongue coating, and floating rapid pulse
	Summer heat injuring Lung	Aversion to cold, heavy sensation of body, body pain, red face, thirst, and hot body without sweats	Red tongue tip, greasy yellow tongue coating, and right pulse bigger than left pulse
	External injury from damp	Aversion to cold, fever, heavy headedness, dizziness, headache, stuffy nose, heavy sensation of body and limbs, fatigue without strength, or irritated and aching joints	White tongue coating and floating pulse
	Dryness injuring Lung	Headache, fever, unproductive cough, or with little sputum that is difficult to expectorate	Dry nose and throat, sore throat, red tongue tip, thin yellow tongue coating, and small rapid pulse
	External injury from wind	Inability to move shoulders and limbs, hemiplegia, inability to retract limbs, sudden loss of consciousness and falling (wind Bi, paraplegia and atrophy, wind rash, and stupor due to wind misting the orifices)	Normal tongue with thin white coating, and wiry tight pulse
	Exterior deficiency not withstanding	Prone to catching colds and spontaneous sweating	Weak immune system, pale tongue, and weak pulse
2. Internal regulation mechanism dysfunction	Upper Jiao heat toxin	Swelling and pain in cheeks and submandibular area, sore and swollen throat, dull chest pain, and coughing; or vomiting pus with foul odor	Fever, red tongue with yellow coating, and rapid pulse
	Extreme heat in Qi level	High fever, sweating, and thirst	Flooding and forceful pulse
	Phlegm–heat obstructing Lung	Cough, dyspnea, and yellow sticky sputum	Fever, red tongue with sticky coating, and rapid pulse
	Cold in Lung stopping fluid metabolism	Cough, dyspnea, and watery sputum	Pale tongue with slippery coating and wiry tight pulse, or together with exterior syndromes

(continued)

Organ System	Pattern	Primary Symptoms	Secondary Symptoms
		Lung System Patterns (continued)	
	Water retention in the three Jiaos	Edema and pleurisy	Acute onset, first presented with exterior syndromes, then edema, normal or slippery yellow tongue coating, and floating rapid pulse; or coughing saliva leading to chest and hypochondriac pain
	Lung deficiency failing to control	Frequent urination, nocturnal enuresis, and incontinence	Spitting saliva and froth, without cough or thirst, pale tongue, and slow pulse; or urinary incontinence triggered by coughing
	Lung Qi deficiency	Dyspnea, cough	Watery sputum, shortness of breath, pale complexion, pale tongue, and weak pulse
	Lung Qi not withholding	Chronic unstoppable cough	Severe cough leading to dyspnea and spontaneous sweating, and deficient and rapid pulse
	Lung Yin depletion	Cough	Dry nose and throat, choking cough with little sputum, tidal fever, night sweats, red cheeks, deep red tongue, and rapid pulse
	Qi and Yin deficiency	Shortness of breath and spontaneous sweating	Fatigue with little desire to talk, dizziness, vertigo, dispiritedness, dry mouth as if thirsty, and deficient and rapid pulse
		Kidney System Patterns	
1. Kidney essence patterns	Kidney Yin deficiency	Lumbar pain, nocturnal emission, priapism (persistent erection of penis without sexual arousal), and intercourse in dreams; uterine bleeding, amenorrhea, painful scanty urination, xiao ke (general symptoms of untreated diabetes such as big thirst, big appetite, and frequent and sweet urination); late in childhood to start teething, walking, speaking, and mental development, nonclosure of fontanel, loose teeth, weakness in legs and feet, tinnitus, and blurry vision	Thin and small tongue body, red tongue color with slight coating, and small and rapid pulse
	Kidney Yang deficiency	Painful and cold lumbar region, aversion to cold, extremely cold hands and feet, impotence, premature ejaculation, azoospermia (lack of sperms), low libido, infertility due to coldness in uterus, difficult urination, clear and copious urine, incontinence, frequent urination, and gradual weight gain	Pale complexion, lips, and nails; pale flabby tongue with thin white coating, and deep slow and weak pulse

(continued)

Organ System	Pattern	Primary Symptoms	Secondary Symptoms
	Kidney System Patterns (continued)		
	Kidney Yin and Yang deficiency	Impotence, premature ejaculation, nocturnal emission, low libido, infertility, tinnitus, blurry vision, and hair loss	Mental fatigue and small and weak pulse
	Gate of essence not consolidated	Nocturnal emission and spontaneous ejaculation	Extreme mental exhaustion, aching lower back and weak knees, tinnitus, blurry vision, limbs without energy, palpitation, shortness of breath, pale tongue with white coating, and deficient and weak pulse
	Chong and Ren channels not consolidated	Miscarriage, uterine bleeding, and vaginal discharge (leukorrhea)	Miscarriage: aching and distended lower back, lower abdomen bearing down, vaginal bleeding, dizziness, tinnitus, weak legs, and frequent urination Uterine bleeding: coldness and pain below navel, weak limbs without energy, and pale tongue with thin coating; or complexion changing from red to pale, shortness of breath, and mental fatigue Leukorrhea: clear, watery, and prolonged vaginal discharge without end; pale complexion, mental fatigue, dizziness, vertigo, blurry vision, palpitation, shortness of breath, lumbar pain as if being broken, frequent urination, and deep and slow pulse
	Kidney not receiving Qi	Dyspnea	Exhalation and more than inhalation, cough, and weak pulse
	Yang deficiency leading to cold stasis	Yin (sunken) carbuncles, lower back pain, and neck and back stiffness and pain	Sunken area of infection; neck and back stiffness and pain without exterior syndromes; lower back pain and inability to bend forward or backward, and normal urination
2. Fluid regulation mechanism dysfunction	Kidney deficiency failing to withhold	Urination incontinence, frequent urination, and involuntary urination (enuresis) at night	Mental exhaustion, aversion to cold, fatigue and weak, dizziness, lumbar ache, weak legs, and deep and thin pulse with chi position being the weakest

(continued)

Organ System	Pattern	Primary Symptoms	Secondary Symptoms
		Kidney System Patterns (continued)	
	Qi-Hua mechanism dysfunction	Xiao ke, edema, clear and watery vaginal discharge, and difficult urination; impotence and involuntary ejaculation; cold body and limbs, aches and pain with heavy sensation, numbness, physical incapacity; abdominal pain, distention, vomiting, diarrhea, and constipation; chest pain and discomfort, palpitation, frequent desire to sleep, abnormal mental status; hypochondriac pain, dizziness, vertigo, blurry vision, twitching of muscles, fatigue of body and limbs, and hiccups; cough, dyspnea; dizziness, headache with heavy sensation, hair loss, gradual loss of memory, runny nose with clear mucus, blurry vision, tinnitus, deafness, and swollen gums; hoarse voice or loss of voice and plum-pit sensation in throat	Pale and flabby tongue with tooth marks, white slippery tongue coating, and slow pulse
	Lower Jiao damp–heat	Frequent, urgent, painful, and difficult urination of the following types: Qi-Lin (Qi type), Xue-Lin (bloody type), Sha-Lin (sandy type), and Gao-Lin (creamy type). Refer to descriptions under Auxiliary Symptoms	Qi-Lin: painful and hesitant urination, fever, bitter taste in mouth, lumbar pain, red tongue with yellow coating, and rapid pulse Xue-Lin: blood in urine Sha-Lin: in addition, lumbar pain or lower abdominal pain, involuntary pauses in urination flow Gao-Lin: cloudy urine or sticky discharge at urethra before and after urination
		Combined Patterns of Two Organ Systems	
1. Lung and Spleen combined disorders	Combined exterior and interior disorder	Chills and fever, aches, and pain of head and body; no sweatimg, fullness in chest and abdomen; or vomiting and diarrhea; or constipation	If cold manifestation: pale tongue with white coating, and moderate pulse If heat manifestation: fever, sweating, red tongue with yellow coating, and rapid pulse
	Lung and Spleen Qi deficiency	Low voice, shortness of breath, low energy, without desire to talk, poor appetite, and diarrhea	Pale complexion, pale tongue, and weak pulse
	Excess heat of Lung and Stomach	High fever with sweats, irritability, thirst, and constipation	Slight red tongue and flooding big pulse or deep, excess, and forceful pulse
	Fluid depletion of Lung and Stomach	Dry cough without sputum, dry mouth and lips, and dry throat	Slight red tongue with slight coating, and thin and rapid pulse
	Qi stagnation leading to fluid stagnation	Cough, dyspnea, with sputum, edema, no appetite, chest fullness, and nausea	Pale tongue with white coating and soggy pulse

(continued)

Organ System	Pattern	Primary Symptoms	Secondary Symptoms
Combined Patterns of Two Organ Systems (continued)			
2. Heart and Lung combined disorders	Disharmony between Ying and Wei (nutritive and protective levels)	Aversion to wind, fever, and sweating	Pale tongue, thirst, and moderate pulse
	Extreme heat in Qi and Ying (nutritive) levels	High fever, sweating, irritability and thirst, delirium, delirious speech	Deep red tongue and rapid pulse
	Knotted phlegm and blood stagnation	Palpitation and productive cough	Cyanosis in nails, lips, and tongue
3. Lung and Kidney combined disorders	Lung and Kidney Yin deficiency	Red cheeks, malar flush, tidal fever, night sweats, and blood in sputum	Red tongue with slight coating and thin and rapid pulse
	Water retention in Lung and Kidney	Phlegm–damp in Lung, edema	Pale swollen tongue and aversion to cold
4. Spleen and Kidney combined disorders	Spleen and Kidney Yang deficiency	Aversion to cold, cold limbs, vomiting, diarrhea, cold and pain in abdomen, diarrhea at dawn, and edema	Pale flabby tongue and deep slow pulse
	Spleen and Kidney Yin deficiency	Constipation	Dry mouth and lips, and red tongue with slight coating
5. Liver and Kidney combined disorders	Liver and Kidney Yin deficiency	Dizziness, vertigo, blurry vision, tinnitus, hot palms and soles, nocturnal emission, ejaculation during dreams, and scanty menstrual flow	Red tongue with slight coating and thin rapid pulse
	Liver and Kidney deficient cold	Cold type hernia, retraction of genitals, irregular menstruation, cold and painful abdomen during menstruation, and cold uterus causing infertility	Pale tongue and wiry and tight pulse
6. Liver and Lung combined disorders	Liver fire invading Lung	Hemoptysis and nosebleed	Agitation, irritability, impatience, hypochondriac pain, bitter taste in mouth, red tongue with slight coating, and wiry and rapid pulse
7. Heart and Liver combined disorders	Heart and Liver blood stagnation	Palpitation, shortness of breath, chest pain, Liver enlargement, and epilepsy	Purple tongue without shine, purple dots on tongue, and extremely weak pulse or wiry pulse
	Blood heat leading to wind	Delirium, convulsions, rash, and hemoptysis	High fever, sweating, deep red tongue, and rapid pulse
8. Heart and Kidney combined disorders	Heart and Kidney disconnection	Insomnia and nocturnal emission	Red tongue with slight coating or pale tongue with white coating
	Heart and Kidney Yin deficiency	Palpitation, forgetfulness, insomnia, and excessive dreams	Dry mouth and throat, red tongue with slight coating, and thin pulse
	Heart and Kidney Yang deficiency	Extremely cold limbs	Pale tongue and weak pulse
9. Heart and Spleen combined disorders	Phlegm–damp invading Heart	Palpitation triggered by fright, insomnia, chest pain, epilepsy, and stiff tongue with speech difficulty	Copious sputum, flabby tongue with sticky coating, and soggy pulse
	Heart and Spleen deficiency	Palpitation, forgetfulness, light menstrual flow, and subcutaneous bleeding	Pale complexion, poor appetite, loose stool, and pale tongue with weak pulse

(continued)

Organ System	Pattern	Primary Symptoms	Secondary Symptoms
	Combined Patterns of Two Organ Systems (continued)		
10. Liver and Spleen combined disorders	Stagnation in Liver and Spleen	Distention and pain in chest and diaphragm, bitter taste, acid reflux, and poor appetite	Depression, poor appetite, and loose stool
	Disharmony between Liver and Stomach	Hypochondriac pain and vomiting	Bitter taste, red-edged tongue with yellow coating, and wiry and rapid pulse
	Disharmony between Liver and Spleen	Abdominal pain	Abdominal pain that desires pressure or diarrhea upon onset of abdominal pain
	Qi and Blood deficiency	Low voice and shortness of breath, low energy with little desire to talk, dizziness, vertigo, blurry vision, and palpitation	Dull complexion and pale tongue with weak pulse

APPENDIX 2

Glossary of Traditional Chinese Medical Terminology

Ben Tun Qi — An upward rushing sensation from the lower abdomen to the chest, throat, or even the vertex of the head. It often occurs with symptoms like a stifling sensation at the chest, rapid breathing heart palpitation, dizziness, and abdominal pain. It is a result of rebellious Qi in Chong channel.

Beng Lou — A group of disorders that would be classified as dysfunctional uterine bleeding in Western medicine. *Beng Lou* refers to uterine bleeding that does not happen during the normal time in the menstrual cycle. *Beng* means excessively heavy uterine bleeding, while *Lou* means continuous spotting that does happen during the normal time in the menstrual cycle.

Bi syndrome — *Bi* literally means "obstruction" in Chinese. Bi syndrome refers to a variety of conditions in which there is a painful obstruction to the normal flow of Qi and Blood. In general, Qi stagnation is less severe, and Blood stagnation is more severe. Bi syndrome can be caused by an external invasion of wind, cold, or damp (see the term *external pathogenic factors*). It can also be caused by a disharmony of the Zang Fu (see below) or by a traumatic injury. Joint and body pains are the conditions most commonly diagnosed as Bi syndromes, but the term applies to a variety of conditions, including chest pain and headaches.

Blood — This word carries a different meaning in the context of Chinese medicine than it does in standard English, often being capitalized. In TCM, blood is defined as a red liquid circulating inside the blood vessels that is a vital substance of the human body. The functions attributed to blood are nourishing and moistening the body's tissues and organs, and providing a basis for mental activities.

Channel — See *meridian.*

Chi section — The third portion of the pulse that is felt with the practitioner's ring finger, which indicates Kidney Yin on the left wrist, Kidney Yang on the right wrist, and their related Fu organs.

Chong (meridian) — One of the *Eight Extra Meridians*. Chong acts as a sea of the blood, as well as the reservoir of all other meridians, and is especially important in women's reproductive functions, such as menstruation, fertility, etc.

Cold — One of the six *External Pathogenic Factors* (see below). Cold is a pathogen that causes contraction of tissues, and stagnation of Qi and blood. It exhausts the body as it consumes body heat (Yang Qi).

Cold and Heat — Two of the Eight Principles (see below). When using the *Eight Principles* to diagnose a patient, a disease is classified as being a "heat" disease or a "cold" disease. This is determined by evaluating characteristics of the tongue and pulse, bowel habits, urination, body temperature, and fatigue/energy level, among other factors.

Cun section — The first portion of the pulse that is felt with the practitioner's index finger, which indicates Heart on the left wrist, Lung on the right wrist, and their related Fu organs.

Dai (meridian) — One of the *Eight Extra Meridians*. Dai goes around the waist like a girdle, and acts as a girdle that holds and restricts the rest of meridians, and regulates the Qi and Blood circulating up and down. Dai is closely related to vaginal discharge.

Damp — One of the six *External Pathogenic Factors* (see below). Dampness is defined as being a heavy and turbid pathogen that lingers and is difficult to get rid of. It causes obstruction and stagnation, and tends to move downward. Dampness can invade the body from the exterior, but it can also be generated in the interior of the body by a disharmony of the Zang Fu (see below).

Deficiency — See *Excess and Deficiency.*

Dian — The manifestations of Dian are depression, melancholy, dull expression, abnormal laughing and crying, hallucination, delusion, and deranged speech. This is a Yin syndrome.

Dry — One of the six *External Pathogenic Factors* (see below). Dryness is mostly seen in chronic diseases. It has the characteristic of drying up body fluids and damaging the lung (see *Lung* below).

Du (meridian) — One of the *Eight Extra Meridians.* Du acts as the "general" of all Yang meridians and controls the distribution of Yang Qi in the body.

Eight Principles — One of several different systems of diagnosis in Chinese medicine. This system can be contrasted with *Six Channel Diagnosis*, *Four Stage Diagnosis*, *San Jiao Diagnosis*, and *Zang Fu Diagnosis.* The Eight Principles form the foundation of all systems of diagnosis in TCM. The principles are cold vs. heat, internal vs. external, excess vs. deficiency, and yin vs. yang. These four pairs of the Eight Principles are defined in more detail individually.

Excess and Deficiency — Two of the *Eight Principles* (see above). When the Eight Principles are used to diagnose a patient, a disease is classified as being a disease of excess or a disease of deficiency. The terms *excess* and *deficiency* refer to an excess of Xie Qi (see below), or a deficiency of Zheng Qi (see below). A variety of symptoms are used to assess whether a condition is from excess or deficiency, including signs from the tongue and pulse.

External — See *Internal and External.*

External Pathogenic Factors — According to Chinese medicine, there are three main ways that the body can become diseased. One way is an internal disharmony of organ functioning, caused by improper diet, overstrain, lack of exercise, or pathological emotions (see *Seven Emotions*). Another way is an invasion of the body by an *external pathogenic factor* (epi), and the third way is trauma and injury. There are six External Pathogenic Factors. They are *wind*, *cold*, *damp*, *dry*, *heat*, *fire*, and *summer heat.* Heat and fire are the same pathogen but differ in degrees of severity. The epis are thought to invade the body through a series of layers, beginning with the exterior (skin) and ending with the interior (internal organs). All of the epis are defined in greater detail individually.

Fei — See *Lung.*

Feng Shui — Categorized as a type of edema caused by external pathogenic wind usually presenting with edema more pronounced on the face and potentially moving throughout the body.

Feng Zhen — Skin rash caused by wind or wind–heat that comes and goes quickly.

Feng Dan — Skin spots and their merged patches that often are pink or red, coming and going as quickly as wind.

Fire — One of the six *External Pathogenic Factors* (see above). Fire is the same pathogen as *heat* (see below), but in a more severe form. Fire has the characteristics of damaging body fluids and consuming Qi. Fire moves upward and disturbs the mind and spirit, causing agitation. It causes the blood to circulate in a disorderly and reckless manner, and can lead to the "stirring of internal wind" (see *wind* below).

Five Elements — An ancient Chinese theory that forms the basic foundation of Chinese medicine, along with the *Yin/Yang* theory (see below). Five Elements theory states that there are five elements—wood, fire, earth, metal, and water—that have specific constructive and destructive relationships with each other. For example, wood generates fire, while water controls fire. Through the relationships among the Five Elements, all the elements control and harmonize each other. It is important to note that the Five Elements are archetypes and not "elements" as we would think of elements in the context of chemistry (hydrogen, helium, etc.). In the human body, the organs (*Zang Fu,* see below) and various body parts are ascribed relationships to the Five Elements.

Four Level diagnosis — One of several different systems of diagnosis in Chinese medicine. This system can be contrasted with *Eight Principles*, *Six Channel diagnosis*, *San Jiao diagnosis*, and *Zang Fu diagnosis.* Four Level diagnosis deals with the invasion of the body by warm-natured *External Pathogenic Factors* (see above). It outlines how a pathogen invades the body from the exterior levels to the deep internal levels, level-by-level; how to diagnose and treat the disease at each level, and how the natural untreated progression of the disease will eventually lead to death.

Gan Ji — This term refers to children who have been malnourished, and have developmental disabilities as a result. They look small for their age, and are slow and listless, without any energy. As they grow up, they have learning disabilities and emotional problems.

Gan — See *Liver.*

Gan Zhuo — A syndrome that results from long-term Qi and blood stagnation in the Liver. Its manifestation includes distended sensation or pain in the hypochondriac region, better with pressure, and a preference for warm drinks.

Gu Qi — *Qi* (see below) is a central concept in Chinese medicine. Qi in the human body is thought to be formed, or derived, from three sources: Jing Qi (see *Jing*), *Qing Qi* (see below), and Gu Qi. Gu Qi is the Qi that is extracted from food by the digestive system (see *Spleen*).

Guan Section — The second portion of the pulse that is felt with the practitioner's middle finger, which indicates Liver on the left wrist, Spleen on the right wrist, and their related Fu organs.

Heart — The Chinese word for Heart is *Xin*. The Heart is one of the five *Zang Fu* (see below). The Heart system is responsible for circulating blood and for providing a place for a person's Shen (see *Shen 2*) to reside. The Heart system is partially responsible for emotionally stability and mental activities, and also has much to do with sleep and sleeping disorders. The tongue, blood vessels, and face are under the influence of the Heart. The Heart's Fu organ "partner" is the Small Intestine. In *Five-Elements* theory (see above), the Heart and Small Intestine are "fire" organs.

Heat — One of the six *External Pathogenic Factors* (see above). Heat is the same pathogen as *Fire* (see above), but in a less severe form. Heat, like Fire, has the characteristics of causing Qi and blood to circulate recklessly, moving upward to disturb the mind and spirit. However, Heat does not consume the body fluids as readily as Fire, and Heat cannot generate internal wind. Heat must first transform into Fire, and then Fire can generate internal wind.

Huang Di — The legendary "Yellow Emperor" and author of the *Nei Jing* (see below). It is unclear as to whether Huang Di was a real person or if his writings are actually a collaboration of many different peoples' writings. The earliest mention of the *Nei Jing* in classical Chinese literature is in a text written around 100 B.C.

Internal and External — Two of the *Eight Principles* (see above). When using the Eight Principles to diagnose a patient, a disease is classified as being situated in either the exterior or the interior. A variety of symptoms are used to assess whether a condition is interior or exterior, including signs from the tongue and pulse. An exterior condition is an invasion of external pathogenic factor (see above), and an interior condition is a disharmony of the *Zang Fu* (see below) or an external pathogenic factor that has invaded the deep levels of the body.

Jiao —"Burner" or "warmer." Any of three specific regions (collectively the San, or three Jiao) of the thorax and its organs: the Upper Jiao is the region above the diaphragm and includes the heart and lungs; the Middle Jiao is the region from the diaphragm to the navel and includes the spleen and stomach; and the Lower Jiao is the region from the navel to the upper border of the pubic bone and includes the liver and kidneys.

Jin Gui Yao Lue — This book was written by *Zhang Zhong Jing* (see below) around 300 A.D. It contains the differentiation and formulas on TCM internal medicine, gynecology, rheumatology, and miscellaneous diseases. Its counterpart is *Shang Han Lun* (see below).

Jin Ye — One of the body's basic materials: body fluids. Usually, *Jin* is the thicker fluid, while *Ye* is the thinner fluid. The pathological counterpart of Jin Ye is called *Tan Yin* or *Shui Yin.*

Jing — Usually translated as "essence," *Jing* can also be called *Jing Qi*. It is thought of as a highly potent essence that is transmitted from parent to child, and is thus also called the "congenital root" or "congenital Qi." A person begins life with a finite amount of Jing and can consume or conserve the Jing through various activities. Jing Qi is one of the three sources for the formation of Qi in the human body, along with *Gu Qi* (see above) and *Qing Qi* (see below). Jing is stored in the *Kidney* (see below).

Jue (syndrome) — A syndrome in which patients manifest sudden cold extremities, especially in the distal ends of the extremities. It is often due to Qi stagnation in the meridians or Qi disturbance with emotional stimulations.

Jue Yin — Usually translated as "Ultimate Yin," this is the last of the *Six Channels* (see *Six Channel Diagnosis*). It is characterized by an imbalance between the body's Yin and Yang, producing heat and cold symptoms. This manifests as thirst, a sensation of air rushing up the chest, with palpitations and chest pain hunger without a desire to eat, cold limbs, vomiting, and diarrhea.

Kidney — The Chinese word for Kidney is Shen (*Shen 1*). The Kidney is one of the five *Zang Fu* (see below). The Kidney system is responsible for storing the *Jing* (see above), regulating water metabolism (forming and excreting urine), and assisting the *Lung* (see below) in receiving Qi from the air. As the Kidney stores the Jing, it is responsible for aging in general. The hair, ears, and bones are all under the influence of the Kidney, and these are all parts of the body that change significantly as people age. The Kidney's Fu organ "partner" is the Urinary Bladder. It is considered a "water" organ in Five Elements theory (see above).

Kuang — The manifestations of Kuang are madness, mania wild behavior, uncontrollable violent behavior, shouting, and screaming obscenities, This is a Yang syndrome.

Li Jie (Feng) — One type of Bi syndrome that can be classified into wandering Bi. Li Jie (Feng) is caused by invasion of wind–cold–damp in the meridians and collaterals leading to Qi and blood blockage. Its manifestations are migrating severe joint pain and swelling, red and hot affected joints that are unable to move smoothly, and pain that is worse at night. Li Jie (Feng) might be seen in conventional medicine classified under diseases such as rheumatic fever or rheumatoid arthritis.

Lin syndrome — This term is often translated as "painful urination syndrome." It refers to a variety of disorders of the urinary tract, including blood in the urine, kidney stones, and cloudy and turbid urine, among other conditions.

Liver — The Chinese word for Liver is Gan. The Liver is one of the five *Zang Fu* (see below). The Liver system is responsible for maintaining the free flow of Qi in the body. When the Qi flows smoothly, emotions are balanced, blood and body fluids circulate smoothly, the digestive tract operates harmoniously, and a female's menstrual cycle is regulated. The Liver also has the function of storing blood. It is thought that blood circulates in the vessels during exertion and returns to the Liver during rest. Because the Liver stores blood, it is also very involved with menstruation. The eyes, nails, and tendons are all under the influence of the Liver, and dysfunction of these body parts can indicate an imbalance in the Liver system. The Liver's Fu organ "partner" is the Gallbladder. It is considered a "wood" organ in *Five Elements* theory (see above).

Lung — The Chinese word for Lung is Fei. The Lung is one of the five *Zang Fu* (see below). The Lung system is responsible for inhaling "good Qi" (see *Qing Qi* below) and exhaling "waste Qi." It is said to "dominate descending and dispersing," which means that the Lung is involved with the dispersing of fluids evenly throughout the body. The Lung also disperses *Wei Qi* (see below) throughout the body, and therefore has functions that Western science attributes to the immune system. The skin, nose, and pores are under the influence of the Lung. The Lung's Fu organ "partner" is the Large Intestine. It is considered a "metal" organ in *Five Elements* theory (see above).

Man — One symptom of Yangming Fu syndrome. *Man* means abdominal bloating.

Meridian — A pathway for the flow of Qi (see below) in the human body. The network of meridians in the human body is analogous to the cardiovascular system; larger vessels branch out into increasingly smaller and smaller vessels. There are 12 major meridians that correspond to the 12 *Zang Fu* (see below), and there are eight "extraordinary" meridians that do not directly correspond to the *Zang Fu*. Most acupuncture points are located along these 20 meridians.

Mingmen — A special place where Kidney Yang or True Fire is stored. The term *Mingmen* fire is usually used when Kidney Yang or True Fire are emphasized.

Nei Jing — The "bible" of TCM. This book is over 2000 years old, and was supposedly written by the legendary "Yellow Emperor," *Huang Di* (see above). Because of this, it is often referred to as the *Huang Di Nei Jing*. The *Nei Jing* discusses Yin/Yang theory and *Five Elements* theory as they apply to the human body; *Zang Fu* theory, *tongue diagnosis*, *pulse diagnosis*, *meridian* theory, disease processes, and acupuncture techniques, among other subjects (the italicized terms are defined in more detail individually in the Glossary).

Pian — When seen in the name of a Chinese herbal prescription, this word usually means that the formula is prepared as a tablet. For example: An Mian *Pian* or Dan Shen *Pian*.

Pi — See *Spleen*.

Pi — One symptom of Yangming Fu syndrome. *Pi* means a congested or "stuck" sensation.

Pinyin — A system of romanizing Chinese words.

Pi Shui — Categorized under skin edema occurring more on the superficial level, usually caused by Lung Qi congestion unable to distribute water.

Plum-Pit Qi — A condition in which a person has the sensation of something being stuck in his/her throat that cannot be coughed up or swallowed down, when there is no actual physical obstruction in the throat. This is caused by a stagnation of Liver Qi due to strong emotions that depress the Liver's function of circulating Qi (see *Liver* above).

Pulse diagnosis — One method through which a practitioner of TCM gains information about his/her patient. The practitioner feels the radial pulse across the length of three fingertips, on both wrists. Each of the *Zang Fu* (see below) are correlated with a specific position in the pulse. A pulse is diagnosed by analyzing its depth, speed, rhythm, strength, and shape. Variations in each of these qualities indicate the characteristics and locations of diseases in the body.

Qi — Pronounced "chee." There is no single English word that can adequately convey the meaning of Qi. Qi is both a physical substance and a nonphysical energy. When Qi is spread out, it is energy, and when it congeals, it is substance. Movements and changes of Qi are used to explain all phenomenons in the universe. In the human body, Qi has six basic functions: warming, protecting (immunity), nourishing, transforming (extracting nutrients from food, filtering out wastes, etc.), holding things in place (keeping blood inside blood vessels, preventing organ prolapse, etc.), and promoting growth.

Qi Gong — A form of meditation in which practitioners strive to control their own body's Qi with their minds. The practice of Qi Gong includes breathing exercises, sitting postures, and visualizations. Contrasted with *Tai Chi* (see below), Qi Gong generally involves much less movement and more mental exercises.

Qi Hua — Hua means transformation or generation. Qi Hua means certain changes are ongoing or completed with involvement of Qi activities. Qi Hua also indicates some functions of Zang Fu organs, such as San Jiao Qi Hua or Urinary Bladder Qi Hua.

Qing Qi — *Qi* (see below) is a central concept to Chinese medicine. Qi in the human body is thought to be formed or derived from three sources: Jing Qi (see *Jing*), *Gu Qi* (see above), and Qing Qi. Qing Qi is the Qi that is extracted from the air by the Lungs (see *Lung*).

Ren (meridian) — One of the *Eight Extra Meridians*. Ren acts as a "general" of all the Yin meridians, and controls conception and nourishment of the fetus for women.

San — When seen in the name of a Chinese herbal formula, this word usually means that the formula is prepared as a powder. For example: Yin Qiao *San* or Dao Chi *San*.

San Jiao — One of the six Fu organs (see *Zang Fu* below). The San Jiao is a series of pathways in the human body that are used to distribute fluids and *Yuan Qi* (see below) throughout the body. The human body is divided into three areas, or Jiaos. They are the Upper Jiao (above the diaphragm), the Middle Jiao (below the diaphragm to the navel), and the Lower Jiao (the lower abdomen below the navel). The San Jiao is a good example of how the *Zang Fu* organs are not the same as the organs in a scientific context. The functions of the San Jiao can be attributed to no single organ in Western medicine. The organs and tissues that share the functions of the San Jiao are the lymphatics, the fascial system, and the hypothalamus, among others.

San Jiao diagnosis — One of several different systems of diagnosis in Chinese medicine. This system can be contrasted with *Six Channels diagnosis*, *Four Levels diagnosis, Eight Principles*, and *Zang Fu diagnosis*. San Jiao diagnosis deals with the invasion of the body by damp *External Pathogenic Factors* (see above). It outlines how a pathogen invades the body from the exterior levels to the deep internal levels, stage by stage, how to diagnose and treat the disease at each stage, and how the natural untreated progression of the disease will eventually lead to death.

Seven Emotions — According to Chinese medicine, there are three main ways that the body can become diseased. One way is an internal disharmony of organ functioning, another way is an invasion of the body by *External Pathogenic Factors* (see above), and the third way is trauma and injury. According to theory, imbalances of the emotions can cause imbalances of organ functioning. Different emotions affect the different organs according to *Five Elements* theory (see above). For example, the Liver is affected by anger, the kidney by fear, and so on.

Shang Han Lun — This book is equally as important as the *Nei Jing* (see above) as a foundation to Chinese medicine. This book was written by *Zhang Zhong Jing* (see below) around 300 A.D. It contains the theory of *Six Channel Diagnosis* (see below), and is the first book that explains differential diagnosis. The counterpart of *Shang Han Lun* is *Jin Gui Yao Lue* (see above).

Shao Yang — Usually translated as "Lesser Yang," this is the third of the Six Channels (see *Six Channel Diagnosis*). In this stage, the pathogen is trapped half-interior and half-exterior, producing symptoms such as alternating chills and fever, vomiting, a bitter taste in the mouth, blurry vision, dizziness, and sweating. The next channel is *Tai Yin*.

Shao Yin — Usually translated as "Lesser Yin," this is the fifth of the Six Channels (see *Six Channel Diagnosis*). This stage is characterized by systemic weakness and dysfunction. A person's constitution determines whether they develop Shao Yin cold or Shao Yin heat syndrome. Shao Yin cold manifests as fatigue and apathy, diarrhea, poor digestion, cold limbs, profuse clear urine, and palpitations. Shao Yin heat manifests as restlessness, insomnia, mania, dark and concentrated urine, and a dry throat. The last stage is *Jue Yin*.

Shen (1) — See *Kidney*. Also, compare with *Shen (2)* below.

Shen (2) — Each of the *Zang Fu* organs (see below) are thought to "house" a spirit. The human body is thought to be animated by five different spirits, each one inhabiting a different *Zang* organ, and each representing a different aspect of human consciousness. The *Shen* is the spirit that resides in the heart. Being awake, aware, and conscious of your surroundings, and interacting with other people are things associated with the Shen. If a person is mentally unsound, it is said that their *Shen* is disturbed.

Shen Zhuo — A syndrome that results from internal accumulation of wind, cold, and damp in the Kidney. Its manifestations include cold and heavy pain in lower back, decreased range of motion due to extreme pain, aggravated by cold and humid weather, and does not improve with rest.

Shi — One symptom of Yangming Fu syndrome. *Shi* means a palpable or visible accumulation or lump in the intestines/abdomen.

Si Qi Wu Wei — Literally, this means, "Four Directions, Five Flavors." It is a fundamental principle to Chinese herbal medicine. Chinese herbs are categorized according to their flavor and direction of action (among other functions). The four directions are up, down, outwards, and inward/down. The five flavors are bitter, sour, salty, sweet, and pungent/acrid. The five flavors are correlated with the five *Zang Fu* (see below) according to *Five Elements* theory (see above).

Six Channel diagnosis — One of several different systems of diagnosis in Chinese medicine. This system can be contrasted with *San Jiao diagnosis*, *Four Level diagnosis*, *Eight Principles*, and *Zang Fu diagnosis*. Six Channel diagnosis deals with invasion of the body by cold *External Pathogenic Factors* (see above). It outlines how a pathogen invades the body from the exterior levels to the deep internal levels, stage by stage, how to diagnose and treat the disease at each stage, and how the natural untreated progression of the disease will eventually lead to death. Six Channel diagnosis was established in the classic book, the *Shang Han Lun* (see above). The Six Channels — *Tai Yang*, *Yang Ming*, *Shao Yang*, *Tai Yin*, *Shao Yin*, and *Jue Yin* — are defined in more detail individually.

Spleen — The Chinese word for Spleen is *Pi*. The Spleen is one of the five *Zang Fu* (see below). The Spleen system is responsible for extracting "good Qi" from food (see *Gu Qi* above) and for distributing this good Qi throughout the body. This function is referred to as "transformation and transportation." The Stomach, Small Intestine, and Large Intestine all extract nutrients from food, and send them up to the Spleen. Additionally, the Spleen is responsible for holding the internal organs in place, and for keeping the blood inside of the blood vessels. The skeletal muscles, lips, and mouth are all under the influence of the Spleen. The Spleen's Fu organ "partner" is the Stomach. It is considered to be an "earth" organ in *Five Elements* theory (see above).

Summer Heat — One of the six *External Pathogenic Factors* (see above). Summer heat is a specific type of pathogen that can only invade the body during the summer. It is characterized by intense heat (high fever, thirst, sweating, etc.) accompanied by sensations that are attributed to dampness (heavy sensations, loose stools, abdominal pain, etc.). It is the same thing as summer flu.

Tai Chi — A form of exercise that involves moving the body very slowly and deliberately through a sequence of poses. All of the poses are done while standing. In some forms of Tai Chi, a sword is incorporated into the poses. Tai Chi is considered a form of "moving meditation," but contrasted with *Qi Gong* (see above), Tai Chi involves much more movement, while still paying attention to the breath and keeping the mind clear.

Tai Yang — Usually translated as "Great Yang," this is the first of the Six Channels (see *Six Channel diagnosis*). This is the initial stage, where febrile disease is caused by an invasion of the exterior parts of the body by cold pathogens. This usually manifests as chills and fever, a stiff neck, headache, and possibly sweating. The next channel is *Yang Ming*.

Tai Yin — Usually translated as "Great Yin," this is the fourth of the Six Channels (see *Six Channel diagnosis*). In this stage, the Spleen Yang has been damaged, causing deficiency cold. This manifests as abdominal fullness, vomiting, diarrhea, poor appetite, and a sensation of coldness. The next channel is *Shao Yin*.

Tang — When seen in the name of a Chinese herbal formula, this word usually means that the formula is prepared as a water decoction. For example: Gui Pi *Tang* or Ma Huang *Tang*.

TCM — An acronym for *Traditional Chinese Medicine*.

Tian Gui — This term refers to a substance that is released in the body to induce sexual maturity. It is equivalent to the hormonal changes that induce puberty.

Tongue diagnosis — One method through which a practitioner of TCM gains information about his/her patient. The tongue is divided into several areas, and each of the organs are represented in these areas. A practitioner will look at the overall appearance of a tongue, including the quality and color of the tongue coating, the color, shape, and size of the tongue body, and the condition of the blood vessels underneath the tongue. The condition of the internal organs is thought to be reflected in the tongue.

Tuo and Bi — *Tuo* means collapse of Qi, which indicates a severe deficient syndrome in TCM, while *Bi* means "locked-up" or severely stagnant Qi, which indicates a severe excessive syndrome in TCM. They are both critical conditions.

Wan — When seen in the name of a Chinese herbal formula, this word usually means that the formula is prepared as a pill. For example: Liu Wei Di Huang *Wan* or Qing Qi Hua Tan *Wan*.

Wei Qi — This type of Qi is often referred to as "defensive Qi." Its functions are equivalent to functions that Western science attributes to the immune system. Wei Qi is derived from *Gu Qi* (see above), and is dispersed throughout the body by the *Lung* (see above). Wei Qi is thought to gather at the surface of the body to fight off *External Pathogenic Factors* (see above). When Wei Qi is weak, External Pathogenic Factors can invade into deeper levels of the body.

Wei syndrome — *Wei* means "atrophy" in Chinese. Wei syndrome refers to a condition in which a muscle or limb has atrophied and has lost function. It is thought to be a natural consequence of untreated *Bi syndrome* (see above). and is due to an obstruction in the flow of Qi and blood. This can cause a muscle or limb to be deprived of nourishment, thus resulting in atrophy.

Wind — One of the six *External Pathogenic Factors* (see above). Wind has the characteristics of moving upward and attacking the exterior areas of the body. It is also a "leading pathogen" that can lead other External Pathogenic Factors into the body, such as cold, heat, or dampness. Most febrile diseases are diagnosed as an invasion of external wind. Wind that is produced internally by a disharmony of the *Zang Fu* is an entirely different phenomenon than external wind. Internal wind refers to convulsions, tremors, seizures, and other such diseases.

Wu Xing — See *Five Elements*.

Xian — Recurrent sudden brief attacks with altered consciousness, loss of spirit, pale complexion, upward staring by the eyes, clenched jaw, foaming at the mouth, and convulsions of the extremities. In some cases, there is squealing like a goat or pig. This is why it is also called Yang Xian Feng (goat epileptic wind).

Xiao Ke — A consumptive condition which manifests as thirst, excessive hunger, profuse urination, and weight loss (similar to diabetes). It can be differentiated into three patterns: Upper Xiao, which mainly presents as thirst; Middle Xiao, which mainly presents as hunger; and Lower Xiao, which mainly presents as profuse urination.

Xie Qi — In very general terms, health and disease is seen in Chinese medicine as being a balance of the body's strength and resistance vs. the strength of pathogenic forces that attack the body. *Xie Qi* is a general term for the pathogenic forces that attack the body. Contrast with *Zheng Qi* (below). There is a saying that there can never be too little Xie Qi or too much Zheng Qi.

Xin — See *Heart*.

Xue — See *Blood*.

Yang Huang — One of two types of jaundice. *Huang* means "yellow." Yang Huang is caused by damp–heat of Liver and Gallbladder and commonly presents with a bright tangerine yellow skin complexion. Additional presentations include dark yellow urine, nausea, vomiting, poor appetite, constipation, abdominal bloating, and pain in the hypochondriac region.

Yang Shui — One of two types of edema. Yang Shui is often acute and has an excessive presentation of heat. It is caused by Lung Qi congestion, San Jiao congestion, and Urinary Bladder Qi stagnation.

Yang Ming — Usually translated as "Bright Yang," this is the second of the Six Channels (see *Six Channel Diagnosis*). In this stage, stagnant heat accumulates in the stomach and intestines if the exterior pathogen is not resolved. This manifests as constipation and abdominal pain, profuse sweating, high fever, and irritability. The next channel is *Shao Yang*.

Yin and Yang — An ancient Chinese philosophical concept. Literally, *Yin* is "the shady side of the mountain," and *Yang* is "the sunny side of the mountain." Yin and Yang represent polar opposites that are observed in all natural phenomenon (for example, fire and water, or night and day.) Yin and Yang are relative to each other, and any phenomenon can be infinitely divided into its Yin and Yang aspects. For example, the skin is considered to be Yang relative to the organs, but among the organs, there are certain Yang organs and certain Yin organs. Yin/Yang theory forms the basic foundation of Chinese medicine, along with the *Five Elements* (see above).

Yin Huang — One of two types of jaundice. *Huang* means yellow. *Yin Huang* is caused by Spleen Yang deficiency leading to cold–damp accumulation and presents with a sallow yellow complexion. Yin Huang may also be caused by the inappropriate use of cold herbs. Additional symptoms may include aversion to cold, fatigue, abdominal distention, poor digestion, and diarrhea.

Ying Qi — Often translated as "nutritive Qi." This Qi is carried by the blood inside the blood vessels, and is distributed to all of the organs. Its functions are to nourish the organs and tissues. Ying Qi is a concentrated form of the highest-quality *Gu Qi* (see above), whereas *Wei Qi* (see above) is formed from the *Gu Qi* that is not high-enough quality to become *Ying Qi*.

Yin Shui — One of two types of edema. *Yin Shui* is often chronic and has a deficient presentation of cold. It is caused by Spleen and Kidney Yang deficiency leading to failure of water transformation and transportation.

Ying Zheng — Also known as Ying Liu, or enlarged neck. It often results from Liver Qi stagnation, Qi, and phlegm accumulation due to Spleen dysfunction or from inadequate iodine intake. *Ying Zheng* is similar to enlargement of the thyroid gland.

Yuan Qi — Often translated as, "original Qi," or "primary Qi." *Yuan Qi* is the most important kind of Qi in the body. It is formed by combining *Gu Qi* (see above) and *Jing Qi* (see above), and is distributed to all of the organs via the *San Jiao* (see above). It has the function of stimulating and maintaining organ functions; when an organ is not functioning properly, there is a problem with its *Yuan Qi*.

Zang Fu — The internal organs of the human body. It is important to understand that the Western scientific concept of the internal organs is entirely different from Zang Fu theory. In Zang Fu theory, there are five main organ systems: Lung, Spleen, Liver, Heart, and Kidney. Each organ system includes a *Zang* organ and its *Fu* organ "partner." In general, Zang organs are solid, and Fu organs are hollow. Each Zang Fu system is responsible for a set of bodily functions that Western science may or may not attribute to those physical organs. For example, the Heart system is responsible for circulating blood (which correlates with Western science), but it is also responsible for memory and thinking (functions that Western science attributes to the brain). Western science and Zang Fu theory are not thought of as antagonistic theories, but more as different ways of viewing the human body. The five organ systems are defined in more detail individually.

Zang Fu diagnosis — One of several different systems of diagnosis in Chinese medicine. This system can be contrasted with *San Jiao Diagnosis*, *Four Level Diagnosis*, *Eight Principles*, and *Six Channel Diagnosis*. Zang Fu diagnosis is basically a synthesis of many of the other systems of diagnosis in TCM. It deals with signs and symptoms associated with dysfunctions of the *Zang Fu* organ systems. Zang Fu diagnosis is the most useful form of diagnosis for diseases of the internal organs, while *San Jiao Diagnosis, Six Channel diagnosis,* and *Four Level diagnosis* are more appropriately used to diagnose febrile diseases.

Zao — One symptom of Yangming Fu syndrome. *Zao* means irritability and dry stools.

Zhang Zhong Jing — The author of the *Shang Han Lun* (see above). He created *Six Channel diagnosis*, which laid the foundation for much of Chinese herbal medicine and for differential diagnosis.

Zheng Qi — In very general terms, health and disease is seen in Chinese medicine as being a balance of the body's strength and resistance vs. the strength of pathogenic forces that attack the body. Zheng Qi is a general term for the body's strength and resistance. Contrast with *Xie Qi* (above).

Zhi Yin — A syndrome that is caused by evil Yin (water) accumulation in the chest and diaphragm leading to Lung Qi rebellion. It presents as chest constriction, shortness of breath, cough, wheezing, inability to lie flat, and edema. It may also present with dizziness, blurred vision, congested focal distention, and hardness in epigastrium, as well as a dark sallow complexion.

APPENDIX 3

Classical Text Listing

Bao Ming Ji also known as Su Wen Bing Ji Qi Ye Bao Ming Ji *Collection of Writings on the Mechanism of Illness, Suitability of Qi, and the Safeguarding of Life as Discussed in the Basic Questions*
Ben Cao Yan Yi *Amplification on Canon of Materia Medica* (1116, Kou Zong-Shi)
Ben Shi Fang *Formulas of Universal Benefit from My Practice* (1132, Xu Shu-Wei)
Cheng Fang Bian Du *Convenient Reader of Established Formulas* (1904, Zhang Bing-Cheng)
Chong Lou Yu Yao *Jade Key to Many Towers* (1838, Zheng Mei-Run)
Ci Shi Nan Zhi *Hard-won Knowledge* (1308, Wang Hao-Gu)
Dan Xi Xin Fa *Dan Xiís Experiential Therapy* (1347, Zhu Dan-Xi)
Fu Qing Zhu Nu Ke *Fu Qing-Zhuís Obstetrics and Gynecology* (1826, Fu Qing-Zhu)
Fu Ren Liang Fang *Fine Formulas for Women* (1237, Chen Zi-Ming)
Gu Fang Ba Zhen *Eight Positions of Ancient Formulas* (Refer to Jing Yue Quan Shu *Complete Work of [Zhang] Jine-Yue*)
Gu Jin Lu Yan *Ancient and Modern Records of Experiential Formulas* (Tang dynasty, Zheng Li Yan)
Gu Jin Yi Tong *Comprehensive Collection of Ancient and Modern Medicine* (1556, Xu Chun-Fu)
Guang Wen Re Lun *Discussion of Widespread Warm febrile Disorders* (1722, Dai Tian-Zhang)
Han Shi Yi Tong *Han's Treatise on General Medicine* (1522, Han Mao)
Han Wen Tiao Bian *Systematic Differentiation of Cold and Warm Disorders* (1784, Yang Xuan)
He Ji Ju Fang *Prescriptions of Peaceful Benevolent Dispensary* (1151, Chen Shi-Wen, et al.)
Huo Luan Lun *Discussion of Sudden Turmoil Disorders* (1862, Wang Shi-Xiong)
Ji Sheng Fang *Formulas to Aid the Living* (1253, Yan Yong-He)
Ji Yin Gang Mu *Synopsis of treating Women's Disease* (1620, Wu Zhi-Wang)
Jin Gui Yao Lue *Essentials from Golden Cabinet* (219, Zhang Zhong-Jing)
Jing Yue Quan Shu *Complete Works of [Zhang] Jing-Yue* (1624, Zhang Jing-Yue)
Lan Shi Mi Cang *Secret Record of the Chamber of Orchids* (1249, Li Dong-Yuan)
Lei Zheng Huo Ren Shu *Book to Safeguard Life Arranged According to Pattern* (1108, Zhu Hong)
Liang Fang Ji Ye *Small Collection of Fine Formulas* (1842, Xie Yuan-Qing)
Liu Zhou Yi Hua *Medical Talks at Liu State* (1838, Wei Zhi-Xiu)
Ma Ke Huo Ren Quan Shu *Complete Book to Safeguard Life in Anesthesiology* (Qing dynasty, Xie Yu-Qiong)
Ming Yi Za Zhu *Miscellaneous Works of Physicians in Ming Dynasty* (1502, Wang Ken-Tang)
Nei Ke Zhai Yao *Summary of Internal Medicine* (1548, Xue Ji)
Nei Wai Shang Bian Huo Lun *Clarifying Doubts about Endogenous and Exogenous Diseases* (1231, Li Dong-Yuan)
Nu Ke Ji Yao *Collected Essentials on Obstetrics and Gynecology* (Qing dynasty, Shen Yao-Feng)
Pi Wei Lun *Discussions on Spleen and Stomach* (1249, Li Dong-Yuan)
Pu Ji Fang *Prescriptions for Universal Benefits* (1406, Chu Su et al.)
Qian Jin Yi Fang Supplement to *Thousand Ducat Formulas* (652, Sun Si-Miao)
Qian Jin Yao Fang *Thousand Golden Prescriptions* (652, Sun Si-Miao)

Quan Guo Zhong Cao Yao Zhi Liao Hui Bian *National Compilation of Chinese Herbal Treatments* (1975, China Academy of TCM)
Quan Guo Zhong Cheng Yao Chu Fang Ji *National Collection of Chinese Herbal Prepared Medicines* (1964, China Academy of TCM)
Quan Sheng Ji (refer to Wai Ke Zheng Zhi Quan Sheng Ji *Complete Collection of Patterns and Treatments in External Medicine*)
Ru Men Shi Qin *Confucians' Duties to Their Parents* (1228, Zhang Cong-Zheng)
San Yin Fang *Discussion of Pathology Based on the Triple-Etiology Doctrine* (1174, Chen Yan)
Shang Han Lun *Discussions of Cold-induced Disorders* (219, Zhang Zhong-Jing)
She Sheng Zhong Miao Fang *Marvelous Formulas for the Health of the Multitudes* (1550, Zhang Shi-Che)
Shen Shi Zun Sheng Shu *Shen's Book for Revering Life* (1773, Shen Jin-Ao)
Sheng Hui Fang *Holy Peaceful Benevolent Prescriptions* (992, Wang Huai-Yin, et al.)
Sheng Ji Zong Lu *Comprehensive Recording of Sage-Like Benefit* (1122, Shen Fu, et al.)
Shi Fang Ge Kuo *Collected Songs about Contemporary Formulas* (1801, Chen Xiu-Yuan)
Shi Yao Shen Shu *Teen Effective Remedies* (1348, Ge Qian-Sun)
Shi Yi De Xiao Fang *Effective Prescriptions for Generations* (1337, Wei Yi-Lin)
Su Wen Bing Ji Qi Yi Bao Ming Ji *Collection of Writings on the Mechanism of Illness, Suitability of Qi, and the Safeguarding of Life as Discussed in the Basic Questions* (1186, Zhang Yuan-Su)
Tang Tou Ge Jue *Prescriptions in Rhyme* (1694, Wang Ang)
Tong Su Shang Han Lun *Popular Guide to the Discussion of Cold-induced Disorders* (Qing dynasty, Yu Gen-Chu)
Wai Ke Quan Sheng Ji *Complete Collection of Patterns and Treatments in External Medicine* (1740, Wang Wei-De)
Wai Ke Zheng Zong *True Lineage of External Medicine* (1617, Chen Shi-Gong)
Wai Tai Mi Yao *Arcane Essentials from the Imperial Library* (752, Wang Tao)
Wei Sheng Bao Jian *Main Rules in Medical and Health Service* (1343, Luo Tian-Yi)
Wen Bing Tiao Bian *Systematic Differentiation of Warm Diseases* (1798, Wu Ju-Tong)
Wen Re Jing Wei *Compendium on Seasonal Febrile Diseases* (1852, Wang Shi-Xiong)
Wen Yi Lun *Discussions on Epidemic Warm Diseases* (1642, Wu You-Xing)
Xian Xing Zhai Yi Xue Guang Bi Ji *Wide-ranging Medical Notes from the First-awakened Studio* (1613, Miao Xi-Yong)
Xiao Er Yao Zheng Zhi Jue *Key to Differentiation and Treatment of Diseases of Children* (1114, Qian Yi)
Yan Fang Xin Bian *New Compilation of Proved Formulas* (1846, Bao Xiang-Ao)
Yang Ke Xin De Ji *Collection of Personal Experiences Concerning Skin Lesions* (1805, Gao Bing-Jun)
Yi Fang Ji Jie *Analytic Collection of Medical Formulas* (1682, Wang Ang)
Yi Fang Kao *Investigations of Medical Formulas* (1584, Wu Kun)
Yi Lin Gai Cuo *Corrections on the Errors of Medical Works* (1830, Wang Qing-Ren)
Yi Men Fa Lu *Law of Medicine* (1856, Yu Chang)
Yi Xue Fa Ming *Medical Innovations* (1315, Li Dong-Yuan)
Yi Xue Ru Men *Elementary Course for Medicine* (1624, Li Yan)
Yi Xue Xin Wu *A Summary on Medicine from Clinical Practice* (1732, Cheng Guo-Peng)
Yi Xue Zheng Zhuan *True Lineage of Medicine* (1515, Yu Tuan)
Yi Xue Zhong Zhong Can Xi Lu *Records of Heart-felt Experiences in Medicine with Reference to the West* (1918–34, Zhang Xi-Chun)
Yi Yuan *Sources of Medicine* (1861, Shi Shou-Tang)
Yi Zhen Yi De *Achievements Regarding Eruptive Epidemic Disorders* (1794, Yu Shi-Yu)
Yi Zong Jin Jian *Golden Mirror for Original Medicine* (1742 Wu Qian, et al.)
You Ke Xin Fa *Experiential Therapy for Pediatrics* (refer to Yi Zong Jin Jian *Golden Mirror for Original Medicine*)
Yuan Ji Qi Wei *Causes and Mechanism of Eye Diseases* (1370, Ni Wei-De)
Zhang Shi Yi Tong *Zhang's Treatise on General Medicine* (1695, Zhang Lu)
Zheng Ti Lei Yao *Classification and Treatment of Traumatic Diseases* (1529, Xue Ji)
Zheng Yin Mai Zhi *Pattern, Cause, Pulse and Treatment* (1702, Qin Jing-Ming)
Zheng Zhi Hui Bu *Complied Supplementation to Standard Patterns and Treatment* (Qing dynasty, Li Yong-Cui)
Zheng Zhi Zhun Sheng *Standards Patterns and Treatment* (1602, Wang Ken-Tang)
Zhi Mi Fang *Clarify the Confusion Formulas* (Song dynasty, Wang Kuang)

Zhi Zhi Fang also known as Ren Zhai Zhi Zhi *Straight Directions of Benevolent Aid* (1264, Yang Shi-Ying)
Zhong Yi Fang Ji Lin Chuang Shou Ce *Clinical Handbook of Chinese Herbal Formulas* (1982, Zhu Hua-De, et al.)
Zhong Yi Fu Ke Zhi Liao Xue *Traditional Chinese Medicine Treatment of Women's Disorders* (1970s, Chengdu College of Traditional Chinese Medicine)
Zhou Hou Fang *Emergency Formulas to Keep Up One's Sleeve* (341, Ge Hong)

APPENDIX 4

Cross-Reference Pinyin–English Herbs

A Wei	*Chinese Giantfennel Resin*
Ai Di Cha	*Japanese Ardisia*
Ai Ye	*Mugwort Leaf*
An Xi Xiang	*Benzoin*
Ba Dou	*Croton Seed*
Ba Ji Tian	*Indian Mulberry Root*
Ba Jiao Feng	*Chinese Alangium Root*
Ba Jiao Hui Xiang	*Truestar Anisetree Fruit*
Ba Qia	*China Root*
Ba Yue Zha	*Akebia Fruit*
Bai Bian Dou	*Hyacinth Bean*
Bai Bu	*Stemona Tuber*
Bai Dou Kou	*Round Cardamom Fruit*
Bai Fu Zi	*Giant Typhonium Tuber*
Bai Guo	*Ginkgo Nut*
Bai Guo Ye	*Ginkgo Leaf*
Bai He	*Lily Bulb*
Bai Hua She	*Agkistrodon Snake*
Bai Hua She She Cao	*Spreading Hedyotis Oldenlandia Plant*
Bai Ji	*Hyacinth Bletilla Tuber*
Bai Ji Li	*Caltrop Fruit*
Bai Jiang Can	*Dead Sick Silkworm*
Bai Jiang Cao	*Patrinia*
Bai Jie Zi	*White Mustard Seed*
Bai Lian	*Ampelopsis*
Bai Mao Gen	*Woolly Grass Rhizome*
Bai Mu Er	*Fruiting Body of Tremella*
Bai Qian	*White Swallowwort*
Bai Shao Yao	*White Peony Root*
Bai Tou Weng	*Chinese Anemone Root*
Bai Wei	*Swallowwort Root*
Bai Xian Pi	*Dictamnus Root Bark*
Bai Ying	*Climbing Nightshade Plant*
Bai Zhi	*Angelica Root*
Bai Zhu	*Large-Headed Atractylodes Rhizome*
Bai Zi Ren	*Arbor-Vitae Seed*
Ban Bian Lian	*Chinese Lobelia*
Ban Lan Gen	*Woad Root*

Ban Mao	*Cantharides*
Ban Xia	*Pinellia Tuber*
Ban Zhi Lian	*Sun Plant*
Bei Xie	*Fish-Poison Yam*
Bei Xing Ren	*Apricot Seed*
Bi Ba	*Long Pepper Fruit*
Bi Cheng Qie	*Litse Fruit*
Bi Hu	*House Lizard*
Bi Qi	*Water Chestnut*
Bian Dou	*Hyacinth Bean*
Bian Dou Hua	*Hyacinth Flower*
Bian Dou Yi	*Hyacinth Bean Peel*
Bian Xu	*Knotweed*
Bie Jia	*Chinese Soft-Shelled Turtle Shell*
Bing Lang	*Betel Nut*
Bing Pian	*Borneol*
Bo He	*Peppermint*
Bu Gu Zhi	*Fruit of Scuffy Pea*
Can Sha	*Silkworm Feces*
Cang Er Zi	*Cocklebur Fruit*
Cang Zhu	*Chinese Atractylodes*
Cao Dou Kou	*Seed of Katsumada's Galangal*
Cao Guo	*Amomum Tsao-Ko*
Cao He Che	*Snakeweed*
Cao Wu	*Kusnezoff Monkshood Root*
Ce Bai Ye	*Oriental Arborvitae Twig*
Cha Shu Gen	*Tea Tree Root*
Chai Hu	*Thorowax Root*
Chan Su	*Toad Venom*
Chan Tui	*Cicada Molting*
Chang Chun Hua	*Madagascar Periwinkle*
Chang Shan	*Dichroa Root*
Che Qian Zi	*Plantago Seeds*
Chen Pi	*Tangerine Peel*
Chen Shi Hui	*Old Lime*
Chen Xiang	*Aloeswood*
Cheng Liu	*Chinese Tamarisk Tops*
Chi Fu Ling	*Red Poria*
Chi Shao Yao	*Red Peony Root*
Chi Shi Zhi	*Halloysite*
Chi Xiao Dou	*Aduki Bean*
Chong Cao	*Cordyceps*
Chong Lou	*Paris Rhizome*
Chong Wei Zi	*Motherwort Fruit*
Chou Wu Tong	*Glorybower Leaf*
Chuan Bei Mu	*Tendrilled Fritillaria Bulb*
Chuan Jiao	*Fruit of Szechuan Pricklyash Pepper*
Chuan Lian Zi	*Chinaberry Fruit*
Chuan Niu Xi	*Sichuan Cyathula Root*
Chuan Shan Jia	*Pangolin Scales*
Chuan Shan Long	*Japanese Yam*
Chuan Wu	*Sichuan Monkshood Root*
Chuan Xin Lian	*Green Chiretta*
Chuan Xiong	*Sichuan Lovage Root*
Chui Pen Cao	*Stringy Stonecrop*
Chun Bai Pi	*Bark of Heaven Tree*

Chun Gen Pi	*Bark of Heaven Tree*
Chun Pi	*Bark of Heaven Tree*
Ci Ji Li	*Caltrop Fruit*
Ci Shi	*Magnetite*
Ci Wei Pi	*Hedgehog Skin*
Ci Wu Jia	*Acanthopanax Root*
Cong Bai	*Scallion*
Da Feng Zi	*Chaulmoogra Seed*
Da Fu Pi	*Betel Husk*
Da Huang	*Rhubarb Root Rhizome*
Da Hui Xiang	*Truestar Anisetree Fruit*
Da Ji	*Peking Spurge Root*
Da Ji	*Japanese Thistle*
Da Pi Chong	*Dung Beetle*
Da Qing Ye	*Woad Leaf*
Da Suan	*Garlic Bulb*
Da Zao	*Chinese Date*
Dai Zhe Shi	*Hematite*
Dan Dou Chi	*Prepared Soybean*
Dan Fan	*Crystallized Blue Vitriol*
Dan Shen	*Salvia Root*
Dan Zhu Ye	*Bamboo Leaves*
Dang Gui	*Chinese Angelica Root*
Dang Shen	*Codonopsis Root*
Dao Dou	*Sword Bean*
Deng Xin Cao	*Rush Pith*
Di Er Cao	*Japanese Saint John's Wort*
Di Fu Zi	*Kochia Fruit*
Di Gu Pi	*Lycium Bark*
Di Long	*Earthworm*
Di Yu	*Burnet-Bloodwort Root*
Dian Di Mei	*Androsace Plant*
Ding Xiang	*Clove Flower Bud*
Dong Chong Xia Cao	*Cordyceps*
Dong Gua Pi	*Chinese Wax Gourd Peel*
Dong Gua Ren	*Chinese Wax Gourd Seed*
Dong Kui Zi	*Musk Mallow Seed*
Dou Chi Jiang	*Aromatic Litsea Root/Stem*
Dou Juan	*Young Soybean Sprout*
Du Huo	*Pubescent Angelica Root*
Du Juan Hua Ye	*Rhododendron Leaf*
Du Zhong	*Eucommia Bark*
E Guan Shi	*Tip of Tubular Stalactites*
E Jiao	*Donkey-Hide Gelatin*
E Zhu	*Zedoary Rhizome*
Er Cha	*Catechu Extract*
Fan Bai Cao	*Cinquefoil*
Fan Mu Bie	*Nux-Vomica Seed*
Fan Xie Ye	*Senna Leaf*
Fang Feng	*Ledebouriella Root*
Fei Zi	*Chinese Torreya Seed*
Fen Sha Shen	*Changium Root*
Feng Mi	*Honey*
Feng Wei Cao	*Bracken*
Fo Shou	*Finger Citron Fruit*
Fo Shou Hua	*Finger Citron Flower*

Fu Hai Shi	*Pumice*
Fu Ling	*White Poria*
Fu Ling Pi	*Poria Skin*
Fu Long Gan	*Yellow Earth from center of cooking stove*
Fu Pen Zi	*Chinese Raspberry*
Fu Ping	*Duckweed*
Fu Shen	*Poria with Wood Root*
Fu Xiao Mai	*Light Wheat Grain*
Fu Zi	*Aconite*
Gan Cao	*Raw Licorice Root*
Gan Jiang	*Dried Ginger*
Gan Qi	*Dry Lacquer Sap*
Gan Sui	*Kan-Sui Root*
Gao Ben	*Chinese Lovage Root*
Gao Liang Jiang	*Lesser Galangal Rhizome*
Ge Gen	*Kudzu Root*
Ge Hua	*Kudzu Flower*
Ge Jie	*Gecko*
Ge Shan Xiao	*Earleaf Cynanchum Root*
Gong Ding Xiang	*Clove Flower Bud*
Gong Fen	*Calomel*
Gou Ji	*Chain Fern Rhizome*
Gou Qi Zi	*Chinese Wolfberry Fruit*
Gou Teng	*Gambir Vine*
Gu Jing Cao	*Pipewort Scapus*
Gu Sui Bu	*Drynaria Rhizome*
Gu Ya	*Rice Sprout*
Gua Di	*Melon Pedicle*
Gua Lou	*Mongolian Snakegourd*
Gua Lou Ren	*Mongolian Snakegourd Seed*
Guan Gui	*Young Cinnamon Bark*
Guan Ye Liao	*Perfoliatum*
Guan Zhong	*Dryopteris Root*
Guang Fang Ji	*Aristolochia Root*
Gui Ban	*Land Tortoise Shell*
Gui Ban Jiao	*Glue of Land Tortoise Shell*
Gui Jiao	*Glue of Land Tortoise Shell*
Gui Yuan Rou	*Flesh of Longan Fruit*
Gui Zhi	*Cinnamon Twig*
Hai Dai	*Common Eelgrass*
Hai Er Shen	*Caryophyllaceous Root*
Hai Feng Teng	*Kadsura Stem*
Hai Ge Ke	*Clam Shell*
Hai Gou Shen	*Male Seal Sexual Organ*
Hai Jin Sha	*Japanese Fern Spore*
Hai Jin Sha Teng	*Japanese Fern*
Hai Long	*Pipe-Fish*
Hai Ma	*Seahorse*
Hai Piao Xiao	*Cuttlefish Bone*
Hai Shen	*Sea Cucumber*
Hai Tong Pi	*Erythrina Bark*
Hai Zao	*Seaweed*
Han Cai	*Nasturtium*
Han Fang Ji	*Stephania Root*
Han Lian Cao	*Eclipta*
Han Shui Shi	*Calcitum*

He Cao Ya	*Bud of Hairyvein*
He Huan Hua	*Mimosa Tree Flower*
He Huan Pi	*Mimosa Tree Bark*
He Li Le	*Myrobalan Fruit*
He Shi	*Starwort Fruit*
He Shou Wu	*Fleeceflower Root*
He Tao Shu Zhi	*Walnut Twig*
He Ye	*Lotus Leaf*
He Zi	*Myrobalan Fruit*
Hei Zhi Ma	*Black Sesame Seeds*
Hong Hua	*Safflower Flower*
Hong Teng	*Sargentodoxa Vine*
Hong Zao	*Chinese Date*
Hou Jiang	*Drynaria Rhizome*
Hou Po	*Magnolia Bark*
Hou Po Hua	*Magnolia Flower*
Hou Zao	*Macaque Gallstone*
Hu Er Cao	*Saxifrage*
Hu Gu	*Tiger Bone*
Hu Huang Lian	*Picrorhiza Rhizome*
Hu Jiao	*Black Pepper*
Hu Lu Ba	*Fenugreek Seed*
Hu Po	*Amber*
Hu Tao Ren	*Walnut Nut*
Hu Tao Rou	*Walnut Nut*
Hu Tui Ye	*Thorny Elaegnus Leaf*
Hu Zhang	*Giant Knotweed Root & Rhizome*
Hua Jiao	*Fruit of Szechuan Pricklyash Pepper*
Hua Rui Shi	*Ophicalcite*
Hua Sheng Yi	*Peanut Peel*
Hua Shi	*Talcum*
Huai Hua Mi	*Pagoda Tree Flower Bud*
Huai Jiao	*Pagoda Tree Fruit*
Huai Niu Xi	*Bidentate Achyranthes Root*
Huang Bai	*Amur Cork-Tree Bark*
Huang Dan	*Lead Oxide* (Pb_3O_4)
Huang Du	*Airpotato Yam*
Huang Jing	*Siberian Solomon Seal Rhizome*
Huang Lian	*Coptis Rhizome*
Huang Qi	*Milk-Vetch Root*
Huang Qin	*Skullcap Root*
Huang Yao Zi	*Air Potato Yam*
Huo Ma Ren	*Cannabis Seeds*
Huo Xiang	*Patchouli or Wrinkled Giant Hyssop*
Ji Cai	*Shepherd's Purse*
Ji Dai	*Human Umbilical Cord*
Ji Guan Hua	*Cockscomb Flower*
Ji Mu	*Strap Flower/Root/Leaf*
Ji Nei Jin	*Chicken Gizzard Lining*
Ji Shi Teng	*Chinese Fevervine*
Ji Xing Zi	*Garden Balsam Seed*
Ji Xue Teng	*Leatherleaf Millettia Root/Vine*
Ji Zi Huang	*Chicken Egg Yolk*
Jiang Can	*Dead Sick Silkworm*
Jiang Huang	*Turmeric Rhizome*
Jiang Xiang	*Center of Rosewood*

Jie Geng	*Balloonflower Root*
Jin Guo Lan	*Tinospora Root*
Jin Qian Cao	*Loosestrife Plant*
Jin Qiao Mai	*Wild Buckwheat*
Jin Yin Hua	*Honeysuckle Flower*
Jin Ying Zi	*Cherokee Rosehip*
Jin Zhi	*Purified Feces*
Jing Jie	*Schizonepeta Stem*
Jing Mao Gou Ji	*Chain Fern Rhizome*
Jing Tian San Qi	*Aizoon Stonecrop Plant*
Jiu Jie Cha	*Sarcandrae Plant*
Jiu Li Ming	*Climbing Groundsel*
Jiu Xiang Chong	*Dung Beetle*
Jiu Zi	*Seeds of Fragrant-Flowered Garlic*
Ju He	*Tangerine Seed*
Ju Hong	*Outer Tangerine Peel*
Ju Hua	*Chrysanthemum Flower*
Ju Luo	*Tangerine Pith*
Ju Ye	*Tangerine Leaf*
Ju Ye San Qi	*Canton Tusanqi*
Jue Ming Zi	*Cassia Seeds*
Kan Qi	*Human Umbilical Cord*
Ku Lian Gen Pi	*China Tree Root Bark*
Ku Shen	*Sophora Root*
Ku Xing Ren	*Apricot Seed*
Kuan Dong Hua	*Coltsfoot Flower*
Kuan Jin Teng	*Chinese Tinospora Stem*
Kun Bu	*Kelp Thallus*
Lai Fu Zi	*Chinese Radish Seed*
Lang Du	*Fischer Euphorbia Root*
Lao Guan Cao	*Heroubill Plant*
Lei Gong Teng	*Tripterygium Wilfordii Hook*
Lei Wan	*Fruiting Body of Stone-like Omphalia*
Li Lu	*Mountain Onion*
Li Pi	*Asian Pear Peel*
Li Zhi He	*Leechee Nut*
Lian Fang	*Mature Lotus Receptacle*
Lian Qiao	*Forsythia Fruit*
Lian Tou Jian	*Anemone*
Lian Xu	*Lotus Stamen*
Lian Zi	*Lotus Seed*
Lian Zi Xin	*Lotus Plumule*
Liang Mian Zhen	*Pricklyash Root*
Ling Jiao	*Water Calptrop*
Ling Xiao Hua	*Chinese Trumpetcreeper Flower*
Ling Yang Jiao	*Antelope Horn*
Ling Zhi	*Glossy Ganoderma*
Liu Huang	*Sulphur*
Liu Ji Nu	*Diverse Wormwood*
Long Dan Cao	*Gentiana Root*
Long Gu	*Fossils of Mammal Vertebrae and Extremity*
Long Kui	*Black Nightshade*
Long Yan Rou	*Flesh of Longan Fruit*
Lou Lu	*Rhaponticum Root*
Lu Dou	*Mung Bean*
Lu Dou Yi	*Mung Bean Skin*

Lu Feng Fang	*Hornet Nest*
Lu Gan Shi	*Smithsonite or Calamine*
Lu Gen	*Reed Rhizome*
Lu Hui	*Juice of Aloe Leaf*
Lu Jiao	*Deer Horn*
Lu Jiao Jiao	*Deer Horn Glue*
Lu Jiao Shuang	*Dregs leftover after making Lu Jiao Jiao*
Lu Lu Tong	*Sweetgum Fruit*
Lu Pi Jiao	*Donkey-Hide Gelatin*
Lu Rong	*Velvet of Young Deer Horn*
Lu Ying	*Chinese Sambucus*
Luo Bu Ma	*Leaf of Apocynum Venetum*
Luo Han Guo	*Round Momordica Fruit*
Luo Han Guo Ye	*Round Momordica Leaf*
Luo Shi Teng	*Star Jasmine Stem*
Ma Bian Cao	*European Verbena*
Ma Bo	*Puffball*
Ma Chi Xian	*Purslane*
Ma Dou Ling	*Birthwort Fruit*
Ma Huang	*Ephedra Stem*
Ma Huang Gen	*Chinese Ephedra Root*
Ma Qian Zi	*Nux-Vomica Seed*
Ma Ti Jin	*Dichondra Plant*
Mai Dong	*Tuber of Dwarf Lilyturf*
Mai Men Dong	*Tuber of Dwarf Lilyturf*
Mai Ya	*Barley Sprout*
Man Jing Zi	*Vitex Fruit*
Man Tuo Luo	*Jamestown Weed Seed/Leaf/Flower*
Mang Chong	*Gadfly*
Mang Xiao	*Glauber's Salt*
Mao Dong Qing	*Hairy Holly Root*
Mao Mei	*Japanese Raspberry*
Mao Zhua Cao	*Buttercup Root*
Mei Gui Hua	*Miniature Chinese Rose*
Meng Shi	*Chlorite*
Mi Ke	*Opium Poppy Husk*
Mi Meng Hua	*Buddleia Flower Bud*
Mi Tuo Seng	*Litharge or Galena*
Mian Hua Gen	*Cotton Plant Root*
Ming Dang Shen	*Changium Root*
Ming Fan	*Purified Alunite*
Mo Yao	*Myrrh*
Mu Bie Zi	*Momordica Seed*
Mu Dan Pi	*Cortex of Tree Peony*
Mu Ding Xiang	*Clove Fruit*
Mu Gua	*Chinese Quince Fruit*
Mu Hu Die	*India Trumpetflower Seed*
Mu Jin Pi	*Shrubalthea Bark*
Mu Li	*Oyster Shell*
Mu Tong	*Five Leaf Akebia Stem*
Mu Xiang	*Aucklandia*
Mu Zei	*Horse Tail*
Nan Gua Zi	*Pumpkin Seed Husk*
Nao Sha	*Sal Ammoniac*
Niu Bang Zi	*Great Burdock Fruit*
Niu Huang	*Cattle Gallstone*

Nong Ji Li	*Crotalaria Plant*
Nu Zhen Zi	*Privet Fruit*
Nuo Dao Gen Xu	*Glutinous Rice Root*
Ou Jie	*Node of Lotus Rhizome*
Pang Da Hai	*Boat Sterculia Seed*
Pao Jiang	*Quick-Fried Ginger Root*
Pei Lan	*Eupatorium*
Peng Sha	*Borax*
Pi Pa Ye	*Loquat Leaf*
Pi Shi	*Arsenolite*
Pi Shuang	*Arsenolite*
Po Gu Zhi	*Fruit of Scuffy Pea*
Pu Gong Ying	*Dandelion*
Pu Huang	*Cattail Pollen*
Pu Xiao	*Less Purfied Mirabilitum*
Qi Cao	*Grub*
Qian Cao Gen	*Madder Root*
Qian Dan	*Lead Oxide* (Pb_3O_4)
Qian Hu	*Hogfennel Root*
Qian Li Guang	*Climbing Groundsel*
Qian Nian Jian	*Homalomena Rhizome*
Qian Niu Zi	*Morning Glory Seed*
Qian Ri Hong	*Globeamaranth Flower*
Qian Shi	*Gordon Euryale Seed*
Qiang Huo	*Notopterygium Root*
Qiang Lang	*Dung Beetle*
Qiang Lang Chong	*Dung Beetle*
Qin Jiao	*Gentiana Macrophylla Root*
Qin Pi	*Bark of Korean Ash Branch*
Qing Dai	*Indigo*
Qing Fen	*Calomel*
Qing Hao	*Wormwood*
Qing Meng Shi	*Chlorite*
Qing Pi	*Immature Tangerine Peel*
Qing Xiang Zi	*Celosia Seeds*
Qu Mai	*Chinese Pink Dianthus*
Quan Shen	*Snakeweed*
Quan Xie	*Scorpion*
Ren Shen	*Ginseng Root*
Ren Shen Ye	*Ginseng Leaf*
Rou Cong Rong	*Fleshy Stem of Broomrape*
Rou Dou Kou	*Nutmeg Seed*
Rou Gui	*Cinnamon Bark*
Rou Guo	*Nutmeg Seed*
Ru Xiang	*Frankincense or Pistaciacense*
San Bai Cao	*Lizardtail*
San Ke Zhen	*Berberis Root*
San Leng	*Bur-Reed Rhizome*
San Qi	*Pseudoginseng Root*
Sang Bai Pi	*Bark of Mulberry Root*
Sang Ji Sheng	*Mulberry Mistletoe*
Sang Piao Xiao	*Praying Mantis Egg-Case laid on Mulberry Leaf*
Sang Shen	*Bud of Mulberry Fruit*
Sang Ye	*Mulberry Leaf*
Sang Zhi	*Mulberry Twig*
Sha Ren	*Seed of Grains-of-Paradise*

Sha Shen	*Root of Beech Silver-Top*
Sha Yuan Ji Li	*Milk-Vetch Seed*
Shan Ci Gu	*Chinese Tulip Bulb*
Shan Dou Gen	*Sophora Root*
Shan Yao	*Chinese Yam*
Shan Yu Rou	*Asiatic Cornelian Cherry Fruit*
Shan Zha	*Hawthorn Fruit*
Shan Zhi Zi	*Gardenia Fruit*
Shan Zhu Yu	*Asiatic Cornelian Cherry Fruit*
Shang Lu	*India Pokeberry Root*
She Chuang Zi	*Cnidium Seed*
She Gan	*Belamcanda Rhizome*
She Mei	*Mock Strawberry Plant*
She Tui	*Snake Skin Slough*
She Xiang	*Musk of Deer*
Shen Jin Cao	*Buck Grass*
Shen Qu	*Medicated Leaven*
Sheng Di Huang	*Fresh Rehmannia*
Sheng Jiang	*Fresh Ginger Rhizome*
Sheng Jiang Pi	*Ginger Peel*
Sheng Jiang Zhi	*Fresh Ginger Juice*
Sheng Ma	*Black Kohosh*
Sheng Tie Luo	*Iron Filings*
Sheng Tou Cao	*Sweet Basil*
Shi Chang Pu	*Sweetflag Rhizome*
Shi Da Gong Lao	*Chinese Mahonia*
Shi Dan	*Crystallized Blue Vitriol*
Shi Di	*Persimmon Calyx*
Shi Gao	*Gypsum*
Shi Hu	*Dendrobium Stem*
Shi Jian Chuan	*Chinese Sage*
Shi Jue Ming	*Abalone Shell*
Shi Jun Zi	*Rangoon Creeper Fruit with Seed*
Shi Liu Gen Pi	*Bark of Pomegranate Tree*
Shi Liu Pi	*Pomegranate Husk*
Shi Nan Teng	*Chinese Photinia Vine*
Shi Shang Bai	*Greater Selaginella*
Shi Wei	*Felt Fern Leaves*
Shou Gong	*House Lizard*
Shou Wu Teng	*Vine of Solomon's Seal*
Shu Di Huang	*Cooked Rehmannia Root*
Shui Hong Hua Zi	*Prince's Feather Fruit*
Shui Niu Jiao	*Water Buffalo Horn*
Shui Yin	*Mercury*
Shui Zhi	*Leech*
Si Gua Luo	*Suakwa Vegetable Sponge*
Si Gua Zi	*Suakwa Seed*
Si Ji Qing	*Chinese Holly*
Song Jie	*Knotty Pine Wood*
Su He Xiang	*Resin of Rose Maloes*
Su Mu	*Sappan Wood*
Suan Jiang	*Winter Cherry Fruit*
Suan Zao Ren	*Sour Jujube Seed*
Suo Yang	*Chinese Cynomorium*
Ta Gan	*Otter Liver*
Tai Pan	*Human Placenta*

Tai Yi	*Human Placenta*
Tai Zi Shen	*Caryophyllaceous Root*
Tan Xiang	*Sandalwood*
Tang Song Cao	*Thalictrum*
Tao Ren	*Peach Kernel*
Teng Li Gen	*Chinese Kiwi Root*
Tian Dong	*Chinese Asparagus Tuber*
Tian Hu Sui	*Hydrocotyle Plant*
Tian Hua Fen	*Trichosantes Root*
Tian Kui	*Muskroot-like Semiaquilegia Root*
Tian Kui Zi	*Muskroot-like Semiaquilegia Root*
Tian Long	*House Lizard*
Tian Ma	*Gastrodia Rhizome*
Tian Men Dong	*Chinese Asparagus Tuber*
Tian Ming Jing	*Carpesium Root*
Tian Nan Xing	*Jack-in-the-Pulpit*
Tian Xian Zi	*Henbane Seed*
Tian Zhu Huang	*Tabasheer*
Tie Shu	*Sago Cycas*
Tie Xian	*Copperleaf*
Ting Li Zi	*Pepperweed Seed or Flixweed Tansymustard Seed*
Tong Bian	*Boy Urine*
Tong Cao	*Rice Paper Pith*
Tou Gu Cao	*Garden Balsam*
Tu Bei Mu	*Paniculate Bolbostemma*
Tu Bie Chong	*Wingless Cockroach*
Tu Fu Ling	*Glabrous Greenbrier Rhizome*
Tu Jin Pi	*Goldenlarch Bark*
Tu Niu Xi	*Prickly Chaff Flower/Root/Twig*
Tu Si Zi	*Chinese Dodder Seeds*
Wa Leng Zi	*Cockle Shell*
Wan Nian Qing	*Japanese Lily Bulb*
Wang Bu Liu Xing	*Cowherb Seed*
Wei Ling Xian	*Chinese Clematis Root*
Wu Bei Zi	*Gallnut of Chinese Sumac*
Wu Gong	*Centipede*
Wu Jia Pi	*Acanthopanax Root Bark*
Wu Ling Zhi	*Flying Squirrel Feces*
Wu Mei	*Mume Fruit*
Wu Shao She	*Zaocys*
Wu Wei Zi	*Schisandra Fruit*
Wu Yao	*Three-Nerved Spicebush Root*
Wu Yi	*Stinky Elm Fruit Paste*
Wu Zei Gu	*Cuttlefish Bone*
Wu Zhu Yu	*Evodia Fruit*
Xi Gua	*Watermelon Fruit*
Xi Gua Cui Yi	*Watermelon Peel*
Xi Gua Shuang	*Frost Powder Prepared by Watermelon Peel and Mirabilite*
Xi He Liu	*Chinese Tamarisk Tops*
Xi Jiao	*Rhinoceros Horn*
Xi Shu	*Happy Tree Fruit Bark*
Xi Xian Cao	*Common Saint Paulswort*
Xi Xin	*Chinese Wild Ginger*
Xi Yang Shen	*American Ginseng Root*
Xia Ku Cao	*Selfheal Spike*
Xia Tian Wu	*Corydalis Rhizome*

Xian He Cao	*Agrimony*
Xian Ling Pi	*Longspur Epimedium*
Xian Mao	*Rhizome of Golden Eye-Grass*
Xiang Fu	*Nutgrass*
Xiang Ru	*Aromatic Madder*
Xiang Yuan	*Citron Fruit*
Xiao Hui Xiang	*Fennel Fruit*
Xiao Ji	*Small Japanese Thistle*
Xiao Mai	*Wheat Grain*
Xie Bai	*Bulb of Chinese Chive*
Xin Yi Hua	*Magnolia Flower*
Xing Ren	*Apricot Seed*
Xiong Dan	*Bear Gallbladder*
Xiong Huang	*Realgar*
Xiu Qiu Hua	*Hydrangea Flower*
Xu Chang Qing	*Chinese Swallowwort*
Xu Duan	*Japanese Teasel Root*
Xu Sui Zi	*Moleplant Seed*
Xuan Fu Hua	*Inula Flower*
Xuan Ming Fen	*Pure Mirabilitum*
Xuan Shen	*Ningpo Figwort Root*
Xue Dan	*Hemsleya Root*
Xue Jian Chou	*Mapleleaf Goosefoot*
Xue Jie	*Dragon's Blood (Resin)*
Xue Shang Yi Zhi Hao	*Brachypodum Monkshood Root*
Xue Yu Tan	*Charred Human Hair*
Xun Gu Feng	*Dutchman's Pipe*
Ya Dan Zi	*Java Brucea Fruit*
Ya Zhi Cao	*Dayflower*
Yan Fu Mu	*Chinese Sumac*
Yan Hu Suo	*Corydalis Rhizome*
Yang Jin Hua	*Jamestown Weed Seed/Leaf/Flower*
Yang Qi Shi	*Actinolite*
Yang Zhi Zhu	*Yellow Azalea*
Ye Jiao Teng	*Vine of Solomon's Seal*
Ye Ju Hua	*Wild Chrysanthemum Flower*
Ye Ming Sha	*Bat Feces*
Yi Mu Cao	*Chinese Motherwort*
Yi Tang	*Barley Malt Sugar*
Yi Yi Ren	*Job's Tears (Coix)*
Yi Zhi Ren	*Galangal Fruit*
Yin Chai Hu	*Stellaria Root*
Yin Chen Hao	*Capillaris*
Yin Xing	*Ginkgo Nut*
Yin Xing Ye	*Ginkgo Leaf*
Yin Yang Huo	*Longspur Epimedium*
Ying Su Ke	*Opium Poppy Husk*
Yu Bai Pi	*Elm Tree Bark*
Yu Jin	*Turmeric Tuber*
Yu Li Ren	*Bush Cherry Pit*
Yu Liang Shi	*Limonite*
Yu Mi Xu	*Cornsilk*
Yu Xing Cao	*Houttuynia*
Yu Yu Liang	*Limonite*
Yu Zhu	*Solomon's Seal Rhizome*
Yuan Hua	*Genkwa Flower*

Yuan Shen	*Ningpo Figwort Root*
Yuan Zhi	*Chinese Senega Root*
Yue Ji Hua	*Chinese Tea Rose*
Zao Jiao	*Chinese Honeylocust Fruit*
Zao Pi	*Asiatic Cornelian Cherry Fruit*
Zao Xiu	*Paris Rhizome*
Ze Lan	*Shiny Bugleweed*
Ze Qi	*Spurge*
Ze Xie	*Water Plantain*
Zhang Nao	*Camphor*
Zhe Bei Mu	*Thunberg Fritillaria Bulb*
Zhe Chong	*Wingless Cockroach*
Zhe Shi Zi	*Paper Mulberry Fruit*
Zhen Zhu	*Pearl*
Zhi Gan Cao	*Prepared Licorice Root*
Zhi Huang Qi	*Prepared Milk-Vetch Root*
Zhi Ke	*Ripe Fruit of Bitter Orange*
Zhi Mu	*Anemarrhena Rhizome*
Zhi Shi	*Immature Fruit of Bitter Orange*
Zhong Jie Feng	*Sarcandrae Plant*
Zhu Jie Xiang Fu	*Anemone*
Zhu Li	*Dried Bamboo Sap*
Zhu Ling	*Umbellate Pore Fungus*
Zhu Ma Gen	*Ramie Root*
Zhu Ru	*Bamboo Shavings*
Zhu Sha Gen	*Coral Ardisia Root*
Zi Bai Pi	*Chinese Catalpa Bark*
Zi Bei Tian Kui	*Begonia*
Zi Cao	*Groomwell Root*
Zi He Che	*Human Placenta*
Zi Hua Di Ding	*Yedeon's Violet*
Zi Jin Niu	*Japanese Ardisia*
Zi Jing Mu	*Chinese Redbud Wood*
Zi Jing Mu Pi	*Chinese Redbud Bark*
Zi Jing Pi	*Chinese Redbud Bark*
Zi Ran Tong	*Pyrite*
Zi Shen	*Snakeweed*
Zi Shi Ying	*Fluorite*
Zi Su Geng	*Perilla Stem*
Zi Su Ye	*Perilla Leaf*
Zi Su Zi	*Purple Perilla Fruit*
Zi Wan	*Purple Aster Root*
Zi Wei	*Chinese Trumpetcreeper Flower*
Zi Zhu	*Beautyberry Leaf*
Zong Lu Pi	*Stiple Fiber of Fortune Windmill Palm*
Zong Lu Tan	*Charred Stiple Fiber of Fortune Windmill Palm*
Zu Shi Ma	*Giraldi Daphne Bark*

APPENDIX 5

Cross-Reference English–Pinyin Herbs

Abalone Shell	**Shi jue Ming**
Acanthopanax Root	**Ci Wu Jia**
Acanthopanax Root Bark	**Wu Jia Pi**
Aconite	**Fu Zi**
Actinolite	**Yang Qi Shi**
Aduki Bean	**Chi Xiao Dou**
Agkistrodon Snake	**Bai Hua She**
Agrimony	**Xian He Cao**
Air Potato Yam	**Huang Du**
Air Potato Yam	**Huang Yao Zi**
Aizoon Stonecrop Plant	**Jing Tian San Qi**
Akebia Fruit	**Ba Yue Zha**
Aloeswood	**Chen Xiang**
Amber	**Hu Po**
American Ginseng Root	**Xi Yang Shen**
Amomum Tsao-Ko	**Cao Guo**
Ampelopsis	**Bai Lian**
Amur Cork-Tree Bark	**Huang Bai**
Androsace Plant	**Dian Di Mei**
Anemarrhena Rhizome	**Zhi Mu**
Anemone	**Lian Tou Jian**
Anemone	**Zhu Jie Xiang Fu**
Angelica Root	**Bai Zhi**
Antelope Horn	**Ling Yang Jiao**
Apricot Seed	**Bei Xing Ren**
Apricot Seed	**Ku Xing Ren**
Apricot Seed	**Xing Ren**
Arbor-Vitae Seed	**Bai Zi Ren**
Aristolochia Root	**Guang Fang Ji**
Aromatic Litsea Root/Stem	**Dou Chi Jiang**
Aromatic Madder	**Xiang Ru**
Arsenolite	**Pi Shi**
Arsenolite	**Pi Shuang**
Asian Pear Peel	**Li Pi**
Asiatic Cornelian Cherry Fruit	**Shan Yu Rou**
Asiatic Cornelian Cherry Fruit	**Shan Zhu Yu**
Asiatic Cornelian Cherry Fruit	**Zao Pi**
Aucklandia	**Mu Xiang**

Balloonflower Root	**Jie Geng**
Bamboo Leaves	**Dan Zhu Ye**
Bamboo Shavings	**Zhu Ru**
Bark of Heaven Tree	**Chun Bai Pi**
Bark of Heaven Tree	**Chun Gen Pi**
Bark of Heaven Tree	**Chun Pi**
Bark of Korean Ash Branch	**Qin Pi**
Bark of Mulberry Root	**Sang Bai Pi**
Bark of Pomegranate Tree	**Shi Liu Gen Pi**
Barley Malt Sugar	**Yi Tang**
Barley Sprout	**Mai Ya**
Bat Feces	**Ye Ming Sha**
Bear Gallbladder	**Xiong Dan**
Beautyberry Leaf	**Zi Zhu**
Begonia	**Zi Bei Tian Kui**
Belamcanda Rhizome	**She Gan**
Benzoin	**An Xi Xiang**
Berberis Root	**San Ke Zhen**
Betel Husk	**Da Fu Pi**
Betel Nut	**Bing Lang**
Bidentate Achyranthes Root	**Huai Niu Xi**
Birthwort Fruit	**Ma Dou Ling**
Black Kohosh	**Sheng Ma**
Black Nightshade	**Long Kui**
Black Pepper	**Hu Jiao**
Black Sesame Seeds	**Hei Zhi Ma**
Boat Sterculia Seed	**Pang Da Hai**
Borax	**Peng Sha**
Borneol	**Bing Pian**
Boy Urine	**Tong Bian**
Brachypodum Monkshood Root	**Xue Shang Yi Zhi Hao**
Bracken	**Feng Wei Cao**
Buck Grass	**Shen Jin Cao**
Bud of Hairyvein	**He Cao Ya**
Bud of Mulberry Fruit	**Sang Shen**
Buddleia Flower Bud	**Mi Meng Hua**
Bulb of Chinese Chive	**Xie Bai**
Burnet-Bloodwort Root	**Di Yu**
Bur-Reed Rhizome	**San Leng**
Bush Cherry Pit	**Yu Li Ren**
Buttercup Root	**Mao Zhua Cao**
Calcitum	**Han Shui Shi**
Calomel	**Gong Fen**
Calomel	**Qing Fen**
Caltrop Fruit	**Bai Ji Li**
Caltrop Fruit	**Ci Ji Li**
Camphor	**Zhang Nao**
Cannabis Seeds	**Huo Ma Ren**
Cantharides	**Ban Mao**
Canton Tusanqi	**Ju Ye San Qi**
Capillaris	**Yin Chen Hao**
Carpesium Root	**Tian Ming Jing**
Caryophyllaceous Root	**Hai Er Shen**
Caryophyllaceous Root	**Tai Zi Shen**
Cassia Seeds	**Jue Ming Zi**
Catechu Extract	**Er Cha**

Cattail Pollen	**Pu Huang**
Cattle Gallstone	**Niu Huang**
Celosia Seeds	**Qing Xiang Zi**
Center of Rosewood	**Jiang Xiang**
Centipede	**Wu Gong**
Chain Fern Rhizome	**Gou Ji**
Chain Fern Rhizome	**Jing Mao Gou Ji**
Changium Root	**Fen Sha Shen**
Changium Root	**Ming Dang Shen**
Charred Human Hair	**Xue Yu Tan**
Charred Stiple Fiber of Fortune Windmill Palm	**Zong Lu Tan**
Chaulmoogra Seed	**Da Feng Zi**
Cherokee Rosehip	**Jin Ying Zi**
Chicken Egg Yolk	**Ji Zi Huang**
Chicken Gizzard Lining	**Ji Nei Jin**
China Root	**Ba Qia**
China Tree Root Bark	**Ku Lian Gen Pi**
Chinaberry Fruit	**Chuan Lian Zi**
Chinese Alangium Root	**Ba Jiao Feng**
Chinese Anemone Root	**Bai Tou Weng**
Chinese Angelica Root	**Dang Gui**
Chinese Asparagus Tuber	**Tian Dong**
Chinese Asparagus Tuber	**Tian Men Dong**
Chinese Atractylodes	**Cang Zhu**
Chinese Catalpa Bark	**Zi Bai Pi**
Chinese Clematis Root	**Wei Ling Xian**
Chinese Cynomorium	**Suo Yang**
Chinese Date	**Da Zao**
Chinese Date	**Hong Zao**
Chinese Dodder Seeds	**Tu Si Zi**
Chinese Ephedra Root	**Ma Huang Gen**
Chinese Fevervine	**Ji Shi Teng**
Chinese Giantfennel Resin	**A Wei**
Chinese Holly	**Si Ji Qing**
Chinese Honeylocust Fruit	**Zao Jiao**
Chinese Kiwi Root	**Teng Li Gen**
Chinese Lobelia	**Ban Bian Lian**
Chinese Lovage Root	**Gao Ben**
Chinese Mahonia	**Shi Da Gong Lao**
Chinese Motherwort	**Yi Mu Cao**
Chinese Photinia Vine	**Shi Nan Teng**
Chinese Pink Dianthus	**Qu Mai**
Chinese Quince Fruit	**Mu Gua**
Chinese Radish Seed	**Lai Fu Zi**
Chinese Raspberry	**Fu Pen Zi**
Chinese Redbud Bark	**Zi Jing Mu Pi**
Chinese Redbud Bark	**Zi Jing Pi**
Chinese Redbud Wood	**Zi Jing Mu**
Chinese Sage	**Shi Jian Chuan**
Chinese Sambucus	**Lu Ying**
Chinese Senega Root	**Yuan Zhi**
Chinese Soft-Shelled Turtle Shell	**Bie Jia**
Chinese Sumac	**Yan Fu Mu**
Chinese Swallowwort	**Xu Chang Qing**
Chinese Tamarisk Tops	**Cheng Liu**
Chinese Tamarisk Tops	**Xi He Liu**

Chinese Tea Rose	**Yue Ji Hua**
Chinese Tinospora Stem	**Kuan Jin Teng**
Chinese Torreya Seed	**Fei Zi**
Chinese Trumpetcreeper Flower	**Ling Xiao Hua**
Chinese Trumpetcreeper Flower	**Zi Wei**
Chinese Tulip Bulb	**Shan Ci Gu**
Chinese Wax Gourd Peel	**Dong Gua Pi**
Chinese Wax Gourd Seed	**Dong Gua Ren**
Chinese Wild Ginger	**Xi Xin**
Chinese Wolfberry Fruit	**Gou Qi Zi**
Chinese Yam	**Shan Yao**
Chlorite	**Meng Shi**
Chlorite	**Qing Meng Shi**
Chrysanthemum Flower	**Ju Hua**
Cicada Molting	**Chan Tui**
Cinnamon Bark	**Rou Gui**
Cinnamon Twig	**Gui Zhi**
Cinquefoil	**Fan Bai Cao**
Citron Fruit	**Xiang Yuan**
Clam Shell	**Hai Ge Ke**
Climbing Groundsel	**Jiu Li Ming**
Climbing Groundsel	**Qian Li Guang**
Climbing Nightshade Plant	**Bai Ying**
Clove Flower Bud	**Ding Xiang**
Clove Flower Bud	**Gong Ding Xiang**
Clove Fruit	**Mu Ding Xiang**
Cnidium Seed	**She Chuang Zi**
Cockle Shell	**Wa Leng Zi**
Cocklebur Fruit	**Cang Er Zi**
Cockscomb Flower	**Ji Guan Hua**
Codonopsis Root	**Dang Shen**
Coltsfoot Flower	**Kuan Dong Hua**
Common Eelgrass	**Hai Dai**
Common Saint Paul's Wort	**Xi Xian Cao**
Cooked Rehmannia Root	**Shu Di Huang**
Copperleaf	**Tie Xian**
Coptis Rhizome	**Huang Lian**
Coral Ardisia Root	**Zhu Sha Gen**
Cordyceps	**Chong Cao**
Cordyceps	**Dong Chong Xia Cao**
Cornsilk	**Yu Mi Xu**
Cortex of Tree Peony	**Mu Dan Pi**
Corydalis Rhizome	**Xia Tian Wu**
Corydalis Rhizome	**Yan Hu Suo**
Cotton Plant Root	**Mian Hua Gen**
Cowherb Seed	**Wang Bu Liu Xing**
Crotalaria Plant	**Nong Ji Li**
Croton Seed	**Ba Dou**
Crystallized Blue Vitriol	**Dan Fan**
Crystallized Blue Vitriol	**Shi Dan**
Cuttlefish Bone	**Hai Piao Xiao**
Cuttlefish Bone	**Wu Zei Gu**
Dandelion	**Pu Gong Ying**
Dayflower	**Ya Zhi Cao**
Dead Sick Silkworm	**Bai Jiang Can**
Dead Sick Silkworm	**Jiang Can**

Deer Horn	**Lu Jiao**
Deer Horn Glue	**Lu Jiao Jiao**
Dendrobium Stem	**Shi Hu**
Dichondra Plant	**Ma Ti Jin**
Dichroa Root	**Chang Shan**
Dictamnus Root Bark	**Bai Xian Pi**
Diverse Wormwood	**Liu Ji Nu**
Donkey-Hide Gelatin	**E Jiao**
Donkey-Hide Gelatin	**Lu Pi Jiao**
Dragon's Blood (Resin)	**Xue Jie**
Dregs leftover after making Lu Jiao Jiao	**Lu Jiao Shuang**
Dried Bamboo Sap	**Zhu Li**
Dried Ginger	**Gan Jiang**
Dry Lacquer Sap	**Gan Qi**
Drynaria Rhizome	**Gu Sui Bu**
Drynaria Rhizome	**Hou Jiang**
Dryopteris Root	**Guan Zhong**
Duckweed	**Fu Ping**
Dung Beetle	**Da Pi Chong**
Dung Beetle	**Jiu Xiang Chong**
Dung Beetle	**Qiang Lang**
Dung Beetle	**Qiang Lang Chong**
Dutchman's Pipe	**Xun Gu Feng**
Earleaf Cynanchum Root	**Ge Shan Xiao**
Earthworm	**Di Long**
Eclipta	**Han Lian Cao**
Elm Tree Bark	**Yu Bai Pi**
Ephedra Stem	**Ma Huang**
Erythrina Bark	**Hai Tong Pi**
Eucommia Bark	**Du Zhong**
Eupatorium	**Pei Lan**
European Verbena	**Ma Bian Cao**
Evodia Fruit	**Wu Zhu Yu**
Felt Fern Leaves	**Shi Wei**
Fennel Fruit	**Xiao Hui Xiang**
Fenugreek Seed	**Hu Lu Ba**
Finger Citron Flower	**Fo Shou Hua**
Finger Citron Fruit	**Fo Shou**
Fischer Euphorbia Root	**Lang Du**
Fish-Poison Yam	**Bei Xie**
Five Leaf Akebia Stem	**Mu Tong**
Fleeceflower Root	**He Shou Wu**
Flesh of Longan Fruit	**Gui Yuan Rou**
Flesh of Longan Fruit	**Long Yan Rou**
Fleshy Stem of Broomrape	**Rou Cong Rong**
Fluorite	**Zi Shi Ying**
Flying Squirrel Feces	**Wu Ling Zhi**
Forsythia Fruit	**Lian Qiao**
Fossils of Mammal Vertebrae and Extremity	**Long Gu**
Frankincense or Pistaciacense	**Ru Xiang**
Fresh Ginger Juice	**Sheng Jiang Zhi**
Fresh Ginger Rhizome	**Sheng Jiang**
Fresh Rehmannia	**Sheng Di Huang**
Frost Powder Prepared by Watermelon Peel and Mirabilite	**Xi Gua Shuang**
Fruit of Scuffy Pea	**Bu Gu Zhi**
Fruit of Scuffy Pea	**Po Gu Zhi**

Fruit of Szechuan Pricklyash Pepper	**Chuan Jiao**
Fruit of Szechuan Pricklyash Pepper	**Hua Jiao**
Fruiting Body of Stone-like Omphalia	**Lei Wan**
Fruiting Body of Tremella	**Bai Mu Er**
Gadfly	**Mang Chong**
Galangal Fruit	**Yi Zhi Ren**
Gallnut of Chinese Sumac	**Wu Bei Zi**
Gambir Vine	**Gou Teng**
Garden Balsam	**Tou Gu Cao**
Garden Balsam Seed	**Ji Xing Zi**
Gardenia Fruit	**Shan Zhi Zi**
Garlic Bulb	**Da Suan**
Gastrodia Rhizome	**Tian Ma**
Gecko	**Ge Jie**
Genkwa Flower	**Yuan Hua**
Gentiana Macrophylla Root	**Qin Jiao**
Gentiana Root	**Long Dan Cao**
Giant Knotweed Root and Rhizome	**Hu Zhang**
Giant Typhonium Tuber	**Bai Fu Zi**
Ginger Peel	**Sheng Jiang Pi**
Ginkgo Leaf	**Bai Guo Ye**
Ginkgo Leaf	**Yin Xing Ye**
Ginkgo Nut	**Bai Guo**
Ginkgo Nut	**Yin Xing**
Ginseng Leaf	**Ren Shen Ye**
Ginseng Root	**Ren Shen**
Giraldi Daphne Bark	**Zu Shi Ma**
Glabrous Greenbrier Rhizome	**Tu Fu Ling**
Glauber's Salt	**Mang Xiao**
Globeamaranth Flower	**Qian Ri Hong**
Glorybower Leaf	**Chou Wu Tong**
Glossy Ganoderma	**Ling Zhi**
Glue of Land Tortoise Shell	**Gui Ban Jiao**
Glue of Land Tortoise Shell	**Gui Jiao**
Glutinous Rice Root	**Nuo Dao Gen Xu**
Goldenlarch Bark	**Tu Jin Pi**
Gordon Euryale Seed	**Qian Shi**
Great Burdock Fruit	**Niu Bang Zi**
Greater Selaginella	**Shi Shang Bai**
Green Chiretta	**Chuan Xin Lian**
Groomwell Root	**Zi Cao**
Grub	**Qi Cao**
Gypsum	**Shi Gao**
Hairy Holly Root	**Mao Dong Qing**
Halloysite	**Chi Shi Zhi**
Happy Tree Fruit Bark	**Xi Shu**
Hawthorn Fruit	**Shan Zha**
Hedgehog Skin	**Ci Wei Pi**
Hematite	**Dai Zhe Shi**
Hemsleya Root	**Xue Dan**
Henbane Seed	**Tian Xian Zi**
Heroubill Plant	**Lao Guan Cao**
Hogfennel Root	**Qian Hu**
Homalomena Rhizome	**Qian Nian Jian**
Honey	**Feng Mi**
Honeysuckle Flower	**Jin Yin Hua**

Hornet Nest	**Lu Feng Fang**
Horse Tail	**Mu Zei**
House Lizard	**Bi Hu**
House Lizard	**Shou Gong**
House Lizard	**Tian Long**
Houttuynia	**Yu Xing Cao**
Human Placenta	**Tai Pan**
Human Placenta	**Tai Yi**
Human Placenta	**Zi He Che**
Human Umbilical Cord	**Ji Dai**
Human Umbilical Cord	**Kan Qi**
Hyacinth Bean	**Bai Bian Dou**
Hyacinth Bean	**Bian Dou**
Hyacinth Bean Peel	**Bian Dou Yi**
Hyacinth Bletilla Tuber	**Bai Ji**
Hyacinth Flower	**Bian Dou Hua**
Hydrangea Flower	**Xiu Qiu Hua**
Hydrocotyle Plant	**Tian Hu Sui**
Immature Fruit of Bitter Orange	**Zhi Shi**
Immature Tangerine Peel	**Qing Pi**
India Pokeberry Root	**Shang Lu**
India Trumpetflower Seed	**Mu Hu Die**
Indian Mulberry Root	**Ba Ji Tian**
Indigo	**Qing Dai**
Inula Flower	**Xuan Fu Hua**
Iron Filings	**Sheng Tie Luo**
Jack-in-the-Pulpit	**Tian Nan Xing**
Jamestown Weed Seed/Leaf/Flower	**Man Tuo Luo**
Jamestown Weed Seed/Leaf/Flower	**Yang Jin Hua**
Japanese Ardisia	**Ai Di Cha**
Japanese Ardisia	**Zi Jin Niu**
Japanese Fern	**Hai Jin Sha Teng**
Japanese Fern Spore	**Hai Jin Sha**
Japanese Lily Bulb	**Wan Nian Qing**
Japanese Raspberry	**Mao Mei**
Japanese Saint John's Wort	**Di Er Cao**
Japanese Teasel Root	**Xu Duan**
Japanese Thistle	**Da Ji**
Japanese Yam	**Chuan Shan Long**
Java Brucea Fruit	**Ya Dan Zi**
Job's Tears (Coix)	**Yi Yi Ren**
Juice of Aloe Leaf	**Lu Hui**
Kadsura Stem	**Hai Feng Teng**
Kan-Sui Root	**Gan Sui**
Kelp Thallus	**Kun Bu**
Knotty Pine Wood	**Song Jie**
Knotweed	**Bian Xu**
Kochia Fruit	**Di Fu Zi**
Kudzu Flower	**Ge Hua**
Kudzu Root	**Ge Gen**
Kusnezoff Monkshood Root	**Cao Wu**
Land Tortoise Shell	**Gui Ban**
Large-headed Atractylodes Rhizome	**Bai Zhu**
Lead Oxide (Pb_3O_4)	**Huang Dan**
Lead Oxide (Pb_3O_4)	**Qian Dan**
Leaf of Apocynum Venetum	**Luo Bu Ma**

Leatherleaf Millettia Root/Vine	**Ji Xue Teng**
Ledebouriella Root	**Fang Feng**
Leech	**Shui Zhi**
Leechee Nut	**Li Zhi He**
Less Purfied Mirabilitum	**Pu Xiao**
Lesser Galangal Rhizome	**Gao Liang Jiang**
Light Wheat Grain	**Fu Xiao Mai**
Lily Bulb	**Bai He**
Limonite	**Yu Liang Shi**
Limonite	**Yu Yu Liang**
Litharge or Galena	**Mi Tuo Seng**
Litse Fruit	**Bi Cheng Qie**
Lizardtail	**San Bai Cao**
Long Pepper Fruit	**Bi Ba**
Longspur Epimedium	**Xian Ling Pi**
Longspur Epimedium	**Yin Yang Huo**
Loosestrife Plant	**Jin Qian Cao**
Loquat Leaf	**Pi Pa Ye**
Lotus Leaf	**He Ye**
Lotus Plumule	**Lian Zi Xin**
Lotus Seed	**Lian Zi**
Lotus Stamen	**Lian Xu**
Lycium Bark	**Di Gu Pi**
Macaque Gallstone	**Hou Zao**
Madagascar Periwinkle	**Chang Chun Hua**
Madder Root	**Qian Cao Gen**
Magnetite	**Ci Shi**
Magnolia Bark	**Hou Po**
Magnolia Flower	**Hou Po Hua**
Magnolia Flower	**Xin Yi Hua**
Male Seal Sexual Organ	**Hai Gou Shen**
Mapleleaf Goosefoot	**Xue Jian Chou**
Mature Lotus Receptacle	**Lian Fang**
Medicated Leaven	**Shen Qu**
Melon Pedicle	**Gua Di**
Mercury	**Shui Yin**
Milk-Vetch Root	**Huang Qi**
Milk-Vetch Seed	**Sha Yuan Ji Li**
Mimosa Tree Bark	**He Huan Pi**
Mimosa Tree Flower	**He Huan Hua**
Miniature Chinese Rose	**Mei Gui Hua**
Mock Strawberry Plant	**She Mei**
Moleplant Seed	**Xu Sui Zi**
Momordica Seed	**Mu Bie Zi**
Mongolian Snakegourd	**Gua Lou**
Mongolian Snakegourd Seed	**Gua Lou Ren**
Morning Glory Seed	**Qian Niu Zi**
Motherwort Fruit	**Chong Wei Zi**
Mountain Onion	**Li Lu**
Mugwort Leaf	**Ai Ye**
Mulberry Leaf	**Sang Ye**
Mulberry Mistletoe	**Sang Ji Sheng**
Mulberry Twig	**Sang Zhi**
Mume Fruit	**Wu Mei**
Mung Bean	**Lu Dou**
Mung Bean Skin	**Lu Dou Yi**

Musk Mallow Seed	**Dong Kui Zi**
Musk of Deer	**She Xiang**
Muskroot-like Semiaquilegia Root	**Tian Kui**
Muskroot-like Semiaquilegia Root	**Tian Kui Zi**
Myrobalan Fruit	**He Li Le**
Myrobalan Fruit	**He Zi**
Myrrh	**Mo Yao**
Nasturtium	**Han Cai**
Ningpo Figwort Root	**Xuan Shen**
Ningpo Figwort Root	**Yuan Shen**
Node of Lotus Rhizome	**Ou Jie**
Notopterygium Root	**Qiang Huo**
Nutgrass	**Xiang Fu**
Nutmeg Seed	**Rou Dou Kou**
Nutmeg Seed	**Rou Guo**
Nux-Vomica Seed	**Fan Mu Bie**
Nux-Vomica Seed	**Ma Qian Zi**
Old Lime	**Chen Shi Hui**
Ophicalcite	**Hua Rui Shi**
Opium Poppy Husk	**Mi Ke**
Opium Poppy Husk	**Ying Su Ke**
Oriental Arborvitae Twig	**Ce Bai Ye**
Otter Liver	**Ta Gan**
Outer Tangerine Peel	**Ju Hong**
Oyster Shell	**Mu Li**
Pagoda Tree Flower Bud	**Huai Hua Mi**
Pagoda Tree Fruit	**Huai Jiao**
Pangolin Scales	**Chuan Shan Jia**
Paniculate Bolbostemma	**Tu Bei Mu**
Paper Mulberry Fruit	**Zhe Shi Zi**
Paris Rhizome	**Chong Lou**
Paris Rhizome	**Zao Xiu**
Patchouli or Wrinkled Giant Hyssop	**Huo Xiang**
Patrinia	**Bai Jiang Cao**
Peach Kernel	**Tao Ren**
Peanut Peel	**Hua Sheng Yi**
Pearl	**Zhen Zhu**
Peking Spurge Root	**Da Ji**
Peppermint	**Bo He**
Pepperweed Seed or Flixweed Tansymustard Seed	**Ting Li Zi**
Perfoliatum	**Guan Ye Liao**
Perilla Leaf	**Zi Su Ye**
Perilla Stem	**Zi Su Geng**
Persimmon Calyx	**Shi Di**
Picrorhiza Rhizome	**Hu Huang Lian**
Pinellia Tuber	**Ban Xia**
Pipe-Fish	**Hai Long**
Pipewort Scapus	**Gu Jing Cao**
Plantago Seeds	**Che Qian Zi**
Pomegranate Husk	**Shi Liu Pi**
Poria Skin	**Fu Ling Pi**
Poria with Wood Root	**Fu Shen**
Praying Mantis Egg-Case laid on Mulberry Leaf	**Sang Piao Xiao**
Prepared Licorice Root	**Zhi Gan Cao**
Prepared Milk-Vetch Root	**Zhi Huang Qi**
Prepared Soybean	**Dan Dou Chi**

Prickly Chaff Flower/Root/Twig	**Tu Niu Xi**
Pricklyash Root	**Liang Mian Zhen**
Prince's Feather Fruit	**Shui Hong Hua Zi**
Privet Fruit	**Nu Zhen Zi**
Pseudoginseng Root	**San Qi**
Pubescent Angelica Root	**Du Huo**
Puffball	**Ma Bo**
Pumice	**Fu Hai Shi**
Pumpkin Seed Husk	**Nan Gua Zi**
Pure Mirabilitum	**Xuan Ming Fen**
Purified Alunite	**Ming Fan**
Purified Feces	**Jin Zhi**
Purple Aster Root	**Zi Wan**
Purple Perilla Fruit	**Zi Su Zi**
Purslane	**Ma Chi Xian**
Pyrite	**Zi Ran Tong**
Quick-Fried Ginger Root	**Pao Jiang**
Ramie Root	**Zhu Ma Gen**
Rangoon Creeper Fruit with Seed	**Shi Jun Zi**
Raw Licorice Root	**Gan Cao**
Realgar	**Xiong Huang**
Red Peony Root	**Chi Shao Yao**
Red Poria	**Chi Fu Ling**
Reed Rhizome	**Lu Gen**
Resin of Rose Maloes	**Su He Xiang**
Rhaponticum Root	**Lou Lu**
Rhinoceros Horn	**Xi Jiao**
Rhizome of Golden Eye-Grass	**Xian Mao**
Rhododendron Leaf	**Du Juan Hua Ye**
Rhubarb Root Rhizome	**Da Huang**
Rice Paper Pith	**Tong Cao**
Rice Sprout	**Gu Ya**
Ripe Fruit of Bitter Orange	**Zhi Ke**
Root of Beech Silver-Top	**Sha Shen**
Round Cardamom Fruit	**Bai Dou Kou**
Round Momordica Fruit	**Luo Han Guo**
Round Momordica Leaf	**Luo Han Guo Ye**
Rush Pith	**Deng Xin Cao**
Safflower Flower	**Hong Hua**
Sago Cycas	**Tie Shu**
Sal Ammoniac	**Nao Sha**
Salvia Root	**Dan Shen**
Sandalwood	**Tan Xiang**
Sappan Wood	**Su Mu**
Sarcandrae Plant	**Jiu Jie Cha**
Sarcandrae Plant	**Zhong Jie Feng**
Sargentodoxa Vine	**Hong Teng**
Saxifrage	**Hu Er Cao**
Scallion	**Cong Bai**
Schisandra Fruit	**Wu Wei Zi**
Schizonepeta Stem	**Jing Jie**
Scorpion	**Quan Xie**
Sea Cucumber	**Hai Shen**
Seahorse	**Hai Ma**
Seaweed	**Hai Zao**
Seed of Grains-of-Paradise	**Sha Ren**

Seed of Katsumada's Galangal	**Cao Dou Kou**
Seeds of Fragrant-Flowered Garlic	**Jiu Zi**
Selfheal Spike	**Xia Ku Cao**
Senna Leaf	**Fan Xie Ye**
Shepherd's Purse	**Ji Cai**
Shiny Bugleweed	**Ze Lan**
Shrubalthea Bark	**Mu Jin Pi**
Siberian Solomon Seal Rhizome	**Huang Jing**
Sichuan Cyathula Root	**Chuan Niu Xi**
Sichuan Lovage Root	**Chuan Xiong**
Sichuan Monkshood Root	**Chuan Wu**
Silkworm Feces	**Can Sha**
Skullcap Root	**Huang Qin**
Small Japanese Thistle	**Xiao Ji**
Smithsonite or Calamine	**Lu Gan Shi**
Snake Skin Slough	**She Tui**
Snakeweed	**Cao He Che**
Snakeweed	**Quan Shen**
Snakeweed	**Zi Shen**
Solomon's Seal Rhizome	**Yu Zhu**
Sophora Root	**Ku Shen**
Sophora Root	**Shan Dou Gen**
Sour Jujube Seed	**Suan Zao Ren**
Spreading Hedyotis Oldenlandia Plant	**Bai Hua She She Cao**
Spurge	**Ze Qi**
Star Jasmine Stem	**Luo Shi Teng**
Starwort Fruit	**He Shi**
Stellaria Root	**Yin Chai Hu**
Stemona Tuber	**Bai Bu**
Stephania Root	**Han Fang Ji**
Stinky Elm Fruit Paste	**Wu Yi**
Stiple Fiber of Fortune Windmill Palm	**Zong Lu Pi**
Strap Flower/Root/Leaf	**Ji Mu**
Stringy Stonecrop	**Chui Pen Cao**
Suakwa Seed	**Si Gua Zi**
Suakwa Vegetable Sponge	**Si Gua Luo**
Sulphur	**Liu Huang**
Sun Plant	**Ban Zhi Lian**
Swallowwort Root	**Bai Wei**
Sweet Basil	**Sheng Tou Cao**
Sweetflag Rhizome	**Shi Chang Pu**
Sweetgum Fruit	**Lu Lu Tong**
Sword Bean	**Dao Dou**
Tabasheer	**Tian Zhu Huang**
Talcum	**Hua Shi**
Tangerine Leaf	**Ju Ye**
Tangerine Peel	**Chen Pi**
Tangerine Pith	**Ju Luo**
Tangerine Seed	**Ju He**
Tea Tree Root	**Cha Shu Gen**
Tendrilled Fritillaria Bulb	**Chuan Bei Mu**
Thalictrum	**Tang Song Cao**
Thorny Elaegnus Leaf	**Hu Tui Ye**
Thorowax Root	**Chai Hu**
Three-Nerved Spicebush Root	**Wu Yao**
Thunberg Fritillaria Bulb	**Zhe Bei Mu**

Tiger Bone	**Hu Gu**
Tinospora Root	**Jin Guo Lan**
Tip of Tubular Stalactites	**E Guan Shi**
Toad Venom	**Chan Su**
Trichosantes Root	**Tian Hua Fen**
Tripterygium Wilfordii Hook	**Lei Gong Teng**
Truestar Anisetree Fruit	**Ba Jiao Hui Xiang**
Truestar Anisetree Fruit	**Da Hui Xiang**
Tuber of Dwarf Lilyturf	**Mai Dong**
Tuber of Dwarf Lilyturf	**Mai Men Dong**
Turmeric Rhizome	**Jiang Huang**
Turmeric Tuber	**Yu Jin**
Umbellate Pore Fungus	**Zhu Ling**
Velvet of Young Deer Horn	**Lu Rong**
Vine of Solomon's Seal	**Shou Wu Teng**
Vine of Solomon's Seal	**Ye Jiao Teng**
Vitex Fruit	**Man Jing Zi**
Walnut Nut	**Hu Tao Ren**
Walnut Nut	**Hu Tao Rou**
Walnut Twig	**He Tao Shu Zhi**
Water Buffalo Horn	**Shui Niu Jiao**
Water Calptrop	**Ling Jiao**
Water Chestnut	**Bi Qi**
Water Plantain	**Ze Xie**
Watermelon Fruit	**Xi Gua**
Watermelon Peel	**Xi Gua Cui Yi**
Wheat Grain	**Xiao Mai**
White Mustard Seed	**Bai Jie Zi**
White Peony Root	**Bai Shao Yao**
White Poria	**Fu Ling**
White Swallowwort	**Bai Qian**
Wild Buckwheat	**Jin Qiao Mai**
Wild Chrysanthemum Flower	**Ye Ju Hua**
Wingless Cockroach	**Tu Bie Chong**
Wingless Cockroach	**Zhe Chong**
Winter Cherry Fruit	**Suan Jiang**
Woad Leaf	**Da Qing Ye**
Woad Root	**Ban Lan Gen**
Woolly Grass Rhizome	**Bai Mao Gen**
Wormwood	**Qing Hao**
Yedeon's Violet	**Zi Hua Di Ding**
Yellow Azalea	**Yang Zhi Zhu**
Yellow Earth from center of cooking stove	**Fu Long Gan**
Young Cinnamon Bark	**Guan Gui**
Young Soybean Sprout	**Dou Juan**
Zaocys	**Wu Shao She**
Zedoary Rhizome	**E Zhu**

APPENDIX 6

Cross-Reference Pinyin–Botanical Latin Herbs

A Wei	*Resina Ferulae*
Ai Di Cha	*Herba Ardisiae Japonicae*
Ai Ye	*Folium Artemisiae Vulgaris/Argyi*
An Xi Xiang	*Benzoinum*
Ba Dou	*Semen Croton Tiglii*
Ba Ji Tian	*Radix Morindae Officinalis*
Ba Jiao Feng	*Radix Alangii*
Ba Jiao Hui Xiang	*Fructus Illicius*
Ba Qia	*Rhizoma et Folium Smilacis Chinensis*
Ba Yue Zha	*Fructus Akebiae*
Bai Bian Dou	*Semen Dolichoris Lablab*
Bai Bu	*Radix Stemonae*
Bai Dou Kou	*Fructus Amomi Cardamomi*
Bai Fu Zi	*Rhizoma Typhonii Gigantei*
Bai Guo	*Semen Ginkgo Bilobae*
Bai Guo Ye	*Folium Ginkgo*
Bai He	*Bulbus Lilii*
Bai Hua She	*Agkistrodon seu Bungarus*
Bai Hua She She Cao	*Herba Hedyotidis Diffusae*
Bai Ji	*Rhizoma Bletillae Striatae*
Bai Ji Li	*Fructus Tribuli Terrestris*
Bai Jiang Can	*Bombyx Batryticatus*
Bai Jiang Cao	*Herba cum Radice Patriniae*
Bai Jie Zi	*Semen Sinapis Albae*
Bai Lian	*Radix Ampelopsis*
Bai Mao Gen	*Rhizoma Imperatae Cylindricae*
Bai Mu Er	*Fructificatio Tremellae Fuciformis*
Bai Qian	*Rhizoma Cynanchii Stautoni*
Bai Shao Yao	*Radix Paeoniae Lactiflorae*
Bai Tou Weng	*Radix Pulsatillae Chinensis*
Bai Wei	*Radix Cynanchi Baiwei*
Bai Xian Pi	*Cortex Dictamni Dasycarpi Radicis*
Bai Ying	*Herba Solani Lyrati*
Bai Zhi	*Radix Angelicae Dahuricae*
Bai Zhu	*Rhizoma Atractylodis Macrocephalae*
Bai Zi Ren	*Semen Biotae Orientalis*
Ban Bian Lian	*Herba Lobeliae Chinensis cun Radice*

Ban Lan Gen	*Radix Isatidis seu Baphicacanthi*
Ban Mao	*Mylabris*
Ban Xia	*Rhizoma Pinelliae Ternatae*
Ban Zhi Lian	*Herba Scutellariae Barbatae*
Bei Xie	*Rhizoma Dioscoreae Hypoglaucae*
Bei Xing Ren	*Semen Pruni Armeniacae*
Bi Ba	*Fructus Piperis Longi*
Bi Cheng Qie	*Fructus Cubebae*
Bi Hu	*Gekko Swinhoana*
Bi Qi	*Cormus Heleochris Plantaginea*
Bian Dou	*Semen Dolichoris Lablab*
Bian Dou Hua	*Flos Dolichoris Lablab*
Bian Dou Yi	*Testa Dolichoris Lablab*
Bian Xu	*Herba Polygoni Avicularis*
Bie Jia	*Carapax Amydae Sinensis*
Bing Lang	*Semen Arecae Catechu*
Bing Pian	*Borneol*
Bo He	*Herba Menthae Haplocalycis*
Bu Gu Zhi	*Fructus Psoraleae Corylifoliae*
Can Sha	*Excrementum Bombycis Mori*
Cang Er Zi	*Fructus Xanthii Sibirici*
Cang Zhu	*Rhizoma Atractylodis Macrocephalae*
Cao Dou Kou	*Semen Alpiniae Katsumadai*
Cao Guo	*Fructus Amomi Tsao-ko*
Cao He Che	*Rhizoma Polygoni Bistortae*
Cao Wu	*Radix Aconiti Kusnezoffii*
Ce Bai Ye	*Cacumen Biotae Orientalis*
Cha Shu Gen	*Radix Camelliae Sinensis*
Chai Hu	*Radix Bupleuri*
Chan Su	*Secretio Bufonis*
Chan Tui	*Periostracum Cicadae*
Chang Chun Hua	*Herba Catharanthi Rosei*
Chang Shan	*Radix Dichroae Febrifugae*
Che Qian Zi	*Semen Plantaginis*
Chen Pi	*Pericarpium Citri Reticulatae*
Chen Shi Hui	*Calyx*
Chen Xiang	*Lignum Aquilariae*
Cheng Liu	*Cacumen Tamaricis*
Chi Fu Ling	*Sclerotium Poriae Cocos Rubrae*
Chi Shao Yao	*Radix Paeoniae Rubrae*
Chi Shi Zhi	*Halloysitum Rubrum*
Chi Xiao Dou	*Semen Phaseoli Calcarati*
Chong Cao	*Cordyceps Sinensis*
Chong Lou	*Rhizoma Paris*
Chong Wei Zi	*Fructus Leonuri*
Chou Wu Tong	*Folium Clerodendri Trichotomi*
Chuan Bei Mu	*Bulbus Fritillariae Cirrhosae*
Chuan Jiao	*Pericarpium Zanthoxyli Bungeani*
Chuan Lian Zi	*Fructus Meliae Toosendan*
Chuan Niu Xi	*Radix Cyathulae Officinalis*
Chuan Shan Jia	*Squama Manitis Pentadactylae*
Chuan Shan Long	*Rhizoma Dioscoreae Nipponicae*
Chuan Wu	*Radix Aconiti Carmichaeli*
Chuan Xin Lian	*Herba Andrographitis Paniculatae*
Chuan Xiong	*Radix Ligustici Chuanxiong*
Chui Pen Cao	*Herba Sedi Sarmentosi*

Chun Bai Pi	*Cortex Ailanthi*
Chun Gen Pi	*Cortex et Radix Ailanthi*
Chun Pi	*Cortex Ailanthi Altissimae*
Ci Ji Li	*Fructus Tribuli Terrestris*
Ci Shi	*Magnetitum*
Ci Wei Pi	*Corium Erinacei*
Ci Wu Jia	*Radix Acanthopanaeis Senticosi*
Cong Bai	*Bulbus Allii Fistulosi*
Da Feng Zi	*Semen Hydnocarpi Anthelminticae*
Da Fu Pi	*Pericarpium Arecae Catechu*
Da Huang	*Radix et Rhizoma Rhei*
Da Hui Xiang	*Fructus Illicii Veri*
Da Ji	*Herba seu Radix Cirsii Japonici*
Da Ji	*Radix Euphorbiae seu Knoxiae*
Da Pi Chong	*Aspongopus Chinensis*
Da Qing Ye	*Folium Daqingye*
Da Suan	*Bulbus Alli Sativi*
Da Zao	*Fructus Zizyphi Jujubae*
Dai Zhe Shi	*Haematitum*
Dan Dou Chi	*Semen Sojae Praeparatum*
Dan Fan	*Chalcanthite*
Dan Shen	*Radix Salviae Miltiorrhizae*
Dan Zhu Ye	*Herba Lophatheri Gracilis*
Dang Gui	*Radix Angelicae Sinensis*
Dang Shen	*Radix Codonopsitis Pilosulae*
Dao Dou	*Semen Canavaliae*
Deng Xin Cao	*Medulla Junci Effusi*
Di Er Cao	*Herba Hyperici Japonici*
Di Fu Zi	*Fructus Kochiae Scopariae*
Di Gu Pi	*Cortex Lycii Radicis*
Di Long	*Lumbricus*
Di Yu	*Radix Sanguisorbae Officinalis*
Dian Di Mei	*Herba Androsace Aizoon*
Ding Xiang	*Flos Caryophylli*
Dong Chong Xia Cao	*Cordyceps Sinensis*
Dong Gua Pi	*Epicarpium Benincasae Hispidae*
Dong Gua Ren	*Semen Benincasae Hispidae*
Dong Kui Zi	*Semen Abutili seu Malvae*
Dou Chi Jiang	*Radix et Ramus Litseae Cubebae*
Dou Juan	*Semen Glycines Germinatum*
Du Huo	*Radix Angelicae Pubescentis*
Du Juan Hua Ye	*Folium Rhododendri Simsii*
Du Zhong	*Cortex Eucommiae Ulmoidis*
E Guan Shi	*Stalactitum*
E Jiao	*Gelatinum Corii Asini*
E Zhu	*Rhizoma Curcumae Zedoariae*
Er Cha	*Acacia Catechu seu Uncaria Gambier*
Fan Bai Cao	*Herba Potenlillae Discoloris*
Fan Mu Bie	*Semen Strychni*
Fan Xie Ye	*Folium Sennae*
Fang Feng	*Radix Ledebouriellae Divaricatae*
Fei Zi	*Semen Torreyae Grandis*
Fen Sha Shen	*Radix Changii*
Feng Mi	*Mel*
Feng Wei Cao	*Herba Pteris*
Fo Shou	*Fructus Citri Sarcodactylis*

Fo Shou Hua	*Flos Citri Sarcodactylis*
Fu Hai Shi	*Pumice*
Fu Ling	*Sclerotium Poriae Cocos*
Fu Ling Pi	*Cortex Poriae Cocos*
Fu Long Gan	*Terra Flava Usta*
Fu Pen Zi	*Fructus Rubi Chingii*
Fu Ping	*Herba Lemnae seu Spirodelae*
Fu Shen	*Sclerotium Poriae Cocos Pararadicis*
Fu Xiao Mai	*Semen Tritici Aestivi Levis*
Fu Zi	*Radix Aconiti Carmichaeli Praeparata*
Gan Cao	*Radix Glycyrrhizae Uralensis*
Gan Jiang	*Rhizom Zingiberis Officinalis*
Gan Qi	*Lacca Sinica Exsiccatae*
Gan Sui	*Radix Euphorbiae Kansui*
Gao Ben	*Rhizoma et Radix Ligustici*
Gao Liang Jiang	*Rhizoma Alpiniae Officinari*
Ge Gen	*Radix Puerariae*
Ge Hua	*Flos Puerariae*
Ge Jie	*Gecko*
Ge Shan Xiao	*Rhizoma Cynanchum Auriculati*
Gong Ding Xiang	*Flos Caryophylli*
Gong Fen	*Calomelas*
Gou Ji	*Rhizoma Cibotii Barometz*
Gou Qi Zi	*Fructus Lycii*
Gou Teng	*Ramulus Uncariae cum Uncis*
Gu Jing Cao	*Scapus Eriocaulonis Buergeriani*
Gu Sui Bu	*Rhizoma Drynariae*
Gu Ya	*Fructus Oryzae Sativae Germinantus*
Gua Di	*Pedicellus Cucumeris*
Gua Lou	*Fructus Trichosanthis*
Gua Lou Ren	*Radix Trichosanthis Kirilowii*
Guan Gui	*Cortex Cinnamomi Cassiae*
Guan Ye Liao	*Herba Polygonum Perfoliati*
Guan Zhong	*Rhizoma Guanzhong*
Guang Fang Ji	*Radix Aristolochiae seu Cocculi*
Gui Ban	*Plastrum Testudinis*
Gui Ban Jiao	*Gelatinum Plastri Testudinis*
Gui Jiao	*Gelatinum Plastri Testudinis*
Gui Yuan Rou	*Arillus Euphoriae Longanae*
Gui Zhi	*Ramulus Cinnamomi Cassiae*
Hai Dai	*Herba Zosterae Marinae*
Hai Er Shen	*Radix Pseudostellariae Heterophyllae*
Hai Feng Teng	*Caulis Piperis Futokadsurae*
Hai Ge Ke	*Concha Cyclinae Sinensis*
Hai Gou Shen	*Testes et Penis Otariae*
Hai Jin Sha	*Spora Lygodii Japonici*
Hai Jin Sha Teng	*Herba Lygodii Japonici*
Hai Long	*Syngnathus*
Hai Ma	*Hippocampus*
Hai Piao Xiao	*Os Sepiae seu Sepiellae*
Hai Shen	*Stichopus Japonicus*
Hai Tong Pi	*Cortex Erythrinae*
Hai Zao	*Herba Sargassii*
Han Cai	*Herba Rorippae*
Han Fang Ji	*Radix Stephaniae Tetrandrae*
Han Lian Cao	*Herba Ecliptae Prostratae*

Han Shui Shi	*Calcitum*
He Cao Ya	*Gemma Agrimoniae*
He Huan Hua	*Albizziae Julibrissin*
He Huan Pi	*Cortex Albizziae Julibrissin*
He Li Le	*Fructus Terminaliae Chebulae*
He Shi	*Fructus Carpesii seu Daucusi*
He Shou Wu	*Radix Polygoni Multiflori*
He Tao Shu Zhi	*Ramulus Juglandis Mandshuricae*
He Ye	*Folium Nelumbinis Nuciferae*
He Zi	*Fructus Terminaliae Chebulae*
Hei Zhi Ma	*Semen Sesami Indici*
Hong Hua	*Flos Carthami Tinctorii*
Hong Teng	*Caulis Sargentodoxae Cuneatae*
Hong Zao	*Fructus Zizyphi Jujubae*
Hou Jiang	*Rhizoma Drynariae*
Hou Po	*Cortex Magnoliae Officinalis*
Hou Po Hua	*Flos Magnoliae Officinalis*
Hou Zao	*Calculus Macacae Mulattae*
Hu Er Cao	*Herba Saxifragae*
Hu Gu	*Os Tigris*
Hu Huang Lian	*Rhizoma Picrorhizae*
Hu Jiao	*Fructus Piperis Nigri*
Hu Lu Ba	*Semen Trigonellae Foeni-graeci*
Hu Po	*Succinum*
Hu Tao Ren	*Semen Juglandis Regiae*
Hu Tao Rou	*Semen Juglandis Regiae*
Hu Tui Ye	*Folium Elaeagni*
Hu Zhang	*Radix et Rhizoma Polygoni Cuspidati*
Hua Jiao	*Pericarpium Zanthoxyli Bungeani*
Hua Rui Shi	*Ophicalcitum*
Hua Sheng Yi	*Testa Arachidis Hypogaeae*
Hua Shi	*Talcum*
Huai Hua Mi	*Flos Sophorae Japonicae Immaturus*
Huai Jiao	*Fructus Sophorae Japonicae*
Huai Niu Xi	*Radix Achyranthis Bidentatae*
Huang Bai	*Cortex Phellodendri*
Huang Dan	*Minium*
Huang Du	*Rhizoma Dioscoreae Bulbiferae*
Huang Jing	*Rhizoma Polygonati*
Huang Lian	*Rhizoma Coptidis*
Huang Qi	*Radix Astragali Membranacei*
Huang Qin	*Radix Scutellariae Baicalensis*
Huang Yao Zi	*Rhizoma Dioscoreae Bulbiferae*
Huo Ma Ren	*Semen Cannabis Sativae*
Huo Xiang	*Herba Agastaches seu Pogostemi*
Ji Cai	*Herba Houttuyniae Cordatae*
Ji Dai	*Funiculus Umbilicalis*
Ji Guan Hua	*Flos Celosiae Cristatae*
Ji Mu	*Folium et Radix Loropetali*
Ji Nei Jin	*Endothelium Corneum Gigeriae Galli*
Ji Shi Teng	*Herba Paederiae*
Ji Xing Zi	*Semen Impatientis*
Ji Xue Teng	*Radix et Caulis Jixueteng*
Ji Zi Huang	*Vitellus Gigeriae Galli*
Jiang Can	*Bombyx Batryticatus*
Jiang Huang	*Rhizoma Curcumae Longae*

Jiang Xiang	*Lignum Dalbergiae Odoriferae*
Jie Geng	*Radix Platycodi Grandiflori*
Jin Guo Lan	*Radix Tinosporae*
Jin Qian Cao	*Herba Lysimachiae*
Jin Qiao Mai	*Rhizoma Fagopyri Cymosi*
Jin Yin Hua	*Flos Lonicerae Japonicae*
Jin Ying Zi	*Fructus Rosae Laevigatae*
Jin Zhi	*Succus Faecalis Aureus*
Jing Jie	*Flos Schizonepetae Tenuifoliae*
Jing Mao Gou Ji	*Rhizoma Cibotii*
Jing Tian San Qi	*Herba Sedi Aizoon*
Jiu Jie Cha	*Herba Sarcandrae Glabrae*
Jiu Li Ming	*Herba Senecionis*
Jiu Xiang Chong	*Aspongopus Chinensis*
Jiu Zi	*Semen Allii Tuberosi*
Ju He	*Semen Citri Reticulatae*
Ju Hong	*Exocarpium Citrus Reticulate Rubrum*
Ju Hua	*Flos Chrysanthemi Morifolii*
Ju Luo	*Fibrum Vascularis Citrus Tangerina*
Ju Ye	*Folium Citrus Tangerina*
Ju Ye San Qi	*Folium et Radix Gynurae Segetii*
Jue Ming Zi	*Semen Cassiae*
Kan Qi	*Funiculus Umbilicalis*
Ku Lian Gen Pi	*Cortex Meliae Radicis*
Ku Shen	*Radix Sophorae Flavescentis*
Ku Xing Ren	*Semen Pruni Armeniacae*
Kuan Dong Hua	*Flos Tussilaginis Farfarae*
Kuan Jin Teng	*Ramus Tinosporae Sinensis*
Kun Bu	*Thallus Algae*
Lai Fu Zi	*Semen Raphani Sativi*
Lang Du	*Radix Euphorbiae et Stellerae*
Lao Guan Cao	*Herba Erodii seu Geranii*
Lei Gong Teng	*Radix Tripterygii Wilfordii*
Lei Wan	*Sclerotium Omphaliae Lapidescens*
Li Lu	*Radix et Rhizoma Veratri*
Li Pi	*Pericarpium Pyri*
Li Zhi He	*Semen Litchi Chinensis*
Lian Fang	*Receptaculum Nelumbinis Nuciferae*
Lian Qiao	*Fructus Forsythiae Suspensae*
Lian Tou Jian	*Rhizoma Anemones Raddeanae*
Lian Xu	*Stamen Nelumbinis Nucifera*
Lian Zi	*Semen Nelumbinis Nucifera*
Lian Zi Xin	*Plumula Nelumbinis Nucifera*
Liang Mian Zhen	*Radix Zanthoxyli*
Ling Jiao	*Fructus Trapae*
Ling Xiao Hua	*Flos Campsis*
Ling Yang Jiao	*Cornu Antelopis*
Ling Zhi	*Excrementum Trogopteri seu Pteromi*
Liu Huang	*Sulphur*
Liu Ji Nu	*Herba Artemisiae Anomalae*
Long Dan Cao	*Radix Gentianae Longdancao*
Long Gu	*Os Draconis*
Long Kui	*Herba Solani Nigri*
Long Yan Rou	*Arillus Euphoriae Longanae*
Lou Lu	*Radix Rhapontici seu Echinops*
Lu Dou	*Semen Phaseoli Radiati*

Lu Dou Yi	*Cortex Phaseoli Radiati*
Lu Feng Fang	*Nidus Vespae*
Lu Gan Shi	*Smithsonitum*
Lu Gen	*Rhizoma Phragmitis Communis*
Lu Hui	*Herba Aloes*
Lu Jiao	*Cornu Cervi*
Lu Jiao Jiao	*Gelatinum Cornu Cervi*
Lu Jiao Shuang	*Cornu Cervi Degelatinatium*
Lu Lu Tong	*Fructus Liquidambaris Taiwanianae*
Lu Pi Jiao	*Gelatinum Corii Asini*
Lu Rong	*Cornu Cervi Parvum*
Lu Ying	*Flos Sambuci Chinensis*
Luo Bu Ma	*Folium Apocyni Veneti*
Luo Han Guo	*Fructus Momordicae Grosvenori*
Luo Han Guo Ye	*Folium Momordicae Grosvenori*
Luo Shi Teng	*Caulis Trachelospermi Jasminoidis*
Ma Bian Cao	*Herba Verbenae*
Ma Bo	*Frucificatio Lasiosphaerae*
Ma Chi Xian	*Herba Portulacae Oleraceae*
Ma Dou Ling	*Fructus Aristolochiae*
Ma Huang	*Herba Ephedrae*
Ma Huang Gen	*Radix Ephedrae*
Ma Qian Zi	*Semen Strychni*
Ma Ti Jin	*Herba Dichondrae*
Mai Dong	*Tuber Ophiopogonis Japonici*
Mai Men Dong	*Tuber Ophiopogonis Japonici*
Mai Ya	*Fructus Hordei Vulgaris Germinantus*
Man Jing Zi	*Fructus Viticis*
Man Tuo Luo	*Flos Daturae*
Mang Chong	*Tabanus Bivittatus*
Mang Xiao	*Mirabilitum*
Mao Dong Qing	*Radix Ilicis Pubescentis*
Mao Mei	*Herba Rubi Parvifolii*
Mao Zhua Cao	*Radix Ranunculus Tematus*
Mei Gui Hua	*Flos Rosae Rugosae*
Meng Shi	*Lapis Micae seu Chloriti*
Mi Ke	*Pericarpium Papaveris Somniferi*
Mi Meng Hua	*Flos Buddleiae Officinalis*
Mi Tuo Seng	*Lithargyrum*
Mian Hua Gen	*Radix Gossypii*
Ming Dang Shen	*Radix Changii*
Ming Fan	*Alumen*
Mo Yao	*Myrrha*
Mu Bie Zi	*Semen Momordicae Cochinchinensis*
Mu Dan Pi	*Cortex Moutan Radicis*
Mu Ding Xiang	*Flos Syzygii Aromatici*
Mu Gua	*Fructus Chaenomelis*
Mu Hu Die	*Semen Oroxyli Indici*
Mu Jin Pi	*Cortex Hibiscus Syriaci*
Mu Li	*Concha Ostreae*
Mu Tong	*Caulis Mutong*
Mu Xiang	*Radix Aucklandiae Lappae*
Mu Zei	*Herba Equiseti Hiemalis*
Nan Gua Zi	*Semen Cucurbitae Moschatae*
Nao Sha	*Sal Ammoniaci*
Niu Bang Zi	*Fructus Arctii Lappae*

Niu Huang	*Calculus Bovis*
Nong Ji Li	*Herba Crotalariae*
Nu Zhen Zi	*Fructus Ligustri Lucidi*
Nuo Dao Gen Xu	*Radix et Rhizoma Oryzae Glutinosae*
Ou Jie	*Nodus Nelumbinis Nuciferae Rhizomatis*
Pang Da Hai	*Semen Sterculiae Scaphigerae*
Pao Jiang	*Quick-Fried Rhizoma Zingiberis Officinalis*
Pei Lan	*Herba Eupatorii Fortunei*
Peng Sha	*Borax*
Pi Pa Ye	*Folium Eriobotryae Japonicae*
Pi Shi	*Arsenolite*
Pi Shuang	*Arsenicum*
Po Gu Zhi	*Fructus Psoraleae Corylifoliae*
Pu Gong Ying	*Herba Taraxaci Mongolici cum Radice*
Pu Huang	*Pollen Typhae*
Pu Xiao	*Mirabilitum Depuratum*
Qi Cao	*Holotrichia Diomphalia*
Qian Cao Gen	*Radix Rubiae Cordifoliae*
Qian Dan	*Minium*
Qian Hu	*Radix Peucedani*
Qian Li Guang	*Herba Senecionis*
Qian Nian Jian	*Rhizoma Homalomenae Occultae*
Qian Niu Zi	*Semen Pharbitidis*
Qian Ri Hong	*Flos Gomphrenae*
Qian Shi	*Semen Euryales Ferocis*
Qiang Huo	*Rhizoma et Radix Notopterygii*
Qiang Lang	*Catharsius*
Qiang Lang Chong	*Catharsius*
Qin Jiao	*Radix Gentianae Qinjiao*
Qin Pi	*Cortex Fraxini*
Qing Dai	*Indigo Pulverata Levis*
Qing Fen	*Calomelas*
Qing Hao	*Herba Artemisiae Annuae*
Qing Meng Shi	*Lapis Micae seu Chloriti*
Qing Pi	*Pericarpium Citri Reticulatae Viride*
Qing Xiang Zi	*Semen Celosiae Argenteae*
Qu Mai	*Herba Dianthi*
Quan Shen	*Rhizoma Polygoni Bistortae*
Quan Xie	*Buthus Martensi*
Ren Shen	*Radix Ginseng*
Ren Shen Ye	*Folium Ginseng*
Rou Cong Rong	*Herba Cistanches Deserticolae*
Rou Dou Kou	*Semen Myristicae Fragrantis*
Rou Gui	*Cortex Cinnamomi Cassiae*
Rou Guo	*Semen Myristicae Fragrantis*
Ru Xiang	*Gummi Olibanum*
San Bai Cao	*Rhizoma Seu Herba Saururii*
San Ke Zhen	*Radix Berberidis*
San Leng	*Rhizoma Sparganii Stoloniferi*
San Qi	*Radix Notoginseng*
Sang Bai Pi	*Cortex Mori Albae Radicis*
Sang Ji Sheng	*Ramulus Sangjisheng*
Sang Piao Xiao	*Ootheca Mantidis*
Sang Shen	*Fructus Mori Albae*
Sang Ye	*Folium Mori Albae*
Sang Zhi	*Ramulus Mori Albae*

Sha Ren	*Fructus Amomi*
Sha Shen	*Radix Adenophorae seu Glehniae*
Sha Yuan Ji Li	*Semen Astragali Complanati*
Shan Ci Gu	*Pseudobulbus Shancigu*
Shan Dou Gen	*Radix Sophorae Tonkinensis*
Shan Yao	*Radix Dioscoreae Oppositae*
Shan Yu Rou	*Fructus Corni Officinalis*
Shan Zha	*Fructus Crataegi*
Shan Zhi Zi	*Fructus Gardeniae Jasminoidis*
Shan Zhu Yu	*Fructus Corni Officinalis*
Shang Lu	*Radix Phytolaccae*
She Chuang Zi	*Fructus Cnidii Monnieri*
She Gan	*Rhizoma Belamcandae Chinensis*
She Mei	*Herba Duchesneae Indicae*
She Tui	*Exuviae Serpentis*
She Xiang	*Secretio Moschus*
Shen Jin Cao	*Herba Lycopodii*
Shen Qu	*Massa Fermentata*
Sheng Di Huang	*Radix Rehmanniae Glutinosae*
Sheng Jiang	*Rhizoma Zingiberis Officinalis Recens*
Sheng Jiang Pi	*Exocarpium Zingiberis Officinalis Recens*
Sheng Jiang Zhi	*Succus Zingiberis*
Sheng Ma	*Rhizoma Cimicifugae*
Sheng Tie Luo	*Frusta Ferri*
Sheng Tou Cao	*Herba Ocimi Pilosi*
Shi Chang Pu	*Rhizoma Acori Graminei*
Shi Da Gong Lao	*Folium Mahoniae*
Shi Dan	*Chalcanthite*
Shi Di	*Calyx Diospyri Kaki*
Shi Gao	*Gypsum*
Shi Hu	*Herba Dendrobii*
Shi Jian Chuan	*Radix Salviae Chinensis*
Shi Jue Ming	*Concha Haliotidis*
Shi Jun Zi	*Fructus Quisqualis Indicae*
Shi Liu Gen Pi	*Cortex Punicae Granati Radicis*
Shi Liu Pi	*Pericarpium Punicae Granati*
Shi Nan Teng	*Ramulus Piper Wallichii*
Shi Shang Bai	*Herba Selagrinellae Doederleinii*
Shi Wei	*Folium Pyrrosiae*
Shou Gong	*Gekko Swinhoana*
Shou Wu Teng	*Caulis Polygoni Multiflori*
Shu Di Huang	*Radix Rehmanniae Glutinosae Conquitae*
Shui Hong Hua Zi	*Fructus Polygoni Orientalis*
Shui Niu Jiao	*Cornu Bubali*
Shui Yin	*Hydrargyrum*
Shui Zhi	*Hirudo seu Whitmania*
Si Gua Luo	*Fasciculus Vascularis Luffae*
Si Gua Zi	*Fructus Vascularis Luffae*
Si Ji Qing	*Folium Ilicis Chinensis*
Song Jie	*Lignum Pini Nodi*
Su He Xiang	*Styrax Liquidis*
Su Mu	*Lignum Sappan*
Suan Jiang	*Rhizoma et Radix Physalitis*
Suan Zao Ren	*Semen Zizyphi Spinosae*
Suo Yang	*Herba Cynomorii Songarici*
Ta Gan	*Jecur Lutrae*

Tai Pan	*Placenta Hominis*
Tai Yi	*Placenta Hominis*
Tai Zi Shen	*Radix Pseudostellariae Heterophyllae*
Tan Xiang	*Lignum Santali Albi*
Tang Song Cao	*Radix et Herba Thalictri Foliolosi*
Tao Ren	*Semen Persicae*
Teng Li Gen	*Radix Actinidiae Argutae*
Tian Dong	*Tuber Asparagi Cochinchinensis*
Tian Hu Sui	*Herba Hydrocotylis*
Tian Hua Fen	*Radix Trichosanthis Kirilowii*
Tian Kui	*Herba Semiaquilegiae Adoxoidis*
Tian Kui Zi	*Radix Semiaquilegiae Adoxoidis*
Tian Long	*Gekko Swinhoana*
Tian Ma	*Rhizoma Gastrodiae Elatae*
Tian Men Dong	*Tuber Asparagi Cochinchinensis*
Tian Ming Jing	*Radix Carpesii*
Tian Nan Xing	*Rhizoma Arisaematis*
Tian Xian Zi	*Semen Hyoscyami*
Tian Zhu Huang	*Concretio Silicea Bambusae Textillis*
Tie Shu	*Folium et Flos Cordylines Fruticosae*
Tie Xian	*Herba Acalyphae*
Ting Li Zi	*Semen Descurainiae seu Lepidii*
Tong Bian	*Urina Puerilis*
Tong Cao	*Medulla Tetrapanacis Papyriferi*
Tou Gu Cao	*Herba Speranskia Tuberculata*
Tu Bei Mu	*Rhizoma Bolbostemmatis*
Tu Bie Chong	*Eupolyphaga seu Opisthoplatia*
Tu Fu Ling	*Rhizoma Smilacis Glabrae*
Tu Jin Pi	*Cortex Pseudolarix Kaempferi*
Tu Niu Xi	*Radix Achyranthis Aspera*
Tu Si Zi	*Semen Cuscutae Chinensis*
Wa Leng Zi	*Concha Arcae*
Wan Nian Qing	*Folium et Radix Rohdeae Japonicae*
Wang Bu Liu Xing	*Semen Vaccariae Segetalis*
Wei Ling Xian	*Radix Clematidis*
Wu Bei Zi	*Galla Rhois Chinensis*
Wu Gong	*Scolopendra Subspinipes*
Wu Jia Pi	*Cortex Acanthopanacis Gracilistyli Radicis*
Wu Ling Zhi	*Excrementum Trogopteri seu Pteromi*
Wu Mei	*Fructus Pruni Mume*
Wu Shao She	*Zaocys Dhumnades*
Wu Wei Zi	*Fructus Schisandrae Chinensis*
Wu Yao	*Radix Linderae Strychnifoliae*
Wu Yi	*Praeparatio Fructus Ulmi Macrocarpi*
Wu Zei Gu	*Os Sepiae seu Sepiellae*
Wu Zhu Yu	*Fructus Evodiae Rutaecarpae*
Xi Gua	*Fructus Citrulli Vulgaris*
Xi Gua Cui Yi	*Exocarpium Citrulli*
Xi Gua Shuang	*Citrullus Lanatus*
Xi He Liu	*Ramulus Tamarix Chinensis*
Xi Jiao	*Cornu Rhinoceri*
Xi Shu	*Fructus Seu Radix Camptothecae*
Xi Xian Cao	*Herba Siegesbeckiae*
Xi Xin	*Herba cum Radice Asari*
Xi Yang Shen	*Radix Panacis Quinquefolii*
Xia Ku Cao	*Spica Prunellae Vulgaris*

Xia Tian Wu	*Rhizoma Corydalis Decumbentis*
Xian He Cao	*Herba Agrimoniae Pilosae*
Xian Ling Pi	*Herba Epimedii*
Xian Mao	*Rhizoma Curculiginis Orchioidis*
Xiang Fu	*Rhizoma Cyperi Rotundi*
Xiang Ru	*Herba Elsholtziae seu Moslae*
Xiang Yuan	*Fructus Citri Medica*
Xiao Hui Xiang	*Fructus Foeniculi Vulgaris*
Xiao Ji	*Herba Cephalanoplos*
Xiao Mai	*Semen Tritici Levis*
Xie Bai	*Bulbus Allii*
Xin Yi Hua	*Flos Magnoliae*
Xing Ren	*Semen Pruni Armeniacae*
Xiong Dan	*Vesica Fellea Ursi*
Xiong Huang	*Realgar*
Xiu Qiu Hua	*Flos Clematidis Ranunculoidis*
Xu Chang Qing	*Radix et Herba Cynanchi Paniculati*
Xu Duan	*Radix Dipsaci Asperi*
Xu Sui Zi	*Semen Euphorbia Lathyridis*
Xuan Fu Hua	*Flos Inulae*
Xuan Ming Fen	*Purum Mirabilitum*
Xuan Shen	*Radix Scrophulariae Ningpoensis*
Xue Dan	*Radix Hemsleyae*
Xue Jian Chou	*Herba Chenopodii Hybridi*
Xue Jie	*Resina Draconis*
Xue Shang Yi Zhi Hao	*Radix Aconiti Brachypodi*
Xue Yu Tan	*Crinis Carbonisatus Hominis*
Xun Gu Feng	*Herba Aristolochiae*
Ya Dan Zi	*Fructus Bruceae Javanicae*
Ya Zhi Cao	*Herba Commelinae*
Yan Fu Mu	*Folium et Cortex Rhois Chinensis*
Yan Hu Suo	*Rhizoma Corydalis Yanhusuo*
Yang Jin Hua	*Flos Daturae Albae*
Yang Qi Shi	*Actinolitum*
Yang Zhi Zhu	*Radix Rhododendri Mollis*
Ye Jiao Teng	*Caulis Polygoni Multiflori*
Ye Ju Hua	*Flos Chrysanthemi Indici*
Ye Ming Sha	*Excrementum Vespertilionis Murini*
Yi Mu Cao	*Herba Leonuri Heterophylli*
Yi Tang	*Saccharum Granorum*
Yi Yi Ren	*Semen Coicis Lachryma-jobi*
Yi Zhi Ren	*Fructus Alpiniae Oxyphyllae*
Yin Chai Hu	*Radix Stellariae Dichotomae*
Yin Chen Hao	*Herba Artemisiae Yinchenhao*
Yin Xing	*Semen Ginkgo Bilobae*
Yin Xing Ye	*Folium Ginkgo*
Yin Yang Huo	*Herba Epimedii*
Ying Su Ke	*Pericarpium Papaveris Somniferi*
Yu Bai Pi	*Cortex Ulmi Pumilae*
Yu Jin	*Tuber Curcumae*
Yu Li Ren	*Semen Pruni*
Yu Liang Shi	*Limonite*
Yu Mi Xu	*Stylus Zeae Mays*
Yu Xing Cao	*Herba cum Radice Houttuyniae Cordatae*
Yu Yu Liang	*Limonite*
Yu Zhu	*Rhizoma Polygonati Odorati*

Yuan Hua	*Flos Daphnes Genkwa*
Yuan Shen	*Radix Scrophulariae Ningpoensis*
Yuan Zhi	*Radix Polygalae Tenuifoliae*
Yue Ji Hua	*Flos et Fructus Rosae Chinensis*
Zao Jiao	*Fructus Gleditsiae Sinensis*
Zao Pi	*Fructus Corni Officinalis*
Zao Xiu	*Rhizoma Paris*
Ze Lan	*Herba Lycopi Lucidi*
Ze Qi	*Herba Euphorbiae Helioscopiae*
Ze Xie	*Rhizoma Alismatis Orientalis*
Zhang Nao	*Camphora*
Zhe Bei Mu	*Bulbus Fritillariae Thunbergii*
Zhe Chong	*Eupolyphaga seu Opisthoplatia*
Zhe Shi Zi	*Fructus Broussonetiae*
Zhen Zhu	*Margarita Usta*
Zhi Gan Cao	*Radix Glycyrrhizae Uralensis Praeparata*
Zhi Huang Qi	*Radix Astragali Membranacei Praeparata*
Zhi Ke	*Fructus Citri Aurantii*
Zhi Mu	*Rhizoma Anemarrhenae Asphodeloidis*
Zhi Shi	*Fructus Immaturus Citri Aurantii*
Zhong Jie Feng	*Herba Sarcandrae*
Zhu Jie Xiang Fu	*Rhizoma Anemones Raddeanae*
Zhu Li	*Succus Bambusae*
Zhu Ling	*Sclerotium Polypori Umbellati*
Zhu Ma Gen	*Radix Boehmeriae*
Zhu Ru	*Caulis Bambusae in Taeniis*
Zhu Sha Gen	*Radix Ardisiae Crenatae*
Zi Bai Pi	*Cortex Catalpae Ovatae*
Zi Bei Tian Kui	*Herba Begoniae Fimbristipulatae*
Zi Cao	*Radix Arnebiae seu Lithospermi*
Zi He Che	*Placenta Hominis*
Zi Hua Di Ding	*Herba cum Radice Violae Yedoensitis*
Zi Jin Niu	*Caulis et Folium Ardisiae Japonicae*
Zi Jing Mu	*Lignum Cercis Chinensis*
Zi Jing Mu Pi	*Cortex Cercis Chinensis*
Zi Jing Pi	*Cortex Cercis Chinensis*
Zi Ran Tong	*Pyritum*
Zi Shen	*Rhizoma Bistortae*
Zi Shi Ying	*Fluoritum*
Zi Su Geng	*Caulis Perillae Frutescentis*
Zi Su Ye	*Folium Perillae Frutescentis*
Zi Su Zi	*Fructus Perillae Frutescentis*
Zi Wan	*Radix Asteris Tatarici*
Zi Wei	*Flos Campsis*
Zi Zhu	*Folium Callicarpae Pedunculatae*
Zong Lu Pi	*Fibra Trachycarpi Vagina*
Zong Lu Tan	*Trachycarpi Carbonisatus*
Zu Shi Ma	*Cortex Daphnes*

APPENDIX 7

Cross-Reference Botanical Latin–Pinyin Herbs

Acacia Catechu seu Uncaria Gambier	**Er Cha**
Actinolitum	**Yang Qi Shi**
Agkistrodon seu Bungarus	**Bai Hua She**
Albizziae Julibrissin	**He Huan Hua**
Alumen	**Ming Fan**
Arillus Euphoriae Longanae	**Gui Yuan Rou**
Arillus Euphoriae Longanae	**Long Yan Rou**
Arsenicum	**Pi Shuang**
Arsenolite	**Pi Shi**
Aspongopus Chinensis	**Da Pi Chong**
Aspongopus Chinensis	**Jiu Xiang Chong**
Benzoinum	**An Xi Xiang**
Bombyx Batryticatus	**Bai Jiang Can**
Bombyx Batryticatus	**Jiang Can**
Borax	**Peng Sha**
Borneol	**Bing Pian**
Bulbus Alli Sativi	**Da Suan**
Bulbus Allii	**Xie Bai**
Bulbus Allii Fistulosi	**Cong Bai**
Bulbus Fritillariae Cirrhosae	**Chuan Bei Mu**
Bulbus Fritillariae Thunbergii	**Zhe Bei Mu**
Bulbus Lilii	**Bai He**
Buthus Martensi	**Quan Xie**
Cacumen Biotae Orientalis	**Ce Bai Ye**
Cacumen Tamaricis	**Cheng Liu**
Calcitum	**Han Shui Shi**
Calculus Bovis	**Niu Huang**
Calculus Macacae Mulattae	**Hou Zao**
Calomelas	**Gong Fen**
Calomelas	**Qing Fen**
Calx	**Chen Shi Hui**
Calyx Diospyri Kaki	**Shi Di**
Camphora	**Zhang Nao**
Carapax Amydae Sinensis	**Bie Jia**
Catharsius	**Qiang Lang**
Catharsius	**Qiang Lang Chong**
Caulis Bambusae in Taeniis	**Zhu Ru**
Caulis et Folium Ardisiae Japonicae	**Zi Jin Niu**

Caulis Mutong	**Mu Tong**
Caulis Perillae Frutescentis	**Zi Su Geng**
Caulis Piperis Futokadsurae	**Hai Feng Teng**
Caulis Polygoni Multiflori	**Shou Wu Teng**
Caulis Polygoni Multiflori	**Ye Jiao Teng**
Caulis Sargentodoxae Cuneatae	**Hong Teng**
Caulis Trachelospermi Jasminoidis	**Luo Shi Teng**
Chalcanthite	**Dan Fan**
Chalcanthite	**Shi Dan**
Citrullus Lanatus	**Xi Gua Shuang**
Concha Arcae	**Wa Leng Zi**
Concha Cyclinae Sinensis	**Hai Ge Ke**
Concha Haliotidis	**Shi Jue Ming**
Concha Ostreae	**Mu Li**
Concretio Silicea Bambusae Textillis	**Tian Zhu Huang**
Cordyceps Sinensis	**Chong Cao**
Cordyceps Sinensis	**Dong Chong Xia Cao**
Corium Erinacei	**Ci Wei Pi**
Cormus Heleochris Plantaginea	**Bi Qi**
Cornu Antelopis	**Ling Yang Jiao**
Cornu Bubali	**Shui Niu Jiao**
Cornu Cervi	**Lu Jiao**
Cornu Cervi Degelatinatium	**Lu Jiao Shuang**
Cornu Cervi Parvum	**Lu Rong**
Cornu Rhinoceri	**Xi Jiao**
Cortex Acanthopanacis Gracilistyli Radicis	**Wu Jia Pi**
Cortex Ailanthi	**Chun Bai Pi**
Cortex Ailanthi Altissimae	**Chun Pi**
Cortex Albizziae Julibrissin	**He Huan Pi**
Cortex Catalpae Ovatae	**Zi Bai Pi**
Cortex Cercis Chinensis	**Zi Jing Mu Pi**
Cortex Cercis Chinensis	**Zi Jing Pi**
Cortex Cinnamomi Cassiae	**Guan Gui**
Cortex Cinnamomi Cassiae	**Rou Gui**
Cortex Daphnes	**Zu Shi Ma**
Cortex Dictamni Dasycarpi Radicis	**Bai Xian Pi**
Cortex Erythrinae	**Hai Tong Pi**
Cortex et Radix Ailanthi	**Chun Gen Pi**
Cortex Eucommiae Ulmoidis	**Du Zhong**
Cortex Fraxini	**Qin Pi**
Cortex Hibiscus Syriaci	**Mu Jin Pi**
Cortex Lycii Radicis	**Di Gu Pi**
Cortex Magnoliae Officinalis	**Hou Po**
Cortex Meliae Radicis	**Ku Lian Gen Pi**
Cortex Mori Albae Radicis	**Sang Bai Pi**
Cortex Moutan Radicis	**Mu Dan Pi**
Cortex Phaseoli Radiati	**Lu Dou Yi**
Cortex Phellodendri	**Huang Bai**
Cortex Poriae Cocos	**Fu Ling Pi**
Cortex Pseudolarix Kaempferi	**Tu Jin Pi**
Cortex Punicae Granati Radicis	**Shi Liu Gen Pi**
Cortex Ulmi Pumilae	**Yu Bai Pi**
Crinis Carbonisatus Hominis	**Xue Yu Tan**
Endothelium Corneum Gigeriae Galli	**Ji Nei Jin**
Epicarpium Benincasae Hispidae	**Dong Gua Pi**
Eupolyphaga seu Opisthoplatia	**Tu Bie Chong**

Eupolyphaga seu Opisthoplatia	**Zhe Chong**
Excrementum Bombycis Mori	**Can Sha**
Excrementum Trogopteri seu Pteromi	**Ling Zhi**
Excrementum Trogopteri seu Pteromi	**Wu Ling Zhi**
Excrementum Vespertilionis Murini	**Ye Ming Sha**
Exocarpium Citrulli	**Xi Gua Cui Yi**
Exocarpium Citrus Reticulate Rubrum	**Ju Hong**
Exocarpium Zingiberis Officinalis Recens	**Sheng Jiang Pi**
Exuviae Serpentis	**She Tui**
Fasciculus Vascularis Luffae	**Si Gua Luo**
Fibra Trachycarpi Vagina	**Zong Lu Pi**
Fibrum Vascularis Citrus Tangerina	**Ju Luo**
Flos Buddleiae Officinalis	**Mi Meng Hua**
Flos Campsis	**Ling Xiao Hua**
Flos Campsis	**Zi Wei**
Flos Carthami Tinctorii	**Hong Hua**
Flos Caryophylli	**Ding Xiang**
Flos Caryophylli	**Gong Ding Xiang**
Flos Celosiae Cristatae	**Ji Guan Hua**
Flos Chrysanthemi Indici	**Ye Ju Hua**
Flos Chrysanthemi Morifolii	**Ju Hua**
Flos Citri Sarcodactylis	**Fo Shou Hua**
Flos Clematidis Ranunculoidis	**Xiu Qiu Hua**
Flos Daphnes Genkwa	**Yuan Hua**
Flos Daturae	**Man Tuo Luo**
Flos Daturae Albae	**Yang Jin Hua**
Flos Dolichoris Lablab	**Bian Dou Hua**
Flos et Fructus Rosae Chinensis	**Yue Ji Hua**
Flos Gomphrenae	**Qian Ri Hong**
Flos Inulae	**Xuan Fu Hua**
Flos Lonicerae Japonicae	**Jin Yin Hua**
Flos Magnoliae	**Xin Yi Hua**
Flos Magnoliae Officinalis	**Hou Po Hua**
Flos Puerariae	**Ge Hua**
Flos Rosae Rugosae	**Mei Gui Hua**
Flos Sambuci Chinensis	**Lu Ying**
Flos Schizonepetae Tenuifoliae	**Jing Jie**
Flos Sophorae Japonicae Immaturus	**Huai Hua Mi**
Flos Syzygii Aromatici	**Mu Ding Xiang**
Flos Tussilaginis Farfarae	**Kuan Dong Hua**
Fluoritum	**Zi Shi Ying**
Folium Apocyni Veneti	**Luo Bu Ma**
Folium Artemisiae Vulgaris/Argyi	**Ai Ye**
Folium Callicarpae Pedunculatae	**Zi Zhu**
Folium Citrus Tangerina	**Ju Ye**
Folium Clerodendri Trichotomi	**Chou Wu Tong**
Folium Daqingye	**Da Qing Ye**
Folium Elaeagni	**Hu Tui Ye**
Folium Eriobotryae Japonicae	**Pi Pa Ye**
Folium et Cortex Rhois Chinensis	**Yan Fu Mu**
Folium et Flos Cordylines Fruticosae	**Tie Shu**
Folium et Radix Gynurae Segetii	**Ju Ye San Qi**
Folium et Radix Loropetali	**Ji Mu**
Folium et Radix Rohdeae Japonicae	**Wan Nian Qing**
Folium Ginkgo	**Bai Guo Ye**
Folium Ginkgo	**Yin Xing Ye**

Folium Ginseng	**Ren Shen Ye**
Folium Ilicis Chinensis	**Si Ji Qing**
Folium Mahoniae	**Shi Da Gong Lao**
Folium Momordicae Grosvenori	**Luo Han Guo Ye**
Folium Mori Albae	**Sang Ye**
Folium Nelumbinis Nuciferae	**He Ye**
Folium Perillae Frutescentis	**Zi Su Ye**
Folium Pyrrosiae	**Shi Wei**
Folium Rhododendri Simsii	**Du Juan Hua Ye**
Folium Sennae	**Fan Xie Ye**
Frucificatio Lasiosphaerae	**Ma Bo**
Fructificatio Tremellae Fuciformis	**Bai Mu Er**
Fructus Akebiae	**Ba Yue Zha**
Fructus Alpiniae Oxyphyllae	**Yi Zhi Ren**
Fructus Amomi	**Sha Ren**
Fructus Amomi Cardamomi	**Bai Dou Kou**
Fructus Amomi Tsao-ko	**Cao Guo**
Fructus Arctii Lappae	**Niu Bang Zi**
Fructus Aristolochiae	**Ma Dou Ling**
Fructus Broussonetiae	**Zhe Shi Zi**
Fructus Bruceae Javanicae	**Ya Dan Zi**
Fructus Carpesii seu Daucusi	**He Shi**
Fructus Chaenomelis	**Mu Gua**
Fructus Citri Aurantii	**Zhi Ke**
Fructus Citri Medica	**Xiang Yuan**
Fructus Citri Sarcodactylis	**Fo Shou**
Fructus Citrulli Vulgaris	**Xi Gua**
Fructus Cnidii Monnieri	**She Chuang Zi**
Fructus Corni Officinalis	**Shan Yu Rou**
Fructus Corni Officinalis	**Shan Zhu Yu**
Fructus Corni Officinalis	**Zao Pi**
Fructus Crataegi	**Shan Zha**
Fructus Cubebae	**Bi Cheng Qie**
Fructus Evodiae Rutaecarpae	**Wu Zhu Yu**
Fructus Foeniculi Vulgaris	**Xiao Hui Xiang**
Fructus Forsythiae Suspensae	**Lian Qiao**
Fructus Gardeniae Jasminoidis	**Shan Zhi Zi**
Fructus Gleditsiae Sinensis	**Zao Jiao**
Fructus Hordei Vulgaris Germinantus	**Mai Ya**
Fructus Illicii Veri	**Da Hui Xiang**
Fructus Illicius	**Ba Jiao Hui Xiang**
Fructus Immaturus Citri Aurantii	**Zhi Shi**
Fructus Kochiae Scopariae	**Di Fu Zi**
Fructus Leonuri	**Chong Wei Zi**
Fructus Ligustri Lucidi	**Nu Zhen Zi**
Fructus Liquidambaris Taiwanianae	**Lu Lu Tong**
Fructus Lycii	**Gou Qi Zi**
Fructus Meliae Toosendan	**Chuan Lian Zi**
Fructus Momordicae Grosvenori	**Luo Han Guo**
Fructus Mori Albae	**Sang Shen**
Fructus Oryzae Sativae Germinantus	**Gu Ya**
Fructus Perillae Frutescentis	**Zi Su Zi**
Fructus Piperis Longi	**Bi Ba**
Fructus Piperis Nigri	**Hu Jiao**
Fructus Polygoni Orientalis	**Shui Hong Hua Zi**
Fructus Pruni Mume	**Wu Mei**

Fructus Psoraleae Corylifoliae	**Bu Gu Zhi**
Fructus Psoraleae Corylifoliae	**Po Gu Zhi**
Fructus Quisqualis Indicae	**Shi Jun Zi**
Fructus Rosae Laevigatae	**Jin Ying Zi**
Fructus Rubi Chingii	**Fu Pen Zi**
Fructus Schisandrae Chinensis	**Wu Wei Zi**
Fructus Seu Radix Camptothecae	**Xi Shu**
Fructus Sophorae Japonicae	**Huai Jiao**
Fructus Terminaliae Chebulae	**He Li Le**
Fructus Terminaliae Chebulae	**He Zi**
Fructus Trapae	**Ling Jiao**
Fructus Tribuli Terrestris	**Bai Ji Li**
Fructus Tribuli Terrestris	**Ci Ji Li**
Fructus Trichosanthis	**Gua Lou**
Fructus Vascularis Luffae	**Si Gua Zi**
Fructus Viticis	**Man Jing Zi**
Fructus Xanthii Sibirici	**Cang Er Zi**
Fructus Zizyphi Jujubae	**Da Zao**
Fructus Zizyphi Jujubae	**Hong Zao**
Frusta Ferri	**Sheng Tie Luo**
Funiculus Umbilicalis	**Ji Dai**
Funiculus Umbilicalis	**Kan Qi**
Galla Rhois Chinensis	**Wu Bei Zi**
Gecko	**Ge Jie**
Gekko Swinhoana	**Bi Hu**
Gekko Swinhoana	**Shou Gong**
Gekko Swinhoana	**Tian Long**
Gelatinum Corii Asini	**E Jiao**
Gelatinum Corii Asini	**Lu Pi Jiao**
Gelatinum Cornu Cervi	**Lu Jiao Jiao**
Gelatinum Plastri Testudinis	**Gui Ban Jiao**
Gelatinum Plastri Testudinis	**Gui Jiao**
Gemma Agrimoniae	**He Cao Ya**
Gummi Olibanum	**Ru Xiang**
Gypsum	**Shi Gao**
Haematitum	**Dai Zhe Shi**
Halloysitum Rubrum	**Chi Shi Zhi**
Herba Acalyphae	**Tie Xian**
Herba Agastaches seu Pogostemi	**Huo Xiang**
Herba Agrimoniae Pilosae	**Xian He Cao**
Herba Aloes	**Lu Hui**
Herba Andrographitis Paniculatae	**Chuan Xin Lian**
Herba Androsace Aizoon	**Dian Di Mei**
Herba Ardisiae Japonicae	**Ai Di Cha**
Herba Aristolochiae	**Xun Gu Feng**
Herba Artemisiae Annuae	**Qing Hao**
Herba Artemisiae Anomalae	**Liu Ji Nu**
Herba Artemisiae Yinchenhao	**Yin Chen Hao**
Herba Begoniae Fimbristipulatae	**Zi Bei Tian Kui**
Herba Catharanthi Rosei	**Chang Chun Hua**
Herba Cephalanoplos	**Xiao Ji**
Herba Chenopodii Hybridi	**Xue Jian Chou**
Herba Cistanches Deserticolae	**Rou Cong Rong**
Herba Commelinae	**Ya Zhi Cao**
Herba Crotalariae	**Nong Ji Li**
Herba cum Radice Asari	**Xi Xin**

Herba cum Radice Houttuyniae Cordatae	**Yu Xing Cao**
Herba cum Radice Patriniae	**Bai Jiang Cao**
Herba cum Radice Violae Yedoensitis	**Zi Hua Di Ding**
Herba Cynomorii Songarici	**Suo Yang**
Herba Dendrobii	**Shi Hu**
Herba Dianthi	**Qu Mai**
Herba Dichondrae	**Ma Ti Jin**
Herba Duchesneae Indicae	**She Mei**
Herba Ecliptae Prostratae	**Han Lian Cao**
Herba Elsholtziae seu Moslae	**Xiang Ru**
Herba Ephedrae	**Ma Huang**
Herba Epimedii	**Xian Ling Pi**
Herba Epimedii	**Yin Yang Huo**
Herba Equiseti Hiemalis	**Mu Zei**
Herba Erodii seu Geranii	**Lao Guan Cao**
Herba Eupatorii Fortunei	**Pei Lan**
Herba Euphorbiae Helioscopiae	**Ze Qi**
Herba Hedyotidis Diffusae	**Bai Hua She She Cao**
Herba Houttuyniae Cordatae	**Ji Cai**
Herba Hydrocotylis	**Tian Hu Sui**
Herba Hyperici Japonici	**Di Er Cao**
Herba Lemnae seu Spirodelae	**Fu Ping**
Herba Leonuri Heterophylli	**Yi Mu Cao**
Herba Lobeliae Chinensis cun Radice	**Ban Bian Lian**
Herba Lophatheri Gracilis	**Dan Zhu Ye**
Herba Lycopi Lucidi	**Ze Lan**
Herba Lycopodii	**Shen Jin Cao**
Herba Lygodii Japonici	**Hai Jin Sha Teng**
Herba Lysimachiae	**Jin Qian Cao**
Herba Menthae Haplocalycis	**Bo He**
Herba Ocimi Pilosi	**Sheng Tou Cao**
Herba Paederiae	**Ji Shi Teng**
Herba Polygoni Avicularis	**Bian Xu**
Herba Polygonum Perfoliati	**Guan Ye Liao**
Herba Portulacae Oleraceae	**Ma Chi Xian**
Herba Potenlillae Discoloris	**Fan Bai Cao**
Herba Pteris	**Feng Wei Cao**
Herba Rorippae	**Han Cai**
Herba Rubi Parvifolii	**Mao Mei**
Herba Sarcandrae	**Zhong Jie Feng**
Herba Sarcandrae Glabrae	**Jiu Jie Cha**
Herba Sargassii	**Hai Zao**
Herba Saxifragae	**Hu Er Cao**
Herba Scutellariae Barbatae	**Ban Zhi Lian**
Herba Sedi Aizoon	**Jing Tian San Qi**
Herba Sedi Sarmentosi	**Chui Pen Cao**
Herba Selagrinellae Doederleinii	**Shi Shang Bai**
Herba Semiaquilegiae Adoxoidis	**Tian Kui**
Herba Senecionis	**Jiu Li Ming**
Herba Senecionis	**Qian Li Guang**
Herba seu Radix Cirsii Japonici	**Da Ji**
Herba Siegesbeckiae	**Xi Xian Cao**
Herba Solani Lyrati	**Bai Ying**
Herba Solani Nigri	**Long Kui**
Herba Speranskia Tuberculata	**Tou Gu Cao**
Herba Taraxaci Mongolici cum Radice	**Pu Gong Ying**

Herba Verbenae	**Ma Bian Cao**
Herba Zosterae Marinae	**Hai Dai**
Hippocampus	**Hai Ma**
Hirudo seu Whitmania	**Shui Zhi**
Holotrichia Diomphalia	**Qi Cao**
Hydrargyrum	**Shui Yin**
Indigo Pulverata Levis	**Qing Dai**
Jecur Lutrae	**Ta Gan**
Lacca Sinica Exsiccatae	**Gan Qi**
Lapis Micae seu Chloriti	**Meng Shi**
Lapis Micae seu Chloriti	**Qing Meng Shi**
Lignum Aquilariae	**Chen Xiang**
Lignum Cercis Chinensis	**Zi Jing Mu**
Lignum Dalbergiae Odoriferae	**Jiang Xiang**
Lignum Pini Nodi	**Song Jie**
Lignum Santali Albi	**Tan Xiang**
Lignum Sappan	**Su Mu**
Limonite	**Yu Liang Shi**
Limonite	**Yu Yu Liang**
Lithargyrum	**Mi Tuo Seng**
Lumbricus	**Di Long**
Magnetitum	**Ci Shi**
Margarita Usta	**Zhen Zhu**
Massa Fermentata	**Shen Qu**
Medulla Junci Effusi	**Deng Xin Cao**
Medulla Tetrapanacis Papyriferi	**Tong Cao**
Mel	**Feng Mi**
Minium	**Huang Dan**
Minium	**Qian Dan**
Mirabilitum	**Mang Xiao**
Mirabilitum Depuratum	**Pu Xiao**
Mylabris	**Ban Mao**
Myrrha	**Mo Yao**
Nidus Vespae	**Lu Feng Fang**
Nodus Nelumbinis Nuciferae Rhizomatis	**Ou Jie**
Ootheca Mantidis	**Sang Piao Xiao**
Ophicalcitum	**Hua Rui Shi**
Os Draconis	**Long Gu**
Os Sepiae seu Sepiellae	**Hai Piao Xiao**
Os Sepiae seu Sepiellae	**Wu Zei Gu**
Os Tigris	**Hu Gu**
Pedicellus Cucumeris	**Gua Di**
Pericarpium Arecae Catechu	**Da Fu Pi**
Pericarpium Citri Reticulatae	**Chen Pi**
Pericarpium Citri Reticulatae Viride	**Qing Pi**
Pericarpium Papaveris Somniferi	**Mi Ke**
Pericarpium Papaveris Somniferi	**Ying Su Ke**
Pericarpium Punicae Granati	**Shi Liu Pi**
Pericarpium Pyri	**Li Pi**
Pericarpium Zanthoxyli Bungeani	**Chuan Jiao**
Pericarpium Zanthoxyli Bungeani	**Hua Jiao**
Periostracum Cicadae	**Chan Tui**
Placenta Hominis	**Tai Pan**
Placenta Hominis	**Tai Yi**
Placenta Hominis	**Zi He Che**
Plastrum Testudinis	**Gui Ban**

Plumula Nelumbinis Nucifera	**Lian Zi Xin**
Pollen Typhae	**Pu Huang**
Praeparatio Fructus Ulmi Macrocarpi	**Wu Yi**
Pseudobulbus Shancigu	**Shan Ci Gu**
Pumice	**Fu Hai Shi**
Purum Mirabilitum	**Xuan Ming Fen**
Pyritum	**Zi Ran Tong**
Quick-Fried Rhizoma Zingiberis Officinalis	**Pao Jiang**
Radix Acanthopanacis Senticosi	**Ci Wu Jia**
Radix Achyranthis	**Tu Niu Xi**
Radix Achyranthis Bidentatae	**Huai Niu Xi**
Radix Aconiti Brachypodi	**Xue Shang Yi Zhi Hao**
Radix Aconiti Carmichaeli	**Chuan Wu**
Radix Aconiti Carmichaeli Praeparata	**Fu Zi**
Radix Aconiti Kusnezoffii	**Cao Wu**
Radix Actinidiae Argutae	**Teng Li Gen**
Radix Adenophorae seu Glehniae	**Sha Shen**
Radix Alangii	**Ba Jiao Feng**
Radix Ampelopsis	**Bai Lian**
Radix Angelicae Dahuricae	**Bai Zhi**
Radix Angelicae Pubescentis	**Du Huo**
Radix Angelicae Sinensis	**Dang Gui**
Radix Ardisiae Crenatae	**Zhu Sha Gen**
Radix Aristolochiae seu Cocculi	**Guang Fang Ji**
Radix Arnebiae seu Lithospermi	**Zi Cao**
Radix Asteris Tatarici	**Zi Wan**
Radix Astragali Membranacei	**Huang Qi**
Radix Astragali Membranacei Praeparata	**Zhi Huang Qi**
Radix Aucklandiae Lappae	**Mu Xiang**
Radix Berberidis	**San Ke Zhen**
Radix Boehmeriae	**Zhu Ma Gen**
Radix Bupleuri	**Chai Hu**
Radix Camelliae Sinensis	**Cha Shu Gen**
Radix Carpesii	**Tian Ming Jing**
Radix Changii	**Fen Sha Shen**
Radix Changii	**Ming Dang Shen**
Radix Clematidis	**Wei Ling Xian**
Radix Codonopsitis Pilosulae	**Dang Shen**
Radix Cyathulae Officinalis	**Chuan Niu Xi**
Radix Cynanchi Baiwei	**Bai Wei**
Radix Dichroae Febrifugae	**Chang Shan**
Radix Dioscoreae Oppositae	**Shan Yao**
Radix Dipsaci Asperi	**Xu Duan**
Radix Ephedrae	**Ma Huang Gen**
Radix et Caulis Jixueteng	**Ji Xue Teng**
Radix et Herba Cynanchi Paniculati	**Xu Chang Qing**
Radix et Herba Thalictri Foliolosi	**Tang Song Cao**
Radix et Ramus Litseae Cubebae	**Dou Chi Jiang**
Radix et Rhizoma Oryzae Glutinosae	**Nuo Dao Gen Xu**
Radix et Rhizoma Polygoni Cuspidati	**Hu Zhang**
Radix et Rhizoma Rhei	**Da Huang**
Radix et Rhizoma Veratri	**Li Lu**
Radix Euphorbiae et Stellerae	**Lang Du**
Radix Euphorbiae Kansui	**Gan Sui**
Radix Euphorbiae seu Knoxiae	**Da Ji**
Radix Gentianae Longdancao	**Long Dan Cao**

Radix Gentianae Qinjiao	**Qin Jiao**
Radix Ginseng	**Ren Shen**
Radix Glycyrrhizae Uralensis	**Gan Cao**
Radix Glycyrrhizae Uralensis Praeparata	**Zhi Gan Cao**
Radix Gossypii	**Mian Hua Gen**
Radix Hemsleyae	**Xue Dan**
Radix Ilicis Pubescentis	**Mao Dong Qing**
Radix Isatidis seu Baphicacanthi	**Ban Lan Gen**
Radix Ledebouriellae Divaricatae	**Fang Feng**
Radix Ligustici Chuanxiong	**Chuan Xiong**
Radix Linderae Strychnifoliae	**Wu Yao**
Radix Morindae Officinalis	**Ba Ji Tian**
Radix Notoginseng	**San Qi**
Radix Paeoniae Lactiflorae	**Bai Shao Yao**
Radix Paeoniae Rubrae	**Chi Shao Yao**
Radix Panacis Quinquefolii	**Xi Yang Shen**
Radix Peucedani	**Qian Hu**
Radix Phytolaccae	**Shang Lu**
Radix Platycodi Grandiflori	**Jie Geng**
Radix Polygalae Tenuifoliae	**Yuan Zhi**
Radix Polygoni Multiflori	**He Shou Wu**
Radix Pseudostellariae Heterophyllae	**Hai Er Shen**
Radix Pseudostellariae Heterophyllae	**Tai Zi Shen**
Radix Puerariae	**Ge Gen**
Radix Pulsatillae Chinensis	**Bai Tou Weng**
Radix Ranunculus Tematus	**Mao Zhua Cao**
Radix Rehmanniae Glutinosae	**Sheng Di Huang**
Radix Rehmanniae Glutinosae Conquitae	**Shu Di Huang**
Radix Rhapontici seu Echinops	**Lou Lu**
Radix Rhododendri Mollis	**Yang Zhi Zhu**
Radix Rubiae Cordifoliae	**Qian Cao Gen**
Radix Salviae Chinensis	**Shi Jian Chuan**
Radix Salviae Miltiorrhizae	**Dan Shen**
Radix Sanguisorbae Officinalis	**Di Yu**
Radix Scrophulariae Ningpoensis	**Xuan Shen**
Radix Scrophulariae Ningpoensis	**Yuan Shen**
Radix Scutellariae Baicalensis	**Huang Qin**
Radix Semiaquilegiae Adoxoidis	**Tian Kui Zi**
Radix Sophorae Flavescentis	**Ku Shen**
Radix Sophorae Tonkinensis	**Shan Dou Gen**
Radix Stellariae Dichotomae	**Yin Chai Hu**
Radix Stemonae	**Bai Bu**
Radix Stephaniae Tetrandrae	**Han Fang Ji**
Radix Tinosporae	**Jin Guo Lan**
Radix Trichosanthis Kirilowii	**Gua Lou Ren**
Radix Trichosanthis Kirilowii	**Tian Hua Fen**
Radix Tripterygii Wilfordii	**Lei Gong Teng**
Radix Zanthoxyli	**Liang Mian Zhen**
Ramulus Cinnamomi Cassiae	**Gui Zhi**
Ramulus Juglandis Mandshuricae	**He Tao Shu Zhi**
Ramulus Mori Albae	**Sang Zhi**
Ramulus Piper Wallichii	**Shi Nan Teng**
Ramulus Sangjisheng	**Sang Ji Sheng**
Ramulus Tamarix Chinensis	**Xi He Liu**
Ramulus Uncariae cum Uncis	**Gou Teng**
Ramus Tinosporae Sinensis	**Kuan Jin Teng**

Realgar	**Xiong Huang**
Receptaculum Nelumbinis Nuciferae	**Lian Fang**
Resina Draconis	**Xue Jie**
Resina Ferulae	**A Wei**
Rhizom Zingiberis Officinalis	**Gan Jiang**
Rhizoma Acori Graminei	**Shi Chang Pu**
Rhizoma Alismatis Orientalis	**Ze Xie**
Rhizoma Alpiniae Officinari	**Gao Liang Jiang**
Rhizoma Anemarrhenae Asphodeloidis	**Zhi Mu**
Rhizoma Anemones Raddeanae	**Lian Tou Jian**
Rhizoma Anemones Raddeanae	**Zhu Jie Xiang Fu**
Rhizoma Arisaematis	**Tian Nan Xing**
Rhizoma Atractylodis Macrocephalae	**Bai Zhu**
Rhizoma Atractylodis Macrocephalae	**Cang Zhu**
Rhizoma Belamcandae Chinensis	**She Gan**
Rhizoma Bistortae	**Zi Shen**
Rhizoma Bletillae Striatae	**Bai Ji**
Rhizoma Bolbostemmatis	**Tu Bei Mu**
Rhizoma Cibotii	**Jing Mao Gou Ji**
Rhizoma Cibotii Barometz	**Gou Ji**
Rhizoma Cimicifugae	**Sheng Ma**
Rhizoma Coptidis	**Huang Lian**
Rhizoma Corydalis Decumbentis	**Xia Tian Wu**
Rhizoma Corydalis Yanhusuo	**Yan Hu Suo**
Rhizoma Curculiginis Orchioidis	**Xian Mao**
Rhizoma Curcumae Longae	**Jiang Huang**
Rhizoma Curcumae Zedoariae	**E Zhu**
Rhizoma Cynanchii Stautoni	**Bai Qian**
Rhizoma Cynanchum Auriculati	**Ge Shan Xiao**
Rhizoma Cyperi Rotundi	**Xiang Fu**
Rhizoma Dioscoreae Bulbiferae	**Huang Du**
Rhizoma Dioscoreae Bulbiferae	**Huang Yao Zi**
Rhizoma Dioscoreae Hypoglaucae	**Bei Xie**
Rhizoma Dioscoreae Nipponicae	**Chuan Shan Long**
Rhizoma Drynariae	**Gu Sui Bu**
Rhizoma Drynariae	**Hou Jiang**
Rhizoma et Folium Smilacis Chinensis	**Ba Qia**
Rhizoma et Radix Ligustici	**Gao Ben**
Rhizoma et Radix Notopterygii	**Qiang Huo**
Rhizoma et Radix Physalitis	**Suan Jiang**
Rhizoma Fagopyri Cymosi	**Jin Qiao Mai**
Rhizoma Gastrodiae Elatae	**Tian Ma**
Rhizoma Guanzhong	**Guan Zhong**
Rhizoma Homalomenae Occultae	**Qian Nian Jian**
Rhizoma Imperatae Cylindricae	**Bai Mao Gen**
Rhizoma Paris	**Chong Lou**
Rhizoma Paris	**Zao Xiu**
Rhizoma Phragmitis Communis	**Lu Gen**
Rhizoma Picrorhizae	**Hu Huang Lian**
Rhizoma Pinelliae Ternatae	**Ban Xia**
Rhizoma Polygonati	**Huang Jing**
Rhizoma Polygonati Odorati	**Yu Zhu**
Rhizoma Polygoni Bistortae	**Cao He Che**
Rhizoma Polygoni Bistortae	**Quan Shen**
Rhizoma Seu Herba Saururii	**San Bai Cao**
Rhizoma Smilacis Glabrae	**Tu Fu Ling**

Rhizoma Sparganii Stoloniferi	**San Leng**
Rhizoma Typhonii Gigantei	**Bai Fu Zi**
Rhizoma Zingiberis Officinalis Recens	**Sheng Jiang**
Saccharum Granorum	**Yi Tang**
Sal Ammoniaci	**Nao Sha**
Scapus Eriocaulonis Buergeriani	**Gu Jing Cao**
Sclerotium Omphaliae Lapidescens	**Lei Wan**
Sclerotium Polypori Umbellati	**Zhu Ling**
Sclerotium Poriae Cocos	**Fu Ling**
Sclerotium Poriae Cocos Pararadicis	**Fu Shen**
Sclerotium Poriae Cocos Rubrae	**Chi Fu Ling**
Scolopendra Subspinipes	**Wu Gong**
Secretio Bufonis	**Chan Su**
Secretio Moschus	**She Xiang**
Semen Abutili seu Malvae	**Dong Kui Zi**
Semen Allii Tuberosi	**Jiu Zi**
Semen Alpiniae Katsumadai	**Cao Dou Kou**
Semen Arecae Catechu	**Bing Lang**
Semen Astragali Complanati	**Sha Yuan Ji Li**
Semen Benincasae Hispidae	**Dong Gua Ren**
Semen Biotae Orientalis	**Bai Zi Ren**
Semen Canavaliae	**Dao Dou**
Semen Cannabis Sativae	**Huo Ma Ren**
Semen Cassiae	**Jue Ming Zi**
Semen Celosiae Argenteae	**Qing Xiang Zi**
Semen Citri Reticulatae	**Ju He**
Semen Coicis Lachryma-jobi	**Yi Yi Ren**
Semen Croton Tiglii	**Ba Dou**
Semen Cucurbitae Moschatae	**Nan Gua Zi**
Semen Cuscutae Chinensis	**Tu Si Zi**
Semen Descurainiae seu Lepidii	**Ting Li Zi**
Semen Dolichoris Lablab	**Bai Bian Dou**
Semen Dolichoris Lablab	**Bian Dou**
Semen Euphorbia Lathyridis	**Xu Sui Zi**
Semen Euryales Ferocis	**Qian Shi**
Semen Ginkgo Bilobae	**Bai Guo**
Semen Ginkgo Bilobae	**Yin Xing**
Semen Glycines Germinatum	**Dou Juan**
Semen Hydnocarpi Anthelminticae	**Da Feng Zi**
Semen Hyoscyami	**Tian Xian Zi**
Semen Impatientis	**Ji Xing Zi**
Semen Juglandis Regiae	**Hu Tao Ren**
Semen Juglandis Regiae	**Hu Tao Rou**
Semen Litchi Chinensis	**Li Zhi He**
Semen Momordicae Cochinchinensis	**Mu Bie Zi**
Semen Myristicae Fragrantis	**Rou Dou Kou**
Semen Myristicae Fragrantis	**Rou Guo**
Semen Nelumbinis Nucifera	**Lian Zi**
Semen Oroxyli Indici	**Mu Hu Die**
Semen Persicae	**Tao Ren**
Semen Pharbitidis	**Qian Niu Zi**
Semen Phaseoli Calcarati	**Chi Xiao Dou**
Semen Phaseoli Radiati	**Lu Dou**
Semen Plantaginis	**Che Qian Zi**
Semen Pruni	**Yu Li Ren**
Semen Pruni Armeniacae	**Bei Xing Ren**

Semen Pruni Armeniacae	**Ku Xing Ren**
Semen Pruni Armeniacae	**Xing Ren**
Semen Raphani Sativi	**Lai Fu Zi**
Semen Sesami Indici	**Hei Zhi Ma**
Semen Sinapis Albae	**Bai Jie Zi**
Semen Sojae Praeparatum	**Dan Dou Chi**
Semen Sterculiae Scaphigerae	**Pang Da Hai**
Semen Strychni	**Fan Mu Bie**
Semen Strychni	**Ma Qian Zi**
Semen Torreyae Grandis	**Fei Zi**
Semen Trigonellae Foeni-graeci	**Hu Lu Ba**
Semen Tritici Aestivi Levis	**Fu Xiao Mai**
Semen Tritici Levis	**Xiao Mai**
Semen Vaccariae Segetalis	**Wang Bu Liu Xing**
Semen Zizyphi Spinosae	**Suan Zao Ren**
Smithsonitum	**Lu Gan Shi**
Spica Prunellae Vulgaris	**Xia Ku Cao**
Spora Lygodii Japonici	**Hai Jin Sha**
Squama Manitis Pentadactylae	**Chuan Shan Jia**
Stalactitum	**E Guan Shi**
Stamen Nelumbinis Nucifera	**Lian Xu**
Stichopus Japonicus	**Hai Shen**
Stylus Zeae Mays	**Yu Mi Xu**
Styrax Liquidis	**Su He Xiang**
Succinum	**Hu Po**
Succus Bambusae	**Zhu Li**
Succus Faecalis Aureus	**Jin Zhi**
Succus Zingiberis	**Sheng Jiang Zhi**
Sulphur	**Liu Huang**
Syngnathus	**Hai Long**
Tabanus Bivittatus	**Mang Chong**
Talcum	**Hua Shi**
Terra Flava Usta	**Fu Long Gan**
Testa Arachidis Hypogaeae	**Hua Sheng Yi**
Testa Dolichoris Lablab	**Bian Dou Yi**
Testes et Penis Otariae	**Hai Gou Shen**
Thallus Algae	**Kun Bu**
Trachycarpi Carbonisatus	**Zong Lu Tan**
Tuber Asparagi Cochinchinensis	**Tian Dong**
Tuber Asparagi Cochinchinensis	**Tian Men Dong**
Tuber Curcumae	**Yu Jin**
Tuber Ophiopogonis Japonici	**Mai Dong**
Tuber Ophiopogonis Japonici	**Mai Men Dong**
Urina Puerilis	**Tong Bian**
Vesica Fellea Ursi	**Xiong Dan**
Vitellus Gigeriae Galli	**Ji Zi Huang**
Zaocys Dhumnades	**Wu Shao She**

APPENDIX 8

Cross-Reference English–Botanical Latin Herbs

English	Botanical Latin
Abalone Shell	*Concha Haliotidis*
Acanthopanax Root	*Radix Acanthopanaeis Senticosi*
Acanthopanax Root Bark	*Cortex Acanthopanacis Gracilistyli Radicis*
Aconite	*Radix Aconiti Carmichaeli Praeparata*
Actinolite	*Actinolitum*
Aduki Bean	*Semen Phaseoli Calcarati*
Agkistrodon Snake	*Agkistrodon seu Bungarus*
Agrimony	*Herba Agrimoniae Pilosae*
Airpotato Yam	*Rhizoma Dioscoreae Bulbiferae*
Aizoon Stonecrop Plant	*Herba Sedi Aizoon*
Akebia Fruit	*Fructus Akebiae*
Aloeswood	*Lignum Aquilariae*
Amber	*Succinum*
American Ginseng Root	*Radix Panacis Quinquefolii*
Amomum Tsao-Ko	*Fructus Amomi Tsao-ko*
Ampelopsis	*Radix Ampelopsis*
Amur Cork-Tree Bark	*Cortex Phellodendri*
Androsace Plant	*Herba Androsace Aizoon*
Anemarrhena Rhizome	*Rhizoma Anemarrhenae Asphodeloidis*
Anemone	*Rhizoma Anemones Raddeanae*
Angelica Root	*Radix Angelicae Dahuricae*
Antelope Horn	*Cornu Antelopis*
Apricot Seed	*Semen Pruni Armeniacae*
Arbor-Vitae Seed	*Semen Biotae Orientalis*
Aristolochia Root	*Radix Aristolochiae seu Cocculi*
Aromatic Litsea Root/Stem	*Radix et Ramus Litseae Cubebae*
Aromatic Madder	*Herba Elsholtziae seu Moslae*
Arsenolite	*Arsenicum*
Asian Pear Peel	*Pericarpium Pyri*
Asiatic Cornelian Cherry Fruit	*Fructus Corni Officinalis*
Aucklandia	*Radix Aucklandiae Lappae*
Balloonflower Root	*Radix Platycodi Grandiflori*
Bamboo Leaves	*Herba Lophatheri Gracilis*
Bamboo Shavings	*Caulis Bambusae in Taeniis*
Bark of Heaven Tree	*Cortex Ailanthi*
Bark of Heaven Tree	*Cortex et Radix Ailanthi*
Bark of Heaven Tree	*Cortex Ailanthi Altissimae*
Bark of Korean Ash Branch	*Cortex Fraxini*
Bark of Mulberry Root	*Cortex Mori Albae Radicis*

Bark of Pomegranate Tree	*Cortex Punicae Granati Radicis*
Barley Malt Sugar	*Saccharum Granorum*
Barley Sprout	*Fructus Hordei Vulgaris Germinantus*
Bat Feces	*Excrementum Vespertilionis Murini*
Bear Gallbladder	*Vesica Fellea Ursi*
Beautyberry Leaf	*Folium Callicarpae Pedunculatae*
Begonia	*Herba Begoniae Fimbristipulatae*
Belamcanda Rhizome	*Rhizoma Belamcandae Chinensis*
Benzoin	*Benzoinum*
Berberis Root	*Radix Berberidis*
Betel Husk	*Pericarpium Arecae Catechu*
Betel Nut	*Semen Arecae Catechu*
Bidentate Achyranthes Root	*Radix Achyranthis Bidentatae*
Birthwort Fruit	*Fructus Aristolochiae*
Black Kohosh	*Rhizoma Cimicifugae*
Black Nightshade	*Herba Solani Nigri*
Black Pepper	*Fructus Piperis Nigri*
Black Sesame Seeds	*Semen Sesami Indici*
Boat Sterculia Seed	*Semen Sterculiae Scaphigerae*
Borax	*Borax*
Borneol	*Borneol*
Boy Urine	*Urina Puerilis*
Brachypodum Monkshood Root	*Radix Aconiti Brachypodi*
Bracken	*Herba Pteris*
Buck Grass	*Herba Lycopodii*
Bud of Hairyvein	*Gemma Agrimoniae*
Bud of Mulberry Fruit	*Fructus Mori Albae*
Buddleia Flower Bud	*Flos Buddleiae Officinalis*
Bulb of Chinese Chive	*Bulbus Allii*
Burnet-Bloodwort Root	*Radix Sanguisorbae Officinalis*
Bur-Reed Rhizome	*Rhizoma Sparganii Stoloniferi*
Bush Cherry Pit	*Semen Pruni*
Buttercup Root	*Radix Ranunculus Tematus*
Calcitum	*Calcitum*
Calomel	*Calomelas*
Caltrop Fruit	*Fructus Tribuli Terrestris*
Camphor	*Camphora*
Cannabis Seeds	*Semen Cannabis Sativae*
Cantharides	*Mylabris*
Canton Tusanqi	*Folium et Radix Gynurae Segetii*
Capillaris	*Herba Artemisiae Yinchenhao*
Carpesium Root	*Radix Carpesii*
Caryophyllaceous Root	*Radix Pseudostellariae Heterophyllae*
Cassia Seeds	*Semen Cassiae*
Catechu Extract	*Acacia Catechu seu Uncaria Gambier*
Cattail Pollen	*Pollen Typhae*
Cattle Gallstone	*Calculus Bovis*
Celosia Seeds	*Semen Celosiae Argenteae*
Center of Rosewood	*Lignum Dalbergiae Odoriferae*
Centipede	*Scolopendra Subspinipes*
Chain Fern Rhizome	*Rhizoma Cibotii Barometz*
Chain Fern Rhizome	*Rhizoma Cibotii*
Changium Root	*Radix Changii*
Charred Human Hair	*Crinis Carbonisatus Hominis*
Charred Stiple Fiber of Fortune Windmill Palm	*Trachycarpi Carbonisatus*
Chaulmoogra Seed	*Semen Hydnocarpi Anthelminticae*

Cherokee Rosehip	*Fructus Rosae Laevigatae*
Chicken Egg Yolk	*Vitellus Gigeriae Galli*
Chicken Gizzard Lining	*Endothelium Corneum Gigeriae Galli*
China Root	*Rhizoma et Folium Smilacis Chinensis*
China Tree Root Bark	*Cortex Meliae Radicis*
Chinaberry Fruit	*Fructus Meliae Toosendan*
Chinese Alangium Root	*Radix Alangii*
Chinese Anemone Root	*Radix Pulsatillae Chinensis*
Chinese Angelica Root	*Radix Angelicae Sinensis*
Chinese Asparagus Tuber	*Tuber Asparagi Cochinchinensis*
Chinese Atractylodes	*Rhizoma Atractylodis Macrocephalae*
Chinese Catalpa Bark	*Cortex Catalpae Ovatae*
Chinese Clematis Root	*Radix Clematidis*
Chinese Cynomorium	*Herba Cynomorii Songarici*
Chinese Date	*Fructus Zizyphi Jujubae*
Chinese Dodder Seeds	*Semen Cuscutae Chinensis*
Chinese Ephedra Root	*Radix Ephedrae*
Chinese Fevervine	*Herba Paederiae*
Chinese Giantfennel Resin	*Resina Ferulae*
Chinese Holly	*Folium Ilicis Chinensis*
Chinese Honeylocust Fruit	*Fructus Gleditsiae Sinensis*
Chinese Kiwi Root	*Radix Actinidiae Argutae*
Chinese Lobelia	*Herba Lobeliae Chinensis cun Radice*
Chinese Lovage Root	*Rhizoma et Radix Ligustici*
Chinese Mahonia	*Folium Mahoniae*
Chinese Motherwort	*Herba Leonuri Heterophylli*
Chinese Photinia Vine	*Ramulus Piper Wallichii*
Chinese Pink Dianthus	*Herba Dianthi*
Chinese Quince Fruit	*Fructus Chaenomelis*
Chinese Radish Seed	*Semen Raphani Sativi*
Chinese Raspberry	*Fructus Rubi Chingii*
Chinese Redbud Bark	*Cortex Cercis Chinensis*
Chinese Redbud Wood	*Lignum Cercis Chinensis*
Chinese Sage	*Radix Salviae Chinensis*
Chinese Sambucus	*Flos Sambuci Chinensis*
Chinese Senega Root	*Radix Polygalae Tenuifoliae*
Chinese Soft-Shelled Turtle Shell	*Carapax Amydae Sinensis*
Chinese Sumac	*Folium et Cortex Rhois Chinensis*
Chinese Swallowwort	*Radix et Herba Cynanchi Paniculati*
Chinese Tamarisk Tops	*Cacumen Tamaricis*
Chinese Tamarisk Tops	*Ramulus Tamarix Chinensis*
Chinese Tea Rose	*Flos et Fructus Rosae Chinensis*
Chinese Tinospora Stem	*Ramus Tinosporae Sinensis*
Chinese Torreya Seed	*Semen Torreyae Grandis*
Chinese Trumpetcreeper Flower	*Flos Campsis*
Chinese Tulip Bulb	*Pseudobulbus Shancigu*
Chinese Wax Gourd Peel	*Epicarpium Benincasae Hispidae*
Chinese Wax Gourd Seed	*Semen Benincasae Hispidae*
Chinese Wild Ginger	*Herba cum Radice Asari*
Chinese Wolfberry Fruit	*Fructus Lycii*
Chinese Yam	*Radix Dioscoreae Oppositae*
Chlorite	*Lapis Micae seu Chloriti*
Chrysanthemum Flower	*Flos Chrysanthemi Morifolii*
Cicada Molting	*Periostracum Cicadae*
Cinnamon Bark	*Cortex Cinnamomi Cassiae*

Cinnamon Twig	*Ramulus Cinnamomi Cassiae*
Cinquefoil	*Herba Potenlillae Discoloris*
Citron Fruit	*Fructus Citri Medica*
Clam Shell	*Concha Cyclinae Sinensis*
Climbing Groundsel	*Herba Senecionis*
Climbing Nightshade Plant	*Herba Solani Lyrati*
Clove Flower Bud	*Flos Caryophylli*
Clove Fruit	*Flos Syzygii Aromatici*
Cnidium Seed	*Fructus Cnidii Monnieri*
Cockle Shell	*Concha Arcae*
Cocklebur Fruit	*Fructus Xanthii Sibirici*
Cockscomb Flower	*Flos Celosiae Cristatae*
Codonopsis Root	*Radix Codonopsitis Pilosulae*
Coltsfoot Flower	*Flos Tussilaginis Farfarae*
Common Eelgrass	*Herba Zosterae Marinae*
Common Saint Paulswort	*Herba Siegesbeckiae*
Cooked Rehmannia Root	*Radix Rehmanniae Glutinosae Conquitae*
Copperleaf	*Herba Acalyphae*
Coptis Rhizome	*Rhizoma Coptidis*
Coral Ardisia Root	*Radix Ardisiae Crenatae*
Cordyceps	*Cordyceps Sinensis*
Cornsilk	*Stylus Zeae Mays*
Cortex of Tree Peony	*Cortex Moutan Radicis*
Corydalis Rhizome	*Rhizoma Corydalis Decumbentis*
Corydalis Rhizome	*Rhizoma Corydalis Yanhusuo*
Cotton Plant Root	*Radix Gossypii*
Cowherb Seed	*Semen Vaccariae Segetalis*
Crotalaria Plant	*Herba Crotalariae*
Croton Seed	*Semen Croton Tiglii*
Crystallized Blue Vitriol	*Chalcanthite*
Cuttlefish Bone	*Os Sepiae seu Sepiellae*
Dandelion	*Herba Taraxaci Mongolici cum Radice*
Dayflower	*Herba Commelinae*
Dead Sick Silkworm	*Bombyx Batryticatus*
Dead Sick Silkworm	*Bombyx Batryticatus*
Deer Horn	*Cornu Cervi*
Deer Horn Glue	*Gelatinum Cornu Cervi*
Dendrobium Stem	*Herba Dendrobii*
Dichondra Plant	*Herba Dichondrae*
Dichroa Root	*Radix Dichroae Febrifugae*
Dictamnus Root Bark	*Cortex Dictamni Dasycarpi Radicis*
Diverse Wormwood	*Herba Artemisiae Anomalae*
Donkey-Hide Gelatin	*Gelatinum Corii Asini*
Dragon's Blood (Resin)	*Resina Draconis*
Dregs leftover after making Lu Jiao Jiao	*Cornu Cervi Degelatinatium*
Dried Bamboo Sap	*Succus Bambusae*
Dried Ginger	*Rhizom Zingiberis Officinalis*
Dry Lacquer Sap	*Lacca Sinica Exsiccatae*
Drynaria Rhizome	*Rhizoma Drynariae*
Dryopteris Root	*Rhizoma Guanzhong*
Duckweed	*Herba Lemnae seu Spirodelae*
Dung Beetle	*Aspongopus Chinensis*
Dung Beetle	*Catharsius*
Dutchman's Pipe	*Herba Aristolochiae*
Earleaf Cynanchum Root	*Rhizoma Cynanchum Auriculati*

Earthworm	*Lumbricus*
Eclipta	*Herba Ecliptae Prostratae*
Elm Tree Bark	*Cortex Ulmi Pumilae*
Ephedra Stem	*Herba Ephedrae*
Erythrina Bark	*Cortex Erythrinae*
Eucommia Bark	*Cortex Eucommiae Ulmoidis*
Eupatorium	*Herba Eupatorii Fortunei*
European Verbena	*Herba Verbenae*
Evodia Fruit	*Fructus Evodiae Rutaecarpae*
Felt Fern Leaves	*Folium Pyrrosiae*
Fennel Fruit	*Fructus Foeniculi Vulgaris*
Fenugreek Seed	*Semen Trigonellae Foeni-graeci*
Finger Citron Flower	*Flos Citri Sarcodactylis*
Finger Citron Fruit	*Fructus Citri Sarcodactylis*
Fischer Euphorbia Root	*Radix Euphorbiae et Stellerae*
Fish-Poison Yam	*Rhizoma Dioscoreae Hypoglaucae*
Five Leaf Akebia Stem	*Caulis Mutong*
Fleeceflower Root	*Radix Polygoni Multiflori*
Flesh of Longan Fruit	*Arillus Euphoriae Longanae*
Fleshy Stem of Broomrape	*Herba Cistanches Deserticolae*
Fluorite	*Fluoritum*
Flying Squirrel Feces	*Excrementum Trogopteri seu Pteromi*
Forsythia Fruit	*Fructus Forsythiae Suspensae*
Fossils of Mammal Vertebrae and Extremity	*Os Draconis*
Frankincense or Pistaciacense	*Gummi Olibanum*
Fresh Ginger Juice	*Succus Zingiberis*
Fresh Ginger Rhizome	*Rhizoma Zingiberis Officinalis Recens*
Fresh Rehmannia	*Radix Rehmanniae Glutinosae*
Frost Pwdr Prep by Watermelon Peel and Mirabilite	*Citrullus Lanatus*
Fruit of Scuffy Pea	*Fructus Psoraleae Corylifoliae*
Fruit of Szechuan Pricklyash Pepper	*Pericarpium Zanthoxyli Bungeani*
Fruiting Body of Stone-like Omphalia	*Sclerotium Omphaliae Lapidescens*
Fruiting Body of Tremella	*Fructificatio Tremellae Fuciformis*
Gadfly	*Tabanus Bivittatus*
Galangal Fruit	*Fructus Alpiniae Oxyphyllae*
Gallnut of Chinese Sumac	*Galla Rhois Chinensis*
Gambir Vine	*Ramulus Uncariae cum Uncis*
Garden Balsam	*Herba Speranskia Tuberculata*
Garden Balsam Seed	*Semen Impatientis*
Gardenia Fruit	*Fructus Gardeniae Jasminoidis*
Garlic Bulb	*Bulbus Alli Sativi*
Gastrodia Rhizome	*Rhizoma Gastrodiae Elatae*
Gecko	*Gecko*
Genkwa Flower	*Flos Daphnes Genkwa*
Gentiana Macrophylla Root	*Radix Gentianae Qinjiao*
Gentiana Root	*Radix Gentianae Longdancao*
Giant Knotweed Root & Rhizome	*Radix et Rhizoma Polygoni Cuspidati*
Giant Typhonium Tuber	*Rhizoma Typhonii Gigantei*
Ginger Peel	*Exocarpium Zingiberis Officinalis Recens*
Ginkgo Leaf	*Folium Ginkgo*
Ginkgo Nut	*Semen Ginkgo Bilobae*
Ginseng Leaf	*Folium Ginseng*
Ginseng Root	*Radix Ginseng*
Giraldi Daphne Bark	*Cortex Daphnes*

Glabrous Greenbrier Rhizome	*Rhizoma Smilacis Glabrae*
Glauber's Salt	*Mirabilitum*
Globeamaranth Flower	*Flos Gomphrenae*
Glorybower Leaf	*Folium Clerodendri Trichotomi*
Glossy Ganoderma	*Excrementum Trogopteri seu Pteromi*
Glue of Land Tortoise Shell	*Gelatinum Plastri Testudinis*
Glutinous Rice Root	*Radix et Rhizoma Oryzae Glutinosae*
Goldenlarch Bark	*Cortex Pseudolarix Kaempferi*
Gordon Euryale Seed	*Semen Euryales Ferocis*
Great Burdock Fruit	*Fructus Arctii Lappae*
Greater Selaginella	*Herba Selagrinellae Doederleinii*
Green Chiretta	*Herba Andrographitis Paniculatae*
Groomwell Root	*Radix Arnebiae seu Lithospermi*
Grub	*Holotrichia Diomphalia*
Gypsum	*Gypsum*
Hairy Holly Root	*Radix Ilicis Pubescentis*
Halloysite	*Halloysitum Rubrum*
Happy Tree Fruit Bark	*Fructus Seu Radix Camptothecae*
Hawthorn Fruit	*Fructus Crataegi*
Hedgehog Skin	*Corium Erinacei*
Hematite	*Haematitum*
Hemsleya Root	*Radix Hemsleyae*
Henbane Seed	*Semen Hyoscyami*
Heroubill Plant	*Herba Erodii seu Geranii*
Hogfennel Root	*Radix Peucedani*
Homalomena Rhizome	*Rhizoma Homalomenae Occultae*
Honey	*Mel*
Honeysuckle Flower	*Flos Lonicerae Japonicae*
Hornet Nest	*Nidus Vespae*
Horse Tail	*Herba Equiseti Hiemalis*
House Lizard	*Gekko Swinhoana*
Houttuynia	*Herba cum Radice Houttuyniae Cordatae*
Human Placenta	*Placenta Hominis*
Human Umbilical Cord	*Funiculus Umbilicalis*
Hyacinth Bean	*Semen Dolichoris Lablab*
Hyacinth Bean Peel	*Testa Dolichoris Lablab*
Hyacinth Bletilla Tuber	*Rhizoma Bletillae Striatae*
Hyacinth Flower	*Flos Dolichoris Lablab*
Hydrangea Flower	*Flos Clematidis Ranunculoidis*
Hydrocotyle Plant	*Herba Hydrocotylis*
Immature Fruit of Bitter Orange	*Fructus Immaturus Citri Aurantii*
Immature Tangerine Peel	*Pericarpium Citri Reticulatae Viride*
India Pokeberry Root	*Radix Phytolaccae*
India Trumpetflower Seed	*Semen Oroxyli Indici*
Indian Mulberry Root	*Radix Morindae Officinalis*
Indigo	*Indigo Pulverata Levis*
Inula Flower	*Flos Inulae*
Iron Filings	*Frusta Ferri*
Jack-in-the-Pulpit	*Rhizoma Arisaematis*
Jamestown Weed Seed/Leaf/Flower	*Flos Daturae*
Jamestown Weed Seed/Leaf/Flower	*Flos Daturae Albae*
Japanese Ardisia	*Herba Ardisiae Japonicae*
Japanese Ardisia	*Caulis et Folium Ardisiae Japonicae*
Japanese Fern	*Herba Lygodii Japonici*
Japanese Fern Spore	*Spora Lygodii Japonici*

Japanese Lily Bulb	*Folium et Radix Rohdeae Japonicae*
Japanese Raspberry	*Herba Rubi Parvifolii*
Japanese Saint Johnswort	*Herba Hyperici Japonici*
Japanese Teasel Root	*Radix Dipsaci Asperi*
Japanese Thistle	*Radix Euphorbiae seu Knoxiae*
Japanese Yam	*Rhizoma Dioscoreae Nipponicae*
Java Brucea Fruit	*Fructus Bruceae Javanicae*
Job's Tears (Coix)	*Semen Coicis Lachryma-jobi*
Juice of Aloe Leaf	*Herba Aloes*
Kadsura Stem	*Caulis Piperis Futokadsurae*
Kan-Sui Root	*Radix Euphorbiae Kansui*
Kelp Thallus	*Thallus Algae*
Knotty Pine Wood	*Lignum Pini Nodi*
Knotweed	*Herba Polygoni Avicularis*
Kochia Fruit	*Fructus Kochiae Scopariae*
Kudzu Flower	*Flos Puerariae*
Kudzu Root	*Radix Puerariae*
Kusnezoff Monkshood Root	*Radix Aconiti Kusnezoffii*
Land Tortoise Shell	*Plastrum Testudinis*
Large-headed Atractylodes Rhizome	*Rhizoma Atractylodis Macrocephalae*
Lead Oxide (Pb_3O_4)	*Minium*
Leaf of Apocynum Venetum	*Folium Apocyni Veneti*
Leatherleaf Millettia Root/Vine	*Radix et Caulis Jixueteng*
Ledebouriella Root	*Radix Ledebouriellae Divaricatae*
Leech	*Hirudo seu Whitmania*
Leechee Nut	*Semen Litchi Chinensis*
Less Purified Mirabilitum	*Mirabilitum Depuratum*
Lesser Galangal Rhizome	*Rhizoma Alpiniae Officinari*
Light Wheat Grain	*Semen Tritici Aestivi Levis*
Lily Bulb	*Bulbus Lilii*
Limonite	*Limonite*
Litharge or Galena	*Lithargyrum*
Litse Fruit	*Fructus Cubebae*
Lizardtail	*Rhizoma Seu Herba Saururii*
Long Pepper Fruit	*Fructus Piperis Longi*
Longspur Epimedium	*Herba Epimedii*
Loosestrife Plant	*Herba Lysimachiae*
Loquat Leaf	*Folium Eriobotryae Japonicae*
Lotus Leaf	*Folium Nelumbinis Nuciferae*
Lotus Plumule	*Plumula Nelumbinis Nucifera*
Lotus Seed	*Semen Nelumbinis Nucifera*
Lotus Stamen	*Stamen Nelumbinis Nucifera*
Lycium Bark	*Cortex Lycii Radicis*
Macaque Gallstone	*Calculus Macacae Mulattae*
Madagascar Periwinkle	*Herba Catharanthi Rosei*
Madder Root	*Radix Rubiae Cordifoliae*
Magnetite	*Magnetitum*
Magnolia Bark	*Cortex Magnoliae Officinalis*
Magnolia Flower	*Flos Magnoliae Officinalis*
Magnolia Flower	*Flos Magnoliae*
Male Seal Sexual Organ	*Testes et Penis Otariae*
Mapleleaf Goosefoot	*Herba Chenopodii Hybridi*
Mature Lotus Receptacle	*Receptaculum Nelumbinis Nuciferae*
Medicated Leaven	*Massa Fermentata*
Melon Pedicle	*Pedicellus Cucumeris*

Mercury	*Hydrargyrum*
Milk-Vetch Root	*Radix Astragali Membranacei*
Milk-Vetch Seed	*Semen Astragali Complanati*
Mimosa Tree Bark	*Cortex Albizziae Julibrissin*
Mimosa Tree Flower	*Albizziae Julibrissin*
Miniature Chinese Rose	*Flos Rosae Rugosae*
Mock Strawberry Plant	*Herba Duchesneae Indicae*
Moleplant Seed	*Semen Euphorbia Lathyridis*
Momordica Seed	*Semen Momordicae Cochinchinensis*
Mongolian Snakegourd	*Fructus Trichosanthis*
Mongolian Snakegourd Seed	*Radix Trichosanthis Kirilowii*
Morning Glory Seed	*Semen Pharbitidis*
Motherwort Fruit	*Fructus Leonuri*
Mountain Onion	*Radix et Rhizoma Veratri*
Mugwort Leaf	*Folium Artemisiae Vulgaris/Argyi*
Mulberry Leaf	*Folium Mori Albae*
Mulberry Mistletoe	*Ramulus Sangjisheng*
Mulberry Twig	*Ramulus Mori Albae*
Mume Fruit	*Fructus Pruni Mume*
Mung Bean	*Semen Phaseoli Radiati*
Mung Bean Skin	*Cortex Phaseoli Radiati*
Musk Mallow Seed	*Semen Abutili seu Malvae*
Musk of Deer	*Secretio Moschus*
Muskroot-like Semiaquilegia Root	*Herba Semiaquilegiae Adoxoidis*
Muskroot-like Semiaquilegia Root	*Radix Semiaquilegiae Adoxoidis*
Myrobalan Fruit	*Fructus Terminaliae Chebulae*
Myrrh	*Myrrha*
Nasturtium	*Herba Rorippae*
Ningpo Figwort Root	*Radix Scrophulariae Ningpoensis*
Node of Lotus Rhizome	*Nodus Nelumbinis Nuciferae Rhizomatis*
Notopterygium Root	*Rhizoma et Radix Notopterygii*
Nutgrass	*Rhizoma Cyperi Rotundi*
Nutmeg Seed	*Semen Myristicae Fragrantis*
Nux-Vomica Seed	*Semen Strychni*
Old Lime	*Calx*
Ophicalcite	*Ophicalcitum*
Opium Poppy Husk	*Pericarpium Papaveris Somniferi*
Opium Poppy Husk	*Pericarpium Papaveris Somniferi*
Oriental Arborvitae Twig	*Cacumen Biotae Orientalis*
Otter Liver	*Jecur Lutrae*
Outer Tangerine Peel	*Exocarpium Citrus Reticulate Rubrum*
Oyster Shell	*Concha Ostreae*
Pagoda Tree Flower Bud	*Flos Sophorae Japonicae Immaturus*
Pagoda Tree Fruit	*Fructus Sophorae Japonicae*
Pangolin Scales	*Squama Manitis Pentadactylae*
Paniculate Bolbostemma	*Rhizoma Bolbostemmatis*
Paper Mulberry Fruit	*Fructus Broussonetiae*
Paris Rhizome	*Rhizoma Paris*
Patchouli or Wrinkled Gianthyssop	*Herba Agastaches seu Pogostemi*
Patrinia	*Herba cum Radice Patriniae*
Peach Kernel	*Semen Persicae*
Peanut Peel	*Testa Arachidis Hypogaeae*
Pearl	*Margarita Usta*
Peking Spurge Root	*Herba seu Radix Cirsii Japonici*
Peppermint	*Herba Menthae Haplocalycis*

Pepperweed Seed or Flixweed Tansymustard Seed	*Semen Descurainiae seu Lepidii*
Perfoliatum	*Herba Polygonum Perfoliati*
Perilla Leaf	*Folium Perillae Frutescentis*
Perilla Stem	*Caulis Perillae Frutescentis*
Persimmon Calyx	*Calyx Diospyri Kaki*
Picrorhiza Rhizome	*Rhizoma Picrorhizae*
Pinellia Tuber	*Rhizoma Pinelliae Ternatae*
Pipe-Fish	*Syngnathus*
Pipewort Scapus	*Scapus Eriocaulonis Buergeriani*
Plantago Seeds	*Semen Plantaginis*
Pomegranate Husk	*Pericarpium Punicae Granati*
Poria Skin	*Cortex Poriae Cocos*
Poria with Wood Root	*Sclerotium Poriae Cocos Pararadicis*
Praying Mantis Egg-Case laid on Mulberry Leaf	*Ootheca Mantidis*
Prepared Licorice Root	*Radix Glycyrrhizae Uralensis Praeparata*
Prepared Milk-Vetch Root	*Radix Astragali Membranacei Praeparata*
Prepared Soybean	*Semen Sojae Praeparatum*
Prickly Chaff Flower/Root/Twig	*Radix Achyranthis Aspera*
Pricklyash Root	*Radix Zanthoxyli*
Prince's Feather Fruit	*Fructus Polygoni Orientalis*
Privet Fruit	*Fructus Ligustri Lucidi*
Pseudoginseng Root	*Radix Notoginseng*
Pubescent Angelica Root	*Radix Angelicae Pubescentis*
Puffball	*Frucificatio Lasiosphaerae*
Pumice	*Pumice*
Pumpkin Seed Husk	*Semen Cucurbitae Moschatae*
Pure Mirabilitum	*Purum Mirabilitum*
Purified Alunite	*Alumen*
Purified Feces	*Succus Faecalis Aureus*
Purple Aster Root	*Radix Asteris Tatarici*
Purple Perilla Fruit	*Fructus Perillae Frutescentis*
Purslane	*Herba Portulacae Oleraceae*
Pyrite	*Pyritum*
Quick-Fried Ginger Root	*Quick-Fried Rhizoma Zingiberis Officinalis*
Ramie Root	*Radix Boehmeriae*
Rangoon Creeper Fruit with Seed	*Fructus Quisqualis Indicae*
Raw Licorice Root	*Radix Glycyrrhizae Uralensis*
Realgar	*Realgar*
Red Peony Root	*Radix Paeoniae Rubrae*
Red Poria	*Sclerotium Poriae Cocos Rubrae*
Reed Rhizome	*Rhizoma Phragmitis Communis*
Resin of Rose Maloes	*Styrax Liquidis*
Rhaponticum Root	*Radix Rhapontici seu Echinops*
Rhinoceros Horn	*Cornu Rhinoceri*
Rhizome of Golden Eye-Grass	*Rhizoma Curculiginis Orchioidis*
Rhododendron Leaf	*Folium Rhododendri Simsii*
Rhubarb Root Rhizome	*Radix et Rhizoma Rhei*
Rice Paper Pith	*Medulla Tetrapanacis Papyriferi*
Rice Sprout	*Fructus Oryzae Sativae Germinantus*
Ripe Fruit of Bitter Orange	*Fructus Citri Aurantii*
Root of Beech Silver-Top	*Radix Adenophorae seu Glehniae*
Round Cardamom Fruit	*Fructus Amomi Cardamomi*
Round Momordica Fruit	*Fructus Momordicae Grosvenori*

Round Momordica Leaf	*Folium Momordicae Grosvenori*
Rush Pith	*Medulla Junci Effusi*
Safflower Flower	*Flos Carthami Tinctorii*
Sago Cycas	*Folium et Flos Cordylines Fruticosae*
Sal Ammoniac	*Sal Ammoniaci*
Salvia Root	*Radix Salviae Miltiorrhizae*
Sandalwood	*Lignum Santali Albi*
Sappan Wood	*Lignum Sappan*
Sarcandrae Plant	*Herba Sarcandrae Glabrae*
Sarcandrae Plant	*Herba Sarcandrae*
Sargentodoxa Vine	*Caulis Sargentodoxae Cuneatae*
Saxifrage	*Herba Saxifragae*
Scallion	*Bulbus Allii Fistulosi*
Schisandra Fruit	*Fructus Schisandrae Chinensis*
Schizonepeta Stem	*Flos Schizonepetae Tenuifoliae*
Scorpion	*Buthus Martensi*
Sea Cucumber	*Stichopus Japonicus*
Seahorse	*Hippocampus*
Seaweed	*Herba Sargassii*
Seed of Grains-of-Paradise	*Fructus Amomi*
Seed of Katsumada's Galangal	*Semen Alpiniae Katsumadai*
Seeds of Fragrant- Flowered Garlic	*Semen Allii Tuberosi*
Selfheal Spike	*Spica Prunellae Vulgaris*
Senna Leaf	*Folium Sennae*
Shepherd's Purse	*Herba Houttuyniae Cordatae*
Shiny Bugleweed	*Herba Lycopi Lucidi*
Shrubalthea Bark	*Cortex Hibiscus Syriaci*
Siberian Solomon Seal Rhizome	*Rhizoma Polygonati*
Sichuan Cyathula Root	*Radix Cyathulae Officinalis*
Sichuan Lovage Root	*Radix Ligustici Chuanxiong*
Sichuan Monkshood Root	*Radix Aconiti Carmichaeli*
Silkworm Feces	*Excrementum Bombycis Mori*
Skullcap Root	*Radix Scutellariae Baicalensis*
Small Japanese Thistle	*Herba Cephalanoplos*
Smithsonite or Calamine	*Smithsonitum*
Snake Skin Slough	*Exuviae Serpentis*
Snakeweed	*Rhizoma Polygoni Bistortae*
Snakeweed	*Rhizoma Bistortae*
Solomon's Seal Rhizome	*Rhizoma Polygonati Odorati*
Sophora Root	*Radix Sophorae Flavescentis*
Sophora Root	*Radix Sophorae Tonkinensis*
Sour Jujube Seed	*Semen Zizyphi Spinosae*
Spreading Hedyotis Oldenlandia Plant	*Herba Hedyotidis Diffusae*
Spurge	*Herba Euphorbiae Helioscopiae*
Star Jasmine Stem	*Caulis Trachelospermi Jasminoidis*
Starwort Fruit	*Fructus Carpesii seu Daucusi*
Stellaria Root	*Radix Stellariae Dichotomae*
Stemona Tuber	*Radix Stemonae*
Stephania Root	*Radix Stephaniae Tetrandrae*
Stinky Elm Fruit Paste	*Praeparatio Fructus Ulmi Macrocarpi*
Stiple Fiber of Fortune Windmill Palm	*Fibra Trachycarpi Vagina*
Strap Flower/Root/Leaf	*Folium et Radix Loropetali*
Stringy Stonecrop	*Herba Sedi Sarmentosi*
Suakwa Seed	*Fructus Vascularis Luffae*
Suakwa Vegetable Sponge	*Fasciculus Vascularis Luffae*

Sulphur	*Sulphur*
Sun Plant	*Herba Scutellariae Barbatae*
Swallowwort Root	*Radix Cynanchi Baiwei*
Sweet Basil	*Herba Ocimi Pilosi*
Sweetflag Rhizome	*Rhizoma Acori Graminei*
Sweetgum Fruit	*Fructus Liquidambaris Taiwanianae*
Sword Bean	*Semen Canavaliae*
Tabasheer	*Concretio Silicea Bambusae Textillis*
Talcum	*Talcum*
Tangerine Leaf	*Folium Citrus Tangerina*
Tangerine Peel	*Pericarpium Citri Reticulatae*
Tangerine Pith	*Fibrum Vascularis Citrus Tangerina*
Tangerine Seed	*Semen Citri Reticulatae*
Tea Tree Root	*Radix Camelliae Sinensis*
Tendrilled Fritillaria Bulb	*Bulbus Fritillariae Cirrhosae*
Thalictrum	*Radix et Herba Thalictri Foliolosi*
Thorny Elaegnus Leaf	*Folium Elaeagni*
Thorowax Root	*Radix Bupleuri*
Three-Nerved Spicebush Root	*Radix Linderae Strychnifoliae*
Thunberg Fritillaria Bulb	*Bulbus Fritillariae Thunbergii*
Tiger Bone	*Os Tigris*
Tinospora Root	*Radix Tinosporae*
Tip of Tubular Stalactites	*Stalactitum*
Toad Venom	*Secretio Bufonis*
Trichosantes Root	*Radix Trichosanthis Kirilowii*
Tripterygium Wilfordii Hook	*Radix Tripterygii Wilfordii*
Truestar Anisetree Fruit	*Fructus Illicius*
Truestar Anisetree Fruit	*Fructus Illicii Veri*
Tuber of Dwarf Lilyturf	*Tuber Ophiopogonis Japonici*
Turmeric Rhizome	*Rhizoma Curcumae Longae*
Turmeric Tuber	*Tuber Curcumae*
Umbellate Pore Fungus	*Sclerotium Polypori Umbellati*
Velvet of Young Deer Horn	*Cornu Cervi Parvum*
Vine of Solomon's Seal	*Caulis Polygoni Multiflori*
Vitex Fruit	*Fructus Viticis*
Walnut Nut	*Semen Juglandis Regiae*
Walnut Twig	*Ramulus Juglandis Mandshuricae*
Water Buffalo Horn	*Cornu Bubali*
Water Calptrop	*Fructus Trapae*
Water Chestnut	*Cormus Heleochris Plantaginea*
Water Plantain	*Rhizoma Alismatis Orientalis*
Watermelon Fruit	*Fructus Citrulli Vulgaris*
Watermelon Peel	*Exocarpium Citrulli*
Wheat Grain	*Semen Tritici Levis*
White Mustard Seed	*Semen Sinapis Albae*
White Peony Root	*Radix Paeoniae Lactiflorae*
White Poria	*Sclerotium Poriae Cocos*
White Swallowwort	*Rhizoma Cynanchii Stautoni*
Wild Buckwheat	*Rhizoma Fagopyri Cymosi*
Wild Chrysanthemum Flower	*Flos Chrysanthemi Indici*
Wingless Cockroach	*Eupolyphaga seu Opisthoplatia*
Winter Cherry Fruit	*Rhizoma et Radix Physalitis*
Woad Leaf	*Folium Daqingye*
Woad Root	*Radix Isatidis seu Baphicacanthi*
Woolly Grass Rhizome	*Rhizoma Imperatae Cylindricae*

Wormwood	*Herba Artemisiae Annuae*
Yedeon's Violet	*Herba cum Radice Violae Yedoensitis*
Yellow Azalea	*Radix Rhododendri Mollis*
Yellow Earth from center of cooking stove	*Terra Flava Usta*
Young Cinnamon Bark	*Cortex Cinnamomi Cassiae*
Young Soybean Sprout	*Semen Glycines Germinatum*
Zaocys	*Zaocys Dhumnades*
Zedoary Rhizome	*Rhizoma Curcumae Zedoariae*

APPENDIX 9

Cross-Reference Botanical Latin–English Herbs

Acacia Catechu seu Uncaria Gambier	*Catechu Extract*
Actinolitum	*Actinolite*
Agkistrodon seu Bungarus	*Agkistrodon Snake*
Albizziae Julibrissin	*Mimosa Tree Flower*
Alumen	*Purified Alunite*
Arillus Euphoriae Longanae	*Flesh of Longan Fruit*
Arsenicum	*Arsenolite*
Arsenolite	*Arsenolite*
Aspongopus Chinensis	*Dung Beetle*
Benzoinum	*Benzoin*
Bombyx Batryticatus	*Dead Sick Silkworm*
Borax	*Borax*
Borneol	*Borneol*
Bulbus Alli Sativi	*Garlic Bulb*
Bulbus Allii	*Bulb of Chinese Chive*
Bulbus Allii Fistulosi	*Scallion*
Bulbus Fritillariae Cirrhosae	*Tendrilled Fritillaria Bulb*
Bulbus Fritillariae Thunbergii	*Thunberg Fritillaria Bulb*
Bulbus Lilii	*Lily Bulb*
Buthus Martensi	*Scorpion*
Cacumen Biotae Orientalis	*Oriental Arborvitae Twig*
Cacumen Tamaricis	*Chinese Tamarisk Tops*
Calcitum	*Calcitum*
Calculus Bovis	*Cattle Gallstone*
Calculus Macacae Mulattae	*Macaque Gallstone*
Calomelas	*Calomel*
Calx	*Old Lime*
Calyx Diospyri Kaki	*Persimmon Calyx*
Camphora	*Camphor*
Carapax Amydae Sinensis	*Chinese Soft-Shelled Turtle Shell*
Catharsius	*Dung Beetle*
Caulis Bambusae in Taeniis	*Bamboo Shavings*
Caulis et Folium Ardisiae Japonicae	*Japanese Ardisia*
Caulis Mutong	*Five Leaf Akebia Stem*
Caulis Perillae Frutescentis	*Perilla Stem*
Caulis Piperis Futokadsurae	*Kadsura Stem*
Caulis Polygoni Multiflori	*Vine of Solomon's Seal*
Caulis Sargentodoxae Cuneatae	*Sargentodoxa Vine*
Caulis Trachelospermi Jasminoidis	*Star Jasmine Stem*

Chalcanthite	*Crystallized Blue Vitriol*
Citrullus Lanatus	*Frost Pwdr Prep by Watermelon Peel and Mirabilite*
Concha Arcae	*Cockle Shell*
Concha Cyclinae Sinensis	*Clam Shell*
Concha Haliotidis	*Abalone Shell*
Concha Ostreae	*Oyster Shell*
Concretio Silicea Bambusae Textillis	*Tabasheer*
Cordyceps Sinensis	*Cordyceps*
Corium Erinacei	*Hedgehog Skin*
Cormus Heleochris Plantaginea	*Water Chestnut*
Cornu Antelopis	*Antelope Horn*
Cornu Bubali	*Water Buffalo Horn*
Cornu Cervi	*Deer Horn*
Cornu Cervi Degelatinatium	*Dregs leftover after making Lu Jiao Jiao*
Cornu Cervi Parvum	*Velvet of Young Deer Horn*
Cornu Rhinoceri	*Rhinoceros Horn*
Cortex Acanthopanacis Gracilistyli Radicis	*Acanthopanax Root Bark*
Cortex Ailanthi	*Bark of Heaven Tree*
Cortex Ailanthi Altissimae	*Bark of Heaven Tree*
Cortex Albizziae Julibrissin	*Mimosa Tree Bark*
Cortex Catalpae Ovatae	*Chinese Catalpa Bark*
Cortex Cercis Chinensis	*Chinese Redbud Bark*
Cortex Cinnamomi Cassiae	*Cinnamon Bark*
Cortex Cinnamomi Cassiae	*Young Cinnamon Bark*
Cortex Daphnes	*Giraldi Daphne Bark*
Cortex Dictamni Dasycarpi Radicis	*Dictamnus Root Bark*
Cortex Erythrinae	*Erythrina Bark*
Cortex et Radix Ailanthi	*Bark of Heaven Tree*
Cortex Eucommiae Ulmoidis	*Eucommia Bark*
Cortex Fraxini	*Bark of Korean Ash Branch*
Cortex Hibiscus Syriaci	*Shrubalthea Bark*
Cortex Lycii Radicis	*Lycium Bark*
Cortex Magnoliae Officinalis	*Magnolia Bark*
Cortex Meliae Radicis	*China Tree Root Bark*
Cortex Mori Albae Radicis	*Bark of Mulberry Root*
Cortex Moutan Radicis	*Cortex of Tree Peony*
Cortex Phaseoli Radiati	*Mung Bean Skin*
Cortex Phellodendri	*Amur Cork-Tree Bark*
Cortex Poriae Cocos	*Poria Skin*
Cortex Pseudolarix Kaempferi	*Goldenlarch Bark*
Cortex Punicae Granati Radicis	*Bark of Pomegranate Tree*
Cortex Ulmi Pumilae	*Elm Tree Bark*
Crinis Carbonisatus Hominis	*Charred Human Hair*
Endothelium Corneum Gigeriae Galli	*Chicken Gizzard Lining*
Epicarpium Benincasae Hispidae	*Chinese Wax Gourd Peel*
Eupolyphaga seu Opisthoplatia	*Wingless Cockroach*
Eupolyphaga seu Opisthoplatia	*Wingless Cockroach*
Excrementum Bombycis Mori	*Silkworm Feces*
Excrementum Trogopteri seu Pteromi	*Flying Squirrel Feces*
Excrementum Trogopteri seu Pteromi	*Glossy Ganoderma*
Excrementum Vespertilionis Murini	*Bat Feces*
Exocarpium Citrulli	*Watermelon Peel*
Exocarpium Citrus Reticulate Rubrum	*Outer Tangerine Peel*
Exocarpium Zingiberis Officinalis Recens	*Ginger Peel*

Exuviae Serpentis	*Snake Skin Slough*
Fasciculus Vascularis Luffae	*Suakwa Vegetable Sponge*
Fibra Trachycarpi Vagina	*Stiple Fiber of Fortune Windmill Palm*
Fibrum Vascularis Citrus Tangerina	*Tangerine Pith*
Flos Buddleiae Officinalis	*Buddleia Flower Bud*
Flos Campsis	*Chinese Trumpetcreeper Flower*
Flos Carthami Tinctorii	*Safflower Flower*
Flos Caryophylli	*Clove Flower Bud*
Flos Celosiae Cristatae	*Cockscomb Flower*
Flos Chrysanthemi Indici	*Wild Chrysanthemum Flower*
Flos Chrysanthemi Morifolii	*Chrysanthemum Flower*
Flos Citri Sarcodactylis	*Finger Citron Flower*
Flos Clematidis Ranunculoidis	*Hydrangea Flower*
Flos Daphnes Genkwa	*Genkwa Flower*
Flos Daturae	*Jamestown Weed Seed/Leaf/Flower*
Flos Daturae Albae	*Jamestown Weed Seed/Leaf/Flower*
Flos Dolichoris Lablab	*Hyacinth Flower*
Flos et Fructus Rosae Chinensis	*Chinese Tea Rose*
Flos Gomphrenae	*Globeamaranth Flower*
Flos Inulae	*Inula Flower*
Flos Lonicerae Japonicae	*Honeysuckle Flower*
Flos Magnoliae	*Magnolia Flower*
Flos Magnoliae Officinalis	*Magnolia Flower*
Flos Puerariae	*Kudzu Flower*
Flos Rosae Rugosae	*Miniature Chinese Rose*
Flos Sambuci Chinensis	*Chinese Sambucus*
Flos Schizonepetae Tenuifoliae	*Schizonepeta Stem*
Flos Sophorae Japonicae Immaturus	*Pagoda Tree Flower Bud*
Flos Syzygii Aromatici	*Clove Fruit*
Flos Tussilaginis Farfarae	*Coltsfoot Flower*
Fluoritum	*Fluorite*
Folium Apocyni Veneti	*Leaf of Apocynum Venetum*
Folium Artemisiae Vulgaris/Argyi	*Mugwort Leaf*
Folium Callicarpae Pedunculatae	*Beautyberry Leaf*
Folium Citrus Tangerina	*Tangerine Leaf*
Folium Clerodendri Trichotomi	*Glorybower Leaf*
Folium Daqingye	*Woad Leaf*
Folium Elaeagni	*Thorny Elaegnus Leaf*
Folium Eriobotryae Japonicae	*Loquat Leaf*
Folium et Cortex Rhois Chinensis	*Chinese Sumac*
Folium et Flos Cordylines Fruticosae	*Sago Cycas*
Folium et Radix Gynurae Segetii	*Canton Tusanqi*
Folium et Radix Loropetali	*Strap Flower/Root/Leaf*
Folium et Radix Rohdeae Japonicae	*Japanese Lily Bulb*
Folium Ginkgo	*Ginkgo Leaf*
Folium Ginseng	*Ginseng Leaf*
Folium Ilicis Chinensis	*Chinese Holly*
Folium Mahoniae	*Chinese Mahonia*
Folium Momordicae Grosvenori	*Round Momordica Leaf*
Folium Mori Albae	*Mulberry Leaf*
Folium Nelumbinis Nuciferae	*Lotus Leaf*
Folium Perillae Frutescentis	*Perilla Leaf*
Folium Pyrrosiae	*Felt Fern Leaves*
Folium Rhododendri Simsii	*Rhododendron Leaf*
Folium Sennae	*Senna Leaf*

Frucificatio Lasiosphaerae	*Puffball*
Fructificatio Tremellae Fuciformis	*Fruiting Body of Tremella*
Fructus Akebiae	*Akebia Fruit*
Fructus Alpiniae Oxyphyllae	*Galangal Fruit*
Fructus Amomi	*Seed of Grains-of-Paradise*
Fructus Amomi Cardamomi	*Round Cardamom Fruit*
Fructus Amomi Tsao-ko	*Amomum Tsao-Ko*
Fructus Arctii Lappae	*Great Burdock Fruit*
Fructus Aristolochiae	*Birthwort Fruit*
Fructus Broussonetiae	*Paper Mulberry Fruit*
Fructus Bruceae Javanicae	*Java Brucea Fruit*
Fructus Carpesii seu Daucusi	*Starwort Fruit*
Fructus Chaenomelis	*Chinese Quince Fruit*
Fructus Citri Aurantii	*Ripe Fruit of Bitter Orange*
Fructus Citri Medica	*Citron Fruit*
Fructus Citri Sarcodactylis	*Finger Citron Fruit*
Fructus Citrulli Vulgaris	*Watermelon Fruit*
Fructus Cnidii Monnieri	*Cnidium Seed*
Fructus Corni Officinalis	*Asiatic Cornelian Cherry Fruit*
Fructus Crataegi	*Hawthorn Fruit*
Fructus Cubebae	*Litse Fruit*
Fructus Evodiae Rutaecarpae	*Evodia Fruit*
Fructus Foeniculi Vulgaris	*Fennel Fruit*
Fructus Forsythiae Suspensae	*Forsythia Fruit*
Fructus Gardeniae Jasminoidis	*Gardenia Fruit*
Fructus Gleditsiae Sinensis	*Chinese Honeylocust Fruit*
Fructus Hordei Vulgaris Germinantus	*Barley Sprout*
Fructus Illicii Veri	*Truestar Anisetree Fruit*
Fructus Illicius	*Truestar Anisetree Fruit*
Fructus Immaturus Citri Aurantii	*Immature Fruit of Bitter Orange*
Fructus Kochiae Scopariae	*Kochia Fruit*
Fructus Leonuri	*Motherwort Fruit*
Fructus Ligustri Lucidi	*Privet Fruit*
Fructus Liquidambaris Taiwanianae	*Sweetgum Fruit*
Fructus Lycii	*Chinese Wolfberry Fruit*
Fructus Meliae Toosendan	*Chinaberry Fruit*
Fructus Momordicae Grosvenori	*Round Momordica Fruit*
Fructus Mori Albae	*Bud of Mulberry Fruit*
Fructus Oryzae Sativae Germinantus	*Rice Sprout*
Fructus Perillae Frutescentis	*Purple Perilla Fruit*
Fructus Piperis Longi	*Long Pepper Fruit*
Fructus Piperis Nigri	*Black Pepper*
Fructus Polygoni Orientalis	*Prince's Feather Fruit*
Fructus Pruni Mume	*Mume Fruit*
Fructus Psoraleae Corylifoliae	*Fruit of Scuffy Pea*
Fructus Quisqualis Indicae	*Rangoon Creeper Fruit with Seed*
Fructus Rosae Laevigatae	*Cherokee Rosehip*
Fructus Rubi Chingii	*Chinese Raspberry*
Fructus Schisandrae Chinensis	*Schisandra Fruit*
Fructus Seu Radix Camptothecae	*Happy Tree Fruit Bark*
Fructus Sophorae Japonicae	*Pagoda Tree Fruit*
Fructus Terminaliae Chebulae	*Myrobalan Fruit*
Fructus Trapae	*Water Calptrop*
Fructus Tribuli Terrestris	*Caltrop Fruit*
Fructus Trichosanthis	*Mongolian Snakegourd*

Fructus Vascularis Luffae	*Suakwa Seed*
Fructus Viticis	*Vitex Fruit*
Fructus Xanthii Sibirici	*Cocklebur Fruit*
Fructus Zizyphi Jujubae	*Chinese Date*
Frusta Ferri	*Iron Filings*
Funiculus Umbilicalis	*Human Umbilical Cord*
Galla Rhois Chinensis	*Gallnut of Chinese Sumac*
Gecko	*Gecko*
Gekko Swinhoana	*House Lizard*
Gelatinum Corii Asini	*Donkey-Hide Gelatin*
Gelatinum Cornu Cervi	*Deer Horn Glue*
Gelatinum Plastri Testudinis	*Glue of Land Tortoise Shell*
Gemma Agrimoniae	*Bud of Hairyvein*
Gummi Olibanum	*Frankincense or Pistaciacense*
Gypsum	*Gypsum*
Haematitum	*Hematite*
Halloysitum Rubrum	*Halloysite*
Herba Acalyphae	*Copperleaf*
Herba Agastaches seu Pogostemi	*Patchouli or Wrinkled Gianthyssop*
Herba Agrimoniae Pilosae	*Agrimony*
Herba Aloes	*Juice of Aloe Leaf*
Herba Andrographitis Paniculatae	*Green Chiretta*
Herba Androsace Aizoon	*Androsace Plant*
Herba Ardisiae Japonicae	*Japanese Ardisia*
Herba Aristolochiae	*Dutchman's Pipe*
Herba Artemisiae Annuae	*Wormwood*
Herba Artemisiae Anomalae	*Diverse Wormwood*
Herba Artemisiae Yinchenhao	*Capillaris*
Herba Begoniae Fimbristipulatae	*Begonia*
Herba Catharanthi Rosei	*Madagascar Periwinkle*
Herba Cephalanoplos	*Small Japanese Thistle*
Herba Chenopodii Hybridi	*Mapleleaf Goosefoot*
Herba Cistanches Deserticolae	*Fleshy Stem of Broomrape*
Herba Commelinae	*Dayflower*
Herba Crotalariae	*Crotalaria Plant*
Herba cum Radice Asari	*Chinese Wild Ginger*
Herba cum Radice Houttuyniae Cordatae	*Houttuynia*
Herba cum Radice Patriniae	*Patrinia*
Herba cum Radice Violae Yedoensitis	*Yedeon's Violet*
Herba Cynomorii Songarici	*Chinese Cynomorium*
Herba Dendrobii	*Dendrobium Stem*
Herba Dianthi	*Chinese Pink Dianthus*
Herba Dichondrae	*Dichondra Plant*
Herba Duchesneae Indicae	*Mock Strawberry Plant*
Herba Ecliptae Prostratae	*Eclipta*
Herba Elsholtziae seu Moslae	*Aromatic Madder*
Herba Ephedrae	*Ephedra Stem*
Herba Epimedii	*Longspur Epimedium*
Herba Equiseti Hiemalis	*Horse Tail*
Herba Erodii seu Geranii	*Heroubill Plant*
Herba Eupatorii Fortunei	*Eupatorium*
Herba Euphorbiae Helioscopiae	*Spurge*
Herba Hedyotidis Diffusae	*Spreading Hedyotis Oldenlandia Plant*
Herba Houttuyniae Cordatae	*Shepherd's Purse*
Herba Hydrocotylis	*Hydrocotyle Plant*

Herba Hyperici Japonici	*Japanese Saint Johnswort*
Herba Lemnae seu Spirodelae	*Duckweed*
Herba Leonuri Heterophylli	*Chinese Motherwort*
Herba Lobeliae Chinensis cun Radice	*Chinese Lobelia*
Herba Lophatheri Gracilis	*Bamboo Leaves*
Herba Lycopi Lucidi	*Shiny Bugleweed*
Herba Lycopodii	*Buck Grass*
Herba Lygodii Japonici	*Japanese Fern*
Herba Lysimachiae	*Loosestrife Plant*
Herba Menthae Haplocalycis	*Peppermint*
Herba Ocimi Pilosi	*Sweet Basil*
Herba Paederiae	*Chinese Fevervine*
Herba Polygoni Avicularis	*Knotweed*
Herba Polygonum Perfoliati	*Perfoliatum*
Herba Portulacae Oleraceae	*Purslane*
Herba Potenlillae Discoloris	*Cinquefoil*
Herba Pteris	*Bracken*
Herba Rorippae	*Nasturtium*
Herba Rubi Parvifolii	*Japanese Raspberry*
Herba Sarcandrae	*Sarcandrae Plant*
Herba Sarcandrae Glabrae	*Sarcandrae Plant*
Herba Sargassii	*Seaweed*
Herba Saxifragae	*Saxifrage*
Herba Scutellariae Barbatae	*Sun Plant*
Herba Sedi Aizoon	*Aizoon Stonecrop Plant*
Herba Sedi Sarmentosi	*Stringy Stonecrop*
Herba Selagrinellae Doederleinii	*Greater Selaginella*
Herba Semiaquilegiae Adoxoidis	*Muskroot-like Semiaquilegia Root*
Herba Senecionis	*Climbing Groundsel*
Herba seu Radix Cirsii Japonici	*Peking Spurge Root*
Herba Siegesbeckiae	*Common Saint Paulswort*
Herba Solani Lyrati	*Climbing Nightshade Plant*
Herba Solani Nigri	*Black Nightshade*
Herba Speranskia Tuberculata	*Garden Balsam*
Herba Taraxaci Mongolici cum Radice	*Dandelion*
Herba Verbenae	*European Verbena*
Herba Zosterae Marinae	*Common Eelgrass*
Hippocampus	*Seahorse*
Hirudo seu Whitmania	*Leech*
Holotrichia Diomphalia	*Grub*
Hydrargyrum	*Mercury*
Indigo Pulverata Levis	*Indigo*
Jecur Lutrae	*Otter Liver*
Lacca Sinica Exsiccatae	*Dry Lacquer Sap*
Lapis Micae seu Chloriti	*Chlorite*
Lignum Aquilariae	*Aloeswood*
Lignum Cercis Chinensis	*Chinese Redbud Wood*
Lignum Dalbergiae Odoriferae	*Center of Rosewood*
Lignum Pini Nodi	*Knotty Pine Wood*
Lignum Santali Albi	*Sandalwood*
Lignum Sappan	*Sappan Wood*
Limonite	*Limonite*
Lithargyrum	*Litharge or Galena*
Lumbricus	*Earthworm*
Magnetitum	*Magnetite*

Margarita Usta	*Pearl*
Massa Fermentata	*Medicated Leaven*
Medulla Junci Effusi	*Rush Pith*
Medulla Tetrapanacis Papyriferi	*Rice Paper Pith*
Mel	*Honey*
Minium	*Lead Oxide (Pb_3O_4)*
Mirabilitum	*Glauber's Salt*
Mirabilitum Depuratum	*Less Purfied Mirabilitum*
Mylabris	*Cantharides*
Myrrha	*Myrrh*
Nidus Vespae	*Hornet Nest*
Nodus Nelumbinis Nuciferae Rhizomatis	*Node of Lotus Rhizome*
Ootheca Mantidis	*Praying Mantis Egg-Case laid on Mulberry Leaf*
Ophicalcitum	*Ophicalcite*
Os Draconis	*Fossils of Mammal Vertebrae and Extremity*
Os Sepiae seu Sepiellae	*Cuttlefish Bone*
Os Tigris	*Tiger Bone*
Pedicellus Cucumeris	*Melon Pedicle*
Pericarpium Arecae Catechu	*Betel Husk*
Pericarpium Citri Reticulatae	*Tangerine Peel*
Pericarpium Citri Reticulatae Viride	*Immature Tangerine Peel*
Pericarpium Papaveris Somniferi	*Opium Poppy Husk*
Pericarpium Punicae Granati	*Pomegranate Husk*
Pericarpium Pyri	*Asian Pear Peel*
Pericarpium Zanthoxyli Bungeani	*Fruit of Szechuan Pricklyash Pepper*
Periostracum Cicadae	*Cicada Molting*
Placenta Hominis	*Human Placenta*
Plastrum Testudinis	*Land Tortoise Shell*
Plumula Nelumbinis Nucifera	*Lotus Plumule*
Pollen Typhae	*Cattail Pollen*
Praeparatio Fructus Ulmi Macrocarpi	*Stinky Elm Fruit Paste*
Pseudobulbus Shancigu	*Chinese Tulip Bulb*
Pumice	*Pumice*
Purum Mirabilitum	*Pure Mirabilitum*
Pyritum	*Pyrite*
Quick-Fried Rhizoma Zingiberis Officinalis	*Quick-Fried Ginger Root*
Radix Acanthopanaeis Senticosi	*Acanthopanax Root*
Radix Achyranthis Aspera	*Prickly Chaff Flower/Root/Twig*
Radix Achyranthis Bidentatae	*Bidentate Achyranthes Root*
Radix Aconiti Brachypodi	*Brachypodum Monkshood Root*
Radix Aconiti Carmichaeli	*Sichuan Monkshood Root*
Radix Aconiti Carmichaeli Praeparata	*Aconite*
Radix Aconiti Kusnezoffii	*Kusnezoff Monkshood Root*
Radix Actinidiae Argutae	*Chinese Kiwi Root*
Radix Adenophorae seu Glehniae	*Root of Beech Silver-Top*
Radix Alangii	*Chinese Alangium Root*
Radix Ampelopsis	*Ampelopsis*
Radix Angelicae Dahuricae	*Angelica Root*
Radix Angelicae Pubescentis	*Pubescent Angelica Root*
Radix Angelicae Sinensis	*Chinese Angelica Root*
Radix Ardisiae Crenatae	*Coral Ardisia Root*
Radix Aristolochiae seu Cocculi	*Aristolochia Root*
Radix Arnebiae seu Lithospermi	*Groomwell Root*
Radix Asteris Tatarici	*Purple Aster Root*
Radix Astragali Membranacei	*Milk-Vetch Root*

Radix Astragali Membranacei Praeparata	*Prepared Milk-Vetch Root*
Radix Aucklandiae Lappae	*Aucklandia*
Radix Berberidis	*Berberis Root*
Radix Boehmeriae	*Ramie Root*
Radix Bupleuri	*Thorowax Root*
Radix Camelliae Sinensis	*Tea Tree Root*
Radix Carpesii	*Carpesium Root*
Radix Changii	*Changium Root*
Radix Clematidis	*Chinese Clematis Root*
Radix Codonopsitis Pilosulae	*Codonopsis Root*
Radix Cyathulae Officinalis	*Sichuan Cyathula Root*
Radix Cynanchi Baiwei	*Swallowwort Root*
Radix Dichroae Febrifugae	*Dichroa Root*
Radix Dioscoreae Oppositae	*Chinese Yam*
Radix Dipsaci Asperi	*Japanese Teasel Root*
Radix Ephedrae	*Chinese Ephedra Root*
Radix et Caulis Jixueteng	*Leatherleaf Millettia Root/Vine*
Radix et Herba Cynanchi Paniculati	*Chinese Swallowwort*
Radix et Herba Thalictri Foliolosi	*Thalictrum*
Radix et Ramus Litseae Cubebae	*Aromatic Litsea Root/Stem*
Radix et Rhizoma Oryzae Glutinosae	*Glutinous Rice Root*
Radix et Rhizoma Polygoni Cuspidati	*Giant Knotweed Root & Rhizome*
Radix et Rhizoma Rhei	*Rhubarb Root Rhizome*
Radix et Rhizoma Veratri	*Mountain Onion*
Radix Euphorbiae et Stellerae	*Fischer Euphorbia Root*
Radix Euphorbiae Kansui	*Kan-Sui Root*
Radix Euphorbiae seu Knoxiae	*Japanese Thistle*
Radix Gentianae Longdancao	*Gentiana Root*
Radix Gentianae Qinjiao	*Gentiana Macrophylla Root*
Radix Ginseng	*Ginseng Root*
Radix Glycyrrhizae Uralensis	*Raw Licorice Root*
Radix Glycyrrhizae Uralensis Praeparata	*Prepared Licorice Root*
Radix Gossypii	*Cotton Plant Root*
Radix Hemsleyae	*Hemsleya Root*
Radix Ilicis Pubescentis	*Hairy Holly Root*
Radix Isatidis seu Baphicacanthi	*Woad Root*
Radix Ledebouriellae Divaricatae	*Ledebouriella Root*
Radix Ligustici Chuanxiong	*Sichuan Lovage Root*
Radix Linderae Strychnifoliae	*Three-Nerved Spicebush Root*
Radix Morindae Officinalis	*Indian Mulberry Root*
Radix Notoginseng	*Pseudoginseng Root*
Radix Paeoniae Lactiflorae	*White Peony Root*
Radix Paeoniae Rubrae	*Red Peony Root*
Radix Panacis Quinquefolii	*American Ginseng Root*
Radix Peucedani	*Hogfennel Root*
Radix Phytolaccae	*India Pokeberry Root*
Radix Platycodi Grandiflori	*Balloonflower Root*
Radix Polygalae Tenuifoliae	*Chinese Senega Root*
Radix Polygoni Multiflori	*Fleeceflower Root*
Radix Pseudostellariae Heterophyllae	*Caryophyllaceous Root*
Radix Puerariae	*Kudzu Root*
Radix Pulsatillae Chinensis	*Chinese Anemone Root*
Radix Ranunculus Tematus	*Buttercup Root*
Radix Rehmanniae Glutinosae	*Fresh Rehmannia*
Radix Rehmanniae Glutinosae Conquitae	*Cooked Rehmannia Root*

Radix Rhapontici seu Echinops	*Rhaponticum Root*
Radix Rhododendri Mollis	*Yellow Azalea*
Radix Rubiae Cordifoliae	*Madder Root*
Radix Salviae Chinensis	*Chinese Sage*
Radix Salviae Miltiorrhizae	*Salvia Root*
Radix Sanguisorbae Officinalis	*Burnet-Bloodwort Root*
Radix Scrophulariae Ningpoensis	*Ningpo Figwort Root*
Radix Scutellariae Baicalensis	*Skullcap Root*
Radix Semiaquilegiae Adoxoidis	*Muskroot-like Semiaquilegia Root*
Radix Sophorae Flavescentis	*Sophora Root*
Radix Sophorae Tonkinensis	*Sophora Root*
Radix Stellariae Dichotomae	*Stellaria Root*
Radix Stemonae	*Stemona Tuber*
Radix Stephaniae Tetrandrae	*Stephania Root*
Radix Tinosporae	*Tinospora Root*
Radix Trichosanthis Kirilowii	*Mongolian Snakegourd Seed*
Radix Trichosanthis Kirilowii	*Trichosantes Root*
Radix Tripterygii Wilfordii	*Tripterygium Wilfordii Hook*
Radix Zanthoxyli	*Pricklyash Root*
Ramulus Cinnamomi Cassiae	*Cinnamon Twig*
Ramulus Juglandis Mandshuricae	*Walnut Twig*
Ramulus Mori Albae	*Mulberry Twig*
Ramulus Piper Wallichii	*Chinese Photinia Vine*
Ramulus Sangjisheng	*Mulberry Mistletoe*
Ramulus Tamarix Chinensis	*Chinese Tamarisk Tops*
Ramulus Uncariae cum Uncis	*Gambir Vine*
Ramus Tinosporae Sinensis	*Chinese Tinospora Stem*
Realgar	*Realgar*
Receptaculum Nelumbinis Nuciferae	*Mature Lotus Receptacle*
Resina Draconis	*Dragon's Blood (Resin)*
Resina Ferulae	*Chinese Giantfennel Resin*
Rhizom Zingiberis Officinalis	*Dried Ginger*
Rhizoma Acori Graminei	*Sweetflag Rhizome*
Rhizoma Alismatis Orientalis	*Water Plantain*
Rhizoma Alpiniae Officinari	*Lesser Galangal Rhizome*
Rhizoma Anemarrhenae Asphodeloidis	*Anemarrhena Rhizome*
Rhizoma Anemones Raddeanae	*Anemone*
Rhizoma Arisaematis	*Jack-in-the-Pulpit*
Rhizoma Atractylodis Macrocephalae	*Chinese Atractylodes*
Rhizoma Atractylodis Macrocephalae	*Large-headed Atractylodes Rhizome*
Rhizoma Belamcandae Chinensis	*Belamcanda Rhizome*
Rhizoma Bistortae	*Snakeweed*
Rhizoma Bletillae Striatae	*Hyacinth Bletilla Tuber*
Rhizoma Bolbostemmatis	*Paniculate Bolbostemma*
Rhizoma Cibotii	*Chain Fern Rhizome*
Rhizoma Cibotii Barometz	*Chain Fern Rhizome*
Rhizoma Cimicifugae	*Black Kohosh*
Rhizoma Coptidis	*Coptis Rhizome*
Rhizoma Corydalis Decumbentis	*Corydalis Rhizome*
Rhizoma Corydalis Yanhusuo	*Corydalis Rhizome*
Rhizoma Curculiginis Orchioidis	*Rhizome of Golden Eye-Grass*
Rhizoma Curcumae Longae	*Turmeric Rhizome*
Rhizoma Curcumae Zedoariae	*Zedoary Rhizome*
Rhizoma Cynanchii Stautoni	*White Swallowwort*
Rhizoma Cynanchum Auriculati	*Earleaf Cynanchum Root*

Rhizoma Cyperi Rotundi	*Nutgrass*
Rhizoma Dioscoreae Bulbiferae	*Airpotato Yam*
Rhizoma Dioscoreae Hypoglaucae	*Fish-Poison Yam*
Rhizoma Dioscoreae Nipponicae	*Japanese Yam*
Rhizoma Drynariae	*Drynaria Rhizome*
Rhizoma et Folium Smilacis Chinensis	*China Root*
Rhizoma et Radix Ligustici	*Chinese Lovage Root*
Rhizoma et Radix Notopterygii	*Notopterygium Root*
Rhizoma et Radix Physalitis	*Winter Cherry Fruit*
Rhizoma Fagopyri Cymosi	*Wild Buckwheat*
Rhizoma Gastrodiae Elatae	*Gastrodia Rhizome*
Rhizoma Guanzhong	*Dryopteris Root*
Rhizoma Homalomenae Occultae	*Homalomena Rhizome*
Rhizoma Imperatae Cylindricae	*Woolly Grass Rhizome*
Rhizoma Paris	*Paris Rhizome*
Rhizoma Phragmitis Communis	*Reed Rhizome*
Rhizoma Picrorhizae	*Picrorhiza Rhizome*
Rhizoma Pinelliae Ternatae	*Pinellia Tuber*
Rhizoma Polygonati	*Siberian Solomon Seal Rhizome*
Rhizoma Polygonati Odorati	*Solomon's Seal Rhizome*
Rhizoma Polygoni Bistortae	*Snakeweed*
Rhizoma Seu Herba Saururii	*Lizardtail*
Rhizoma Smilacis Glabrae	*Glabrous Greenbrier Rhizome*
Rhizoma Sparganii Stoloniferi	*Bur-Reed Rhizome*
Rhizoma Typhonii Gigantei	*Giant Typhonium Tuber*
Rhizoma Zingiberis Officinalis Recens	*Fresh Ginger Rhizome*
Saccharum Granorum	*Barley Malt Sugar*
Sal Ammoniaci	*Sal Ammoniac*
Scapus Eriocaulonis Buergeriani	*Pipewort Scapus*
Sclerotium Omphaliae Lapidescens	*Fruiting Body of Stone-like Omphalia*
Sclerotium Polypori Umbellati	*Umbellate Pore Fungus*
Sclerotium Poriae Cocos	*White Poria*
Sclerotium Poriae Cocos Pararadicis	*Poria with Wood Root*
Sclerotium Poriae Cocos Rubrae	*Red Poria*
Scolopendra Subspinipes	*Centipede*
Secretio Bufonis	*Toad Venom*
Secretio Moschus	*Musk of Deer*
Semen Abutili seu Malvae	*Musk Mallow Seed*
Semen Allii Tuberosi	*Seeds of Fragrant- Flowered Garlic*
Semen Alpiniae Katsumadai	*Seed of Katsumada's Galangal*
Semen Arecae Catechu	*Betel Nut*
Semen Astragali Complanati	*Milk-Vetch Seed*
Semen Benincasae Hispidae	*Chinese Wax Gourd Seed*
Semen Biotae Orientalis	*Arbor-Vitae Seed*
Semen Canavaliae	*Sword Bean*
Semen Cannabis Sativae	*Cannabis Seeds*
Semen Cassiae	*Cassia Seeds*
Semen Celosiae Argenteae	*Celosia Seeds*
Semen Citri Reticulatae	*Tangerine Seed*
Semen Coicis Lachryma-jobi	*Job's Tears (Coix)*
Semen Croton Tiglii	*Croton Seed*
Semen Cucurbitae Moschatae	*Pumpkin Seed Husk*
Semen Cuscutae Chinensis	*Chinese Dodder Seeds*
Semen Descurainiae seu Lepidii	*Pepperweed Seed or Flixweed Tansymustard Seed*
Semen Dolichoris Lablab	*Hyacinth Bean*

Semen Euphorbia Lathyridis	*Moleplant Seed*
Semen Euryales Ferocis	*Gordon Euryale Seed*
Semen Ginkgo Bilobae	*Ginkgo Nut*
Semen Ginkgo Bilobae	*Ginkgo Nut*
Semen Glycines Germinatum	*Young Soybean Sprout*
Semen Hydnocarpi Anthelminticae	*Chaulmoogra Seed*
Semen Hyoscyami	*Henbane Seed*
Semen Impatientis	*Garden Balsam Seed*
Semen Juglandis Regiae	*Walnut Nut*
Semen Litchi Chinensis	*Leechee Nut*
Semen Momordicae Cochinchinensis	*Momordica Seed*
Semen Myristicae Fragrantis	*Nutmeg Seed*
Semen Nelumbinis Nucifera	*Lotus Seed*
Semen Oroxyli Indici	*India Trumpetflower Seed*
Semen Persicae	*Peach Kernel*
Semen Pharbitidis	*Morning Glory Seed*
Semen Phaseoli Calcarati	*Aduki Bean*
Semen Phaseoli Radiati	*Mung Bean*
Semen Plantaginis	*Plantago Seeds*
Semen Pruni	*Bush Cherry Pit*
Semen Pruni Armeniacae	*Apricot Seed*
Semen Raphani Sativi	*Chinese Radish Seed*
Semen Sesami Indici	*Black Sesame Seeds*
Semen Sinapis Albae	*White Mustard Seed*
Semen Sojae Praeparatum	*Prepared Soybean*
Semen Sterculiae Scaphigerae	*Boat Sterculia Seed*
Semen Strychni	*Nux-Vomica Seed*
Semen Torreyae Grandis	*Chinese Torreya Seed*
Semen Trigonellae Foeni-graeci	*Fenugreek Seed*
Semen Tritici Aestivi Levis	*Light Wheat Grain*
Semen Tritici Levis	*Wheat Grain*
Semen Vaccariae Segetalis	*Cowherb Seed*
Semen Zizyphi Spinosae	*Sour Jujube Seed*
Smithsonitum	*Smithsonite or Calamine*
Spica Prunellae Vulgaris	*Selfheal Spike*
Spora Lygodii Japonici	*Japanese Fern Spore*
Squama Manitis Pentadactylae	*Pangolin Scales*
Stalactitum	*Tip of Tubular Stalactites*
Stamen Nelumbinis Nucifera	*Lotus Stamen*
Stichopus Japonicus	*Sea Cucumber*
Stylus Zeae Mays	*Cornsilk*
Styrax Liquidis	*Resin of Rose Maloes*
Succinum	*Amber*
Succus Bambusae	*Dried Bamboo Sap*
Succus Faecalis Aureus	*Purified Feces*
Succus Zingiberis	*Fresh Ginger Juice*
Sulphur	*Sulphur*
Syngnathus	*Pipe-Fish*
Tabanus Bivittatus	*Gadfly*
Talcum	*Talcum*
Terra Flava Usta	*Yellow Earth from center of cooking stove*
Testa Arachidis Hypogaeae	*Peanut Peel*
Testa Dolichoris Lablab	*Hyacinth Bean Peel*
Testes et Penis Otariae	*Male Seal Sexual Organ*
Thallus Algae	*Kelp Thallus*

Trachycarpi Carbonisatus	*Charred Stiple Fiber of Fortune Windmill Palm*
Tuber Asparagi Cochinchinensis	*Chinese Asparagus Tuber*
Tuber Curcumae	*Turmeric Tuber*
Tuber Ophiopogonis Japonici	*Tuber of Dwarf Lilyturf*
Urina Puerilis	*Boy Urine*
Vesica Fellea Ursi	*Bear Gallbladder*
Vitellus Gigeriae Galli	*Chicken Egg Yolk*
Zaocys Dhumnades	*Zaocys*

APPENDIX 10

Cross-Reference Pinyin–English Herbal Formulas

Ai Fu Nuan Gong Wan [Mugwort and Cyperi Rhizome Pill for Warming the Womb]
An Gong Niu Huang Wan [Calm the Palace Pill with Cattle Gallstone]
An Shen Ding Zhi Wan [Calm the Spirit and Settle the Will Pill]
An Shen Wan [Calm the Spirit Pill]
Ba Li San [Eight-Thousandths of a Tael Powder]
Ba Xian Chang Shou Wan [Eight-Immortal Pill for Longevity]
Ba Zhen Tang [Eight-Treasure Decoction]
Ba Zheng Jie Du Tang [Eight-Herb Decoction for Rectification and Relieving Toxicity]
Ba Zheng San [Eight-Herb Powder for Rectification]
Bai Du San [Overcome Pathogenic Influences Powder]
Bai He Gu Jin Tang [Lily Bulb Decoction to Preserve the Metal]
Bai He Pian [Lily Bulb Tablet]
Bai Hu Cheng Qi Tang [White Tiger and Order the Qi Decoction]
Bai Hu He Huang Lian Jie Du Tang [White Tiger with Coptis Decoction to Relieve Toxicity]
Bai Hu Jia Gui Zhi Tang [White Tiger plus Cinnamon Twig Decoction]
Bai Hu Jia Ren Shen Tang [White Tiger plus Ginseng Decoction]
Bai Hu Tang [White Tiger Decoction]
Bai Ji Pi Pa Wan [Bletilla and Loquat Leaf Pill]
Bai Tou Weng Jia Gan Cao E Jiao Tang [Pulsatilla Decoction plus Licorice and Ass-Hide Gelatin]
Bai Tou Weng Tang [Pulsatilla Decoction]
Bai Ye Tang [Biota Leaves Decoction]
Bai Zhu Fu Zi Tang [Atractylodes and Prepared Aconite Decoction]
Bai Zhu San [Atractylodes Powder]
Bai Zi Yang Xin Wan [Biota Seed Pill to Nourish the Heart]
Ban Xia Bai Zhu Tian Ma Tang [Pinellia, Atractylodes Macrocephala, and Gastrodia Decoction]
Ban Xia Fu Ling Wan [Pinellia and Poria Pill]
Ban Xia Gan Jiang San [Pinellia and Dry Ginger Powder]
Ban Xia Hou Po Tang [Pinellia and Magnolia Bark Decoction]
Ban Xia Tang [Pinellia Decoction]
Ban Xia Xie Xin Tang Qu Gan Jiang, Gan Cao, Jia Zhi Shi Xing Ren Fang [Pinellia Decoction to Drain the Epigastrium without Dry Ginger, Licorice and with the Addition of Immature Orange and Apricot Seed]
Ban Xia Xie Xin Tang [Pinellia Decoction to Drain the Epigastrium]
Bao He Wan [Preserve Harmony Pill]
Bao Long Wan [Hold the Dragon Pill]
Bao Long Wan [Hold the Dragon Pill]
Bao Sheng Wu You San [Protect the Fetus without Any Worries Powder]

Bao Zhen Tang [Preserve-True Decoction]
Bei Mu Gua Lou San [Fritillaria and Trichosanthes Fruit Powder]
Bei Xie Fen Qing Yin [Dioscorea Hypoglauca Decoction to Separate the Clear]
Bi Yu San [Jasper Powder]
Bian Que Ding Xiang San [Clove Powder of Bian Que]
Bian Tong Bai Tou Weng Tang [Adapted Pulsatilla Decoction]
Bian Tong Shi Wei Wen Dan Tang [Adapted Ten Flavored Warm the Gallbladder Decoction]
Bie Jia Jian Wan [Decocted Turtle Shell Pill]
Bie Jia Qing Hao Yin [Soft-shelled Turtle Shell and Artemisa Annua Decoction]
Bie Jia Yang Yin Jian [Nourish Yin Decoction with Soft-shelled Turtle Shell]
Bu Dai Wan [Cloth Sack Pill]
Bu Fei E Jiao Tang [Tonify the Lungs Decoction with Ass-Hide Gelatin]
Bu Fei Tang [Tonify the Lung Decoction]
Bu Gan Tang [Tonify the Liver Decoction]
Bu Huan Jin Zheng Qi San [Rectify the Qi Powder Worth More than Gold]
Bu Yang Huan Wu Tang [Tonify the Yang to Restore Five-Tenths Decoction]
Bu Yi Wan [Tonify and Augment Pill]
Bu Zhong Yi Qi Tang [Tonify the Middle and Augment the Qi Decoction]
Can Shi Tang [Silk Worm Excrement Decoction]
Cang Er Zi San [Xanthium Powder]
Cang Zhu Bai Hu Tang [Atractylodes White Tiger Decoction]
Chai Hu Bai Hu Tang [Bupleurum White Tiger Decoction]
Chai Hu Da Yuan Yin [Bupleurum Decoction to Reach the Membrane Source]
Chai Hu Gui Jiang Tang [Bupleurum, Cinnamon and Ginger Decoction]
Chai Hu Gui Zhi Tang [Bupleurum and Cinnamon Twig Decoction]
Chai Hu Jia Long Gu Mu Li Tang [Bupleurum Plus Dragon Bone and Oyster Shell Decoction]
Chai Hu Jia Lu Gen Tang [Bupleurum Decoction plus Reed]
Chai Hu Jia Mang Xiao Tang [Bupleurum Decoction plus Mirabilite]
Chai Hu Shu Gan San [Bupleurum Powder to Spread the Liver]
Chai Hu Shu Gan San [Bupleurum Powder to Spread the Liver] *Jing Yue Quan Shu*
Chai Hu Si Wu Tang [Bupleurum and Four-Substance Decoction]
Chai Hu Xian Xiong Tang [Bupleurum Decoction for Sinking into the Chest]
Chai Hu Zhi Jie Tang [Bupleurum, Bitter Orange, and Platycodon Decoction]
Chai Ping Tang [Bupleurum and Calm the Stomach Decoction]
Chai Shao Liu Jun Zi Tang [Six-Gentleman Decoction with Bupleurum and Peony]
Chan Su Wan [Toad Cake Pill]
Chen Xiang Jiang Qi San [Aquilaria to Direct Qi Downward Powder]
Chen Xiang San [Aquilaria Powder]
Chi Fu Ling Tang [Red Poria Decoction]
Chi Shi Zhi Yu Yu Liang Tang [Halloysitum and Limonite Decoction]
Chuan Xiong Cha Tiao San [Ligusticum Chuanxiong Powder to be Taken with Green Tea]
Chun Gen Pi Wan [Ailanhti Bark Pill]
Chun Ze Tang [Spring Pond Decoction]
Cong Chi Bai Hu Tang [Scallion, Prepared Soybean, and White Tiger Decoction]
Cong Chi Bo Bang Tang [Scallion, Prepared Soybean, Peppermint, and Arctii Decoction]
Cong Chi Jie Geng Tang [Scallion, Prepared Soybean, and Platycodon Decoction]
Cong Chi Tang [Scallion, and Prepared Soybean Decoction]
Cui Sheng Yin [Induce Labor Decoction]
Da An Wan [Great Peace Pill]
Da Ban Xia Tang [Major Pinellia Decoction]
Da Bu Wan [Great Tonify Pill]
Da Bu Yin Wan [Great Tonify the Yin Pill]
Da Chai Hu Tang [Major Bupleurum Decoction]
Da Cheng Qi Tang [Major Order the Qi Decoction]
Da Ding Feng Zhu [Major Arrest Wind Pearl]
Da Huang Fu Zi Tang [Rhubarb and Prepared Aconite Decoction]

Da Huang Gan Cao Tang [Rhubarb and Licorice Decoction]
Da Huang Mu Dan Tang [Rhubarb and Moutan Decoction]
Da Huang Xiao Shi Tang [Rhubarb and Mirabilite Decoction]
Da Huang Zhe Chong Wan [Rhubarb and Eupolyphaga Pill]
Da Huo Luo Dan [Major Invigorate the Collaterals Pill]
Da Jian Zhong Tang [Major Construct the Middle Decoction]
Da Jiao Ai Tang [Major Ass-Hide Gelatin and Mugwort Decoction]
Da Ju Pi Tang [Big Tangerine Peel Decoction from Wai Tai Mi Yao]
Da Qin Jiao Tang [Major Gentiana Qinjiao Decoction]
Da Qing Long Tang [Major Bluegreen Dragon Decoction]
Da Xian Xiong Tang [Major Sinking into the Chest Decoction]
Da Xian Xiong Wan [Major Sinking into the Chest Pill]
Da Yuan Yin [Reach the Membrane Source Decoction]
Da Zao Wan [Great Creation Pill]
Da Zao Wan [Great Creation Pill]
Dan Bai Si Ni San [Moutan and Phellodendron Powder for Frigid Extremities]
Dan Dao Pai Shi Tang [Expelling Stones from Biliary Tract Decoction]
Dan Nang Yan Da Chai Hu Tang [Major Bupleurum Decoction for Cholecystitis]
Dan Shen Yin [Salvia Decoction]
Dan Xi Bei Xie Fen Qing Yin [Dioscorea Hypoglauca Decoction to Separate the Clear] *Dan Xi Xin Fa*]
Dan Zhi Xiao Yao San [Relax Wanderer Powder with Moutan and Gardenia]
Dang Gui Bu Xue Tang [Tangkuei Decoction to Tonify Blood]
Dang Gui Cheng Qi Tang [Order of Qi Decoction with Tangkuei]
Dang Gui Jian Zhong Tang [Tangkuei Decoction to Construct the Middle]
Dang Gui Liu Huang Tang [Tangkuei and Six-Yellow Decoction]
Dang Gui Long Hui Wan [Tangkuei, Gentiana Longdancao and Aloe Pill]
Dang Gui San [Tangkuei Powder]
Dang Gui Shao Yao San [Tangkuei and Peony Powder]
Dang Gui Sheng Jiang Yang Rou Tang [Mutton Stew with Tangkuei, Fresh Ginger]
Dang Gui Si Ni Jia Wu Zhu Yu Sheng Jiang Tang [Tangkuei Decoction for Frigid Extremities plus Evodia and Fresh Ginger]
Dang Gui Si Ni Tang [Tangkuei Decoction for Frigid Extremities]
Dang Gui Tang [Tangkuei Decoction]
Dao Chi San [Guide Out the Red Powder]
Dao Chi San He Jia Wei Hu Zhang San [Guide Out the Red and Augmented Polygoni Cuspidati Powder]
Dao Qi Tang [Guide the Qi Decoction]
Dao Tan Tang [Guide Out Phlegm Decoction]
Di Gu Pi Yin [Lycium Root Bark Decoction]
Di Gu Yang Yin Jian [Lycium Root Bark to Nourish Yin Decoction]
Di Huang Wan [Rehmannia Pill]
Di Tan Tang [Scour Phlegm Decoction]
Di Yu Wan [Sanguisorba Pill]
Die Da Wan [Trauma Pill]
Ding Chuan Tang [Arrest Wheezing Decoction]
Ding Xiang Jiao Ai Tang [Clove, Ass-Hide Gelatin, and Mugwort Decoction]
Ding Xiang Shi Di Tang [Clove and Persimmon Calyx Decoction]
Ding Xiang Shi Di Zhu Ru Tang [Clove, Persimmon Calyx and Bamboo Shaving Decoction]
Ding Yu Li Zhong Tang [Clove and Evodia Decoction to Regulate the Middle]
Ding Zhi Wan [Settle the Emotion Pill]
Dong Yuan Long Dan Xie Gan Tang [Gentiana Longdancao Decoction to Drain the Liver by Li Dong Yuan]
Du Huo Ji Sheng Tang [Angelica Pubescens and Sangjisheng Decoction]
Du Qi Wan [Capital Qi Pill]
Du Shen Tang [Unaccompanied Ginseng Decoction]
Duo Ming San [Seize the Life Powder]

E Jiao Huang Lian Tang [Ass-Hide Gelatin and Coptis Decoction]
E Jiao Ji Zi Huang Tang [Ass-Hide Gelatin and Chicken Egg Yolk Decoction] *Guang Wen Re Lun*
E Jiao Ji Zi Huang Tang [Ass-Hide Gelatin and Egg Yolk Decoction]
E Jiao Yi Gong San [Extraordinary Merit Powder with Ass-Hide Gelatin]
Er Chen Tang [Two-Cured Decoction]
Er Jia Fu Mai Tang [Two-Shell Decoction to Restore the Pulse]
Er Jia Jian Zheng Qi San [Second Modification of Rectify the Qi Powder]
Er Long Zuo Ci Wan [Pill for Deafness that is Kind to the Left (Kidney)]
Er Miao San [Two-Marvel Powder]
Er Xian Tang [Two-Immortal Decoction]
Er Zhi Wan [Two-Ultimate Pill]
Er Zhu Er Chen Tang [Two-Cured Decoction with Two Atractylodes]
Fang Feng Tong Sheng San [Ledebouriella Powder that Sagely Unblocks]
Fang Ji Fu Ling Tang [Stepania and Poria Decoction]
Fang Ji Huang Qi Tang [Stepania and Astragalus Decoction]
Fei Er Wan [Strengthen Baby Pill]
Fu Bao Dan [Women's Treasure Pill]
Fu Fang Da Cheng Qi Tang [Revised Major Order of Qi Decoction]
Fu Fang Da Huang Mu Dan Tang [Revised Rhubarb and Moutan Decoction]
Fu Fang Hong Teng Jian [Revised Millettia Decoction]
Fu Ling Pi Tang [Poria Skin Decoction]
Fu Ling Wan [Poria Pill]
Fu Mai Tang [Restore the Pulse Decoction]
Fu Pi Shu Gan Tang [Assist Spleen and Spread the Liver Decoction]
Fu Sang Wan [Assist with Mulberry Leaf Pill]
Fu Yuan Huo Xue Tang [Revive Health by Invigorating the Blood Decoction]
Fu Zi Li Zhong Wan [Prepared Aconite Pill to Regulate the Middle]
Fu Zi Tang [Prepared Aconite Decoction]
Gan Cao Fen Mi Tang [Licorice, Lead and Honey Decoction]
Gan Cao Gan Jiang Fu Ling Bai Zhu Tang [Licorice, Ginger, Poria and Atractylodes Macrocephalae Decoction]
Gan Cao Gan Jiang Tang [Licorice and Ginger Decoction]
Gan Cao Xie Xin Tang [Licorice Decoction to Drain the Epigastrium]
Gan Dan Guan Jie Shi Fang [Formula for Stones in the Hepatic Duct and Biliary Duct]
Gan Jiang Ren Shen Ban Xia Wan [Ginger, Ginseng and Pinellia Pill]
Gan Lu Xiao Du Dan [Sweet Dew Special Pill to Eliminate Toxin]
Gan Lu Yin [Sweet Dew Decoction]
Gan Sui Tong Jie Tang [Open the Blockage Decoction with Euphorbia]
Ge Gen Huang Qin Huang Lian Tang [Kudzu, Scutellaria and Coptis Decoction]
Ge Gen Tang [Kudzu Decoction]
Ge Xia Zhu Yu Tang [Drive Out Blood Stasis Below the Diaphragm Decoction]
Gong Wai Yun Er Hao Fang [No. 2 Extrauterine Pregnancy Formula]
Gong Wai Yun Yi Hao Fang [No. 1 Extrauterine Pregnancy Formula]
Gong Xiao He Jie Ruan Jian Tang [Attack, Reduce, Harmonize, Resolve, Soften Hardness Decoction]
Gou Chong Wan [Hook Worm Pill]
Gou Duan Luo Xuan Ti Bing Fang [Leptospirosis Formula]
Gou Teng Yin [Uncaria Decoction]
Gu Ben Zhi Beng San [Stabilize the Root and Stop Excessive Uterine Bleeding Powder]
Gu Jin Lu Yan Xu Ming Tang [Prolong Life Decoction] *Gu Jin Lu Yan*
Gu Jing Wan [Stabilize the Essence Pill] *Ji Sheng Fang*
Gu Jing Wan [Stabilize the Menses Pill] *Yi Xue Ru Men*
Gu Yin Jian [Stabilize the Yin Decoction]
Gua Lou Xie Bai Bai Jiu Tang [Trichosanthes Fruit, Chinese Chive, and Wine Decoction]
Gua Lou Xie Bai Ban Xia Tang [Trichosanthes Fruit, Chinese Chive and Pinellia Decoction]
Guan Xin Er Hao Fang [No.2 Formula for Coronary Artery Disease]
Guan Xin Su He Xiang Wan [Liquid Styrax Pill for Coronary Heart Conditions]

Gui Fu Er Chen Tang [Two-Cured Decoction with Cinnamon and Aconite]
Gui Fu Li Zhong Tang [Cinnamon and Prepared Aconite Decoction to Regulate the Middle]
Gui Fu Si Wu Tang [Four-Substance Decoction with Cinnamon and Prepared Aconite]
Gui Lu Er Xian Jiao [Tortoise Shell and Deer Antler Syrup]
Gui Pi Tang [Restore the Spleen Decoction]
Gui Qi Jian Zhong Tang [Tangkuei and Astragalus Decoction to Construct the Middle]
Gui Zhi Fu Ling Wan [Cinnamon Twigs and Poria Pill]
Gui Zhi Fu Zi Tang [Cinnamon Twig and Prepared Aconite Decoction]
Gui Zhi Jia Cang Er Xin Yi Tang [Cinnamon Twig Decoction plus Xanthium and Magnolia Flower]
Gui Zhi Jia Fu Zi Tang [Cinnamon Twig Decoction plus Prepared Aconite]
Gui Zhi Jia Ge Gen Tang [Cinnamon Twig Decoction plus Kudzu]
Gui Zhi Jia Gui Tang [Cinnamon Twig Decoction with Increased Dosage of Cinnamon Twig]
Gui Zhi Jia Hou Po Xing Zi Tang [Cinnamon Twig Decoction plus Magnolia Bark and Apricot Kernel]
Gui Zhi Jia Huang Qi Tang [Cinnamon Twig Decoction plus Astragalus]
Gui Zhi Jia Long Gu Mu Li Tang [Cinnamon Twig Decoction plus Dragon Bone and Oyster Shell]
Gui Zhi Jia Shao Yao Tang [Cinnamon Twig Decoction plus Peony]
Gui Zhi Shao Yao Zhi Mu Tang [Cinnamon Twig, Peony, and Anemarrhena Decoction]
Gui Zhi Tang [Cinnamon Twig Decoction]
Gun Tan Wan [Vaporize Phlegm Pill]
Guo Qi Yin [Delayed Menstruation Decoction]
Hai Ge Er Chen Tang [Two-Cured Decoction with Pumice and Clam Shell]
Hai Zao Yu Hu Tang [Sargassum Decoction for the Jade Flask]
Hao Qin Qing Dan Tang [Artemisia Annua and Scutellaria Decoction to Clear the Gallbaldder] °n
He Che Ba Wei Wan [Eight-Flavor Pill with Placenta]
He Wei Er Chen Tang [Two-Cured Decoction to Harmonize the Stomach]
He Xiong Huang San [Combined Ligusticum and Rhubarb Powder]
Hei Bai San [Black and White Powder]
Hei Xi Dan [Lead Special Pill]
Hei Xiao Yao San [Black Relax Wanderer Powder]
Hong Teng Yu Li Ren Jian [Millettia and Prune Seed Decoction]
Hou Po Ma Huang Tang [Magnolia Bark and Ephedra Decoction]
Hou Po Qi Wu Tang [Seven-Substance Decoction with Magnolia Bark]
Hou Po Wen Zhong Tang [Magnolia Bark Decoction to Warm the Middle]
Hou Po Yin [Magnolia Bark Decoction]
Hu Po Bao Long Wan [Hold the Dragon Pill with Amber]
Hu Qian Wan [Hidden Tiger Pill]
Hua Ban Tang [Transform Blotches Decoction]
Hua Chong Wan [Dissolve Parasites Pill]
Hua Gai San [Canopy Powder]
Hua Shi Huo Xiang Tang [Talcum and Agastache Decoction]
Hua Shi San [Transform the Stone Powder]
Huai Hua San [Sophora Japonica Flower Powder]
Huai Jiao Wan [Sophora Japonica Fruit Pill]
Huan Gan Li Pi Tang [Moderate the Liver, Regulate the Spleen Decoction]
Huang Lian E Jiao Tang [Coptis and Ass-Hide Gelatin Decoction]
Huang Lian Liu Jun Zi Tang [Six-Gentleman Decoction with Coptis]
Huang Lian Tang [Coptis Decoction]
Huang Lian Wen Dan Tang [Coptis Decoction to Warm the Gallbladder]
Huang Lian Xiang Ru Yin [Elsholtzia Decoction with Coptis]
Huang Lian Xiao Pi Wan [Reduce the Focal Distention Pill with Coptis]
Huang Long Tang [Yellow Dragon Decoction]
Huang Qi Gui Zhi Wu Wu Tang [Astragalus and Cinnamon Twig Five-Substance Decoction]
Huang Qi Jian Zhong Tang [Astragalus Decoction to Construct the Middle]
Huang Qin Hua Shi Tang [Scutellaria and Talcum Decoctionl]
Huang Qin Jia Ban Xia Sheng Jiang Tang [Scutellaria Decoction with Pinellia and Ginger]
Huang Qin Shao Yao Tang [Scutellaria and Peony Decoction]

Huang Qin Tang [Scutellaria Decoction]
Huang Tu Tang [Yellow Earth Decoction]
Hui Xiang San [Fennel Powder]
Hui Yang Ji Jiu Tang [Restore and Revive the Yang Decoction]
Huo Luo Dan [Invigorate the Collaterals Special Pill]
Huo Luo Xiao Ling Dan [Fantastically Effective Pill to Invigorate the Collaterals]
Huo Po Xia Ling Tang [Agastache, Magnolia Bark, Pinellia and Poria Decoction]
Huo Ren Cong Chi Tang [Scallion, and Prepared Soybean Decoction from the Book to Safeguard Life]
Huo Ren Ju Pi Zhu Ru Tang [Tangerine Peel and Bamboo Shaving Decoction from the Book to Safeguard Life]
Huo Xiang Zheng Qi San [Agastache Powder to Rectify the Qi]
Huo Xue Jiu [Invigorate the Blood Wine]
Huo Xue Shu Jin Tang [Invigorate the Blood, and Relax the Tendon Decoction]
Ji Chuan Jian [Benefit the River(Flow) Decoction]
Ji Jiao Li Huang Tang [Stephania, Zanthoxylum, Descurainia and Rhubarb Pill]
Ji Sheng Dang Gui Yang Rou Tang [Tangkuei and Mutton Decoction from the Formulas that Aid the Living]
Ji Sheng Shen Qi Wan [Kidney Qi Pill from Formulas that Aid the Living]
Ji Sheng Tong Mai Si Ni Tang [Unblock the Channel for Frigid Extremities Decoction from Formulas that Aid the Living]
Jia Jian Bai Tou Weng Tang [Modified Pulsatilla Decoction]
Jia Jian Ban Xia Xie Xin Tang [Modified Pinellia Decoction to Drain the Epigastrium]
Jia Jian Bu Zhong Yi Qi Tang [Modified Tonify the Middle and Augment Qi Decoction]
Jia Jian Cang Sha Yin [Modified Atractylodes and Cyperi Decoction]
Jia Jian Chai Ling Tang [Modified Bupleurum and Poria Decoction]
Jia Jian Da Chai Hu Tang [Modified Major Bupleurum Decoction]
Jia Jian Dan Dao Pai Shi Tang [Modified Expelling Stone from Biliary Tract Decoction]
Jia Jian Fu Mai Tang [Modified Restore the Pulse Decoction]
Jia Jian Gua Lou Xie Bai Tang [Modified Trichosanthes Fruit and Chinese Chive Decoction]
Jia Jian Huo Luo Xiao Ling Dan [Modified Fantastically Effective Pill to Invigorate the Collaterals]
Jia Jian Liang Di Tang [Modified Rehmannia and Lycium Root Bark Decoction]
Jia Jian Long Dan Xie Gan Tang [Modified Gentiana Longdancao Decoction to Drain the Liver]
Jia Jian Mu Fang Ji Tang [Modified Cocculus Decoction]
Jia Jian Niu Xi Tang [Modified Achyranthis Decoction]
Jia Jian Pu Ji Xiao Du Yin #1 [Modified Universal Benefit Decoction to Eliminate Toxin #1]
Jia Jian Pu Ji Xiao Du Yin #2 [Modified Universal Benefit Decoction to Eliminate Toxin #2]
Jia Jian Pu Ji Xiao Du Yin [Modified Universal Benefit Decoction to Eliminate Toxin]
Jia Jian Qing Hao Bie Jia Tang [Modified Artemisia Annua and Soft-shelled Turtle Shell Decoction]
Jia Jian Qing Jing Tang [Modified Clear the Menses Decoction]
Jia Jian Qing Wen Bai Du Yin [Modified Clear Epidemics and Overcome Toxin Decoction]
Jia Jian San Bi Tang [Modified Three Painful-Obstruction Decoction]
Jia Jian Shao Yao Tang [Modified Peony Decoction]
Jia Jian Shen Ling Bai Zhu San [Modified Ginseng, Poria, and Atractylodes Macrocephala Powder]
Jia Jian Sheng Mai San [Modified Generate the Pulse Powder]
Jia Jian Wan Dai Tang [Modified End Discharge Decoction]
Jia Jian Wei Rui Tang [Modified Polygonatum Odoratum]
Jia Jian Wu Pi Yin [Modified Five-Peel Decoction]
Jia Jian Xiao Chai Hu Tang [Modified Minor Bupleurum Decoction]
Jia Jian Xiao Yao San [Modified Relax Wanderer Powder]
Jia Jian Xiao Yao San [Modified Relax Wanderer Powder] *Nei Ke Zhai Yao*
Jia Jian Xie Xin Tang [Modified Drain the Epigastrium Decoction]
Jia Jian Yi Yin Jian [Modified One Yin Decoction]
Jia Jian Yin Chen Da Chai Hu Tang [Modified Major Bupleurum Decoction with Artemesia Yinchenhao]
Jia Jian Yin Qiao Bai Hu Tang [Modified Honeysuckle, Forsythia and White Tiger Decoction]
Jia Jian Yin Qiao San [Modified Honeysuckle and Forsythia Powder]

Jia Jian Yu Shi Qing Zao Jiu Fei Tang [Modified Yu's Eliminate Dryness and Rescue the Lung Decoction]
Jia Jian Zhu Ye Shi Gao Tang [Modified Lophatherus and Gypsum Decoction]
Jia Wei Bai Tou Weng Tang [Augmented Pulsatilla Decoction]
Jia Wei Bao He Wan [Augmented Reserve the Harmony Pill]
Jia Wei Cang Er Zi San [Augmented Xanthium Powder]
Jia Wei Chai Hu Shu Gan San [Augmented Bupleurum Powder to Spread the Liver]
Jia Wei Cui Sheng Xiong Gui Tang [Augmented Inducing Labor Decoction with Ligusticum and Tangkuei]
Jia Wei Da Chai Hu Tang [Augmented Major Bupleurum Decoction]
Jia Wei Dan Gui Shao Yao San [Augmented Tangkuei and Peony Powder]
Jia Wei Dan Zhi Xiao Yao San [Augmented Relax Wanderer Powder with Moutan and Gardenia]
Jia Wei Dang Gui Bu Xue Tang I [Augmented Tangkuei Decoction to Tonify Blood I]
Jia Wei Dang Gui Bu Xue Tang II [Augmented Tangkuei Decoction to Tonify Blood II]
Jia Wei Dao Chi San [Augmented Guide Out the Red Powder]
Jia Wei Dao Tan Tang [Augmented Guide Out Phlegm Decoction]
Jia Wei Er Chen Tang [Augmented Two-Cured Decoction]
Jia Wei Ge Gen Huang Qin Huang Lian Tang [Augmented Kudzu, Coptis, and Scutellaria Decoction]
Jia Wei Gui Zhi Tang [Augmented Cinnamon Twig Decoction]
Jia Wei Huang Lian Su Ye Tang [Augmented Coptis and Perilla Decoction]
Jia Wei Liu Jun Zi Tang [Augmented Six-Gentleman Decoction]
Jia Wei Long Dan Xie Gan Tang [Augmented Gentiana Longdancao Decoction to Drain the Liver]
Jia Wei Mai Men Dong Tang [Augmented Ophiopogonis Decoction]
Jia Wei Qian Zheng San [Augmented Lead to Symmetry Powder]
Jia Wei Qing Wei San [Augmented Clear the Stomach Powder]
Jia Wei Shao Gan Tang [Augmented Peony and Licorice Decoction]
Jia Wei Shen Qi Wan [Augmented Kidney Qi Pill]
Jia Wei Shi Xiao San [Augmented Sudden Smile Powder]
Jia Wei Si Qi Tang [Augmented Four Seven Decoction]
Jia Wei Si Wu Er Chen Tang [Augmented Four-Substance and Two-Cured Decoction]
Jia Wei Tian Ma Tang [Augmented Gastrodia Decoction]
Jia Wei Ting Li Da Zao Xie Fei Tang [Augmented Descurainia and Jujube Decoction to Drain the Lungs]
Jia Wei Wen Dan Tang [Augmented Warm the Gallbladder Decoction]
Jia Wei Wu Ling San [Augmented Five-Ingredient Powder with Poria]
Jia Wei Wu Yao Tang [Augmented Lindera Decoction]
Jia Wei Xiang Ru Tang [Augmented Elsholtzia Decoction]
Jia Wei Xiao Yao San [Augmented Relax Wanderer Powder]
Jia Wei Zhi Jing San [Augmented Stop Spasms Powder]
Jia Wei Zhi Zi Chi Tang [Augmented Gardenia and Prepared Soybean Decoction]
Jian Ling Tang [Construct Roof Tiles Decoction]
Jian Pi Wan [Strengthen the Spleen Pill]
Jian Wei Zhu Ye Shi Gao Tang [Reduced Lophatherus and Gypsum Decoction]
Jiang Ni Zhi E Tang [Direct the Rebellious Qi Downward and Stop Hiccup Decoction]
Jiao Ai Tang [Ass-Hide Gelatin and Mugwort Decoction] *Jin Gui Yao Lue*
Jiao Mei Tang [Pepper and Mume Decoction]
Jiao Tai Wan [Grand Communication Pill]
Jie Geng Tang [Platycodon Decoction]
Jie Gu Wan [Set the Bone Pill]
Jin Gui Shen Qi Wan [Kidney Qi Pill from the Golden Cabinet]
Jin Jian Qing Wei Yin [Clear the Stomach Decoction from Yi Zong Jin Jian]
Jin Ling San [Melia Toosendan Powder]
Jin Ling Zi San [Melia Toosendan Powder]
Jin San Bao Long Wan [Hold the Dragon Pill]
Jin She Zhi Jing San [Stop Spasms Powder with Viper]

Jin Suo Gu Jin Wan [Metal Lock Pills to Stabilize the Essence]
Jin Zhu Wan [Melia Toosendan and Evodia Pill]
Jing Fang Bai Du San [Schizonepeta and Ledebouriella Powder to Overcome Pathogenic Influences]
Jiu Ni Tang [Rescue the Critical Condition Decoction]
Jiu Wei Qiang Huo Tang [Nine-Herb Decoction with Notopterygium]
Jiu Xian San [Nine-Immortal Powder]
Jiu Zhi Er Chen Tang [Two-Cured Decoction with Chinese Chive Juice]
Ju Ban Gui Ling Zhi Jiang Tang [Tangerine, Pinellia, Cinnamon Twig, Immatured Bitter Orange, and Ginger Decoction]
Ju Ban Zhi Zhu Wan [Orange Peel, Pinellia, Immature Bitter Orange and Atractylodes Macrocephala Pill]
Ju Fang Wen Jing Tang [Warm the Menses Decoction from Imperial Grace Formula]
Ju Fang Xi Jiao Di Huang Wan [Rhinoceros Horn and Rehmannia Pill]
Ju He Wan [Tangerine Seed Pill]
Ju Hua Cha Tiao San [Chrysanthemum Powder to be Taken with Green Tea]
Ju Pi Tang [Tangerine Peel Decoction from the Golden Cabinet]
Ju Pi Zhu Ru Tang [Tangerine Peel and Bamboo Shaving Decoction]
Ju Pi Zhu Ru Tang [Tangerine Peel and Bamboo Shaving Decoction from Formulas that Aid the Living]
Ju Yuan Jian [Lift the Source Decoction]
Juan Bi Tang [Remove Painful Obstruction Decoction]
Juan Yin Liu Shen Tang [Six-Miracle Decoction to Expel Congested Fluid]
Kai Wei Jian Pi Wan [Stimulating Appetite and Strengthen the Spleen Pill]
Kang Bai Hou He Ji [Anti-Diphtheria Mixture]
Ke Xue Fang [Coughing of Blood Formula]
Kong Xian Dan [Control Mucous Special Pill]
Ku Shen Zhi Li Jian [Sophorae to Stop Diarrhea Decoction]
Lai Po Tong Jie Tang [Open the Blockage Decoction with Radish Seeds and Magnolia Bark]
Lan Wei Hua Yu Tang [Transform Blood Stasis Decoction for Appendicitis]
Lan Wei Qing Hua Tang [Clear and Transform the Appendicitis Decoction]
Lan Wei Qing Jie Tang [Clear and Relieve the Appendicitis Decoction]
Li Dan Wan [Promote Gallbladder Pill]
Li Ling Tang [Regulate the Middle Decoction with Five-Ingredient Powder with Poria]
Li Yu Tang [Carp Decoction]
Li Zhong An Hui Wan [Regulate the Middle and Calm Roundworms Pill]
Li Zhong Hua Tan Wan [Regulate the Middle and Transform Phlegm Pill]
Li Zhong Jiang Tan Wan [Regulate the Middle and Direct Phlegm Decoction]
Li Zhong Wan [Regulate the Middle Pill]
Lian Li Tang [Coptis Decoction to Regulate the Middle]
Lian Mei An Hui Tang [Picrorhizae and Mume Decoction to Calm Roundworms]
Lian Po Yin [Coptis and Magnolia Bark Decoction]
Lian Ru Er Chen Tang [Two-Cured Decoction with Coptis and Bamboo Shaving]
Lian Yu Li Zhong Tang [Coptis and Evodia Decoction to Regulate the Middle]
Liang Di Tang [Rehmannia and Lycium Root Bark Decoction]
Liang Fu Wan [Galangal and Cyperus Pill]
Liang Ge San [Cool the Diaphragm Powder]
Liang Ying Qing Qi Tang [Cool Ying and Clear the Qi Decoction]
Ling Gui Zhu Gan Tang [Poria, Cinnamon Twig, Atractylodes Macrocephalae, and Licorice Decoction]
Ling Ma Bai Hu Tang [Antelope, Gastrodia White Tiger Decoction]
Ling Yang Gou Teng Yin [Antelope Horn and Uncaria Decoction]
Liu He Tang [Harmonize the Six Decoction]
Liu Jun Zi Tang [Six-Gentleman Decoction]
Liu Mo Tang [Six Milled-Herb Decoction]
Liu Nao He Ji [Compound Mixture for Epidemic Encephalitis]
Liu Wei Di Huang Wan [Six-Ingredient Pill with Rehmannia]
Liu Yi San [Six-to-One Powder]
Long Dan Xie Gan Tang [Gentiana Longdancao Decoction to Drain the Liver]

Long Gu Wan [Dragon Bone Pill]
Long Hu Ji Yin Dan [Dragon and Tiger to Benefit Yin Special Pill]
Lou Bei Er Chen Tang [Two-Cured Decoction with Trichosanthis and Fritillaria]
Lu Jiao Tu Si Wan [Deer Antler and Cuscuta Seed Pill]
Ma Huang Fu Zi Gan Cao Tang [Ephedra, Prepared Aconite, and Licorice Decoction]
Ma Huang Fu Zi Xi Xin Tang [Ephedra, Asarum, and Prepared Aconite Decoction]
Ma Huang Jia Zhu Tang [Ephedra Decoction plus Atractylodes]
Ma Huang Lian Qiao Chi Xiao Dou Tang [Ephedra, Forsythia and Azuki Bean Decoction]
Ma Huang Tang [Ephedra Decoction]
Ma Xing Er Chen Tang [Two-Cured Decoction with Ephedra and Apricot Kernel]
Ma Xing Shi Gan Tang [Ephedra, Apricot Kernel, Gypsum and Licorice Decoction]
Ma Xing Yi Gan Tang [Ephedra, Apricot Kernel, Coicis, and Licorice Decoction]
Ma Zi Ren Wan [Hemp Seed Pill]
Mai Men Dong Tang [Ophiopogonis Decoction]
Mai Wei Di Huang Wan [Ophiopogonis, Schisandra and Rehmannia Pill]
Mei Ren Tang [Prune Seed Decoction]
Meng Shi Gun Tan Wan [Vaporize Phlegm Pill with Chlorite-Mica]
Mi Niao Dao Pai Shi Fang [Expel Stone from the Urinary Tract Formula]
Mi Yan Qing Wei San [Secret Clear the Stomach Powder from Experiences]
Miao Yin Wan [Beautiful Voice Pill]
Mu Fang Ji Jia Fu Ling Mang Xiao Tang [Cocculus with Mirabilitum and Poria Decoction]
Mu Fang Ji Tang [Cocculus Decoction]
Mu Li San [Oyster Shell Powder]
Mu Xiang Bing Lang Wan [Aucklandia and Betel Nut Pill]
Mu Xiang Wu Gong San [Aucklandia and Centipede Powder]
Nan Gua Zi Fen Bing Lang Jian [Pumpkin Seed Powder and Betel Nut Decoction]
Nei Bu Wan [Internal Tonify Pill]
Nian Tou San [Twist the Head Powder]
Niu Huang Bao Long Wan [Hold the Dragon Pill with Cattle Gallstone]
Niu Huang Bao Long Wan [Hold the Dragon Pill with Cattle Gallstone] *Ming Yi Za Zhu*
Niu Huang Qing Xin Wan [Cattle Gallstone Decoction to Clear the Heart]
Nuan Gan Jian [Warm the Liver Decoction]
Nuan Gong Wan [Mugwort and Cyperi Rhizome Pill for Warming the Womb]
Pai Shi Tang [Expelling Stone Decoction]
Pi Pa Qing Fei Yin [Loquat Leaf Decoction to Clear the Lung]
Pi Yue Ma Ren Wan [Constraint Spleen Hemp Seed Pill]
Ping Wei San [Calm and Stomach Powder]
Pu Ji Xiao Du Yin [Universal Benefit Decoction to Eliminate Toxin]
Qi Bao Mei Ran Dan [Seven-Treasure Special Pill for Beautiful Whiskers]
Qi Fu Tang [Astragalus and Prepared Aconite Decoction]
Qi Ju Di Huang Wan [Lycium Fruit, Chrysanthemum, and Rehmannia Pill]
Qi Li San [Seven-Thousandths of a Tael Powder]
Qi Pi Yin [Seven-Peel Decoction]
Qi Wei Bai Zhu San [Seven-Ingredient Actractylodes Macrocephala Decocotion]
Qi Wei Cong Bai Yin [Seven-Ingredient Scallion Decoction]
Qi Xiao Si Wu Tang [Miraculous Effective Four-Substance Deccotion]
Qian Jin Dang Bao Tang [Cleanse the Uterus Decoction from Thousand Ducat Formula]
Qian Jin Dang Gui Tang [Tangkuei Decoction from Thousand Ducat Formula]
Qian Jin Fu Shen Tang [Poria Cocos Pararadicis Decoction from Thousand Ducat Formula]
Qian Jin Gui Zhi Qu Shao Yao Jia Zao Jia Tang [Cinnamon Decoction minus Peony plus Gleditsiae from Thousang Ducat Formula]
Qian Jin Wei Jing Tang Jia Xing Ren Hua Shi Tang [Thousand Gold Reed Decoction with Apricot Kernel and Talcum Decoction]
Qian Zheng San [Lead to Symmetry Powder]
Qiang Huo Sheng Shi Tang [Notopterygium Decoction to Overcome Dampness]
Qin Jiao Bie Jia San [Gentiana Qinjiao and Soft-shelled Turtle Shell Powder]

Qin Lian Ju Ru Tang [Scutellaria, Coptis, Tangerine Peel and Bamboo Shaving Decoction]
Qin Lian Si Wu Tang [Four-Substance Decoction with Scutellaria and Coptis]
Qing Dan Li Shi Tang [Clear the Gallbladder and Drain Damp Decoction]
Qing Dan Xie Huo Tang [Clear the Gallbladder and Drain Fire Decoction]
Qing Dan Xing Qi Tang [Clear the Gallbladder and Move Qi Decoction]
Qing Dan Xuan Bi Tang [Clear the Gallbladder and Disband Painful Obstruction Decoction]
Qing E Wan [Young Maiden Pill]
Qing Fei Jie Du Tang [Clear the Lung and Relieve Toxicity Decoction]
Qing Fei Yin [Clear the Lung Decoction]
Qing Gan Da Yu Tang [Clear the Liver to Reach Stagnation Decoction]
Qing Gong Tang [Clear the Palace Decoction]
Qing Gu San [Cool the Bones Powder] *Zheng Zhi Zhun Sheng*
Qing Hao Bie Jia Tang [Artemisia Annua and Soft-shelled Turtle Shell Decoction]
Qing Jin Hua Tan Tang [Clear the Metal and Transform Phlegm Decoction]
Qing Jing Tang [Clear the Menses Decoction]
Qing Jing Zhi Xue Tang [Cool the Menses and Stop Bleeding Decoction]
Qing Pi Yin [Clear the Spleen Decoction]
Qing Qi Hua Tan Wan [Clear the Qi and Transform Phlegm Pill]
Qing Re Xi Feng Tang [Clear Heat and Extinguish Wind Decoction]
Qing Re Zhi Beng Tang [Clear Heat and Stop Excessive Uterine Bleeding Decoction]
Qing Re Zhi Dai Tang [Clear Heat and Stop Discharge Decoction]
Qing Shu Yi Qi Tang [Clear Summerheat and Augment Qi Decoction]
Qing Wei San [Clear the Stomach Powder]
Qing Wen Bai Du Yin [Clear Epidemics and Overcome Toxin Decoction]
Qing Xin Lian Zi Yin [Lotus Seed Decoction to Clear the Heart]
Qing Xin Liang Ge San [Clear the Heart and Cool the Diaphragm Powder]
Qing Yi Tang [Clear the Pancreas Decoction]
Qing Ying Tang [Clear the Nutritive Level Decoction]
Qing Zao Jiu Fei Tang [Eliminate Dryness and Rescue the Lungs Decoction]
Qing Zhen Tang [Clear the Rash Decoction]
Qing Zhong Tang [Clear the Middle Decoction]
Qu Hui Tang [Expel Roundworm Decoction]
Qu Mai Zhi Zhu Wan [Medicated Leaven, Barley Sprout, Immature Bitter Orange and Atractylodes Macrocephala Pill]
Quan Sheng Bai Zhu San [Atractylodes Macrocephala Powder] *Quan Sheng Ji*
Ren Ma Ping An San [Safe Powder for People and Horses]
Ren Shen Bai Du San [Ginseng Powder to Overcome Pathogenic Influences]
Ren Shen Ge Jie San [Ginseng and Gecko Powder]
Ren Shen Ge Jie San [Ginseng and Gecko Powder] *Wei Sheng Bao Jian*
Ren Shen Gu Ben Wan [Ginseng Pill to Reserve the Source]
Ren Shen Hu Tao Tang [Ginseng and Walnut Decoction]
Ren Shen Ju Pi Zhu Ru Tang [Ginseng, Tangerine Peel and Bamboo Shaving Decoction]
Ren Shen Xi Jiao Tang [Ginseng and Rhinoceros Horn Decoction]
Ren Shen Xie Xin Tang [Ginseng Decoction to Drain the Epigastrium]
Ren Shen Yang Ying Tang [Ginseng Decoction to Nourish the Nutritive Qi]
Ren Shen Zhu Ru Tang [Ginseng and Bamboo Shaving Decoction]
Ru Fang Zhang Tong Jing Yan Fang [Empirical Formula for Breast Distention and Pain]
Ru Xian Zeng Sheng Fang [Mastoplasia Formula]
Ruan Jian San Jie Tang [Soften the Hardness and Scatter the Accumulation Decoction]
Run Chang Wan [Moisten the Intestines Pill]
San Ao Tang [Three-Unbinding Decoction]
San Bi Tang [Three-Painful Obstruction Decoction]
San Ceng Hui Xiang Wan [Three Layers Fennel Pill]
San Hua Shen You Wan [Protected by Deities Pill with Three Flowers]
San Huang Si Wu Tang [Three-Yellow and Four-Substance Decoction]
San Huang Zhi Zhu Wan [Three-Yellow, Immature Bitter Orange and Atractylodes Macrocephala Pill]

San Jai Fu Mai Tang [Three-Shell Decoction to Restore the Pulse]
San Jia Jian Zheng Qi San [Third Modification of Rectify the Qi Powder]
San Miao Wan [Three-Marvel Pill]
San Ren Tang [Three-Nut Decoction]
San Wu Bei Ji Wan [Three-Substance Pill Prepared for Emergency]
San Xiang Tang [Three-Fragrant Decoction]
San Zi Yang Qin Tang [Three-Seed Decoction to Nourish One's Parents]
Sang Dan Xie Bai San [Mulberry Leaf, Moutan Powder to Drain the White]
Sang Dan Xie Bai Tang [Mulberry Leaf, Moutan Decoction to Drain the White]
Sang Ju Yin [Mulberry Leaf and Chrysanthemum Decoction]
Sang Ma Wan [Mulberry Leaf and Sesame Seed Pill]
Sang Piao Xiao San [Mantis Egg-Case Powder]
Sang Xing Tang [Mulberry Leaf and Apricot Kernel Decoction]
Sha Ban Li Zhong Tang [Amomum and Pinellia Decoction to Regulate the Middle]
Sha Lin Wan [Pill for Sandy Lin Syndrome]
Sha Shen Mai Dong Tang [Glenhnia and Ophiopogonis Decoction]
Shang Jiao Xuan Bi Tang [Disperse Painful Obstruction at the Upper Jiao Decoction]
Shao Fu Zhu Yu Tang [Drive Out Blood Stasis in the Lower Abdomen Decoction]
Shao Gan Zhi Jing Tang [Peony and Licorice Decoction to Stop Spasms]
Shao Yao Gan Cao Fu Zi Tang [Peony, Licorice, and Prepared Aconite Decoction]
Shao Yao Gan Cao Tang [Peony and Licorice Decoction]
Shao Yao Tang [Peony Decoction]
She Gan Ma Huang Tang [Belamcanda and Ephedra Decoction]
Shen Fu Long Mu Tang [Ginseng, Prepared Aconite, Dragon Bone and Oyster Shell Decoction]
Shen Fu Tang [Ginseng and Prepared Aconite Decoction]
Shen Ling Bai Zhu San [Ginseng, Poria, and Atractylodes Macrocephala Powder]
Shen Su San [Ginseng and Perilla Leaf Powder]
Shen Su Yin [Ginseng and Perilla Leaf Decoction]
Shen Tong Zhu Yu Tang [Drive Out Blood Stasis from a Painful Body Decoction]
Shen Xi Dan [Magical Rhinoceros Special Pill]
Shen Ying San [Magical Response Powder]
Shen Zhu San [Marvelous Atractylodis Powder]
Shen Zhuo Tang [Fixed at Kidney Decoction]
Sheng Chai Liu Jun Zi Tang [Six-Gentleman Decoction with Cimicifuga and Bupleurum]
Sheng Hua Tang [Generation and Transformation Decoction]
Sheng Hua Tong Jing Tang [Generation and Transformation Decoction to Unblock the Menses]
Sheng Jiang Ban Xia Tang [Fresh Ginger and Pinellia Decoction]
Sheng Jiang San [Ascending Descending Powder]
Sheng Jiang Xie Xin Tang [Fresh Ginger Decoction to Drain the Epigastrium]
Sheng Ma Huang Qi Tang [Cimicifuga and Astragalus Decoction]
Sheng Mai San [Generate the Pulse Powder]
Sheng Pu Huang Tang [Fresh Cattail Pollen Decoction]
Sheng Tie Luo Yin [Iron Filings Decoction]
Sheng Xian Tang [Raise the Sinking Decoction]
Sheng Yu Tang [Sage-Like Healing Decoction]
Shi Bu Wan [Ten-Tonic Pill]
Shi Di Tang [Persimmon Calyx Decoction]
Shi Hu Ye Guang Wan [Dendrobium Pill for Night Vision]
Shi Hui San [Ten Partially-Charred Substance Powder]
Shi Jue Ming San [Haliotidis Shell Powder]
Shi Pi San [Bolster the Spleen Powder]
Shi Pi Yin [Bolster the Spleen Decoction]
Shi Qian San [Persimmon Coin Powder]
Shi Quan Da Bu Tang [All-Inclusive Great Tonifying Decoction]
Shi Quan Ku Han Jiu Bu Tang [All-Inclusive Bitter and Cold Decoction to Rescue and Tonify]
Shi Wei San [Pyrrosiae Powder]

Shi Wei San [Pyrrosiae Powder] *Pu Ji Fang*
Shi Wei San [Pyrrosiae Powder] *Zheng Zhi Hui Bu*
Shi Wei Wen Dan Tang [Ten-Ingredient Decoction to Warm the Gallbladder]
Shi Wei Xiang Ru Yin [Ten-Ingredient Elsholtzia Decoction]
Shi Xiao San [Sudden Smile Powder]
Shi Zao Tang [Ten-Jujube Decoction]
Shou Se Zhi Dai Tang [Astringe and Stop Discharge Decoction]
Shu Gan Jie Yu Tang [Spread the Liver and Resolve the Stagnation Decoction]
Shu Jin Huo Luo Tang [Relax the Tendon and Invigorate Collaterals Decoction]
Shu Yu Qing Gan Tang [Spread the Depression and Clear the Liver Decoction]
Shu Yu Qing Gan Yin [Spread the Depression and Clear the Liver Beverage]
Shu Zao Yin Zi [Unblock and Dig Formula]
Shui Lu Er Xian Dan [Water and Earth Immortals Special Pill]
Shun Qi San [Smooth the Qi Powder]
Si Jia Jian Zheng Qi San [Fourth Modification of Rectify the Qi Powder]
Si Jun Zi Tang [Four-Gentleman Decoction]
Si Ling San [Four-Ingredient Powder with Poria]
Si Ling San Jia Mu Gua Cao Guo Ban Xia Tang [Four-Ingredient Powder with Poria plus Chaenomelis Fruit, Amomi Tsaoko and Pinellia]
Si Ling Wan [Four Order Pill]
Si Miao Tang [Four-Marvel Decoction]
Si Miao Wan [Four-Marvel Pill]
Si Miao Yong An Tang [Four-Valiant Decoction for Well-being]
Si Mo Tang [Four Milled-Herb Decoction]
Si Ni Jia Ren Shen Tang [Ginseng Decoction for Frigid Extremities]
Si Ni San [Frigid Extremities Powder]
Si Ni Tang [Frigid Extremities Decoction]
Si Qi Tang [Four Seven Decoction]
Si Shen Wan [Four-Miracle Pill]
Si Shen Wan [Four-Miracle Pill] *Zheng Zhi Zhun Sheng*
Si Sheng Wan [Four-Fresh Pill]
Si Shou Yin [Four Beasts Decoction]
Si Wei Xiang Ru Yin [Four-Ingredient Elsholtzia Decoction]
Si Wu Tang [Four-Substance with Lidera Decoction]
Si Wu Tang [Fou-Substance Decoction]
Su He Xiang Wan [Liquid Styrax Pill]
Su Ting Ma Xing Shi Gan Tang [Perilla Fruit, Lepidii Seeds, Ephedra, Apricot Kernel, Gypsum, and Licorice Decoction]
Su Xing Er Chen Tang [Two-Cured Decoction with Perilla and Apricot Kernel]
Su Zi Jiang Qi Tang [Perilla Fruit Decoction for Directing Qi Downward]
Suan Zao Ren Tang [Sour Jujube Decoction]
Suo Quan Wan [Shut the Sluice Pill]
Tao He Cheng Qi Tang [Peach Pit Decoction to Order the Qi]
Tao Hong Si Wu Tang [Four-Substance Decoction with Peach Pit and Safflower]
Tao Hua Tang [Peach Blossom Decoction]
Tao Ren Cheng Qi Tang [Order of Qi Decoction with Peach Kernel]
Tian Tai Wu Yao San [Top-Quality Lindera Powder]
Tian Wang Bu Xin Dan [Emperor of Heaven's Special Pill to Tonify the Heart]
Tiao Wei Cheng Qi Tang [Regulate the Stomach and Order the Qi Decoction]
Ting Li Da Zao Xie Fei Tang [Descurainia and Jujube Decoction to Drain the Lungs]
Tong Jing Huo Luo Tang [Unblock the Channel and Invigorate the Collateral Decoction]
Tong Jing Zhu Yu Tang [Drive Out Blood Stasis and Unblock the Channel Decoction]
Tong Mai Si Ni Jia Zhu Dan Zhi Tang [Unblock the Pulse Decoction with Pig Bile for Frigid Extremities]
Tong Mai Si Ni Tang [Unblock the Pulse Decoction for Frigid Extremities]
Tong Qiao Huo Xue Tang [Unblock the Orifices and Invigorate the Blood Decoction]

Tong Ru San Jie Tang [Unblock Breast Milk and Scatter Accumulations Decoction]
Tong Xie Yao Fang [Important Formula for Painful Diarrhea]
Tong Yu Jian [Unblock the Stasis Decoction]
Tu Si Zi Wan [Cuscuta Seed Pill]
Tuo Hua Jian [Release from Flower Decoction]
Wan Dai Tang [End Discharge Decoction]
Wei Jing Tang [Reed Decoction]
Wei Ling Tang [Calm the Stomach and Poria Decoction]
Wen Dan Tang [Warm the Gallbladder Decoction]
Wen Jing She Xue Tang [Warm the Menses, Control the Blood Decoction]
Wen Jing Tang [Warm the Menses Decoction]
Wen Pi San [Warm the Spleen Powder]
Wen Pi Tang [Warm the Spleen Decoction] *Ben Shi Fang*
Wen Pi Tang [Warm the Spleen Decoction] *Qian Jin Yao Fang*
Wu Hu Zhui Feng Tang [Five-Tiger Decoction to Pursue Wind]
Wu Ji Wan [Fifth and Sixth Heavenly Stem Pill]
Wu Ji Wan [Infinity Pill]
Wu Jia Jian Zheng Qi San [Fifth Modification of Rectify the Qi Powder]
Wu Lin San [Five-Ingredient Powder for Painful Urinary Dysfunction]
Wu Ling San [Five-Ingredient Powder with Poria]
Wu Mei Wan [Mume Pill]
Wu Mo Tang [Five Milled-Herb Decoction]
Wu Pi San [Five-Peel Powder]
Wu Pi Yin [Five-Peel Decoction]
Wu Ren Wan [Five-Seed Pill]
Wu Tou Tang [Aconite Decoction]
Wu Wei Mai Dong Tang [Five-Ingredient Ophiopogonis Decoction]
Wu Wei Xiao Du Yin [Five-Ingredient Decoction to Eliminate Toxin]
Wu Wei Zi Tang [Schisandra Decoction]
Wu Xie Liu Jun Zi Tang [Six-Gentleman Decoction with Aconite and Scorpion]
Wu Xie Si Jun Zi Tang [Four-Gentleman Decoction with Aconite and Scorpion]
Wu Yao Tang [Lindera Decoction]
Wu Yu Mu Gua Tang [Evodia and Chinese Quince Fruit Decoction]
Wu Zhi Yin [Five-Juice Decoction]
Wu Zhu Yu Jia Fu Zi Tang [Evodia Decoction with Prepared Aconite]
Wu Zhu Yu Tang [Evodia Decoction]
Xi Di Qing Luo Yin [Clear the Collaterals Decoction with Rhinoceros Horn and Rehmannia]
Xi Huang Wan [Rhinoceros Gallstone Pill]
Xi Jiao Di Huang He Xie Xin Tang [Rhinoceros Horn and Rehmannia, and Drain the Epigastrium Decoction]
Xi Jiao Di Huang Tang [Rhinoceros Horn and Rehmannia Decoction]
Xi Ling Bai Hu Tang [Rhinoceros Horn, Antelope and White Tiger Decoction]
Xia Si Tai Fang [Purge the Dead Fetus Formula]
Xia Yu Xue Tang [Purge the Stagnant Blood Decoction]
Xian Fang Huo Ming Yin [Sublime Formula for Sustaining Life]
Zhen Ren Huo Ming Yin [True Man's Formula for Sustaining Life]
Xian Xiong Cheng Qi Tang [Sinking into the Chest and Order of Qi Decoction]
Xiang Fu Xuan Fu Hua Tang [Cyperus and Inula Decoction]
Xiang Lian Wan [Aucklandia and Coptis Pill]
Xiang Ru San [Elsholtzia Powder]
Xiang Sha Liu Jun Zi Tang [Six-Gentleman Decoction with Aucklandia and Amomum]
Xiang Sha Ping Wei San [Calm the Stomach Powder with Aucklandia and Amomum]
Xiang Sha Zhi Zhu Wan [Aucklandia, Amomum, Immature Bitter Orange and Atractylodes Macrocephala Pill]
Xiang Su Cong Chi Tang [Cyperum, Perilla Leaf, Scallion, and Prepared Soybean Decoction]
Xiang Su San [Cyperum and Perilla Leaf Decoction]

Xiao Ban Qing Dai Yin [Indigo Decoction to Clear Rash]
Xiao Ban Xia Jia Fu Ling Tang [Minor Pinellia Decoction with Poria]
Xiao Ban Xia Tang [Minor Pinellia Decoction]
Xiao Chai Hu Tang [Minor Bupleurum Decoction]
Xiao Cheng Qi Tang [Minor Order of Qi Decoction]
Xiao Ding Feng Zhu [Minor Arrest the Wind Pearl]
Xiao Er Hui Chun Dan [Return of Spring Special Pill]
Xiao Feng San [Eliminate Wind Powder] *Zhang Shi Yi Tong*
Xiao Feng San [Eliminate Wind Powder]
Xiao Fu Tang [Mirabilitum and Turnip Juice Mixture]
Xiao Huang Bai Du San [Mirabilitum and Rhubarb Powder to Overcome Pathogenic Influences]
Xiao Huo Luo Dan [Minor Invigorate the Collaterals Special Pill]
Xiao Ji Yin Zi [Cephalanoplos Decoction] *Ji Sheng Fang*
Xiao Jian Zhong Tang [Minor Construct the Middle Decoction]
Xiao Pi Wan [Reduce the Focal Distention Pill]
Xiao Pin Sheng Di Huang Tang [Small Rehmannia Decoction]
Xiao Qing Long Jia Shi Gao Tang [Minor Bluegreen Dragon Decoction with Gypsum]
Xiao Qing Long Tang [Minor Bluegreen Dragon Decoction]
Xiao Shi Fan Shi San [Niter and Alum Powder]
Xiao Shui San [Reduce the Water Powder]
Xiao Xian Xiong Tang [Minor Sinking into the Chest Decoction]
Xiao Xu Ming Tang [Minor Prolong Life Decoction]
Xiao Yao San [Relax Wanderer Powder]
Xiao Ying Jian [Minor Nourishing Decoction]
Xie Bai San [Drain the White Powder]
Xie Qing Wan [Drain the Green Pill]
Xie Xin Dao Zhi Tang [Drain the Epigastrium and Guide Out the Red Decoction]
Xin Ding Da Yuan Yin [Newly-Settled Reach the Membrane Source Decoction]
Xin Jia Ba Wei Di Huang Wan [Newly-Augmented Eight-Ingredient Pill with Rehmannia]
Xin Jia Bai Hu Tang [Newly Augmented White Tiger Decoction]
Xin Jia Huang Long Tang [Newly Augmented Yellow Dragon Decoction]
Xin Jia San Ao Tang [Newly Augmented Three-Unbinding Decoction]
Xin Jia Xiang Ru Yin [Newly Augmented Elsholtzia Decoction]
Xin Jia Yu Nu Jian [Newly Augmented Jade Woman Decoction]
Xin Pi Shuang Bu Wan [Tonify Both Heart and Spleen Pill]
Xin Shen Jiao Tai Tang [Grand Communication Decoction for Heart and Kidney]
Xin Zhi Ju Pi Zhu Ru Tang [Newly Formulated Tangerine Peel and Bamboo Shaving Decoction]
Xing Fu Liu Jun Zi Tang [Six-Gentleman Decoction with Arisaemae and Typhonii]
Xing Jun San [Marching Powder]
Xing Pi San [Awaken the Spleen Powder]
Xing Ren Hua Shi Tang [Apricot Kernel and Talcum Decoction]
Xing Su San [Apricot Kernel and Perilla Leaf Decoction]
Xing Xiao Wan [Awaken and Eliminate Pill]
Xiong Gui Er Chen Tang [Two–Cured Decoction with Ligusticum Chuanxiong and Tangkuei]
Xiong Zhi Shi Gao Tang [Ligusticum Chuanxiong, Angelica Dahurica, and Gypsum Decoction]
Xiong Zhu Wan [Ligusticum and Atractylodes Macrocephala Pill] *Dan Xi Xin Fa*
Xuan Bai Cheng Qi Tang [Order of Qi Decoction to Diffuse the White]
Xuan Bi Tang [Disband Painful Obstruction Decoction]
Xuan Fu Dai Zhe Tang [Inula and Haematite Decoction]
Xuan Fu Hua Tang [Inula Decoction]
Xuan Hu Ku Liang Tang [Corydalis and Bitter Melia Toosendan Decoction]
Xuan Hu Tang [Corydalis Decoction]
Xuan Qing Dao Zhuo Tang [Disperse, Clear and Guide the Turbid Decoction]
Xuan Tou Mo Yuan Fa [Disperse and Vent the Membrane Source Formula]
Xue Fu Zhu Yu Tang [Drive Out Stasis in the Mansion of Blood Decoction]
Yan Hu Suo San [Corydalis Powder]

Yang He Tang [Yang-Heartening Decoction]
Yang Xin Tang [Nourish the Heart Decoction]
Yang Yin Qing Fei Tang [Nourish the Yin and Clear the Lungs Decoction]
Yang Zang Tang [Nourishing the Organs Decoction]
Ye Shi Yi Wei Tang [Ye's Benefit the Stomach Decoction]
Yi Gong San [Extraordinary Merit Powder]
Yi Guan Jian [Linking Decoction]
Yi Huang Tang [Change Yellow Decoction]
Yi Jia Fu Mai Tang [One-Shell Decoction to Restore the Pulse]
Yi Jia Jian Zheng Qi San [First Modification of Rectify the Qi Powder]
Yi Nian Jin [One Pinch of Gold Decoction]
Yi Qing Wan [Suppress the Green Pill]
Yi Wei Tang [Benefit the Stomach Decoction]
Yi Yi Zhu Ye San [Coicis and Lophatherus Powder]
Yin Chen Hao Tang [Artemisia Yinchenhao Decoction]
Yin Chen Si Ni Tang [Artemisia Yinchenhao Decoction for Frigid Extremities]
Yin Chen Wei Ling Tang [Calm the Stomach and Poria Decoction with Artemisia Yinchenhao]
Yin Chen Wu Ling San [Artemisia Yinchenhao and Five-Ingredient Powder with Poria]
Yin Chen Zhu Fu Tang [Arstemisia Yinchenhao, Atractylodes and Prepared Aconite Decoction]
Yin Hua Jie Du Tang [Honeysuckle Decoction to Relieve Toxicity]
Yin Jia Wan [Silver Armour Pill]
Yin Qiao Bai Du San [Honeysuckle and Forsythia Powder to Overcome Pathogenic Influences]
Yin Qiao Bai Hu Tang [Honeysuckle, Forsythia and White Tiger Decoction]
Yin Qiao Ma Bo San [Honeysuckle, Forsythia, and Puffball Powder]
Yin Qiao Ma Xing Shi Gan Tang [Honeysuckle, Forsythia, Ephedra, Apricot Kernel, Gypsum, and Licorice Decoction]
Yin Qiao San [Honeysuckle and Forsythia Powder]
You Gui Wan [Restore the Right (Kidney) Pill]
You Gui Yin [Restore the Right (Kidney) Decoction]
Yu Dai Wan [Cure Discharge Pill]
Yu Ping Feng San [Jade Windscreen Powder]
Yu Zhen San [True Jade Powder]
Yu Zhu San [Jade Candle Decoction]
Yuan Zhi Wan [Polygalae Pill]
Yue Bi Jia Ban Xia Tang [Maidservant from Yue Decoction plus Pinellia]
Yue Bi Jia Zhu Fu Tang [Maidservant from Yue Decoction plus Atractylodes and Prepared Aconite]
Yue Bi Jia Zhu Tang [Maidservant from Yue Decoction plus Atractylodes]
Yue Bi Tang [Maidservant from Yue Decoction]
Yue Hua Wan [Moonlight Pill]
Yue Ju Wan [Escape Restraint Pill]
Zan Yu Dan [Special Pill to Aid Fertility]
Zao Fan Wan [Red Date and Alum Pill]
Zao Li Er Chen Tang [Two-Cured Decoction with Gleditsia and Bamboo Sap]
Zao Shi Wan [Chisel the Stone Pill]
Zeng Ye Cheng Qi Tang [Increase the Fluids and Order the Qi Decoction]
Zeng Ye Tang [Increase the Fluids Decoction]
Zha Qu Er Chen Tang [Two-Cured Decoction with Hawthorn and Medicated Leaven]
Zha Qu Liu Jun Zi Tang [Six-Gentleman Decoction with Hawthorn Berry and Medicated Leaven]
Zha Qu Ping Wei San [Calm the Stomach Powder with Hawthorn and Fermented Leaven]
Zhen Gan Xi Feng Tang [Sedate the Liver and Extinguish the Wind Decoction]
Zhen Gan Xi Feng Tang Jia Jian [Modified Sedate the Liver Extinguish Wind Decoction]
Zhen Ni Bai Hu Tang [White Tiger Decoction to Suppress Rebellion]
Zhen Wu Tang [True Warrior Decoction]
Zhen Zhu Mu Wan [Mother of Pearl Pill]
Zhen Zhu Mu Wan [True Mother of Pearl Pill]
Zheng Gu Zi Jin Dan [Purple Gold Pill for Correcting the Bone]

Zheng She San [Correct the Tongue Powder]
Zhi Bai Ba Wei Wan [Eight-Ingredient Pill with Anemarrhena and Phellodendron]
Zhi Bai Di Huang Wan [Anemarrhena, Phellodendron and Rehmannia Pill]
Zhi Bao Dan [Greatest Treasure Special Pill]
Zhi Beng Zheng Ji Yan Fang [Extreme Effective Formula in Treating Beng Syndromes]
Zhi Gan Cao Tang [Honey-Fried Licorice Decoction]
Zhi Jing San [Stop Spasms Powder]
Zhi Mi Fu Ling Wan [Poria Pill from *Zhi Mi Fang*]
Zhi Qian Lie Xian Yan Fang [Formula that Treats Prostatitis]
Zhi Ru Yong Yan Fang [Empirical Formula in Treating Mastitis]
Zhi Shi Dao Zhi Wan [Immature Bitter Orange Pill to Guide the Stagnation]
Zhi Shi Li Zhong Tang [Immature Bitter Orange Decoction to Regulate the Middle]
Zhi Shi Shao Yao San [Immature Bitter Orange and Peony Powder]
Zhi Shi Xiao Pi Wan [Immature Bitter Orange Pill to Reduce Focal Distention]
Zhi Shi Xie Bai Gui Zhi Tang [Immature Bitter Orange, Chinese Chive, and Cinnamon Twig Decoction]
Zhi Suo San [Stop Coughing Powder]
Zhi Xue San [Stop Bleeding Powder]
Zhi Yu Wan [Gardenia and Evodia Pill]
Zhi Zhong Tang [Treats the Middle Decoction]
Zhi Zhu Pi Wei San [Calm the Stomach Powder with Immature Bitter Orange and Atractylodes Macrocephala]
Zhi Zhu Wan [Immature Bitter Orange and Atractylodes Macrocephala Pill]
Zhi Zi Bai Pi Tang [Gardenia, Phellodendron Decoction]
Zhi Zi Chi Tang [Gardenia and Prepared Soybean Decoction]
Zhi Zi Da Huang Tang [Gardenia and Rhubarb Decoction]
Zhi Zi Gan Cao Chi Tang [Gardenia, Licorice, and Prepared Soybean Decoction]
Zhi Zi Gan Jiang Tang [Gardenia and Ginger Decoction]
Zhi Zi Hou Po Tang [Gardenia and Magnolia Bark Decoction]
Zhi Zi Sheng Jiang Chi Tang [Gardenia, Fresh Ginger and Prepared Soybean Decoction]
Zhong He Tang [Middle-Heartening Decoction]
Zhou Che Wan [Vessel and Vehicle Pill]
Zhu Che Wan [Stationed Vehicle Pill]
Zhu Fu Tang [Atractylodes and Prepared Aconite Decoction]]
Zhu Jing Wan [Preserve Vistas Pill]
Zhu Li Yin Zi [Bamboo Sap Decoction]
Zhu Ling Tang [Polyporus Decoction]
Zhu Mai Liu Jun Zi Tang [Six-Gentleman Decoction with Lophatherus and Ophiopogonis]
Zhu Ru Tang [Bamboo Shaving Decoction]
Zhu Sha An Shen Wan [Cinnabar Pill to Calm the Spirit]
Zhu Ye Liu Bang Tang [Lophatherus, Willow and Burdock Seeds Decoction]
Zhu Ye Shi Gao Tang [Lophatherus and Gypsum Decoction]
Zi Long Wan [Purple Dragon Pill]
Zi Shen Tong Guan Wan [Enrich the Kidneys and Open the Gates Pill]
Zi Shui Qing Gan Yin [Nourish Water Clear the Liver Decoction]
Zi Su San [Perilla Powder]
Zi Xue Dan [Purple Snow Special Pill]
Zuo Gui Wan [Restore the Left (Kidney) Pill] *Jing Yue Quan Shu*
Zuo Gui Yin [Restore the Left (Kidney) Decoction]
Zuo Jin Wan [Left Metal Pill]

APPENDIX 11

Cross-Reference English–Pinyin Herbal Formulas

Aconite Decoction [Wu Tou Tang]
Adapted Pulsatilla Decoction [Bian Tong Bai Tou Weng Tang]
Adapted Ten Flavored Warm the Gallbladder Decoction [Bian Tong Shi Wei Wen Dan Tang]
Agastache Powder to Rectify the Qi [Huo Xiang Zheng Qi San]
Agastache, Magnolia Bark, Pinellia and Poria Decoction [Huo Po Xia Ling Tang]
Ailanhti Bark Pill [Chun Gen Pi Wan]
All-Inclusive Bitter and Cold Decoction to Rescue and Tonify [Shi Quan Ku Han Jiu Bu Tang]
All-Inclusive Great Tonifying Decoction [Shi Quan Da Bu Tang]
Amomum and Pinellia Decoction to Regulate the Middle [Sha Ban Li Zhong Tang]
Anemarrhena, Phellodendron and Rehmannia Pill [Zhi Bai Di Huang Wan]
Angelica Pubescens and Sangjisheng Decoction [Du Huo Ji Sheng Tang]
Antelope Horn and Uncaria Decoction [Ling Yang Gou Teng Yin]
Antelope, Gastrodia White Tiger Decoction [Ling Ma Bai Hu Tang]
Anti-Diphtheria Mixture [Kang Bai Hou He Ji]
Apricot Kernel and Perilla Leaf Decoction [Xing Su San]
Apricot Kernel and Talcum Decoction [Xing Ren Hua Shi Tang]
Aquilaria Powder [Chen Xiang San]
Aquilaria to Direct Qi Downward Powder [Chen Xiang Jiang Qi San]
Arrest Wheezing Decoction [Ding Chuan Tang]
Artemisia Annua and Scutellaria Decoction to Clear the Gallbladder [Hao Qin Qing Dan Tang]
Artemisia Annua and Soft-Shelled Turtle Shell Decoction [Qing Hao Bie Jia Tang]
Artemisia Yinchenhao and Five-Ingredient Powder with Poria [Yin Chen Wu Ling San]
Artemisia Yinchenhao Decoction [Yin Chen Hao Tang]
Artemisia Yinchenhao Decoction for Frigid Extremities [Yin Chen Si Ni Tang]
Artemisia Yinchenhao, Atractylodes and Prepared Aconite Decoction [Yin Chen Zhu Fu Tang]
Ascending Descending Powder [Sheng Jiang San]
Ass-Hide Gelatin and Chicken Egg Yolk Decoction [E Jiao Ji Zi Huang Tang] *Guang Wen Re Lun*
Ass-Hide Gelatin and Coptis Decoction [E Jiao Huang Lian Tang]
Ass-Hide Gelatin and Egg Yolk Decoction [E Jiao Ji Zi Huang Tang] *Tong Su Shang Han Lun*
Ass-Hide Gelatin and Mugwort Decoction [Jiao Ai Tang]
Assist Spleen and Spread the Liver Decoction [Fu Pi Shu Gan Tang]
Assist with Mulberry Leaf Pill [Fu Sang Wan]
Astragalus and Cinnamon Twig Five-Substance Decoction [Huang Qi Gui Zhi Wu Wu Tang]
Astragalus and Prepared Aconite Decoction [Qi Fu Tang]
Astragalus Decoction to Construct the Middle [Huang Qi Jian Zhong Tang]
Astringe and Stop Discharge Decoction [Shou Se Zhi Dai Tang]

Atractylodes and Prepared Aconite Decoction [Bai Zhu Fu Zi Tang]
Atractylodes and Prepared Aconite Decoction [Zhu Fu Tang]
Atractylodes Macrocephala Powder [Quan Sheng Bai Zhu San] *Quan Sheng Ji*
Atractylodes Powder [Bai Zhu San]
Atractylodes White Tiger Decoction [Cang Zhu Bai Hu Tang]
Attack, Reduce, Harmonize, Resolve, Soften Hardness Decoction [Gong Xiao He Jie Ruan Jian Tang]
Aucklandia and Betel Nut Pill [Mu Xiang Bing Lang Wan]
Aucklandia and Centipede Powder [Mu Xiang Wu Gong San]
Aucklandia and Coptis Pill [Xiang Lian Wan]
Aucklandia, Amomum, Immature Bitter Orange and Atractylodes Macrocephala Pill [Xiang Sha Zhi Zhu Wan]
Augmented Bupleurum Powder to Spread the Liver [Jia Wei Chai Hu Shu Gan San]
Augmented Cinnamon Twig Decoction [Jia Wei Gui Zhi Tang]
Augmented Clear the Stomach Powder [Jia Wei Qing Wei San]
Augmented Coptis and Perilla Decoction [Jia Wei Huang Lian Su Ye Tang]
Augmented Descurainia and Jujube Decoction to Drain the Lungs [Jia Wei Ting Li Da Zao Xie Fei Tang]
Augmented Elsholtzia Decoction [Jia Wei Xiang Ru Tang]
Augmented Five-Ingredient Powder with Poria [Jia Wei Wu Ling San]
Augmented Four Seven Decoction [Jia Wei Si Qi Tang]
Augmented Four-Substance and Two-Cured Decoction [Jia Wei Si Wu Er Chen Tang]
Augmented Gardenia and Prepared Soybean Decoction [Jia Wei Zhi Zi Chi Tang]
Augmented Gastrodia Decoction [Jia Wei Tian Ma Tang]
Augmented Gentiana Longdancao Decoction to Drain the Liver [Jia Wei Long Dan Xie Gan Tang]
Augmented Guide Out the Red Powder [Jia Wei Dao Chi San]
Augmented Guide Out the Red Powder [Jai Wei Dao Tan Tang]
Augmented Inducing Labor Decoction with Ligusticum and Tangkuei [Jia Wei Cui Sheng Xiong Gui Tang]
Augmented Kidney Qi Pill [Jia Wei Shen Qi Wan]
Augmented Kudzu, Coptis, and Scutellaria Decoction [Jia Wei Ge Gen Huang Qin Huang Lian Tang]
Augmented Lead to Symmetry Powder [Jia Wei Qian Zheng San]
Augmented Lindera Decoction [Jia Wei Wu Yao Tang]
Augmented Major Bupleurum Decoction [Jia Wei Da Chai Hu Tang]
Augmented Ophiopogonis Decoction [Jia Wei Mai Men Dong Tang]
Augmented Peony and Licorice Decoction [Jia Wei Shao Gan Tang]
Augmented Pulsatilla Decoction [Jia Wei Bai Tou Weng Tang]
Augmented Relax Wanderer Powder [Jia Wei Xiao Yao San]
Augmented Relax Wanderer Powder with Moutan and Gardenia [Jia Wei Dan Zhi Xiao Yao San]
Augmented Reserve the Harmony Pill [Jia Wei Bao He Wan]
Augmented Six-Gentleman Decoction [Jia Wei Liu Jun Zi Tang]
Augmented Stop Spasms Powder [Jia Wei Zhi Jing San]
Augmented Sudden Smile Powder [Jia Wei Shi Xiao San]
Augmented Tangkuei and Peony Powder [Jia Wei Dan Gui Shao Yao San]
Augmented Tangkuei Decoction to Tonify Blood I [Jia Wei Dang Gui Bu Xue Tang I]
Augmented Tangkuei Decoction to Tonify Blood II [Jia Wei Dang Gui Bu Xue Tang II]
Augmented Two-Cured Decoction [Jia Wei Er Chen Tang]
Augmented Warm the Gallbladder Decoction [Jia Wei Wen Dan Tang]
Augmented Xanthium Powder [Jia Wei Can Er Zi San]
Awaken and Eliminate Pill [Xing Xiao Wan]
Awaken the Spleen Powder [Xing Pi San]
Bamboo Sap Decoction [Zhu Li Yin Zi]
Bamboo Shaving Decoction [Zhu Ru Tang]
Beautiful Voice Pill [Miao Yin Wan]
Belamcanda and Ephedra Decoction [She Gan Ma Huang Tang]

Benefit the River [Flow] Decoction [Ji Chuan Jian]
Benefit the Stomach Decoction [Yi Wei Tang]
Big Tangerine Peel Decoction [Da Ju Pi Tang] Wai Tai Mi Yao
Biota Leaves Decoction [Bai Ye Tang]
Biota Seed Pill to Nourish the Heart [Bai Zi Yang Xin Wan]
Black and White Powder [Hei Bai San]
Black Relax Wanderer Powder [Hei Xiao Yao San]
Bletilla and Loquat Leaf Pill [Bai Ji Pi Pa Wan]
Bolster the Spleen Decoction [Shi Pi Yin]
Bolster the Spleen Powder [Shi Pi San]
Bupleurum and Calm the Stomach Decoction [Chai Ping Tang]
Bupleurum and Cinnamon Twig Decoction] [Chai Hu Gui Zhi Tang]
Bupleurum and Four-Substance Decoction [Chai Hu Si Wu Tang]
Bupleurum Decoction for Sinking into Chest [Chai Hu Xian Xiong Tang]
Bupleurum Decoction plus Mirabilite [Chai Hu Jia Mang Xiao Tang]
Bupleurum Decoction plus Reed [Chai Hu Jia Lu Gen Tang]
Bupleurum Decoction to Reach the Membrane Source [Chai Hu Da Yuan Yin]
Bupleurum plus Dragon Bone and Oyster Shell Decoction [Chai Hu Jia Long Gu Mu Li Tang]
Bupleurum Powder to Spread the Liver [Chai Hu Shu Gan San]
Bupleurum Powder to Spread the Liver [Chai Hu Shu Gan San] *Jing Yue Quan Shu*
Bupleurum White Tiger Decoction [Chai Hu Bai Hu Tang]
Bupleurum, Bitter Orange, and Platycodon Decoction [Chai Hu Zhi Jie Tang]
Bupleurum, Cinnamon and Ginger Decoction [Chai Hu Gui Jiang Tang]
Calm the Palace Pill with Cattle Gallstone [An Gong Niu Huang Wan]
Calm the Spirit and Settle the Will Pill [An Shen Ding Zhi Wan]
Calm the Spirit Pill [An Shen Wan]
Calm the Stomach and Poria Decoction [Wei Ling Tang]
Calm the Stomach and Poria Decoction with Artemisia Yinchenhao [Yin Chen Wei Ling Tang]
Calm the Stomach Powder [Ping Wei San]
Calm the Stomach Powder with Aucklandia and Amomum [Xiang Sha Ping Wei San]
Calm the Stomach Powder with Hawthorn and Medicated Leaven [Zha Qu Ping Wei San]
Calm the Stomach Powder with Immature Bitter Orange and Atractylodes Macrocephala [Zhi Zhu Pi Wei San]
Canopy Powder [Hua Gai San]
Capital Qi Pill [Du Qi Wan]
Carp Decoction [Li Yu Tang]
Cattle Gallstone Decoction to Clear the Heart [Niu Huang Qing Xin Wan]
Cephalanoplos Decoction [Xiao Ji Yin Zi]
Change Yellow Decoction [Yi Huang Tang]
Chisel the Stone Pill [Zao Shi Wan]
Chrysanthemum Powder to be Taken with Green Tea [Ju Hua Cha Tiao San]
Cimicifuga and Astragalus Decoction [Sheng Ma Huang Qi Tang]
Cinnabar Pill to Calm the Spirit [Zhu Sha An Shen Wan]
Cinnamon and Prepared Aconite Decoction to Regulate the Middle [Gui Fu Li Zhong Tang]
Cinnamon Decoction minus Peony plus Gleditsiae from Thousand Ducat Formula [Qian Jin Gui Zhi Qu Shao Yao Jia Zao Jia Tang]
Cinnamon Twig and Prepared Aconite Decoction [Gui Zhi Fu Zi Tang]
Cinnamon Twig Decoction [Gui Zhi Tang]
Cinnamon Twig Decoction plus Astragalus [Gui Zhi Jia Huang Qi Tang]
Cinnamon Twig Decoction plus Dragon Bone and Oyster Shell [Gui Zhi Jia Long Gu Mu Li Tang]
Cinnamon Twig Decoction plus Kudzu [Gui Zhi Jia Ge Gen Tang]
Cinnamon Twig Decoction plus Magnolia Bark and Apricot Kernel [Gui Zhi Jia Hou Po Xing Zi Tang]
Cinnamon Twig Decoction plus Peony [Gui Zhi Jia Shao Yao Tang]
Cinnamon Twig Decoction plus Prepared Aconite [Gui Zhi Jia Fu Zi Tang]

Cinnamon Twig Decoction Plus Xanthium and Magnolia Flower [Gui Zhi Jia Cang Er Xin Yi Tang]
Cinnamon Twig Decoction with Increased Dosage of Cinnamon Twig [Gui Zhi Jia Gui Tang]
Cinnamon Twig, Peony, and Anemarrhena Decoction [Gui Zhi Shao Yao Zhi Mu Tang]
Cinnamon Twigs and Poria Pill [Gui Zhi Fu Ling Wan]
Cleanse the Uterus Decoction from Thousand Ducat Formula [Qian Jin Dang Bao Tang]
Clear and Relieve the Appendicitis Decoction [Lan Wei Qing Jie Tang]
Clear and Transform the Appendicitis Decoction [Lan Wei Qing Hua Tang]
Clear Epidemics and Overcome Toxin Decoction [Qing Wen Bai Du Yin]
Clear Heat and Extiguish Wind Decoction [Qing Re Xi Feng Tang]
Clear Heat and Stop Discharge Decoction [Qing Re Zhi Dai Tang]
Clear Heat and Stop Excessive Uterine Bleeding Decoction [Qing Re Zhi Beng Tang]
Clear Summerheat and Augment Qi Decoction [Qing Shu Yi Qi Tang]
Clear the Collaterals Decoction with Rhinoceros Horn and Rehmannia [Xi Di Qing Luo Yi]
Clear the Gallbladder and Disband Painful Obstruction Decoction [Qing Dan Xuan Bi Tang]
Clear the Gallbladder and Drain Damp Decoction [Qing Dan Li Shi Tang]
Clear the Gallbladder and Drain Fire Decoction [Qing Dan Xie Huo Tang]
Clear the Gallbladder and Move Qi Decoction [Qing Dan Xing Qi Tang]
Clear the Heart and Cool the Diaphragm Powder [Qing Xin Liang Ge San]
Clear the Liver to Reach Stagnation Decoction [Qing Gan Da Yu Tang]
Clear the Lung and Relieve Toxicity Decoction [Qing Fei Jie Du Tang]
Clear the Lung Decoction [Qing Fei Yin]
Clear the Menses Decoction [Qing Jing Tang]
Clear the Metal and Transform Phlegm Decoction [Qing Jin Hua Tan Tang]
Clear the Middle Decoction [Qing Zhong Tang]
Clear the Nutritive Level Decoction [Qing Ying Tang]
Clear the Palace Decoction [Qing Gong Tang]
Clear the Pancreas Decoction [Qing Yi Tang]
Clear the Qi and Transform Phlegm Pill [Qing Qi Hua Tan Wan]
Clear the Rash Decoction [Qing Zhen Tang]
Clear the Spleen Decoction [Qing Pi Yin]
Clear the Stomach Decoction from Yi Zong Jin Jian [Jin Jian Qing Wei Yin]
Clear the Stomach Powder [Qing Wei San]
Cloth Sack Pill [Bu Dai Wan]
Clove and Evodia Decoction to Regulate the Middle [Ding Yu Li Zhong Tang]
Clove and Persimmon Calyx Decoction [Ding Xiang Shi Di Tang]
Clove Powder of Bian Que [Bian Que Ding Xiang San]
Clove, Ass-Hide Gelatin, and Mugwort Decoction [Ding Xiang Jiao Ai Tang]
Clove, Persimmon Calyx and Bamboo Shaving Decoction [Ding Xiang Shi Di Zhu Ru Tang]
Cocculus Decoction [Mu Fang Ji Tang]
Cocculus with Mirabilitum and Poria Decoction [Mu Fang Ji Jia Fu Ling Mang Xiao Tang]
Coicis and Lophatherus Powder [Yi Yi Zhu Ye San]
Combined Ligusticum and Rhubarb Powder [He Xiong Huang San]
Compound Mixture for Epidemic Encephalitis [Liu Nao He Ji]
Constraint Spleen Hemp Seed Pill [Pi Yue Ma Ren Wan]
Construct Roof Tiles Decoction [Jian Ling Tang]
Control Mucus Special Pill [Kong Xian Dan]
Cool the Bones Powder [Qing Gu San]
Cool the Diaphragm Powder [Liang Ge San]
Cool the Menses and Stop Bleeding Decoction [Qing Jing Zhi Xue Tang]
Cool Ying and Clear the Qi Decoction [Liang Ying Qing Qi Tang]
Coptis and Ass-Hide Gelatin Decoction [Huang Lian E Jiao Tang]
Coptis and Evodia Decoction to Regulate the Middle [Lian Yu Li Zhong Tang]
Coptis and Magnolia Bark Decoction [Lian Po Yin]
Coptis Decoction [Huang Lian Tang]
Coptis Decoction to Regulate the Middle [Lian Li Tang]

Coptis Decoction to Warm the Gallbladder [Huang Lian Wen Dan Tang]
Correct the Tongue Powder [Zheng She San]
Corydalis and Bitter Melia Toosendan Decoction [Xuan Hu Ku Liang Tang]
Corydalis Decoction [Xuan Hu Tang]
Corydalis Powder [Yan Hu Suo San]
Coughing of Blood Formula [Ke Xue Fang]
Cure Discharge Pill [Yu Dai Wan]
Cuscuta Seed Pill [Tu Si Zi Wan]
Cyperum and Perilla Leaf Decoction [Xiang Su San]
Cyperus and Inula Decoction [Xiang Fu Xuan Fu Hua Tang]
Cyperus, Perilla Leaf, Scallion, and Prepared Soybean Decoction [Xiang Su Cong Chi Tang]
Decocted Turtle Shell Pill [Bie Jia Jian Wan]
Deer Antler and Cuscuta Seed Pill [Lu Jiao Tu Si Wan]
Delayed Menstruation Decoction [Guo Qi Yin]
Dendrobium Pill for Night Vision [Shi Hu Ye Guang Wan]
Descurainia and Jujube Decoction to Drain the Lungs [Ting Li Da Zao Xie Fei Tang]
Dioscorea Hypoglauca Decoction to Separate the Clear [Bei Xie Fen Qing Yin]
Dioscorea Hypoglauca Decoction to Separate the Clear from *Dan Xi Xin Fa* [Dan Xi Bei Xie Fen Qing Yin]
Direct the Rebellious Qi Downward and Stop Hiccup Decoction [Jiang Ni Zhi E Tang]
Disband Painful Obstruction Decoction [Xuan Bi Tang]
Disperse and Vent the Membrane Source Formula [Xuan Tou Mo Yuan Fa]
Disperse Painful Obstruction at the Upper Jiao Decoction [Shang Jiao Xuan Bi Tang]
Disperse, Clear and Guide the Turbid Decoction [Xuan Qing Dao Zhuo Tang]
Dissolve Parasites Pill [Hua Chong Wan]
Dmelia Toosendan and Evodia Pill [Jin Zhu Wan]
Dragon and Tiger to Benefit Yin Special Pill [Long Hu Ji Yin Dan]
Dragon Bone Pill [Long Gu Wan]
Drain the Epigastrium and Guide Out the Red Decoction [Xie Xin Dao Zhi Tang]
Drain the Green Pill [Xie Qing Wan]
Drain the White Powder [Xie Bai San]
Drive Out Blood Stasis and Unblock the Channel Decoction [Tong Jing Zhu Yu Tang]
Drive Out Blood Stasis below the Diaphragm Decoction [Ge Xia Zhu Yu Tang]
Drive Out Blood Stasis from a Painful Body Decoction [Shen Tong Zhu Yu Tang]
Drive Out Blood Stasis in the Lower Abdomen Decoction [Shao Fu Zhu Yu Tang]
Drive Out Stasis in the Mansion of Blood Decoction [Xue Fu Zhu Yu Tang]
Eight-Flavor Pill with Placenta [He Che Ba Wei Wan]
Eight-Herb Decoction for Rectification and Relieving Toxicity [Ba Zheng Jie Du Tang]
Eight-Herb Powder for Rectification [Ba Zheng San]
Eight-Immortal Pill for Longevity [Ba Xian Chang Shou Wan]
Eight-Ingredient Pill with Anemarrhena and Phellodendron [Zhi Bai Ba Wei Wan]
Eight-Thousandths of a Tael Powder [Ba Li San]
Eight-Treasure Decoction [Ba Zhen Tang]
Eliminate Dryness and Rescue the Lungs Decoction [Qing Zao Jiu Fei Tang]
Eliminate Wind Powder [Xiao Feng San]
Eliminate Wind Powder from *Zhang Shi Yi Tong* [Xiao Feng San]
Elsholtzia Decoction with Coptis [Huang Lian Xiang Ru Yin]
Elsholtzia Powder [Xiang Ru San]
Emperor of Heaven's Special Pill to Tonify the Heart [Tian Wang Bu Xin Dan]
Empirical Formula for Breast Distention and Pain [Ru Fang Zhang Tong Jing Yan Fang]
Empirical Formula in Treating Mastitis [Zhi Ru Yong Yan Fang]
End Discharge Decoction [Wan Dai Tang]
Enrich the Kidney and Open the Gates Pill [Zi Shen Tong Guan Wan]
Ephedra Decoction [Ma Huang Tang]
Ephedra Decoction plus Atractylodes [Ma Huang Jia Zhu Tang]

Ephedra, Apricot Kernel, Coicis, and Licorice Decoction [Ma Xing Yi Gan Tang]
Ephedra, Apricot Kernel, Gypsum and Licorice Decoction [Ma Xing Shi Gan Tang]
Ephedra, Asarum, and Prepared Aconite Decoction [Ma Huang Fu Zi Xi Xin Tang]
Ephedra, Forsythia and Azuki Bean Decoction [Ma Huang Lian Qiao Chi Xiao Dou Tang]
Ephedra, Prepared Aconite and Licorice Decoction [Ma Huang Fu Zi Gan Cao Tang]
Escape Restraint Pill [Yue Ju Wan]
Evodia and Chinese Quince Fruit Decoction [Wu Yu Mu Gua Tang]
Evodia Decoction [Wu Zhu Yu Tang]
Evodia Decoction with Prepared Aconite [Wu Zhu Y Jia Fu Zi Tang]
Expel Roundworm Decoction [Qu Hui Tang]
Expel Stone from the Urinary Tract Formula [Mi Niao Dao Pai Shi Fang]
Expelling Stone Decoction [Pai Shi Tang]
Expelling Stones from Biliary Tract Decoction [Da Dao Pai Shi Tang]
Extraordinary Merit Powder [Yi Gong San]
Extraordinary Merit Powder with Ass-Hide Gelatin [E Jiao Yi Gong San]
Extreme Effective Formula in Treating Beng Syndrome [Zhi Beng Zheng Ji Yan Fang]
Fantastically Effective Pill to Invigorate the Collaterals [Huo Luo Xiao Ling Dan]
Fennel Powder [Hui Xiang San]
Fifth and Sixth Heavenly Stem Pill [Wu Ji Wan]
Fifth Modification of Rectify the Qi Powder [Wu Jia Jian Zheng Qi San]
First Modification of Rectify the Qi Powder [Yi Jia Jian Zheng Qi San]
Five Milled-Herb Decoction [Wu Mo Tang]
Five-Ingredient Decoction to Eliminate Toxin [Wu Wei Xiao Du Yin]
Five-Ingredient Ophiopogonis Decoction [Wu Wei Mai Dong Tang]
Five-Ingredient Powder for Painful Urinary Dysfunction [Wu Lin San]
Five-Ingredient Powder with Poria [Wu Ling San]
Five-Juice Decoction [Wu Zhi Yin]
Five-Peel Decoction [Wu Pi Yin]
Five-Peel Powder [Wu Pi San]
Five-Seed Pill [Wu Ren Wan]
Five-Tiger Decoction to Pursue Wind [Wu Hu Zhui Feng Tang]
Fixed at Kidney Decoction [Shen Zhuo Tang]
Formula for Stones in the Hepatic Duct and Biliary Duct [Gan Dan Guan Jie Shi Fang]
Formula that Treats Prostatitis [Zhi Qian Lie Xian Yan Fang]
Four Beasts Decoction [Si Shou Yin]
Four Milled-Herb Decoction [Si Mo Tang]
Four Order Pill [Si Ling Wan]
Four Seven Decoction [Si Qi Tang]
Four-Gentleman Decoction [Si Jun Zi Tang]
Four-Gentleman Decoction with Aconite and Scorpion [Wu Xie Si Jun Zi Tang]
Four-Ingredient Elsholtzia Decoction [Si Wei Xiang Ru Yin]
Four-Ingredient Powder with Poria [Si Ling San]
Four-Ingredient Powder with Poria plus Chaenomelis Fruit, Amomi Tsaoko and Pinellia [Si Ling San Jia Mu Gua Cao Guo Ban Xia Tang]
Four-Marvel Decoction [Si Miao Tang]
Four-Marvel Pill [Si Miao Wan]
Four-Miracle Pill [Si Shen Wan]
Four-Miracle Pill [Si Shen Wan]
Four-Miracle Pill [Si Sheng Wan]
Four-Substance Decoction [Si Wu Tang]
Four-Substance Decoction with Cinnamon and Prepared Aconite [Gui Fu Si Wu Tang]
Four-Substance Decoction with Peach Pit and Safflower [Tao Hong Si Wu Tang]
Four-Substance Decoction with Scutellaria and Coptis [Qin Lian Si Wu Tang]
Four-Substance with Lidera Decoction [Si Wu Tang]
Fourth Modification of Rectify the Qi Powder [Si Jia Jian Zheng Qi San]

Four-Valiant Decoction for Well-Being [Si Miao Yong An Tang]
Fresh Cattail Pollen Decoction [Sheng Pu Huang Tang]
Fresh Ginger and Pinellia Decoction [Sheng Jiang Ban Xia Tang]
Fresh Ginger Decoction to Drain the Epigastrium [Sheng Jiang Xie Xin Tang]
Frigid Extremities Decoction [Si Ni Tang]
Frigid Extremities Powder [Si Ni San]
Fritillaria and Trichosanthes Fruit Powder [Bei Mu Gua Lou San]
Galangal and Cyperus Pill [Liang Fu Wan]
Gardenia and Evodia Pill [Zhi Yu Wan]
Gardenia and Ginger Decoction [Zhi Zi Gan Jiang Tang]
Gardenia and Magnolia Bark Decoction [Zhi Zi Hou Po Tang]
Gardenia and Prepared Soybean Decoction [Zhi Zi Chi Tang]
Gardenia and Rhubarb Decoction [Zhi Zi Da Huang Tang]
Gardenia, Fresh Ginger and Prepared Soybean Decoction [Zhi Zi Sheng Jiang Chi Tang]
Gardenia, Licorice and Prepared Soybean Decoction [Zhi Zi Gan Cao Chi Tang]
Gardenia, Phellodendron Decoction [Zhi Zi Bai Pi Tang]
Generate the Pulse Powder [Sheng Mai San]
Generation and Transformation Decoction [Sheng Hua Tang]
Generation and Transformation Decoction to Unblock the Menses [Sheng Hua Tong Jing Tang]
Gentiana Longdancao Decoction to Drain the Liver [Long Dan Xie Gan Tang]
Gentiana Longdancao Decoction to Drain the Liver by Li Dong Yuan [Dong Yuan Long Dan Xie Gan Tang]
Gentiana Qinjiao and Soft-shelled Turtle Shell Powder [Qin Jiao Bie Jia San]
Ginger, Ginseng and Pinellia Pill [Gan Jiang Ren Shen Ban Xia Wan]
Ginseng and Bamboo Shaving Decoction [Ren She Zhu Ru Tang]
Ginseng and Gecko Powder [Ren Shen Ge Jie San]
Ginseng and Gecko Powder [Ren Shen Ge Jie San] *Wei Sheng Bao Jian*
Ginseng and Perilla Leaf Decoction [Shen Su Yin]
Ginseng and Perilla Leaf Powder [Shen Su San]
Ginseng and Prepared Aconite Decoction [Shen Fu Tang]
Ginseng and Rhinoceros Horn Decoction [Ren Shen Xi Jiao Tang]
Ginseng and Walnut Decoction [Ren Shen Hu Tao Tang]
Ginseng Decoction for Frigid Extremities [Si Ni Jia Ren Shen Tang]
Ginseng Decoction to Drain the Epigastrium [Ren Shen Xie Xin Tang]
Ginseng Decoction to Nourish the Nutritive Qi [Ren Shen Yang Ying Tang]
Ginseng Pill to Reserve the Source [Ren Shen Gu Ben Wan]
Ginseng Powder to Overcome Pathogenic Influences [Ren Shen Bai Du San]
Ginseng, Poria, and Atractylodes Macrocephala Powder [Shen Ling Bai Zhu San]
Ginseng, Prepared Aconite, Dragon Bone and Oyster Shell Decoction [Shen Fu Long Mu Tang]
Ginseng, Tangerine Peel and Bamboo Shaving Decoction [Ren Shen Ju Pi Zhu Ru Tang]
Glenhnia and Ophiopogonis Decoction [Sha Shen Mai Dong Tang]
Grand Communication Decoction for Heart and Kidney [Xin Shen Jiao Tai Tang]
Grand Communication Pill [Jiao Tai Wan]
Great Creation Pill [Da Zao Wan]
Great Creation Pill [Da Zao Wan] *Fu Shou Jing Fang*
Great Peace Pill [Da An Wan]
Great Tonify Pill [Da Bu Wan]
Great Tonify the Yin Pill [Da Bu Yin Wan]
Greatest Treasure Special Pill [Zhi Bao Dan]
Guide Out Phlegm Decoction [Dao Tan Tang]
Guide Out the Red and Augmented Polygoni Cuspidati Powder [Dao Chi San He Jia Wei Hu Zhang San]
Guide Out the Red Powder [Dao Chi San]
Guide the Qi Decoction [Dao Qi Tang]
Haliotidis Shell Powder [Shi Jue Ming San]

Halloysitum and Limonite Decoction [Chi Shi Zhi Yu Yu Liang Tang]
Harmonize the Six Decoction [Liu He Tang]
Hemp Seed Pill [Ma Zi Ren Wan]
Hidden Tiger Pill [Hu Qian Wan]
Hold the Dragon Pill [Bao Long Wan] *He Ji Ju Fang*
Hold the Dragon Pill [Bao Long Wan] *Xiao Er Yao Zheng Zhi Jue*
Hold the Dragon Pill [Jin San Bao Long Wan]
Hold the Dragon Pill with Amber [Hu Po Bao Long Wan]
Hold the Dragon Pill with Cattle Gallstone [Niu Huang Bao Long Wan]
Hold the Dragon Pill with Cattle Gallstone [Niu Huang Bao Long Wan] *Ming Yi Za Zhu*
Honey-Fried Licorice Decoction [Zhi Gan Cao Tang]
Honeysuckle and Forsythia and White Tiger Decoction [Yin Qiao Bai Hu Tang]
Honeysuckle and Forsythia Powder [Yin Qiao San]
Honeysuckle and Forsythia Powder to Overcome Pathogenic Influences [Yin Qiao Bai Du San]
Honeysuckle Decoction to Relieve Toxicity [Yin Hua Jie Du Tang]
Honeysuckle, Forsythia, and Puffball Powder [Yin Qiao Ma Bo San]
Honeysuckle, Forsythia, Ephedra, Apricot Kernel, Gypsum and Licorice Decoction [Yin Qiao Ma Xing Shi Gan Tang]
Hook Worm Pill [Gou Chong Wan]
Immature Bitter Orange and Atractylodes Macrocephala Pill [Zhi Zhu Wan]
Immature Bitter Orange and Peony Powder [Zhi Shi Shao Yao San]
Immature Bitter Orange Decoction to Regulate the Middle [Zhi Shi Li Zhong Tang]
Immature Bitter Orange Pill to Guide the Stagnation [Zhi Shi Dao Zhi Wan]
Immature Bitter Orange Pill to Reduce Focal Distention [Zhi Shi Xiao Pi Wan]
Immature Bitter Orange, Chinese Chive, and Cinnamon Twig Decoction [Zhi Shi Xie Bai Gui Zhi Tang]
Important Formula for Painful Diarrhea [Tong Xie Yao Fang]
Increase the Fluids and Order the Qi Decoction [Zeng Ye Cheng Qi Tang]
Increase the Fluids Decoction [Zeng Ye Tang]
Indigo Decoction to Clear Rash [Xiao Ban Qing Dai Yin]
Induce Labor Decoction [Cui Sheng Yin]
Infinity Pill [Wu Ji Wan]
Internal Tonify Pill [Nei Bu Wan]
Inula and Haematite Decoction [Xuan Fu Dai Zhe Tang]
Inula Decoction [Xuan Fu Hua Tang]
Invigorate the Blood Wine [Huo Xue Jiu]
Invigorate the Blood, and Relax the Tendon Decoction [Huo Xue Shu Jin Tang]
Invigorate the Collaterals Special Pill [Huo Luo Dan]
Iron Filings Decoction [Sheng Tie Luo Yin]
Jade Candle Decoction [Yu Zhu San]
Jade Windscreen Powder [Yu Ping Feng San]
Jasper Powder [Bi Yu San]
Kidney Qi Pill from Formulas that Aid the Living [Ji Sheng Shen Qi Wan]
Kidney Qi Pill from the Golden Cabinet [Jin Gui Shen Qi Wan]
Kudzu Decoction [Ge Gen Tang]
Kudzu, Scutellaria and Coptis Decoction [Ge Gen Huang Qin Huang Lian Tang]
Lead Special Pill [Hei Xi Dan]
Lead to Symmetry Powder [Qian Zheng San]
Ledebouriella Powder that Sagely Unblocks [Fang Feng Tong Sheng San]
Left Metal Pill [Zuo Jin Wan]
Leptospirosis Formula [Gou Duan Luo Xuan Ti Bing Fang]
Licorice and Ginger Decoction [Gan Cao Gan Jiang Tang]
Licorice Decoction to Drain the Epigastrium [Gan Cao Xie Xin Tang]
Licorice, Ginger, Poria and Atractylodes Macrocephalae Decoction] [Gan Cao Gan Jiang Fu Ling Bai Zhu Tang]

Licorice, Lead and Honey Decoction [Gan Cao Fen Mi Tang]
Lift the Source Decoction [Ju Yuan Jian]
Ligusicum Chuanxiong, Angelica Dahurica, and Gypsum Decoction [Xiong Zhi Shi Gao Tang]
Ligusticum and Atractylodes Macrocephala Pill [Xiong Zhu Wan] *Dan Xi Xin Fa*
Lily Bulb Decoction to Preserve the Metal [Bai He Gu Jin Tang]
LiLy Bulb Table [Bai He Pian]
Lindera Decoction [Wu Yao Tang]
Linking Decoction [Yi Guan Jian]
Liquid Styrax Pill [Su He Xiang Wan]
Liquid Styrax Pill for Coronary Heart Conditions [Guan Xin Su He Xiang Wan]
Liqusticum Chuanxiong Powder to be Taken with Green Tea [Chuan Xiong Cha Tiao San]
Lophatherus and Gypsum Decoction [Zhu Ye Shi Gao Tang]
Lophatherus, Willow and Burdock Seeds Decoction [Zhu Ye Liu Bang Tang]
Loquat Leaf Decoction to Clear the Lung [Pi Pa Qing Fei Yin]
Lotus Seed Decoction to Clear the Heart [Qing Xin Lian Zi Yin]
Lycium Fruit, Chrysanthemum, and Rehmannia Pill [Qi Ju Di Huang Wan]
Lycium Root Bark Decoction [Di Gu Pi Yin]
Lycium Root Bark to Nourish Yin Decoction [Di Gu Yang Yin Jian]
Magical Response Powder [Shen Ying San]
Magical Rhinoceros Special Pill [Shen Xi Dan]
Magnolia Bark and Ephedra Decoction [Hou Pu Ma Huang Tang]
Magnolia Bark Decoction [Hou Po Yin]
Magnolia Bark Decoction to Warm the Middle [Hou Po Wen Zhong Tang]
Maidservant from Yue Decoction [Yue Bi Tang]
Maidservant from Yue Decoction plus Atractylodes [Yue Bi Jia Zhu Tang]
Maidservant from Yue Decoction plus Atractylodes and Prepared Aconite [Yue Bi Jia Zhu Fu Tang]
Maidservant from Yue Decoction plus Pinellia [Yue Bi Jia Ban Xia Tang]
Major Arrest Wind Pearl [Da Ding Feng Zhu]
Major Ass-Hide Gelatin and Mugwort Decoction [Da Jiao Ai Tang]
Major Bluegreen Dragon Decoction [Da Qing Long Tang]
Major Bupleurum Decoction [Da Chai Hu Tang]
Major Bupleurum Decoction for Cholecystitis [Dan Nang Yan Da Chai Hu Tang]
Major Construct the Middle Decoction [Da Jian Zhong Tang]
Major Gentiana Qinjiao Decoction [Da Qin Jiao Tang]
Major Invigorate the Collaterals Pill [Da Huo Luo Dan]
Major Order the Qi Decoction [Da Cheng Qi Tang]
Major Pinellia Decoction [Da Ban Xia Tang]
Major Sinking into the Chest Decoction [Da Xian Xiong Tang]
Major Sinking into the Chest Pill [Da Xian Xiong Wan]
Mantis Egg-Case Powder [Sang Piao Xiao San]
Marching Powder [Xing Jun San]
Marvelous Atractylodis Powder [Shen Zhu San]
Mastoplasia Formula [Ru Xian Zeng Sheng Fang
Medicated Leaven, Barley Sprout, Immature Bitter Orange and Atractylodes Macrocephala Pill [Qu Mai Zhi Zhu Wan]
Melia Toosendan Powder [Jin Ling San]
Melia Toosendan Powder [Jin Ling Zi San]
Metal Lock Pills to Stabilize the Essence [Jin Suo Gu Jin Wan]
Middle-Heartening Decoction [Zhong He Tang]
Millettia and Prune Seed Decoction [Hong Teng Yu Li Ren Jian]
Minor Arrest the Wind Pearl [Xiao Ding Feng Zhu]
Minor Bluegreen Dragon Decoction [Xiao Qing Long Tang]
Minor Bluegreen Dragon Decoction with Gypsum [Xiao Qing Long Dan Tang]
Minor Bupleurum Decoction [Xiao Chai Hu Tang]
Minor Construct the Middle [Xiao Jian Zhong Tang]

Minor Invigorate the Collaterals Special Pill [Xiao Huo Luo Dan]
Minor Nourishing Decoction [Xiao Ying Jian]
Minor Order of Qi Decoction [Xiao Cheng Qi Tang]
Minor Pinellia Decoction [Xiao Ban Xia Tang]
Minor Pinellia Decoction with Poria [Xiao Ban Xia Jia Fu Ling Tang]
Minor Prolong Life Decoction [Xiao Xu Ming Tang]
Minor Sinking into the Chest Decoction [Xiao Xian Xiong Tang]
Mirabilitum and Rhubarb Powder to Overcome Pathogenic Influences [Xiao Huang Bai Du San]
Mirabilitum and Turnip Juice Mixture [Xiao Fu Tang]
Miraculous Effective Four-Substance Decoction [Qi Xiao Si Wu Tang]
Moderate the Liver, Regulate the Spleen Decoction [Huan Gan Li Pi Tang]
Modfied Honeysuckle, Forsythia and White Tiger Decoction [Jia Jian Yin Qiao Bai Hu Tang]
Modified Achyranthis Decoction [Jia Jian Niu Xi Tang]
Modified Artemisia Annua and Soft-shelled Turtle Shell Decoction [Jia Jian Qing Hao Bie Jia Tang]
Modified Atractylodes and Cyperi Decoction [Jia Jian Cang Sha Yin]
Modified Bupleurum and Poria Decoction [Jia Jian Chai Ling Tang]
Modified Clear Epidemics and Overcome Toxin Decoction [Jia Jian Qing Wen Bai Du Yin]
Modified Clear the Menses Decoction [Jia Jian Qing Jing Tang]
Modified Cocculus Decoction [Jia Jian Mu Fang Ji Tang]
Modified Drain the Epigastrium Decoction [Jia Jian Xie Xin Tang]
Modified End Discharge Decoction [Jia Jian Wan Dai Tang]
Modified Expelling Stone from Biliary Tract Decoction [Jia Jian Dan Dao Pai Shi Tang]
Modified Fantastically Effective Pill to Invigorate the Collaterals [Jia Jian Huo Luo Xiao Ling Dan]
Modified Five-Peel Decoction [Jia Jian Wu Pi Yin]
Modified Generate the Pulse Powder [Jia Jian Sheng Mai San]
Modified Gentiana Longedancao Decoction to Drain the Liver [Jia Jian Long Dan Xie Gan Tang]
Modified Ginseng, Poria and Atractylodes Macrocephala Powder [Jia Jian Shen Ling Bai Zhu San]
Modified Honeysuckle and Forsythia Powder [Jia Jian Yin Qiao San]
Modified Lophatherus and Gypsum Decoction [Jia Jian Zhu Ye Shi Gao Tang]
Modified Major Bupleurum Decoction [Jia Jian Da Chai Hu Tang]
Modified Major Bupleurum Decoction with Artemesia Yinchenhao [Jia Jian Yin Chen Da Chai Hu Tang]
Modified Minor Bupleurum Decoction [Jia Jian Xiao Chai Hu Tang]
Modified One Yin Decoction [Jia Jian Yi Yin Jian]
Modified Peony Decoction [Jia Jian Shao Yao Tang]
Modified Pinellia Decoction to Drain the Epigastrium [Jia Jian Ban Xia Xie Xin Tang]
Modified Polygonatum Odoratum [Jia Jian Wei Rui Tang]
Modified Pulsatilla Decoction [Jia Jian Bai Tou Weng Tang]
Modified Rehmannia and Lycium Root Bark Decoction [Jia Jian Liang Di Tang]
Modified Relax Wanderer Powder [Jia Jian Xiao Yao San]
Modified Relax Wanderer Powder [Jia Jian Xiao Yao San]
Modified Restore the Pulse Decoction [Jia Jian Fu Mai Tang]
Modified Sedate the Liver and Extinguish the Wind Decoction [Zhen Gan Xi Feng Tang Jia Jian]
Modified Three Painful-Obstruction Decoction [Jia Jian San Bi Tang]
Modified Tonify the Middle and Augment Qi Decoction [Jia Jian Bu Zhong Yi Qi Tang]
Modified Trichosanthes Fruit and Chinese Chive Decoction [Jia Jian Gua Lou Xie Bai Tang]
Modified Univeral Benefit Decoction to Eliminate Toxin #1 [Jian Jian Pu Ji Xiao Du Yin #1]
Modified Univeral Benefit Decoction to Eliminate Toxin #2 [Jia Jian Pu Ji Xiao Du Yin #2]
Modified Universal Benefit Decoction to Eliminate Toxin [Jia Jian Pu Ji Xiao Du Yin]
Modified Yu's Eliminate Dryness and Rescue the Lung Decoction [Jia Jian Yu Shi Qing Zao Jiu Fei Tang]
Moisten the Intestines Pill [Run Chang Wan]
Moonlight Pill [Yue Hua Wan]
Mother of Pearl Pill [Zhen Zhu Mu Wan]
Moutan and Phellodendron Powder for Frigid Extremities [Dan Bai Si Ni San]

Mugwort and Cyperi Rhizome Pill for Warming the Womb [Ai Fu Nuan Gong Wan]
Mugwort and Cyperi Rhizome Pill for Warming the Womb [Nuan Gong Wan]
Mulberry Leaf and Apricot Kernel Decoction [Sang Xing Tang]
Mulberry Leaf and Chrysanthemum Decoction [Sang Ju Yin]
Mulberry Leaf and Sesame Seed Pill [Sang Ma Wan]
Mulberry Leaf, Moutan Decoction to Drain the White [Sang Dan Xie Bai Tang]
Mulberry Leaf, Moutan Powder to Drain the White [Sang Dan Xie Bai San]
Mume Pill [Wu Mei Wan]
Mutton Stew with Tangkuei, Fresh Ginger [Dang Gui Sheng Jiang Yang Rou Tang]
Newly Formulated Tangerine Peel and Bamboo Shaving Decoction [Xin Zhi Ju Pi Zhu Ru Tang]
Newly Augmented Eight-Ingredient Pill with Rehmannia [Xin Jia Ba Wei Di Huang Wan]
Newly Augmented Elsholtzia Decoction [Xin Jia Xiang Ru Yin]
Newly Augmented Jade Woman Decoction [Xin Jia Yu Nu Jian]
Newly Augmented Three-Unbinding Decoction [Xin Jia San Ao Tang]
Newly Augmented White Tiger Decoction [Xin Jia Bai Hu Tang]
Newly Augmented Yellow Dragon Decoction [Xin Jia Huang Long Tang]
Newly Settled Reach the Membrane Source Decoction [Xin Ding Dan Yuan Yin]
Nine-Herb Decoction with Notopterygium [Jiu Wei Qiang Huo Tang]
Nine-Immortal Powder [Jiu Xian San]
Niter and Alum Powder [Xiao Shi Fan Shi San]
No. 1 Extrauterine Pregnancy Formula [Gong Wai Yun Yi Hao Fang]
No. 2 Extrauterine Pregnancy Formula [Gong Wai Yun Er Hao Fang]
No. 2 Formula for Coronary Artery Disease [Guan Xin Er Hao Fang]
Notopterygium Decoction to Overcome Dampness [Qiang Huo Sheng Shi Tang]
Nourish the Heart Decoction [Yang Xin Tang]
Nourish the Yin and Clear the Lungs Decoction [Yang Yin Qing Fei Tang]
Nourish Water Clear the Liver Decoction [Zi Shui Qing Gan Yin]
Nourish Yin Decoction with Soft-shelled Turtle Shell [Bie Jia Yang Yin Jian]
Nourishing the Organs Decoction [Yang Zang Tang]
One Pinch of Gold Decoction [Yi Nian Jin]
One-Shell Decoction to Restore the Pulse [Yi Jia Fu Mai Tang]
Open the Blockage Decoction with Euphorbia [Gan Sui Tong Jie Tang]
Open the Blockage Decoction with Radish Seeds and Magnolia [Lai Po Tong Jie Tang]
Ophiopogonis Decoction [Mai Men Dong Tang]
Ophiopogonis, Schisandra and Rehmannia Pill [Mai Wei Di Huang Wan]
Orange Peel, Pinellia, Immature Bitter Orange and Atractylodes Macrocephala Pill [Ju Ban Zhi Zhu Wan]
Order of Qi Decociton with Tangkuei [Dang Gui Cheng Qi Tang]
Order of Qi Decoction to Diffuse the White [Xuan Bai Cheng Qi Tang]
Order of Qi Decoction with Peach Kernel [Tao Ren Cheng Qi Tang]
Overcome Pathogenic Influences Powder [Bai Du San]
Oyster Shell Powder [Mu Li San]
Peach Blossom Decoction [Tao Hua Tang]
Peach Pit Decoction to Order the Qi [Tao He Cheng Qi Tang]
Peony and Licorice Decoction [Shao Yao Gan cao Tang]
Peony and Licorice Decoction to Stop Spasms [Shao Gan Zhi Ling Tang]
Peony Decoction [Shao Yao Tang]
Peony, Licorice, and Prepared Aconite Decoction [Shao Yao Gan Cao Fu Zi Tang]
Pepper and Mume Decoction [Jiao Mei Tang]
Perilla Fruit Decoction for Directing Qi Downward [Su Zi Jiang Qi Tang]
Perilla Fruit, Lepidii Seeds, Ephedra, Apricot Kernel, Gypsum and Licorice Decoction [Su Ting Ma Xing Shi Gan Tang]
Perilla Powder [Zi Su San]
Persimmon Calyx Decoction [Shi Di Tang]
Persimmon Coin Powder [Shi Qian San]

Picrorhizae and Mume Decoction to Calm Roundworms [Lian Mei An Hui Tang]
Pill for Deafness that is Kind to the Left [Kidney] [Er Long Zuo Ci Wan]
Pill for Sandy Lin Syndrome [Sha Lin Wan]
Pinellia and Dry Ginger Powder [Ban Xia Gan Jiang San]
Pinellia and Magnolia Bark Decoction [Ban Xia Hou Po Tang]
Pinellia and Poria Pill [Ban Xia Fu Ling Wan]
Pinellia Decociton to Drain the Epigastrium [Ban Xia Xie Xin Tang]
Pinellia Decoction [Ban Xia Tang]
Pinellia Decoction to Drain the Epigastrium without Dry Ginger, Licorice and with the Addition of Immature Orange and Apricot Seed [Ban Xia Xie Xin Tang Qu Gan Jiang, Gan Cao, Jia Zhi Shi Xing Ren Fang]
Pinellia, Atractylodes Macrocephala, and Gastrodia Decoction [Ban Xia Bai Zhu Tian Ma Tang]
Platycodon Decoction [Jie Geng Tang]
Polygalae Pill [Yuan Zhi Wan]
Polyporus Decoction [Zhu Ling Tang]
Poria Cocos Pararadicis Decoction from Thousand Ducat Formula [Qian Jin Fu Shen Tang]
Poria Pill [Fu Ling Wan]
Poria Pill from *Zhi Mi Fang* [Zhi Mi Fu Ling Wan]
Poria Skin Decoction [Fu Ling Pi Tang]
Poria, Cinnamon Twig, Atractylodes Macrocephalae, and Licorice Decoction [Ling Gui Zhu Gan Tang]
Prepared Aconite Decoction [Fu Zi Tang]
Prepared Aconite Pill to Regulate the Middle [Fu Zi Li Zhong Wan]
Preserve Harmony Pill [Bao He Wan]
Preserved Vistas Pill [Zhu Jing Wan]
Preserve-True Decoction [Bao Zhen Tang]
Prolong Life Decoction from *Gu Jin Lu Yan* [Gu Jin Lu Yan Xu Ming Tang]
Promote Gallbladder Pill [Li Dan Wan]
Protect the Fetus without Any Worries Powder [Bao Sheng Wu You San]
Protected by Deities Pill with Three Flowers [San Hua Shen You Wan]
Prune Seed Decoction [Mei Ren Tang]
Pulsatilla Decoction [Bai Tou Weng Tang]
Pulsatilla Decoction plus Licorice and Ass-Hide Gelatin [Bai Tou Weng Jia Gan Cao E Jiao Tang]
Pumpkin Seed Powder and Betel Nut Decoction [Nan Gua Zi Fen Bing Lang Jian]
Purge the Dead Fetus Formula [Xia Si Tai Fang]
Purge the Stagnant Blood Decoction [Xia Yu Xue Tang]
Purple Dragon Pill [Zi Long Wan]
Purple Gold Pill for Correcting the Bone [Zheng Gu Zi Jin Dan]
Purple Snow Special Pill [Zi Xue Dan]
Pyrrosiae Powder [Shi Wei San] *Ben Shi Fang*
Pyrrosiae Powder [Shi Wei San] *Pu Ji Fang*
Pyrrosiae Powder [Shi Wei San] *Zhen Zhi Hui Bu*
Raise the Sinking Decoction [Sheng Xian Tang]
Reach the Membrane Source Decoction [Da Yuan Yin]
Rectify the Qi Powder Worth More than Gold [Bu Huan Jin Zheng Qi San]
Red Date and Alum Pill [Zao Fan Wan]
Red Poria Decoction [Chi Fu Ling Tang]
Reduce the Focal Distention Pill with Coptis [Huang Lian Xiao Pi Wan]
Reduce the Food Distention Pill [Xiao Pi Wan]
Reduce the Water Powder [Xiao Shui San]
Reduced Lophatherus and Gypsum Decoction [Jian Wei Zhu Ye Shi Gao Tang]
Reed Decoction [Wei Jing Tang]
Regualte the Middle and Transform Phlegm Pill [Li Zhong Hua Tan Wan]
Regulate the Middle and Calm Roundworms Pill [Li Zhong An Hui Wan]
Regulate the Middle and Direct Phlegm Decoction [Li Zhong Jiang Tan Wan]

Regulate the Middle Decoction with Five-Ingredient Powder with Poria [Li Ling Tang]
Regulate the Middle Pill [Li Zhong Wan]
Regulate the Stomach and Order the Qi Decoction [Tiao Wei Cheng Qi Tang]
Rehamannia Pill [Di Huang Wan]
Rehmannia and Lycium Root Bark Decoction [Liang Di Tang]
Relax the Tendon and Invigorate Collaterals Decoction [Shu Jin Huo Luo Tang]
Relax Wanderer Powder [Xiao Yao San]
Relax Wanderer Powder with Moutan and Gardenia [Dan Zhi Xiao Yao San]
Release from Flower Decoction [Tuo Hua Jian]
Remove Painful Obstruction Decoction [Juan Bi Tang]
Rescue the Critical Condition Decoction [Jiu Ni Tang]
Restore and Revive the Yang Decoction [Hui Yang Ji Jiu Tang]
Restore the Left [Kidney] Decoction [Zuo Gui Yin]
Restore the Left [Kidney] Pill [Zuo Gui Wan]
Restore the Pulse Decoction [Fu Mai Tang]
Restore the Right [Kidney] Decoction [You Gui Yin]
Restore the Right [Kidney] Pill [You Gui Wan]
Restore the Spleen Decoction [Gui Pi Tang]
Return of Spring Special Pill [Xiao Er Hui Chun Dan]
Revised Millettia Decoction [Fu Fang Hong Teng Jian]
Revised Major Order of Qi Decoction [Fu Fang Da Cheng Qi Tang]
Revised Rhubarb and Moutan Decoction [Fu Fang Da Huang Mu Dan Tang]
Revive Health by Invigorating the Blood Decoction [Fu Yuan Huo Xue Tang]
Rhinoceros Gallstone Pill [Xi Huang Wan]
Rhinoceros Horn and Rehmannia Decoction [Xi Jiao Di Huang Tang]
Rhinoceros Horn and Rehmannia Pill [Ju Fang Xi Jiao Di Huang Wan]
Rhinoceros Horn and Rehmannia, and Drain the Epigastrium Decoction [Xi Jiao Di Huang He Xie Xin Tang]
Rhinoceros Horn, Antelope and Whtie Tiger Decoction [Xi Ling Bai Hu Tang]
Rhubarb and Eupolyphaga Pill [Da Huang Zhe Chong Wan]
Rhubarb and Licorice Decoction [Da Huang Gan Cao Tang]
Rhubarb and Mirabilite Decoction [Da Huang Xiao Shi Tang]
Rhubarb and Moutan Decoction [Da Huang Mu Dan Tang]
Rhubarb and Prepared Aconite Decoction [Da Huang Fu Zi Tang]
Safe Powder for People and Horses [Ren Ma Ping An San]
Sage-Like Healing Decoction [Sheng Yu Tang]
Salvia Decoction [Dan Shen Yin]
Sanguisorba Pill [Di Yu Wan]
Sargassum Decoction for the Jade Flask [Hai Zao Yu Hu Tang]
Scallion, and Prepared Soybean Decoction [Cong Chi Tang]
Scallion, and Prepared Soybean Decoction from the Book to Safeguard Life [Huo Ren Cong Chi Tang]
Scallion, Prepared Soybean, and Platycodon Decoction [Cong Chi Jie Geng Tang]
Scallion, Prepared Soybean, and White Tiger Decoction [Cong Chi Bai Hu Tang]
Scallion, Prepared Soybean, Peppermint and Arctii Decoction [Cong Chi Bo Bang Tang]
Schisandra Decoction [Wu Wei Zi Tang]
Schizonepeta and Ledebouriella Powder to Overcome Pathogenic Influences [Jing Fang Bai Du San]
Scutellaria and Peony Decoction [Huang Qin Shao Yao Tang]
Scutellaria and Talcum Decoction [Huang Qin Hua Shi Tang]
Scutellaria Decoction [Huang Qin Tang]
Scutellaria Decoction with Pinellia and Ginger [Huang Qin Jia Ban Xia Sheng Jiang Tang]
Scutellaria, Coptis, Tangerine Peel and Bamboo Shaving Decoction [Qin Lian Ju Ru Tang]
Second Modification of Rectify the Qi Powder [Er Jia Jian Zheng Qi San]
Secret Clear the Stomach Powder from Experiences [Mi Yan Qing Wei San]
Sedate the Liver and Extinguish the Wind Decoction [Zhen Gan Xi Feng Tang]

Seize the Life Powder [Duo Ming San]
Set the Bone Pill [Jie Gu Wan]
Settle the Emotion Pill [Ding Zhi Wan]
Seven-Ingredient Actractylodes Macrocephala Decoction [Qi Wei Bai Zhu San]
Seven-Ingredient Scallion Decoction [Qi Wei Cong Bai Yin]
Seven-Peel Decoction [Qi Pi Yin]
Seven-Substance Decoction with Magnolia Bark [Hou Po Qi Wu Tang]
Seven-Thousands of a Tael Powder [Qi Li San]
Seven-Treasure Special Pill for Beautiful Whiskers [Qi Bao Mei Ran Dan]
Shut the Sluice Pill [Suo Quan Wan]
Silk Worm Excrement Decoction [Can Shi Tang]
Silver Armour Pill [Yin Jia Wan]
Sinking into the Chest and Order of Qi Decoction [Xian Xiong Cheng Qi Tang]
Six Milled-Herb Decoction [Liu Mo Tang]
Six-Gemtleman Decoction with Lophatherus and Ophiopogonis [Zhu Mai Liu Jun Zi Tang]
Six-Gentleman Decoction [Liu Jun Zi Tang]
Six-Gentleman Decoction with Aconite and Scorpion [Wu Xie Liu Jun Zi Tang]
Six-Gentleman Decoction with Arisaemae and Typhonii [Xing Fu Liu Jun Zi Tang]
Six-Gentleman Decoction with Aucklandia and Amomum [Xiang Sha Liu Jun Zi Tang]
Six-Gentleman Decoction with Bupleurum and Peony [Chai Shao Liu Jun Zi Tang]
Six-Gentleman Decoction with Cimicifuga and Bupleurum [Sheng Chai Liu Jun Zi Tang]
Six-Gentleman Decoction with Coptis [Huang Lian Liu Jun Zi Tang]
Six-Gentleman Decoction with Hawthorn Berry and Medicated Leaven [Zha Qu Liu Jun Zi Tang]
Six-Ingredient Pill with Rehmannia [Liu Wei Di Huang Wan]
Six-Miracle Decoction to Expel Congested Fluid [Juan Yin Liu Shen Tang]
Six-to-One Powder [Liu Yi San]
Small Rehamnnia Decoction [Xiao Pin Sheng Di Huang Tang]
Smooth the Qi Powder [Shun Qi San]
Soften the Hardness and Scatter the Accumulation Decoction [Ruan Jian San Jie Tang]
Soft-shelled Turtle Shell and Artemisia Annua Decoction [Bie Jia Qing Hao Yin]
Sophora Japonica Flower Powder [Huai Hua San]
Sophora Japonica Fruit Pill [Huai Jiao Wan]
Sophorae to Stop Diarrhea Decoction [Ku Shen Zhi Li Jian]
Sour Jujube Decoction [Suan Zao Ren Tang]
Sour Phlegm Decoction [Di Tan Tang]
Special Pill to Aid Fertility [Zan Yu Dan]
Spread the Depression and Clear the Liver Beverage [Shu Yu Qing Gan Yin]
Spread the Depression and Clear the Liver Decoction [Shu Yu Qing Gan Tang]
Spread the Liver and Resolve the Stagnation Decoction [Shu Gan Jie Yu Tang]
Spring Pond Decoction [Chun Ze Tang]
Stabilize the Essence Pill [Gu Jing Wan]
Stabilize the Essence Pill [Gu Jing Wan]
Stabilize the Root and Stop Excessive Uterine Bleeding Powder [Gu Ben Zhi Beng San]
Stabilize the Yin Decoction [Gu Yin Jian]
Stationed Vehicle Pill [Zhu Che Wan]
Stepania and Astragalus Decoction [Fang Ji Huang Qi Tang]
Stepania and Poria Decoction [Fang Ji Fu Ling Tang]
Stephania, Zanthoxylum, Descurainia and Rhubarb Pill [Ji Jiao Li Huang Tang]
Stimulating Appetite and Strengthen the Spleen Pill [Kai Wei Jian Pi Wan]
Stop Bleeding Powder [Zhi Xue San]
Stop Coughing Powder [Zhi Suo San]
Stop Spasms Powder [Zhi Jing San]
Stop Spasms Powder with Viper [Jin She Zhi Jing San]
Strengthen Baby Pill [Fei Er Wan]
Strengthen the Spleen Pill [Jian Pi Wan]

Sublime Formula for Sustaining Life [Xian Fang Huo Ming Yin]
Sudden Smile Powder [Shi Xiao San]
Supress the Green Pill [Yi Qing Wan]
Sweet Dew Decoction [Gan Lu Yin]
Sweet Dew Special Pill to Eliminate Toxin [Gan Lu Xiao Du Dan]
Talcum and Agastache Decoction [Hua Shi Huo Xiang Tang]
Tangerine Peel and Bamboo Shaving Decoction [Ju Pi Zhu Ru Tang]
Tangerine Peel and Bamboo Shaving Decoction from Formulas that Aid the Living [Ju Pi Zhu Ru Tang] *Ji Sheng Fang*
Tangerine Peel and Bamboo Shaving Decoction from the Book to Safeguard Life [Huo Ren Ju Pi Zhu Ru Tang]
Tangerine Peel Decoction from the Golden Cabinet [Ju Pi Tang] *Jin Gui Yao Lue*
Tangerine Seed Pill [Ju He Wan]
Tangerine, Pinellia, Cinnamon Twig, Immatured Bitter Orange and Ginger Decoction [Ju Ban Gui Ling Zhi Jiang Tang]
Tangkuei and Astragalus Decoction to Construct the Middle [Gui Qi Jian Zhong Tang]
Tangkuei and Mutton Decoction from the Formulas that Aid the Living [Ji Sheng Dang Gui Yang Rou Tang]
Tangkuei and Peony Powder [Dang Gui Shao Yao San]
Tangkuei and Six-Yellow Decoction [Dang Gui Liu Huang Tang]
Tangkuei Decoction [Dang Gui Tang]
Tangkuei Decoction for Frigid Extremities [Dang Gui Si Ni Tang]
Tangkuei Decoction for Frigid Extremities plus Evodia and Fresh Ginger [Dang Gui Si Ni Jia Wu Zhu Yu Sheng Jiang Tang]
Tangkuei Decoction from Thousand Ducat Formula [Qian Jin Dang Gui Tang]
Tangkuei Decoction to Construct the Middle [Dang Gui Jian Zhong Tang]
Tangkuei Decoction to Tonify Blood [Dang Gui Bu Xue Tang]
Tangkuei Powder [Dang Gui San]
Tangkuei, Gentiana Londancao and Aloe Pill [Dang Gui Long Hui Wan]
Ten Partially Charred Substance Powder [Shi Hui San]
Ten-Ingredient Decoction to Warm the Gallbladder [Shi Wei Wen Dan Tang]
Ten-Ingredient Elsholtzia Decoction [Shi Wei Xiang Ru Yin]
Ten-Jujube Decoction [Shi Zao Tang]
Ten-Tonic Pill [Shi Bu Wan]
Third Modification of Rectify the Qi Powder [San Jia Jian Zheng Qi San]
Thousand Gold Reed Decoction with Apricot Kernel and Talcum Decoction [Qian Jin Wei Jing Tang Jia Xing Ren Hua Shi Tang]
Three Layers Fennel Pill [San Ceng Hui Xiang Wan]
Three-Fragrant Decoction [San Xiang Tang]
Three-Marvel Pill [San Miao Wan]
Three-Nut Decoction [San Ren Tang]
Three-Painful Obstruction Decoction [San Bi Tang]
Three-Seed Decoction to Nourish One's Parents [San Zi Yang Qin Tang]
Three-Shell Decoction to Restore the Pulse [San Jia Fu Mai Tang]
Three-Substance Pill Prepared fro Emergency [San Wu Bei Ji Wan]
Three-Unbinding Decoction [San Ao Tang]
Three-Yellow and Four-Substance Decoction [San Huang Si Wu Tang]
Three-Yellow, Immature Bitter Orange and Atractylodes Macrocephala Pill [San Huang Zhi Zhu Wan]
Toad Cake Pill [Chan Su Wan]
Tonify and Augment Pill [Bu Yi Wan]
Tonify Both Heart and Spleen Pill [Xin Pi Shuang Bu Wan]
Tonify the Liver Decoction [Bu Gan Tang]
Tonify the Lung Decoction [Bu Fei Tang]
Tonify the Lungs Decoction with Ass-Hide Gelatin [Bu Fei E Jiao Tang]

Tonify the Middle and Augment Qi Decoction [Bu Zhong Yi Qi Tang]
Tonify the Yang to Restore Five-Tenths Decoction [Bu Yang Huan Wu Tang]
Top-Quality Lindera Powder [Tian Tai Wu Yao San]
Tortoise Shell and Deer Antler Syrup [Gui Lu Er Xian Jaio]
Transform Blood Stasis Decoction for Appendicitis [Lan Wei Hua Yu Tang]
Transform Blotches Decoction [Hua Ban Tang]
Transform the Stone Powder [Hua Shi San]
Trauma Pill [Die Da Wan]
Treats the Middle Decoction [Zhi Zhong Tang]
Trichosanthes Fruit, Chinese Chive, and Pinellia Decoction [Gua Lou Xie Bai Ban Xia Tang]
Trichosanthes Fruit, Chinese Chive, and Wine Decoction [Gua Lou Xie Bai Bai Jiu Tang]
True Jade Powder [Yu Zhen San]
True Man's Formula for Sustaining Life [Zhen Ren Huo Ming Yin]
True Mother of Pearl Pill [Zhen Zhu Mu Wan]
True Warrior Decoction [Zhen Wu Tang]
Twist the Head Powder [Nian Tou San]
Two-Cured Decoction [Er Chen Tang]
Two-Cured Decoction to Harmonize the Stomach [He Wei Er Chen Tang]
Two-Cured Decoction with Chinese Chive Juice [Jiu Zhi Er Chen Tang]
Two-Cured Decoction with Cinnamon and Aconite [Gui Fu Er Chen Tang]
Two-Cured Decoction with Coptis and Bamboo Shaving [Lian Ru Er Chen Tang]
Two-Cured Decoction with Ephedra and Apricot Kernel [Ma Xing Er Chen Tang]
Two-Cured Decoction with Gleditsia and Bamboo Sap [Zao Li Er Chen Tang]
Two-Cured Decoction with Hawthorn and Medicated Leaven [Zha Qu Er Chen Tang]
Two-Cured Decoction with Ligusticum Chuanxiong and Tangkuei [Xiong Gui Er Chen Tang]
Two-Cured Decoction with Perilla and Apricot Kernel [Su Xing Er Chen Tang]
Two-Cured Decoction with Pumice and Clam Shell [Hai Ge Er Chen Tang]
Two-Cured Decoction with Trichosanthis and Fritillaria [Lou Bei Er Chen Tang]
Two-Cured Decoction with Two Atractylodes [Er Zhu Er Chen Tang]
Two-Immortal Decoction [Er Xian Tang]
Two-Marvel Powder [Er Miao San]
Two-Shell Decoction to Restore the Pulse [Er Jia Fu Mai Tang]
Two-Ultimate Pill [Er Zhi Wan]
Unaccompanied Ginseng Decoction [Du Shen Tang]
Unblock and Dig Formula [Shu Zao Yin Zi]
Unblock Breast Milk and Scatter Accumulation Decoction [Tong Ru San Jie Tang]
Unblock the Channel and Invigorate the Collateral Decoction [Tong Jing Huo Luo Tang]
Unblock the Channel for Frigid Extremities Decoction from the Formulas that Aid the Living [Ji Sheng Tong Mai Si Ni Tang]
Unblock the Orifices and Invigorate the Blood Decoction [Tong Qiao Huo Xue Tang]
Unblock the Pulse Decoction for Frigid Extremities [Tong Mai Si Ni Tang]
Unblock the Pulse Decoction with Pig Bile for Frigid Extremities [Tong Mai Si Ni Jia Zhu Dan Zhi Tang]
Unblock the Stasis Decoction [Tong Yu Jian]
Uncaria Decoction [Gou Teng Yin]
Universal Benefit Decoction to Eliminate Toxin [Pu Ji Xiao Du Yin]
Vaporize Phlegm Pill [Gun Tan Wan]
Vaporize Phlegm Pill with Chlorite-Mica [Meng Shi Gun Tan Wan]
Vessel and Vehicle Pill [Zhou Che Wan]
Warm the Gallbladder Decoction [Wen Dan Tang]
Warm the Liver Decoction [Nuan Gan Jian]
Warm the Menses Decoction [Wen Jing Tang]
Warm the Menses Decoction from Imperial Grace Formula [Ju Fang Wen Jing Tang]
Warm the Menses, Control the Blood Decoction [Wen Jing She Xue Tang]
Warm the Spleen Decoction [Wen Pi Tang]°nBen Shi Fang°n

Warm the Spleen Decoction [Wen Pi Tang]°nQian Jin Yao Fang°n
Warm the Spleen Powder [Wen Pi San]
Water and Earth Immortals Special Pill [Shui Lu Er Xian Dan]
White Tiger and Order the Qi Decoction [Bai He Cheng Qi Tang]
White Tiger Decoction [Bai Hu Tang]
White Tiger Decoction to Suppress Rebellion [Zhen Ni Bai Hu Tang]
White Tiger plus Cinnamon Twig Decoction [Bai Hu Jia Ren Shen Tang]
White Tiger plus Ginseng Decoction [Bai Hu Jia Gui Zhi Tang]
White Tiger with Coptis Decoction to Relieve Toxicity [Bai Hu He Huang Lian Jie Du Tang]
Women's Treasure Pill [Fu Bao Dan]
Xanthium Powder [Cang Er Zi San]
Yang-Heartening Decoction [Yang He Tang]
Yello Dragon Decoction [Huang Long Tang]
Yellow Earth Decoction [Huang Tu Tang]
Ye's Benefit the Stomach Decoction [Ye Shi Yi Wei Tang]
Young Maiden Pill [Qing E Wan]

Chinese Herbs Index*

* Page numbers in bold type indicate more extensive information regarding specific herbs.

C

D

E

F

G

H

J

M

N

O

P

Q

R

S

T

Y

Z

Index

B

C

D

E

F

G

M

N

Q

R

T

V

W

X

Y

Z